The
FEEDING *of* NATIONS
REDEFINING FOOD SECURITY
FOR THE 21ST CENTURY

The FEEDING *of* NATIONS

REDEFINING FOOD SECURITY FOR THE 21ST CENTURY

Mark Gibson

CRC Press
Taylor & Francis Group
Boca Raton   London   New York

CRC Press is an imprint of the
Taylor & Francis Group, an **informa** business

CRC Press
Taylor & Francis Group
6000 Broken Sound Parkway NW, Suite 300
Boca Raton, FL 33487-2742

First issued in paperback 2016

© 2012 by Taylor & Francis Group, LLC
CRC Press is an imprint of Taylor & Francis Group, an Informa business

Version Date: 20111213

ISBN 13: 978-1-138-19851-7 (pbk)
ISBN 13: 978-1-4398-3950-8 (hbk)

Library of Congress Cataloging-in-Publication Data

Gibson, Mark.
  The feeding of nations : re-defining food security for the 21st century / Mark Gibson.
    p. cm.
  Includes bibliographical references and index.
  ISBN 978-1-4398-3950-8 (hardcover : alk. paper)
  1. Food security. 2. Hunger--Prevention. 3. Food industry and trade. 4. Food relief. 5. International cooperation. I. Title.

HD9000.9.A1G557 2012
338.1'9--dc23   2011046937

Visit the Taylor & Francis Web site at
http://www.taylorandfrancis.com

and the CRC Press Web site at
http://www.crcpress.com

To Hannah

To see the world in a grain of sand,

and to see heaven in a wild flower,

hold infinity in the palm of your hands,

and eternity in an hour.

William Bake, *Auguries of Innocence*

Contents

Part I   Food Security: What Is It, How and Who Does It Affect?

Part V   The Final Analysis—Food Security

Appendices

List of Figures

List of Tables

Preface

The Paradox of Hunger

"Hunger has increased as the world has grown richer and produced more food than ever in the last decade." This statement in the FAO's State of Food Insecurity 2008 (SOFI) report succinctly sums up the ethical dilemma that as of 2011*, globally sees over 1.6 billion people overweight while at the same time a further 925 million are underfed and suffering malnutrition. Of these an estimated 10 million this year will die directly as a result of hunger and hunger-related diseases. In a world where we produce enough food to feed the population comfortably, there still exist confusion and lack of consensus over conceptualising and dealing with the multiple components of food security. As a result, the topic becomes ambiguous and emotive and often leads to strong convictions and much blame, infighting and mudslinging.

The quest to end hunger and malnutrition, an endeavour most notably championed in the twentieth century has its roots firmly embedded in science, philosophy and historical altruism. However, there is no clear or linear path that leads us directly from the past to present-day ideologies; instead an amalgam of inter-related events, ideas, notions and philosophies have collectively emerged under a single banner of food security. Continually propelled by moral and ethical righteousness, the concept itself enforces healthy periodic introspection and reflection, yet in the ensuing endeavour, and despite attempts to the contrary, there is no getting away from the sheer complexity and disunity of food security issues.

At its very simplest, the history of food security is a history of cultural and political civilisation itself. From the earliest days, man has had to contend with feeding himself and his fellow beings, and while the earliest hunter–gatherer, diversification into agriculture aided the growth of complex interdependent societal civilisations: with it developed new problems of food insecurity. Population growth and famine worries spurred innovation, discussion and scientific experimentation, and for millennia people sought to work the land to its maximum potential in order to feed the ever growing masses. Consequentially, over the centuries, multiple diverse threads from philosophy to agricultural advancements began to converge around the time of the 'Age of Enlightenment'. This was a seminal period around the eighteenth century when Western intellectualism, the sciences and social currents all came together espousing open fora on societal issues. From this, critical ideas emerged that promoted rationality, freedom democracy and ultimately the rights of individuals. This was a time, too, that witnessed public discursive debate influence political action, affect cultural change and set the foundation of present-day Western intellectual and political philosophy. Meanwhile, national identity and self-interests sat awkwardly alongside a growing internationalism of the eighteenth and nineteenth centuries whilst various revolutions—agricultural, industrial, and scientific—paved the way for the widely disseminated knowledge revolution. Up to this point, food security was still very much a local, or at best a national concern. However, at the end of the nineteenth century and with the further growth of globalisation, the advancing media and travel industries helped cultivate a new enthusiasm of moral enlightenment and the beginnings of a new global cultural revolution. About this time, social awakenings and the growth of humanitarianism also led to many new international instruments, agreements, treaties and conventions aimed at improving the lot of mankind.

From here, it was not long before awareness and growing social distaste over the national and international suffering of hunger and famine percolated into the political arena with the resultant and inevitable politicisation of hunger. Later, after the two world wars had sharpened the public focus on health and nutrition and with science and the green revolution providing further impetus, ambitious goals were

---

* The latest 2011 State of Food Insecurity Report (SOFI) makes no estimate of figures for 2011 nor does it revise estimates of the previous two years due to a transition in the methodological approach to calculating these figures. In lieu of this, figures for 2010 are quoted.

elucidated and formalised. In the midst of this, the concept of food security was becoming ever more entangled with wider social issues of public health, sustainability and national agendas. For some, hunger in the third world* offered an opportunity for growing exports, while for others, it was the chance to cement the role of the UN and other international organisations in world matters. A period of conflict over national versus international governance ensued that, to this day, has had no real or clear resolution. Instead, at times, an uneasy collaboration is effected that witnesses periodic trade-offs between domestic and foreign policies. Also from around the 1970s onwards, the already increasingly complex food security phenomenon was fast becoming redefined by yet more apparently disparate driving forces. World population, for instance, began increasing at an alarming rate whilst at the same time world energy needs were being hampered by higher oil prices, placing further pressure on agricultural inputs and economies. Furthermore, climate change, land and soil degradation, urbanisation and sustainability, all, too, began to be closely associated with food security. All the while, as globalisation became ever more entrenched, an increasing share of the global food chain was being placed in the hands of fewer larger multinationals whose core philosophies of profit and growth were often seen to be fundamentally at odds with the political and humanitarian notions of equity and access. In fact, these many influences came together and colluded to transform not only regional and global food production and consumption patterns, but the very nature and foundation of the philosophies concerning food security itself. Analysing and interpreting these seemingly exponential emerging trends and underlying philosophies is a constant challenge.

Continually throughout this time and largely as a result of this fluxial nature, new targets and initiatives were created, usually dictated by new theories and food security models which were then altered, absorbed and redefined. This testing ground has had some marginal successes and it has also offered new insights into a complex phenomenon. However, despite some modest successes, our accumulated knowledge has ultimately made very little difference to the overall numbers of hungry people in the last four decades.

Aims of the Book

As mentioned earlier, the concept of food security itself, despite its already inherent complexity, is becoming increasingly difficult to comprehend with any degree of clarity. Moreover, any such attempt to do so threatens to open the Pandora's box. The aim of this book is to do just that: by opening up the box and offering a glimpse into the full spectrum of food security issues, the aetiology, the workings, the pitfalls and successes; we see the concept in its entirety—only then can definitions be clarified, goals articulated, consensus sought and, ultimately, solutions found. Unfortunately these days with food security being wrapped up inside a cacophonous melee of confusion and inclusion, it might seem to the uninitiated that the idea has become a hook onto which wider societal issues are being hung. With this in mind, there is no skinny version of food security and any attempt to cherry-pick certain of the ideas for discussion runs counter to the goal of total appreciation of the subject. Having said that, there have to be limits, after all there is only so much that can be included in 700 or so pages. Moreover, if such an attempt were made, and the many ideas taken to their logical conclusions, then this book would probably be sitting on the shelves of social philosophy rather than laid open as a working tool of food security practitioners.

As becomes evident throughout this book, behind the deceptively simple idea lies a far more complex and often misunderstood notion. It is this fundamental assertion that drives the narrative. All too often these days, the hunger and malnutrition literature refers to this complex and multifaceted aspect of food security. Yet only a few, if any, expand or articulate on this often casual statement beyond a sentence or two regarding the difficulties of exploring the subject in any great detail. As a result, it is the view of the author that any real progress needs a rethink on just how the subject is approached, one that encompasses all aspects of food security—an holistic view. However, as is becoming the new call to arms of late, lip

---

* The third world is an obsolete term referring to today's developing nations. The term was thought derogatory in some quarters; however, it is a term that is still in use. As such the terms 'third world' and 'developing' countries are used interchangeably throughout this book and no derogatory inferences are implied.

service to this type of approach has been paid by more than a few—yet there seems to be scant evidence of its realisation. This book aims to address this gap. A difficult and daunting task for sure, but one that I feel would only benefit future research and action to alleviate the problems involved. With this in mind, and tapping into the exceptionally rich resources of nutrition, health, social, scientific and political history, the book aims to explore just how the simple aspiration of global food security has evolved and unfolded, particularly at the international level to become a goal replete with complex, interwoven, sometimes contradictory and counterproductive policies, agendas and ideologies. The explicit aim therefore is to identify and draw all the threads of food security together, all the disparate, seemingly unconnected disciplines, events, breakthroughs, people and policies that have contributed to and shaped current ideology, policy and understanding.

In some ways, the focus of the book is conceptual where emphasis on the exploration of contemporary philosophical currents and evolutionary aspirations sets the undertone of prevailing ideology. Looking at the book from another view, the work is empirically oriented and grounded in the practicalities of food security issues. This juxtaposition is deliberate and acts as a sort of symposium bringing into the fold the relevant origins, issues, concepts and underlying philosophies both from a contemporary and historical perspective. This work has been two years in the making and has involved thousands of documents, texts, archival research, books and databases. Where possible, original and primary sources of information were sought in support of the default research stance of the author; this often meant reinterpreting or analysing such material. This book also made use of meta-analysis and secondary research. This helped illuminate and strengthen threads, connections and analyses although I have endeavoured to keep this to a minimum in order that a fresh perspective can be obtained.

The principal focus of this book then is to unravel as much of the complexities of the subject as a book of this length allows. In doing so, I hope to assimilate into one volume enough incongruent information which will strengthen existing understanding and credibility of food security analyses as well as engender a new lasting collaborative paradigm of cooperation among disciplines. Also in writing what is perhaps the first comprehensive historical work on the evolution and diversity of the food security concept itself, it is hoped the book will provide sufficient foundational knowledge of an oft misunderstood concept that benefits current and future students, historians and policy makers alike.

What This Book Is Not ...

While it is important to outline the nature and scope of this book, important caveats are needed in recognising what this book is not. While the core of the text focuses on many key players, events and underpinnings in nutrition and food security history seen as central to the unfolding narrative, it is limiting in a book of this size to offer anything other than an overview of many such issues. Therefore, it is natural that there will perhaps be omissions that some will see as significant. For this, I can only apologise and state that the chosen threads of history that have been included are in the opinion of the author sufficient to give as wide a breadth of understanding in the current debate as is possible without transforming the work into an historical tome. In attempting to reconnect the past with the present I have endeavoured to present or analyse key ideas from the same contemporary viewpoint as far as possible without reinterpretation from the modern perspective. Moreover, it was essential that I not undertake to offer a polemic or diatribe, instead to present analyses and at times critical reviews that are grounded in fact and where interpretation is required I have undertaken to do so unencumbered by bias or prejudice. One last important caveat is that the focus of much of this work regards food security issues predominantly from a Western perspective; again this is deliberate as research suggests much of the international dimension of food security today can trace its genesis in one way or another to Western economic and philosophical ideology.

This book also does not necessarily attempt to provide solutions to current problems. It only attempts simply to unravel as much of the complexity surrounding current conceptions and to properly elucidate perhaps for the first time the many considerations involved in a topic that affects every man, woman and child on the planet. High hopes indeed. Yet I am dogged in my pursuit with singular focus of mind that this undertaking will be carried out to the best of my abilities and my critics will decide whether I have skilfully navigated a linear path through a non-linear concept.

Structure of the Book

Before we can begin to address the notion of food for all, we need to properly understand the concept itself. In this respect, the book starts in the present to establish a brief overview of current understanding and assemble a picture of the prevalence of hunger and malnutrition. From here, the book proceeds to work backwards in time with a cursory look from the historical perspective to understand how we arrived at such a point. Issues and philosophies that gave impetus either directly or indirectly to current orthodoxy are examined and given context. Once many of these key ideas have been explored, consideration is then given to the interaction of such issues and how such seemingly diverse threads came together at the beginning of the twentieth century. The twentieth century is then analysed in detail. The aim is to help shed light on contemporary motivations and chronicle the flow of events that ultimately led to the creation of the modern concept of food security. Once such key events and thinking are in place, the book then takes an overarching perspective to view the concept in relation to the many different sectors, which include, among others, agriculture, environment and policy. By doing so, the book also uncovers and to some extent unravels some of the many complex issues of food security. After this there follows a penultimate and detailed review of the concept itself taking in everything that has been uncovered thus far. Lastly, the final part of the book examines many of the more important considerations with regard to addressing the issues. At this point, the book will explore the goals society has set itself, the means by which these are to be achieved and commentary on some of the current thoughts on solutions.

This book is intended to be fully comprehensive and inclusive and introduces numerous concepts and ideas. In covering such a vast field, it becomes necessary to dip in and out of detail. That is to say in some instances, seemingly important events might deserve only a cursory examination while perhaps apparently not-so-important topics might require more attention. While it is my intention to furnish as much relevant information as possible, it is beyond reason for the author to be authoritative on all aspects. In this respect, some ideas are treated sparingly and act solely as an introduction to the topic. The danger here then is that dense technical narrative might disrupt the flow or crowd out key ideas and concepts. To avoid this, the book relies on a broad set of appendices, allowing the reader to better follow the narrative as it unfolds.

Operational Concepts

Because of the diverse, complex and often intertwined issues of food security, such expansive research cannot be studied in isolation. Instead, this study crosses multiple disciplines which often entail the use of vernacular or esoteric language that may require supplementary clarification. Lastly and more worryingly, in much literature certain terminologies and concepts can and are often used interchangeably to the confusion of the reader. To this end, all concepts, nomenclature or terminology, unless otherwise defined in the main text, relate specifically to operational definitions detailed in the glossary provided.

Acknowledgements

Writing this book has been a tremendous journey of discovery. After having undertaken two studies of the food security phenomenon for both my masters and PhD, I thought I was particularly well read on the subject. Yet after another 18 months and many many hundreds of hours later, I am still left with the feeling there is so much more to learn. Of course, this has a lot to do with the sheer breadth of the topic and the many various disciplines with which I have had to become familiar. However, it also has as much to do with the nature, the complexity and the interaction of the multitude of disciplines involved. In particular, the emergence of food security as a field (or discipline if you like) in its own right is something new and one that has yet to articulate a navigable path through the dense forest of confusion and misunderstanding. I have been guided by many in this challenge. In this respect, I have been very lucky in working alongside some enthusiastic individuals who have provided both encouragement and the occasional epiphany. In particular I would like to acknowledge Dr. Tim Knowles (Manchester Metropolitan University) whom I initially, and regrettably, underestimated and whose strength of character turned out to be a great source of inspiration. I would also like to thank Dr. Wayne Martindale (Sheffield Hallam University) who was particularly generous with his time, especially at the beginning of the project. Indeed it was Wayne who provided me the first opportunity to speak at an international conference on food security and sustainability. Another individual whom I would like to give recognition to is Professor Maurice Moloney of Rothamsted Agricultural Research Centre. Despite his busy schedule, he always made himself accessible for advice, as a mentor or simply for a chat; for this I am truly grateful. There are, of course, numerous others in different fields with whom I have corresponded from the United Nations to the USDA and many more. Collectively these people have provided me with insight, clarification and direction in an effort to navigate such a difficult field without which this project would have been unworkable.

On a personal note I would especially like to thank my life-long friend Patrick Newsham who kindly donated his time and skills to provide some wonderful illustrations throughout the book. Several other close friends and loved ones have also constantly been supportive and inspirational, and these are John Hobson, Ian Aniscough, Gary Lindsay, Amanda Beedon, Alan Novak, Maureen Twoomey, David (fatboy) Gibson, Helen, Mikey, Jamie-Leigh, James, Olly, Paul Parsons, Gopi, Tao, Stewart, Ieva, Kathy, Simeone and Bart. Lastly there are two people in particular without whose encouragement and support this project would have been a non-starter: Kaur and my daughter Dalvina, both of whom I adore and who never cease to surprise me or fill me with joy.

Introduction

Hunger and malnutrition are greater than AIDS, malaria and tuberculosis combined and present the number one risk to public health worldwide. By attempting to shed light on some of the issues involved this work raises many many questions, perhaps more so than can be answered in a book such as this. Many of the questions, too, are of a fundamental nature and of such magnitude as to be of significant social and political import. In light of this, coming to grips with the constantly shifting social, scientific and political tectonics of food security is of paramount importance.

Starting with the central theme of food security and pondering on the notion for any length of time, it becomes immediately evident that it is indeed complex and multi-faceted. Having enough food means *regularly* having enough food and not only for today or tomorrow, but next month and next year. It entails ensuring people have adequate access, both physical and economic: access to food through growing it; individual subsistence farming, pastoralism or community cooperative agricultural ventures; or through purchasing it, bartering or trading for it etc. This requires thought to be given in respect of market mechanisms that facilitate trade, whether at government level, through legislative quota's, taxes, incentives and the like, or maybe at the local level with more informal arrangements. Also needed in place are the infrastructures to process, store and move such commodities whether imports or exports, regionally, nationally or globally. Further considerations, too, arise as to how the needy are to fund their food purchases? And when it fails does the humanitarian sector step in and if so to what end? To provide food and shelter or to perhaps fund longer-term development programmes with the aim of raising personal wealth and the economic health of a country? On top of this, the FAO definition (discussed in Part I) encompasses "sufficient, safe and nutritious food for an active and healthy life". Who or how are we to determine the sufficient needs of individuals, of populations and what purports to be an "active and healthy life"? Later discussion, too, shows that food and nutrition have become inextricably linked with the environment and sustainability. This has wide implications for the agricultural community at large ranging from land management and degradation to the use and range of agricultural inputs such as agrochemicals, labour and mechanisation. That's not all, the growing needs of increased populations require biotechnologies and other scientific disciplines to work closely with the agrarian community to maximise yields and the resilience of plant and crop breeds. Further ethical questions are then brought in to the debate regarding natural versus genetically modified organisms? And what of public perception and acceptance? Then there is the non-food aspect of good nutrition of among others, water, shelter, sanitation and education. And what of health? Links have been made between poor health, malnutrition and disease. This puts primary health care into the mix with everything else. What of those who cannot provide for themselves, do we once again turn to the humanitarian sector or perhaps offer myriad social safety nets?

The questions are ad infinitum, the subject is as broad as it is diverse, yet solutions raise as many questions as the problems themselves. Perhaps the most important consideration that comes to the fore is who should take responsibility: the individual, governments or institutions perhaps? Are we indeed to work collaboratively as nations or a global collective, or are the intrinsic self interests of nation states to be put before those of the wider community. In the wider debate do we provide solutions, or simply the means to solutions? Amidst all this we have to ask ourselves, at what point do we stop with the broader issues and refocus on food for the hungry? Do we give in confronting hunger head on and concentrate on the wider social context? Raise economic standards; attack poverty and gender issues; introduce extensive social welfare programmes? Or do we dilute our efforts, so to speak, and find a simpler more direct way to address the core central issue, that of hunger?

At what point do we stop projecting theoretical solutions from the comfort of the textbook and start to engage? The answer I fear is not that simple. Many of these questions are political and philosophical in nature and are to be addressed by others. Other scientific questions might perhaps be addressed in the course of time. Yet more are sociological, physiological and environmental, but perhaps the most impor-tant consideration, however, regards the ethical and moral question of doing nothing.

Concerning the nature and type of approach used in coming to grips with the holistic concept of food security and its plethora of associated phenomena, some might criticise the author for trying to apply linear logic to a non-linear problem. In my defense it can be argued that given sufficient mind mapping, paper trails and the like, all non-linear problems will eventually find a linearity of connection that can be explored and elucidated. Nor am I against the non-linear model either, on the contrary, instead this research employs both linear (logical and progressive) and non-linear thinking (creative and outside the box). In other words, both reductionist or mechanistic if you like, as well as a systems thinking approach. Far from being contradictive, this approach lends itself well to the food security phenomenon. This I feel is justified for the simple reason that food security itself is not a deterministic phenomenon, rather it is dynamic system more akin to complexity and chaos theory, where complex, heterogeneous variables involve a multitude of linear and non-linear causality and feedback mechanisms. The logic, therefore, requires that the phenomenon be studied as a whole, yet this is clearly not without difficulty. Thus a dual-istic approach is adopted. It is in the marriage of these two apparently dichotomous approaches that this book aims to offer an overarching view of a concept that continually defies comprehension.

Author

Mark Gibson has always taken an interest in the way food has been treated, not just locally but in the global context, too. For Mark there was an elemental desire to understand more of the social, political and economic tectonics of food culture, particularly in relation to issues of food security. After having trained initially in the culinary arts, Mark remained in the industry for two decades before finally dipping his toe in the academic world. Mark now lectures about many aspects related to food culture from governance to sustainability issues as well as keeping his hand-in, in the kitchen. After having completed his PhD on the issues surrounding the comprehension of the subject of food security, Mark undertook to share his knowledge in the present book.

List of Abbreviations

APEC	Asia Pacific Economic Cooperation
ASEAN	Association of South-East Asian Nations
AU	African Union
BMI	Body mass index
BMR	Basal metabolic rate
CAP	Common Agricultural Policy (of the European Union)
CCD	Convention to Combat Desertification
CDC	Centers for Disease Control and Prevention
CDIAC	Carbon Dioxide Information Analysis Centre
CEC	Commission of the European Community
CGIAR	Consultative Group on International Agriculture Research
CH$_4$	Methane
CIS	Commonwealth of Independent States: A community of States and economic union composed of 12 former constituent republics of the Soviet Union
CITES	Convention on International Trade in Endangered Species of Wild Fauna and Flora Adopted in 1973, and entered into force in 1975
CO$_2$	Carbon dioxide
CPI	Consumer price index
CSD	Commission on Sustainable Development Called for in Agenda 21 and established by ECOSOC
CSSD	Consultative Sub-Committee on Surplus Disposal
CST	Committee on Science and Technology; subsidiary body established under the UN Convention to combat desertification
DEFRA	UK Department for Environment, Food and Rural Affairs
DES	Dietary energy supply
DESA	Department of Economic and Social Affairs
EBRD	European Bank for Reconstruction and Development
EC	European Community
ECA	Economic Commission for Africa
ECCAS	Economic Community of Central African States
ECE	Economic Commission for Europe
ECOSOC	UN Economic and Social Council; one of the principal organs of the UN
EDF	European Development Fund
EEA	European Environment Agency
EFTA	European free trade area
EIA	Environmental impact assessment
ENSO	El niño southern oscillation
EPA	Environmental Protection Agency
ESA	European Space Agency
EU	European Union
FAC	Food Aid Convention
FAO	UN Food and Agriculture Organisation; the UN specialised organisation for agriculture, forestry, fisheries and rural development established in 1945

FAOSTAT	FAO Statistical Databases
FBS	Food Balance Sheets
FEWSNET	USAIDs Food Early Warning System Network
FIVIMS	Food Insecurity and Vulnerability Information and Mapping Systems
GATT	General Agreement on Tariffs and Trade
GDP	Gross domestic product
GEC	Global environmental change
GEO	Global Environment Outlook
GHG	Greenhouse gases
GIEWS	Global Information and Early Warning System on Food and Agriculture
GLP	Global Land Project
GMO	Genetically modified organism
GNI	Gross national income
GRID	Global Resource Information Database
GWP	Global warming potential
HDI	Human development index
IBRD	International Bank for Reconstruction and Development; also known as World Bank
ICCP	International Climate Change Partnership
IDA	International Development Agency
IEA	International Energy Agency
IFAD	International Fund for Agricultural Development
IFPRI	International Food Policy Research Institute
IGBP	International Geosphere Biosphere Programme
IGC	International Grains Council
IGO	Intergovernmental Organisation
IMF	International Monetary Fund; established in 1945
IMR	Infant mortality rate
IPCC	Intergovernmental Panel on Climate Change
IRC	International Rice Commission
LDCs	Least developed countries
LIFDC	Low-Income Food-Deficit Country
MDGs	Millennium Development Goals
MENA	Middle East and North Africa
NAFTA	North American Free Trade Agreement
NASA	National Aeronautics and Space Administration
NEC	Not elsewhere classified; used in the UN FAO food balance sheet reporting
NGO	Non-governmental organisation
OCHA	Office for the Coordination of Humanitarian Affairs
ODA	Official Development Assistance
OECD	Organisation of Economic and Cooperative Development
OHCHR	Office of the United Nations High Commissioner for Human Rights
OPEC	Organisation of the Petroleum Exporting Countries of eleven developing countries whose economies rely on oil export revenues
POPs	Persistent organic pollutants
PPP	Purchasing power parity
SAC	Scientific advisory committee
SAP	Structural adjustment program
SCN	United Nations System Standing Committee on Nutrition

SOFI	State of Food Insecurity
UN	United Nations
UNCCD	UN Convention to Combat Desertification
UNCTAD	UN Conference on Trade and Development
UNDG	United Nations Development Group
UNDP	United Nations Development Programme
UNEP	United Nations Environment Programme
UNESCO	UN Educational, Scientific and Cultural Organisation
UNFCCC	UN Framework Convention on Climate Change
UNFPA	United Nations Population Fund
UNHCHR	United Nations High Commissioner for Human Rights
UNHCR	United Nations High Commissioner for Refugees
UNICEF	United Nations Children's Fund
UNU	United Nations University
USAID	US Agency for International Development
USDA	United States Department of Agriculture
WFP	World Food Programme
WFS	World Food Summit
WHO	World Health Organisation; UN specialised agency
WMO	World Meteorological Organisation; UN specialised agency
WRI	World Resources Institute
WTO	World Trade Organisation

Part I

Food Security: What Is It, How and Who Does It Affect?

How little we know of what there is to know.

Ernest Hemingway
For Whom the Bell Tolls

The aim of Part I is to present a global picture of food insecurity as it exists today. However, as becomes increasingly evident as the book unfolds the issues of food security are vast. In light of this, the goal at this point is not to overwhelm the reader with all of the issues involved, contentious or otherwise, but instead to lay a bedrock upon which fuller later analysis can be embedded. Therefore, in presenting the notion in its narrowest form at this point the benefits are twofold; first it helps to introduce the essential ideas that underpin the concept for later discussions, and second, it aids in the overview of the current situational analysis without overly clouding the picture with food security's own baggage. Moreover, in proffering a blunt slice of reality as it exists, we offer insights into some of the difficulties in defining, quantifying and measuring food security. Furthermore, by equipping the reader with an understanding of the more vulnerable individuals, groups and regions and by examining the effects of poor security on individual's health, mental capacity and future prospects, it is hoped that a fuller, more overarching perspective of the difficulties in addressing hunger and malnutrition issues will become apparent. As such and as far as possible, the following section importantly tries to be more descriptive rather than analytical. In this undertaking the information is also presented where practical, value-free: that is to say without opinion or inference. This is in an effort not to detract from the stark and essential reality of the facts; there is plenty of room for analysis later. Also an initial cursory examination of the broader aspects of the phenomenon introduces several key ideas, concepts and inherent issues that collude to undermine universal consensus on the problem and its solutions. These are introduced here in Part I simply as they arise and solely in an introductory capacity.

Furthermore, in placing the problems of food security and general nutrition in the wider context of global mortality and morbidity, it helps place emphasis not only on the problems of hunger itself but also on associated diseases resulting from poor nutrition practices such as overnutrition and many ischemic heart diseases (IHD). This is because IHDs are becoming an increasing global dilemma closely associated with problems of reduced blood circulation resulting from poor diets and physical inactivity, among other things. Generally speaking then overall modalities of death can be summarised

in three categories: communicable diseases (CD), non-communicable diseases (NCD), and injuries. The short list is presented in Table I.1, and the full list can be seen in Appendix B.

In the mortality figures, as with many other statistics, the latest, most up-to-date data are often lagging behind current dates by several years. The latest available mortality figures for 2011 for instance are the 2004 figures last updated in 2008 and applied retrospectively. In this way, data need to be carefully considered in the light of estimations, predictions and suppositions.

With this in mind in 2004 while total global mortality was over 58 million, over the last few years, this figure has consistently hovered around 52–58 million deaths per year. It has also been estimated that a good proportion of those (between 10 and 11 million) is related to hunger. Moreover, of these 10–11 million, up to 92% were specifically related to chronic hunger and malnutrition, while only 8% were attributable to 'emergencies'. Importantly too, 60% of this annual figure is usually found to be women (Fleshman 2006; ACF 2010; Rajaratnam et al. 2010; WHO/GHO 2010). These figures are summarised in Table I.2.

As will be seen throughout the book, hunger and malnutrition affect, and are affected by, many conditions and diseases, so much so that an estimated 17%–21% of all annual deaths worldwide are attributable to hunger-related deaths (HRD) (calculated from the above figures). Presently each year, of the

TABLE I.1

WHO Mortality Categories

World Health Organization Categories of Mortality		
All causes		58,676.50
	Lower uncertainty bound	54,095.90
	Upper uncertainty bound	64,433.00
Communicable, maternal, perinatal and nutritional conditions (CD)		17,951.30
A.	Infectious and parasitic diseases	9,508.70
B.	Respiratory infections	4,254.00
C.	Maternal conditions	525.8
D.	Perinatal conditions	3,176.90
E.	Nutritional deficiencies	485.9
Non-communicable diseases (NCD)		34,950.90
A.	Malignant neoplasms	7,411.60
B.	Other neoplasms	162.5
C.	Diabetes mellitus	1,135.90
D.	Endocrine disorders	301.8
E.	Neuropsychiatric conditions	1,259.50
F.	Sense organ diseases	4.3
G.	Cardiovascular diseases	17,043.10
H.	Respiratory diseases	4,030.50
I.	Digestive diseases	2,040.80
J.	Genitourinary diseases	924.9
K.	Skin diseases	67.2
L.	Musculoskeletal diseases	126.4
M.	Congenital anomalies	439.3
N.	Oral conditions	3.2
Injuries		5,774.30
A.	Unintentional injuries	3,899.90
B.	Intentional injuries	1,639.10

Source: Reprinted from the *Global Health Observatory*, World Health Organization. http://www.who.int/healthinfo/statistics/en/, 2010. With permission.

TABLE 1.2

Approximate Recent Annual Mortality Characteristics

Deaths	Approx (millions)	Notes
Total annual	52–58[a]	
Hunger-related (malnutrition, undernutrition, associated disease)	10–11	92% due to chronic issues and 8% due to emergencies (WHO 2010); 50% of these are children (UNICEF 2008; WHO/GHO 2010) and 60% are women (ACF 2010).
Total child mortality (under 5 years)	8–9	Undernutrition contributes to more than one-third of child deaths (UNICEF 2009; WHO/GHO 2010). Malnutrition and hunger-related diseases cause 60% of the deaths in developing countries (UNICEF 2007) or 53% of child deaths worldwide (UNICEF 2006)

[a] Depending on year.

10 million or so people that die of HRD, over 50% of these are children. Furthermore while over one-third of these deaths are exacerbated by undernutrition, collectively malnutrition and hunger-related conditions actually account for over 53% of worldwide child deaths (increasing to 60% in developing countries) every year (Caulfield et al. 2004). This equates to one child's death every 5–6 seconds (UNICEF 2008; WHO/GHO 2010).

The good news is these trends are set to change while the bad news however is that other causes of food-related deaths are predicted to take their place. For instance, WHO forecasts that by 2030 tuberculosis and malaria are set to decline along with other infectious diseases; however, among the increases, HIV/AIDS, stroke and IHD are expected to be significant. This has consequences for the many annual nutrition-related deaths for it is unclear how the forecast trends will affect the 'hungry' numbers the projected increase in strokes and ischaemic diseases will be a reflection of, among other things, increasing poor nutritional diet, overnutrition and lack of exercise.

1

Food Security: What Is It?

So what is food security? At its very basic, food security deals with the notion that everyone has enough food to eat, not just today but every day. However as is discovered later, there is ample food being produced annually, so surely the simplest solution would be to just hand out the surplus food to those in need; would it not be? Well, in fact, in emergency situations, this is in essence what actually happens; therefore, it is a necessary and valuable tool short-term measure in such circumstances. Unfortunately, in non-emergency situations, only limited assistance is given and this is where things become problematical. There are several issues involved here, and a brief glimpse is given in the following discussion. There are economic issues concerned with the effects of introducing free or subsidised food into the local markets that could adversely affect local trading conditions and prices. Such action could destabilise the local markets and have knock-on effects on local livelihoods and future production, further aggravating a delicate or unstable situation. There are also policy considerations where if this practice were a regular occurrence, domestic governance might be tempted to cede responsibility to the international community, or perhaps more likely, the local community would come to depend on such handouts, reducing motivation to fend for themselves. This is particularly important as the prevailing development paradigm is not about handouts rather it is about empowering people and nations with the tools to provide for themselves. Other reasons too are discussed later however, suffice to say that long-term free or subsidised food in non-emergency situations is clearly not a solution.

The challenge then is twofold: a duality of purpose that motivates all stakeholders from the policymaker to the frontline warrior. First, there is a clear need to facilitate a position of emergency fallback where the infrastructure and food are in place in times of emergencies or crises. This entails an enabling political and institutional environment as well as the logistical capabilities on the ground. It also requires monitoring and an understanding of how events develop into emergencies. In this situation further understanding too needs to be sought with reference to appropriate and adequate foods as well as health care for the worst affected. This requires collaboration of effort, consensus of methodology and expediency of action. Emergency situations aside the second important condition that must be strived for is prevention: prevention of food insecurity through the creation and improvement of conditions conducive to empowering the disaffected. In short we need to help people achieve their own security of food through investment and development. This is where it gets tricky. Apart from the main protagonists of poverty, political will, conflict and disasters, there are so many other factors involved that collude to further exacerbate the problem or maintain the status quo. Moreover, not enough is currently known about how all these variables combine or interact to promote food insecurity. Some like climate change might have obvious implications on the pattern of food production, or so we might think but what of the other variables and how do they manifest exactly? How does the philosophy of economic development affect poverty alleviation and do we fully understand it? What of migration or urbanisation? How about natural resources or biofuels and what else might we be missing? These questions and more need to be considered and addressed if any real progress is to be made in allowing people to determine their own futures in respect of the security of food.

Food security then is about understanding food *in*security; by reverse engineering the social construct, students of the phenomenon aim to break down the problem into its component parts and to figure out how to put them back, only better. An unforgivable oversimplification perhaps but sufficiently descriptive. It is important to note too that when literature talks of food security, it is invariably directed at the developing nations yet food insecurity is not just a problem of the poorer countries. Many developed economies suffer the same inequalities of nutritional distribution and poor diets as their poorer counterparts, although usually to a much lesser extent. Nevertheless, this in turn creates enormous public health problems that need to be addressed by both developing and developed countries alike. Thus, the notion of adequate food for all today is a global phenomena impinging on many aspects of every human being's daily life. Not surprisingly, then the idea encourages many interpretations and draws on just as many disparate scientific disciplines from biotechnology to political ideology and from economic development philosophy to environmentalism.

In summary then from the above introduction it can be seen that the seemingly simple notion of food security is very quickly complicated by difficult or uncomfortable realities. This reality also leads to further questions of governance and the niggling question of who is ultimately responsible for the people's security of food: the individual, the state or one of the many multilateral institutions?

1.1 By Whose Standards?

This is probably the hardest and singularly most frustrating aspect in understanding the phenomenon. For, while the United Nations (UN) for many assume the lead role in global governance, the United States, European Union (EU) agencies and various small but influential think tanks, policy analysts and charitable organisations all have major inputs which cannot, nor should not, be ignored. Indeed, much vitally needed input with regards to action, research and oversight can and do often originate from outside the UN family. In fact, in this regard, just as many others see the UN not so much as the global figurehead of the fight against hunger but instead a much-needed clearing house of research and current thinking; although even this role has been called into question of late.

Such issues are discussed later, but the point being for now is that without global consensus and without a single source of stewardship, it becomes very difficult to educate ourselves in the discipline of food security. Moreover, without an agreed single clearing house of information, it can be just as arduous to stay abreast of the latest research or understanding with regard to food security. Publications, journal articles, studies, consultations all take time to filter into general discussion. Then there are the conferences, the symposia and technical briefings which debate everything from definitions to methodologies and from concepts to targets. Objective and subjective agreements are made; multilaterally and unilaterally, decisions are questioned and policy strategised, all the while the latest ideologies fronted for general acceptance might include or exclude previously contended issues. It should come as no surprise then that given this melee it becomes increasingly difficult to grasp the extant level of comprehension of the subject or to fully understand by what standards we are measuring, analysing or comparing data, ideas or current thinking. The issue of consensus or competing analyses in the face of what sometimes might seem overwhelmingly voluminous, complex and contradictory information sets needs to be addressed to a good degree of satisfaction before any real progress in the full understanding of food security can take place. That said, we must start somewhere and in this endeavour we now take a look at the many different understandings of the concept. The modern concept is, and as has been repeatedly alluded to, complex and beyond the brief aim of this section; however, as mentioned, the previous overview is important and will aid in answering some fundamental questions. The next task is then is to identify what the modern concept of food security is, and from that we can examine what it entails and how this knowledge is then used.

1.2 Popular Definitions

The notion of food security for many is replete with confusion and subjectivity (McCalla and Revoredo 2001). It is a broad and compound concept, the outcome of which is determined by the interaction of many political, social and economic considerations which may include geophysical, socioeconomic and biological factors among others (Riely et al. 1999). With this in mind there is an inherent difficulty in precisely defining the nature and scope of the food security construct, and depending on who one reads, there are numerous definitions, interpretations and permutations of the ideas involved (Maxwell and Frankenberger 1992; Riely and Mock 1995; Hoddinott 1999). In this respect, perhaps the most often quoted, and for very good reason, was the extensive study of household food security definitions by Maxwell and Frankenberger in 1992. In an attempt to collate and clarify the concept, the study drew on all the disparate definitions and permutations of the time. Aided by Smith and Pointing, Maxwell and Frankenberger managed to collate a staggering near 200 (194) unique definitions. In their study's summary however it was stated that while many definitions shared common strands, many other definitions were being specifically tailored to suit individual needs (Maxwell and Frankenberger 1992). Maxwell and Frankenbergers's was not the only attempt to pin down the concept either and over the years many other attempts have aimed to qualify the concept either directly or through its measurement. While these are too numerous to go into here suffice to say that Maxwell's findings were not unique (Hoddinott 1999). Indeed, more and more of the literature reviewed in the undertaking of this book revealed a continual trend to redefine food security in the researcher's own image. A little unfair perhaps but once again illuminating. Indeed such continual relentless attempts at defining the concept from a less than objective viewpoint has had the unintended effect of creating weary readers who look for motive in many definitions rather than substance. From an academic standpoint too this cannot be good. Although that said its not all bad news as nowadays common trend shows that many of the important studies are beginning to converge on a small number of widely accepted definitions on offer. With that in mind and for introductory purposes I offer a few of the most widely quoted and most up-to-date definitions that appear throughout current literature. This is not meant to be a definitive list; it is simply a convenient place to start. Of the many definitions that exist perhaps the two most widely recognised definitions are those of the UN and the United States.

1.2.1 FAO Definition

The Food and Agriculture Organization (FAO)'s food security definition is based on the 1996 UN FAO 'Declaration on World Food Security and World Summit Plan for Action' definition, which was then further expanded in 2001. Currently the FAO definition reads as follows:

> Food security [is] a situation that exists when all people, at all times, have physical, social and economic access to sufficient, safe and nutritious food that meets their dietary needs and food preferences for an active and healthy life. (FAO 1996; SOFI 2001)

This definition is one of the most commonly used or quoted. Users include the World Food Programme (WFP), the World Bank, the European Union as well as the International Fund for Agricultural Development (IFAD) and numerous others (EU 2006; IFAD 2009; EU 2010).

1.2.2 US Definitions

The United States on the other hand uses several definitions. Both the US Department of Agriculture (USDA) and the US Agency for International Development (USAID) define food security depending on its particular purpose. In general, the USDA focuses on national hunger issues while USAID mainly operates on the international front.

1.2.2.1 USDA Definition

The currently used USDA definition was originally defined in 1990 by Anderson of the US Life Sciences Research Office, which reads as follows:

> Access by all people at all times to enough food for an active, healthy life. Food security includes at a minimum: (1) the ready availability of nutritionally adequate and safe foods, and (2) an assured ability to acquire acceptable foods in socially acceptable ways (e.g., without resorting to emergency food supplies, scavenging, stealing, or other coping strategies). (Andersen 1990; USDA 2009)

1.2.2.2 USAID Definition

USAID's current general classification is based on the USAID Policy Determination #19 from 1992:

> When all people at all times have both physical and economic access to sufficient food to meet their dietary needs in order to lead a healthy and productive life. (USAID 1992a, 2010a)

1.2.2.3 PL 480: Food for Peace

Officially called the Agricultural Trade Development and Assistance Act, Public Law 480 (PL 480) offers a more flexible definition, again based on Policy Determination #19 and necessary in their view to allow for a range of possible interventions:

> Access by all people at all times to sufficient food and nutrition for a healthy and productive life. (USAID 1995)

With regard to the above definitions of the United States, the USDA Economic Research Service (ERS) also defines the food insecure as those consuming less than 2100 kcal per day (USDA/ERS 2010).

Not wholly dissimilar, the above group represents the more convergent of the many available definitions. As is evident, the definitions of both the FAO and the United States share strong commonalities. One interesting aside to all this was seen in 2003 with a rare introspective analysis by the FAO on the use of such definitions. In their view there was a growing propensity for people to apply contrasting significance to such definitions suggesting on the one hand that they have been used as ' ... little more than a proxy for chronic poverty ... ", while on the other hand there has been a tendency for international committees to apply

> ... an all-encompassing definition, which ensures that the concept is morally unimpeachable and politically acceptable, but unrealistically broad. (FAO 2003)

This is not surprising as the definitions by themselves, once stripped of their simplistic façade, can in fact be seen to be complex and unwieldy and open to subjective interpretation.

1.3 Understanding Food Security

1.3.1 The Four-Pillars Model

In an effort to lay the foundation of the concept without getting drawn in at this stage into the nuances and complexities of the subject, we introduce a simple framework used by the FAO (Figure 1.1). However, when it comes to modelling the food security phenomenon, such an exercise, can also be just as confusing with erroneous linkages, unclear causality and implicit rather than explicit complexities. This is because over the years as the paradigms, or the way the food security concept has been viewed, have

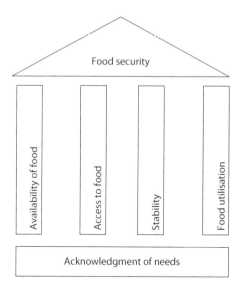

FIGURE 1.1 The FAO four-pillars model of food security.

changed, different approaches to understanding the ideas have been innovated. As the discussions in the book unfold, these will become evident; suffice to say for the moment that food security paradigms are in constant shift and evolve or co-evolve, leaving a legacy of alternative viewpoints such as the livelihoods approach; the household perspective; rights-based approaches; entitlement mapping; food sovereignty and many more (WFP 2009b). For now though the simple FAO four-pillar framework is sufficient for introductory purposes and helps to break down the concept into its various component parts. Moreover the model itself is it is also sufficiently well defined and descriptive for the organisation of ideas within this book and as such will be used throughout. The interpretative focus of this model divides the definition or the idea of food security into a set of distinct but inter-related constituent dimensions: the four pillars of food availability, food access, stability or vulnerability and food utilisation. So fundamental to the ideas of food security that these four pillars have come to be accepted across institutional divides; although this was not always the case as up until recently the USAID only explicitly offered only three dimensions in their definition waiting until 2007 to re-define it to include the fourth pillar of stability (Riely et al. 1999; CFS 2000; Devereux et al. 2004; Scaramozzino 2006a).

The various pillars or dimensions offered by the FAO are now outlined in a little more detail.

1. *Availability*: This dimension tackles the availability of food in sufficient quantity and of appropriate quality. Food can be supplied through various means from domestic production and/ or imports together with any donations or food aid. This dimension also implies variety and physical availability of food at farms and in local markets as well as adequate transport infrastructure through to storage and processing technologies. Moreover, all such foods for successful food security are needed on a consistent basis and on that point the USAID also suggests that such availability requires that this food needs to be within reasonable proximity too (Riely et al. 1999; FAO 2006a; Maunder 2006b; CFS 2007; FANTA2 2010).

2. *Access*: In interpreting the access component of the definition, agreement is more or less universal. Access involves a package of entitlements used to acquire and maintain appropriate foods for an adequate diet and nutritional level. Such entitlements include all those measures that an individual has command over: whether directly, through adequate income, barter and exchange; or indirectly via social arrangements, either at the community or national levels such as kinship, welfare systems, traditional rights, access to common resources and once again food aid (Sen 1981; Riely et al. 1999; FAO 2006a; USAID 2010b). In short, access refers to the ability of individuals or households to purchase or produce sufficient food for their needs. This also entails the adequate integration and functioning of both local and international markets to effectively supply the food. (Maunder 2006a).

3. *Utilisation*: As far as the FAO, the United States and the EU are concerned, utilisation is implied in their respective definitions and is explicitly documented as a dimension of importance. Utilisation is interpreted as ensuring the efficient and maximisation of food to its fullest potential and directly refers to people's ability to absorb the nutrients from the food they eat. Research shows this is closely related to factors such as a person's health status as well as their access to health services. Issues concerned with utilisation then include quality and safety of food and processing and storage techniques (IFAD 2009). Just as importantly all three institutions emphasise the importance of non-food inputs whereby optimum utilisation necessitates the need for proper health care, clean water and sanitation services. On top of this there is also a need to consider adequate knowledge of nutritional and physiological needs and not solely at the individual level either but more increasingly at the level of provision too (Sen 1981; Riely et al. 1999; FAO 2006a; USAID 2010b).

4. *Stability/Vulnerability*: Becoming more and more entrenched in food security literature is the notion of stability or vulnerability. Although not a new idea, it is gaining popularity as a risk management tool in the fight against hunger. To be food secure means to have access to adequate food at 'all times'. This implies that food security can be lost as well as gained, in this respect consideration must be given to vulnerability within all dimensions of availability, access and utilisation. Stability addresses the risks inherent, impending or conditional that impact negatively on the availability, access or utilisation of food. This might refer to

the risk of sudden shocks, for example economic or climatic crisis or cyclical events such as seasonal or labour fluctuations and their subsequent impacts on food security. Vulnerability also can be thought of as the inability to manage risk, and in this sense the time element also becomes a critical factor (FAO 2006a; Maunder 2006a; CFS 2007; USAID 2007).

The above 4-pillars framework briefly highlights the breadth and extent of the multi-dimensionality of food security. This inclusivity was even touched upon back in 1992 in Maxwell and Frankenberger's superb study when in conclusion they suggested that in and of itself food security cannot be seen in isolation of its much wider societal context (Maxwell and Frankenberger 1992). Indeed on this point there is a growing convergence of ideology whereby security of food is being viewed more and more as a constituent of the broader concept of the social welfare construct including inter-alia nutrition security,* health care, poverty alleviation, education and human rights (Ruxin 1996; IFAD 2009). As a result and of late, agreement is being sought on the need to define food security in respect of these various dimensions (FANTA2 2010).

Before we take a closer look at the different types of food security, it is perhaps worth examining the different levels of the concept first.

1.3.2 Different Levels

In the conceptual framework of food security drawn up by the Inter-agency Working Group (IAWG) at FIVIMS, they identify '15 information domains' as they call them. These aid in identifying and understanding the causes of poor food consumption and nutritional status. The Various levels can be seen in Figure 1.2 and range from the individual to the household to the national. These in turn are underpinned by several contextual factors whose understanding and interplay are the subject of much research (FIVIMS 2008).

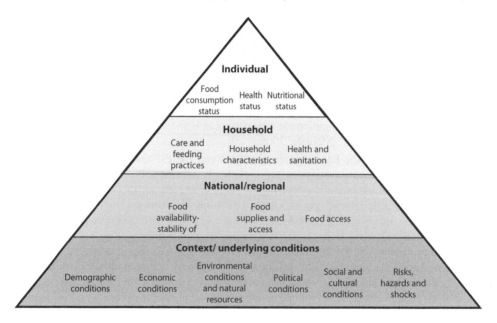

FIGURE 1.2 Different levels of food security. (Based on ideas from the Committee on Food Security and the FIVIMS initiative; CFS [1998]. Guidelines for National Food Insecurity and Vulnerability Information and Mapping Systems [FIVIMS]: Background and Principles. Committee on World Food Security: Twenty-fourth Session, Rome, Committee on World Food Security. FIVIMS, *The FIVIMS Initiative: Food Insecurity and Vulnerability Information and Mapping Systems: Tools and Tips*, FIVIMS, Rome, 2008.)

---

* Food security is necessary but often insufficient for ensuring nutrition security, as adequate food does not necessarily equate with adequate nutritional content (IFAD 2009).

Whatsoever the policy goals or interventions, understanding of the macro and micro aspects of the dynamically changing food security groups or levels is vital to future progress in this field. With this in mind. research into the types of food security suggest that on the ground it can exist in several distinct typologies. This was brought about largely due to the highly influential World Bank report, 'Poverty and Hunger' in 1986 and which was further elaborated on in 1999 by the Food Information and Vulnerability Mapping Systems (FIVIMS) then subsequently endorsed by the World Committee on Food Security (CFS). As a result, food insecurity can now be categorised as taking one of three forms: temporary (temporal), chronic (continuous) or cyclical (seasonal).

1.3.3 Chronic, Temporal, Temporary or Cyclical

This widely accepted distinction allows policy makers to distinguish the endemic or structural aspect of insecurity from the more transient or unexpected shocks often associated with the sudden onset of emergencies and allows the better targeting of more appropriate programmatic policies (World Bank 1986; USAID 1992b; FAO 2000).

1.3.3.1 Chronic (Continuous)

When an individual or household is persistently unable to meet their food requirements over a long or protracted period of time or where such provision is constantly marred by continuous, temporary good or bad incidents of shortfalls, it is known as chronic food insecurity. Such chronic insecurity is often endemic or structural in nature and closely associated with poverty or people with low income. Importantly chronic food security by its structural nature can often be foreseen or predicted (World Bank 1986; USAID 1992b; FAO 2000; ActionAid 2003; Barrett and Maxwell 2006; EC/FAO 2008; IFAD 2009).

1.3.3.2 Temporal (Temporary or Transitory)

Transitory food insecurity on the other hand concerns shocks that cause a temporary decrease in access or availability of food below requirements. Such periods of intensified pressure can be caused by any one or a combination of many factors including, but not exclusively, exposure to high incidence of natural disasters, economic collapse, conflict, production shortfalls or crop failures; flooding and drought; price changes; or loss of income through illness or unemployment. Urban dwellers in this instance are particularly susceptible to highly unstable markets and agricultural producers (World Bank 1986; McHarry et al. 2002; FAO 2006b). Such transitory phenomena may also, regardless of any pre-existing long-term chronic problem, affect households without warning (IFAD 2009). Temporary sharp reductions too, it has been suggested, undermine long-term development strategies which can take years to recover (FAO 2003).

1.3.3.3 Cyclic (Seasonal)

The third consideration is seasonal food insecurity, which generally sits between or alongside chronic and transitory. It is chronic or predictable in the sense that it is inherent or inbuilt into existing patterns of endemic hunger and transitory in that it can often be associated with seasonal fluctuations in cropping patterns or employment trends. In rural communities it is not uncommon to talk of 'hungry periods' and 'periods of plenty' in the agricultural calendar (FAO 2000; EC/FAO 2008; Gerster-Bentaya and Maunder 2008).

Note that the above causal factors do not necessarily fall exclusively in one camp or the other, conflict for instance can be the cause of both sudden flares of temporal food insecurities or of a more long-lasting attritional nature. Lastly, whether chronic, temporal or seasonal, individuals, households, communities and nations can only be considered food secure if they have protection against all kinds of insecurity (IFAD 2009).

1.3.4 Different Things to Different People

As has been suggested, food security is a multi-faceted phenomenon involving many variables and existing in several typologies, which can and often do originate from a number of possible causes. Further, compounding a full understanding of the subject is the fact that security is often studied from a particular specialist perspective. That is to say, with regard to food security, the following is worth noting:

- Agriculturalists might concentrate on increasing production through better land management and/or maximising yield potential with improved crop varieties; cross breeding and hybridisation, disease and pest control measures, etc.
- Politicians and economists look to implement policy and to understand the economics of food security. As food security issues wax and wane, certain politically popular aspects of the wider construct might take prominence; therefore, understanding the economic and social trade-offs between policies is expected to balance needs with objectives and reflect the prevailing economic and political sentiments of the day.
- Sociologists might look for causality by studying the cultural and societal influences of inse-curity; population growth and how poverty affects malnourishment; the psychology of income growth and changing dietary habits; as well as the dynamics of rural-urban change among others.

This is just a snapshot of a few perspectives, and the difficulty lies with gaining an overarching per-spective as you would not expect an economist to study the effective nutrient uptake of hybrid plants, nor would he be asked to comprehend the subtleties of culturally acceptable food aid. The point here is that while researchers specialising in particular disciplines can and often do dip into other disciplines in the context of their research, food security analysis is such that a cursory dip so to speak would never be sufficient to properly comprehend its many facets. In this regard, food security tends to suffer (FAO 2000). While this might seem a little flippant or generalised, it is not to denigrate the work of these vitally important disciplines. Instead it is illustrative of the often insular dimension in which food secu-rity research is undertaken and also serves to further highlight the lack of an appropriate overarching perspective. However there are some exceptions and it seems the few who do take a holistic approach to understanding the phenomenon are institutions, think tanks and agencies such as the UN, the Overseas Development Institute (ODI), the International Food Policy Research Institute (IFPRI) and various humanitarian agencies or foundations like OXFAM, CARE, the Ford Foundation and the Red Cross (Darcy and Hofmann 2003). However, even many of these tend to have priority areas focusing on one aspect to the detriment of others. So with such widely differing perspectives, it is perhaps not surprising then that opinions of such things as causality and solutions come just as often from as many different directions as there are viewpoints (FAO 2002). For some, food security represents the ability to trade, supply or simply purchase food in a global marketplace unimpeded by barriers such as preferential trade agreements, import quotas or export tariffs. Others refer to food security as the right of a country's food sovereignty or its ability to directly or indirectly exercise control over their food supplies; to determine what is produced and under what conditions. Yet others still see hunger and nutrition issues as central to an individual's basic human rights. In the end and irrespective of one's personal or professional view-point the salient point here is that food insecurity is a multi-headed beast with many different masters (Earth Summit 2002; ActionAid 2003; FIAN 2010).

As mentioned earlier, there are different ways in which the concept can be viewed. A few of the more common approaches are looked at in a little more detail.

1.3.4.1 Sustainable Livelihoods and Asset-Based Models

An important viewpoint within the food security concept is the increasingly popular sustainable liveli-hoods approach (SLA). A livelihoods framework is people-centric and stresses the relationship between people's vulnerability to food security and their subsequent coping mechanisms. By extension, this aspect focuses on the means by which people or groups access the food they need and is often used as an alternative to traditional poverty analysis (DFID 1999; Frankenberger et al. 2002; Devereux et al. 2004; Shoham and Lopriore 2007). Livelihoods as a development strategy was introduced around the 1970s to indicate a more holistic view of access to food security and was in stark contrast to the contem-porary dominant thinking, which up till that point had treated livelihoods as merely synonymous with paid employment or cash generation (Chambers and Conway 1991). In this way, livelihoods became much more than just wages and money it encompassed the multifarious activities and sources of support (including cash) that made up the whole livelihood approach (Chambers 2005). This perspective in effect deals with the 'means of making a living' or in the words of the Committee on Food Security

> Encompass the capabilities, material and social resources, and activities required for a particular means of living. (CFS 2000)

Furthermore, this approach effectively places the individual back at the centre of the issue and focuses on supporting access, developing and enabling policies (DFID 1999). It also involves understanding trends; resources—economic, natural, physical, human, social and political capital; institutional and organisational processes such as government, civil society and private sector bodies; and access strategies whether productive, exchange and coping. Essentially,

> The sustainable livelihoods approach provides an analytical framework that promotes systematic analysis of the underlying processes and causes of poverty … [providing a] framework that helps provide structure to debate and analysis, … . (DFID 1999)

Ultimately it can be seen that this approach has gained much momentum since the mid-1970s and it is one that has increasingly come to dominate poverty and food insecurity analysis contributing greatly to the design of poverty and famine prevention policies today (Devereux et al. 2004).

1.3.4.2 SLA, Endowment Sets and Entitlement Mapping

Entitlement mapping refers to the asset base (or the endowment set) which can be converted into food. This concept builds directly on the work of Amartya Sen (discussed later) whose seminal work in the 1980s formally introduced the notion of entitlements. This view also marked a departure away from food production as the main focus of food security and shifted the emphasis towards the affordability of food or the exchangeability of assets. Entitlement mapping comprises various input–output ratios based on three endowment sets: a production component; an exchange component; and a transfer component (IFAD 2009). These can be likened to the food access construct of the FAO, which suggests that people can grow or produce their own food (production); buy or trade for their food through employment and other remittances (exchange); and through social networks, family or institutional aid (transfer). Collectively, this endowment set determines a household's ability to acquire food which has important implications in today's environment where the growth of urban neighbourhoods places extra emphasis on access and incomes as fewer people tend to grow their own food (Ericksen 2008). A good illustration of the entitlement mapping process in action is the use of 'capital assets' in the livelihood's approach. In this model, SLA identifies the various 'capital assets' that combine to make up the entire individual or household's means at their disposal (Table 1.1). This in short provides a more forward-looking proactive assessment of an individual's or household's potential food security situation.

The livelihoods, asset base and entitlement mapping approaches have all gained increasing importance over the years and have provided much valuable insight into food security's associated concepts of

TABLE 1.1

Capital Assets of the Livelihoods Strategy

Natural capital	Boasts the natural capital base from which livelihoods are derived: environmental resources such as land and water, livestock or wildlife.
Financial capital	These can be liquid assets such as savings, credit, remittances, pensions or assets that can be sold.
Physical capital	Physical capital refers to the infrastructure of transport, energy and water as well as needs of shelter and communication, etc.
Human capital	Consist of a person's skills, their knowledge or ability and considerations of health.
Social capital	Is the access to social resources such as networks, relations and the wider supportive societal institutions whether, health, financial or other.

poverty and vulnerability. Of particular note, according to Ericksen (2008) and others, is the notion that households have multiple objectives beyond the immediacy of food security (Ericksen 2008; Gerster-Dentaya and Maunder 2008). That is to say individuals or households might go hungry or sacrifice short-term food stability in order to preserve other household assets. This is in contrast to the traditional hierarchical 'food first' view where food security was seen as primacy and which superseded all other needs. This idea that food security is only one of many objectives is, according to Gerster-Bentaya and Maunder (2008), perhaps the main reason why the livelihoods approach has gained such a great following over the last decade or so (Gerster-Bentaya and Maunder 2008).

1.3.4.3 Advantages and Limitations

The livelihoods approach to food security analysis has several obvious advantages and limitations. Perhaps its greatest advantage is its people-centric perspective that focuses attention on an individual's or household's various assets on which they rely to achieve their personal objectives. The concept further recognises multiple influences at work and seeks to understand these relationships and their impact upon livelihoods. It is also multi-levelled in that it crosses traditional geographic and social divides; that is to say, it is just as applicable in the deserts of the Sahara as in the cities of India and it is just as relevant between rural and urban areas and also in different economic sectors (DFID 1999; ELDIS 2011). Perhaps its greatest advantage however might also be its greatest weakness. This is because countries are essentially heterogeneous in which many alternative livelihoods co-exist. Although having said that, in some aspects this approach can be applied in broad strokes to certain sections of the communities, for instance with regards to fishermen, factory workers or pastoralists, etc. A further limitation in this approach is that it makes the constant demand for aggregated statistical analysis at the national level difficult to fulfil (Devereux et al. 2004). Others however reject this limitation and assert that SLA approach is a well-placed alternative to traditional poverty analysis and has recognisable potential for scaling up to the macro. Ultimately however, despite being a sub-national metric, SLA allows for greater insight into causality of food insecurity over time and allows for better understanding of people's coping strategies in times of need. Furthermore, the FAO sees livelihoods as a more personal or tailored assessment over which needs can be better ascertained rather than simply

> … importing blueprint development models that often ignore or even undermine these positive features. (FAO 2006a)

1.3.4.4 Household and Community Food Security

Similar to SLA, the household perspective is another way of looking at food security at the level of the individual and household unit. Popularised by Maxwell and Frankenberger in the 1980s and 1990s, this move also shifted focus away from the simple food supply paradigm and concentrated on the series of entitlements that allowed them to pursue food either through purchase, trade, barter, grow themselves or acquire through family connections. As previously discussed, this came to be known as a household's asset base or its endowment set however, although similar to SLA in principle the household perspective goes further and also examines things such as the intra-household dynamic of food security. This is an important addition to the SLA approach in that it recognises that households are not equitable places. Gender issues for instance still prevail in many parts of the world where women of the household do not necessarily command an equal share of the incoming food; this might equally apply to children who only are fed after the males of the unit are fully sated. On top of this, there are also the non-food issues of security dealing with things such as water, sanitation, education, health services and care practices of the household, which are all important contributions to overall household food security (Maxwell and Frankenberger 1992; IFAD 2009).

1.3.4.5 Vulnerability

Another perspective in which the phenomenon of food security is increasingly viewed is from the standpoint of vulnerability. This exposure to risk can be mirrored by an individual or a group's ability to cope and lends itself well as a tool to be used in conjunction with things like entitlement mapping or livelihoods analysis (FAO 2000; WFP 2009b). Although while entitlement mapping is a good example of this type of perspective, vulnerability analysis in general is much wider in breadth and more inclusive of external influences outside of the control of the SLA or the capital asset model. This might involve identifying vulnerable areas to frequent droughts or floods; to the political stability of a region; or perhaps to the instability of financial infrastructure. The biggest advantage of this viewpoint according to Løvendal and Knowles (2005) is that whereas traditional food security analysis concentrates on retrospective measures and is constantly pre-occupied with the rearview, vulnerability analysis ensures a more forward-looking proactive measure that is more likely to identify causality and illuminate strategies for dealing with risks in uncertain environments (Løvendal and Knowles 2005). This in Scaramozzino's view, however, will only be effective if the methodology is employed appropriately. Indeed, Scaramozzino suggests that many failures in food security programmes and policies directly relate to ignoring the specific needs and interests of the differing socio-cultural groups in favour of treating large groups of people as homogeneous (Scaramozzino 2006b). Another criticism charged by Løvendal and Knowles also suggests that vulnerability is not consistently applied nor always adequately operationalised. In this way, it is offered that vulnerability might at various times refer to hunger or food insecurity in general or to specifics such as malnutrition, famine, food production or even used to explicitly identify indirect risks. In answer, many advocate the use of specifically targeted risk analysis and management tools to adequately and precisely map the vulnerabilities of various populations and groups. In this, it is recognised that the overall food security policy response could be greatly improved (Maxwell and Frankenberger 1992; Gerster-Bentaya and Maunder 2008).

1.4 Confusion, Consternation and Contradiction

It can be seen then that there are numerous ways in which food security is defined, viewed and addressed. Howsoever the notion is viewed, whether from a holistic perspective at national or academic levels or in its narrower more specific grass roots perspective, it is important that it is properly understood. However, it is this very complexity and cross-discipline or sectoral nature that continues to hinder progress as much today as in the past. It is indeed, according to some, this complexity in essence that is both the cause of much misunderstanding and the barrier to any real consensual solution. Unfortunately, the extent of the confusion is widespread and almost seems to be endemic as the following illustrates:

1. 'Malnutrition is poorly understood and has not benefited from expertise in communications and advocacy.' (WEF 2009)
2. 'An analysis of the current literature on vulnerability make it apparent that there is ... no consensus on ... how to define and measure vulnerability.' (Scaramozzino 2006a)
3. 'Urban poverty and food insecurity—Urban livelihoods and coping strategies remain poorly understood.' (Floro and Swain 2010)
4. 'What are the water and food challenges faced by the world? Why are they so poorly understood?' (Rijsberman 2010)
5. 'Food security and nutrition have not been given adequate attention [with regard to AIDS] despite broad agreement on their significance as an essential component of a comprehensive social protection package.' (Greenway 2008)

These are not isolated views either; a quick scan of the available literature soon identifies numerous feelings of misunderstanding and misapprehension (Budge and Slade 2009; Rijsberman 2010). That's not all, on top of this general lack of comprehension, there is also subjectivity, misquotes and misrepresentation to deal with while interpretation of primary and secondary research sometimes subjective and sometimes without recourse to fact continues to causes confusion, frustration and consternation. This collectively fosters an environment beset with difficult analyses and divergence of opinion, and it is not just the fault of the individuals or media either; errors are made at the institutional level too. By way of example, one of the biggest failures in food security analysis of late, and by their own admission concerns one of omission. It was not until 2006 that USAID Food for Peace (FFP) recognised a serious flaw in their conceptology of food security analysis, one that failed to take into account peoples', communities' and countries' vulnerability to risk of food security. Up until this point, USAID had only elucidated three dimensions to the concept: availability, access and utilisation. The fourth, vulnerability, although retrospectively stated as being implicit in its official definition with the usage of the phrase 'at all times' it did not in-fact became explicit until the official PL 480 Title II proposed programme policy update for 2008, Annex C (published in 2007) (USAID 2007). Such an oversight in contemporary understanding on USAID's part almost defies belief. However, as is becoming increasingly clear though, the dynamic complexity of the subject is such that, this will probably not be the last admission, redefinition or reclassification of its kind that will continue to contribute to the three 'C's of food security: confusion, consternation and contradiction.

In Section 1.4, we take a snapshot of global prevalence of food insecurity but before we can gauge the extent of the problem we must have some way of assessing who the hungry are.

1.5 Measuring and Monitoring the Food Insecure

Determining just who the food insecure are is a difficult and complex task and as is characteristic of the nature of the food security debate in general it is also one that enjoys far-from-universal agreement (Svedberg 2000). The need to quantify the extent of the problem is self-evident and not in question. Policy makers, analysts, humanitarian agencies and a wider public regularly require updated analyses on the big questions of the who's, the why's and the where's of the disaffected. Indeed an expert analysis can give a much needed snapshot-in-time of the overall extent of food insecurity of a region; it can also determine perhaps the nature of a vulnerability within a particular group and with adequate statistical evidence can even ascertain the degree to which the hungry are hungry (Scaramozzino 2006a; EC/FAO 2008). Food security analysis therefore offers the rationale behind much policy response whether identifying the appropriate type and scale of emergency or protracted relief and recovery responses; the extent of any food aid, school feeding, cash or voucher programmes; the required support to re-establish livelihoods or just as importantly the monitoring of evolving or recurrent crises (Hoddinott 1999; WFP/VAM 2010).

1.5.1 No Single Measure

All this however is predicated on the assumption that there are adequate measures of food security. The sad glaring truth however is that there is no extant measure, no yardstick by which the food insecure can be gauged (Riely et al. 1999). This has been the case from the start although it was once again recently reaffirmed at the International Scientific Symposium on Measurement and Assessment of Food Deprivation and Undernutrition in 2002 (FAO 2002). It was also noted at the same conference that the absence of such a measure has perhaps impeded progress particularly in the 'political advocacy' of more integrated approaches (Weisell 2002). The aspiration of a single indicator has not gone away though and it is one that continues to be the goal of many stakeholders who see the potential benefit of promoting quick and easy analysis and more timely policy responses (FAO 2003).

So the question then arises, how do we measure food security or its adversary, food insecurity? The answer is by proxy. By triangulating outcomes or variables closely associated with the concept, we are able to build a picture, piece by piece of those affected and to what extent. Once again, such laudable aims are racked with disunity and inconsistency. Take for example the construct of 'hunger'. This is closely related to food security yet how can this be defined or quantified? Other notions too such as risk, vulnerability, adequate food and optimum diet all too become difficult to define and in this context even more challenging to quantify (Mason 2002). Yet, despite this there has to be a starting point. Currently, food security measures can be viewed loosely as consisting of three varieties or sorts: those measures that address the various dimensions such as access, availability, stability and utilisation; those that look at the associated influences such as poverty, water, sanitation and health and perhaps other wider societal issues; and those that measure outcomes of food insecurity, which by their very definition are ex-post or retrospective. Of course in practice, such clear demarcation is not always easy, practical or even necessary.

In this way, it is widely acknowledged that several of the most quoted indicators are: malnourishment figures, via various methods discussed later; poverty, via those living on less than one dollar-a-day and anthropometric measures such as wasting, stunting and underweight. Importantly too, while malnutrition garners a measure of general worldwide acceptance as the most ostensible manifestation of food insecurity on an individual's health, there is less agreement over the best method of calculating this (Svedberg 2000, 2002a, 2002b; World Bank 2004; Nubé and Sonneveld 2005; Black et al. 2008; Wesenbeeck et al. 2009). These caveats are explored in detail in the measurements of malnourishment in Appendix F. That said, restricting ourselves to just one or two indicators limits the scope of food security to those elements alone and misses by a wide margin many of the contributing factors that taken together build up the concept to develop a more fuller, comprehensive picture.

1.5.2 Suite of Indicators

In response, the method of understanding the extent of the problem is achieved using the notion of a suite of indicators. This was also reaffirmed and fully sanctioned at the 2002 FAO conference. A suite of indicators is oftentimes made up of the different dimensions of the food security concept; access, availability, utilisation and vulnerability as well as a second underlying set that is used to monitor associated influences (FAO 2002). But how many is enough? Almost every major government or organisation involved in food security collects, analyses, uses and disseminates data based on their own and/or internationally accepted standards. Once again depending on who you read, there are anywhere between 25 broadly defined indicators to over 450 (Riely and Mock 1995; Hoddinott 1999; Maire and Delpeuch 2005; Scaramozzino 2006a). The Institut de Recherche pour le Développement (IRD), for instance, published the nutrition indicators for development reference guide in 2005. In the report, there was a list of over 450 indicators, 110 of which were designed around three core dimensions of food security and the rest were devoted to closely associated issues of nutritional status; care and caring capacity; health and demography; socio-economic; agro-ecological and sustainable environment indicators. Although in fairness this list was never designed to be used in full, instead it is a sort of pick 'n' mix of indicators that can be used depending on situation and circumstance (Maire and Delpeuch 2005; FIVIMS 2008). It is also worth noting that while the list might seem long, in Maire and Delpeuch words, it is certainly not exhaustive. Moreover, this list is set to increase, since the book was written, more indices covering dietary diversity, hunger scale and a coping strategies are also being considered (personal communication, Claude 2010).

On a national or sub-national level, these large databases are probably fine as the full compliment can be used for myriad purposes. However, for cross-border comparisons at the international level, very few measures are truly comparable. Consequently, there has been a consensual shift towards deriving a core set of indicators that can be compared at national levels and above; this was led in the first instance by the United Nations World Food Summit 1996. As a result and after much consultation and collaboration between the Food Insecurity and Vulnerability Information and Mapping Systems (FIVIMS); the Key Indicators Database System (KIDS); the Inter-Agency Working Group (IAWG) as well as select others, the Committee on Food Security (CFS) in 2001 endorsed a suite of seven core food security indicators (Table 1.2) and a further suite of indicators monitoring the underlying food economy.

Since that time however there has been growing recognition that the concept incorporates far more broader societal issues such as education, water and sanitation, morbidity and many others. As a result, the CFS now routinely monitors 20 or so other contextual indicators while the FAO now regularly monitors over 80 indicators based around six core groupings for their food security country profile series (FAO 2011c). USAID too employs up to 41 generic indicators from six categories for their Title II projects outlined by the Food and Nutrition Technical Assistance II Project (FANTA-2) (Swindale and Ohri-Vachaspati 2005). The World Food Program's (WFP) Comprehensive Food Security and Vulnerability Analysis (CFSVA) guidelines on the other hand suggest up to 47 generic indicators while the WHO also monitors national indices around food security and associated concepts covering nearly 40 separate variables around six core groupings (Mock et al. 2006; WFP 2009a). Table 1.3 highlights the diversity of monitored indicators.

TABLE 1.2

Indicators Endorsed by CFS/26

Core Food Security Indicators

Average per person dietary energy supply (DES)
Cereals, roots and tubers as % of DES
Percentage of population undernourished
Life expectancy at birth
Under-5 mortality rate
Proportion of children under 5 that are underweight
Percentage of adults with body mass index (BMI) <18.5

Possible Indicators for Monitoring Underlying Conditions Proposed for Inclusion in Future CFS Assessment Documents

Economic conditions	GNP per capita
	Growth rate in GNP per capita
	GNP per capita at purchasing power parity
Risks, hazards, shocks	Number of countries facing food emergencies
Food availability	Volume of production, food use, trade and stock changes for major food commodities, by commodity group and by country groupings
	Ratio of five major grain exporters' supplies to requirements
	Food production index
Food access	Gini-index of income distribution
	People living below national poverty line
	People living on less than $1 per day
Stability of food supplies and access	Changes in cereal production in LIFDCs with and without China and India
	Export price movements for wheat, maize and rice
	Variability of food prices
	Index of variability of food production

Source: FAO, Indicators from the assessment of the world food security situation, FAO Committee on Food Security, Twenty-Seventh Session, Food and Agriculture Organization, Rome, 2001; CFS, Assessment of the world food security situation, FAO Committee on Food Security, Twenty-Seventh Session, Food and Agriculture Organization, Rome, 2001; FIVIMS, *The FIVIMS Initiative: Food Insecurity and Vulnerability Information and Mapping Systems: Tools and Tips*, FIVIMS, Rome, 2008.

In the final analysis, the suite of seven core indicators first endorsed by the CFS in 2000/2001 remains the central nucleus of indicators used by FIVIMS and others. Today, however, between the agencies, a total of over 300 separate indices are collated and monitored regularly by many agencies, which combine to form the backbone of central data sets for analyses. These indicators are not definitive by a wide margin, but many more literally running into the hundreds if not thousands and are associated with the wider aspects of food security, risk, access and development goals in general, etc.

Such an approach allows individuals to look at a suite of linked indicators and analyse them in a sort of cross-comparative fashion of triangulation. This gives policy makers much needed choice and flexibility although, by the same token, the absence of a single definitive indicator means that such analysis will always be open to charges of bias or subjectivity. Another interesting development in the focus of analysis these days places less emphasis on absolute numbers but rather on trends over time. This trend better reflects progress or otherwise and limits the short-sightedness of one-off metrics being taken out of context (Mason 2002). Also in recent times, while a single measure of food security still eludes us, there has been a concerted move towards compiling composite indicators to classify degrees of food security (WFP 2009b). These are usually formulaic and work on a similar principle as the suite of indicator approach. By combining several indices using various methodologies (Appendix F), it becomes possible to grade and compare populations accordingly. There has been some encouraging developments in this regard although as yet there is no international standard. Despite that there are some leading contenders though; in particular, the International Food Policy Research Institute's (IFPRI's) Global Hunger Index (GHI) and the collaborative Integrated Food Security Phase Classification (IPC), both of which are examined in the appendices. The propensity to integrate such data based on linkages from empirical research is a welcome approach although the big questions still remain: what is linked and to what degree?

TABLE 1.3

A Selection of Institutionally Monitored Indicator Groups

USAID Title II Programmes	CFS	FAO Country Profile Indicators	WHO Country Profile Indicators
Health, Nutrition and Maternal and Child Health	Food Deprivation and Child Malnutrition	Food consumption and deprivation	Malnutrition in children
Water and Sanitation	Food Consumption and Diet Diversification	Food production	Malnutrition in women
Household Food Consumption		Food trade	Food security
Agricultural Productivity	Economic Growth, Poverty and Employment	Macro and socioeconomic	Commitment
Natural Resource Management		Agriculture	Capacity
Food-for-Work/CFW roads	Education and Gender Equality	Health, nutritional and sanitation	Caring practices
	Health and Sanitation		Health services
	Agricultural Development		Meta-indicators
	Water, Natural Resources and Infrastructure		
	Trade and National Debt		
	Development Assistance		

Sources: CFS, Assessment of the World Food Security Situation: Suggested core indicators for monitoring food security status, Committee on World Food Security: Twenty-sixth Session, Rome, 2000; Swindale, A. and Ohri-Vachaspati, P., *Fanta-2: Measuring Household Food Consumption: A Technical Guide*, US Agency for International Development, Washington, DC, 2005; CFS, Committee on World Food Security, 2011. http://www.fao.org/economic/cfs09/cfs-home/en/ (accessed 15 January 2011); Website of the Food and Agriculture Organisation of the United Nations. FAO. http://www.fao.org/hunger/faqs-on-hunger/en/ (accessed 15 February 2011). FAOSTAT, *Food and Agriculture Statistics*, Food and Agriculture Organization, Rome, 2011; WHO, *WHO Trade, Foreign Policy, Diplomacy and Health: Food Security*, World Health Organization, Geneva, Switzerland 2011. http://www.who.int/trade/glossary/story028/en/.

1.6 Identifying the Vulnerable

As the four-pillar framework dictates, it has become essential not just to look at current incidence of food security but also to identify potential individuals, households or the communities that are at risk of suffering in the future. The concept of vulnerability analysis has for a long time been implicit in the various definitions particularly by the UN, but it is increasingly being acknowledged by the wider community as an important dimension of the analysis of food insecurity as a whole (Scaramozzino 2006a). In respect of food security, the World Food Programme (WFP) defines vulnerability as 'exposure to risk, mitigated by the ability to cope' or more specifically, the 'probability of an acute decline in food access or consumption' (WFP 2009b). Vulnerability also involves the need to understand, quantify and mange risk, and there is much tacit agreement that suggests 'food security can be lost as well as gained' (USAID 2007). Such risks might take the form of droughts or floods that affect many hundreds even thousands or conversely, the death of a household's income provider might only affect a few. The risk might also be at the national or government level too, that is to say failing, failed or fragile states undermined by political, social and economic factors might affect a country's ability to address such vulnerabilities as they arise. In this respect, the measure of a country's economic development impacts on its ability to cope. Weak institutions and insufficient institutional oversight or governance also contributes to a region's vulnerability, as does armed conflict. In this way, vulnerability is concerned with the causes of and the risks to food security as well as an understanding of the many coping mechanisms involved. This is perfectly summed up by USAID, which suggests that

> Unmanaged risk leads to food insecurity, while managing risks can protect and enhance food security. (USAID 2007)

For now though we are interested in the flip side of the equation, the typology of the vulnerable, who they are and where might they be. At this juncture it is worth remembering that while food insecurity in a temporal sense can strike with little or no warning as in acute onset emergencies, chronic suffering can be brought on by either sudden or slow onset shocks. We have mentioned that each year about 10 million people die of hunger and hunger-related diseases while only 8% of these are considered a direct result of emergencies and the rest are from chronic or foreseeable transitory situations (UN 2007; WFP 2011). As a result, it becomes important from both policy and intervention points of view to accurately identify exactly who these vulnerable are, in what temporal sense the suffering occurs and to ultimately understand how best such populations can be helped. Such understanding facilitates a more proactive or preventative approach which in turn can help reduce the risks or mitigate the worst effects of such vulnerabilities (FAO 2000; Scaramozzino 2006a; USAID 2007). One place to start is with vulnerability analysis.

Vulnerability can be grouped or categorised in several ways. The National Food Insecurity and Vulnerability Information and Mapping Systems (FIVIMS) for instance includes groupings such as occupational—fishermen and pastoralists; spatial—marginal urban or forest dwellers; cultural—bushmen and nomads; demographic—children under 5; social status—refugees and the displaced and physical condition—the handicapped and severely ill. (CFS 2000). The difficulty here lies with the complexity of the myriad sub-categories that may arise. For instance, the sedentary indigenous under 5 urban dwellers in Mauritania for example have a completely different set of food security issues from those faced by the elderly relatives of the busmen of the Kalahari.

However, we must start somewhere and to simplify the basics, the following gives a brief overview of some of the most vulnerable categories starting from the macro to the micro. These are further elaborated in Part 5: The Sectoral Analysis.

- *Vulnerable Areas*: These include regions or areas like those of fragile balanced ecosystems or regions heavily entrenched in the production and supply of food and food materials. These might become vulnerable to extreme weather events such as droughts or floods, degradation and mismanagement of resources. While predominantly rural, such vulnerable areas can just as easily be urban, and this impacts particularly where there are large numbers of individuals, families or whole communities reliant on a particular area or region. Therefore, any significant variations in production and supply or sustainability of the land will have adverse effects on such groups.
- *Vulnerable Groups*: These groups tend to be homogenous such as women, the under-5s, the poor, seasonal agricultural workers, new immigrants into an area, nomadic or transhumant populations (seasonal migration of herders typically between lower valleys in the winter and high pastures in the summer). These groups rarely command strong financial or physical continuity over food security.
- *Vulnerable Households*: This term is used to describe family groupings often categorised by vulnerable typologies including: large or female-headed households, or those with prevalences of disease like AIDS or tuberculosis.
- *Vulnerable Individuals*: These tend to be individuals of the community who are especially vulnerable to food insecurity, those that need extra attention or consideration. These include infants and children under 5, pregnant or lactating mothers, the ill, the handicapped and the elderly, etc. (FAO 2000).

Among all the above, there are perhaps a few vulnerable groups who stand out from the rest and who are widely agreed to be at most risk than others: these are generally considered to be women, especially pregnant women; the poor, in particular the rural poor; and children.

1.6.1 Women

Women play an important and key role in the family unit. It is often women who are responsible for growing or purchasing the food the family eats. They are also the primary health givers, particularly of the young, and are especially vulnerable during pregnancy and lactation when their additional nutritional demands are greatest. Yet it is an unfortunate truth that many women becoming pregnant do so from the position of suboptimal nutritional status. This exacerbates the need for immediate and continual nutritional support for both the mother and her baby (SD 2001; Morley 2007; Black et al. 2008; Muller and Jahn 2009; Freedom from Hunger 2010). Also of difficulty here is the persistent cultural values in which women continually lack equal access to most things such as food, health care, support and education that continue to undermine their food security and perpetuate the status quo (UNICEF 2007; Dewan 2008). Such gender subordination also ensures that women persistently make up the majority of the global poor and more than 60% of the chronically hungry (SOFI 2006).

1.6.2 Infants and Children

When it comes to vulnerability, because of their reliance on others, few are more vulnerable to the effects of hunger than children. This reflects the child's dependency on the adult for all its nutritional needs. It also reflects children's growing needs that require constant and increased energy and essential nutrients over time. Children younger than two years are especially vulnerable as poor foetal growth combined with poor nutrition in the first two years, particularly during the transition period from breastfeeding to meat, cereal and plant-based diets (weaning), may result in reduced mental development, stunting or chronic malnutrition. This is further complicated if malnutrition is missed or left untreated or unless circumstances do not drastically improve for malnourished youngsters, catch-up growth might prove especially difficult, if not impossible. In such cases, the effects are largely irreversible and could possibly adversely affect the individual for the rest of their lives (Frongillo 1999; WFP 2000; Theron et al. 2006; Morley 2007; WFP 2008; DFID 2009; MacAuslan 2009; Muller and Jahn 2009; Nnakwe 2009; World Vision 2009; Save the Children 2010; WFP 2011).

1.6.3 Poor

A strong correlation between poverty, health and by extension undernutrition and food security is well documented (Louria 2000; Arcand 2001; UN 2002/3; Cohen 2005; Dittoh et al. 2007). Globally, there are around 1.2 billion people who live on less than $1 a day while a further 2.1 billion get by on less than $2 a day (World Bank 2009). This has a profound impact on their ability to provide food security for themselves or their families. Where richer countries normally spend around 10%–20% of their income on food, the poorer countries regularly have to expend between 60% and 80% (USDA/ERS 2008). Such large proportional expenditure on food makes the poor especially vulnerable to economic downturns, food price rises or financial volatility, particularly in the current ongoing financial crisis (WFP 2011). This is even more devastating for the rural poor, which in themselves make up three-quarters of all poor people in developing countries (World Bank 2008). Further of these three quarters, over 85%, with little or no alternative source of income, are overwhelmingly dependent on agriculture for food and income, which makes them particularly vulnerable to natural or man-made crisis. Moreover, while the majority of the remaining 25% of the global poor reside in shantytowns or on the urban city peripheries, the changing demographic will see more growth in this particular group. Being more detached from the source or the growing of their own foods a different set of challenges then face those people. This group tends to be more wage- and employment-reliant for their security than those in the agricultural sector, and as such future social safety nets for these people are of particular importance (SOFI 2005).

References

ActionAid (2003). Food Aid: An Actionaid Briefing Paper. ActionAid.

Andersen, S. A. (1990). Core Indicators of Nutritional State for Difficult to Sample Populations. *The Journal of Nutrition* 120 (Suppl 11): 1559–1600.

Arcand, J. L. (2001). Undernourishment and Economic Growth: The Efficiency Cost of Hunger. European Media Seminar on Global Food Security, Royal Swedish Academy of Agriculture and Forestry, Stockholm, Food and Agriculture Organization.

Barrett, C. and D. Maxwell (2006). Towards a Global Food Aid Compact. *Food Policy* 31(2): 105–118.

Black, R. E., L. H. Allen, Z. A. Bhutta, L. E. Caulfield, M. D. Onis, M. Ezzati, C. Mathers and J. Rivera (2008). Maternal and Child Undernutrition: Global and Regional Exposures and Health Consequences. *Lancet* 371(9608): 243–260.

Budge, T. and C. Slade (2009). Integrating Land Use Planning and Community Food Security. Victoria, Australia: Victorian Local Governance Association 76.

CFS (1998). Guidelines for National Food Insecurity and Vulnerability Information and Mapping Systems (Fivims): Background and Principles. Committee on World Food Security: Twenty-fourth Session, Rome, Committee on World Food Security.

CFS (2000). Assessment of the World Food Security Situation: Suggested Core Indicators for Monitoring Food Security Status. Committee on World Food Security: Twenty-Sixth Session, Rome, Committee on World Food Security.

CFS (2001). Assessment of the World Food Security Situation. FAO Committee on Food Security, Twenty-Seventh Session. Rome: Food and Agriculture Organization.

CFS (2007). Assessment of the World Food Security Situation. Committee On World Food Security: Thirty-third Session, Rome, Food and Agriculture Organisation.

CFS (2011). FAO Committee on World Food Security (CFS). http://www.fao.org/economic/cfs09/cfs-home/ en/ (accessed 15 January 2011).

Chambers, R. (2005). *Ideas for Development*. Brighton: Earthscan Ltd: Institute of Development Studies.

Chambers, R. and G. R. Conway (1991). Sustainable Rural Livelihoods: Practical Concepts for the 21st Century. *Discussion Paper 296*. Brighton: Institute of Development Studies.

Claude, M. (2010). *Indicators of Nutrition and Food Security*. Rome: Nutrient Requirements and Assessment Group (AGNA).

Cohen, D. (2005). Achieving Food Security in Vulnerable Populations. *British Medical Journal* 331: 775–777.

Darcy, J. and C. Hofmann (2003). *According to Need? Needs Assessment and Decision Making in the Humanitarian Sector*. London: Overseas Development Institute.

Devereux, S., B. Baulch, K. Hussein, J. Shoham, H. Sida and D. Wilcock (2004). *Improving the Analysis of Food Insecurity: Food Insecurity Measurement, Livelihoods Approaches and Policy: Applications in Fivims*. Rome: FAO-Food Insecurity and Vulnerability Information and Mapping Systems (FIVIMS) 52.

Dewan, M. (2008). Malnutrition in Women. *The Journal Studies of Home and Community Science* 2(1): 7–10.

DFID (1999). Sustainable Livelihoods Guidance Sheets: Overview. *Sustainable livelihoods Guidance Sheets*. London: Department for International Development DFID.

DFID (2009). Draft Terms of Reference Tackling the Neglected Crisis of Undernutrition Research Programme Consortia. London: Department for International Development (DFID).

Dittoh, S., A.-R. Abizari and M. Akuriba (2007). Agriculture for Food and Nutrition Security: A Must for Achieving the Millennium Development Goals in Africa. Second Annual African Association of Agricultural Economists (AAAE), Accra, Ghana AAAE.

Earth Summit (2002). Decision Making: Briefing Sheet. United Nations. http://www.earthsummit2002.org/es/ life/Decision-making.pdf (accessed 20 July 2011).

EC/FAO (2008). An Introduction to the Basic Concepts of Food Security. Rome: EC–FAO Food Security Programme 3.

ELDIS (2011). What Are Livelihoods Approaches? Institute of Development Studies (IDS). http://www.eldis. org/go/topics/dossiers/livelihoods-connect/what-are-livelihoods-approaches (accessed 12 February 2011).

Ericksen, P. J. (2008). Conceptualizing Food Systems for Global Environmental Change Research *Global Environmental Change* 18(1): 234–245.

EU (2006). *European Union Regulation (Ec) No 1905/2006*: European Union.

EU (2010). Eu Development Policy: Food Security. Council of the European Union. http://www.consilium. europa.eu/showPage.aspx?id=1662&lang=EN (accessed 30 March 2011).

FANTA2 (2010). About Food and Nutrition Technical Assistance II Project (FANTA-2). US Agency for International Development. http://www.fantaproject.org/ (accessed 15 July 2011).

FAO (1996). Rome Declaration on World Food Security and World Food Summit Plan of Action. Rome: Food and Agriculture Organization of the United Nations.

FAO (2000). *Handbook for Defining and Setting up a Food Security Information and Early Warning System (FSIEWS)*. Rome: Food and Agriculture Organization.

FAO (2002). Measurement and Assessment of Food Deprivation and Undernutrition: International Scientific Symposium. International Scientific Symposium: FIVIMS, An Inter-Agency Initiative to Promote Information and Mapping Systems on Food Insecurity and Vulnerability. Rome: Food and Agriculture Organization.

FAO (2003). *Trade Reforms and Food Security: Conceptualizing the Linkages*. Rome: Food and Agriculture of the United Nations.

FAO (2006a). *Policy Brief: Food Security*. Rome: Food and Agriculture Organisation 4.

FAO (2006b). *The State of Food and Agriculture 2006: Food Aid for Food Security?* Rome: U.N. Food and Agriculture Organisation 183pp.

FAO (2011a). *Food Security Statistics: Country Profiles*. Food and Agriculture Organization.

FAO (2011b). Website of the Food and Agriculture Organization of the United Nations. FAO. http://www.fao.org/hunger/faqs-on-hunger/en/ (accessed 15 February 2011).

FAO (2011c). Website of the Food and Agriculture Organisation of the United Nations. FAO. http://www.fao.org/hunger/faqs-on-hunger/en/ (accessed 15 February 2011).

FAOSTAT (2011). *Food and Agriculture Statistics*: Food and Agriculture Organization

FIAN (2010). Website of the Foodfirst Information and Action Network. FoodFirst Information and Action Network. http://www.fian.org/ (accessed 12 July 2010).

FIVIMS (2008). The FIVIMS Initiative: Food Insecurity and Vulnerability Information and Mapping Systems: Tools and Tips.

Floro, M. and R. Swain (2010). Food Security, Gender and Occupational Choice among Urban Low-Income Households. *Working Papers, American University, Department of Economics*.

Frankenberger, T. R., K. Luther, J. Becht and M. K. McCaston (2002). *Household Livelihood Security Assessments: A Toolkit for Practitioners*. Tuscon: CARE USA.

Freedom from Hunger (2010). Website of the Freedom from Hunger. Freedom from Hunger. http://www.freedomfromhunger.org/ (accessed 15 June 2010).

Frongillo, E. A. (1999). Symposium: Causes and Etiology of Stunting. *Journal of Nutrition*(129): 529S–530S.

Gerster-Bentaya, M. and N. Maunder (2008). *Food Security Concepts and Frameworks: What Is Food Security?*: EC-FAO.

Greenway, K. (2008). Inter-Agency Task Team on Children and HIV and AIDS: Working Group on Food Security and Nutrition. Paris: IATT Food Security and Nutrition Working Group.

Hoddinott, J. (1999). *Choosing Outcome Indicators of Household Food Security*. Washington, DC: International Food Policy Research Institute 29.

IFAD (2009). Food Security: A Conceptual Framework. International Fund for Agricultural Development. http://www.ifad.org/hfs/thematic/rural/rural_2.htm (accessed 15 August 2009).

Løvendal, C. R. and M. Knowles (2005). Esa Working Paper No. 05-07: Tomorrow's Hunger: A Framework for Analysing Vulnerability to Food Insecurity. *ESA Working Papers*. Rome: FAO: Agricultural and Development Economics Division.

Louria, D. B. (2000). Emerging and Re-Emerging Infections: The Societal Determinants. *Futures* 32(6): 581–594.

MacAuslan, I. (2009). Hunger, Discourse and the Policy Process: How Do Conceptualizations of the Problem of 'Hunger' Affect Its Measurement and Solution? *European Journal of Development Research* 21: 397–418.

Maire, B. and F. Delpeuch (2005). *Nutrition Indicators for Development: Reference Guide*. Montpellier, France: FAO & IRD: Institut de Recherche pour le Développement.

Mason, J. B. (2002). Measurement and Assessment of Food Deprivation and Undernutrition: Keynote Paper: Measuring Hunger and Malnutrition. International Scientific Symposium. Rome: FAO.

Maxwell, S. and T. Frankenberger (1992). *Household Food Security: Concepts, Indicators and Measurements: A Technical Review*. New York, Rome: UNICEF and IFAD.

Maunder, N. (2006a). *Food Security Information Systems and Networks: Lesson 1 - Food Security Information Systems*. Rome: EC FAO.

Maunder, N. (2006b). *Food Security Information Systems and Networks: Lesson 3 - Improving Food Security Information Systems*. Rome: EC FAO.

McCalla, A. F. (2007). FAO in the Changing Global Landscape. *UCD. ARE Working Papers, vol. Paper 07-006*. Dublin: University College Dublin.

McHarry, J., F. Scott and J. Green (2002). Towards Global Food Security: Fighting against Hunger. *Towards Earth Summit 2002*.

Mock, N., N. Morrow, S. Aguiari, X. Chen, S. Chotard, Y. Lin, A. Papendieck and D. Rose (2006). Comprehensive Food Security and Vulnerability Analysis (CFSVA): An External Review of WFP Guidance and Practice. Rome: Development Information Services International.

Morley, J. E. (2007). Nutritional Disorders. In *The Merck Manual Online Medical Library*. R. S. Porter and J. L. Kaplan, eds. Whitehouse Station, NJ: Merck Sharp & Dohme Corporation.

Muller, O. and A. Jahn (2009). *Malnutrition and Maternal and Child Care. Maternal and Child Health* (Part 3): 287–310.

Nnakwe, N. E. (2009). *Community Nutrition: Planning Health Promotion and Disease Prevention.* Sudbury, MA: Jones and Bartlett.

Nubé, M. and B. G. J. S. Sonneveld (2005). The Geographical Distribution of Underweight Children in Africa. *Bulletin of the World Health Organization* 83(10): 764–770.

Riely, F. and N. Mock (1995). Inventory of Food Security Impact Indicators: Food Security Indicators and Framework. In *A Handbook for Monitoring and Evaluation of Food Aid Programs.* Arlington, VA: IMPACT.

Riely, F., Nancy Mock, B. Cogill, L. Bailey and E. Kenefick (1999). Food Security Indicators and Framework for Use in the Monitoring and Evaluation of Food Aid Programs. Washington, DC: US Agency for International Development 45.

Rijsberman, F. (2010). Energy & Climate: Water and Food Security. *America.Gov* March: http://www.america.gov/st/energy-english/2010/March/20100310143830fsyelkaew0.4131739.html. (accessed 15 June 2011).

Ruxin, J. N. (1996). *Hunger, Science, and Politics: FAO, WHO, and UNICEF Nutrition Policies, 1945–1978, Chapter II: The Backdrop of UN Nutrition Agencies, by Joshua Nalibow Ruxin.* London: University College London. PhD.

Save the Children (2010). Website of Save the Children Alliance. http://www.savethechildren.net/alliance/index.html (accessed 12 May 2011).

Scaramozzino, P. (2006a). Measuring Vulnerability to Food Insecurity. Rome: FAO Agricultural and Development Economics Division (ESA) 26.

Scaramozzino, P. (2006b). Measuring Vulnerability to Food Insecurity: Esa Working Paper No. 06–12. Rome: FAO: The Agricultural Development Economics Division (ESA) 24.

SD (2001). Gender and Nutrition: Gender and Development Fact Sheets. Rome: FAO/Sustainable Development Department.

Sen, A. (1981). *Poverty and Famines: An Essay on Entitlement and Deprivation.* Oxford: Clarendon Press.

Shoham, J. and C. Lopriore (2007). *Livelihoods Assessment and Analysis: Introduction to Livelihoods*: EC FAO.

SOFI (2001). The State of Food Insecurity in the World 2001. Rome: Food and Agriculture Organization.

SOFI (2005). *State of Food Insecurity.* Rome: Food and Agricultural Organization.

SOFI (2006). *State of Food Insecurity: Eradicating World Hunger—Taking Stock Ten Years after the World Food Summit.* Rome: Food and Agricultural Organization

Svedberg, P. (2000). *Poverty and Undernutrition: Theory, Measurement, and Policy.* Oxford: Oxford University Press.

Svedberg, P. (2002a). Measurement and Assessment of Food Deprivation and Undernutrition: Part III: Parallel Contributed Papers Sessions Fallacies in—and Ways of Improving—The FAO Methodology for Estimating Prevalence of Undernutrition. International Scientific Symposium: FIVIMS—An Inter-Agency Initiative to Promote Information and Mapping Systems on Food Insecurity and Vulnerability, Rome: FAO.

Svedberg, P. (2002b). Undernutrition Overestimated. *Economic Development and Cultural Change* 51(1): 5–36.

Swindale, A. and P. Ohri-Vachaspati (2005). Fanta-2: Measuring Household Food Consumption: A Technical Guide Washington, DC: US Agency for International Development.

Theron, M., A. Amissah, I. C. Kleynhans, E. Albertse and U. E. MacIntyre (2006). Inadequate Dietary Intake Is Not the Cause of Stunting Amongst Young Children Living in an Informal Settlement in Gauteng and Rural Limpopo Province in South Africa: The Nutrigro Study. *Journal of Public Health Nutrition* 10(4): 379–389.

UN (2002/3). Fighting Poverty through Better Health and Education. *Bulletin of the Eradication of Poverty* 2002/3 Edition: 14.

UN (2007). Millennium Development Goals: Eradicate Extreme Poverty and Hunger United Nations. http://www.mdgmonitor.org/goal1.cfm (accessed 15 July 2010).

UNICEF (2006). Progress for Children. New York: United Nations Children's Fund.

UNICEF (2007). The State of the World's Children 2007: Women and Children—The Double Dividend of Gender Equality New York City: United Nations Children's Fund.

UNICEF (2008). The State of the World's Children 2008: Women and Children—Child Survival. New York City: United Nations Children's Fund.

UNICEF (2009). Tracking Progress on Child and Maternal Nutrition: Survival and Development Priority. New York: United Nations Children's Fund.

USAID (1992a). *Policy Determination: Definition of Food Security*: US Agency for International Development.

USAID (1992b). *USAID Policy Determination: Definition of Food Security*. Washington, DC: US Agency for International Development.

USAID (1995). *Policy Determination: Food Aid and Food Security*: US Agency for International Development.

USAID (2007). Food for Peace: Fy 2008 P.L. 480 Title II Program Policies and Proposal Guidelines Washington, DC: United States Agency for International Development.

USAID (2010a). Our Work: A Better Future for All. US Agency for International Development. http://www.usaid.gov/our_work/ (accessed 12 October 2011).

USAID (2010b). What Is the Famine Early Warning Systems Network (FEWS NET)? USAID. http://www.fews.net/ml/en/info/Pages/default.aspx?l=en (accessed 30 June 2011).

USDA/ERS (2008). Food and Consumer Price Index (CPI) and Expenditures. *Food Security Assessment*. Washington, DC: USDA/Economic Research Service.

USDA (2009). Food Security in the United States: Measuring Household Food Security. United States Department of Agriculture. http://www.ers.usda.gov/Briefing/FoodSecurity/measurement.htm (accessed 15 October 2011).

USDA/ERS (2010). Food Security Assessment, 2010-20. *GFA: Food Security Assessment*. Washington: United States Department of Agriculture.

WEF (2009) Nutrition. *Summit on the Global Agenda 2009 Council Reports* 2009: http://www.weforum.org/pdf/GAC09/council/nutrition/proposal.htm.

Weisell, R. (2002). Measurement and Assessment of Food Deprivation and Undernutrition: Part 3. Food Security Information Systems Using Combined Methods—An International Nutrition Index: Concept and Analyses of Food Insecurity and Undernutrition at Country Levels. International Scientific Symposium, Rome, FAO.

Wesenbeeck, C. F. v., M. A. Keyzer and M. Nubé (2009). Estimation of Undernutrition and Mean Calorie Intake in Africa: Methodology, Findings and Implications. *International Journal of Health Geographics* 8(37): 1–18.

WFP (2000). *Food and Nutrition Handbook*. Rome: World Food Programme (WFP).

WFP (2008). *Ten Minutes to Learn About Improving the Nutritional Quality of WFP's Food Basket—An Overview of Nutrition Issues, Commodity Options and Programming Choices*: World Food Programme. 2008 (S).

WFP (2009a). *Comprehensive Food Security & Vulnerability Analysis Guidelines*. Rome: World Food Programme.

WFP (2009b). *Emergency Food Security Assessment Handbook*. Rome: World Food Program.

WFP (2011). Website of the World Food Program. World Food Programme. http://www.wfp.org/ (accessed 25 February 2011).

WFP/VAM (2010). *Vam Understanding Vulnerability Food Security Analysis*. Rome: World Food Program.

WHO (2010). Website of the World Health Organization. World Health Organization. http://www.who.int/ (accessed 2 April 2010).

WHO/GHO (2010). Global Health Observatory. World Health Statistics. http://www.who.int/gho/publications/world_health_statistics/en/index.html (accessed 13 February 2011).

WHO (2011). *WHO Trade, Foreign Policy, Diplomacy and Health: Food Security*. Geneva, Switzerland: World Health Organization. http://www.who.int/trade/glossary/story028/en/ (accessed 1 January 2011).

World Bank (1986). *Poverty and Hunger: Issues and Options for Food Security in Developing Countries*. Washington DC: World Bank Group.

World Bank (2004). Millennium Development Goals: Proportion of the Population Below Minimum Level of Dietary Energy Consumption. World Bank. http://ddp-ext.worldbank.org/ext/GMIS/gdmis.do?siteId=2&contentId=Content_t5&menuId=LNAV01HOME1 (accessed 12 October 2010).

World Bank (2008). World Development Report 2008 Agriculture for Development. Washington, DC: World Bank.

World Bank (2009). Understanding Poverty. World Bank. http://web.worldbank.org/WBSITE/EXTERNAL/TOPICS/EXTPOVERTY/0,,contentMDK:20153855~menuPK:373757~pagePK:148956~piPK:216618~theSitePK:336992,00.html (accessed 5 June 2009).

World Vision (2009). Website of World Vision. World Vision. http://www.worldvision.org.uk/ (accessed 23 May 2010).

2

Good Nutrition: A Basic Introduction

One of the most important generally accepted proxies of inadequate food security is in its outcome: the prevalence of undernourishment. Thus, the concept of food security is predicated on and underpinned by the notion of nutrition, whether good or bad. Indeed, diet and nutrition as components of maintenance and good health is so important throughout an individual's lifetime that education and understanding of nutritional issues as well as public health activities are taking increasingly prominent positions in food security policies (WFP 2011). Therefore, before we are able to discuss food security and malnutrition in any kind of detail, we must first understand the basics of nutrition itself. Questions such as what our nutritional needs are and what is considered sufficient to maintain good health need to be considered. For this reason, a perfunctory glance at some of the more fundamental nutritional issues at this point helps ease the narrative along. The following section looks at current nutritional knowledge and understanding particularly in respect to good or healthy nutrition before any further exploration of the various aspects of hunger and malnutrition is undertaken.

While it is recognised that every cell in the body requires energy, it is just as important and sometimes overlooked that the human body also requires many more nutrients beyond just those with energy-giving properties (WHO 1997). In this endeavour, we derive all our needs from the nutrients present in the food and water we consume without which we risk malnourishment and hunger, which if prolonged causes starvation and death (WFP 2011). In this way a healthy balanced diet aims to satisfy not only all the human energy needs but also all essential nutrients needed for growth and maintenance of the body (FAO/WHO/UNU 2004). In this sense, while food security is clearly necessary, in itself it is not necessarily a sufficient condition to ensure good nutrition. For this reason, food security is more than just about the provision of food; it is about the quality, the bioavailability, the supply and preservation of adequate nutrients (Maxwell and Frankenberger 1992; Latham 1997; WFP 2009b; USAID 2010).

So what is good nutrition? At its very basic level, good nutrition is the foundation necessary for the support and continued well-being, health and development of life. In the words of the World Health Organization (WHO), nutrition is defined as follows:

> An adequate supply of foods ... needed to maintain all the functions of the body ... ensuring [good] healthy living. (Beaton and Patwardhan 1976).

Or more expansively good nutrition aims to achieve a desirable balance between the essential intake of macronutrients and micronutrients vital for physical and mental energy expenditure whilst maintaining a desirable body weight for height or body mass index (BMI) so that the body is neither over- nor undernourished (Wilson 2007; Elamin 2008; Müller and Jahn 2009; WHO 2011).

2.1 Nutrients

Nutrition is concerned with nutrients. The body is made of nutrients derived from all the food we eat, and without exception everything we eat is derived from six major nutrient classes: carbohydrates, proteins, fats, vitamins, minerals and water (WHO/FAO 1996; Lipp et al. 1999; CFNI 2004; Inyang 2007; Kalman 2010). Consequently, health and good nutrition are both concerned with obtaining the right balance of the right nutrients (EUFIC 1998). By way of illustration, the different proportions of the main nutrients that make up the bulk of the human body mass can be seen from Table 2.1 and Figure 2.1.

For ease of categorisation, nutrients can be divided into two distinct groups of dietary components: macronutrients and micronutrients. As will be revealed, each nutrient, whether micro or macro, is important for the vitality of a healthy functioning body (EUFIC 1998). While the remainder of this section explores the various nutrients, their roles and our needs, an examination of the body's metabolic processes in respect of the different nutrients can be seen in Appendix E.

TABLE 2.1

Average Lean Man Body Composition Percent by Weight

	Rowett Institute	FAO	EUFIC[a]	Gibney
Water	60	61.6	60	60 (26%[b], 34%[c])
Proteins	17	17	17	19.1
Fats	17	13.8	14	14.4
Carbohydrate		1.5		0.6
Minerals		6.1		5.3
Remaining	6% including carbohydrates			

Source: Compiled from Latham, M. C., *Human Nutrition in the Developing World. Food and Nutrition Series No. 29.* Food and Agriculture Organization of the United Nations, Rome, 1997; EUFIC, *The European Food Information Centre: What Do We Mean by Nutrition?* The European Food Information Council, 1998. http://www.eufic.org/article/en/health-and-lifestyle/healthy-eating/artid/nutrition-2/; Gibney, M. J., Vortex, H. H., and Kok, F. J. *Introduction to Human Nutrition* Blackwell Publishing, Oxford, 2009; Fact Sheets: Body Composition. The Rowett Institute. http://www.rowett.ac.uk/edu_web/.

[a] Based on proteins and fats constituting 44% and 36% of the dry weight of the body, respectively.
[b] Extracellular water.
[c] Intracellular water.

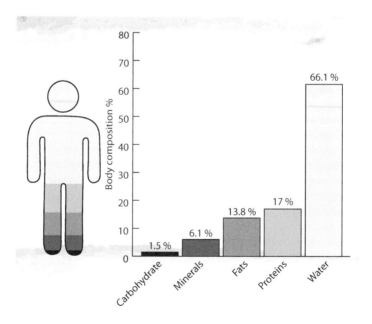

FIGURE 2.1   Nutrient composition of the body by weight. (Courtesy of Patrick Newsham, © 2011.)

2.1.1 Macronutrients and Micronutrients

The two fundamental components of nutrition, macronutrients and micronutrients, are examined in a little more detail.

Macronutrients are carbohydrates, lipids (fats and oils), protein and water and collectively these macronutrients are needed for energy, cell regeneration and repair.

Micronutrients include vitamins, minerals and trace elements which enable the body to produce enzymes, hormones and other substances essential for the metabolic processes of proper growth and development (IoM 2005; Wilson 2008; WHO 2010).

2.1.1.1 Macronutrients

Macronutrients are the nutrients needed in large quantities; these are proteins, lipids (fats), carbohydrates and water. Importantly, although we do not receive either energy or large quantities of nutrients from water, it is often overlooked as a macronutrient, yet as it becomes clear, water is a vital and necessary component of the macronutrients (Wilson 2008; EUFIC 2009). With the exception of water, the macronutrients are energy-providing nutrients, and a person must take in enough of each macronutrient in the right proportions to provide not only the essential 'building blocks' of the body but also a balanced energy profile (WHO 1997, 2006).

2.1.1.1.1 Carbohydrates

While collectively each macronutrient is vital for health and development, carbohydrates are particularly important. Throughout history, grain has constantly been the major source of calories for much of the world (Roorbach 1917; Johnson 2000). Indeed, this continues even today with nearly half the annual cereal/grain supply providing about two-thirds of all energy consumed by humans (Eswaran et al. 2006). However, more importantly for most of the world's poorest, carbohydrates are the main source of food, in some cases providing as much as 80% of their total diet (USDA/ERS 1997; WFP 2000). Carbohydrates mostly originate from plant foods and are the cheapest and most plentiful of all the nutrients. They consist of carbon, hydrogen and oxygen molecules and can be categorised in several ways. At a chemical level, they are often referred to by their molecular structure (or degree of polymerisation) and grouped into two main categories, simple and complex carbohydrates. Carbohydrates can also be grouped according to their absorptive capacity. In other words, how readily the body can metabolise and utilise different carbohydrates in blood sugar (glucose) compared to pure glucose. This is called the glycemic index (GI); it is essentially a scalar measurement with glucose (at 100) representing the highest GI score, medium scores (60–70) are represented by whole wheat and some rice and table sugar while low GI scores (below 55) include most fruits and vegetables (except potatoes), legumes and pulses as well as milk and yogurt. High GI foods cause rapid rises in blood sugar (glucose) levels and is good for recovery after exercise; however, low GI foods are recommended in a balanced diet as glucose is released more slowly and steadily. Low GI foods have also been linked to reduced risk of heart disease and diabetes (FAO 1997; Foster-Powell et al. 2002; Wilson 2008; GI Database 2010; HSPH 2010; WHO/EMRO 2010).

As mentioned, it is also orthodox to categorise or group carbohydrates along the lines of polymerisation. Building on this convention, another orthodoxy allows for similar groupings only this time in ways that further identify the food groupings; sugar, starch and fibre. This helps articulate carbohydrates from the other perspectives and one that is more recognisable to the wider public.

Sugar: Collectively sugars are simple carbohydrates (or simple sugars) and include both monosaccharides (e.g. glucose, fructose, and galactose) and disaccharides (e.g. sucrose, lactose, and maltose).

Starch: Both starch and fibre are complex carbohydrates; a polysaccharide (or complex-sugar polymers with three or more monosaccharides). Starch however differs from nearly all fibre in that starch is readily digestible. Glycogen is digestible starch from animal origin.

Fibre: Fibre is the other complex carbohydrate. In humans, fibre is generally* indigestible. It is sometimes referred to as roughage or dietary fibre although the US Food and Nutrition Board (FNB), among others, makes a further distinction of 'functional fibre', that is beneficial fibre (natural or artificial) which is added to the diet. Fibre does not provide energy directly (see note 1) but is important for the proper functioning of the body. Traditionally, dietary fibre solely referred to those of plant origin nowadays though this term refers to plant, animal and artificial origins. Dietary fibres can also be soluble or insoluble; soluble fibres dissolve in water and ferments while insoluble fibres swell absorbing waste products as soft and bulky faecal matter. Combined, both types help to ease the passage of food through the gastrointestinal tract. The most common fibres include insoluble fibres such as cellulose, hemicelluloses and other substances that offer rigidity to plant material and soluble fibres such as pectins, gums and mucilages (Kelsay et al. 1978; FAO 1997; Latham 1997; IoM 2001, 2005; Brown 2007; HSPH 2010; UoArkansas 2010).

2.1.1.1.2 Proteins

Proteins are the most satiating macronutrient. They are made up of amino acids which are the main building blocks of the body, and like carbohydrates, proteins contain carbon, hydrogen and oxygen; however they also contain nitrogen (WHO/FAO/UNU 2007; Rodriguez and Garlick 2008). As discussed in metabolism (Appendix D), although proteins can supply energy their primary importance is that they are the key component of most cells including muscle, connective tissues and skin cells. As such the nitrogenous substances of proteins are particularly important with respect to the growth and repair of these cells (Latham 1997). Of the many amino acids in existence the body requires 20: of these, 8 amino acids (isoleucine, leucine, lysine, methionine, phenylalanine, threonine, tryptophan and valine) plus a ninth one (histidine for infants) cannot be synthesised by the body. These are called the essential amino acids and must be consumed as part of the diet (Beaton and Patwardhan 1976; Latham 1997; WHO 1997; Wilson 2008). The other amino acids either ingested or synthesised in the body include glycine, alanine, serine, cystine, tyrosine, aspartic acid, glutamic acid, proline, hydroxyproline, citrulline and arginine (Latham 1997). Proteins can be further classified as either complete or incomplete.

Complete proteins contain all the essential amino acids. As noted, essential amino acids cannot be chemically converted or replicated in the body. As a result proteins from meat or animal products like milk and eggs are by far the best source of complete proteins from one source (WFP 2000). However, while animal proteins offer a complete one-stop shop for essential amino acids, they are not always the best source, as they often contain large amounts of cholesterol and saturated fat.

Incomplete proteins are generally of fruit and vegetable origin. Proteins from vegetables, nuts, seeds, legumes and grains all vary in their amino acid makeup and are often combined to complete the full needed compliment. There is one notable exception however and that is soy. Soybean's amino acid profile means that it is one of the few vegetable sources considered to be a complete protein in that it supplies all of the essential amino acids in the right quantities (Erdman 2000; WFP 2000; Tovar et al. 2002; Velasquez and Bhathena 2007).

2.1.1.1.3 Lipids/Fats

Lipids are an important nutrient that the body cannot produce for itself. Often used synonymously with oils and fats, lipids are in fact fat-like chemical compounds that cover a wide variety of naturally occurring molecules, which do indeed include fats but also include waxes, sterols (including cholesterol), fat-soluble vitamins (A, D, E and K) and phospholipids, among others. Fats and oils are collectively known as triglycerides or triglycerols (derived from glycerol and three fatty acids) and are commonly referred to as dietary fat. These are obtained from both animal and vegetable sources and make up the bulk of human lipid intake. They are primarily used for energy while any reserves are stored for future use as

---

* Unlike ruminants that have more of the right enzymes, humans generally find it difficult to digest fibre. Some fibre however is anaerobically fermented by bacteria in the intestinal tract producing short-chain fatty acid by-products that become available as an energy source. The amount amount digested and metabolised varies and depends on many factors and nutritional status of the individual and current understanding suggests that available energy from these sources ranges between 1.5 and 2.5 kcal per gram of fibre.

fat or adipose tissue. On the other hand, phospholipids constitute a major component of cell membranes, while sterols, in the form of cholesterol, are used by the body to regulate various physiological functions (Latham 1997; WFP 2000; WKU 2010). In common with general usage, the term 'fat' or 'fats' in the context of this discussion is used to represent all edible fats and oils (triglycerides). It also follows from the above that the term 'dietary fats', strictly speaking, is distinct from 'dietary lipids' however, and again following common usage, 'dietary fats' in the course of this book is used to represent all lipids (with the exception of the fat-soluble vitamins) used in the body. Not only do humans ingest fats but as discussed in the section on metabolism these can be made from excess carbohydrate intake as well. Most lipids are composed of some sort of fatty acid* arrangement; triglycerides are no different and are made of glycerol (a type of alcohol) and three fatty acids. This separation is important as the nature of the fat depends on the type of its fatty acids, which can be divided into two groups: saturated fatty acids and unsaturated fatty acids, abbreviated to saturated and unsaturated fats, respectively. While in truth all fats contain both saturated and unsaturated fatty acids, they are usually identified according to the predominant fatty acid present (Latham 1997; Ortega 2007; WKU 2010).

Saturated fats are fats, broadly speaking, predominantly of animal origin, for example meat, eggs, milk, cream and butter, although exceptions from the vegetable kingdom include coconut and palm. Saturated fats are distinguished from their unsaturated cousins in that they have no carbon double bonds and are instead saturated with hydrogen atoms. Diets high in saturated fats tend to raise blood cholesterol levels, leading to possible heart disease.

Unsaturated fats originate mainly from plants and can be further subdivided into monounsaturated and polyunsaturated fats. While both help in lowering low-density lipoproteins (LDL cholesterol) and reducing heart disease, some consider polyunsaturates marginally better in this respect by providing more membrane fluidity than monounsaturated fats. Unsaturated fats contain carbon to carbon double bonds, and it is the number and configuration of these bonds that ultimately determine their health properties in the human diet. Monounsaturated fats contains one such bond while polyunsaturated fats contain more than one. With regard to the configuration, the healthy fatty acids are those with the *cis-* as opposed to the *trans-*carbon arrangements (EFSA 2008a; WHO/EMRO 2010; WKU 2010). Both the following, and unless otherwise stated, unsaturated fats in this book are of the *cis-*configuration.

Monounsaturated fats (MUFA) include oils from olive, peanut, sesame seed, canola and avocados. Monounsaturated fats are also high in vitamin E, which helps to develop and maintain cells in the body.

Polyunsaturated fats (PUFA) include oils from corn, cotton seed, safflower, sunflower and soybean. Fish oil too is also polyunsaturated as are omega-3 and omega-6 fatty acids, which have been suggested to be particularly good for brain development.

Although polyunsaturated fats are the preferred choice for some, they are more vulnerable to lipid peroxidation (rancidity). In an attempt to combat this condition and prolong the shelf life of the fats, scientists introduced hydrogenation†; unfortunately however, partial hydrogenation of polyunsaturated fats creates *trans-*fatty acids or *trans* fats (see earlier footnote) (EFSA 2008a, WHO/EMRO 2010; WKU 2010).

*Trans-*fats can either be *trans-*PUFA or *trans-*MUFA, and while these can naturally occur in nature, they are not common; instead, by far the biggest intake of human *trans-*fats is derived from artificial sources. *Trans-*fats are often found in processed foods such as soft or semi-hard margarines and some industrially manufactured cakes, pastries. Trans-fats are particularly harmful in that they act like

---

* Fatty acids are the building blocks of fats. They are considered 'good fats' and can be used for energy by most types of cells. Fatty acids are long hydrocarbon chains capped by a carboxyl group (COOH) at one end and a methyl group (the fatty part) at the other. Fatty acid chains can be saturated (contain no double bonds connecting their carbons and as many hydrogen atoms as possible) or unsaturated (with one (monounsaturated) or more (polyunsaturated) double bonds and fewer hydrogen atoms) (IoM 2005; Ortega 2007).

† Trans-Fats and Hydrogenation. In their natural state exposed unsaturated fats oxidize to create rancid, stale or unpleasant smells or flavours. By saturating the fat, that is fixing more hydrogen to natural unsaturated fats to decrease the number of double bonds in a process called hydrogenation; the process of rancidity is slowed or eliminated altogether. Manufacturers only partially hydrogenate oils as full saturated fats are too solid to use as a food additive. The process of partial hydrogenation of unsaturated fats thus creates *trans-*fats, although they do also exist in fully hydrogenated fat it is not to the same degree, this is purely based on the less-than 100% chemical efficiency of the process.

saturated fats (see note on hydrogenation) and raise levels of cholesterol in the blood contributing to coronary health problems. It has also been found that these fats contribute to membrane loss of fatty acids in the brain with implications for neurodegenerative disorders such as multiple sclerosis (MS), Parkinson's disease and Alzheimer's disease (EFSA 2008a; WHO/EMRO 2010; WKU 2010).

2.1.1.1.4 Water

As mentioned earlier, water is often not seen, or is overlooked, as a macronutrient, yet it is perhaps one of the most important of the macro nutrients both quantitatively and qualitatively as it affects almost every part of the process of nutrition (Lipp et al. 1999; CFNI 2004; Matthys et al. 2007; Wilson 2008; EUFIC 2009). Water variously accounts for around 60%–70% of our total body weight (EUFIC 1998) and losing as little as about 8% can bring about serious consequences while losing 10% can prove to be fatal (EFSA 2008a). Water is the chief component of all body fluids and regulates much of the body's activity. It regulates the body's temperature through perspiration; aids in the digestion of the foods we eat; facilitates the transportation of nutrients around the body; and is also involved at the chemical level helping the metabolisation of foods. It even promotes efficient waste management through the faeces and urine (CFNI 2004; WHO/EMRO 2010). Yet for all its importance, water intake requirements were still up until recently primarily dictated solely by the need to prevent such things as dehydration (Grandjean 2004). In fact, it has been calculated that just over 2.5 litres is lost from the average person through respiration, perspiration, urination and defecation every day. Based on such calculations, average water requirements are in the region of 2.2–2.9 litres per person per day (Howard and Bartram 2003). In short, without food, we can survive many weeks, yet without water we can survive for perhaps a few days (CFNI 2004). The main sources are directly through suitable drinking water or indirectly through the food we eat, particularly in fruits and vegetables.

2.1.1.2 Micronutrients

Unlike macronutrients, micronutrients provide virtually no energy and are only required in small quantities, hence the name. It is a volumetric term only and should not be used to relegate micronutrients to a position of secondary importance; micronutrients are essential in many processes of the body without which, even in these small amounts, severe problems would arise. Such nutrients promote biochemical reactions such as enzyme production and metabolisation, and they also produce hormones as well as many other substances essential for the proper regulation of growth and development (WHO 1997; EUFIC 2009; WHO 2010). Micronutrients consist primarily of vitamins, minerals and trace elements.

2.1.1.2.1 Vitamins

For clarification, the fat-soluble vitamins (A, D, E and K) were earlier categorised to belong to the lipid category associated with macronutrient fats. While they are indeed lipids, they are not considered macronutrients but rather micronutrients due to their negligible energy-giving properties and the low intake requirements. That said, everyone requires vitamins: the organic substances found in various foods and in varying quantities (Latham 1997; WHO 1997). They are either water- or fat-soluble, and many are essential as metabolic catalysts while others are involved in the diverse biochemical functions that regulate every chemical reaction within the body. Also very few vitamins can be manufactured by, and many cannot be retained in, the body; therefore, they have to be ingested regularly as part of the diet. Appendix E is a useful reference that outlines the main sources, uses and characteristics of each vitamin.

2.1.1.2.2 Minerals

Minerals are naturally occurring inorganic chemical elements that are not produced or synthesised in living organisms. Plants obtain these from the soil, whereas animals obtain them from plants or other animals. Minerals can also be carried in the water supply, which varies greatly depending on geographic location (LPI 2010). The human body requires certain mineral elements for proper functioning. Some are collaborative and work with other nutrients, minerals or vitamins regulating many of the body's processes such as metabolisation and energy production while others have specific functions such as providing structure in bones and teeth (WHO/FAO 2004; FSA 2010; MIT 2010; WHO/EMRO 2010).

Depending on utility, minerals can be categorised by importance or function, as in essential elements or electrolytes*; by their structural compounds such as metals, salts and so on; or by groupings like macro (or major) minerals, micro minerals and trace minerals depending on the amount required by the body each day (WHO/FAO 2002; MIT 2010). This book follows the latter convention. Once again however there is no universal agreement over the separation of macro- and micro-minerals. That said, some form of classification is required, and one common orthodoxy is to denote macro-mineral nutrients as those needed in quantities above 100 mg and micro-minerals as those between 1 and 100 mg. The convention followed throughout this discussion will be to aggregate macro- and micro-minerals as minerals needed by the body in amounts from 1 mg upwards. Trace elements for the purposes of this book then are those that fall below this 1 mg threshold (Britannica 2009; MIT 2010). Importantly and to reiterate this categorisation does not reflect the relative importance of such elements but merely their quantitative requirements. The minerals such as potassium, chloride, sodium, calcium, phosphorus, magnesium, manganese and iron and trace elements such as copper, iodine, selenium and chromium are discussed in more detail in Appendix E.

As can be seen from the examination of the various nutrients, the body requires a constant supply of a good balance of nutritional foods for health maintenance and growth. These are required in various amounts depending on age, sex, life stages (pregnancy, lactation, retirement) and lifestyles (sedentary or active, and it is important that these micronutrients (overlooked for a long time) are considered alongside the energy macronutrients (FAO/WHO 2001). In summary, there are good and bad foods, complimentary and cooperative foods and foods that work against each other. As a result, food security is not only about availability, access or risk but just as importantly the utilisation of food. Following on from this questions are then naturally raised as to the determinants of the right foods and the right quantities. This is where the dietary guidelines assume importance.

2.2 Dietary Guidelines

As yet there is no single international standard encompassing overall dietary recommendations although nowadays most countries have established their own nutrient guidelines in order to asses and plan appropriate dietary policies. Consequently there are many differing opinions from as many different countries and institutions. Variously described as Dietary Guidelines (DG), Dietary Reference Intakes (DRI), Dietary Reference Values (DRV) and Dietary Recommendations (DR) among others, the dietary guidelines vary to some degree or another, and determining appropriate nutritional requirements can be both confusing and at times contradictory. It is a situation that needs interpretation and as such is often subject to 'mis'-interpretation (IoM 2001; Wilson 2007; Doets et al. 2008; Rodriguez and Garlick 2008; BNF 2009; USDA 2009; WHO 2010). Ultimately, the goal of all reference intakes is to provide recommendations for an optimum balance of the various energy-giving nutrients. On top of this are the recommendations for the essential minerals and vitamins required for the good health and vitality of the individual, consistent with a low prevalence of diet-related deficiencies or diseases (WFP 2009b). From a user's point of view, dietary guidelines are viewed as prescriptive. However, many guidelines specifically state that such recommendations are to be applied to populations rather than individuals per se. These guidelines are also used by policy makers for assessing, among other things, the adequacy of food supplies in relation to nutritional planning and populations' needs at national and regional levels (Weisell 2002; Brooks et al. 2004). For whichever reason, accurate and up-to-date information must be available. Unfortunately once again, on this matter as with many others in the food security arena, universal agreement is conspicuously lacking. Although to be fair, there is convergence of opinion on many of the important nutrients where in many cases it is just the degree of agreement that, it seems, needs to be fine-tuned. Not in contention however is the notion that there is rarely a single 'best value' for any one

---

* Electrolytes are a category of minerals that exist in body fluids such as blood, plasma and interstitial fluid (fluid between cells). In their capacity, electrolytes exist as acids, bases, and salts, and their main functions are to regulate such body fluids as well as to facilitate electrical impulses between cells (nerve impulses and muscle contractions). Among the many: sodium, calcium, potassium, chlorine, magnesium and bicarbonate are considered to be the most important (MedlinePlus 2010).

nutrient; instead, it is widely accepted that there exist a range of acceptable daily intakes that can achieve equilibrium in different individuals (WHO 2010). This is because an intake level that creates balance in one person might not do so in another. This could be for any number of reasons including age or gender; an individual's basal metabolic rate (BMR); the bioavailability of any particular nutrient and inter alia, the nutrient's interactions with other nutrients. In recognition of this, there is a growing movement away from defining requirements in terms of single figure balances and in favour of a range to encompass the above variables. That said, supporting scientific evidence for such ranges is often scant, and the default fall-back position is that of the status quo, etc. (WFP 2000; WHO/FAO 2004).

Methods used to calculate requirements have been and are continually changing over time, and although a detailed analysis of the relative methods and their merits is beyond the scope of this book, we do look at those used to estimate energy requirements in a little detail later. It must, however, be noted that prescribing daily requirements is not easy, and distinctions are also made between references for consumption of macronutrients (energy) and those of the macronutrients; by way of example, three reference value groupings—those of the UN, the United States and the European Union (EU)—are examined.

2.2.1 UN (WHO/FAO): Dietary Recommendations/Nutritional Requirements

Many countries rely on the WHO and the Food and Agriculture Organization (FAO) recommendations either adopted as national dietary allowances in their entirety or as valuable reference points in their own guidelines. The FAO has been compiling recommendations since its inception, and every 10–15 years sufficient new evidence accumulates that allows revisions and recalculations to be made (WHO 2010). In the case of their most recent recommendations (as of 2011), vitamin, mineral and energy requirements are all based on the latest 2004 reports, while some trace elements in the latest report remains that of 1996 (WHO 1996; FAO/WHO/UNU 2004; WHO/FAO 2004).

In the WHO's vernacular, the following terminologies are used.

2.2.1.1 Macronutrients (Energy)

Energy requirements (ER): Often referred to as daily requirements or recommended daily intakes, these are calculations of energy balance based on the equilibrium between energy input (food) and energy expenditure (BMR + PAL) plus any allowances for growth, pregnancy or lactation. Estimated needs for groups or populations are based on the mean energy requirement of the individuals who constitute that group (FAO/WHO/UNU 2004). These are also broken down into their various constituent components (carbohydrates, proteins and fats) and given a range of values as a percentage of total energy.

2.2.1.2 Micronutrients

Estimated average requirement (EAR) is an estimate of the average daily nutrient intake needed to meet the requirement of half the healthy population of a given group (WHO/FAO 2004).

Recommended nutrient intake (RNI) is set at the EAR values plus two standard deviations (SD), or in the absence of such data, a value based on EAR plus about 10%–12.5% based on the particular nutrient's physiology. This value is expected to meet 97.5% of a healthy population's requirements.

Protective nutrient intake (PNI): The idea of protective nutrients are those intakes above the RNI used to guard against public health risks, for example vitamin C intake of 25 mg to 'enhance iron absorption and prevent anaemia' (WHO/FAO 2004).

Upper tolerable limits (ULs) are used to denote the maximum intake of micronutrients unlikely to pose of adverse health risks in almost all (97.5%) apparently healthy individuals. These figures are based on long-term usage; also of note is that in the absence of adverse health risks, a UL of 10 times the RNI is often used.

2.2.2 The United States: Dietary Reference Intakes

The US dietary guidelines are those proposed by the US FNB of the Institute of Medicine of the National Academies (IoM), which provides 'authoritative judgment[s]' on the relationships between food, health and nutrition (USDHHS/USDA 2005; IoM 2010). The research is extensive and as with other guidelines is being continually assessed as new research comes to the fore.

Originally called the Recommended Dietary Allowances (RDAs), they were renamed the Dietary Reference Intakes (DRIs), the latest of which (as of 2011) are those for 2005 (IoM 2005). The DRIs are a collection of references for both macro- and micronutrients.

In the US vernacular, the following terminologies are used.

2.2.2.1 Macronutrients (Energy)

In the case of energy nutrients (carbohydrates, fats and proteins), recommendations are based on average requirements.

Estimated energy requirement (EER) is the average dietary intake expected to maintain energy equilibrium in healthy individuals of specifically defined groupings with any additions needed for growth, pregnancy or lactation. Within the IoM, macronutrient recommendations are calculated solely based on EERs with no RDAs above and beyond this as, according to the IoM, EERs are sufficient, beyond which unhealthy weight gain is likely to occur (IoM 2005). As with the WHO/FAO, total energy values are expressed in joules or kilocalories. These too are also broken down into their constituent components of carbohydrates, proteins, fats, etc. with a corresponding range of values.

2.2.2.2 Micronutrients

In reference to micronutrient allowances, the US offers four reference values:

Estimated average requirement (EAR) refers to the average daily intake of nutrients expected to fulfil half a healthy population's requirements.

Recommended dietary allowance (RDA) is the average daily intake sufficient to meet the nutrient requirement of approximately 97%–98% of healthy group's individuals.

Adequate intake (AI): In the absence of a determined RDA, an AI recommendation based on empirical data is offered.

Tolerable upper intake level (UL): This is the highest average daily intake level likely to pose no adverse health risks to almost all individuals (97.5) of the population (IoM 2005).

2.2.3 EU Dietary Reference Values

The EU, through the European Food Safety Authority (EFSA) (preceded by the Scientific Committee on Food [SCF]), is responsible for overall dietary guidelines within the EU. That said, many countries produce their own guidelines independent of or based on these recommendations.

The latest reference data (as of 2011) from the EU are due to be fully released in 2010 after a period of public and scientific consultation. In this respect, much of the following data are taken from their draft opinions of 2008/10 (EFSA 2008b, 2009, 2010a). Within the EU's DRVs catchall, there are seven reference values.

2.2.3.1 Macronutrients (Energy)

Recommended intake ranges for macronutrients (RI): The recommended energy requirements are expressed as a range of adequate values based on a percentage of overall energy intake (EFSA 2008b).

2.2.3.2 Micronutrients

The EU micronutrient allowances offer four reference values:

Lowest threshold intake (LTI): This is described as the intake below which most people will not be able to maintain metabolic integrity.

TABLE 2.2

Comparative Nomenclature of Energy and Micronutrient Guidelines

Institution	Energy	Micronutrients				
		Low		Ideal	Upper Safe/ Protective Levels	
			50%	97.5%		
FAO/WHO	ER		EAR	RNI	PNI	UL
US	EER		EAR	RDA	AI[a]	UL
EU	RI	LTI	AR	PRI	AI	UL

[a] Adequate intake: used when no ideal recommendation is present but is not the same as a recommendation.

The average requirement (AR) is the average, given normal distribution, of nutrient requirements that satisfy half groups' population.

Population reference intake (PRI) is a level of intake that aims to reach 97.5% of healthy people's needs.

Adequate intake (AI): This value is used when no PRI is available or cannot be reasonably established. It is based on empirical studies and assumed to be adequate.

Tolerable upper intake level (UL) is considered to be the maximum continual level of daily intake unlikely to pose any adverse health risks (EFSA 2008b).

It can be seen from the above examination that there are numerous different categorisations of dietary reference values, which at first glance might be seen to be a great source of confusion. However, as the previous summary table shows, there are sufficient similarities as to be able to make objective comparisons. Of particular note is the convergence of opinion that aims to satisfy between 97% and 98% of the population with recommended intakes. These are the UN's RNI, the US's RDA and the EU's PRI as evidenced in Table 2.2.

Determining adequate energy requirements for individuals and groups or populations is an involved process that, as will be discovered later, draws directly on over a century of experiential research. For now, a brief evaluation of the challenges, measurements and calculations of the various components of energy requirements is given.

2.2.4 Energy Requirement Guidelines

Most dietary guidelines include an energy intake component (carbohydrates, fats, proteins) often referred to in variations of 'average energy requirement'; this is not to be confused with *minimum* requirements used in the estimation of undernourishment figures. These are usually summed as a total calorie intake. Generally, the fundamental goal of such energy requirement guidelines is to achieve balance. That is, total energy expenditure of an individual is balanced with the total energy intake (FAO/WHO/UNU 2004; IoM 2005). Importantly, such recommendations are generally predicated on average healthy individuals of defined groupings (gender, age, etc.) looking to maintain long-term good health. Special considerations are given to the overweight or those with extra energy needs, such as infants and growing children, pregnant and lactating mothers as well as the overweight or obese (Shetty 2005).

The problem starts with how to determine and measure the critical energy balance below which people are considered to be nutritionally or energy deficient. The main controversy here is whether there is one single state of good nourishment or indeed a range of different states (body weights, physical activity levels, etc.) at which point the individual can be regarded as being well nourished (Seckler 1984; Osmani 1987). Svedberg, in his book, echoes Payne and Cutler's notion of two opposing paradigms on the determinates of the nutritional status of an individual. In the first paradigm, the genetic potential paradigm (GP), it is suggested that in individuals there exists a single unique optimum or preferred state which the body aims to achieve (Payne and Cutler 1984; Svedberg 2000). The second paradigm, the adjustment and adaptation paradigm, proposes that within all individuals there exist a range of possible pseudo-optimal states in which an individual

> … can adjust his energy requirements through changes in his body weight and by varying his PAL in accordance with changes in his external environment without any harmful effects on health or functions. (Svedberg 2000, p. 19)

In practice, empirical evidence borne out by the many BMI studies show that in fact there does exist a range of values within which a person can be said to be in equilibrium. This is also the view taken by the UN and many other institutions.

2.2.4.1 Calculating Total Energy Expenditure

Of the many factors involved, the biggest determinants of energy needs are age and weight as well as the different work and social activities. However, in order to calculate these energy needs often referred to as Total Energy Expenditure (TEE) the solution is to divide these activities into its three component parts: basal metabolic rate (BMR), the physical activity level (PAL) and thermic effect of food (TEF) (FAO/WHO/UNU 2004; IoM 2005; Shetty 2005; WFP 2009b).

Basal metabolic rate (BMR) has essentially been described as 'the minimal rate of energy expenditure compatible with life'. This is the basic energy needed to sustain life at rest and is a major component of total energy needs reaching as high as 60%–70% (FAO/WHO/UNU 2004; Shetty 2005). Energy in this regard is the minimal required for essential bodily functions like maintaining core body temperature, eating, breathing and sleeping, etc. This does not include any component of physical activity beyond such basic needs.

Physical activity level (PAL), although the second largest component of TEE, is by far the most variable among individuals and over time. In sedentary individuals, one-third of TEE might be used for light physical activity, while in very active individuals this can rise to as much as twice the basal energy expenditure (FAO/WHO/UNU 2004; Shetty 2005). Physical activity energy needs are usually calculated for activity types based on increasing degrees of energy expenditure from low to high. Commonly these are broken down into three different habitual or 'average' day-types; the WHO/FAO describes three levels of activity—sedentary or light activity, active or moderately active and vigorous or vigorously active lifestyles—while the United States in their dietary guidelines uses four levels—sedentary, low active, active and very active (FAO/WHO/UNU 2004; IoM 2005). Energy values can then be assigned to each category depending on each day's specific physical needs. Typical energy requirements for each level of physical activity (PAL) is commonly described in terms of multiples of BMR and is calculated as a ratio of TEE/BMR (see Section 2.2.4.1.2 Factorial Approach).

2.2.4.1.1 Thermic Effect of Food

The body also requires caloric energy to consume and digest the food it eats. This energy expended (calories burned) is called thermic effect of food (TEF) and can be between 6% and 15% of total energy expenditure.

Many methods have been used to calculate energy needs over the years. Previously the most accurate was considered to be the use of calorimetry.* Traditionally too, calculating energy needs also focused on estimating intakes that sufficiently maintained a person's equilibrium, that is to say enough intake that neither allowed an individual to gain or lose weight. Conventional contemporary wisdom however now dictates that energy requirements are calculated from TEE of individuals as it is a more reliable estimate of needs than the former (Warwick et al. 1988; Brooks et al. 2004; FAO/WHO/UNU 2004; IoM 2005).

---

* Direct calorimetry is the direct measurement of the heat produced during a chemical process that provides important information about metabolic energy expenditure. It is measured by placing the organism inside a sealed unit. This has many limitations, not least of which was the restrictive nature of the experiment. In response 'indirect' calorimetry methods have been developed, which include the measuring of respiratory gases. That is translating the oxygen consumption and carbon dioxide released into their caloric heat equivalent for living organisms (Nicholson et al. 1996).

TABLE 2.3

Factorial Calculations of Total Energy Expenditure for Sedentary or Light Activity Lifestyle

Main Daily Activities	Time Allocation (hours)	Energy Cost PAR	Time × Energy Cost	Mean PAL Multiple of 24-Hour BMR
Sedentary or light activity lifestyle				
Sleeping	8	1	8.0	
Personal care (dressing, showering)	1	2.3	2.3	
Eating	1	1.5	1.5	
Cooking	1	2.1	2.1	
Sitting (office work, selling produce, tending shop)	8	1.5	12.0	
General household work	1	2.8	2.8	
Driving car to/from work	1	2.0	2.0	
Walking at varying paces without a load	1	3.2	3.2	
Light leisure activities (watching TV, chatting)	2	1.4	2.8	
Total	24		36.7	36.7/24 = 1.53

Source: Adapted from FAO/WHO/UNU, *Human Energy Requirements: Report of a Joint FAO/WHO/UNU Expert Consultation: Rome, 17–24 October 2001*, FAO/WHO/UNU, Rome, 2004. With permission.

Besides the methods of indirect calorimetry still used today there are currently two other methods that are in common use. The first, the doubly labelled water method (DLW) is a relatively new technique while the second is the more traditional factorial analysis approach (the study of the independent variables or factors). Collectively these are perhaps the preferred choices in calculating energy expenditure/needs today.

2.2.4.1.2 Factorial Approach

Nowadays despite the plethora of offered alternative equations which has arisen over the years, it is still more common to mathematically calculate estimates of BMR using the predictive equations based on weight and height,* which were refined by Schofield in 1985 (Schofield et al. 1985; Wong et al. 1996; Woodward 2003; Shetty 2005). In this way, the factorial approach becomes a

> ... convenient way of controlling for age, sex, body weight and body composition and for expressing the energy needs of a wide range of people in a shorthand form. (Shetty 2005)

The factorial approach essentially separates out the BMR and the PAL parts of the Schofield equation and breaks the PAL into component parts of the day. An example can be seen in Table 2.3 taken from the FAO's light sedentary group of activities during the course of one day. Each activity is allocated a physical activity ratio (PAR) component calculated per unit of time; the two are thus combined to give a final PAL which in turn is multiplied by the BMR to give total daily energy expenditure (TEE).

2.2.4.1.3 DLW Approach

The DLW method uses a novel approach adapted from Lifson's animals studies in 1966 and is the most accurate means of testing energy needs in free living individuals (Weisell 2002; Brooks et al. 2004; Shetty 2005). It starts with the introduction into the body of two stable (non-radioactive) isotopic forms of water (D_2O and $H_2^{18}O$). The disappearance or dilution of these isotopes from body fluids (i.e. urine or blood) is equivalent by way of analogy to a half-life of a radioactive equivalent isotope of nearly 8 days for an average man (Wolfe and Chinkes 2005). In this way, as enrichment of the stable isotopes of oxygen and hydrogen in body water declines, a measure of the CO_2 production

---

* Schofield's meta-analysis in 1985 refined BMR's predictive equations works using factors of height, weight and age and is considered a suitable proxy in the absence of calorimetric measures. It is also now the standard calculation used by FAO in predictive BMR calculations.

can be gauged. Because CO_2 is a result of the oxidative metabolisation of fat, carbohydrate and protein, a measure of energy expenditure can then be made. Being both non-invasive and extremely accurate, the DLW methodology is now considered by many as the 'gold standard' of measuring techniques and far more accurate and applicable than the factorial analysis. Others however, like Prakash Shetty of the Food and Nutrition Division of the FAO, while acknowledging the many benefits of DLW, see DLW as complimenting rather than replacing existing methodologies. The DLW is so widely accepted now too that for the first time the United States via the IoM, in their latest dietary energy guidelines, abandoned the traditional factorial approach for all physiologic groups in favour of the DLW methodology (Brooks et al. 2004; IoM 2005). Conversely, the WHO/FAO only uses DLW methods in their calculations for infants, children and adolescents. For adults (both men and non-pregnant, non-lactating women) however, they continue to use the customary method of calculating energy requirements using the factorial approach (FAO/WHO/UNU 2004). This, they cite, is because despite the growing database of DLW figures generally, there is still insufficient data from the developing world to allow universal application.

Adults and children: Energy recommendations for adults, based either on the DLW or factorial approach, commonly utilise the BMR and PAL allowance. For infants (up to 12 months) because of the close correlation between weight and TEE irrespective of the child's age, gender and length, weight is generally used as the energy predictor. Guidelines are based on body weight and include components of TEE plus extra energy for growth (deposition of tissue) and are expressed as energy requirements per kg per day (FAO/WHO/UNU 2004). For children and adolescents (1–18 years), energy totals are calculated on similar lines (using weight as a predictor of energy needs) for the same reasons. Recommendations are then adjusted on a weight-for-age basis and are expressed as both: total daily requirements in milli joules (MJ) or kilocalories (kcal) as well as on a per kg basis MJ (kcal)/kg/day.

2.2.4.2 Average and Minimum Guidelines

It is important to note that daily energy guidelines are different from micronutrient guidelines in that total daily energy recommendations (averaged over several days) are based on the lower end of daily intake levels. The distinction is that the average daily intake is not as with most micronutrients based on EER plus two standard deviations (SD) to cover a safe level for 97%–98% of a group population. Instead, it is noted that energy intake above requirements based on EER or equivalent would result in excess intake of energy that would be deposited as stored energy in the body (fat). Therefore, a safe level (EER+2 SD) is not sought; instead an 'average' (EER) to meet the needs of half the population* forms the basis of such recommendations (FAO/WHO/UNU 2004).

Another important consideration when discussing about energy requirements is the notion of a cutoff point, a point below the recommended average that can be used to quantify the undernourished. This is variously known as among other things the minimum dietary energy requirement (MDER) and the minimum per capita calorie requirement (MPCCR), etc. and is used by the FAO and the United States in quantifying hunger. In establishing a minimum baseline, both the FAO and the ERS in this regard differ slightly in their methodology. The FAO have calculated minimum daily food requirements for each country based on their specific demographic and physiological make-up; these range from 1990 kcal in Croatia, Czech, Republic Estonia and Slovakia to as low as 1680 kcal in Eritrea, Ethiopia and the Occupied Palestinian Territory (FAO 2011a). Although as a general and convenient standard for usage across all countries, the FAO states that on average, a person needs about a minimum 1800 kcal per day (FAO 2011b), which incidentally is also used as the basis for their definition of hunger (SOFI 2008; EarthTrends 2010). The USDA's ERS on the other hand uses a flat rate higher level of around 2100 kcal per person per day (Svedberg 2002; Blössner and Onis 2005; FAO 2008; USDA/ERS 2008).

---

* Average, the use of 'average requirements to meet half the populations needs' often found in the literature can be misleading. This is not to imply the aim of such recommendations are to meet only half of a groups' needs, simply that, 'average' by definition is a mid-point metric that implies 50% fall above and 50% fall below this point.

TABLE 2.4

Varying Institutional Average and Minimum Energy Requirements

Body	Reference	Average Intake	Minimum Intake	
WHO/FAO				
Industrialised countries	Men (age 18–29.9, 70 kg)	2550–2800[a]		
	Women (age 18–29.9, 60 kg)	2000–2200[a]		
Developing countries			Year	
		1996[b]	1996[b]	2009[c]
Sub-Saharan Africa		2100	1800	1760
Middle East[d] and North Africa		2150	1840	1835
East and Southeast Asia		2220	1880	1812
South Asia		2110	1790	1768
Latin America and the Caribbean		2200	1870	1833
	Regional Average	2156	1836	1802
WFP Full Food Rations		2100[e]		
FAO all country minimum requirement + hunger definition[f]			1800	
Europe EFSA[g]	Men	2500		
	Women	2000		
USDA[h]				
Domestic	Men (age 19–30, 70 kg)	2400		
	Women (age 19–30, 57 kg)	2000		
USDA[i]			Approximately 2100	
All developing countries			calories	

Sources: Naiken, L., Measurement and assessment of food deprivation and undernutrition: FAO methodology for estimating the prevalence of undernourishment, International Scientific Symposium, Rome, 2002; WHO/FAO, *Vitamin and Mineral Requirements in Human Nutrition (Draft Version)*, WHO/FAO, Geneva, Switzerland, 2004; USDA/ERS, *GFA-19: Food Security Assessments Reports. Agriculture and Trade Reports*. USDA Economic Research Service, Washington, DC, 2008; EFSA, Scientific opinion of the panel on dietetic products, nutrition and allergies on a request from the commission related to the review of labelling reference intake values for selected nutritional elements, European Food Safety Authority, Parma, 2009; USAID, *Food for Peace*, United States Agency for International Development, Washington, DC, 2009. http://www.usaid.gov/our_work/humanitarian_assistance/ffp/history.html; Meade, B., *Calorie Requirements*, ERS/USDA, Washington, DC, 2010.

[a] Based on Sedentary PAL (FAO/WHO/UNU 2004).

[b] Regional figures taken from the FAO Sixth World Food Survey, 1996 (FAO 1996b).

[c] While updated minimum requirement figures on individual countries are available as of 2009 ranging from 1680 to 1980 kcal per person, regional aggregates have not been published, however, approximate calculations based on these data show similarities with average decreases of just under 2% for each region (FAO 2011a).

[d] Middle East, formerly Near East.

[e] According to FAO, a person needs on average, a minimum energy intake about 1800 kcal per day—consumption of fewer than this figure also as FAO's definition of hunger (EarthTrends 2010; FAO 2011b).

[f] A full World Food Programme (WFP) emergency ration is 2100 kcal (WFP 2009b, 2011).

[g] European Food Safety Authority (EFSA) figures are based on SCF 1992 and CIAA guidelines (SCF, 1993; CIAA 2006).

[h] Based on Sedentary PAL of reference man 70 kg and woman 57 kg, respectively (IoM 2005; USDHHS/USDA 2005).

[i] The USDA including the Economic Research Service (ERS), and USAID use an average 'minimum' energy requirement for each country which averages about 2 100 kcal/person/day in all developing countries studied (67).

Table 2.4 highlights the most up-to-date average and minimum requirements that the various organisations use in their analysis of food security issues.

2.2.4.2.1 Joules or Calories?

A brief note on joules and calories. Technically, in the International System of Units (SI) energy is
expressed in joules (J); however, it is also customary due more to habit and familiarity than anything else
to also use pre-SI thermal energy units (kilocalories [kcal]). An important distinction and some cause
of confusion are the differences in usage of 'calories', 'Calories' and 'kilocalories'. They are explained
below to provide clarity.

Gram calorie is the amount of energy required to raise one gram of water's temperature by 1°C. This
is known as a 'small' calorie with a small 'c', sometimes written as 'cal'.

Kilogram Calorie is the amount of energy required to raise one kilogram of water's temperature by
1°C. This is known as a 'large' calorie with a capital 'C' or kcal. The confusion arises when quite commonly, 'calorie', 'Calorie' and 'kcal' are all used interchangeably however with this in mind the author's
intentions can usually be determined from the context of the piece. It is also worth mentioning too that in
many instances the use of calorie or Calorie is employed where kcal is meant and it has become accepted
convention to do so. In other words calories, Calories and kilocalories are often used interchangeably to
mean the same thing.

Joules: As for joules, a calorie conversion is based on one calorie being equal to 4.184 joules; conversely, 1 J = 0.239 calories (IoM 2005; Hargrove 2006; EUFIC 2010).

TABLE 2.5
Energy Balance Portfolio

Dietary Factor	WHO/FAO	USDA	EU[a]
Total carbohydrate	55%–75%	45%–65%	45%–60%
Total fat	15%–30%	20%–35%	20%–35%
Saturated fatty acids	<10%	As low as possible while ensuring a nutritionally adequate diet	As low as possible while ensuring a nutritionally adequate diet
Polyunsaturated fatty acids (PUFAs)	6%–10%	5.6%–11.2%	No DRV[b]
n–6 Polyunsaturated fatty acids (PUFAs) (linoleic acid)	5%–8%	5%–10%	No DRV[b] (AI[c]=4%)
n–3 Polyunsaturated fatty acids (PUFAs) (α-linolenic acid)	1%–2%	0.6%–1.2%	No DRV[b] (AI[c]=0.5%)
Trans-fatty acids	<1%	As low as possible while ensuring a nutritionally adequate diet	As low as possible while ensuring a nutritionally adequate diet
Monounsaturated fatty acids (MUFAs)	By difference[a]		No DRV[b]
Protein	10%–15%	10%–35%	15%
Cholesterol	<300 mg per day	As low as possible while ensuring a nutritionally adequate diet	No DRV[b]

Sources: WHO/FAO, *Diet, Nutrition and the Prevention of Chronic Diseases, WHO Technical Report Series 916,* WHO/
FAO, Switzerland, 2003; IoM, *Dietary Reference Intakes for Energy, Carbohydrate, Fiber, Fat, Fatty Acids,
Cholesterol, Protein, and Amino Acids,* National Academy Press, Washington, DC, 2005; WHO/FAO/UNU, *Joint
FAO/WHO/UNU Expert Consultation on Protein and Amino Acid Requirements in Human Nutrition. WHO
Technical Report Series 935,* World Health Organization, Geneva, Switzerland, 2007; WHO, Website of the World
Health Organization. http://www.who.int/.

[a] EU figures are based on European Food Safety Authority (EFSA) Panel on Dietetic Products, Nutrition and Allergies draft
proposals 2009, undergoing consultation and subject to change (SCF 1993; EFSA 2008b, 2010b).

[b] No Dietary Recommended Values set.

[c] AI when there is deemed insufficient evidence for a DRV the EU can offer an AI value (adequate intake). This is not to be
seen as a recommendation.

TABLE 2.6

Nutrient Guidelines of the WHO, the United States and the EU

			Energy					
			Carbohydrate			Fat		
	Water	Daily Calories	Total	Sugar	Fiber	Total	Saturated Fat	*n*–6 PUFA Linoleic Acid
Body	**(L/d)**	**Kcal**	**(g/d)**		**(g/d)**	**(g/d)**	**g/d**	**(g/d)**
USDA[a]	2.7–3.7[l]*	W: 2000[j] M: 2400[j]	>130[c] W: 225–325 M: 270–390		25–38*	ND W: 44–67 M: 53–93		12–17*
FAO/ WHO[b]	1.4–2 (3[l])	W: 2000[j] M: 2550[j]	W: 275–375 M: 350–480			W:33–66 M:42–84		
EU	2–2.5[a]*[c]	W: 2000[k] M: 2500[k]	230[d] W: 230–270[f] M: 300–340[f]	90[f] W: –90[f] M: 110–120[f]		70[d] W:65–70[f] M:80–95[f]	20[d] W: 20[f] M: 30[f]	

Body	Vitamin C	Vitamin D[e]	Vitamin E	Vitamin K	Calcium	Chloride	Choline	Chromium
	(mg/d)	**(μg/d)**	**(mg/d)**	**(μg/d)**	**(mg/d)**	**(g/d)**	**(mg/d)**	**(μg/d)**
USDA[a]	75–90	5*	15	90–120*	1000*	2.3*	425–550*	25–35*
FAO/ WHO[b]	45	5	7.5–10	55–65	1000		750*	–
EU	80	5	12	75	800	0.8	425–550*	40

Sources: SCF, Nutrient and Energy Intakes for the European Community. Reports of the Scientific Committee for Food, Thirty-first Series, European Commission Scientific Committee for Food, Luxembourg, 1993; FAO/WHO, Report of the Joint FAO/WHO Expert Consultation on Human Vitamin and Mineral Requirements. FAO/WHO 303, Bangkok, Thailand, 2001; Europa, Opinion of the Scientific Committee on Cosmetic Products and Non-Food Products Intended for Consumers Concerning Chloride, Europa 9, Parma, 2003; WHO/FAO, Diet, Nutrition and the Prevention of Chronic Diseases. *WHO Technical Report Series 916*, WHO/FAO, Switzerland, 2003; IoM, *Dietary Reference Intakes for Energy, Carbohydrate, Fiber, Fat, Fatty Acids, Cholesterol, Protein, and Amino Acids*, National Academy Press, Washington, DC, 2005; EUFIC (2006). Minerals: What They Do and Where to Find Them. The European Food Information Council. http://www.eufic.org/article/en/page/MARCHIVE/expid/miniguide-minerals/#6; WHO/FAO/UNU, Joint FAO/WHO/UNU Expert Consultation on Protein and Amino Acid Requirements in Human Nutrition. *WHO Technical Report Series 935*, World Health Organization, Geneva, Switzerland, 2007; EFSA, Draft: Dietary Reference Values for Water. *Scientific Opinion of the Panel on Dietetic Products, Nutrition and Allergies*, European Food Safety Authority, Parma, 2008a; EFSA, Scientific Opinion of the Panel on Dietetic Products, Nutrition and Allergies on a Request from the Commission Related to the Review of Labelling Reference Intake Values for Selected Nutritional Elements. European Food Safety Authority, Parma, 2009; CIAA (2010). GDAs: Guideline Daily Amounts. Confederation of the Food and Drink Industries of the EU. http://gda.ciaa.eu/asp3/gda_faq/gda_faq.asp; IoM (2010). Website of the Institute of Medicine. National Academy of Sciences Institute of Medicine. http://www.iom.edu/About-IOM/Leadership-Staff/Boards/Food-and-Nutrition-Board.aspx.

[a] USDA values for adults, 19–30 years, based on Institute of Medicine tables;

[b] Adults 19–50 years (approx);

TABLE 2.6 (Continued)

Nutrient Guidelines of the WHO, the United States and the EU

Body	n–3 PUFA α-Linolenic Acid (g/d)	Protein[m] Total (g/d)	Vitamin A[n] (μg/d)	B₁: Thiamine (mg/d)	B₂: Riboflavin (mg/d)	B₃: Niacin[o] (mg/d)[p]	B₅: Pantothenic Acid (mg/d)	B₇: Biotin (μg/d)	Vit B₆: Pyridoxine (mg/d)	B₉: Folate (μg/d)	Vit B₁₂: Cobalamin (μg/d)
USDA[d]	1.1–1.6*	46–56 W: 50–175 M: 60–210	7–900	1.1–1.2	1.1–1.3	14–16	5*	30*	1.3	400	2.4
FAO/ WHO[b]		54–58[g] W: 50–75 M: 64–96	5 –600	1.1–1.2	1.1–1.3	14–16	5	30	1.3	400	2.4
EU		W : – M:	800	1.1	1.4	16	6	50	1.4	200	2.5

Body	Copper (μg/d)	Fluoride (mg/d)	Iodine (μg/d)	Iron (mg/d)	Magnesium (mg/d)	Manganese (mg/d)	Phosphorous (mg/d)	Potassium (g/d)	Selenium (μg/d)	Sodium (g/d)	Zinc (mg/d)
USDA[a]	900	3–4*	150	8–18	310–400	43–45	700	4.7*	55	1.5*	8–11
FAO/ WHO[b]	1100	3–4*	150	9–59[h]	220–260	–	–	2*	26–34	0.5*	3–14[h]
EU	1000	3.5	150	14	375	2	700	2	55	0.575 –3.5	10

[c] The RDA for carbohydrate is set at 130 g/d for both adults and children based on the average minimum needed to keep brain activities functional. This level is typically exceeded with median intakes of 220 to 330 g/d for men and 180 to 230 g/d for women;

[d] Taken from the labeling Reference Intake at 2000 kcal values and based on scientific opinion undergoing consultation and subject to change;

[e] European Food Safety Authority (EFSA) water intakes are preliminary based on scientific opinion; undergoing consultation and subject to change;

[f] Based on EU's current labeling Intake Values;

[g] Based on a 65–70 kg adult;

[h] Depending on Bioavailability;

[i] USDA's domestic reference energy intakes of sedentary adults;

[j] FAO's sedentary adult requirements (for average body weights of 70 kg);

[k] Based on EU labeling Reference Intake guides; Red 'Type' is calculated using stated domestic or industrialised calorie requirements for each institution and recommended energy balance portfolios (Table 2.5).

[l] *Total* water includes all drinking water and water contained in food and beverages.

[m] Based adults 0.8 g/kg body weight for the reference body weight.

[n] USDA figures=*Retinol activity equivalents (RAEs)*.

[o] *Niacin equivalents.*

[p] In the absence of adequate sunlight.

* In the absence of firm recommendations in the form of RDAs, RNIs and PRIs the relevant institutions own Adequate Intakes (AIs) or equivalent are used based on estimations from balance studies, factorial analysis, daily intakes and/or biochemical indicators.

2.2.4.3 Balanced Energy Portfolio

As we have seen, it is not sufficient simply to ingest adequate calories. Instead because of the body's variable physiological needs and the preferential way it treats the different energy macronutrients (carbohydrates, fats, proteins), a balanced energy portfolio is recommended by all major dietary guidelines (Table 2.5). Deciding adequate protein, carbohydrate and fat balances is a complicated process but necessary too as such imbalances can be costly from both a personal and from a public health perspective (WHO/FAO/UNU 2007; WFP 2009a). It has already been mentioned that carbohydrates can provide as much as 80% of total energy needs in some regions; however, it can also be shown that in some industrialised countries, daily fat intake can be as high as 36% (the United States) while others get by on a more reasonable 8%–10% (Latham 1997). This can particularly skew energy intake as was seen in the discussion on metabolism (Appendix D). Recalling too that the energy content of fat at 9 kcal per gram is more than twice that of protein and carbohydrate (both at 4 kcal per gram), any excess intake beyond expenditure is deposited as stored fat usually as adipose tissue. So what are the recommended balances and who decides this? As can be seen from Table 2.5 three of the top institutional and governmental guidelines highlight the variations among the different parties.

Obtaining a relative dietary balance of macronutrients has taken on more significance in recent times and as has been mentioned it is becoming commonplace to recommend a range of values rather than a single numerical requirement. Table 2.5 shows that despite the specific differences there does appear to be general convergence on the relative importance of each of the macronutrients.

2.2.5 Macronutrient and Micronutrient Tables

Having looked at the macronutrients in detail, it just remains to clarify the available micronutrient guidelines in this section. As with previous guidelines, while there are some guidelines more acceptable at the international level such as those of the WHO, the United States and to some degree the EU, many countries have their own set of standards. Sometimes these are based loosely or in large part on some of these more familiar standards while others are wholly the work of the country's host governments or institutions themselves. Table 2.6 summarises the main nutrient guidelines covering the EU, the US and WHO/FAO's recommendations.

References

Beaton, G. H. and V. H. Patwardhan (1976). Physiological and Practical Considerations of Nutrient Function and Requirement. In *Nutrition in Preventative Medicine: The Major Deficiency Syndromes, Epidemiology and Approaches to Control*. G. H. Beaton and J. M., eds. Bengoa. Geneva, Switzerland: World Health Organization.

Blössner, M. and M. D. Onis (2005). Malnutrition: Quantifying the Health Impact at National and Local Levels. *Environmental Burden of Disease Series, No. 12*. Geneva, Switzerland: World Health Organization.

BNF (2009). Nutrient Requirements. British Nutrition Foundation. http://www.nutrition.org.uk/nutrition-science/nutrients/nutrient-requirements (accessed 30 June 2010).

Britannica (2009). Encyclopædia Britannica Online. 2009.

Brooks, G. A., N. F. Butte, W. M. Rand, J.-P. Flatt and B. Caballero (2004). Chronicle of the Institute of Medicine Physical Activity Recommendation: How a Physical Activity Recommendation Came to Be among Dietary Recommendations. *American Journal of Clinical Nutrition* 79(supplement 5): 921S–930S.

Brown, A. C. (2007). *Understanding Food: Principles and Preparation*. Belmont, CA: Wadsworth Publishing Company.

CFNI (2004). Water the Forgotten but Essential Nutrient. Jamaica: Carribean Food and Nutrition Institute.

CIAA (2006). CIAA Recommendation for a Common Nutrition Labelling Scheme. Brussels: Confederation of the Food and Drink Industries of the EU.

CIAA (2010). GDAs: Guideline Daily Amounts. Confederation of the Food and Drink Industries of the EU. http://gda.ciaa.eu/asp3/gda_faq/gda_faq.asp (accessed 25 March 2011).

Doets E. L., L. S. deWit, R. A. Dhonukshe-Rutten, A. E. Cavelaars, M. M. Raats, L. Timotijevic, A. Brzozowska, T. M. Wijnhoven, M. Pavlovic, T. H. Totland, L. F. Andersen, J. Ruprich, L. T. Pijls, M. Ashwell, J. P. Lambert, P. van 't Veer and L. C. deGroot (2008). Current Micronutrient Recommendations in Europe: Towards Understanding Their Differences and Similarities. *European Journal of Mamoun* 47(Suppl 1): 17–40.

EarthTrends (2010). Website of Earthtrends Environmental Information. World Resources Institute. http://earthtrends.wri.org/ (accessed 13 February 2011).

Eduweb (2010). Fact Sheets: Body Composition. The Rowett Institute. http://www.rowett.ac.uk/edu_web/ (accessed 15 October 2010).

EFSA (2008a). Draft: Dietary Reference Values for Water. *Scientific Opinion of the Panel on Dietetic Products, Nutrition and Allergies*. Parma: European Food Safety Authority.

EFSA (2008b). Draft: Principles for Deriving and Applying Dietary Reference Values. *Opinion of the Scientific Panel on Dietetic Products, Nutrition and Allergies* xxx: 1–28.

EFSA (2009). Scientific Opinion of the Panel on Dietetic Products, Nutrition and Allergies on a Request from the Commission Related to the Review of Labelling Reference Intake Values for Selected Nutritional Elements. Parma: European Food Safety Authority.

EFSA (2010a). Dietary Reference Values and Dietary Guidelines. European Food Safety Authority. http://www.efsa.europa.eu/en/ndatopics/topic/drv.htm (accessed 21 March 2011).

EFSA (2010b). Scientific Opinion: Scientific Opinion on Dietary Reference Values for Fats, Including Saturated Fatty Acids, Polyunsaturated Fatty Acids, Monounsaturated Fatty Acids, Trans Fatty Acids, and Cholesterol. Parma: European Food Safety Authority 5.

Elamin, A. (2008). *Protein Energy Malnutrition*. Pittsburg, PA: University of Pittsburgh.

Erdman, J. W. (2000). Aha Science Advisory: Soy Protein and Cardiovascular Disease—A Statement for Healthcare Professionals from the Nutrition Committee of the AHA. *Circulation* 102: 2555–2559.

Eswaran, H., P. Reich and F. Beinroth (2006). Land Degradation: An Assessment of the Human Impact on Global Land Resources. At the 18th World Congress on Soil Science.

EUFIC (1998). The European Food Information Centre: What Do We Mean by Nutrition? The European Food Information Council. http://www.eufic.org/article/en/health-and-lifestyle/healthy-eating/artid/nutrition-2/ (accessed 11 November 2010).

EUFIC (2006). Minerals: What They Do and Where to Find Them. The European Food Information Council. http://www.eufic.org/article/en/page/MARCHIVE/expid/miniguide-minerals/#6 (accessed 2 November 2010).

EUFIC (2009). The European Food Information Centre: The Basics: Nutrition. The European Food Information Council. http://www.eufic.org/article/en/page/BARCHIVE/expid/basics-nutrition/ (accessed 15 November 2010).

EUFIC (2010). The European Food Information Centre: Energy. The European Food Information Council. http://www.eufic.org/page/en/page/what-is-energy (accessed 15 November 2010).

Europa (2003). Opinion of the Scientific Committee on Cosmetic Products and Non-Food Products Intended for Consumers Concerning Chloride. Parma: Europa 9.

FAO (1996). The Sixth World Food Survey. Rome: Food and Agiculture Orgainsation

FAO (1997). Carbohydrates in Human Nutrition: Report of a Joint Fao/Who Expert Consultation. Rome: Food and Agiculture Orgainzation.

FAO (2008). FAO Methodology for the Measurement and Assessment of Food Deprivation: Updating the Minimum Dietary Energy Requirements. Rome: FAO Statistics Division.

FAO (2011a). *Food Security Statistics*. Rome: Food and Agriculture Organization.

FAO (2011b). Website of the Food and Agriculture Organization of the United Nations. FAO. http://www.fao.org/hunger/faqs-on-hunger/en/ (accessed 15 February 2011).

FAO/WHO (2001). Report of the Joint FAO/WHO Expert Consultation on Human Vitamin and Mineral Requirements. Bangkok, Thailand: FAO/WHO 303.

FAO/WHO/UNU (2004). Human Energy Requirements Report of a Joint FAO/WHO/UNU Expert Consultation: Rome, 17–24 October 2001. Rome: UNU/WHO/FAO.

Foster-Powell, K., S. H. Holt and J. C. Brand-Miller (2002). International Table of Glycemic Index and Glycemic Load Values: 2002. *American Journal of Clinical Nutrition* 76(1): 5–56.

FSA (2010). Healthy Diet. UK Food Standards Agency. http://www.eatwell.gov.uk/healthydiet/ (accessed).

GI Database (2010). The Official Website of the Glycemic Index and GI Database. Human Nutrition Unit, School of Molecular and Microbial Biosciences, University of Sydney. http://www.glycemicindex.com/ (accessed 15 October 2010).

Gibney, M. J., H. H. Vortex and F. J. Kok (2009). *Introduction to Human Nutrition.* Oxford: Blackwell Publishing.

Grandjean, A. (2004). *Water Requirements: Impinging Factors and Recommended Intakes.* Geneva, Switzerland: World Health Organization.

Hargrove, J. L. (2006). History of the Calorie in Nutrition. *Journal of Nutrition* 136: 2957–2961.

Howard, G. and J. Bartram (2003). Domestic Water Quantity, Service Level and Health. WHO/SDE/WSH 39.

HSPH (2010). The Nutrition Source. Harvard School of Public Health. http://www.hsph.harvard.edu/ (accessed 5 October 2010).

Inyang, M. P. (2007). Food Intake Pattern of Public Primary School Children in Eleme Local Government Area, Rivers State, Nigerai *Port Harcourt Medical Journal* 1: 190–196.

IoM (2001). *Institute of Medicine: Dietary Reference Intakes—Proposed Definition of Dietary Fiber.* Washington, DC: National Academy Press.

IoM (2005). *Dietary Reference Intakes for Energy, Carbohydrate, Fiber, Fat, Fatty Acids, Cholesterol, Protein, and Amino Acids.* Washington, DC: National Academy Press.

IoM (2010). Website of the Institute of Medicine. National Academy of Sciences Institute of Medicine. http://www.iom.edu/About-IOM/Leadership-Staff/Boards/Food-and-Nutrition-Board.aspx (accessed 13 February 2011).

Johnson, D. G. (2000). Population, Food, and Knowledge. *The American Economic Review* 90(1): 1–14.

Kalman, D. S. a. L., A (2010). A Review of Hydration. *Strength & Conditioning Journal.* Published ahead of print.

Kelsay, J. L., K. M. Behall and E. S. Prather (1978). Effect of Fiber from Fruits and Vegetables on Metabolic Responses of Human Subjects I. Bowel Transit Time, Number of Defecations, Fecal Weight, Urinary Excretions of Energy and Nitrogen and Apparent Digestibilities of Energy, Nitrogen, and Fat. *American Journal of Clinical Nutrition* 31: 1149–1153.

Latham, M. C. (1997). Human Nutrition in the Developing World. *Food and Nutrition Series No. 29.* Rome: Food and Agriculture Organization.

Lipp, J., L. M. Lord and L. H. Scholer (1999). Techniques and Procedures; Fluid Management in Enteral Nutrition. *Nutrition in Clinical Practice* 14(5): 232–237.

LPI (2010). Micronutrient Information Center. Oregon State University: Linus Pauling Institute. http://lpi.oregonstate.edu/ (accessed 12 September 2010).

Matthys, C., S. D. Henauw, M. Bellemans and M. D. Maeyer (2007). Beverage Consumption in Belgian Adolescents. *Asia Pacific Journal of Clinical Nutrition* 16 (Suppl 3): S58.

Maxwell, S. and T. Frankenberger (1992). *Household Food Security: Concepts, Indicators and Measurements: A Technical Review.* New York, Rome: UNICEF and IFAD.

Meade, B. (2010). *Calorie Requirements.* Washington, DC: ERS/USDA.

MedlinePlus (2010). *Medline/Pubmed Database*: US National Library of Medicine/National Institute of Health.

MIT (2010). Optimizing Your Diet. Massachusetts Institute of Technology. http://web.mit.edu/athletics/sports-medicine/wcrminerals.html (accessed 21 March 2011).

Müller, O. and A. Jahn (2009). Malnutrition and Maternal and Child Health.In *Maternal and Child Health: Global Challenges, Programs, and Policies.* J. Ehiri, ed. New York: Springer.

Naiken, L. (2002). Measurement and Assessment of Food Deprivation and Undernutrition: FAO Methodology for Estimating the Prevalence of Undernourishment, International Scientific Symposium, Rome.

Nicholson, M. J., J. Holton, A. P. Bradley, P. C. W. Beatty and I. T. Campbell (1996). The Performance of a Variable-Flow Indirect Calorimeter. *Physiological Measurement* 17(1): 17–43.

Ortega, J. B. (2007). *Polyunsaturated Fatty Acid Metabolism in Broiler Chickens: Effects of Maternal Diet.* Oregon: Oregon State University. Master of Science.

Osmani, S. R. (1987). Controversies in Nutrition and Their Implications for the Economics of Food. Helsinki. UNU-WDIER—The World Institue for Development Economics Research 133.

Payne, P. and P. Cutler (1984). Measuring Malnutrition: Technical Problems and Ideological Perspectives. *Economic and Political Weekly* 19(34): 1485–91.

Rodriguez, N. R. and P. J. Garlick (2008). Introduction to Protein Summit 2007: Exploring the Impact of High-Quality Protein on Optimal Health. *American Journal of Clinical Nutrition* 87 (suppl): 1551S–1553S.

Roorbach, G. B. (1917). The World's Food Supply. *The ANNALS of the American Academy of Political and Social Science* 74: 1–33.

SCF (1993). Nutrient and Energy Intakes for the European Community. Reports of the Scientific Committee for Food, Thirty-first Series. Luxembourg: European Commission. Scientific Committee for Food.

Schofield, W. N., C. Schofield and W. P. T. James (1985). Basal Metabolic Rate—Review and Prediction, Together with an Annotated Bibliography of Source Material. *Human Nutrition Clinical Nutrition* 39C Suppl 1: 5–96.

Seckler, D. (1984). The 'Small but Healthy?' Hypothesis: A Reply to Critics. *Economic and Political Weekly* 19(44): 1886–1888.

Shetty, P. (2005). Energy Requirements of Adults. *Public Health Nutrition* 8(7A): 994–1009.

SOFI (2008). The State of Food Insecurity in the World 2008. Rome: Food and Agriculture Organization 59.

Svedberg, P. (2000). *Poverty and Undernutrition: Theory, Measurement, and Policy*. Oxford: Oxford University Press.

Svedberg, P. (2002). Measurement and Assessment of Food Deprivation and Undernutrition: Part III: Parallel Contributed Papers Sessions: Fallacies in and Ways of Improving the FAO Methodology for Estimating Prevalence of Undernutrition. International Scientific Symposium: FIVIMS—An Inter-Agency Initiative to Promote Information and Mapping Systems on Food Insecurity and Vulnerability, Rome, FAO.

Tovar, A. R., F. Murguia, C. Cruz, R. Hernandez-Pando, C. A. Aguilar-Salinas, J. Pedraza-Chaverri, R. Correa-Rotter and N. Torres (2002). A Soy Protein Diet Alters Hepatic Lipid Metabolism Gene Expression and Reduces Serum Lipids and Renal Fibrogenic Cytokines in Rats with Chronic Nephrotic Syndrome. *The Journal of Nutrition* 132: 2562–2569.

UoArkansas (2010). Starch Versus Highly Digestible Fiber as a Supplemental Energy Source. Arkansas: USDA/University of Arkansas.

USAID (2009). *Food for Peace*. Washington, DC: United States Agency for International Development. http://www.usaid.gov/our_work/humanitarian_assistance/ffp/history.html (accessed 4 January 2010).

USAID (2010). Our Work: A Better Future for All. US Agency for International Development. United States Agency for International Development. http://www.usaid.gov/our_work/ (accessed 12th Jan 2010).

USDA (2009). Food and Nutrition Information Centre: Dietary Guidelines. US Department of Agriculture. http://fnic.nal.usda.gov/nal_display/index.php?info_center=4&tax_level=2&tax_subject=256&topic_id=1342 (accessed 20 November 2010).

USDA/ERS (1997). Food Security Assessment Gfa-9. *Situation and Outlook Series*. Washington, DC: USDA Economic Research Service.

USDA/ERS (2008). GFA-19: Food Security Assessments Reports. *Agriculture and Trade Reports*. Washington, DC: USDA Economic Research Service.

USDHHS/USDA (2005). *Dietary Guidelines for Americans 2005*; US Department of Health and Human Services/USDA.

Velasquez, M. T. and S. J. Bhathena (2007). Role of Dietary Soy Protein in Obesity. *International Journal of Medical Sciences* 4(2): 72–82.

Warwick, P. M., H. M. Edmundson and E. S. Thomson (1988). Prediction of Energy Expenditure: Simplified FAO/WHO/UNU Factorial Method. *The American Journal of Clinical Nutrition* 48: 1188–1196.

Weisell, R. (2002). Measurement and Assessment of Food Deprivation and Undernutrition: Part IV—Summary of the Draft Findings of the Joint FAO/WIIO/UNU Expert Consultation on Human Energy Requirements. International Scientific Symposium, Rome, FAO.

WFP (2000). *Food and Nutrition Handbook*. Rome: World Food Programme (WFP).

WFP (2009a). *Emergency Food Security Assessment Handbook*. Rome: World Food Program.

WFP (2009b). The World Food Programme: Food Quality Control. World Food Programme. http://foodquality.wfp.org/Home/tabid/36/Default.aspx (accessed 2 April 2010).

WFP (2011). Website of the World Food Program. World Food Programme. http://www.wfp.org/ (accessed 25 February 2011).

WHO (1996). *Trace Elements in Human Nutrition and Health*. Geneva, Switzerland: World Health Organization.

WHO (1997). Nursing Care of the Sick: A Guide for Nurses Working in Small Rural Hospitals. *Western Pacific Education in Action Series No. 12*. Manila, Philippines: World Health Organization.

WHO (2006). *Neurological Disorders: Public Health Challenges.* Geneva, Switzerland: World Health Organization.

WHO (2010). Website of the World Health Organization. http://www.who.int/ (accessed 2nd April 2010).

WHO (2011). Who Trade, Foreign Policy, Diplomacy and Health: Food Security. World Health Organization. http://www.who.int/trade/glossary/story028/en/ (accessed 1 January 2011)

WHO/EMRO (2010). *You Are What You Eat*: World Health Organization: Eastern Mediterranean Regional Office.

WHO/FAO (1996). Report of a Joint FAO/WHO Consultation Nicosia, Cyprus: Preparation and Use of Food-Based Dietary Guidelines: Annex Three: the Scientific Basis for Diet, Nutrition and Health Relationships. Geneva, Switzerland: WHO/FAO.

WHO/FAO (2002). *Vitamin and Mineral Requirements.* Rome: WHO/FAO.

WHO/FAO (2003). Diet, Nutrition and the Prevention of Chronic Diseases. *WHO Technical Report Series 916.* Switzerland: WHO/FAO.

WHO/FAO (2004). *Vitamin and Mineral Requirements in Human Nutrition (Draft Version).* Switzerland: WHO/FAO.

WHO/FAO/UNU (2007). Joint FAO/WHO/UNU Expert Consultation on Protein and Amino Acid Requirements in Human Nutrition. *WHO Technical Report Series 935.* Geneva, Switzerland: World Health Organization.

Wilson, M. M. G. (2007). Nutritional Requirements.In *The Merck Manual Online Medical Library.* R. S. Porter and J. L. Kaplan. New Jersey: Merck Sharp & Dohme Corperation.

Wilson, M. M. G. (2008). Disorders of Nutrition and Metabolism.In *The Merck Manual Online Medical Library.* R. S. Porter and J. L. Kaplan. New Jersey: Merck Sharp & Dohme Corperation.

WKU (2010). Bio 113- Lipids. Western Kentucky University. http://bioweb.wku.edu/courses/biol115/Wyatt/Biochem/Lipid/lipid1.htm (accessed 15th July 2010).

Wolfe, R. R. and D. L. Chinkes (2005). *Isotope Traces in Metabolic Research: Principles and Practice of Kinetic Anlaysis.* Hoboken, NJ: John Wiley.

Wong, W. W., N. F. Butte, A. C. Hergenroeder, R. B. Hill, J. E. Stuff and E. O. B. Smith (1996). Are Basal Metabolic Rate Prediction Equations Appropriate for Female Children and Adolescents? *Journal of Applied Physiology* 81(6): 2407–2414.

Woodward, S. (2003). *Biomes of Earth, Terrestrial, Aquatic, and Human-Dominated.* New York: Greenwood Press.

3

Bad 'Mal'-nutrition: The Physiology of Hunger

Having identified some of the fundamentals of nutrition and good nutrition in particular and before we go on to explore the current state of food insecurity in the world today in Chapter 4, it becomes prudent to dissect some of the ideas of malnutrition.

As one of the major proxies of the outcomes of food insecurity, the next big question arises as to how exactly do hunger and malnutrition affect the individual both physiologically and mentally? How does hunger manifest? It has been suggested earlier that much literature concerns itself with undernourishment rather than wider aspects of malnutrition per se. However, until the recent past for many chroniclers of the hungry, when people referred to hunger or malnutrition they often referred solely to the figures of undernourished, which concentrated predominantly on issues of caloric intake. For a long time, this was the norm to the point where undernourished figures tended to overshadow micronutrient disorders leading to an unhealthy preoccupation with the former (Udall et al. 2002). This de-emphasis on the multitude of micronutrient disorders (pre-1980s) was in danger of relegating the problem to a position of secondary importance and indeed for a while the situation was referred to as the *hidden* or *silent* hunger. Thankfully this position is changing, and the salient point today is that malnutrition is now more widely understood to be more than just hunger or undernourishment it represents the spectrum of poor or bad nutrition ranging from the under-nourished to the over-nourished.

3.1 Malnutrition and Nutritional Disorders

According to the UN's Standing Committee on Nutrition (SCN) as well as the World Health Organization (WHO) and others, malnutrition today remains one of the most serious worldwide public health problems and a major contributor to the total annual global disease burden (DFID 2009; WHO/EMRO 2010; WFP 2011). At the same time, undernutrition is generally considered the most pervasive form of malnutrition in developing countries. However, taking into account the accelerated worldwide nutrition transition, then obesity and overweight (overnutrition) are fast becoming comparable public health problems (WHO 2010b) (Figure 3.1). Overnutrition too then is a form of malnutrition, yet despite its less-than-threatening terminology, it can in fact be just as dangerous as undernutrition. This form of malnutrition a trend that is alarmingly forecast to increase, is particularly prevalent in the industrialised countries and the urbanised areas of developing regions characterised by inexpensive, energy-dense foods (Muller and Jahn 2009).

But what of all these terms—undernutrition, overnutrition, malnutrition and hunger? How are they used and more importantly how do they differ?

3.1.1 Nomenclature

Whether malnutrition takes the form of undernutrition, overnutrition, chronic or acute, it is a incumbent to find adequate ways to define or assess the extent of its prevalence in both individuals and groups. A marked difficulty in achieving this aim lies with the nomenclature itself. Familiar languages like hunger and malnutrition need to be properly defined before they can be measured; yet, there seems to be indecision and imprecision over many such attempts. Moreover enormous interchangeability of terminology associated with hunger and malnutrition leads to many terms being misused (WHO 2010c). While in general circulation this may be of little consequence, in a book such as this it becomes necessary to differentiate such matters.

Regional nutrition %

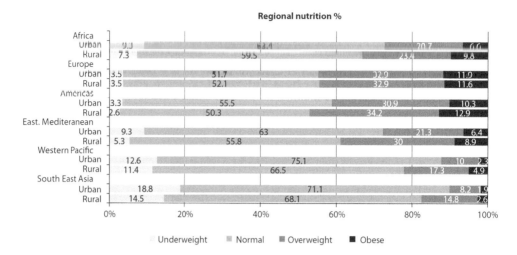

FIGURE 3.1   Regional nutrition: Prevalence of underweight and overweight 2002/2003. (From Moore et al. [2010]. *J. Obesity*, Article ID 514674. With permission.)

3.1.1.1 Hunger

Take for example the term hunger. It is a word often used synonymously with both malnutrition and undernutrition, yet it is also perhaps more importantly an emotive word which both the FAO and WHO use, according to MacAuslan's novel analysis, in subtly different ways: hunger as lack of food and hunger as malnutrition (MacAuslan 2009). While this may seem of little consequence we have already noted that malnutrition is a sort of catch-all that includes incidences of poor, under- and over-nutrition: in this way 'hunger's' usage needs to be tempered or at least qualified. The Oxford English Dictionary (OED) suggests in its proper usage that hunger is the body's physiological response (discomfort or weakness) owing to a lack of food (OED 1989). On this point, because protein is often considered the satiating macronutrient an absence or a lack of this nutrient in the diet can cause feelings of hunger. This comes about whether the lack of protein is a result of undernutrition or through poor nutrition. Therefore, strictly speaking, hunger may adequately describe many states of malnutrition, and as such no separation or distinction is made in this book to the contrary. However, throughout this book the term hunger is used sparingly and carefully in contexts where its meaning is made clear.

3.1.1.2 Malnutrition

Following the same vein, malnutrition by itself refers to the overarching problems of 'mal' or bad nutrition, that is, malnutrition refers to all deviations from adequate or optimal nutritional status. This importantly encompasses not only deficiencies as is commonly cited but also inappropriate nutritional combinations, poor nutrient absorption and/or poor biological use of nutrients (bioavailability*) and, as is becoming increasingly common, excessive intake too (Thomas 2007; DFID 2009).

Broadly speaking, malnutrition whether undernutrition or overnutrition can be thought of as taking two forms: chronic and acute (Cleaver et al. 2006).

- *Chronic*: Chronic malnutrition as with food security itself is thought of as a prolonged or protracted crisis where people constantly fall short of adequate and appropriate nutritional intake. This continuous lack of dietary intake means people, especially children, do not obtain regular or sufficient key vitamins (such as iron, zinc, vitamin A and iodine) that consequently hinder development. For this reason, chronic malnutrition is often associated with stunting (too short

---

* Bioavailability is the availability to the body of usable nutrients; this is, irrespective of the quantity consumed as not all ingested nutrients are available through perhaps infection and disease (Shetty 2002).

for their age) and is ultimately responsible for most hunger-related deaths (USAID 2007; World Vision 2009). In addition, chronic undernutrition affecting children early in life (between 6 and 24 months) reduces their ability to cope as adults. This further traps them in a cycle of vulnerability impairing their mental development and making them more vulnerable to chronic illnesses throughout their lives (Checchi et al. 2007; USAID 2007).

- *Acute*: Acute malnutrition is brought on quickly and often, although not always, without warning. Sudden food shortages usually through emergencies of weather, poor harvests or conflict mean that once available supplies are no longer obtainable in adequate amounts. This results in rapid weight loss or wasted (too thin for their height) individuals with weakened body's. Acute malnutrition as opposed to chronic or slow-onset malnutrition has a more immediate and profound effect on a person's immune system. This is particularly so in children under 5, resulting in a close association between acute malnutrition and mortality (Checchi et al. 2007). Thus, the swift deterioration associated with acute malnutrition kills rapidly, and as a result in such emergencies, it is the acutely malnourished who are to be dealt with first (USAID 2007; World Vision 2009).

3.1.1.3 Undernourishment/Undernutrition

Undernourishment/undernutrition is primarily the result of insufficient intake or inadequate utilisation of ingested nutrients (Shetty 2006; Thomas 2007; DFID 2009). This was historically associated with overall caloric (energy) intake, but strictly speaking it also includes micronutrients, as when calories are deficient, micronutrients (vitamins and minerals) are also likely to be deficient (Thomas 2007). In this way, undernutrition can be seen as the imbalance between what the body needs and what is has biologically available (WFP 2000; Shetty 2002; FAO 2011).

3.1.1.4 Overnutrition

Overnutrition, as is the case of undernutrition, is a component of malnutrition too. Far from being the once-labelled 'disease of affluence', overnutrition in the forms of overweight (excessive weight relative to height) and obesity (excessive body fat content) is becoming increasingly prevalent in developing countries (Cleaver et al. 2006; WHO 2010b). Sweeping dietary changes at the global level, initially in industrial regions but more recently in developing countries see largely plant-based diets being quickly replaced by energy-dense, high-fat, high-sugar, processed diets (WHO 2010b). The health implications in turn are being further exacerbated by physical inactivity as the propensity for a more sedentary lifestyle (which some have suggested is associative of the dietary changes) is becoming the norm in many regions of developing countries too. Combined, these changes are drastically reshaping mortality dynamics, so much so in-fact that when the World Health Reports over the next few years are fully analysed the WHO fully expect to see chronic diseases such as cardiovascular diseases (CVD) (heart disease, hypertension, stroke) and its attendant co-morbid diseases such as diabetes to be the leading cause of death in all developing countries (ibid).

In sum, the use of appropriate language when considering nutrition and food security is an important consideration that is often overlooked. Malnutrition involves a multitude of considerations and can be thought of as existing along a continuum that ranges from undernutrition (hunger and famine) to overnutrition (overweight and obesity) and everything in-between.

Bearing all this in mind we can now consider the actual effects of hunger and malnutrition, and it should not be of any great surprise by now that this is not as straightforward as one might imagine.

3.1.2 Dying of Hunger

As previously hinted at, to answer this question, we have to consider the meaning of hunger in this context. Is its meaning synonymous with starvation? Or, perhaps a wider more inclusive definition that encompasses a broader range of clinical malnutritional disorders? Either way, it would be hoped that mortality attributable to hunger or starvation would be readily identifiable. In practice, it is not that easy. In itself, the mortality numbers resulting directly from hunger or famine have been and still are notoriously difficult to estimate.

Historically, the interaction between hunger, malnutrition and disease was not particularly well understood, and many people in the past invariably attributed such deaths simply to the catchall of hunger and malnutrition (Hionidou 2002). On top of this, poor methodologies or errors in reporting have resulted in a number of problems. Among many issues, for instance, was the lack of official or unreliable records for some of the poorer countries where average crude mortality rates in non-famine years often formed the basis of mortality calculations in years of famine with the difference being attributed to famine. This is a crude simplification but one that illustrates the point. All this however, according to Devereux, has resulted in a lack of confidence in such historical estimates (Devereux 2000).

But what of the present? Nowadays, in practice, it has become customary to talk of mortality due to mal- or undernutrition as 'hunger and hunger-related deaths'. However, this does not address the definitional issue of hunger; indeed, the terminology is still as vague as the numbers it represents. Moreover, it does nothing to answer the question of the actual numbers. Leaving aside the term hunger for a moment and substituting in its place 'starvation' to represent the prolonged and severe absence of adequate vitamins, nutrients and energy intake, then the definition becomes more readily identifiable. Unfortunately the figures do not. In truth, it can be said death directly from starvation or systemic atrophy as it is known is not common, as the following demonstrates.

The main cause of death in cases of undernutrition or instances of famine is not presently nor has it ever been through actual starvation (Mokyr and Ó Gráda 2002). This apparent and long-standing paradox led researchers long ago to look elsewhere for alternative or coincidental factors that might interact with malnutrition in mortality (Ó Gráda 2008). After many studies and much research, there is now widespread consensus that this dichotomy exists as a result of the interaction between malnutrition and morbidity or disease (Pelletier 1994a; Frongillo Jr. 1999; Caulfield et al. 2004). This has a lot to do with the reduced immunity that malnutrition affords the body and the imbalance of the many nutrients that work in harmony to protect against infection and disease. This increased susceptibility has been implicated in many normally non-fatal morbidities, such as diarrhoea, dysentery, respiratory infections (including tuberculosis and pneumonia.), dropsy (oedema, sometimes spelt edema) and to a lesser extent fever (influenza) as well as other opportunistic diseases such as cholera. (Appendix F) (Watkins and Menken 1985; Hionidou 2002; Mokyr and Ó Gráda 2002; Caulfield et al. 2004; WHO 2010b). Of course this works in reverse too; certain diseases such as those considered in secondary protein energy undernutrition (PEU) (Section 3.1.3.3) work against the body's ability to properly absorb or utilise ingested nutrients. Thus, a synergistic relationship can be said to exist between hunger and disease (Elamin 2008). In short, starvation is rarely the terminal event but rather it often leads to other fatal opportunistic diseases.

This aside, taking starvation in adults to its logical conclusion and depending on a person's existing condition and weight, etc., starvation is fatal in approximately 7–12 weeks (Altun et al. 2004; Morley 2007). The process is fairly straightforward and as with the body's metabolic processes which convert food into energy, total withdrawal of food causes the body to catabolise* its own stores of energy. This is analogous to the body ingesting itself. First, the body's glycogen reserves in the liver and indirectly from the muscles are converted into glucose[†] (via glycogenolysis and gluconeogenesis) (Ophardt 2003); this might take up to a few days and is used to balance the blood sugar (glucose) levels (Blakemore and Jennett 2001). Once depleted, the body then looks to the fatty deposits (the adipose cells); for every pound of fat stored in the body, this is equivalent to about 4086 kcal of energy. This process could last for several weeks depending on the individual's initial condition (and assuming sufficient water). Meanwhile, as the individual maintains essential bodily functions, it might adapt by decreasing its basal metabolic rate; however, even in spite of this, fat stores ultimately begin to dwindle and the body becomes emaciated (Altun et al. 2004). Fat depletion causes the cheeks to hollow and eyes to appear sunken, bones protrude and all the while the skin becomes pallid, dry and inelastic (Thomas 2007). At this point, once all the body's fat is used up,

---

* Catabolism is a metabolic process that breaks down large polymer food 'nutrient' molecules such as polysaccharides, lipids, nucleic acids and proteins into the more easily usable monomers of monosaccharides, fatty acids, nucleotides and amino acids, respectively. These can then be further degraded for energy use or recombined to form cell structures in a process called anabolic reactions (King 2010).

† During starvation, the body uses up the liver glycogen to balance blood sugar (glucose) levels within about 24 hours. After this, muscle glycogen is used. Muscle glycogen cannot be directly converted into glucose; however, through glycogenolysis, glucose can be secreted into the blood in an attempt to restore balance (Blakemore and Jennett 2001).

the body initiates the final and catastrophic process of protein catabolism. This process metabolises the body's muscle mass into useable energy. This muscle atrophy is a further attempt to keep vital organs and the immune system functioning. Without immediate nutrient intake, a person becomes counterproductive to his or her own needs, becoming lethargic, apathetic and stuporous. By now, the heart muscle and other organs (except the brain) are shrinking and the skin starts to hang loosely off the body. The pulse is already slowing and blood pressure is dropping. Anaemia commonly develops and is sometimes accompanied by signs of oedema (extracellular fluid collection), and all the while the sufferer's sense of thirst is impaired. Further, malabsorption of water and nutrients manifests as diarrhoea,* which may be ongoing and thus further exacerbating dehydration. All the while concurrent symptoms of vitamin and mineral deficiencies (particularly vitamins B and C) might be also present however, by this time the patient is usually suffering psychological and mental disorders. About now, perhaps 7–12 weeks into starvation if left untreated, then death might result directly from multiple organ failure, severe sepsis, vascular collapse or ventricular fibrillation (Udall et al. 2002; Altun et al. 2004; MedlinePlus 2010). More likely however, even in such extreme circumstances it is well recognised that if diarrhoea (often associated with extreme starvation) has not already become the terminal event, then due to the person's much weakened state, they are more liable to succumb to associated nutritional co-morbidities such as pneumonia, tuberculosis or some other opportunistic disease long before atrophy or catabolism takes its toll (FAO/WHO 1951; Latham 1997). Therefore, in answer to the question of how many people die of hunger (in this case, starvation) the truth is not that many; rarely do people actually starve to death (Checchi et al. 2007; Walker et al. 2007).

If death by starvation is not the greatest killer of the hungry, then it follows that by assessing the types and causes of malnutrition and their contribution to mortality, a fuller more complete picture would be helpful in our pursuit of unravelling morbidity and mortality in the overall understanding of food insecurity. As an aid, we start by separating malnutrition into those of its macro- and micro-nutrient components; this classification helps to logically progress through the many typologies. Commonly however in nutritional literature, there is a predilection of solely referring to deficiency disorders. This misses the point of malnutrition and serves to reinforce existing perceptions of its one dimensionality; although to be fair, the majority of the disorders of malnutrition are indeed brought on by deficiency rather than excesses. However, despite this much of the literature nowadays is beginning to recognise that excesses can be just as problematic as deficiencies. For this reason, the following is labelled *contra*-convention: macronutrient and micronutrient malnutrition.

3.1.3 Macronutrient Malnutrition

Traditionally, macronutrient malnutrition generally focuses on undernutrition in the form of PEU. Overnutrition, however, while not new is increasingly becoming an alarming phenomenon in developing countries, and while the focus of this book is concentrated on malnutrition associated with undernutrition, it is worth including in this section for the sake of completeness.

3.1.3.1 Diet and Non-Communicable Diseases

The last few years has cemented the role of food and food security as central to wider issues of public health. Food for life, food for diet and food for health are at once separate and inextricably linked (WHO/FAO 2003). Later analysis will show how this came about for now though it can be seen that both globalisation and urbanisation have brought major changes in people's physical and dietary lifestyles that are in turn becoming important determinants of non-communicable diseases (NCDs) (WHO/FAO 2003; Sen n.d.). A detailed discussion in relation to food, diet, health and the rise of chronic illnesses is beyond the remit of this book although a brief look at the overall impact would be complimentary.

Increased mortality and morbidity due to NCDs such as cancer, cardiovascular, respiratory obstructive diseases and other chronic illness share many common factors not least of which is diet and physical inactivity. As a result, malnutrition in both forms, undernutrition and overnutrition, occupy prominent positions

---

* Diarrhoea can result from several sources: infections (bacteria, viral and parasitic); malabsorption through intolerance as in celiac disease and lactose intolerance or pancreatic problems associated with cystic fibrosis and short bowel syndrome; or finally inflammatory problems such as irritable bowel or other disorders.

in prevention activities (WHO/FAO 2003). Of these NCDs, it has been stated by WHO that in 2003 cardiovascular diseases alone killed nearly 17 million people; that was nearly 30% of all global deaths. CVDs include coronary (or ischaemic–restricted blood) heart disease; cerebrovascular disease (stroke); hypertension (high blood pressure); heart failure and rheumatic heart disease (ibid). This large death toll, the WHO states, reflects significant changes, inter alia, physical inactivity and improperly balanced dietary habits. Further, this growing trend of overnutrition (the one-time disease of affluence) and inactivity are major risk factors that are no longer simply confined to the so-called developed countries (USDA/ERS 2008b; WHO 2010b). Instead, as previously mentioned, WHO fully expected CVDs to become the leading cause of death in developing countries (WHO 2010b). Accordingly, further research in this field is much warranted and worthy of future consideration, especially in how it relates to the continuum of malnutrition.

Presently however, overnutrition aside and continuing with the focus of hunger and undernutrition, many more individuals within the macronutrient field in fact suffer from deficiencies of calorie and protein intake called PEU (Smith et al. 2000). PEU, of which starvation is the extreme example, has been for a long time and continues to be the most pervasive form of undernutrition in the world (DeMaeyer 1976; DeRose and Millman 1998).

3.1.3.2 Protein Energy Undernutrition

Protein energy malnutrition (PEM) as it is still commonly referred to is a term adopted by the FAO in 1976 (Elamin 2008). It was later changed by the wider community to PEU, possibly reflecting a more focused or specific approach (Morley 2007). That said, both are in common usage and are often used interchangeably. PEU is a deficiency resulting predominantly from the insufficient intake of macronutrients. However, deficiency in macronutrients is often accompanied by deficiency in micronutrients too. This also works in reverse where micronutrient deficiencies themselves can increase the risk of protein-energy malnutrition (Udall et al. 2002; Checchi et al. 2007). However, convention has it that these two issues are dealt with separately and as such the same will be followed here.

Protein-energy malnutrition can be of two causes (primary and secondary), take several forms (kwashiorkor, marasmus, etc.) and be assessed or graded by severity (mild, moderate, or severe) (Jolliffe et al. 1950).

3.1.3.2.1 PEU/PEM Classification

While primary PEU is caused by the inadequate consumption of the energy nutrients, secondary PEU is caused as the result of other disorders, infections and diseases (including micronutrient deficiencies) that interfere with the proper body's utilisation of nutrients (Jolliffe et al. 1950; Udall et al. 2002; Morley 2007). Anyone can be affected regardless of age although it tends to be very common in malnourished children (Thomas 2007). PEU can be sudden and total as in starvation or gradual ranging in severity from sub-clinical at one end of the spectrum where mild PEU manifests simply as poor physical growth in children to clinical kwashiorkor and marasmus (Latham 1997; Morley 2007).

PEU is so called because of the classical theory that suggests problems of dietary protein and energy malnutrition usually occur together. Classical theory also dictates that the form PEU can take depends intimately on the relative intensity and balance of both deficiencies. In milder forms of protein-energy intake deficiency, little distinction can be made but in more severe cases and if one predominates, then certain typologies can be distinguished (Udall et al. 2002). Following on from this, it is said that kwashiorkor (oedematous PEM) is the condition in which the prevailing deficiency is protein. Marasmus, on the other hand, indicates a dominant deficiency of energy while a third condition marasmic-kwashiorkor is prevalent where a combination of energy and protein deficiencies occurs more or less in equal measure (Udall et al. 2002). Although as with many things to do with food security, there are caveats to these simplifications, the following explains.

3.1.3.2.2 Kwashiorkor (Oedematous PEU)

Of all the PEU disorders, kwashiorkor, although not the most pervasive, still remains extremely common and is also the most lethal form of malnutrition with fatality rates at 25% or more in many instances (Amadi et al. 2009). While age is not a barrier to developing kwashiorkor, it is more often seen in younger children perhaps 1- to 4-year-olds (Williams and Oxon 1935). For many, this was thought to be the result of the child-bearing practice whereby older breastfed children were displaced by newborns resulting in the main source of protein from mother's milk being replaced by inferior weaning foods deficient in adequate protein. For many years the syndrome had been known by a multitude of different names including Mehlnahrschaden, carential syndrome, infantile pellagra and sugar baby; this was so both prior to and since the naming of the condition by Cicely Williams in 1933 (Williams 1933; Williams and Oxon 1935; Golden 1998; Bengoa 2000). Cicely originally named the condition after the local Ga dialect from the Gold Coast (now Ghana) Kwashiorkor has been variously translated as 'the sickness of the weaning' and 'first child–second child' however, it was not really until the 1950s that worldwide usage of the term kwashiorkor became standard (Williams 1933; Williams and Oxon 1935; Bengoa 2000; Udall et al. 2002; Thomas 2007; Grover and Ee 2009).

As mentioned, for a long time primary kwashiorkor PEU has been, and is still widely considered to result from an overall deficiency of calories exacerbated by a specific lack of protein in the diet (particularly when the prevailing diet consists of a large proportion of carbohydrates). As such many continue to cite deficient protein as causation in kwashiorkor (Udall et al. 2002; Heird 2007; Morley 2007; Thomas 2007; Al-Mubarak et al. 2010; MedlinePlus 2010). This is in spite of an increasing body of evidence that questions this long-held assumption. One of the first to suggest that deficient protein was perhaps not the actual cause was Gopalan when in 1968 he hypothesised that potential genetic failures in the adaptation of reduced protein intake might in fact be the underlying cause rather than deficiency per se (Gopalan 1968; Golden 1998). Other theories too have emerged over the years; some relate kwashiorkor to multiple dietary deficiencies of amino acids, vitamins and minerals (Black et al. 1998) while others determine that free radicals and their safe disposal compromise cell integrity (Golden 1988). More recent studies too indicate that children suffering from oedematous PEU show a slowing of the protein breakdown in response to food deprivation than those with non-oedematous malnutrition (Jahoor et al. 2005; Amadi et al. 2009). Clearly, consensus is divided on the exact cause of kwashiorkor however, it is looking increasingly unlikely to be the result of low-protein intake as such and more to do with how the body utilises it (Ahmed et al. 2009; Amadi et al. 2009). Although these and other studies are currently challenging the accepted status quo, it seems that literature has been slow to adapt to these and other findings. This is perhaps because, as Golden suggested, there has been no universally agreed alternative paradigm to fill the void (Golden 1998). If the syndrome however turns out after further research to be a symptom of genetic or other non-dietary origin, then as in cachexia (next section) kwashiorkor would need to be reclassified as causal and relegated to secondary PEU. At the moment however, in the absence of any definitive evidence one way or the other, this is a redundant or moot point but one, nonetheless, that should be kept in mind. Although less common than marasmus, kwashiorkor tends to be geo-specific occurring simultaneously where local staple diets that are high in carbohydrate and low in protein (yams, cassavas, sweet potatoes and green bananas, etc.) predominate; these areas tend to be rural Africa, the Caribbean and the Pacific islands (Thomas 2007). Again further research will decide if this is significant or coincidental.

Pathophysiological and clinical symptoms: Fat and muscle wasting is common in all forms of PEU although kwashiorkor in particular is often referred to as the wet, swollen or oedematous form of PEU. This is to do with the kwashiorkor's associated symptomatic characteristic oedema which occurs as a result of sodium and potassium imbalances. Fluid from the lymphatic system accumulates in the extracellular tissue space either through poor drainage and/or increased cell membrane permeability (leakage), resulting in retention and swelling. This is perhaps the single overt distinguishing feature of kwashiorkor, and it is this retention that gives the characteristic 'moon face' or periorbital oedema: a combination of swollen cheeks and sunken eyes. Also common in severe cases is the distended belly (ascites) (Golden 1998; LRF 2006; Morley 2007; Ahmed et al. 2009; NHS 2010). Further symptoms are varied and can be quite dramatic. With acute or chronic PEU, heart size decreases, the pulse slows and blood pressure falls while reduced muscle mass contributes to slow respiratory rates. Affected children can become profoundly apathetic but irritable when moved or held. Petechial haemorrhaging causes blood spots to appear on the surface of the skin while the skin itself can also become dry, thin and wrinkled, which may eventually rupture causing skin rashes and sores. The hair becomes hypopigmented and wispy, falling out easily and eventually becoming sparse. In addition, low plasma protein concentrations (hypoalbuminemia) and hepatomegaly, (the condition of an enlarged liver caused by intense fatty infiltration), might present alongside anaemia and jaundice. The immune system is also compromised, increasing the body's susceptibility to bacterial and other infections such as pneumonia, gastroenteritis, otitis media, urinary tract infections (UTIs), sepsis or diarrhoea. The body's defence communication system as well (the cytokines) can interfere with metabolism and promote further wasting. A suppressed appetite becomes counterproductive, and the body temperature is difficult to regulate possibly resulting in hypothermia or hyperthermia which can also contribute to death. At this point, any number of infections or complications and co-morbidities such as liver, kidney or heart failure may occur, eventually resulting in death (Golden 1998; Maleta 2003; Checchi et al. 2007; Morley 2007; Thomas 2007; Duggan et al. 2008; Roubenoff 2008; Ahmed et al. 2009; Amadi et al. 2009; Rubin and Reisner 2009; MedlinePlus 2010; NHS 2010).

The other two major forms of PEU are marasmus and kwashiorkor-marasmus.

3.1.3.2.3　Marasmus (Non-Oedematous PEU)

It is thought that marasmus occurs when the diet is insufficient in both calories and protein, and unlike kwashiorkor, marasmus has a low mortality rate and also tends to be easier to treat successfully (Jahoor et al. 2005). Although just why severe PEU progresses to either marasmus or kwashiorkor is not clearly understood and as has been suggested cannot be solely explained by the composition of the diet alone (Rabinowitz et al. 2009b). Pathophysiologically, marasmus is described as the body's evolving or adaptive response of those facing insufficient energy intake as opposed to kwashiorkor, which might be considered the body's maladaptive response (Elamin 2008; Rabinowitz et al. 2009b). Derived from the Greek *marasmus*, it means withering or wasting and tends to be prevalent among young vulnerable infants in the first year of life; particularly at about the time of breastfeeding. As a result of insufficient energy intake, marasmus results in progressive emaciation (WFP 2000; Elamin 2008).

Pathophysiological and clinical symptoms: Referred to as the dry form of PEU, marasmus is the most common form of PEU found in children (WFP 2000; Morley 2007; Elamin 2008). Resulting from a partial or total lack of nutrients, the body reacts in the same way as starvation. Faced with reduced food supply, the primary goal of the body is to preserve sufficient energy for the brain and other vital organs (Rabinowitz et al. 2009b). As marasmus progresses, the body's fat stores are depleted before glycogenolysis and gluconeogenesis begins the muscle protein breakdown (atrophy) (WFP 2000; Morley 2007; Elamin 2008; Rabinowitz et al. 2009b). Weight loss and wasting of the fat and muscle as well as emaciated limbs and buttocks makes ribs and other bones protrude and appear prominent. Decreased muscle tone (hypotonia) also makes the muscles appear floppy or rag-doll like while thin, loose skin hangs in folds. Over time, basal metabolic rate slows adapting to decreased intake; this makes sufferers more vulnerable to hypothermia and increases the likelihood of fever during infection. Reduced heart size might also induce hypotension, a lowering of the blood pressure. All the while the more metabolically active organs like the brain are protected for as long as possible (Rabinowitz et al. 2009b). Simultaneously malnutrition compromises mucosal membrane integrity, affecting absorption and secretion, which can commonly aggravate mutually reinforcing bouts of diarrhoea (Brown 2003). During this time further nutrient deficiency weakens the body, making it more vulnerable to disease and infection. This might be compounded by the liver's reduced ability to detoxify the blood, further compromising the immune system. In severe marasmus, cognitive functions and brain development might be impaired while potassium deficiency may result in listlessness and apathy and may contribute to irritability. Lastly in chronic rather than acute episodes of marasmus, reduced linear growth results in permanent stunting (Maleta 2003; Morley 2007; Thomas 2007; Elamin 2008; Rabinowitz et al. 2009b; Rubin and Reisner 2009).

3.1.3.2.4　Marasmic-kwashiorkor

Marasmic-kwashiorkor is a mixed form of PEU with marasmic-type symptoms including severe wasting and/or stunting as well as some signs of kwashiorkor, particularly oedema. It may be that children currently suffering with marasmus deteriorate and present with oedema to become marasmic-kwashiorkor. With regard to the still widely used Wellcome classification of PEM (Appendix F), marasmic-kwashiorkor is defined as those whose weight is below 60% of the median (50th centile) who also present with oedema yet the exact process of how or why PEU develops into one form of malnutrition as opposed to another is not yet fully understood (Jahoor et al. 2006).

3.1.3.3 Secondary Causes of PEU

While primary PEU results from an insufficient diet of protein and/or energy, secondary PEU can occur as a result of, or complication of, other infections, diseases or illnesses; Table 3.1 illustrates some of these secondary causes.

Although Table 3.1 is not exhaustive, it can be seen that two of the most common or devastating complications of secondary PEU are malabsorption and wasting diseases. Failure of the gastrointestinal (GI) tract to absorb nutrients from the diet is often the result of defective or damaged mucosal lining. The most common symptoms of this malabsorption are diarrhoea, bloating, flatulence, cramping and weight loss. In this sense, diarrhoea is particularly noteworthy as not only can it be directly caused by disease or infection, it is also symptomatic of the breakdown of the body's ability to absorb nutrients when in malnutrition. This promotes a synergistic vicious cycle whereby malnutrition is initially predisposed to diarrhoea while in and of itself, diarrhoea aggravates malnutrition (Reddy 1985; Guerrant et al. 1992; Fagundes-Neto and Andrade 1999; Brown 2003). As a result, diarrhoea turns out not only to be a common feature of malnutrition but also one of the major causes of child mortality in developing countries and the second most common cause of death in infants globally, especially in the under 5s (Fagundes-Neto and Andrade 1999; Ashworth et al. 2003).

3.1.3.3.1 Similarities and Differences

Kwashiorkor and marasmic-kwashiorkor are associated with higher morbidity and mortality rates and are easier to treat than marasmus, which tends to be more straightforward and associated with lower mortality. While similarities of all forms of PEU include wasting and lethargy, there are certain important differences present in kwashiorkor and marasmic-kwashiorkor. Most notable of these is the presence of oedema. Others include a higher incidence of disruptive liver functioning resulting in fatty liver (hepatic steatosis). This allows triglycerides to build up, and in worst cases the liver can grow to over three times its original size, be tender and very painful. Signs of skin and hair deterioration tend to be more common in kwashiorkor and marasmic-kwashiorkor and include dermatitis, hypopigmented skin and hair as well as hair thinning and brittleness. Marasmus sufferers also tend to retain a strong appetite whereas non-marasmus patients lose the feelings of hunger. Furthermore, in energy-deficient malnutrition, decreased sodium-potassium pump (Na/K ATPase) activity,* because of its important and multiple functions, can affect cellular nutrient uptake as well as adversely affecting the cell's fluid balance. Other differences involve lowered amino acid and nutrient transport proteins within plasma. Further, decreased levels of cortisol, which aids in the metabolism of protein, are noted in oedematous PEU sufferer's plasma. Lastly, neurological problems in kwashiorkor and marasmic-kwashiorkor manifest as apathy and irritability (McGrath and Goldspink 1982; Jahoor et al. 2006; Norbury et al. 2008).

3.1.4 Micronutrient Malnutrition

While traditionally international malnutrition focus tended to veer towards PEU, more and more current literature is recognising micronutrient malnutrition as being as big a challenge as that of PEU. The numbers themselves are staggering (following chapter) but briefly, in one form or another, the World Food Programme suggested in 2009 that roughly 2 billion people globally suffered some kind of micronutrient deficiency (WFP 2009). Micronutrients are important for a whole host of bodily processes (Appendix E), ranging from aiding metabolism as metabolic co-factors to being the main component of certain of the cellular structures. Timing of micronutrient deficiencies are important too, for while adults generally do not suffer neurological damage due to insufficient micronutrients children certainly do (WHO 2006). This is connected firstly to the fact that the child's nervous system is developed *in utero* and is very susceptible

---

* The sodium-potassium pump (Na/K ATPase activity) is an electrogenic responsible for controlling the cell membrane's electrical potential. It also controls the cell volume through ionic osmosis as well as acting as a cell signal transducer. The sodium-potassium pump also acts as an intercellular nutrient transporter through the use of the sodium gradient (potential difference).

TABLE 3.1

Secondary Causes of PEU

Causes	Examples
Malabsorption or secretory disorders might include gastroenteritic problems (viral and bacterial[a]), as well as pancreatic or lymphatic[b] problems.	*Cholera* caused by the bacterium vibrio cholera often through contaminated drinking water causes extreme diarrhoea and severe vomiting.
	Dysentery usually as a result of parasitic worms, which causes extreme bloody diarrhoea.
	Celiac sprue (celiac disease) is gluten-induced enteropathy of the digestive tract interfering with digestion and absorption of nutrients; the lining (mucosa) of the intestine is damaged, leading to malabsorption.
	Tropical sprue of unknown cause is endemic in tropical regions and causes malabsorptive problems.
	Lymphatic problems in the intestines cause malabsorption of fats from food, which can lead to severe malnutrition (secondary PEU).
	Others include: congenital familial lymphedema (Milroy's disease), cystic fibrosis, Shwachman-Diamond syndrome, giardiasis and other diarrhoea-producing infections and numerous other ailments.
Wasting disorders like those of cancer or infectious diseases	*Anorexia* and *Cachexia* are wasting syndromes. Cachexia is primarily a muscle wasting phenomenon whereby the body produces excess cytokines which appear to interfere with the body's normal metabolic process by promoting the catabolic breakdown of proteins. Importantly and contrary to some literature, cachexia is not the result of starvation or malnutrition, instead it is a wasting away of the body thought to be a syndrome of a metabolic disorder. As a corollary to hunger victims' in which body fat is first consumed in times of deficiency, cachectic patients initially burn off proteins first followed by fats afterwards. For some there is a close resemblance between the processes in cachexia and the 'cytokine storm'.[*] It has been noted that 'more than 80% of patients with cancer or AIDS develop cachexia before death'. It is also associated with conditions such as tuberculosis, parasitic and autoimmune disorders as well chronic heart failure, trauma and major surgery.
Increased metabolic demands	The body's need for energy might surpass usual regular intake of nutrients. These increased metabolic needs are often self-regulated as in eating more due to the hunger response brought about by strenuous physical activity, etc. Extra needs however can also be required in the absence of such physical cues as in such cases of infection, fever, stress, trauma, surgery as well as things like hyperthyroidism and other endocrine disorders.

Sources: Bruera, E., *Br. Med. J.*, 315, 1219–1222, 1997; WFP, *Food and Nutrition Handbook*, World Food Programme, Rome, 2000; Gupte, S., *Recent Advances in Pediatrics: Special Volume 8—Emergency Pediatrics*, Jitendar P. Vij, New Delhi, 2001; Ashworth et al., *Guidelines for the Inpatient Treatment of Severely Malnourished Children*, WHO/UNICEF, Geneva, Switzerland, 2003; Ramos et al., *Curr. Opin. Clin. Nutr. Metab. Care*, 7, 427–434, 2004; Merck Manual, *Merck and the Merck Manuals*, Merck Sharp & Dohme Corporation, New Jersey, 2005; LRF, Website of the Lymphatic Research Foundation. Lymphatic Research Foundation, http://www.lymphaticresearch.org/main.php?menu=about&content=lymphsys, 2006; Luft, F.C., *J. Mol. Med.*, 85, 783–785, 2007; Morley, *The Merck Manual Online Medical Library*. R.S. Porter and J.L. Kaplan, eds, Merck Sharp & Dohme Corporation, New Jersey, 2007; Bauer et al., *Dtsch Med Wochenschr* 133, 305–310, 2008 ; Nnakwe, N.E., *Community Nutrition: Planning Health Promotion and Disease Prevention*, Jones and Bartlett, Sudbury, MA, 2009; Illingworth, J., *Nutrition and Energy: Lecture 25: Cytokines, Thyroid Hormones, Thermoregulation and Basal Metabolic Rate—Anorexia, Bulimia and Cachexia*, School of Medicine, University of Leeds, 2010; MedlinePlus, *Medline/Pubmed Database*, US National Library of Medicine/National Institute of Health, 2010; NHS, Website of the NHS Direct, National Health Service UK, http://www.cks.nhs.uk/, 2010.

[a] Viral causes of enteric disorders belong to the Norovirus, Rotavirus Adenovirus and Astrovirus families while bacterial causes of enteric problems include Salmonella, Shigella, Staphylococcus, *Campylobacter jejuni*, Clostridium, Escherichia coli, Yersinia, Vibrio cholera.

[b] The lymphatic system is a collection of lymphatic organs such as bone marrow, tonsils the spleen as well as lymph nodes (or glands) which are connected by a network of small lymphatic vessels or capillaries. The job of the system is mainly threefold: to transport lymph, a fluid that contains a type of white blood cells for the immune-defence called lymphocytes around the body; secondly to regulate the control of extra-cellular fluid (interstitial and plasma) and to carry away the waste flotsam and jetsam of cellular activity: and lastly the lymphatic system absorbs lipids from foods in the intestine and transports them to the blood (LRF 2006; Macmillan 2007).

[*] Cytokines are proteins released by cells that effect interactions and communications between or on the behaviour of cells. A storm occurs when infection-induced cytokine release can go into overdrive, causing a positive feedback loop resulting in a cytokine storm or hypercytokinaemia (Illingworth 2010).

to deficiencies; second and just as importantly evidence is mounting confirming the correlation between malnutrition in early childhood and long-term adverse mental and cognitive capacity (WHO 2006). When it comes to micronutrient deficiencies, some are more common and more severe than others (WFP 2009). Table 3.2 briefly outlines symptoms and manifestations associated with micronutrient deficiencies while following this some of the more severe deficiencies are examined in a little more detail.

TABLE 3.2

Micronutrient Deficiencies and Their Symptoms

Name	Deficiency Symptoms
Vitamin A (Retinol)	Vitamin A deficiency (VAD), malfunction of the tear glands, drying of the eyes. Xerophthalmia (night blindness, Bitot's spots, etc.), blindness, increased morbidity and mortality and increased risk of anaemia.
Vitamin B_1 (Thiamin)	Beriberi, polyneuritis and Wernicke-Korsakoff syndrome, cardiac failure, polyneuropathy and oedema.
Vitamin B_2 (Riboflavin)	Cheilosis, angular stomatitis, dermatitis, ariboflavinosis, itchy eyes, migraines, peripheral neuropathy.
Vitamin B_3 (Niacin)	Pellagra with diarrhoea, dermatitis, and dementia.
Vitamin B_5 (Pantothenic acid)	Deficiency is uncommon but if it does occur symptoms include fatigue, sleep disturbances, irritability, impaired coordination, nausea, increased heart rate on exertion, numbness and tingling of the hands and feet ('burning feet' syndrome—paresthesia).
Vitamin B_6 (Pyridoxine)	There is no clinical deficient state of vitamin B_6 although sub-clinical deficiency presents with nasolateral seborrhoea, glossitis and peripheral neuropathy and anaemia.
Vitamin B_7 (Biotin)	Fatigue, depression, nausea, dermatitis, enteritis and muscular pains, conjunctivitis, alopecia, central nervous disorders (hypotonia) numbness or tingling (paresthesia) and sometimes hallucinations.
Vitamin B_9 (Folate)	Megaloblastic anaemia. In pregnancy, neural tube defects (NTD) such as anencephaly and spina bifida. Also the feeling of tiredness, breathlessness after little exercise, palpitations and depression. Cognitive dysfunction in children.
Vitamin B_{12} (Cobalamin)	Megaloblastic anaemia, paresthesia, weakness, loss of balance, appetite loss and constipation. Memory loss, disorientation and dementia, also known to damage the sheath (myelin) that covers the cranial, spinal and peripheral nerves.
Vitamin C (Ascorbic acid)	Scurvy, dry and splitting hair, bleeding gums, rough, dry, scaly skin, slow wound-healing, bruise easily, nosebleeds, weakened tooth enamel, swollen and painful joints, anaemia, weakened immune system and slowed metabolism. If untreated can cause death.
Vitamin D (Ergocalciferol, cholecalciferol)	Deficiency leads to rickets in children and osteomalacia in adults (soft bones), also osteoporosis (reduced bone density).
Vitamin E (Tocopherols)	Vitamin E deficiency is common in developing countries. Infants are particularly at risk as their stores at birth are quite low. Main symptoms include hemolytic anaemia, damage to nerves of the peripheral nervous system (neuropathy), impaired reflexes and coordination, difficulty walking, weak muscles, also damage to cell membranes can lead to cell leakage.
Vitamin K_1 (Phylloquinone), K_2 (Menaquinone)	Although rare, vitamin K deficiency in infants up to around 6 months old is a significant public health problem. Because of vitamin E's role in coagulation, deficiency can cause excessive bruising and bleeding known as haemorrhagic disease, this can result in excessive blood accumulation outside the blood vessels (haematoma) in either small purplish spots on the skin (petechia) or largish patches (1cm+) called ecchymosis.

Sources: FAO/WHO, *Report of the Joint FAO/WHO Expert Consultation on Human Vitamin and Mineral Requirements*, FAO/WHO, Bangkok, Thailand, 2001; WHO/FAO, *Vitamin and Mineral Requirements in Human Nutrition (Draft Version)*, WHO/FAO, 2004; WFP, *The World Food Programme: The First 45 Years*, World Food Programme, Rome, 2008; Britannica, Encyclopædia Britannica Online, 2009; Gibney et al., *Introduction to Human Nutrition*, Blackwell Publishing, Oxford, 2009; Nguyen-Khoa, D.-T., Cope, D. W., Busschots, G. V., and Vallee, P. A., *Beriberi (Thiamine Deficiency)*, 2009; Rabinowitz et al., *Beriberi*, WebMD Professional 2009a; Rabinowitz et al., *Marasmus*, WebMD Professional, 2009b; Rubin, E. and Reisner, H. M. *Essentials of Rubin's Pathology*, Lippincott, Williams and Wilkins, Baltimore, MD, 2009; WHO, *WHO Global Database on Vitamin A Deficiency Report: Global Prevalence of Vitamin A Deficiency in Populations at Risk 1995–2005*, World Health Organization, Geneva, Switzerland 2009; LPI, Micronutrient Information Center. Oregon State University: Linus Pauling Institute, 2010; MedlinePlus, Medline/Pubmed Database, US National Library of Medicine/National Institute of Health, 2010.

3.1.4.1 Vitamin Deficiencies

3.1.4.1.1 Vitamin A Deficiency (VAD/Xerophthalmia)

VAD is a prevalent form of global micronutrient deficiency and is widely acknowledged as the single biggest cause of preventable blindness in the world. While vitamin A occurs naturally as retinoids or as a precursor β-carotene, its deficiency can commonly lead to xerophthalmia (Rubin and Reisner 2009). Xerophthalmia is a term that encompasses the clinical spectrum of visual disorders associated with VAD. That said, the effects of deficiency are not limited to visual disorders as VAD also affects multiple other processes. In VAD sufferers, cases of infection appear to worsen, and in severe illnesses such as diarrheal disease and measles can even lead to a significantly increased risk of death. It is also particularly prevalent in times of fat malabsorption in the body, as in bouts of diarrhoea, etc. Lastly, vitamin A also plays a role supporting the transport of iron around the body and deficiency tends to promote anaemia (WHO 2010a). There are two ways commonly used to detect the prevalence of VAD and xerophthalmia; these are clinical eye examinations and biochemically determined plasma concentrations of retinol (AKA serum retinol). Table 3.3 shows the WHO standards of serum retinol, which determine levels of VAD while Table 3.4 is taken from the WHO report (WHO 2009). In Table 3.4 the WHO indicates the gradient classification from mild eye disorders of night blindness and Bitot spots (which still represent moderate-to-severe VAD) to potentially blinding later stages.

The earliest sign of VAD is diminished vision in low light, a condition known as night blindness, and if left untreated, it can progress by degrees to eventual blindness. Several interventions can reduce the prevalence of VAD conditions. First, dietary diversification with better access to vitamin A or provitamin A-rich foods as in mangoes, papaya and dark green leafy vegetables. Second, nutrition education can help to highlight the problem and mitigate the worst effects through dietary change. Third, vitamin A's fortification of staple foods, supplements or even condiments to under-5s has been shown to be effective in reducing the risk of mortality by about 23%–30% (Rubin and Reisner 2009; WHO 2009; LPI 2010; WHO 2010b).

TABLE 3.3

Vitamin A Thresholds in Micromoles

<1.05 µmol/L	Proposed International Standard reflecting low vitamin A among pregnant and lactating women
<0.70 µmol/L	Vitamin A deficiency (VAD)
<0.35 µmol/L	Severe vitamin A deficiency (VAD)

Source: Based on the WHO Global Database, *Vitamin A Deficiency Report: Global Prevalence of Vitamin A Deficiency in Populations at Risk 1995–2005*, World Health Organization, Geneva, Switzerland, 2009.

TABLE 3.4

Classifying Xerophthalmia

XN	Night blindness
X1A	Conjunctival xerosis
X1B	Bitot's spot
X2	Corneal xerosis
X3A	Corneal ulceration/keratomalacia (<1/3 corneal surface)
X3B	Corneal ulceration/keratomalacia (≥1/3 corneal surface)
XS	Corneal scar
XF	Xerophthalmic fundus

Source: Based on the WHO Global Database, *Vitamin A Deficiency Report: Global Prevalence of Vitamin A Deficiency in Populations at Risk 1995–2005*, World Health Organization, Geneva, 2009.

3.1.4.1.2 Vitamin B Deficiencies

Vitamin B deficiencies encompass a few of the more familiar and historically widespread conditions including beriberi and pellagra. The more clinically severe manifestations seem to have declined in many parts of the world in recent years due in part to dietary changes, education and availability of improved food sources. Some common ailments of vitamin B deficiencies include dermatological, neurological and immunological effects (FAO/WHO 2001; LPI 2010).

Thiamine (vitamin B_1) deficiency generally affects the cardiovascular, muscular, nervous and GI systems. Its deficiency can also lead to two types of beriberi: wet and dry. Wet beriberi affects the cardiovascular system while dry beriberi and Wernicke–Korsakoff syndrome disrupt both the neuromuscular and nervous systems (WHO/FAO 2004; Rubin and Reisner 2009).

- Wet beriberi (oedematous) affects the cardiovascular system and can present as acute or chronic. Both acute wet beriberi (Shoshin beriberi) and chronic wet beriberi are high-output cardiac failure-type conditions which ultimately result from the heart's inability to maintain proper function. They are commonly characterised by increased heart rate (tachycardia), swelling of the lower legs (peripheral oedema) and shortness of breath. The condition can be successfully treated with thiamine if caught early (FAO/WHO 2001; WHO/FAO 2004; Britannica 2009; Nguyen-Khoa et al. 2009; Rabinowitz et al. 2009a; Rubin and Reisner 2009).
- Dry beriberi (paralytic) affects the neurological system. A gradual and progressive degeneration of the peripheral nervous system (polyneuropathy) begins; first the sensory (tactile) system is compromised, leading to numbness and tingling (paresthesia) and then pain. This is followed by the degeneration of the motor and autonomic systems, resulting in slow tendon reflexes, foot and wrist drop and eventual complete paralysis (ibid).

Thiamine-deficient mortality is rare but if undiagnosed for long enough, then wet beriberi can be fatal. Morbidity in dry beriberi is also rare and occurs again only if diagnosis is late. Early detection ensures a good prognosis (ibid).

Infantile thiamine deficiency typically occurs in infants whose mothers have previously suffered from beriberi. It is recognised in three forms: cardiologic, aphonic and pseudomeningitic forms with varying degrees of morbidity and mortality (ibid).

Wernicke-Korsakoff syndrome is a predominantly ethnic European genetic disease with weak thiamine-pyrophosphate bonding, which means that thiamine-deficient symptoms present in situations with much less severe thiamine depletion.

Riboflavin (vitamin B_2): Clinical riboflavin deficiency is called ariboflavinosis. It expresses certain similarities to niacin deficiency pellagra, such as typical inflammation and skin lesions especially around the lips and corners of the mouth (*cheilosis and stomatitis*). It tends to be widespread around the world with high concentrations in Africa and Asia. It is rarely fatal it although it has been implicated in preeclampsia, a condition in pregnant women of increased blood pressure, protein discharge in the urine and significant oedema. It is also often seen in combination with other B vitamin deficiencies and becomes common when gastrointestinal infections are commonplace. As well as the skin lesions riboflavin deficiency also presents with oedema of the pharyngeal and oral mucous membranes; accumulation of blood in certain parts of the body (hyperaemia); and disruption of red blood cell production in the bone marrow (normochromic or normocytic anaemia), all the while eyes may become bloodshot and itchy. Riboflavin deficiency has also been associated with developmental abnormalities including cleft lip. Riboflavin replenishment in consideration with other B-complex deficiencies is the preferred course of treatment (Wacker et al. 2000; FAO/WHO 2001; Allee and Baker 2009; Rubin and Reisner 2009; LPI 2010).

Niacin (vitamin B_3) clinical deficiency is known as pellagra, so named by Frappoli in 1771 from the combination of 'pella' meaning skin and 'agra' meaning rough. It is chronic wasting condition which is also aggravated by infection or disease in weakened patients (FAO/WHO 2001). Although on the decline, it is still endemic in poorer areas of India, Africa and China as well as in areas where corn tends to be the staple food. This is because the niacin in corn is not particularly bio-available. Cases of pellagra tend to present with diarrhoea, dermatitis, and dementia. Skin lesions are common and can also be found in the mucous membrane, which when coupled with oedema can see the tongue become inflamed and fissured. In extreme cases vitamin B_6 deficiency can also induce dementia sometimes, bordering on psychosis accompanied by ganglion cell degeneration within the cortex; other neurological changes include insomnia and apathy. Lastly pellagra also accompanies chronic alcoholism and has also been found in people suffering from Crohn's disease (FAO/WHO 2001; Rubin and Reisner 2009).

Folate (vitamin B_9): Although found in abundance in fruit and vegetables, deficiency is not uncommon particularly in regions where fruit and vegetables do not constitute a significant proportion of the diet. B_9 is similar to B_{12} in that deficiency can introduce immature red blood cell precursors into the blood stream (megaloblastic anaemia). In pregnancy common birth defects where spinal cord and/or brain are exposed at birth are known as neural tube defects (NTD) and can include various complications such as anencephaly and spina bifida. Studies have shown that folate intake can considerably reduce this risk (Gibney et al. 2009; Rubin and Reisner 2009).

Cobalamin (vitamin B_{12}): People with mild vitamin B_{12} deficiency may not experience any symptoms. Deficiency itself can be caused by the autoimmune disease pernicious anaemia, which results in the introduction of immature red blood cell precursors into the blood (megaloblastic anaemia). This can further interfere with B_{12} uptake from the intestinal tract. Low levels of B_{12} can also cause numbness or tingling (paresthesia), weakness, disorientation, dementia (with or without mood changes) and loss of balance. Lastly deficiency is also known to damage the sheath (myelin) that covers the cranial, spinal and peripheral nerves. Not synthesised by plants and mainly found in meats, vegans not taking supplements tend to be at most risk. It has been estimated that B_{12} deficiency affects between 10% and 15% of individuals older than 60 years of age (FAO/WHO 2001; Gibney et al. 2009; MedlinePlus 2010).

3.1.4.1.3 Vitamin C Deficiency

Ascorbic acid (vitamin C) is found in abundance in fruit and vegetables and fresh milk. General vitamin C deficiency can be found in regions of scarcity or where vitamin rich foods do not form part of the main diet. General deficiency causes tiredness, weakness and irritation. Vitamin C also aids in the iron absorption, and deficiency can contribute towards anaemia. It is also an antioxidant which protects against naturally occurring free radicals while its prolonged (months) deficiency can lead to scurvy.

Scurvy: Vitamin C deficiency interferes with the synthesis of the body's tissues (collagen). Deterioration occurs as a result with a range of symptoms including inflamed and bleeding gums (gingivitis) and mucous membranes, muscle pain and joint pain. The skin can become rough, dry and scaly while wounds can take longer to heal. Frequent nosebleeds and an overall weakened immune system are coupled with slowed metabolism. If untreated, scurvy can cause haemorrhaging into various tissues or cause cardiac arrest and both can be fatal (WHO/FAO 2004; MedlinePlus 2010).

3.1.4.1.4 Vitamin D Deficiency

Vitamin D deficiency is common, particularly in children and breastfed babies because of their increased needs associated with skeletal growth and the natural low levels of vitamin D in human milk. Deficiency is caused by lack of sunlight (skin synthesis) or inadequate dietary intake. Continued deficiency in children leads to a condition known as rickets while in adults it is known as osteomalacia. Rickets is very common in many developing countries, and both are disorders of the bone where the bones are softened, leading to potential fractures and deformity. Vitamin D deficiency can also contribute (with calcium) to osteoporosis where the bone mineral density is reduced, leading to increased risk of fractures (WHO/FAO 2004; Rubin and Reisner 2009; MedlinePlus 2010).

3.1.4.2 Mineral Deficiencies

As with vitamins, mineral deficiencies can be just as prevalent. Of all the minerals, the body requires deficiencies of only four or so are particularly pervasive; these are iron, iodine, selenium and zinc. These are summarised in Table 3.5 with iron and iodine deficiencies subsequently examined in more detail.

TABLE 3.5

Mineral Deficiencies and Their Symptoms

Name	Deficiency Symptoms
Calcium	Hypocalcaemia (low blood calcium levels) can lead to osteoporosis, rickets or tetany. If calcium levels fall too low, it can result in nerve and muscle impairments and perhaps fatal heart arrhythmia.
Iron	Iron deficiency anaemia can lead to delayed mental development in children, tiredness, shortness of breath, palpitations, chest pain, heart problems or infections.
Iodine	Iodine deficiency syndromes include goitre; hypothyroidism and Cretin disease, culminating in increased foetal mortality; miscarriages; stillbirths; congenital abnormalities; severe mental and growth retardation; deaf-mutism and other physical disabilities.
Selenium	Selenium deficiency can result in the potentially fatal form of cardiomyopathy (Keshan disease), adverse mood states, cancer, liver necrosis, and can also affect the immune system too.
Zinc	Zinc deficiency can lead to delayed motor development in children, depression, growth retardation, skeletal immaturity, neurological disturbances, dermatitis, alopecia, diarrhoea, susceptibility to infection and loss of appetite.

Sources: Beck et al., *J. Nutr.*, 133, 1463S–1467S, 2003; Saito et al., *J. Biol. Chem.*, 278, 39428–39434, 2003; WHO, *Neurological Disorders: Public Health Challenges*, World Health Organization, Geneva, 2006; Zimmermann et al., *Lancet*, 372, 1251–1262, 2008; Gibney et al., *Introduction to Human Nutrition*, Blackwell Publishing, Oxford, 2009.

3.1.4.2.1 Iron

The term 'anaemia' refers to conditions that reduce the amount of oxygen carried by the blood in the body. This can be either as a result of a lowered red blood cell count or if the existing red blood cells do not contain enough haemoglobin.* Iron deficiency anaemia results from reduced iron being available to the haemoglobin protein. Sufferers feel tired, shortness of breath, palpitations and chest pain and if untreated severe deficiency can cause heart problems, infections, problems with growth and development (WFP 2000; Gibney et al. 2009; WHO 2010b).

3.1.4.2.2 Iodine

Iodine is needed for the production of thyroid hormones, and its deficiency causes several conditions collectively known as iodine deficiency syndromes. These can involve wide ranging symptoms and contribute to increased mortality of foetuses and children through miscarriages and stillbirths. Without adequate iodine, the thyroid can also progressively enlarge, causing goitre. It can also lead to decreased production of thyroid hormones (hypothyroidism). Lastly, iodine deficiency can contribute to perinatal mortality, congenital abnormalities and Cretin disease (WHO 2006). Cretinism results when insufficient iodine leads to severe mental and growth retardation, which can also lead to deaf-mutism and other physical disabilities (WHO 2006). Both Goitre and Cretin disease are often concentrated in areas of iodine-poor soils, typically these are mountainous or glaciated regions and other areas prone to flooding or heavy rainfall which tends to wash the iodine from the soils (WFP 2000; Zimmermann et al. 2008).

3.1.4.3 Treatment and Rehabilitation

Millions of under-5-year-old children are malnourished and of those admitted to hospital, face fatality rates of between 30% and 50%, which according to the WHO is a huge failing of management. In fact, the WHO cites evidence of proper palliative care as incontrovertible proof that improved management could effectively reduce this number to less than 5% (Ashworth et al. 2003, 2005). So how are the clinically malnourished treated? In the initial stages any attempts to intravenously rehydrate or increase weight gain before addressing other underlying problems can be fatal, as can manipulating abnormal blood chemistry or the use of diuretics to address oedema or perhaps prescribing iron in the treatment

---

* Haemoglobin: Red blood cells (erythrocytes) are important in the body for several reasons chief of which is that they carry oxygen around and remove carbon dioxide from the body. They achieve this via the biomolecule haemoglobin, which is an iron-rich protein pigment that binds iron with oxygen to fuel the cells and binds carbon dioxide in its waste management cycle.

TABLE 3.6

Ten Steps in Addressing Malnutrition

General Principles for Routine Care

1. Treat/prevent hypoglycaemia
2. Treat/prevent hypothermia
3. Treat/prevent dehydration
4. Correct electrolyte imbalance
5. Treat/prevent infection
6. Correct micronutrient deficiencies
7. Start cautious feeding
8. Achieve catch-up growth
9. Provide sensory stimulation and emotional support
10. Prepare for follow-up after recovery

Sources: Based on Ashworth et al., *Guidelines for the Inpatient Treatment of Severely Malnourished Children*, WHO/UNICEF, Geneva, Switzerland, 2003; Ashworth et al., *Guidelines for the Inpatient Treatment of Severely Malnourished Children*, London School of Hygiene & Tropical Medicine, London, 2005.

of anaemia. In fact, these can all be dangerously fatal in the early stages of treatment. In truth while it might seem to go against common sense or common decency, taking a severely malnourished child and attempting to aggressively re-feed or rehydrate them without consideration of the bigger picture will actually cost lives, not save them. This has a lot to do with the fact that the malnourished do not respond to treatment in the same way as the well-nourished. Successful and proper management of care therefore means addressing the symptoms methodically in an explicit manner and in a specific order (WHO 1999; Ashworth et al. 2003, 2005).

Malnourishment affects every cell, organ and system in the body and in treatment, the WHO/UNICEF guidelines recommend a 10-step process of treatment for both marasmus and kwashiorkor sufferers (Table 3.6). This consists of a two-phase approach: stabilisation and rehabilitation.

Hypoglycaemia and hypothermia often occur together as a result of infection. By treating these together with frequent glucose or sucrose feeding either orally or intravenously (IV) as well as simultaneously warming the child helps to stabilise blood sugar levels and raise the core body temperature. Next, rehydration with rehydration solution for malnutrition (ReSoMal) rather than the usual rehydration salts* should be administered either orally, nasogastrically or through IV. This is continued until respiration and pulse rates slow down (assuming no co-existing infection or over-hydration) and the child begins to pass urine. Of note here is that for rehydrating oedematous patients, the IV route should be avoided if possible as oedema often presents with low blood volume and IV in such cases might result in danger of flooding the circulation. These first three steps might take between one and two days to rectify (WHO 1999; Ashworth et al. 2003, 2005). In step 4, correcting the electrolyte imbalance, although the child's plasma sodium often shows low levels, can be misleading as all severely malnourished children have excess body sodium and low levels of potassium and magnesium (see sodium-potassium pump). This can be rectified by adding extra potassium and magnesium prepared in a liquid form directly to existing feeds. This imbalance can take up to two weeks to redress. As for treating or preventing infection, the WHO suggests that the usual tell-tale signs of fever might be absent and prescribes a broad spectrum of antibiotics as cautionary and preventative, plus specific drugs in the event of specific identifiable ailments. By addressing micronutrient deficiency in step 6, a multivitamin supplement is recommended although with the exclusion of iron for the first week as this could worsen infections. If the child responds favourably, then cautious feeding can commence in step 7, this involves small frequent milk-based

---

* Standard rehydration salts contain sodium and potassium. However, for severely malnutritioned children, these salts contain too much sodium and not enough potassium (Ashworth et al. 2003, 2005).

formulas and can begin as soon as the child becomes stable (perhaps in the first 1 or 2 days). Volume is increased and frequency is decreased over time. Lastly, all going well and if the appetite is returned, then steps 8 through 10 are concerned with: catch up growth (vigorous weight gain feeding); stimulation and support; and lastly follow-up monitoring ensuring that the child remains within above the 90% weight-for-length (equivalent to -1 SD) target (WHO 1999; Ashworth et al. 2003, 2005). Even despite treatment however, in children at least, the condition may continue to be long-lasting (perhaps life-long), leading to mental impairment or digestive difficulties, while in adults, treatment usually means full recovery.

It can be seen from the above that malnutrition and health are clearly intertwined and from previous discussions in this section, it has also been alluded to that disease and morbidity are more often the cause of death than actual starvation. The question then remains, to what degree does malnutrition potentiate or increase the risk of death? To answer this more specifically, the following explores this question through one of the more vulnerable population subgroups, children. Children are chosen partly because of their increased vulnerability and representability and partly because much of the work in this area for the same reasons tend to be aimed at this group.

3.2 Malnutrition, Disease and Child Mortality

The close relationship between child mortality and malnutrition has been observed in numerous studies over the years. However, it was not until the advent of a series of major studies some two decades apart that the first real formal attempts to qualify and estimate the impact of malnutrition as a major risk factor in child mortality were undertaken. Three of these path-breaking epidemiological studies were conducted by Leonardo Mata, then the chief of the INCAP Division of Environmental Biology in 1963, and later by Dr. Ruth Puffer, a biostatistician and public health professional in the United States, in 1967 and 1973. The first study in 1963 investigated child health in the Guatemalan villages and went on for nearly 10 years and overwhelmingly illustrated that infection indeed played a major role in the onset of malnutrition. A similarly motivated study by Dr. Puffer in 1967 began with an investigation into urban childhood mortality while the second followed in 1973 with the Pan American Health Study which spanned some 10 countries within the America's (developed and developing) surveying over 35,000 child deaths between 1968 and 1972 (Colley 1973; PAHO). Using interviews and death certificates as primary sources of data, the outcome of the studies determined that malnutrition was responsible for a staggering 55% of childhood (under 5s) deaths.

As groundbreaking as these and similar studies were though, general limitations in assumptions, according to Pelletier et al. (1993) saw many such studies treat malnutrition almost as a disease in its own right instead of the synergistic adjunct that it is. Pelletier and colleagues responded in 1993 by creating a theoretical formula to describe the synergism between malnutrition and morbidity in six studies in developing countries. Essentially his theory proposed that while exposure to disease and infection was identical for the whole population, risk of fatality was directly correlated with malnutrition (Pelletier et al. 1993). This remarkable study applied the model to previous empirical studies with profound results and implications for policy. As a result, rather than reinforcing previous practices that seemingly treated malnutrition as an additive, something that could almost be separated out and treated independently, Pelletier's study confirmed the potentiating effect of malnutrition. That is malnutrition was empirically seen to be in synergy with disease and infection, encouraging and exacerbating the infections pathology. Furthermore, Pelletier importantly suggested that mild-to-moderate malnutrition was responsible for the majority of mortality on account of the sheer numbers in this category compared to the lesser but more severely malnourished numbers per se (Pelletier et al. 1993; Blössner and Onis 2005). Several important outcomes of this study are worth highlighting, not least of which was the realisation of the need to address the mechanisms by which malnutrition exacerbated infection. This had implications for policy responses too which suggested a dual approach to tackling mortality by treating the mild-to-moderate malnourished as well as the diseased and infected. However, perhaps the most important outcome of Pelletier's study was the practical implications that elevated the concept of the malnutrition-disease synergy out of the clinic to that of real-world applications (Pelletier et al. 1993, 1994a; DeRose and Millman 1998; WFP 2000; Blössner and Onis 2005).

TABLE 3.7

Relative Risks of Mortality Associated with a Weight-for-Age Lower than 1 SD from the Median Value

Diarrhoea	60.7%
Measles	44.8%
Malaria	57.3%
Acute respiratory infections	52.3%
Other infectious diseases	53.1%
All deaths in young children attributable to undernutrition	52.5%

Sources: Caulfield et al., *Am. J. Clin. Nutr.*, 80, 193–198, 2004; Blössner, M. and Onis, M.D., *Malnutrition: Quantifying the Health Impact at National and Local Levels. Environmental Burden of Disease Series, No. 12*, World Health Organization, Geneva, 2005.

Further studies, in particular a joint WHO/Johns Hopkins University working group, built on Pelletier's work and attempted, as with Puffer's 1973 study, to quantify the effect of malnutrition on disease-specific mortality in children (Pelletier et al. 1993; Pelletier 1994b). Using a formulaic approach based on Pelletier's risk exposure (1993), the severity of malnutrition in populations (10 sub-Saharan Africa and Asian studies) were first determined to which relative risks of mortality were then calculated for each condition. Table 3.7 details the specific outcomes of this study. Of note is the total figure attributed to malnutrition of 52.5%, this turns out not to be too dissimilar to the 55% calculated by Puffer in 1973.

From this point, it could confidently be said that the co-interaction between malnutrition and disease was such that if left untreated, mutual synergy would ensure a downward spiral of health (Caulfield et al. 2004). Moreover, that one exacerbates the other is now so widely taken for granted that it is said that many (some say most) malnutrition-related deaths in developing countries are thought to be related to mild-or-moderate forms of malnutrition interacting with other common infectious diseases (Watkins and Menken 1985; Tomkins and Watson 1989; Sachdev 1995; DeRose and Millman 1998; Hionidou 2002; TOI 2010). In summary then, according to the joint WHO/Johns Hopkins study in 2004, malnutrition leaves people, particularly children, debilitated and vulnerable to illnesses that by themselves should not otherwise be fatal. In the study they also suggested that properly nourished children could prevent a staggering 1,000,000 deaths from pneumonia, 800,000 from diarrhoea, 500,000 from malaria, and 250,000 from smallpox annually (Kliksberg 2004).

From the above examination, it becomes clear that now, not only can malnutrition and health be linked but also by ascribing a large proportion of morbidity and disease to nutritional issues, the concept of food security collaborates to solidify the trinity of the public health burden, that of: malnutrition, health and disease. This reciprocal and synergistic relationship enforces the circular dichotomy that sees the malnourished predisposed to infection and disease at a time when they are at their most vulnerable.

However, this leads to problems of its own, particularly in the complication of measurable statistics. Picking up on the confusion of the attribution of 'hunger-related' deaths mentioned earlier, it now becomes clear just why it remains difficult to be precise about such causes of death; especially when malnutrition is involved. Indeed this is not a new problem. Pelletier and others back in 1994 observed that health statistics tended to ascribe hunger-related deaths to either malnutrition or infectious disease and further offered that such a distinction might often be too closely linked to be separated (Pelletier 1994a; DeRose and Millman 1998; WFP 2000).

3.3 Measurement and Assessment

So bearing all this in mind, it is not hard to imagine the difficulties inherent in trying to measure or quantify the extent of the problem of malnutrition, remembering that malnutrition is only one aspect of food insecurity and that food insecurity itself cannot be measured directly. For now, with regard to malnutrition then what is needed are close proxies for representing malnutritional outcomes. Once again this is not that straightforward either, as malnutrition to some includes overnutrition while to others it is inclusive of just the collective PEU disorders. Of course, these are calculated separately but how are they used to detail the bigger picture? The process of assessing and measuring malnutrition is explored in detail in Appendix F. Briefly, there are several methods used to gather and analyse nutritional data (Senauer and Sur 2001). For convenience, these are categorised as direct and indirect. Direct methods include clinical, laboratory and anthropometric, whereas indirect methods include surveys and statistical analysis.

3.3.1 Direct

Clinical: As an aid, doctors can diagnose certain effects of undernutrition from a person's appearance and examination of the hair, mouth, gums, nails, skin, eyes, tongue, muscles, bones and thyroid glands. These might be supplemented with an oral history to confirm or strengthen any diagnosis (Thomas 2007). This tends to be a quick, easy and relatively cheap method of assessment however in the absence of any definitive conclusions, further studies might also be necessary.

Biochemical or laboratory tests: These are good objective tests that may include biochemical, haematological or microbiology analysis. Blood plasma or serum tests have been particularly well used to date, although their reliability as appropriate and sensitive measures of nutritional status are now being brought into question (Appendix F). Other tests specific to individual nutrient shortages might test for vitamin or mineral deficiencies while other nutritional antagonists such as parasites might, depending on symptoms, be tested for in stool samples. Collectively, though while laboratory testing can draw a convincing picture, alone they are insufficient to determine nutritional status. In this way, they are best used to accompany a clinical examination and/or anthropometric measurements (Bowers 1999).

Anthropometry: Anthropometrics are measurements of the body. In the nutritional field, anthropometrics uses body morphology and certain bodily interrelationships to determine nutritional status of individuals and groups. At present, there is no generally agreed single suitable anthropometric measure, nor is their consensus as to the best method for anthropometric classification of undernutrition in groups or populations. In single cases of adults and children however, the best method is that of a suite of body measurements. These include individual measures like skin fold thicknesses or multiple relational measures such as the body mass index (BMI); waist and hip ratio (WHR); weight-for-age (W/A); height-for-age (H/A); etc. The difficulty lies with scaling up these tests on representative groups that characterise the national prevalence. In adults, the most common method of determining nutritional status in groups or populations despite its limitations remains the BMI, although in children the case is not so clear. In the case of children while the WHO attempts to establish a worldwide standard of growth charts with universal cut-off Z scores (Appendix F) some like the Indian health sector continues to use other classifications out of familiarity and an erroneous fear of the complicated Z score system (Ghosh and Benerjee 2006).

After measuring malnutrition there is then the question of classification; the degree or extent of the problem. Several attempts have been made to classify malnutrition over the years (DeMaeyer 1976; Gopalan and Kamala 1984; Brown 2000). These are discussed in detail in Appendix F however, for now a brief introduction suffices. One of the earliest and still widely used measures used for classification in children is the weight-for-age metric. Indeed this measure forms the basis of several popular classification systems currently still in use including the Gomez, the Indian Academy of Paediatrics (IAP) and the Wellcome classifications. As with other measures the weight-for-age index of undernutrition has certain shortcomings, the most notable of which is that it is unable to distinguish typology. Yet, even in spite of this misgiving it is also a measure that still enjoys widespread use with both the Gomez and the IAP classifications still enjoying popular use in India and other countries (Ghosh and Benerjee 2006; Seetharaman et al. 2007). Furthermore, the weight-for-age metric is also the indicator of choice used in the Millennium Development Goals (MDGs) target of halving the prevalence of underweight in children under-5s (UNDP/MDG 2010).

However, while its use is widespread, it is not considered the best measure by many, and in response to perceived shortcomings others have attempted to introduce replacement composite measures consisting of multiple anthropometric data sets. Specifically, Waterlow's indices was designed using the weight-for-height and height-for-age measures and along with the WHO's Z score cut-off point system and it was a classification that was endorsed by the WHO in the early 1980s. Ultimately this led to its widespread adoption and which in turn allowed for more direct clinical measures giving its users more flexibility and more specificity to gradients of classification. Nowadays however, the WHO advocates the use of Z scores against three metrics: the weight-for-height index to represent stunting; the height-for-age to signify wasting; and the weight-for-age to represent underweight. In each case, a Z score of −2 standard deviations (SD) indicates moderate malnutrition while −3 SD indicates the severely undernourished. Lastly, as a result of the previously mentioned limitations within these classifications, Svedberg proposed (2000) a new classification called the composite index of anthropometric failure (CIAF). This further disaggregated Waterlow's original categories of wasted, stunted, stunted and wasted, etc. into what Svedberg saw as their six component parts (Svedberg 2000). These were then summed formulaically to furnish a single more 'inclusive' index of undernourishment (Bhattacharyya 2006).

However, even these leaps, though popular and assuredly a step in the right direction, fail in their basic validity. The reasons are investigated in Appendix F, but briefly despite being grounded in empirical and experiential studies, the basic problem of cut-off points remain arbitrary in that they have no direct correlation and by extension comparison to actual physiological or clinical measures of undernutrition. In other words, the equations of where the lines are drawn at mild, moderate and severely undernourished are still formally not determined (WHO 1999; Avencena and Cleghorn 2001; FAO 2002). While few contest the general appropriateness of clinical, laboratory or anthropometric assessments of undernutrition on the individual, the use of such data at a global and regional levels by its very nature is difficult and attracts much scrutiny and criticism. As such, this is a hotly debated area with many attempting over the years to define, redefine and challenge prevailing concepts and methodologies. As a result several means of indirect assessment or measurements have grown up alongside those of the more direct nature.

3.3.2 Indirect

Two indirect methods of collating and further nutritional data involves the use of surveys and statistical analysis.

Surveys: There are numerous types and sub-types of surveys conducted with regard to food security. These are also explored in the appendices although those that directly relate to food consumption data might include household food surveys, 24-hours dietary recall, food diaries and food frequency questionnaires. While each has their merits and disadvantages, the main drawback is the inherent errors of respondents' memory recall, the problems of interpretation as well as the specialist and expensive nature of this data collecting method (FAO 2002).

Statistical analysis: Sometimes direct methods are inappropriate or unfeasible for developing figures on malnutrition. This might be the case at the national or global levels. In this case, there must be methods that facilitate the extrapolation of existing data that allows for such up-scaling or more commonly, the interpretation of compiled nationally representative data. In this endeavour several methodologies that attempt to indirectly measure undernutrition come to the fore. Those of the FAO, the WHO, USAID and the International Food Policy Research Institute (IFPRI) are worth a closer look.

For the FAO, determining energy requirements and applying these to estimates of global undernourishment has been one of FAO's longest running activities (Weisell 2002). They are the only people to provide global estimates of the undernourished in their annual assessment 'State of Food Insecurity' reports (SOFI), which in turn are used by numerous others. The FAO's measure of food deprivation referred to as the 'prevalence of undernourishment' is the estimation of those that fall below the energy requirement norm. This is based on known energy intake requirements as well as food consumption data determined by production, imports, exports and wastage. Distribution variances are then calculated to represent economic access to such foods, the results of which are then compared against the minimum requirement to determine the numbers of undernourished (Naiken 2002). A detailed explanation along with analysis of the FAO's methodology is outlined in the Section F.3. A caveat concerning the use of prevalence of undernourishment methods as representative of hunger attracts a certain amount of criticism. For many the main argument relates to the notion that the real issue of hunger relates more to malnutrition than mere food availability. In this thinking, the wider malnourishment issues including those once referred to as the 'silent' hunger are considered implicit whereas the prevalence of undernourishment tends to focus solely on energy deficits (MacAuslan 2009).

The WHO on the other hand take a different approach to their calculations of the prevalence of under-nourishment. Overall the WHO sign up to the ideology that young children especially 6 to 59 months old are ' ... a good proxy for the general health of a population' (WHO 2010b). This is because with children being more reliant on others and because of their increased growth needs at this age, it makes them particularly susceptible to changes in their diet, and as a result, nutritional deficits tend to manifest that much earlier than in adults (USDA 1999; CE-DAT 2010). Building on this, the WHO advocates the use of anthropometric measurements in their calculations of undernourishment and given the fact that child malnutrition is an important internationally recognised public health problem, the WHO feels that these measures are robust enough that inferences can be confidentially extrapolated to reflect the wider national nutritional status (Onis et al. 2004; Britannica 2009; WHO 2010d). Methodologically, the WHO collects anthropometric survey information that fits specific data profiles and apply their Z score cut-off formula to determine those that fall below the required stages compared to an international reference standard (which is also set by the WHO). Once again more detailed analysis can be seen in Appendix F.

The United States is another complier of statistics of undernourishment. In the United States, these are calculated by the US Department Of Agriculture's Economic Research Service (ERS) and comprise two measures, one relating to the food gap and another comprising the distribution gap. In a similar methodology to the FAO, the United States in their food gap figures aim to determine caloric shortfall from estimations of total food availability/consumption. This is adjusted for income and distribution and then compared against set calorie requirements, which, unlike the FAO, is set at a standard 2100 kcal. A second metric allows for the disaggregation of these data into quintile (5) income groups. This allows for the determination of shortfalls based on income groups rather than as a country as a whole. Data for both measures is presented as calorie equivalent tons of grain (Reutlinger and Selowsky 1978; USDA/ERS 1997; Senauer and Sur 2001; USDA/ERS 2008b).

Lastly, the International Food Policy Research Institute's (IFPRI) Global Hunger Index (GHI) is a multi-dimensional approach to measuring hunger. It is a composite index of three key dimensions of undernutrition: the FAO's proportion of undernourishment; the WHO's prevalence of underweight and the UNICEF's under-5 mortality rates. The index very simply gives equal weighting to the sum of these three indices expressed in percentages and presented as numerical values between 0 and 100; 100 being the highest (Wiesmann 2006).

In summary then, it can be seen that malnutrition is a complicated business. Thus far the literature has suggested that while widespread hunger and undernutrition is the result of extreme malnutrition and obesity the blight of overnutrition, both extremes are occurring simultaneously in developing countries. Moreover it can also be seen that the effects of energy and micronutrient disorders are pervasive, complex and often fatal. Further, complicating the problem is the synergistic relationship that sees malnutrition and health combine to exacerbate the worst effects of each. On top of this there is also less than universal agreement over the best methods to measure and quantify malnutrition at both the individual and global levels. All in all this makes identifying those affected and determining the extent of the problem that much more difficult.

References

Ahmed, T., S. Rahman and A. Cravioto (2009). Oedematous Malnutrition. *Indian Journal of Medical Research* 130: 651–654.

Al-Mubarak, L., S. Al-Khenaizan and T. A. Goufi (2010). Cutaneous Presentation of Kwashiorkor Due to Infantile Crohn's Disease. *European Journal of Pediatrics* 169(1): 117–119.

Allee, M. R. and M. Z. Baker (2009). *Riboflavin Deficiency*. WebMD Professional.

Altun, G., B. Akansu, B. U. Altun, D. Azmak and A. Yilmaz (2004). Deaths Due to Hunger Strike: Post-Mortem Findings. *Forensic Science International* 146(1): 35–38.

Amadi, B., A. O. Fagbemi, P. Kelly, M. Mwiya, F. Torrente, C. Salvestrini, R. Day, M. H. Golden, E. A. Eklund, H. H. Freeze and S. H. Murch (2009). Reduced Production of Sulfated Glycosaminoglycans Occurs in Zambian Children with Kwashiorkor but Not Marasmus. *American Journal of Clinical Nutrition* 89: 592–560.

Ashworth, A., S. Khanum, A. Jackson and ClaireSchofield (2003). *Guidelines for the Inpatient Treatment of Severely Malnourished Children*. Geneva: WHO/UNICEF.

Ashworth, A., S. Khanum, A. Jackson and Claire Schofield (2005). *Guidelines for the Inpatient Treatment of Severely Malnourished Children*. London: London School of Hygiene & Tropical Medicine.

Avencena, I. T. and G. Cleghorn (2001). The Nature and Extent of Malnutrition in Children. In *Nutrition in the Infant: Problems and Practical Procedures* V. R. Preedy and R. Watson, eds. London: Greenwich Medical Media.

Bauer, J. M., R. Wirth, D. Volkert, H. Werner and C. Sieber (2008). Malnutrition, Sarcopenia and Cachexia in the Elderly: From Pathophysiology to Treatment. Conclusions of an International Meeting of Experts, Sponsored by the Banss Foundation. *Dtsch Med Wochenschr* 133(7): 305–310.

Beck, M. A., O. A. Levander and J. Handy (2003). Selenium Deficiency and Viral Infection. *The Journal of Nutrition* 133(Supplement): 1463S–1467S.

Bengoa, J. (2000). From Kwashiorkor to Chronic Pluricarential Syndrome. *Nutrition Bulletin* 16(7/8): 642–644.

Bhattacharyya, A. K. (2006). Composite Index of Anthropometric Failure (Ciaf) Classification: Is It More Useful? *Bulletin of the World Health Organization* 84(4): 335.

Black, M. M., D. J. Gawkrodger, C. A. Seymour and K. Weismann (1998). Metabolic and Nutritional Disorders. In *Textbook of Dermatology*. R. H. Champion, J. L. Burton, D. A. Burns and S. M. Breathnach, eds. Oxford: Blackwell-Science.

Blakemore, C. and S. Jennett (2001). *The Oxford Companion to the Body*. New York: Oxford University Press.

Blössner, M. and M. d. Onis (2005). Malnutrition: Quantifying the Health Impact at National and Local Levels. *Environmental Burden of Disease Series, No. 12*. Geneva, Switzerland: World Health Organization.

Bowers, L. J. (1999). Assessment of Nutritional Status. In *Best Practices in Clinical Chiropractic*. R. D. Mootz and H. T. Vernon, eds. Frederick, MD: Aspen Publishers.

Britannica (2009). Encyclopaedia Britannica Online. 2009.

Brown, K. H. (2000). Nutrition and Assessment: Application and Interpretation of Commonly Used Nutritional Assessment Techniques. In *Nutrition and Immunology: Principles and Practice*. E. Gershwin, J. B. German and C. L. Kee, eds. New Jersey: Humana Press Inc.

Brown, K. H. (2003). Symposium: Nutrition and Infection, Prologue and Progress since 1968: Diarrhea and Malnutrition. *Journal of Nutrition*(133): 328S–332S.

Bruera, E. (1997). ABC of Palliative Care: Anorexia, Cachexia, and Nutrition. *British Medical Journal* 315: 1219–1222.

Caulfield, L. E., M. d. Onis, M. Blössner and R. E. Black (2004). Undernutrition as an Underlying Cause of Child Deaths Associated with Diarrhea, Pneumonia, Malaria, and Measles. *American Journal of Clinical Nutrition* 80(1): 193–198.

CE-DAT (2010). *The Complex Emergency Database (Ce-Dat)*: Centre for Research on the Epidemiology of Disasters (CRED).

Checchi, F., M. Gayer, R. F. Grais and E. J. Mills (2007). Public Health in Crisis Affected Populations a Practical Guide for Decision-Makers. London: Humanitarian Practice Network (HPN) at ODI.

Cleaver, K., N. Okidegbe and E. D. Nys (2006). Agriculture and Rural Development: Hunger and Malnutrition. *World Bank Seminar Series: Global Issues Facing Humanity*. Washington DC: The World Bank.

Colley, J. R. T. (1973). Inter-American Paediatric Investigation. *British Medical Journal* 4: 363.

DeMaeyer, E. M. (1976). Protein Energy Malnutrition. In *Nutrition in Preventative Medicine: The Major Deficiency Syndromes, Epidemiology and Approaches to Control*. G. H. Beaton and J. M. Bengoa. Geneva, Switzerland: World Health Organization.

DeRose, L. and S. Millman (1998). Framework: Food Shortage, Food Poverty, Food Deprivation. In *Who's Hungry? And How Do We Know? Food Shortage, Poverty, and Deprivation*. L. DeRose, E. Messer, and S. Millman, eds. Tokyo, New York, Paris: United Nations University Press.

Devereux, S. (2000). *Famine in the Twentieth Century*. Brighton: Institute for Development Studies.

DFID (2009). Draft Terms of Reference Tackling the Neglected Crisis of Undernutrition Research Programme Consortia. London: Department for International Development (DFID).

Duggan, C., J. B. Watkins and W. A. Walker (2008). *Nutrition in Pediatrics: Basic Science, Clinical Applications*. Ontario: BC Decker Inc.

Elamin, A. (2008). *Protein Energy Malnutrition*. Pittsburg, PA: University of Pittsburgh.

Fagundes-Neto, U. and J. A. B. d. Andrade (1999). Acute Diarrhea and Malnutrition: Lethality Risk in Hospitalized Infants. *Journal of the American College of Nutrition* 18(4): 303–308.

FAO (2002). Measurement and Assessment of Food Deprivation and Undernutrition: International Scientific Symposium. International Scientific Symposium: FIVIMS—An Inter-Agency Initiative to Promote Information and Mapping Systems on Food Insecurity and Vulnerability, Rome, Food and Agriculture Organization.

FAO (2011). Website of the Food and Agriculture Organization of the United Nations. FAO. http://www.fao.org/hunger/faqs-on-hunger/en/ (accessed 15 February 2011).

FAO/WHO (1951). Prevention and Treatment of Severe Malnutrition in Times of Disaster. *Technical Report Series. No. 45*. Rome/Geneva: FAO/WHO 56.

FAO/WHO (2001). Report of the Joint FAO/WHO Expert Consultation on Human Vitamin and Mineral Requirements. Bangkok, Thailand: FAO/WHO 303.

Frongillo Jr., E. A. (1999). Validation of Measures of Food Insecurity and Hunger. *Journal of Nutrition* 129: 506–509.

Ghosh, G. N. and M. Benerjee (2006). Indices for Measuring Nutritional Status in Children. *Solution Exchange for the Food and Nutrition Security Community*. New Delhi: UN FAO Food and Nutrition Security Community

Gibney, M. J., H. H. Vortex and F. J. Kok (2009). *Introduction to Human Nutrition*. Oxford: Blackwell Publishing.

Golden, M. H. N. (1988). The Effects of Malnutrition in the Metabolismof Children. *Transactions of the Royal Society of Tropical Medicine and Hygiene* 82: 3–6.

Golden, M. H. N. (1998). Oedematous Malnutrition. *British medical Journal* 54(2): 433–444.

Gopalan, C. (1968). Kwashiorkor and Marasmus: Evolution and Distinguishing Features. In *Calorie Deficiencies and Protein Deficiencies*. M. R. A. and W. E. M. eds. London: Churchill.

Gopalan, C. and S. J. R. Kamala (1984). Classifications of Undernutrition: Their Limitations and Fallacies. *Journal of Tropical Pediatrics* 30(1): 7–10.

Grover, Z. and L. C. Ee (2009). Protein Energy Malnutrition. *Pediatric Clinics of North America* 56(5): 1055–1068.

Guerrant, R. L., J. B. Schorling, J. F. McAuliffe and M. A. D. Souza (1992). Diarrhea as a Cause and an Effect of Malnutrition: Diarrhea Prevents Catch-up Growth and Malnutrition Increases Diarrhea Frequency and Duration. *The American Journal of Tropical Medicine and Hygiene* 47(Supplement 1): 28–35.

Gupte, S. (2001). *Recent Advances in Pediatrics: Special Volume 8—Emergency Pediatrics*. New Delhi: Jitendar P. Vij.

Heird, W. C. (2007). Food Insecurity, Hunger, and Undernutrition In *Nelson Textbook of Pediatrics*. B. R. Kliegman RM, Jenson HB, Stanton BF, eds. Philadelphia: Saunders.

Hionidou, V. (2002). Why Do People Die in Famines? Evidence from Three Island Populations. *Population Studies* 56(1): 65–80.

Illingworth, J. (2010). *Nutrition and Energy: Lecture 25: Cytokines, Thyroid Hormones, Thermoregulation and Basal Metabolic Rate—Anorexia, Bulimia and Cachexia*: University of Leeds: School of Medicine.

Jahoor, F., A. Badaloo, M. Reid and T. Forrester (2005). Protein Kinetic Differences between Children with Edematous and Nonedematous Severe Childhood Undernutrition in the Fed and Postabsorptive States *American Journal of Clinical Nutrition* 82(4): 792–800.

Jahoor, F., A. Badaloo, M. Reid and T. Forrester (2006). Unique Metabolic Characteristics of the Major Syndromes of Severe Childhood Malnutrition. In *The House That John Built: The Tropical Metabolism Research Unit, the University of the West Indies Jamaica 1956–2006*. D. Picou, T. Forrester and S. P. Walker, eds. Kingston: Ian Randle Publishers.

Jolliffe, N., F. Tisdall and P. Cannon (1950). *Clinical Nutrition*. New York: Paul B. Hoeber.

King, M. W. (2010). The Medical Biochemistry Page, IU Center for Regenerative Biology and Medicine. http://themedicalbiochemistrypage.org/home.html (accessed 6 November 2010).

Kliksberg, B. (2004). Iasia Symposium on Public Administration: Challenges of Poverty and Exclusion Introduction: New Expectation for the Role of the State. *International Review of Administrative Sciences* 70(4): 645–647.

Latham, M. C. (1997). Human Nutrition in the Developing World. *Food and Nutrition Series, No. 29*. Rome: Food and Agriculture Organization of the United Nations.

LPI (2010). Micronutrient Information Center. Oregon State University: Linus Pauling Institute. http://lpi.oregonstate.edu/ (accessed 12 September 2010).

LRF (2006). Website of the Lymphatic Research Foundation. Lymphatic Research Foundation. http://www.lymphaticresearch.org/main.php?menu=about&content=lymphsys (accessed 15 July 2010).

Luft, F. C. (2007). Cachexia Has Only One Meaning. *Journal of Molecular Medicine* 85(8): 783–785.

MacAuslan, I. (2009). Hunger, Discourse and the Policy Process: How Do Conceptualizations of the Problem of 'Hunger' Affect Its Measurement and Solution? *European Journal of Development Research* 21: 397–418.

Macmillan (2007). The Lymphatic System Macmillan Cancer Support. http://www.macmillan.org.uk/ Cancerinformation/Cancertypes/Lymphomanon-Hodgkin/AboutNHL/Thelymphaticsystem.aspx (accessed 12 June 2010).

Maleta, K. (2003). *Growth in Undernutrition in Rural Malawi*. Medisiinarinkatu: University of Tampere.

McGrath, J. A. and D. F. Goldspink (1982). Glucocorticoid Action on Protein Synthesis and Protein Breakdown in Isolated Skeletal Muscles. *Biochemical Journal* 206(3): 641–645.

MedlinePlus (2010). *Medline/Pubmed Database*. US National Library of Medicine/National Institute of Health.

Merck Manual (2005). *Merck and the Merck Manuals*. Whitehouse Station, NJ: Merck Sharp & Dohme Corporation.

Mokyr, J. and C. Ó Gráda (2002). What Do People Die of During Famines: The Great Irish Famine in Comparative Perspective. *European Review of Economic History* 6(1): 339–363.

Moore, S., J. N. Hall, S. Harper and J. W. Lynch (2010). *Global and National Socioeconomic Disparities in Obesity, Overweight, and Underweight Status*, *Journal of Obesity*, Article ID 514674, 2010.

Morley, J. E. (2007). Nutritional Disorders. In *The Merck Manual Online Medical Library*. R. S. Porter and J. L. Kaplan, eds. Whitehouse Station, NJ: Merck Sharp & Dohme Corporation.

Muller, O. and A. Jahn (2009). Malnutrition and Maternal and Child Care. *Maternal and Child Health* (Part 3): 287–310.

Naiken, L. (2002). Measurement and Assessment of Food Deprivation and Undernutrition: FAO Methodology for Estimating the Prevalence of Undernourishment International Scientific Symposium, Rome.

Nguyen-Khoa, D.-T., D. W. Cope, G. V. Busschots and P. A. Vallee (2009). *Beriberi (Thiamine Deficiency)*. WebMD Professional.

NHS (2010). Website of the NHS Direct. National Health Service UK. http://www.cks.nhs.uk/ (accessed 12 June 2010).

Nnakwe, N. E. (2009). *Community Nutrition: Planning Health Promotion and Disease Prevention*. Sudbury, MA: Jones and Bartlett.

Norbury, W. B., D. N. Herndon, L. K. Branski, D. L. Chinkes and M. G. Jeschke (2008). Urinary Cortisol and Catecholamine Excretion after Burn Injury in Children. *Journal of Clinical Endocrinology & Metabolism* 93(4): 1270–1275.

Ó Grada, C. (2008). The Ripple That Drowns? Twentieth-Century Famines in China and India as Economic History. *The Economic History Review* 61(Supplement 1): 5–37.

OED (1989). *The Oxford English Dictionary*. Oxford Clarendon Press.

Onis, M. d., M. Blössner, E. Borghi, R. Morris and E. A. Frongillo (2004). Methodology for Estimating Regional and Global Trends of Child Malnutrition. *International Journal of Epidemiology* 33: 1–11.

Ophardt, C. E. (2003). *Virtual Chembook: Glycogenesis, Glycogenolysis and Gluconeogenesis*. Elmhurst College.

PAHO (2010). Public Health Heroes. The Pan American Health Organization (PAHO). http://www.paho.org/ English/DPI/100/heroes.htm (accessed 15 November 2010).

Pelletier, D. L., E. A. Frongillo, Jr and J.-P. Habicht (1993). Epidemiologic Evidence for a Potentiating Effect of Malnutrition on Child Mortality. *American Journal of Public Health* 83(8): 1130–1133.

Pelletier, D. (1994a). The Potentiating Effects of Malnutrition on Child Mortality: Epidemiologic Evidence and Policy Implications. *Nutrition Reviews* 52(12): 409–415.

Pelletier, D. L. (1994b). The Relationship between Child Anthropometry and Mortality in Developing Countries. *The Journal of Nutrition* 124: 2047S–2081S.

Rabinowitz, S. S., L. A. Batres and S. Moorthy (2009a). *Beriberi*. WebMD Professional.

Rabinowitz, S. S., M. Gehri, E. R. D. Paolo and N. M. Wetterer (2009b). *Marasmus*. WebMD Professional.

Ramos, E. J., S. Suzuki, D. Marks, A. Inui, A. Asakawa and M. M. Meguid (2004). Cancer Anorexia-Cachexia Syndrome: Cytokines and Neuropeptides. *Current Opinion in Clinical Nutrition & Metabolic Care* 7(4): 427–434.

Reddy, V. (1985). Relationship between Diarrhea and Malnutrition. *Indian Journal of Pediatrics* 52(5): 463–467.

Reutlinger, S. and M. Selowsky (1978). *Malnutrition and Poverty; Magnitude and Policy Options*. Baltimore, MD: Johns Hopkins University Press.

Roubenoff, R. (2008). Molecular Basis of Inflammation: Relationships between Catabolic Cytokines, Hormones, Energy Balance, and Muscle. *Journal of Parenteral and Enteral Nutrition* 32(6): 630–632.

Rubin, E. and H. M. Reisner (2009). *Essentials of Rubin's Pathology*. Baltimore, MD: Lippincott, Williams and Wilkins.

Sachdev, H. P. S. (1995). Assessing Child Malnutrition: Some Basic Issues. *NFI*. http://nutritionfoundationofindia. res.in/archives.asp?archiveid=97&back=bydate.asp.

Saito, Y., Y. Yoshida, T. Akazawa, K. Takahashi and E. Niki (2003). Cell Death Caused by Selenium Deficiency and Protective Effect of Antioxidants. *The Journal of Biological Chemistry* 278: 39428–39434.

Seetharaman, N., T. V. Chacko, S. L. R. Shankar and A. C. Mathew (2007). Measuring Malnutrition—The Role of Z Scores and the Composite Index of Anthropometric Failure (Ciaf). *Indian Journal of Community Medicine* 32, No. (2007–01—2007–03)(1).

Sen, A. (n.d.). Non-Communicable Diseases and Achieving the Millennium Development Goals. Bangkok: UNESCAP/ HDS: Emerging Social Issues Division.

Senauer, B. and M. Sur (2001). Ending Global Hunger in the 21st Century: Projections of the Number of Food Insecure People. *Review of Agricultural Economics* 23(1): 68–81.

Shetty, P. (2002). Measurement and Assessment of Food Deprivation and Undernutrition: Keynote Paper: Measures of Nutritional Status from Anthropometric Survey Data. International Scientific Symposium, Rome.

Shetty, P. (2006). Malnutrition and Undernutrition. *Medicine* 31(4): 18–22.

Smith, L. C., A. E. E. Obeid and H. H. Jensen (2000). The Geography and Causes of Food Insecurity in Developing Countries. *Agricultural Economics* 22(2): 199–215.

Svedberg, P. (2000). *Poverty and Undernutrition: Theory, Measurement, and Policy*. Oxford: Oxford University Press.

Thomas, D. R. (2007). Disorders of Nutrition and Metabolism: Undernutrition. In *The Merck Manual Online Medical Library*. R. S. Porter and J. L. Kaplan, eds. Whitehouse Station, NJ: Merck Sharp & Dohme Corporation.

TOI (2010) India: 45000 Child Malnutrition Deaths Every Yr in Maharashtra. *The Times of India* 2010, (4 February 2010).

Tomkins, A. and F. Watson (1989). Malnutrition and Infection—A Review—Nutrition Policy Discussion Paper No. 5. Geneva, Switzerland : UN Administrative Committee on Coordination—Subcommittee on Nutrition.

Udall, Jr, J. N., Z. A. Bhutta, A. Firmansyah, P. Goyens, M. J. Lentze and C. Lifschitz (2002). Malnutrition and Diarrhea: Working Group Report of the First World Congress of Pediatric Gastroenterology, Hepatology, and Nutrition. *Journal of Pediatric Gastroenterology and Nutrition* 35: S173–S179.

UNDP/MDG (2010). Millennium Development Goals (MDGs). United Nations Development Programme. http://www.undp.org/mdg/ (accessed 24 June 2010).

USAID (2007). Food for Peace: Fy 2008 P.L. 480 Title II Program Policies and Proposal Guidelines. Washington, DC: United States Agency for International Development

USDA (1999). US Action Plan on Food Security: Solutions to Hunger. US Department of Agriculture (USDA).

USDA/ERS (1997). Food Security Assessment Gfa-9. *Situation and Outlook Series*. Washington, DC: USDA Economic Research Service.

USDA/ERS (2008a). Food Security Assessment, 2008–09: Overview—Food-Security Impact of the Financial Downturn, 2008–18. *Food Security Assessment*. Washington, DC: USDA/Economic Research Service GFA-20.

USDA/ERS (2008b). GFA-19: Food Security Assessments Reports. *Agriculture and Trade Reports*. USDA Economic Research Service.

Wacker, J., J. Frühauf, M. Schulz, F. M. Chiwora, J. Volz and K. Becker (2000). Riboflavin Deficiency and Preeclampsia. *Journal of Obstetrics & Gynecology* 96(1): 38–44.

Walker, N., J. Bryce and R. E. Black (2007). Interpreting Health Statistics for Policymaking: The Story Behind the Headlines. *The Lancet*(369): 956–963.

Watkins, S. C. and J. Menken (1985). Famines in Historical Perspective. *Population and Development Review* 11(4): 647–675.

Weisell, R. (2002). Measurement and Assessment of Food Deprivation and Undernutrition: Part IV—Summary of the Draft Findings of the Joint FAO/WHO/UNU Expert Consultation on Human Energy Requirements. International Scientific Symposium, Rome, FAO.

WFP (2000). *Food and Nutrition Handbook*. Rome: World Food Programme (WFP).

WFP (2008). The World Food Programme. The First 45 Years. World Food Programme. World Food Programme. http://www.wfp.org/aboutwfp/history/index.asp?section=1&sub section=2 (accessed 25 April 2010).

WFP (2009). The World Food Programme: Food Quality Control. World Food Programme. http://foodquality.wfp.org/Home/tabid/36/Default.aspx (accessed 2 April 2010).

WFP (2011). Website of the World Food Program. World Food Programme. http://www.wfp.org/ (accessed 25 February 2011).

WHO (1999). Management of Severe Malnutrition: A Manual for Physicians and Other Senior Health Workers. Geneva, Switzerland: World Health Organization.

WHO (2006). *Neurological Disorders: Public Health Challenges*. Geneva, Switzerland: World Health Organization.

WHO (2009). *WHO Global Database on Vitamin a Deficiency Report: Global Prevalence of Vitamin a Deficiency in Populations at Risk 1995–2005*. Geneva: World Health Organization.

WHO (2010a). *Global Database on Child Growth and Malnutrition*. Geneva, Switzerland: World Health Organization.

WHO (2010b). Website of the World Health Organization. World Health Organization. http://www.who.int/ (accessed 2 April 2010).

WHO (2010c). World Health Organization: Children's Environmental Health Indicators. World Health Authority. http://www.who.int/ceh/indicators/malnourishedwomen.pdf (accessed 2 April 2010).

WHO (2010d). World Health Statistics. Geneva, Switzerland: World Health Organization 149.

WHO/EMRO (2010). *You Are What You Eat*: World Health Organization: Eastern Mediterranean Regional Office.

WHO/FAO (2003). Diet, Nutrition and the Prevention of Chronic Diseases. *WHO Technical Report Series 916*. Geneva: WHO/FAO.

WHO/FAO (2004). *Vitamin and Mineral Requirements in Human Nutrition (Draft Version)*: WHO/FAO.

Wiesmann, D. (2006). A Global Hunger Index: Measurement Concept, Ranking of Countries, and Trends. Washington, DC: International Food Policy Institute.

Williams, C. D. (1933). A Nutritional Disease of Childhood Associated with a Maize Diet. *Archives of Disease in Childhood* 8: 423–433.

Williams, C. D. and B. M. Oxon (1935). Kwashiorkor: A Nutritional Disease of Children Fed with a Maaize Diet. *The Lancet*: 1151.

World Vision (2009). Website of World Vision. World Vision. http://www.worldvision.org.uk/ (accessed 23 May 2010).

Zimmermann, M. B., P. L. Jooste and C. S. Pandav (2008). Iodine-Deficiency Disorders. *The Lancet* 372(9645): 1251–1262.

4

Food Security: The Global Picture

The previous chapters have outlined some of the minutiae of the food security concept—how it is defined and understood as well as something of the role of governance. Importantly too, ideas of good and bad nutrition were explored along with the effects of disease and the problems of classification. At this juncture, then it is worth briefly looking at the current global situation of food insecurity. It is worth mentioning too that this book is not intended to identify those affected in any great detail or analysis, there are other publications that cover this topic in much more detail. Instead this book's aim, as mentioned and worth reiterating here is simply to identify the concepts and areas of confusion and to get to the bottom of how we can improve our comprehension of the subject in its entirety. By understanding what type of data are studied, where it comes from and how it is used can only be beneficial. In this way, future in-depth analysis of the concept is ultimately improved. However, before we can explore the global picture in a little more detail, it is incumbent on the reader to understand something of the various country classifications and how these might potentially confuse or complicate global, cross-institutional comparisons.

4.1 Boundaries, Classifications and Geo-Spatial Demarcation

When data are presented, there is a clear need to demarcate certain boundaries, least of all to frame and contextualise the information. In the field of food security and malnutrition, there is much need for data to be presented geo-spatially; that is, with reference to particular areas or regions. It is also important, especially in statistical analysis, that such data are compared on a like-for-like basis. In classifications concerning area-specific data, this might be as simple as using geographical boundaries although as with much else to do with food security, it is rarely that simple. Questions arise as to which boundaries to be used whether geo-specific, as in continental, regional or political country borders or perhaps socio-cultural separations as in the under-5s or the agricultural workforce. Other conventions might also involve presenting data based on regional, politically affiliated or geo-political groupings such as the European Union (EU) or the Commonwealth of Independent States (CIS). Also, trading blocs, common markets and custom unions like the Organisation of Eastern Caribbean States (OECS) and the South African Customs Union (SACU) might all be employed. Economic or industrial groupings are not uncommon either and data might be presented reflecting regions based on the G8 group of industrialised nations. Lastly, although by no means least are the conceptual boundaries which include the UN's 'developed' and 'developing' country classifications.

While many of these diverse classifications are indeed necessary and useful for clarity sake, any discussion on global food security requires a brief introduction on the current conventions in use today. Numerous institutions and bodies between them have created a plethora of diverse informational needs. In turn, the framing of the data also depends to a large extent on the needs and usage of its stakeholders as well as the wider public. This is just as prevalent in the field of nutrition where policy makers might want regional data disaggregated at a sub-national level, or where the UN perhaps needs global and regional reports for cross-border analysis. Other information like the spread of a disease might be associated with eating habits in the Pan Asian area and would need to be framed as such. Also cropping cycles and harvesting or failure of crops might more closely follow weather patterns than any political boundaries; a shift in the El Niño-Southern Oscillation for example could easily influence Africa's Sub-Saharan food production.

The need then for a variety of demarcational constructs is evident; the problem is one of standardisation. In 1969 the UN, recognising this very real need for consistency of coding and aggregations,

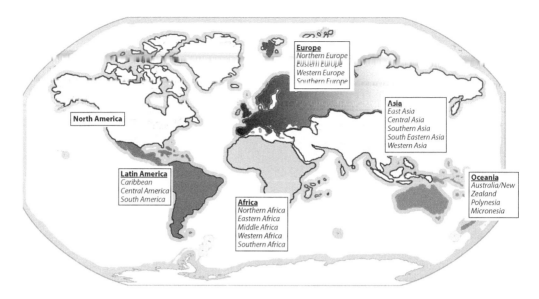

FIGURE 4.1   Series 'M49' UN regional classification. (Based on data from the United Nations Website, *Note on Definition of Regions for Statistical Analysis*, UNSD: Committee for the Coordination of Statistical Activities, Montreal, 2006b. Reprinted with courtesy of Patrick Newsham ©.)

introduced the Standard Country or Area Codes for Statistical Use revised in 1999 (Series M, No.49/ Rev 4) (UN 2010). These were based on the International Organization for Standardization (ISO) standards and are generally referred to as the 'M49' classification (Figure 4.1). It is a dynamic list that needs to be updated and changed periodically to represent country or regional changes (UN 2010). This classification formed the foundation on which regional and other aggregates could be based. However, even after such efforts, classifications are still not homogenous and vary among agencies and sometimes even within different bodies under the same umbrella (see Appendix A). There are also particular issues with reports using different aggregations that supposedly represent similar regions, particularly when it comes to groupings like sub-Saharan Africa (SSA) and the Near East. This has led to a system of proprietary groupings where different agencies employ different conventions, which is tolerable if one is familiar with that particular grouping but becomes unworkable if the user is unfamiliar with the convention and drawing on several reports from different agencies.

In this respect, different UN agency remits cover different regional composites: the UN Office for the Co-ordination of Humanitarian Affairs (OCHA), for instance, incorporates a regional office covering the Middle East, North Africa, Iran and Afghanistan (OCHA 2006), while the UN's regional office, the Economic and Social Commission for Western Asia (ESCWA), covers the Middle East, Sudan and Egypt and if it refers to the Middle East alone, it incorporates the Islamic Republic of Iran (ESCWA 2009). Lastly on this note is the way that Sudan is treated in the UN regional map. In this map for instance Sudan falls in northern Africa, unless the UN talks of SSA in which case Sudan is re-classified as part of SSA, all the while the WHO pops Sudan under the Eastern Mediterranean Region. Many of these disparate collections become clearer in Appendix A (UN 2010; World Bank 2010a).

In recognition of this and directly as a result of the implementation of the Millennium Development Goals (MDG), the UN needed a level-playing field in which all MDG data from all UN agencies could be presented in the same format. After several meetings of UN inter-agency and expert groups, it was decided in 2002 that an appropriate global aggregation of data for MDG purposes was best achieved using a combination of geographical and developmental criteria. For this reason, a new geo-developmental map was introduced (see Figure 4.3) (UNSD 2006a; UN 2010). Although this is a step in the right direction, it only applies to aggregated data for the MDGs and institutional reporting continues to use proprietary aggregations, which while undoubtedly useful in context, can and does lead to confusion when attempting to cross-reference or tease out other useful information (UNSD 2006a,b; UN 2010).

With reference to classifications used in this book then, a vast body of official UN and other agency reports and statistics referenced throughout this work use proprietary country classifications as well as sub-regional aggregates. As far as is possible we rely on three key groupings—those of the regional classification systems of the UN and the World Health Organization (WHO) as well as the geographical and developmental classification used by the MDGs. Having said that, where references are made to other groupings, these will be referred to and identified by their issuing agency.

The following three maps detail the most commonly used referenced data throughout this book.

1. The United Nations introduced the 'M49' country classification series, which divides the world into five regions: the Americas; Europe; Asia; Africa and Oceana (Figure 4.1). These are also divided into various sub-regions and a varying assortment of sub-regional and economic aggregates. Most aggregations in this classification are self-explanatory except the use of the terms 'developing' and 'developed' regions (following discussion).

2. *WHO Regions*: WHO member states are grouped into six geographical areas or regions. These are AFRO (Africa); AMRO (Americas); EMRO (Eastern Mediterranean); EURO (Europe); SEARO (South-East Asia) and WPRO (Western Pacific) (Figure 4.2). When needed, the WHO also creates a seventh grouping based on the World Bank's high-income category leaving the remaining six regions populated solely by low-middle-income countries (LMIC) (WHO 2008). Lastly, for further detailed breakdown, the WHO disaggregates the original six regions into sub-regions on the basis of levels of child and adult mortality (WHO 2003; Boschi-Pinto et al. 2008).

3. The Millennium Development Goals (MDG) groupings offer a combination of geographical and level of development criteria. The MDG groups consist of developed countries, Commonwealth of Independent States, northern Africa, sub-Saharan Africa, Latin America and the Caribbean, Eastern Asia, Southern Asia, South-Eastern Asia, Western Asia and Oceania (Figure 4.3).

As can be seen from this brief discussion caution must be exercised when interpreting or comparing regional or other data from these organisations particularly in the areas of Africa and Asia. As mentioned, a comprehensive table detailing the various conventions can be found in Appendix A.

Perhaps one more institutional groupings is worth a mention here as this organisation often underpins much economic data.

The World Bank (WB): WB countries are separated into six regions: East Asia and Pacific (EAP), Europe and Central Asia (ECA), Latin America and the Caribbean (LAC), Middle East and North Africa

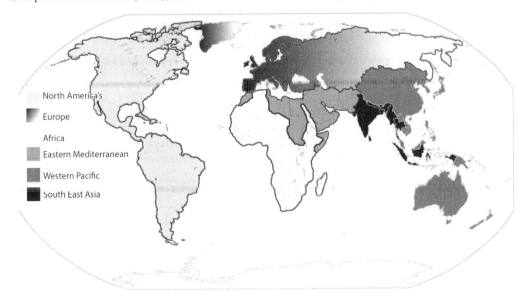

FIGURE 4.2 World Health Organization regional classification. (Based on regions of the World Health Organization, Website of the World Health Organization, 2010b. http://www.who.int/. Reprinted with courtesy of Patrick Newsham ©.)

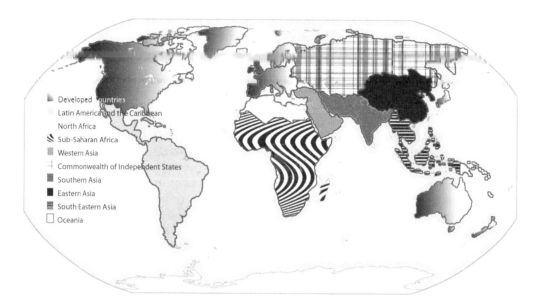

FIGURE 4.3 Millennium Development Goals regional aggregations. (Based on Millennium Development Goals data at
http://unstats.un.org/unsd/mdg/Host.aspx?Content=Data/RegionalGroupings.htm.)

(MENA), South Asia (SA) and sub-Saharan Africa (SSA). The WB also divides its member states into
four income groups currently based on gross national income (GNI) per capita figures. Of the four income
groups, low, lower-middle, upper-middle and high, all but the high income (which is deemed to be developed economies) are classed as 'developing' economies (World Bank 2010a). Lastly the International
Monetary Fund (IMF), part of the World Bank Group and publishers of the World Economic Outlook
(WEO), also divides the globe into two groups: 'advanced' economies and 'emerging' (or developing)
economies. This grouping however rather than being rigidly based on economic or other criteria is one
that has emerged over time with a view to facilitating the organisation of data (IMF 2010).

On the use of the terms 'developing' and 'developed', there is not surprisingly much confusion in their
express definitions and actual usage by a multitude of different stakeholders. The UN states their use of
the term is not bound by any established economic or other conventions and is in fact a conceptual grouping. A developed country for instance has been described as one that allows

> ... all its citizens to enjoy a free and healthy life ... And a genuinely developing country is one
> in which civil society is able to insist, not only on material wellbeing, but on improving standards
> of human rights and environmental protection as well. (ESCAP 2000)

Another UN website however suggests that 'developed' countries includes the high-income groupings
as well as the economies in transition (Cyberschoolbus 2010). On this note, the MDG also the base their
developed and developing regions on a variation of the UN M49 classification, the main difference being
that all transition countries in Europe are included whilst excluding those CIS countries re-classified as
CIS-Europe (UNSD 2006a). By contrast, the United Nations Development Programme (UNDP), responsible for the Human Development Index (HDI), utilises both designations—'developed' and 'developing'
in very specific contexts. The index determines its HDI rankings based on a tripartite of metrics including healthy lives, access to knowledge and a good standard of living. In this way, the index awards countries achieving 0.9 or above as 'developed'. All those not reaching this level are considered developing
(HDR 2009). Finally, each of the above three separate groupings are not to be confused with the World
Bank's income determined 'developed' and 'developing' country income classifications.

Considering the country classifications, the proprietary usage and terminological contradictions
comparing cross-agency, cross-country and cross-institutional data is a potential minefield that has to
be navigated with care. The next and last consideration before we can take a look at the global picture
of the undernourished lies in the relevance and age of the latest available data sets.

4.2 Data Relevance

Much data collected for many reports, books (this one included) and statistical analysis will invariably aim to use the most up-to-date information available. However, all too often, many of the important data sets that are collected are out of date by several years. This results from several reasons, but predominantly it can be seen that many countries, institutions and others collect and collate information over different time periods and with different frequencies. This tends to leave us with two options and that is we can quote from these latest outdated sources or we can use estimates often based on empirical data to estimate present and future statistics. Both are used with equal frequency and the determination of which to use depends on many factors inter alia: relevance of data and skill levels of analytical forecasters as well as time constraints. Institutionally, for estimates such as population figures, for instance, the latest figures are calculated using past growth rates whilst taking into consideration factors such as death rates, birth trends and life expectancy among others. In this way, in some scenario, forecasting can become very complex. As a result, much of the data in this book as well as other major reports tend to utilise the latest published figures whilst quoting where appropriate or available any estimates from reputable sources. This has implications for the way people look at and analyse data sets and it must be borne in mind when also referencing single data sets which give different years for different sources, the often cited as 'latest available data'.

4.3 Food Security Outlook

Having established that there were no single measures of food security, the question then arose as to how people's food security situations were determined. The answer as we discovered was by proxy; by measuring variables closely associated with the concept it then becomes possible through the use of multiple indicators to reflect the various dimensions of a person's, region's or nation's food security status. The following is based loosely around the core Committee on Food Security (CFS) and Food Information and Vulnerability Mapping Systems (FIVIMS) recommended indicators identified in Table 1.2. Although I have added a couple of indices and recategorised them, this does not represent a personal ideal; it is merely seen as easing the narrative along. The section below then is based on the following groups:

Availability
- Average per person dietary energy supply (DES)
- Cereals, roots and tubers as percentage of DES

Access
- People living on less than $1 per day

Utilisation
- Proportion of population using improved drinking water sources
- Proportion of population using improved sanitation facilities

Stability
- Number of countries facing food emergencies—drought, famine, conflict

Health and Development
- Life expectancy at birth
- Under-5 mortality rate

Nutritional Status
- Percentage of population undernourished
- Proportion of children under-5 that are underweight
- Percentage of adults with BMI < 18.5
- Macronutrient deficiency

4.3.1 Availability

Food availability is measured as the amount of food available to the individual or the nation. Food production is influenced by many factors (Chapter 17) and can be measured in many ways. A country's food production indices for instance can determine relative productivity while food use, trade and stock changes can measure the relative health of the food economy. Added to this are measures of stability and dependency such as the ratio of the five major grain exporters' supplies to requirements. However, one of the main measures that is often used as a general barometer of availability is the dietary energy supply (DES).

4.3.1.1 Dietary Energy Supply

DES or consumption, as it is sometimes confusingly labelled, is expressed in kilocalories (kcal) per person per day and refers to the amount of available food for each individual during the reference period (usually one year). Calculated by national institutions and centrally for global comparison via the FAO from a country's food balance sheets, the per capita DES is also based on the country's demographic weighting (Hoddinott 1999). In practice, the FBS takes into account overall agricultural production as well as the trade of food commodities while further analysis takes account of seed use, waste coefficients, country stock changes and uses other than direct consumption such as animal feed, food processing and others. What results is a supply/utilisation account from which the available food commodities for human consumption are calculated (FAOSTAT 2011). Figure 4.4 illustrates the availability of per capita DES for the world as a whole and the least developed countries (LDC) to 2007. What is striking about this is the growing gap between the LDC and the rest of the world.

This widening gap can be further illustrated by examining regional DES (Figure 4.5). It can be seen that North America, Europe and Oceana enjoy considerably more availability than Africa and Asia, and as can be seen, it is a condition that has not changed much in the last four decades.

This disparity is further highlighted when we look at the most recent regional Food and Agriculture Statistics (FAOSTAT) of 2007 shown in Tables 4.1 and 4.2. The first table clearly shows that much of Africa, South and South-Eastern Asia bear the brunt of this regional inequality while the second highlights those countries with the worst DES availability of 2007. Once again, it is shown that all but one of the worst affected countries belong to the African continent.

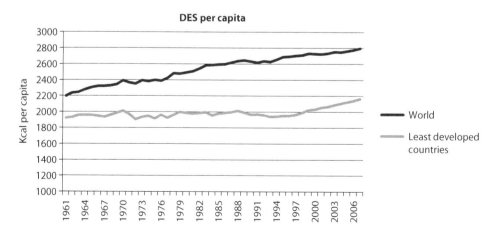

FIGURE 4.4 Dietary energy consumption per capita. (From FAOSTAT and UN definition of LDC [UN/DPI, What are the least developed countries (LDCs)? Fact-Sheet Number 20, 2004; FAOSTAT, *Food and Agriculture Statistics*, Food and Agriculture Organization, Rome, 2011; UNCTAD, UN list of LDCs after the 2006 triennial review, United Nations Conference on Trade and Development, 2011].)

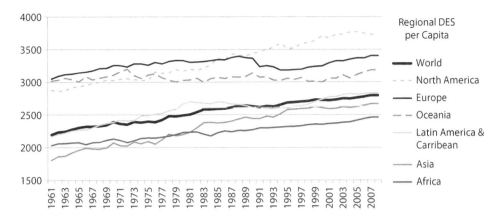

FIGURE 4.5 Regional DES 1961–2007. (From *FAOSTAT Food and Agriculture Statistics*, Food and Agriculture Organization, Rome, 2011. http://faostat.fao.org/.)

TABLE 4.1

Per Capita Dietary Energy Supply by World and by Region, 2007

World (kcal) 2798			
Africa	*2462*	*Asia cont.*	
Eastern Africa	2054	Southern Asia	2370
Middle Africa	1860	South-Eastern Asia	2586
Northern Africa	3016	Western Asia	3104
Southern Africa	2918	*Europe*	*3406*
Western Africa	2649	Eastern Europe	3323
Americas	3216	Northern Europe	3401
Northern America	3727	Southern Europe	3410
Central America	3043	Western Europe	3536
Caribbean	2561	*Oceana*	*3182*
South America	2886	Australia and New Zealand	3216
Asia	*2668*	*Melanesia*	*2800*
Central Asia	2783	Micronesia	2899
Eastern Asia	2955	Polynesia	2919

Source: FAOSTAT Food and Agriculture Statistics, Food and Agriculture Organization. http://faostat.fao.org/, 2011.

4.3.1.2 Depth of Hunger Stats 2004–06

The FAO also calculates the depth of hunger as the difference in kcal between the minimum requirements and consumption of the undernourished population (FAO 2011b). In their view those above 300 kcal shortfall (Table 4.3) suffer a high intensity of food deprivation. In the latest figures available, from the data of 2004–2006, a further 66 countries suffered medium intensity deprivation and as in the previous table among many of those affected, the majority were also made up of African and Asian countries.

4.3.1.3 USDA/Economic Research Service

The Economic Research Service's (ERS's) nutritional and distribution gap metrics are along the same concept as the FAO's depth of hunger measure. There are a few differences, the main difference being

TABLE 4.2

Countries with Per Capita DES below 2000 kcal

Below 2000 kcal		Below 2000 kcal	
Democratic Republic of the Congo	1605.08		
Eritrea	1605.4	Comoros	1884.49
Burundi	1684.72	Angola	1973.29
Haiti	1869.77	Ethiopia	1979.68
Zambia	1873.04	Central African Republic	1985.84

Source: FAOSTAT Food and Agriculture Statistics, Food and Agriculture Organization. http://faostat.fao.org/, 2011.

TABLE 4.3

Intensity of Food Deprivation: Depth of Hunger 2004–2006

Countries with High Intensity of Deprivation	Shortfall in kcal	Countries with High Intensity of Deprivation	Shortfall in kcal
Bermuda	920	Belarus	340
Belize	430	Comoros	340
Democratic Republic of the Congo	430	Rwanda	330
Haiti	430	Zambia	330
Sierra Leone	390	Nicaragua	320
United Arab Emirates	390	Ethiopia	310
Burundi	360	Liberia	310
Eritrea	350	Zimbabwe	310
		Myanmar	300

Source: Food Security Statistics, Food and Agriculture Organization. http://www.fao.org/economic/ess/food-security-statistics/en/, 2011.

that the FAO utilises a regional or country average whereas the ERS uses a higher cut-off point of 2100 kcal (Table 2.4). A second difference concerns ERS's two specific measures of calorie gaps: nutritional and distribution, both of which are calculated in grain equivalent metric tonnes. The nutritional gap is the national average and is more akin to FAO's depth of hunger in which the gap between available food and that which is needed to support minimum national requirements is calculated. The distribution gap on the other hand takes account of economic and income conditions and is the amount of grain equivalent of food needed '… to raise consumption in each income quintile to [meet] the nutritional requirement' (USDA/ERS 2010). Of the 70 developing countries surveyed by the ERS, the total nutritional gap for 2010 was estimated to be 11,553 tonnes while the distribution gap was 24,230 tonnes of grain equivalent.

4.3.1.4 Cereals, Roots and Tubers as Percentage of DES

A further measure worth considering in terms of availability is the weighting of cereals, roots and tubers as a percentage of total DES. The main reason for this is that being the cheapest form of readily accessible foods, commodities such as wheat, rice, potatoes and cassava, etc. provide a substantial contribution to many of the poorer nations' overall diet. This indicator can be regarded as measuring the availability of supply in meeting these needs as well as a measure of risk, or reliance on a smaller variety of diet. It would be wrong to describe such a diet as poor man's food for the dichotomy here is that as we proceed in the book the original concept of this poor man's diet has in fact come full circle and now indeed turns out to be a healthy (assuming nutritional balance) and sustainable diet. Having said that, we cannot get away from the fact that in poorer countries total percentage of cereals, roots and tubers make up increasingly higher proportions of the diet. This can be seen in the FAO country profiles from FAOSTAT where

TABLE 4.4

Cereals, Roots and Tubers as Percentage of Total DES

Bottom Decile of CRT as % DES	%	Top Decile of CRT as % DES	%
Grenada	22.2	Solomon Islands	68.6
Serbia	22.6	Mali	70.2
Iceland	23.1	Timor-Leste	70.8
Switzerland	24.0	Benin	70.8
Cyprus	24.6	Togo	72.7
United States of America	24.7	Nepal	73.0
Netherlands	24.8	Malawi	73.0
Australia	25.0	Burkina Faso	73.5
Bermuda	25.0	Lao People's Democratic Republic	73.6
Belgium	25.6	Cambodia	75.1
Spain	26.3	Zambia	76.3
Antigua and Barbuda	26.6	Madagascar	76.9
Denmark	26.6	Mozambique	76.9
Luxembourg	26.7	Democratic Republic of the Congo	77.1
Austria	27.7	Eritrea	79.2
New Zealand	27.8	Ethiopia	79.2
Canada	28.2	Bangladesh	80.5
Sweden	28.3	Lesotho	80.9
Decile Average	**25.5**	**Decile Average**	**75.3**

Source: *FAOSTAT Food and Agriculture Statistics,* Food and Agriculture Organization. http://faostat. fao.org/, 2011.

shares of cereals roots and tubers in countries of the developed world are a lesser percentage of the total diet; for example, the United States consumes 26%, the UK 28% whilst countries like Lesotho and Bangladesh consume over 80% each (FAO 2010a; FAOSTAT 2011). In fact, it can be seen from Table 4.4 that the top decile rely on cereals roots and tubers (CRT) to the tune of 75.3% and the bottom decile average only 25.5%. Averaged out globally, the per capita population still consume over 51% of all the cereal produced annually. It is also worth noting that the pattern of dependency on such staples has not changed considerable in regional terms in the last 70 years at least, and more likely much longer than that (Bennett 1941).

4.3.2 Access

When considering access to food, it is common to think of access as being both physical and economic. On the physical side, this might include growing it or perhaps the location of markets from which to purchase food. It might also mean the ability of individuals to travel or non-financial bartering, among others. Economically however, access implies the financial ability to pay for or trade for goods in the marketplace. This is true at both the national level via imports and a healthy national current account balance as well as at the individual level through employment and income. Measures of total direct personal income can be compared against national poverty lines while nationally the Gini coefficient or index of income distribution can determine the spread of wealth among populations. More often, a good measure of cross-border comparable economic access to adequate food is made using the international $1-a-day poverty line.

4.3.2.1 People Living on Less Than $1 per Day

Within countries, the idea of poverty is a subjective consideration that does not translate well across borders. Individual national poverty thresholds for instance will vary widely across countries, and it is often the case that richer countries often adopt higher thresholds. This makes comparing the poor in one

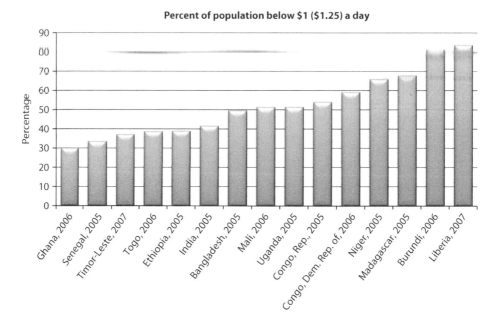

FIGURE 4.6   Percentage of population living on less than $1 a day. (From latest figures from the World Bank, *Worldbank: World Development Indices*, World Bank Group, 2010b.)

country to another difficult. As a result, when comparing the cross-border cost of living or the poverty threshold for instance a single metric is needed. While market-led exchange rate systems offers one solution, cross-border poverty lines building on the principle of purchasing power parity (PPP) is widely accepted as a better measure of the comparative cost of living across countries. Based on the PPP, the World Bank's measure of $1-a-day aims to address this.

Starting in 1968, the World Bank's International Comparison Programme (ICP) examined 10 countries and costed a vast range of goods and services that people spent their incomes on and compared it with similar goods bought in the United States. This comparison allowed different countries to be compared with each other on an absolute scale. While the methods used and the range of data have increased and improved over the years, the idea itself is very simple and ultimately underpins international comparisons for all economic indicators (Shaohua and Ravallion 2001; Sillers 2008). From this point, it was a case of determining an acceptable cross-border level of poverty. Originally based on Ravallion, Datt and van de Walle's work in 1990, the international poverty line was deemed to be $1 presented in 1983 prices at PPP. In light of updated research the World Bank later upped the poverty line to $1.08 adjusted to 1993 PPP figures. A further revision in 2005 found that previous estimates were indeed under-estimated and that the actual cost of living of a basket of goods necessary to meet the needs of the poor were much higher, standing at $1.25 a day. The measure however was and is still abbreviated to $1-a-day and ultimately means that many more people now fall below the poverty threshold (Chen and Ravallion 2007). The important point here is when establishing trends or comparing data, particularly historical data, the same $1-a-day measure is used. Figure 4.6 highlights those countries whose populations live on less than $1 a day rise to 30% and over.

Overridingly, from this figure, it can be seen that the majority of the poorest earning populations are from countries on the African continent.

4.3.3 Utilisation

Utilisation of food in the security concept refers to the notion of sufficient and appropriate food supplies that meet the dietary needs of individuals. Specifically, the idea encapsulates the proper biological use of food. Implicit in most institutional definitions is that appropriate diets contain essential nutrients as well as non-food inputs such as clean water and adequate sanitation. All these factors in turn depend to a large extent on immediate knowledge of individuals or households as to the basic principles of nutrition; food storage and processing techniques; suitable child care practices; and general healthcare management (Riely et al. 1999). This notion also in turn effectively draws into the food security mix the wider societal support mechanisms such as adequate health care services in terms of both availability and accessibility and the subsequent education and the investment in both, among other things. With regard to indices covering such measures then, there are numerous ones to choose from any of which might include expenditure on health services; education enrolment rates; access to adequate water and sanitation as well as morbidity rates of closely associated disease, etc. The two illustrated here (Figures 4.7 and 4.8) are for the proportion of population using improved drinking water sources and improved sanitation facilities.

These figures illustrate the worst-affected countries and serve to highlight the extent to which some of the lesser developed countries, particularly in Africa and parts of Asia, suffer from lack of access to improved water and sanitation. Of note in both instances is the weighting of the rural versus urban situation with the rural areas undoubtedly faring worse.

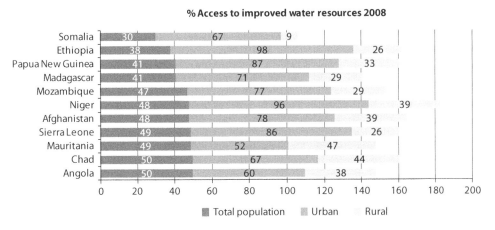

FIGURE 4.7 Proportion of population using improved drinking water sources. (From data at the World Health Organization's Global Health Observatory, *Global Health Observatory*, World Health Organization, Geneva, Switzerland, 2010.)

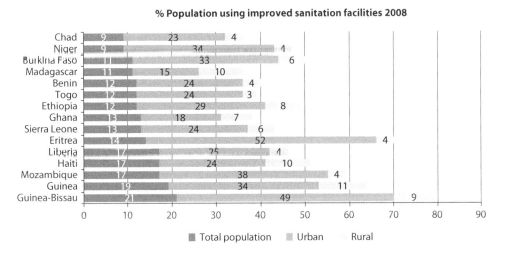

FIGURE 4.8 Proportion of population using improved sanitation facilities. (From data at the World Health Organization's Global Health Observatory, *Global Health Observatory*, World Health Organization, Geneva, Switzerland, 2010.)

4.3.4 Stability

When discussing risk, vulnerability or stability of the food situation, there are a number of variables that can be monitored. Such measures as the variability of food production index allows us to examine the ups and downs of each growing season while the index of food prices explores the market response to supply and demand. Further indices like the changes in cereal production in low-income food-deficit countries (LIFDCs) allows us to assess the susceptibility or risk factors affecting those most food vulnerable. Another common measure is that of the of number of natural disasters affecting a country. This is monitored by the Centre for Research on the Epidemiology of Disasters (CRED) which has since 1900 compiled a database—EM DAT—of over 18,000 natural disasters based on five categories: geophysical, meteorological, hydrological, climatological and biological (CRED 2009). While natural disasters like drought and floods have a direct bearing on food security issues so too do human-mader disasters such as conflict and war. These are monitored by the Office of the United Nations High Commissioner for Refugees (UNHCR).

4.3.4.1 Number of Countries Facing Food Emergencies

Although there were a reported 335 worldwide natural disasters in 2009, the upward trend over the last few years appears to be slowing (Vos et al. 2009). Of the 2009 figures, there were a total 26 reported droughts and a further 150 major floods. It can be seen from Table 4.5 that beyond the immediate death tolls, both of these disaster types carried with them huge humanitarian consequences in terms of displaced or affected people together with an equally large financial burden.

Of note here is that while the Americas suffered a large number of flood disasters during this period, it is interesting how well some of these countries coped with this risk compared to some others. In fact, this notion that some are more prepared than others is echoed in the 2009 Annual Disaster Statistical Review when Vos et al. suggest that

> Countries need to be better prepared for the destructive impact of natural disasters. Underlying factors and preconditions that make human populations vulnerable to disasters need to be addressed in order to mitigate impacts and create resilient and sustainable societies. (Vos et al. 2009)

4.3.4.2 Conflict, War and the Displaced

Whether through war, conflict or political reasons, every year millions of people are involuntarily moved both within and across borders (UNHCR 2009). Taking people out of their natural environment and away from familiar and recognisable means of food access and availability, the displaced form a major

TABLE 4.5

Natural Disasters 2009

Region	Droughts				Floods			
	Number of Incidents	Number Killed	Number Affected (thousands)	Est. Damage ($ million)	Number of Incidents	Number Killed	Number Affected (thousands)	Est. Damage ($ million)
World	26	6	90,000	563,000	150	3487	57, 000	8,004,000
Asia	6	0	55,000	234,000	47	2242	53,000	Incomplete
Africa	10	6	31,000	No data	27	416	1600	No data
The Americas	9	0	3000	Incomplete	33	275	2200	Incomplete
Europe	No data	No data	No data	No data	15	61	38	Incomplete
Oceana	No data	No data	No data	No data	5	40	28	Incomplete

Source: Courtesy of EM-DAT: The OFDA/CRED International Disaster Database (www.emdat.be) (Université Catholique de Louvain, Brussels, Belgium).

TABLE 4.6

Refugees, Asylum Seekers and Internally Displaced Persons

Category of Displaced Population	2008 Protected/ Assisted by UNHCR or UNRWA Mandates	2008 Total	2009 Protected/Assisted by UNHCR or UNRWA Mandates	2009 Total
Refugees	10.5	15.2	10.4	15.2
Asylum-seekers (pending cases)	0.2	0.8	0.2	1.0
Conflict-generated IDPs	14.4	26.0	15.6	27.1
Total	25.1	42.0	26.2	43.3

Source: *Global Trends: Refugees, Asylum-Seekers, Returnees, Internally Displaced and Stateless Persons,* Office of the
United Nations High Commissioner for Refugees, Geneva, Switzerland, 2009.

Note: Figures do not include natural disaster–related displacement.

vulnerable group in the food security stakes. Adding to the problem is the protracted nature of many of these crises. Prevailing political and/or humanitarian emergencies not only continue to uproot people but it makes it very difficult to return or resettle existing refugees and internally displaced persons (IDPs). In total in 2009, there were over 43 million forcibly displaced people worldwide, 27 million of which were IDPs (see Table 4.6).

4.3.5 Health and Development

Health and development indices allow the reader to gauge the development stage of a country or to explore metrics like the standard of living through health care and the rate of trained professionals within the sector, etc. There are numerous metrics to choose from and in keeping with the CFS and FIVIMS loose framework, we will examine two of the metrics: life expectancy and under-5 mortality.

4.3.5.1 Life Expectancy at Birth

The life expectancy metric summarises mortality rates at all ages. Many considerations such as good health care as well as nutritional and financial stability combine to improve longevity. In this way, it can be regarded as a good measure of the development of a country and by extension the quality of life. Table 4.7 shows that the average life expectancy for the bottom decile of countries at 48 years is considerably less than the 81 plus years of the top decile. Once again, comparing these data with those of Figure 4.6, you will notice that several of the countries in the low life expectancy range also appear in the poorer $1-a-day category.

4.3.5.2 Under 5 Mortality Rate

Another measure of the health of the population is that of mortality rates of under-5-year-olds. This is because it is felt that children in this age group are more vulnerable than adults and are more likely to be among the first of the vulnerable groups to succumb to shocks, shortages and emergencies (Figure 4.9).

Put in perspective, Figure 4.10 indicates the mortality trends over the last 20 years or so. Once again, the majority of the affected fall in two major regions: Africa, in particular sub-Saharan Africa, and Asia. While it can be seen that major progress has already been achieved in overall numbers within the developing world, most of these gains have been via the progress made in the Asian countries. By contrast it can be seen that sub-Saharan Africa has made little progress over the last few years to address this problem.

Further analysis conducted by UNICEF also indicates that mortality in under-5s varies considerably between rural and urban populations with the rural areas generally suffering higher mortality (UNICEF 2006). Furthermore the breakdown of causes in Figure 4.11 shows that a massive 31.8% of childhood deaths are perinatal while the other major causes include pneumonia, malaria and diarrhoeal. This once again highlights the extent to which nutrition is involved in health and outcomes and as previously discussed, it is worth remembering that sometimes even just the simple, inexpensive preventative measures can reap great returns.

TABLE 4.7

Life Expectancy at Birth: Top and Bottom Deciles

Country	Age	Country	Age
Afghanistan	44	Malta	80
Zimbabwe	44	Netherlands, The	80
Lesotho	45	New Zealand	80
Zambia	45	United Kingdom	80
Swaziland	46	Australia	81
Angola	47	Canada	81
Central African Republic	47	Israel	81
Congo, Dem. Rep. of	48	Luxembourg	81
Guinea-Bissau	48	Macao SAR, China	81
Mali	48	Norway	81
Mozambique	48	Singapore	81
Nigeria	48	Spain	81
Sierra Leone	48	Sweden	81
Chad	49	France	82
Burundi	50	Hong Kong SAR, China	82
Equatorial Guinea	50	Iceland	82
Rwanda	50	Italy	82
Somalia	50	Switzerland	82
Cameroon	51	Japan	83
Niger	51	Liechtenstein	83
Average	47.85	Average	81.25

Source: World Bank, *World Bank: World Development Indices.* World Bank Group. http://databank.worldbank.org/ddp/home.do, 2010b.

Mortality rate under-5s (per 1000 live births)

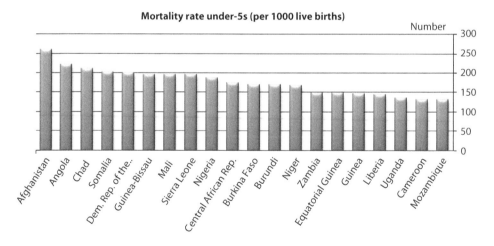

FIGURE 4.9 Mortality rate of under-5-year-olds: 2008 (per 1000 live births). (From data available at the WHO Global Observatory Database, *Global Database on Child Growth and Malnutrition,* World Health Organization, Geneva, Switzerland, 2010a.)

A caveat here highlights the continuing difficulties in assembling reliable and up-to-date data. In calculating mortality figures, the University of Washington's Institute for Health Metrics and Evaluation (IMHE) shows a faster rate of decline than the UN figures. According to the UN, under-5 mortality decreased from 12.6 million in 1990 to 8.8 million annual deaths in 2008, while the IMHE offers lower estimates from 11.9 million in 1990 to 7.7 million in 2010 (Rajaratnam et al. 2010). This comes about,

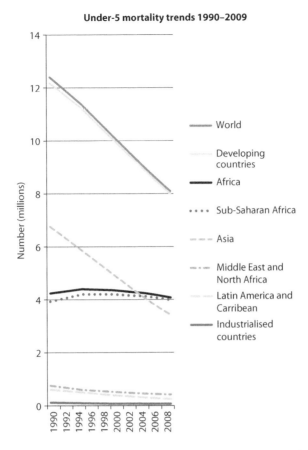

FIGURE 4.10 UN mortality trends 1990–2009. (From data available at the UNICEF, *Trends in Under-Five Mortality Rates (1960–2009)*, UNICEF, New York, 2010. ChildInfo.org.)

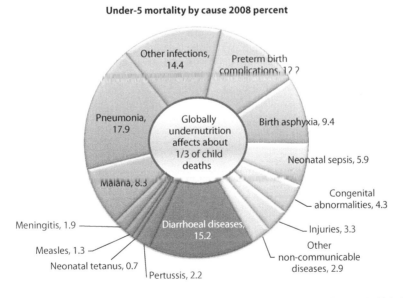

FIGURE 4.11 Under-5 mortality by cause in 2008 (millions). (From data available at the WHO Global Observatory Database, Website of the World Health Organization. World Health Organization, 2010b. http://www.who.int.)

according to the Eric Zuehlke, an editor at the Population Reference Bureau, through variations in methodology (Zuehlke 2010). According to Zuehlke, while the IMHE and the UN both use the same data, variations arise from the different statistical models and weight given to particular data sources. These are then extrapolated to the national level, which invariably leads to different interpretation of the results. Once again this is a small point but one that importantly highlights the different ways in which data is viewed by different groups.

4.3.6 Nutritional Status

4.3.6.1 Undernourished and Underweight

Perhaps some of the more often quoted of the indices are the food security outcome measures. These are useful tools in understanding the extent of the problem as well as assessing the success or otherwise of any intervention programmes. Choosing the right indicator however for many reasons already mentioned is not so easy. There are as many critics of each measure as there are advocates and many institutions are beginning to regularly use several alternative measures to strengthen or compliment their own. One of these widely used measures is the FAO's Prevalence of Undernourishment figures.

4.3.6.1.1 FAO Percentage of Population Undernourished

The percentage of a population suffering from undernutrition is calculated on the basis of three key parameters: the average food consumption (availability) per person; inequality of access and the minimum calories requirement for the average person weighted by the population demographic (MDG, 2000). The FAO categorises different levels or prevalences of undernourishment according to the percentage levels of each country or region shown in Table 4.8.

Figure 4.12 shows the prevalence of undernourishment. From this it can be seen that in 2010 there were an estimated 925 million undernourished people in the world, and although this was considerably less than the 2009 figures of 1.02 billion. This downturn however, between these two years has been attributed more to the improved economic conditions than any great advances in hunger alleviation (FAO 2010b). Yet even despite the downturn this figure, based on 2009 mid-year population estimates, still represented around 11% of the global population (CIA Factbook 2010; FAO 2011a). In fact, the latest State of Food Insecurity in the World (SOFI) 2009 report highlighted that even before the food price and economic crises of 2008/10, the number of hungry people had been steadily, albeit slowly, increasing before reaching its present level. A level incidentally which is higher than at any time since 1970 (see Table 4.9 and Figure 4.13) (ibid). This is deeply suggestive, even by FAO's recognition '… that present solutions are insufficient …' (SOFI 2009). This is further reinforced by the graph in Figure 4.13 which graph helps to identify the trends over time. While the data are erratic, the graph clearly shows a troubling trend. In the three decades from 1970 to late 1990s, there was a minor but definite gradual decrease

TABLE 4.8

FAO's Prevalence of
Undernourishment Categories

Prevalence	Category
<2.5%	Extremely low
2.5 to <5%	Very low
5 to <20%	Moderately low
20 to <35%	Moderately high
35% +	Very high

Source: Adapted from Maire and
Delpeuch, *Nutrition Indicators for
Development: Reference Guide*,
Institut de Recherche pour le
Développement, FAO & IRD,
Montpellier, France, 2005. With
permission.

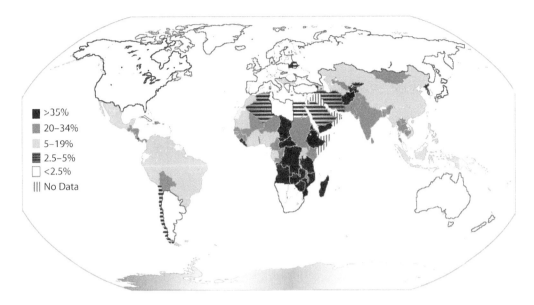

FIGURE 4.12 Prevalence of undernourishment. (Re-created from FAO Global Hunger Map, http://www.fao.org/fileadmin/templates/es/Hunger_Portal/Hunger_Map_2010b.pdf. With Permission.)

in both absolute numbers together with the prevalence or percentage of globally undernourished people in developing countries. This trend was reversed however over the last 10 years or so, but even so it can be seen that overall global figures have remained worryingly high over the last four decades, with the last 5-year prevalence of undernourishment hovering at around 14%–15% of the total world population (CFS 2008; SOFI 2008; FAO 2009; SOFI 2009; WFP 2011).

4.3.6.1.2 Regional Trends

An overview of global figures by themselves, whilst useful, do little to highlight any distributional disparities. To this end, a foray into regional variations is quite illuminating. Much of the previous data along with much literature attest to the fact that food insecurity is predominantly a phenomenon of the less developed countries. So while developing countries account for approximately 70% of the global population, they continually represent in excess of 95% of the overall global prevalence of hunger figures annually (Figure 4.14). Considering too the growth rates of these countries are more than twice that of the developed countries, it is perhaps not surprising then that much attention is focused on these regions (USDA/ERS 2008).

In 2009, the undernourished represented about 19% of the developing world (SOFI 2009). While it seems almost customary to bewail the efforts of the international community for their many failures, on a positive note and for the sake of perspective, it is perhaps worth examining the figures from another perspective. Today's figure of 19% for instance is actually considerably less than the 1969 developing world proportion of nearly 35%, as can be seen in Figure 4.15. In fact, the period from 1970 to early 2000 actually saw great gains in food security throughout the developing world. One way to look at the advances over the last 40 years is to take the near-35% hunger prevalence in the developing world of the 1969–1971 period and assume no progress by projecting these numbers onto today's figures. At an estimated population of 5.67 for the whole developing region in 2010 (DESA/POP 2010), this would represent undernourished figures of nearly 1.99 billion in today's terms. Compared to actual figures of 906 million and the reduction of over 1 billion is quite an achievement; of course this might not enough by many standards but it is definitely a step in the right direction.

This is not to sound trite, indeed for all the criticism this and other works levelled at the food security industry, much of this book is aimed at improving knowledge and engendering a sense of purpose and urgency in the fight against hunger and malnutrition. By the same token and in the interest of a balanced critique it is also worth remembering the many frontline workers, both today and in the past, fighting and succeeding in making a marked difference.

TABLE 4.9

Malnourishment Trends 1970–2009

Year	1969–1971	1979–1981	1990–1992	1995–1997	2000–2002	2004–2006	2008	2009	2010
World	878	853	845.3	824.9	856.8	872.9	915	1020[a]	925
Developed world			19.1	21.4	18.7	15.2		15	19
Developing world	864.2*	774.2*	826.2	803.5	838.0	857.7		1005	
Asia and the Pacific	**689.7***	**591.5***	**585.7**	**528.5**	**552.1**	**566.2**		**642**	**578**
East Asia	464.7*[b]	336.1*[b]	183.3	152.0	141.7	136.3			
Southeast Asia			105.7	88.6	93.9	84.7			
South Asia	225*	255.3*	286.1	278.3	302.8	336.6			
Central Asia			4.0	4.7	9.3	5.8			
Western Asia			6.1	4.4	3.5	2.1			
Latin America and the Caribbean	**49.6***	**43.2***	**52.6**	**51.8**	**49.4**	**45.3**		**53**	**53**
Central And North America			9.3	10.2	9.3	9.0			
The Caribbean			7.5	8.6	7.2	7.8			
South America			35.8	33.0	32.9	28.5			
Near East and North Africa	**38.6***	**21.1***	**19.1**	**29.6**	**31.6**	**33.8**		**42**	**37**
Near East			15.0	25.3	27.1	29.0			
North Africa			4.0	4.3	4.5	4.9			
Sub-Saharan Africa	**86.3***	**118.5***	**168.8**	**193.6**	**205.0**	**212.3**		**265**	**239**
Central Africa			22.0	38.4	47.3	54.3			
East Africa			77.2	85.7	83.4	86.5			
Southern Africa			32.4	35.8	36.5	36.7			
West Africa			37.3	33.8	37.7	34.7			

Sources: Alexandratos, N., *World Agriculture: Towards 2010*, FAO & John Wiley, Chichester, 1995; SOFI, *The State of Food Insecurity in the World 2008*, Food and Agriculture Organization, Rome, 2008; FAO, Address by Jacques Diouf, Director-General of the Food and Agriculture Organization of the United Nations (FAO), Food Security for All, Food and Agriculture Organization, Madrid, 2009; SOFI, *The State of Food Insecurity in the World 2009*, Food and Agriculture Organization, Rome, 2009; DESA/CSD, Achieving Sustainable Food Security: New Trends and Emerging Agenda. Multistakeholder Dialogue on Implementing Sustainable Development, United Nations Headquarters, New York, DESA: Division for Sustainable Development, CSD Secretariat, New York, 2010; DESA/POP, *World Population Prospects*, UN Department of Economic and Social Affairs, New York, 2010; FAO, Global Hunger Declining, but still Unacceptably High: International Hunger Targets Difficult to Reach, *News Brief*, Food and Agriculture Organization, Rome, 2010b, 2011a.

Notes: (1) Based on SOFI country compositions. Several historic figures have been retrospectively revised by the FAO based on the new FAO standards for human energy requirements of 2004 the new WHO Body Mass Index standards of 2006. (2) * indicates author's estimations based on 1996 FAO's Sixth World Food Survey figures and FAO's Agriculture: Towards 2010, which are then adjusted based on FAO's downward revision. These figures are intended as approximations only. (3) The numbers in bold represent the consolidated values for the regions represented in bold.

[a] While (DESA/CSD 2010)+ (SOFI 2009) suggest that the totals are 1020 million, discrepancies in their sub-aggregates equate to 1035 million (SOFI 2009).

[b] Includes Southeast Asia.

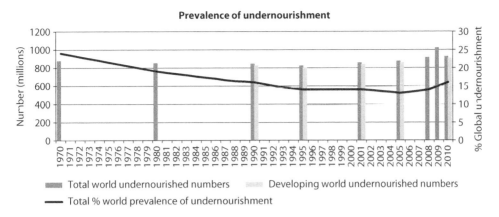

FIGURE 4.13 World undernourishment trend 1969/71–2009. (Compiled from FAO statistics; Alexandratos, N., *World Agriculture: Towards 2010*, FAO & John Wiley, Chichester, 1995; SOFI, *The State of Food Insecurity in the World 2008*, Food and Agriculture Organization, Rome, 2008; FAO 2009; SOFI, *The State of Food Insecurity in the World 2009*, Food and Agriculture Organization, Rome, 2009; DESA/CSD, Achieving Sustainable Food Security: New Trends and Emerging Agenda. Multistakeholder Dialogue on Implementing Sustainable Development, United Nations Headquarters, Division for Sustainable Development (DESA) CSD Secretariat, New York, 2010; DESA/POP, *World Population Prospects*, UN Department of Economic and Social Affairs, New York, 2010; FAO, Global Hunger Declining, but still Unacceptably High: International Hunger Targets Difficult to Reach. *News Brief*, Food and Agriculture Organization, Rome 2010b.)

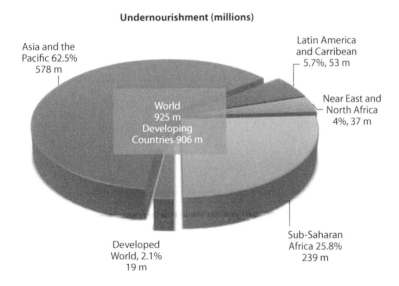

FIGURE 4.14 Number and prevalence of undernourished by region 2010. (Compiled from data available at FAO, 2011a.)

Taking and combining the above two figures of numbers and prevalence, a slightly different picture emerges. However, rather than using the 2010 estimates above, the following figure employs the most recent regional fully disaggregated figures as of 2010, which are from the rolling period 2005/7 (FAO 2011c). As can be seen, Figure 4.16 presents the full grim reality. Of the 847.5 million undernourished in 2005/7, the vast majority (over 98%) were from the developing countries. Notably, in the Asia and Pacific Regions, one in nearly five people suffered from undernourishment. Most of these numbers were made up of Indian (South Asia) and Chinese people (East Asia), which when combined (251 million and 127 million, respectively) represented nearly 67% of undernourished people in the region. The sub-Saharan Africa countries too suffered an alarming incidence of undernourishment at 30% of the overall

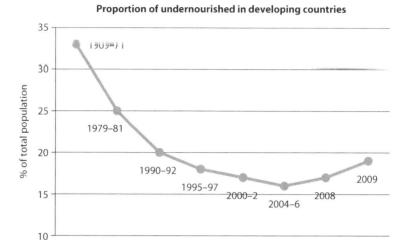

FIGURE 4.15 Proportion of undernourished in developing countries 1969–71 to 2009. (Compiled from FAO statistics; SOFI, *The State of Food Insecurity in the World 2009*, Food and Agriculture Organization, Rome, 2009; FAO, *Food Security Statistics: Country Profiles*, Food and Agriculture Organization, Rome, 2011b.)

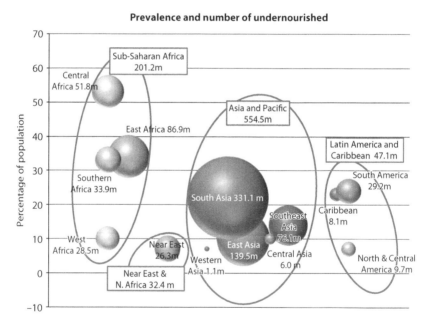

FIGURE 4.16 Undernourishment and prevalence combined 2005/7. Prevalence is represented by the percent of the population and number is represented by the size of sphere. (From FAO Statistics; FAO, Global Hunger Declining, but still Unacceptably High: International Hunger Targets Difficult to Reach, *News Brief*, Food and Agriculture Organization, Rome, 2010b.)

population. However, while Asia might have the majority of overall numbers of hungry, individually the central, east and southern African regions make up the highest prevalence of undernourishment anywhere in the world (Figure 4.16).

4.3.6.1.3 USDA/ERS

In a similar vein, the United States Department of Agriculture's Economic Research Service (USDA/ERS) also calculates an estimate of the prevalence of undernourishment figures; these are discussed in detail in the Section F.3. The two main differences between the UN's and the USDA/ERS's figures in essence are first that the ERS uses a different methodology relying greatly on a universal cut-off point of 2100 kcal per person and secondly the data is restricted to 70 developing countries. With this in mind the last year (2010) the prevalence of undernourishment in the ERS's 70 countries was estimated to have declined from 953 million in 2009 to 882 million in 2010 (Rosen 2010; USDA/ERS 2010).

4.3.6.1.4 WHO

The WHO also collects and collates indices of undernourishment, particularly where child measurements are concerned. However, unlike the measures of other institutions, which are subject to complex statistical manipulation, the WHO collates unadulterated anthropometric indices. The three main child measures in their undernourishment portfolio are: low height-for-age (stunting); low weight-for-height (wasting); and low weight-for-age (underweight). These are also looked at in more detail in Appendix F. For now, we are interested in the children's low weight-for-age metric, the underweight. The main adult anthropometric measure of the WHO in which we are also interested at this moment is the body mass index (BMI).

Proportion of Children Under-5 That Are Underweight: In terms of undernourishment, the WHO adopts the principle that under-5s are more susceptible to any reduction in food intake and would therefore be the first to manifest any signs of deficiency. In this way, they argue that such measures act as a barometer of the wider population at large. Figure 4.17 indicates widespread underweight children in Africa, Asia and the Indian subcontinent.

In terms of actual numbers, the UN Children's Fund (UNICEF) estimated that the total developing world numbers of children (under-5s) underweight in 2009 to be 129 million while in 2008 stunting affected a total of 195 million under-5s (UNICEF 2009; WFP 2011).

Percentage of Adults with Body Mass Index (BMI) <18.5: When it comes to adults the WHO utilises the body mass index matrix, a simple weight measure divided by the square of the height. The data itself are taken from information gathered from the household survey data and is a simple but effective measure that is both widely measured and readily accessible. Figure 4.18 lists the latest available figures; the caveat here is that these recent figures can range from anywhere between the 1990s through to 2008 (WHO 2010c). In this way, these data can only be used as a guide. Take Pakistan for instance. The most recent national statistics for underweight as determined by the BMI calculation are from 1992 and are

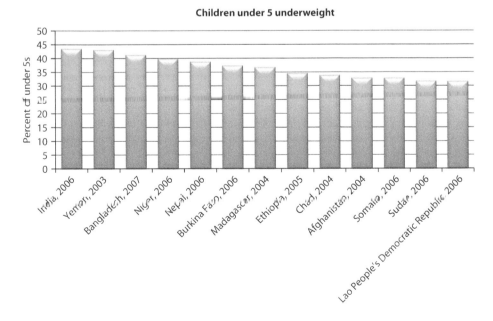

FIGURE 4.17 Proportion of children under 5 who are underweight. (From the WHO's Global Health Observatory, http://www.who.int/healthinfo/statistics/en/).

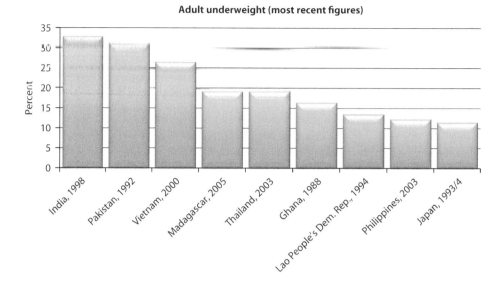

FIGURE 4.18   Body mass index. (From the WHO's Global Health Observatory, http://www.who.int/healthinfo/statistics/en/).

nearly 20 years out of date. That said and as is often the case, the measure of past performance is utilised as a guide for present and future predictions. With this in mind, it can be seen from the graph that India, Pakistan and Vietnam carry the lion's share of the regions' underweight population.

4.3.6.2 Micronutrient Deficiency

Having looked at the general global picture, it is perhaps worth taking a look at the wider nutritional status as dictated by the major micronutrient issues. It was estimated in 2009 that up to 2 billion people globally suffered from some kind of micronutrient deficiency. This in turn caused an increasing assortment of disorders that further increased the risk of death, disease and disability (WFP 2009b). This has a bearing on both the 'utilisation of food' aspect of food security as well as on the perception of food security in general. This is because simple measures of undernourished alone tend to focus on macronutrient or energy deficiencies. Moreover, even when an individual is receiving their full complement of calories, they still might be consuming an insufficient amount of the micronutrients. Micronutrient measures, on the other hand, when taken in conjunction with those of macronutrients, offer a fuller picture as to the burden and extent of nutritional deficiency (MacAuslan 2009). Epidemiologically speaking, it is also important to understand the prevalence of these deficiencies as oftentimes the effects of such deficiencies can be grave while effective treatment can sometimes be relatively cheap and simple. Table 4.10 outlines the criteria for establishing clinical levels of deficiency for iron, iodine and vitamin A.

4.3.6.2.1 Anaemia

Unlike many other nutrient deficiencies, anaemia or iron deficiency anaemia (IDA) is prevalent worldwide, occurring in both developed and developing countries alike. It affects up to 25% of the global population, mostly women and pre-school children and is characterised by tiredness (fatigue) and breathlessness as well as impaired mental development (UNICEF 2009; WFP 2009b). Combating or eradicating this most prevalent form of malnutrition has been posited to improve national productivity levels by as much as 20% (WHO 2009a).

4.3.6.2.2 Iodine

As with anaemia, iodine too affects both the developing and developed world. While many are protected through fortified foods such as salt, there still remained considerable prevalence of deficiencies

TABLE 4.10

Epidemiological Criteria for Assessing the Extent of Micronutrient Deficiencies

Deficiency	Population Group	Indicator	Situation of Deficiency		
			Mild	Moderate	Severe
Iron	All groups	Anaemia (%)	5.0–19.9	20.0–39.9	40
Iodine	Children 6–12 years	Total goitre rate (%)	5.0–19.9	20.0–29.9	30
	Children 6–12 years	Median urinary iodine (mg/L)	50–99	20–49	<20
Vitamin A	Pregnant women	Night blindness, XN (%)	>5		
	Children 2–5 years	Night blindness, XN (or Bitot spots X1B) (%)	>1 (ou >0,5)		
	Children 2–5 years	Low serum retinol (%)	>15		
	Children <5 years	Mortality (%)	>50 (ou 20–50[a])		

Source: Re-created from Maire and Delpeuch, *Nutrition Indicators for Development: Reference Guide,* Institut de Recherche pour le Développement, FAO & IRD, Montpellier, France, 2005. With permission.

[a] Indicates a problem could exist and that additional investigations are necessary.

in 2009, particularly in Europe (52%) and Africa (42%) (UNICEF 2009). Both the WHO and the UN Sub-Committee on Nutrition (UN/SCN) in 2006 suggested that between 1.6 and 1.9 billion people are at risk annually of iodine deficiency, which is especially harmful during early pregnancy and childhood (ChildInfo 2009). In its severest form, iodine deficiency disorders (IDD) include cretinism, stillbirth and miscarriage while milder deficiencies manifest other symptoms such as goitre. In fact, iodine deficiencies continue to be the single greatest obstacle to full mental development in infants and young children (SCN 2004; WHO 2006; ChildInfo 2009).

4.3.6.2.3 Vitamin A

As we have discovered earlier, vitamin A is essential for both eye health and the proper functioning of the body's immune system. It was estimated that vitamin A deficiency worldwide could account for anywhere from between 1 and 3 million overall deaths a year while a further 684,000 childhood deaths could be prevented by increasing access to vitamin A and zinc (Reinhardt 2004; SCN 2004; WFP 2009a). Besides mortality figures vitamin A deficiency (VAD) continues to be a significant global public health problem and the single leading cause of preventable blindness (UNICEF 2009; WHO 2009b). The WHO calculated in their 2009 report that 5.2 million pre-school children worldwide were affected by night blindness and an estimated 190 million (one-third of the global pre-school population) displayed low serum retinol concentrations of <0.70 µmol/L (Figure 4.19). Low serum retinol numbers were also highest in South-East Asia, with 91.5 million representing a staggering 49.9% of pre-schoolers. This was closely followed by Africa with 56.4 million pre-schoolers (44.4%) suffering from low serum retinol.

It was also calculated in the same report that a further 9.8 million pregnant women were affected by night blindness and 19.1 million displayed low serum retinol concentrations globally. Again the highest concentrations were in Africa and South-East Asia with both regions showing just under 10% of pregnant population with low serum retinol concentrations.

4.4 Sufficient Food

It can be clearly seen from any of the above metrics that the vast majority of hunger, malnutrition, burden of disease and poverty, among others, falls in the regions of Africa, Asia, the Caribbean and parts of South America. This inequality is not a recent phenomenon and marks a long-time trend that has its genesis in many aspects of economic and social development spanning decades if not centuries (coming sections). This is reflected too by extension in the regional availability of food and food security with many lesser developed countries receiving a less than equal share of world food availability. Moreover, given the large food deficits in many countries, it is not surprising then that perceived worldwide food

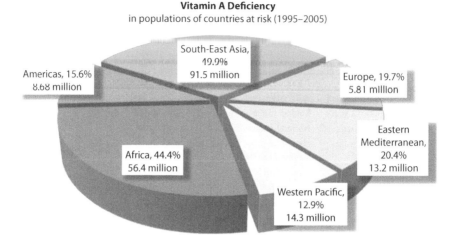

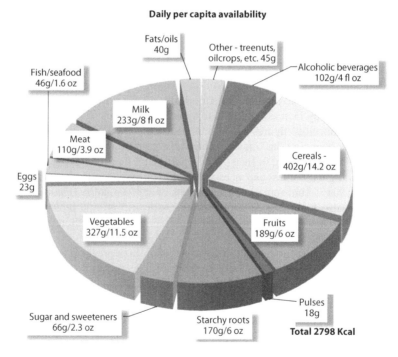

FIGURE 4.19   Global vitamin A deficiency in under-5s. (From WHO, *Global Database on Vitamin A Deficiency Report: Global Prevalence of Vitamin A Deficiency in Populations at Risk 1995–2005*, World Health Organization, Geneva, Switzerland.) *Notes:* WHO calculations are based on inclusive member states each having GDPs as at 2005 less than or equal to $15,000. The remaining 37 member states above this threshold are assumed to be free of public health significant VAD (WHO 2009c).

FIGURE 4.20   Per capita dietary energy supply 2007 estimates. (Figures calculated from the FAO food balance sheets; FAOSTAT, *Food and Agriculture Statistics*, Food and Agriculture Organization, Rome, 2011.)

shortages continue to persist (Charles 2008; Sachs 2008). The reality in fact is very different as evidenced in Figure 4.5. Disregarding regional food inequality for the moment, the global food supply continues to provide an adequate diet consisting of nearly 2800 kcal for every single man, woman and child on the planet (FAO 2011a; 2011c). This can be better illustrated in the pie chart (Figure 4.20) that clearly shows the makings of an enviable balanced daily diet (FAO 2011c).

Put simply, the world has for a while and continues to produce enough food for everybody. This has come about thanks in no small part to the green revolution, which has seen food production continuously outpace population growth over the last few decades (FAOSTAT 2011). It can therefore be shown that while food shortages, hunger and starvation are still prevalent in many regions of the world, insufficient aggregate global food production is not the cause (DeRose et al. 1998; Freedom 21, 2008). This highlights the fundamental and underlying truth that while there is no global food shortage as such, there does in fact exist vast regional disparities in availability, access and ultimately consumption patterns. Further exploring how this might have come about, the following chapters deal with the origins of food security and associated concepts.

4.5 Different Methodologies, Different Pictures

Some important methodological and conceptual differences in understanding or calculating the many variables or indices have been and will continue to be highlighted as they arise both throughout the book and in the appendices. In the above examples, I have highlighted some of the main differences in the various undernutrition metrics. These can sometimes lead to marked differences in results. Strikingly for example, Svedberg, a long time reviewer of the FAO methodology, argues that the FAO approach is not sufficient to give accurate estimates of malnutrition and illustrated this at the 2002 International Symposium on the Measurement and Assessment of Food Deprivation and Undernutrition. At this meeting Svedberg pointed out that while the FAO's results of that year placed the sub-Saharan African countries as having the highest prevalence of undernourishment, by using anthropometric indices South Asia was instead identified as having the highest incidence of underweight and stunting (indications of malnutrition) by far (Svedberg 2000). This 'most challenging observation', he explained, was drawn from the two different methodologies involved. Whereas the FAO uses a top-down approach to estimate the proportion of the population below a certain nutritional cut-off point based on food availability and the distribution coefficient, WHO and the Demographic and Health Survey (DHS) conversely use a bottoms-up approach in which they use representative survey data's anthropometric measures to estimate the number of people who have actual body weights below the norms. The large differences then, Svedberg suggests, can possibly be explained by measurement errors and biases inherent in both methodologies (Svedberg 2002). Others however disagree in principle, suggesting that in both the FAO and WHO methodologies, the lack of geographic concordance was of little importance as the two methodologies actually measured different things (Kennedy 2002). Stephan Klasen also agrees and suggests that the two measures ought to be seen as complimentary and simplifying his observations, suggests as an alternative using anthropometry in a clinical sense at the grass roots level and the FAO methodology at the national level (Klasen 2002).

Further criticising extant methodologies, in 2009 the Centre for World Food Studies calculated the prevalence of hunger based on anthropometric data lower than presented in the Millennium Development reports (17.3% against 27.8%; Wesenbeeck et al. 2009). While another paper by Nandy et al., exploring similar methods, suggested that the standard indices of stunting, wasting and underweight may actually be critically underestimating the problem (Nandy et al. 2005). Furthermore, when it comes to calorie requirements, it can be seen that the FAO's minimum daily food requirements range from 1990 calories in Croatia, Czech, Republic Estonia and Slovakia to as low as 1680 calories in Eritrea, Ethiopia and Occupied Palestinian Territory (FAO 2011b). On the other hand, the US Department of Agriculture's ERS uses a similar method to FAO's however, the ERS sets a higher caloric benchmark than FAO ranging from 2000 to 2100 calories per day. Also, whereas FAO's calculation factors in calories for physical activity level, the ERS does not allow for anything other than gathering food (USDA 1999). In conclusion, an interesting and telling result of such methodological differences showed up in 1999 when the USDA ERS used their methodology to retrospectively recalculate 58 of the FAO's 93 countries studied and found over one billion people by ERS standards fell short of adequate nutritional requirements compared to under 850 million by FAO criteria (USDA 1999). This has further implications for the 925 million undernourished today for the USDA's figures would surely revise this figure upwards.

References

Alexandratos, N., Ed. (1995). *World Agriculture: Towards 2010.* Chichester, UK: FAO & John Wiley

Bennett, M. K. (1941). International Contrasts in Food Consumption. *Geographical Review* 31(3): 365–376.

Boschi-Pinto, C., L. Velebit and K. Shibuya (2008). Estimating Child Mortality Due to Diarrhoea in Developing Countries. *Bulletin of the World Health Organization* 86(9): 657–736.

CFS (2008). Agenda Item II: Assessment of the World Food Security and Nutrition Situation. Committee On World Food Security: Thirty-fourth Session, Rome, Food and Agriculture Organization.

Charles, C. (2008) Food Shortages: How Will We Feed the World? *Telegraph* online: http://www.telegraph.co.uk/Earth/main.jhtml?xml=/Earth/2008/04/22/scifood122.xml.

Chen, S. and M. Ravallion (2007). The Developing World Is Poorer Than We Thought, but No Less Successful in the Fight against Poverty. *Policy Research Working Paper 4703*. Washington, DC: World Bank.

ChildInfo (2009). A World Fit for Children. *Child Nutrition*. United Nations Children's Fund (UNICEF). http://www.childinfo.org/idd.html (accessed 12 February 2011).

CIA Factbook (2010). The CIA World Factbook. https://www.cia.gov/library/publications/the-world-factbook/geos/xx.html (accessed 12 October 2010).

CRED (2009). The Centre for Research on the Epidemiology of Disasters: History. The Centre for Research on the Epidemiology of Disasters. http://www.cred.be/history (accessed 12 June 2010).

Cyberschoolbus, U. (2010). Un Global Teaching and Learning Project. UN Global Teaching and Learning Project. http://cyberschoolbus.un.org (accessed 13 February 2011).

DeRose, L., E. Messer and S. Millman (1998). Case Study: The Importance of Non-Market Entitlements. In *Who's Hungry? And How Do We Know? Food Shortage, Poverty, and Deprivation,* Derose, L., Messer, E., and Millman, S., eds. Tokyo, New York, Paris: United Nations University Press.

DESA/CSD (2010). Achieving Sustainable Food Security: New Trends and Emerging Agenda. Multistakeholder Dialogue on Implementing Sustainable Development, United Nations Headquarters, New York, DESA: Division For Sustainable Development CSD Secretariat.

DESA/POP (2010). *World Population Prospects*. UN Department of Economic and Social Affairs.

ESCAP (2000) Press Release No. G/05/2000. *UN department of Information Services*: http://www.unescap.org/unis/press/G_05_00.htm.

ESCWA (2009). The Impacts of the International Financial and Economic Crisis on ESCWA Member Countries: Challenges and Opportunities United Nations: UN Economic and Social Commission for Western ASIA (ESCWA).

FAO (2009). Address by Jacques Diouf, Director-General of the Food and Agriculture Organization of the United Nations (FAO). Food Security for All. Madrid: Food and Agriculture Organization.

FAO (2010a). FAO Country Profiles. UN Food and Agriculture Organization. http://www.fao.org/countryprofiles/selectcountry.asp?lang=en (accessed 30 November 2010).

FAO (2010b). Global Hunger Declining, but still Unacceptably High: International Hunger Targets Difficult to Reach. *News Brief*. Rome: Food and Agriculture Organization.

FAO (2011a). *Food Security Statistics*. Rome: Food and Agriculture Organization.

FAO (2011b). *Food Security Statistics: Country Profiles*. Rome: Food and Agriculture Organization.

FAO (2011c). Website of the Food and Agriculture Organization of the United Nations. FAO. http://www.fao.org/hunger/faqs-on-hunger/en/ (accessed 15 February 2011).

FAOSTAT (2011). *Food and Agriculture Statistics*. Food and Agriculture Organization.

Freedom 21 (2008). Meeting Essential Human Needs. 9th Annual National Conference, Dallas, Freedom 21.

Grebmer, K. v., B. Nestorova, A. Quisumbing, R. Fertziger, H. Fritschel, R. Pandya-Lorch and Y. Yohannes (2009). 2009 Global Hunger Index the Challenge of Hunger: Focus on Financial Crisis and Gender Inequality. *The Global Hunger Index*. Washington DC: International Food Policy Research Institute .

HDR (2009). Human Development Report 2009 Overcoming Barriers: Human Mobility and Development. New York: United Nations Development Programme (UNDP).

Hoddinott, J. (1999). Choosing Outcome Indicators of Household Food Security. Washington, DC: International Food Policy Research Institute 29.

IMF (2010). International Montetary Fund Website. International Monetary Fund. http://www.imf.org/external/ (accessed 4 January 2010).

Kennedy, G. (2002). Measurement and Assessment of Food Deprivation and Undernutrition: Part III: Parallel Contributed Papers Sessions—Discussion Group Report—Anthropometric Survey Methods. International Scientific Symposium, Rome, FAO.

Klasen, S. (2002). Measurement and Assessment of Food Deprivation and Undernutrition: Part III: Parallel Contributed Papers Sessions—Discussion Opener—Anthropometric Survey Methods. International Scientific Symposium: FIVIMS—An Inter-Agency Initiative to Promote Information and Mapping Systems on Food Insecurity and Vulnerability, Rome, FAO.

MacAuslan, I. (2009). Hunger, Discourse and the Policy Process: How Do Conceptualizations of the Problem of 'Hunger' Affect Its Measurement and Solution? *European Journal of Development Research* 21: 397–418.

MDG (2000). Proportion of Population Below Minimum Level of Dietary Energy Consumption. Millennium Development Goals. http://mdgs.un.org/unsd/mdg/Metadata.aspx?IndicatorId=5 (accessed 10 July 2011).

Nandy, S., M. Irving, D. Gordon, S. V. Subramanian and G. D. Smith (2005). Poverty, Child Undernutrition and Morbidity: New Evidence from India. *Bulletin of the World Health Organization* 83(3): 161–240.

OCHA (2006). Annual Report of the OCHA. UN Office for the Coordination of Humanitarian Affairs (OCHA).

Rajaratnam, J. K., J. R. Marcus, A. D. Flaxman, H. Wang, A. Levin-Rector, L. Dwyer, M. Costa, A. D. Lopez and C. J. Murray (2010). Neonatal, Postneonatal, Childhood, and under-5 Mortality for 187 Countries, 1970—2010: A Systematic Analysis of Progress Towards Millennium Development Goal 4. *The Lancet* 375(9730): 1988–2008.

Reinhardt, E. (2004). Vitamin and Mineral Deficiency, a Global Progress Report. *UN Chronicle* Sept.-Nov.

Riely, F., N. Mock, B. Cogill, L. Bailey and E. Kenefick (1999). Food Security Indicators and Framework for Use in the Monitoring and Evaluation of Food Aid Programs. Washington, DC: Food and Nutrition Technical Assistance Project (FANTA) 50.

Rosen, S. (2010). *Global Food Security Assesment*. Washington, DC: USDA/ERS.

Sachs, J. (2008) The Power of One: How to End the Global Food Shortage. *Time Magazine Online*, http://www.time.com/time/magazine/article/0,9171,1734834,00.html

SCN (2004). The Fifth Report on the World Nutrition Situation: Nutrition for Improved Development Outcomes. New York: UN Standing Committee on Nutrition.

Shaohua, C. and M. Ravallion (2001). How Did the World's Poorest Fare in the 1990s? *Review of Income and Wealth* 47: 283–300.

Sillers, D. (2008). National and International Poverty Lines: An Overview. US: USAID/EGAT/PR.

SOFI (2008). The State of Food Insecurity in the World 2008. Rome: Food and Agriculture Organization 59.

SOFI (2009). The State of Food Insecurity in the World 2009. Rome: Food and Agriculture Organization.

Svedberg, P. (2000). *Poverty and Undernutrition: Theory, Measurement, and Policy*. Oxford: Oxford University Press.

Svedberg, P. (2002). Measurement and Assessment of Food Deprivation and Undernutrition: Part III: Parallel Contributed Papers Sessions—Fallacies in—and Ways of Improving—the FAO Methodology for Estimating Prevalence of Undernutrition. International Scientific Symposium: FIVIMS—An Inter-Agency Initiative to Promote Information and Mapping Systems on Food Insecurity and Vulnerability. Rome: FAO.

UN (2010). *United Nations Statistical Database*. United Nations.

UN/DPI (2004) What Are the Least Developed Countries (LDCs)? *Fact-Sheet Number 20* (20): http://www.un.org/geninfo/faq/factsheets/FS20.htm.

UNCTAD (2011). Un List of LDCs after the 2006 Triennial Review. United Nations Conference on Trade and Development. http://www.unctad.org/Templates/Page.asp?intItemID=3641&lang=1 (accessed 23 February 2010).

UNHCR (2009). Global Trends: Refugees, Asylum-Seekers, Returnees, Internally Displaced and Stateless Persons. Geneva: Office of the United Nations High Commissioner for Refugees.

UNICEF (2006). Progress for Children. New York: United Nations Children's Fund.

UNICEF (2009). Tracking Progress on Child and Maternal Nutrition: Survival and Development Priority. New York: United Nations Children's Fund.

UNICEF (2010). *Trends in Under-Five Mortality Rates (1960–2009)*. New York: UNICEF: ChildInfo.org

UNSD (2006a). Committee for the Coordination of Statistical Activities: Note on Definition of Regions for Statistical Analysis. Montreal: UN Statistical Department.

UNSD (2006b). Note on Definition of Regions for Statistical Analysis. Montreal: UNSD: Committee for the Coordination of Statistical Activities.

USDA (1999). US Action Plan on Food Security: Solutions to Hunger. US Department of Agriculture (USDA).

USDA/ERS (2008). GFA-19: Food Security Assessments Reports. *Agriculture and Trade Reports*. USDA Economic Research Service.

USDA/ERS (2010). Food Security Assessment 2010–20. *GFA: Food Security Assessment*. Washington, DC: US Department of Agriculture.

Vos, F., J. Rodrlguez, R. Below and D. Guha-Sapir (2009). Annual Disaster Statistical Review 2009: The Numbers and Trends. Brussels, Belgium: CRED: Centre for Research on the Epidemiology of Disasters 46.

Wesenbeeck, C. F. v., M. A. Keyzer and M. Nubé (2009). Estimation of Undernutrition and Mean Calorie Intake in Africa: Methodology, Findings and Implications. *International Journal of Health Geographics* 8(37): 1–18.

WFP (2009a). World Food Program Annual Report. Rome: UN/WFP.

WFP (2009b). The World Food Programme: Food Quality Control. World Food Programme. http://foodquality.wfp.org/Home/tabid/36/Default.aspx (accessed 2 April 2010).

WFP (2011). Website of the World Food Program. World Food Programme. http://www.wfp.org/ (accessed 25 February 2011).

WHO (2003). The World Health Report 2003—Shaping the Future. Geneva: World Health Organization.

WHO (2006). *Neurological Disorders: Public Health Challenges*. Geneva, Switzerland: World Health Organization.

WHO (2008). The Global Burden of Disease: 2004 Update. Geneva, Switzerland: World Health Organization.

WHO (2009a). *WHO Global Database on Anaemia*. Geneva, Switzerland: World Health Organization.

WHO (2009b). WHO Global Database on Vitamin A Deficiency Report: Global Prevalence of Vitamin A Deficiency in Populations at Risk 1995–2005. Geneva, Switzerland: World Health Organization.

WHO (2010a). *Global Database on Child Growth and Malnutrition*. Geneva, Switzerland: World Health Organization.

WHO (2010b). Website of the World Health Organization. World Health Organization. http://www.who.int/ (accessed 2 April 2010).

WHO (2010c). *The WHO Global Database on Body Mass Index (Bmi)*. Geneva, Switzerland: World Health Organization. 2010

WHO/GHO (2010). *Global Health Observatory*. Geneva, Switzerland: World Health Organization.

World Bank (2010a). Website of the Worldbank. World Bank. http://www.worldbank.org/ (accessed 13 July 2010).

World Bank (2010b). *Worldbank: World Development Indices*. World Bank Group.

Zuehlke, E. (2010). New Estimates Reassess Progress toward Reducing Maternal and under-5 Mortality. Washington, DC: Population Reference Bureau.

Summary of Part I

Nutrition is the cornerstone of survival, good health and development can be thought of as being on a continuum with overnutrition at one end and undernutrition at the other. Malnutrition, however, can and does occur at any point along this plane and together with overnutrition and undernutrition collectively accounts for the major health issues worldwide. It is also important to emphasise that unlike in many infectious diseases there is no clear point of onset of undernutrition, no easily discernible line between normalcy and clinical morbidity. Despite perceptions to the contrary too, nowadays chronic disease burdens such as overnutrition are not solely limited to the developed regions of the world. Instead, traditional diets and lifestyles are changing and developing countries are increasingly suffering from just as high levels of overnutrition as their developed counterparts. This creates the dichotomy where in some countries both under and overnutrition are occurring simultaneously.

Energy or macro-nutrients along with micro-nutrients and water make up the entire ingredients of the diet. These are the basic materials from which the body is made and achieving the right balance of the right nutrients is the goal of good nutrition. For a long too while dietary guidelines were preoccupied with providing specific targets for this achievement; however, of late it is increasingly recognised that a healthy balanced diet can be achieved in more ways than one using a variety of foods in a given ecologic and population setting. In this way, the development of food-based dietary guidelines (FBDG) by the FAO, the WHO and others recognise that rather than focusing on how each specific nutrient might be supplied in full, the focus instead is on the combination of foods that collectively meet nutrient requirements. This approach highlights the importance of dietary diversification and local food culture as well as addressing regional or local nutrient issues. In departing from the entrenched recommended daily allowance ranges, the FBDG is region- or country-specific and effectively aims to bridge the gap between daily recommended requirements and local food availability and culture.

This optimum balance of both macronutrients and micronutrients has been the pre-occupation of many researchers and scientists for centuries (more on this in the following sections). It has been noted too that producing enough food does not translate into sated populations as regional inequality prevails on an enormous scale. In this way it can be seen that most hungry people live in the developing world; they live in rural areas and are often the poorest in society. We have also discovered thus far that in women and children, nutrition is especially important. Women tend to be the care-givers and are responsible for the nutrition of most children and in many cases the household too. Women also make up 60% of the malnourished and are often relegated to secondary importance due to continued inequality and other archaic practices. Infants and young children too are also particularly vulnerable because of their nutritional requirement needs for growth and development. Children suffering from undernourishment demonstrate lower resistance to infection and consequentially are more likely to die from common complaints such as diarrhea and respiratory infections. Furthermore, frequent illness has been shown to deplete existing nutrients and further hinder the absorption of digested nutrients. This effectively locks people into a vicious cycle of sickness and poor development both physically and mentally. In this regard, well-nourished women are stronger and experience fewer problems during pregnancy and childbirth, leading to better development potential for their children.

What has also been established is that actual physical starvation is a rare occurrence and that under-nutrition and co-morbidity mutually exacerbate and magnify the worst effects of each. In short, the physiological effects of hunger vary and are dependant in large part on an individual's age, health and reproductive status. It also depends to a greater extent on the particular type of nutrient deficiency within the diet. Beyond these specifics, some generalisations can be made. Hunger affects everything from a person's physical and intellectual development as well as their ability to fight disease. A healthy brain uses 20% or so of the body's energy and as a result, malnutrition, particularly in children, can affect intellect, awareness and sociability. Once acute or chronic hunger has taken hold, a person's sensory and motor skills deteriorate while the internal organs begin to break down. In its bleakest, a total lack of food or starvation is fatal in 7–12 weeks. However, more likely hunger sufferers die from opportunistic diseases like diarrhea or measles or infectious diseases. Furthermore, while chronic malnutrition kills more people in terms of numbers, the onset of acute malnutrition by its nature is traditionally quick and kills more rapidly. In this way, severe acute malnutrition is a sensitive indicator of an unfolding emergency. These differences also have implications for policy and intervention and must be understood and addressed differently.

When talking of measuring in general and food security in particular it can be seen that there is a plethora of issues to consider. Short of polling every person on the planet to determine their nutritional status it is clear some assumptions have to be made; however, the essential aspect here is to ensure the accuracy and reliability of such postulations. Considerations need to account for the different concepts: hunger, malnutrition, recommended allowances, etc., and how these are defined before they can be measured. Once consensus is reached on the various metrics the next difficulty lies in choosing the right indicator, which measurements to take or which one best suits which purpose. Appropriate metrics also need to be considered with the end use in mind. If for instance the goal is to address undernutrition in a health clinic then anthropometrics are by far the better metric, if on the other hand, the end purpose is to gauge the enormity of undernutrition in the world in a comparative study, then anthropometrics alone has its limits. In terms of indicators themselves, what has been clearly determined is that there still does not exist, despite some headway, a single measure of food security. Single measures of associated concepts too such as malnutrition, poverty and GNP for instance, while useful do not give the reader a full picture of the food insecurity either. Instead, a suite of indicators is chosen over a single indicator which when combined are ultimately strengthened by their complimentarity. The difficulty then lies in choosing the right proxies through which triangulation can deliver a picture of the nutritional and food security status of individuals, regions and countries. Indices can be of national importance or designed for cross-border comparison. Cross-border metrics have proven to be powerful tools for advocacy and when used in international rankings can promote a healthy sense of competition among countries which in turn help to promote good policies. It is also shown that when it comes to measuring undernutrition in particular, the FAO's prevalence of undernourishment and the WHO's use of child anthropometrics despite many criticisms remain popular options. This highlights the fact that there is often more than one way of measuring any single indicator, and while there might be multiple measures of the same indicator, each in turn may have multiple methodologies. As such, some indices that purport to measure the same variable cannot always be directly compared. Furthermore, inconsistency in reporting and selective or inappropriate usage of statistics compounds the problems of understanding an already complex phenomenon. New composite indices like the Global Hunger Index and the Integrated Food Security Phase Classification (IPC) are steps in the right direction; however, once again there is little widespread consensus over accepted or suitable indices and methodologies.

With regard to measuring energy requirements, generally two approaches are employed. The first is largely discredited or rather rendered obsolete by advances in understanding. This is based on the principle of equilibrium, that is, a person who is neither gaining nor losing weight is considered a healthy person and all that remains would be to measure the food intake of that person. The second is based on energy expenditure and attempts to measure the energy losses of individuals, and in this way the goal is to replace this lost energy in order to meet the ongoing energy requirements. Similarly, when it comes to measuring the output of nutrition in children and adults, two approaches are made use of. In children, anthropometric measurements are based on growth with measures of height/age, weight/height and weight/age, while in adults the body mass index (BMI) (weight divided by the square of the height) is considered to be a better measurement of nutrition.

Country-specific data too are at risk of being misinterpreted as institutional classifications are proprietary and oftentimes non-comparable. Novices and the unsuspecting might easily be comparing two data sets of seemingly identical regions yet in reality are actually comparing apples with oranges. For instance, if reading two documents on developing countries—one by the World Bank and one by the UN—you could be forgiven for thinking they were talking about the same aspects. In this way, it becomes difficult to compare like for like across institutions and regions.

Lastly, there exists a problem with reported statistics. It is not always clear from the literature, even when scanning the references cited, just how old the data that is being quoted is. Also there are many instances, sometimes within the same organisations, in which older reports are quoting newer statistics than more recent reports. Statistics then become confusing at best and contradictory at worst. Compounding this confusion is the time lag differentials between data collection periods (which can be intermittent in themselves) to the reporting of such statistics. At the moment, it is perhaps necessary, although arguably unhelpful, that in some instances, the latest available statistics of 2010 are those of 2000. In several instances however, it can be said that the time lag between the reference collection period and the publication of some data might take on up to one to two years. It is not as if some of these organisations have not attempted to update their data either. Many notables in the food security sector use websites to keep their publics informed and up-to-date yet all too often these attempts are seemingly badly organised and poorly maintained. In light of this fact, it should not be surprising then that data is far too often contradictory, being misused, misquoted or plainly abused. This can lead to a state of statistical blindness where data begins to challenge the readers' trust where people become overwhelmed and switch off to statistics or dismiss them completely; tune them out so to speak. Conversely readers might also just use such statistics at face value without considering the questions of comparability, or the origin of the data or perhaps its age. Moreover, it would seem the current system of 'publish-and-be-damned' encourages the seemingly unregulated practice of just throwing out an overwhelming amount of raw data without proper regard to analysis or context. In this regard there is a propensity with literature that aims to make a point to cherry pick stats that promote or enforce a particular viewpoint. After all, a child dying every 3.6 seconds has more impact than one dying every 6 seconds.

The terminology of food security can be difficult too. Hunger means one thing to one person and something completely different to another. It is a word that has taken on emotive connotations and one that should be used with caution in academic literature. That said the author sees no problem in its usage, if the literature so chooses and if used appropriately the emotive overtones of hunger can effectively be used as a sort of rallying cry, or a call to arms as it were. Another example of difficult terminology sees the US agencies refer to their annual Global Food Assessment calculations as 'food security' when they are in fact measures of undernutrition. A small point but one that perpetuates a simple confusion that sees too many people believing that food security is solely about undernutrition. The FAO is also at fault in this area when it uses 'consumption' to mean supply. It was a practice that was highlighted by Svedberg way back in 1987 yet it continues even today. Lastly, it is seen that undernutrition and malnutrition are often used interchangeably and once again demarcation is not always clear.

Part I then is an assimilation of current knowledge and understanding. It is a brief introductory outline of the food security concept as well as a look at both good and bad nutrition. We have also examined how the concept is viewed, measured and monitored and while we have superficially teased out some of the core precepts; how we got here and how this affects or is affected by wider societal issues are examined in later sections. For now an initial analysis of food security shows that the concept is about adequate provision and access to food for all peoples at all times. Moreover, while many attempts have been made and are continuing to be made regarding the implicit or explicit inclusion of its many facets, suffice to say for now that it is a complex phenomena that at this juncture belies precise definition beyond the touched upon pillars of availability, access, utilisation and stability. Nevertheless from whichever perspective one chooses to view the concept, the emotive and continued relevance of current hunger issues ensures that it is set to remain relevant and topical for many years to come.

Part II

History: A Fledgling Construct

Tis education forms the common mind,
Just as the twig is bent, the tree's inclined.

Alexander Pope
Moral Essays

Over the last three to four decades, little real progress has been made in reducing the numbers of the globally malnourished. It has been suggested that this is due largely to the multi-disciplinary nature and the complexity of the debate itself while many more offer that a fundamental flaw in our understanding is more likely to blame. In answer to both charges, the following sections aim to present a re-evaluation of our current understanding. In doing so, we seek to re-examine food security from the historical perspective to the present and in due course perhaps shed some light on the questions of: what is the essence of food security? Where does it come from and where is it going?

Despite what has been suggested in other literature, the modern notion of food security did not begin in the 1970s; instead it borrows, regurgitates and builds on numerous age-old ideological, political, economic and philosophical foundations. Furthermore, as should be becoming clear by now, modern issues of food security are increasingly being linked to wider economic and societal goals and as such, the subject requires that it be explored from a variety of perspectives (Ruxin 1996; Farnworth 2003). Building on this, while Part I aimed at giving a snapshot or a cursory examination of the current food security concept, the next sections delve into a little history in an attempt to underpin the re-evaluation of the subject.

With this in mind any study of the subject needs to be conducted with one foot in many disciplinary camps including those of social historian, food scientist and technologist as well as those of government, politics and current philosophical ideology, etc. Furthermore, by acknowledging some of the relevant ideological and social movements, more contextual settings are created and motives better elucidated. Towards this goal, Part II delineates the historical context in which key ideas took hold and began to converge on the health of nations. In doing so, this foray both challenges and tests some of today's long-held and prevailing wisdom's that underpin modern food security's fundamentals. After which time the narrative conveniently leads us up to the twentieth century (Part III) and a period that lays claim to the popularisation of the concept of food for all.

5

Sociocultural Evolution

Populations can only grow when they have sufficient means to do so. This involves a plentiful supply of natural resources, food and water and the collective organisation and governance of its people. In this way, social evolution and collective ideology continually seek to reshape humanity for the better. This chapter aims to answer some basic questions of how the idea of food security emerged and under what contemporary conditions. This section also explores the growing international dimension of food and ultimately the multilateral nature of governance and responsibility. For much of this, events are explored chronologically so as to better inform the reader of contemporary motivational forces. Beginning with social collectivisation, the research then goes on to explore agriculture, science and technologies' efforts to adapt to the growing needs of the population before looking at conflict, economic disparity and the growing food security gap and the subsequent introduction of global governance and humanitarianism.

Firstly, though ever since civilisations first began to collectively participate in sedentary cultivation and share the fortunes and failures of their labours, the concept of food security has emerged, adapted and evolved. As such and with this in mind, I include a brief overview of the agricultural progression of civilisation.

5.1 The Beginnings and Growth of Agricultural Civilisation

Many scholarly debates abound with regards to the specific origins of, and the impetus behind, the transition from hunter-gatherer to the birth of the agricultural evolution. Many such theories include the oasis, demographic, evolutionary or socio-economic hypotheses (Childe 1936; Sauer 1952; Binford 1968; Rindos 1987; Hayden 1992, 1995; Weisdorf 2005; Rosen 2007). However, specific underlying theories aside it is generally widely acknowledged that the origins of farming evolved independently about 10,000 years ago at several sites around the world. With freshwater acting as an overriding geologic influence, the previous mobile bands of hunter-gatherers came to settle close to rivers, springs and lakes (Miller 1980; Gopher et al. 2001; Weisdorf 2005; Guisepi 2009; UOR 2009). Widespread adoption of this agrarian revolution, coined by philologist-turned-archaeologist, V. Gordon Childe in 1935, as the 'neolithic revolution' was the primary driving force in the growth of early sedentary cultivation and subsequent early population increases (Greene 1999; Guisepi 2009). So important was this transition from hunter-gatherer to agriculturalist that it has been universally recognised as one of the most pivotal moments in human civilisation (Weisdorf 2005). It resulted in the establishment and growth of many early civilisations: the Mehrgarh and Harappan civilisations along the Indus; the Chinese Empire along the Yangtze, Huang and Yellow Rivers (UOR 2009) as well as many Mesopotamian and Near East cultures that finally settled along the Tigris, Euphrates and the Nile Delta regions (an area later known as the 'Fertile Crescent') (*Time* 1936). The transition from hunter-gatherer was also probably a slow one with early settlers more likely supplementing crop plantations with foods gathered in the wild before the full transition to pure domestic agriculture-based economies was finally achieved (UOR 2009). Such animal and plant domestication also marked a pivotal evolutionary time for mankind and one in which humans became increasingly dependent on and constrained by the environment (Furon 1958; Weisdorf 2005; UOR 2009). It was these practices that also effectively set the agriculturalists apart from the hunter-gathers of the previous epoch.

During this period farming tools were made of wood and stone. These included the stone adz (an ax-like tool); the sickle or reaping knife (used to gather grain); the digging stick (later adapted as a spade or hoe); and a rudimentary plow (later adapted to be pulled by oxen). As civilisation progressed, new and improved practices and metal tools of bronze and iron greatly improved cultivation with things like the cast-iron mold bar plow. New irrigation systems too powered by windmills and water mills allowed hitherto unyielding land to be brought into cultivation. Productivity was also increased through the introduction of animal manure fertilisers and the practice of crop rotation and land left to fallow. Storage methods were also refined in the form of granaries where jars, dry cisterns, silos and bins containing stored grains began to crop up, providing the needed food for increasing populations.

5.1.1 The Rise of Modern Agriculture: Standing on the Shoulders of Giants

The next few millennia witnessed the socio-cultural dynamic transformation of the small, mobile egalitarian band of hunter-gatherers into sedentary societies with increased social, political and technological complexity (Johnson 1997; Guisepi 2009). This complex societal and political re-organisation allowed for the diversification and development of non-agricultural trades such as craftsmen, politicians and priests and in some places the introduction of social class systems (Bender 1975; Price and Gebauer 1995; Weisdorf 2005; Agropolis Museum 2009; Guisepi 2009). This marked change in the way food and non-food goods were produced formed the fledgling basis of modern economic civilisation, however it also brought with it fears of sufficient food provision and access (Cavalli-Sforza et al. 1993; Agropolis Museum 2009).

These fears aside however, the modern agricultural economy owes a great deal to the collective knowledge of our forebears. The 'Muslim Agricultural Revolution' of the eighth century for instance witnessed measurable advancements and by the Middle Ages were the first to disseminate this knowledge through detailed written agricultural techniques and technologies across the Islamic world (Glick 1977; Al-Hassani et al. 2007). The Romans too built on techniques pioneered by the Sumerians and cultivated crops whose emphasis was in trade and export. Collectively, as a consequence of these advances, an economy across the Old World was established, which, over time, further facilitated the diffusion of farming techniques plants and crops.

5.1.2 1492: The Columbian Exchange

The Columbian Exchange, often cited as the second great food revolution, was a period that started with Christopher Columbus and continued with his successors. This involved increased trade between the 'Old' and the 'New Worlds' of crops, livestock and disease. New crops such as American corn (maize), the potato and tomato among others were introduced into Western Europe, while North America benefited from hitherto new livestock such as cattle, pigs and sheep. Grains too, such as wheat, found their way into the Americas while from Africa and Asia, via Western traders and slave labour, rice, onions, coffee beans, olives, grapes, bananas and sugarcane all put down roots in the New World (Agropolis Museum 2009). Agriculture also became a key element in the Atlantic slave trade with the introduction of the plantation economy. Large agricultural estates producing cocoa, cotton, sisal, sugarcane, coffee, banana, citrus trees, palm trees and indigo were introduced and while heavily dependent upon slave labour, allowed populations to grow through greatly increasing productivity (ibid).

5.1.3 Agricultural Colonialism

Around about the fifteenth- and sixteenth-century Europe, new economic theories emerged predicated on wealth creation. Furthering, or perhaps fuelling, this was the new period of global exploration and colonisation providing food and wealth for the home countries. Up to this point, agricultural knowledge was passed on from generation to generation, from culture to culture becoming the sum collective of civilisations' forebears (Danhof 1949; Johnson 1997). The substantial flow of knowledge across time and space started before the Neolithic period and subsequently included the Arab agricultural revolution as well as the Columbian exchange (Crosby 1972; Watson 1974; Glick 1977; Salvaggio 1992; McNeill 2003; Al-Hassani et al. 2007; Agropolis Museum 2009). This flow of knowledge cannot be understated. Indeed, Justus Von Liebig eloquently remarked in his book of the period that

> One of the most remarkable features of modern times is the combination of large numbers of individuals representing the whole intelligence of nations, for the express purpose of advancing science by their united efforts, of learning its progress and communicating new discoveries ... (Playfair 1847, p. 3)

And so it was this accumulation of knowledge that set the backdrop from which the agricultural and industrial revolutions grew.

5.2 The Revolutions: 1650 Onwards

5.2.1 Agricultural Evolution

Indeed, true pioneering agricultural research has often been cited as beginning from around this period (Johnson 1997). It began in England and spread to Europe and the United States in the mid-seventeenth century before continuing with the industrial revolution of the early to mid-eighteenth century (Frey 1996; UNEP 1996; Johnson 2000). Along with the ongoing transfer of crops from their lands of origin as exploration and trade flourished between the old and new worlds, progress was also made in the widespread selective breeding of plants and livestock and new cropping rotations. This period also saw the first systematic attempts at pest control management with poisons and other biological controls. Significant increases in production output and rising yields also aided in the reclamation of land and the heavy use of fertilisers. Moreover, extensive irrigation, drainage, improved and increased animal husbandry and the use of hormones and antibiotics further aided the beginning of the industrialisation of agriculture (Johnson 1997, 2000). All in all, such developments led to the increased specialisation of agricultural businesses. As a result, it was not long before the great cities of North Africa, the Near East, Europe, Asia and the Americas were all supporting technologically advanced agricultural systems of crop rotation, soil fertilisation, irrigation and pest control measures (Cohen 1995; UOR 2009). This period particularly in Europe also witnessed the widespread practice of feudalism in the agricultural sector where the lords of the manor (or liege) presided over their vassals. These systems along with the introduction of new crops improved agricultural efficiency and the subsequent increase in the carrying capacity of the land (UOR 2009). Much was expected and indeed anticipated from such scientific advancement and what effectively started with advances in agricultural practices, experimentation and scientific application continued with leaps in transportation, mechanisation and knowledge transfer (Durand 1916). This had unintended consequences however and in turn led to competition and the displacement of hitherto long-established trading partners as well as changes in the location of suppliers. Another outcome of this global agricultural revolution had the effect of increasing food production per unit area thus allowing for populations to be fed much easier from the same area of land (UOR 2009). From these advances, this period saw populations begin to grow considerably (Figure 5.1). This was also accompanied by a general increase in the standard of living in the industrialised nations and marked changes in among other things, local economies, income growth and distribution of the labour force (Watson 1974, 1983; Gardner 2002).

5.2.2 Industrial Revolution

Hot on the heels of these agricultural improvements was the Industrial Revolution. It was a period of rapid industrial growth which began in England during the early part of the eighteenth century (1725–1750) and spread throughout Europe and the United States. The beginning of this period witnessed many leaps, inventions and improvements in both work practices and mechanical innovation. Advances in the wooden plough, new horse-drawn threshers, cultivators, grain and grass cutters, rakes and corn shellers followed in quick succession. Many in turn were superseded or improved upon with the invention of steam power in the nineteenth century. This essentially led to the mechanisation and industrialisation of agriculture and ultimately to the beginning of the food-processing industries (UOR 2009). Table 5.1 highlights some of the seminal moments and key inventions of both the industrial and agricultural revolutions however; undoubtedly the real coup was the invention of the internal combustion engine in the 1850s. This had the effect of freeing up much labour and for the first time allowed for the large migration of workforces out of agriculture, releasing millions for urban employment (Johnson 1997, 2000).

In the twentieth century, steam, gasoline, diesel and electric power became widely accessible. Improvements in the transportation infrastructure of roads, canals, ships and rail affected agriculture and empowered farmers. Remote supplies could now be economically accessed and produce could be

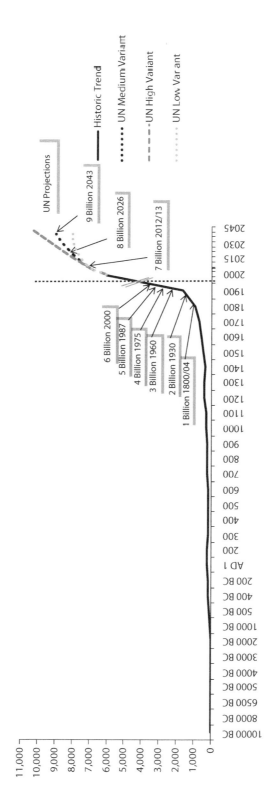

FIGURE 5.1 Population trends: Historical and projected. (Based on mean averages of past trends as offered by UN. *The Determinants and Consequences of Population Trends: New Summary of Findings on Interaction of Demographic, Economic and Social Factors*, Vol. I, Population Studies, No. 50. Sales No. E.71.XIII.5, United Nations, New York, 1973; McEvedy, C. and Jones, R., *Atlas of World Population History*, Facts on File, New York 1974; Tomlinson, H., *Int. J. Consumer Stud.*, 2, 15–26, 1978; Biraben, J.-N., *J. Hum. Evol.*, 9, 655–663, 1980; Johnson, D.G., *Am. Econ. Rev.*, 90, 1–14, 2000; Haub, C., *Populat. Ref. Bur.*, 30, 3–4, 2002; USCB, *Historical Estimates of World Population*, US Census Bureau, Washington, D.C., 2008a; Zhang, W., *Environ. Dev. Sustain.*, 10, 717–730, 2008 and future figure based on UN medium, high and low variant projections [UNPP, *World Population Projects: The 2008 Revision Population Database*, United Nations Population Fund, New York, 2009]).

TABLE 5.1

Key Innovations of the Agricultural and Industrial Revolutions

Date	Innovation/Invention/Progress
1698: First Steam Engine	The first practical steam powered engine (a water pump) was developed by Thomas Savery was inefficient and prone to explosions.
1701: Seed Drill	Invented by agrarian Jethro Tull, the seed drill allowed more efficient seeding.
1712: Improved Steam Powered Engines	Thomas Newcomen developed a steam engine more robust and reliable operating at atmospheric pressure.
1730: The (Rotherham) Iron plough	The first successful commercial iron plough was the Rotherham plough introduced by Joseph Foljambe in 1730.
1732–86: The First Threshing Machines	Building on Michael Menzies unsuccessful machine of 1732 and Mr Stirling's machine of 1758 (which only threshed wheat) Andrew Meikle in 1786 devised the first successful mechanised threshing machine.
1775: James Watt Steam Engine	Matthew Boulton and James Watt partnership improved upon previous engines with a 75% reduction in fuel usage.
1794–98: Plough	Improvements in the mouldboard design of the plough by Thomas Jefferson allowed deeper and more efficient pulling of the plough.
1799: High-Pressure Steam Engine	Around 1800, the first engines using high-pressure steam were introduced by Richard Trevithick; these were more powerful and smaller in design than previous atmospheric engines.
1800–31: Mechanical Reapers	After unsuccessful attempts between 1800 and 1831, the first practical mechanical reapers were introduced in 1830–1834 by McCormack and Hussey.
1804–10: Sealed containers and canning	Hermetically sealed foods for preservation in 1804 by Francois Appert and canning by Peter Durand in 1810.
1837: Steel plough	Steel plough invented by John Deere in 1837.
1840s: Fertiliser manufacture	Introduction of manufactured chemical fertilisers by Baron Justis Von Liebig in the 1840s.
1841: First Portable steam Threshers	Ransomes introduced the first portable steam threshing machines.
1850s to 1878: Internal Combustion Engines	The first successful gas-fired internal combustion engines were developed by Etienne Lenoir 1859 and refined by Nikolaus Otto in 1878.
1871: Pasteurisation	Louis Pasteur invents pasteurisation.
1890s to 1910: Tractors	Engine technology moved forward with Benjamin Holts early steam traction engines of the 1900s, and the internal combustion engines of the 1850s, these paved the way for the first internal combustion tractor of 1910.
1888/95: Pneumatic Tyres	In 1888, John Dunlop invented the first air-filled or pneumatic tyres for his son's bicycle. In 1895, André Michelin was the first to use pneumatic tires on automobiles.
1895: Refrigeration	Although refrigeration had been around for 40 years or so Carl Von Linde developed the first safe domestic refrigerators in 1895.
1899: Artificial Insemination (AI)	Pioneering efforts building on previous work by Spallanzani (1784), Heape (1897) and others allowed E. I. Ivanow to establish AI as a practical procedure in Russia.

Sources: Tull, J., *Horse-Hoeing Husbandry or, an Essay on the Principles of Vegetation and Tillage*, printed for A. Millar London, 1762; de Graffigny, *Gas and Petroleum Engines*, Whittaker and Co., London, 1898; Fouts, L.X., *J. Patent Trademark Off. Soc.*, 4, 1921; Ogburn, W.F. and Thomas, D., *Polit. Sci. Q.*, 37, 83–98 1922; Morris, T.N., *Principles of Fruit Preservation Jam Making, Canning and Drying*, Chapman and Hall, London, 1933; Kuo-Chün, C., *Agric. Hist.*, 32, 25–31, 1958; Olmstead, A.L., *J. Econ. Hist.*, 35, 327–352, 1975; Rasmussen, W.D., *Agriculture in the United States: A Documentary History*, Greenwood Press, 1977; Powell, *Scottish Agricultural Implements*, C.I. Thomas and Sons, UK, 1988; Hills, R.L., *Power from Steam: A History of the Stationary Steam Engine*, Cambridge University Press, Cambridge, 1989; Martin, J.H., *Oceanography*, 4, 52–55, 1991; Fox, P.F., *Cheese: Chemistry, Physics and Microbiology*, Chapman and Hall, New York, 1993; McMichael, P., *Food and Agrarian Orders in the World Economy*, Greenwood Publishing Group, Oxford, 1995; Brunt, L., *Econ. Hist. Rev.*, 56, 444–477, 2003; *Heldman, Encyclopedia of Agricultural, Food, and Biological Engineering*, Marcel Dekker, New York, 2003; Kauffman, K.D., *Advances in Agricultural Economic History*, Elsevier Science, Oxford, 2003; Nuvolar, A., *Cambridge J. Econ.*, 28, 347–363, 2004; Elliott, A.G., *Gas and Petroleum Engines*, BiblioBazaar, Charleston, 2008; Britannica, Encyclopaedia Britannica Online, 2009.

sown, reaped and harvested in much less time. In essence, the Industrial Revolution was a period characterised by the use of fossil fuels and the increased use of synthetic fertilisers and pesticides; scientific
agricultural research; refrigeration; mechanisation; increased yields and labour productivity; extensive
irrigation; large-scale animal husbandry; the use of hormones and antibiotics; the growth of agribusiness (food-processing industries); the decline of family farming; new rotations with leguminous and root
crops; and creating jobs in fledgling industries of agriculture (agricultural chemistry and biotechnology,
etc.) among other things (UOR 2009). In fact, the Industrial Revolution had such a profound effect on
the whole agricultural industry that it facilitated the wholesale transport of food commodities for exports
effectively expanding markets and ultimately ushering in an era of the globalisation of agriculture. This
in turn saw the beginning of the re-alignment of world trade and by extension, it also had the effect of
lowering the cost of production and commodity prices in general.

5.2.3 The Productivity Curve

Previous to this period, greater agricultural productivity was often achieved through the reclamation or
acquisition of land, further increasing cultivatable acreage and thus aggregate yield. Moreover, up to this
point, another of the more serious barriers to increased productivity remained the labour-intensive harvesting of grain. By way of example, up to the beginning of the nineteenth century for instance, harvesting
was still largely achieved using the same methods as was used in the fourteenth century with little or no
change. However, by the 1900s, advances were such that in America and the United Kingdom at least,
growth through land expansion was on the decline, and subsequent yield increases were now achieved
through innovations in technology and improved farming practices (Johnson 2000; Gardner 2002). Also
as a result of intensification and the industrialisation of agriculture, the resultant productivity gains meant
that more farm workers were free to pursue other employment in industry and service sectors. This migration out of agriculture gave rise to more urbanised living and also further fuelled and reinforced the industrialisation process. Effectively, millions of workers were now released for urban employment, which
subsequently and collectively contributed to the growing wealth of nations (Johnson 1997, 2000; Weisdorf
2005). Not surprisingly, it was about this time that population levels began to increase significantly (Cohen
1995; Zhang 2008). This led to marked changes in societal dynamics evident in among other things, the
changing local economies as well as income growth and the distribution of the labour force. Such growth
also contributed to the subsequent decline of family farming, although discoveries and inventions at the
time allowed those farmers to become more productive (Watson 1974, 1983).

   Simultaneously as the population numbers increased so it piqued the minds of those concerned about
the balance of such numbers against the earth's ability to provide.

5.3 Population Pressures: Growth, Capacity and Sustainability

5.3.1 Growth

Concerns over the deleterious effects of unchecked population increases go back millennia (Freen 1996).
In his paper exploring growth of agriculture and population, Johnson quotes Quintus Septimus Florence
Tertillianus as suggesting that populations had, even back in Rome, AD 200, become 'burdensome to
the world' (Johnson 2000, p. 1). The insightful Tertillianus, according to Johnson, also touched on many
of the complaints of overpopulation relevant in today's collective conscience: that of deforestation, biodiversity, farming marginal or unsuitable lands and urbanisation (ibid). In the context of the day, this
was really quite a secular and prophetic insight at a time when most were pre-occupied at the insular
level. Indeed to place this in further context, global population in Tertillianus' time has been estimated
at around 200 million, which by today's standards is remarkable. With regards to population numbers,
Johnson's revised figures of 2000 stipulate that from that point until about AD 1000 the rate of growth
was about 0.04% per annum increasing to 0.12% annually up until about AD 1700 After this point and
for the next 100 years, there was a marked annual increase equalling 0.41% by which time population
levels had reached the first one billion (AD 1800) (Zhang 2008). From this point onwards, different
regions grew at differing rates, dependant on many factors and in the 130 years to 1930 world population

doubled adding another one billion to its numbers (ibid). After this, a further one billion inhabitants were added in ever shorter cycles, reaching 3 billion in 1960, 4 billion in 1975, 5 billion in 1987 and 6 billion in 2000. That is an increase of one billion inhabitants in periods of 130, 30, 15, 12 and 13 years, respectively (Zhang 2008). Zhang (2008) further extrapolates future population growth on past rises and offers a population of between 7.94 and 8.33 billion by 2030 and 9.5 billion by 2050. This is in line with the US Census Bureau's and the UN's estimates by 2050 (see Figure 5.1; UNDP 2006; USCB 2008b).

5.3.2 Trends

Pondering this initial slow growth before AD 1800, Johnson (2000) suggested it was due in part to restricted available calories per capita of between 1600 and 2000 kcal. He based these estimated figures on calorific availability in England and France before the rise of the Industrial Revolution (assumed to be a period of underdevelopment); and those of developing countries between 1934 and 1938 (the earliest available records) (Johnson 2000). Johnson also contends that life expectancy throughout history remained at a constant average 25–30 years until after about AD 1650 at which time longevity increased markedly, likely again in no small part to increased nutrition (Johnson 2000).

As population began to increase after about the seventeenth century (Figure 5.1), mounting concern grew over perceived threats from excessive population pressures to both the social fabric and the sustainable resources of the day (Malthus 1798; Chalmers 1852). Over time, two theories emerged, interlinked and running parallel to each other. The first was Thomas Malthus's theory that food resources would ultimately limit population growth and similarly the second concept was of the optimum population, or carrying capacity of the earth. Together these occupied the minds of many of the leading thinkers of the day and indeed became the dominant discourse for much of the population debate for centuries thereafter.

5.3.3 The Reverend Thomas Robert Malthus: The Malthusian Debate

Most notable and most often quoted on the issue of population explosion is the influential political economist and demographer Thomas Robert Malthus whose theory posited the fundamental inability of the earth to sustain unchecked population growth. His work in 1798 'Essay on the Principle of Population', which was further revised and elaborated in 1803, compellingly argued against the 'perfectibility of man' (Malthus 1798, 1803). Between them, the two editions combined to formally outline, for perhaps the first time, one of the lasting and well-debated theories on population pressures and the earth's ability to provide (Grigg 1982; Abramitzky and Braggion 2009).

Building on the central theme that population growth adversely affects society, Malthus's central tenet was that population growth has the capacity to grow geometrically or exponentially, which is to say growth at regular annual percentage increases while the food supply only has the capacity to grow arithmetically or by a fixed amount annually. He further offers that if left unchecked, the geometric growth of the population would eventually overtake food growth with predictable disastrous results (Malthus 1798, 1803). This fundamental concept is illustrated in Figure 5.2, which shows how the exponential curve (growth in population numbers) quickly overwhelms the ability of the food supply to cater for the masses (the arithmetic line). In Malthus's view, the food supply, unable to keep up with population growth, would depart at point A, leaving many hungry and at risk of famine.

While this formed the backbone of Malthus's central theme, he also recognised that since population seemed to be in equilibrium and was not running away vis-à-vis the food supply that several checks had to be in play regulating the situation. Malthus proposed two types of checks: preventative and positive. The preventative check in Malthus's opinion was simply moral restraint in things like deferment of marriage or putting off having kids until they could be properly fed. Second he considered positive checks, events that effectively shortened life and principally in Malthus's view, these took the form of war, famine and disease (Malthus 1798, 1803). Malthus's theory led to what subsequently became known as the Malthusian cycle or oscillating population numbers. Simply put, populations increased as incomes and increased this allowed for earlier marriages and more children but was then decreased as the food supply became restricted, leading once again to deferred marriages and lower fertility rates (Grigg 1982). Fundamentally, in Malthus's view, population could and would not increase without the proper means of nutritional subsistence, and the two forces, he contested, must at all times be kept in equilibrium.

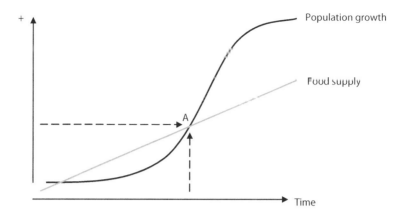

FIGURE 5.2   Malthus's population and food growth curves. (Based on Malthus, T., *An Essay on the Principle of Population,* printed for J. Johnson, St. Paul's Church Yard, London, 1798.)

Malthus also expanded on his ideas and offered certain corollaries to his argument, notable of which was that much of the fault of overpopulation lay at the feet of the 'reproductive and productive' nature of the poor. This, he suggested, was often in excess of their ability to provide for themselves (Vorzimmer 1969; Ross 2003). He also regarded famine, mortality and environmental degradation as a natural occurrence of overpopulation and in his view an almost necessary means of keeping the population numbers in check (Malthus 1798, 1803; Ross 2003; Ó Gráda 2009).

Malthus not only offered a theory of population pressure, he offered some solutions too. These were many and included moral restraint through deferred marriages and birth control; agriculture over manufacture by not allowing manufacturing industries to squeeze out agricultural industries; tillage over pasturage by growing grains and legumes to increase land efficiency; free labour market—the freedom to take work wherever it was available; the abolition of parish laws and the establishment of workhouses, thus negating the need for public assistance for the poor, which, in Malthus's view, only encouraged the poor to have children they could not feed. In much of this, excluding the workhouses, Malthus might have been ahead of his time; yet fundamentally he was vehemently against the idea of the redistribution of wealth in the belief that increasing incomes would only encourage the poor to have more children (Malthus 1798, 1803). However, while there are many proponents of Malthusian theory, there are equally as many critics. Early detractors included Karl Marx and Friedrich Engels, separately they were fierce critics of what they saw as Malthus's bourgeois attempts to cement or reify the social strata by ignoring the exploitation of the wage workers and the '... antagonistic relations between the landed and the industrial interests ...' (Gimenez 2008). Furthermore, Engels was particularly distasteful of the denigration of the poor and hapless as inevitable outcomes of universal natural laws. He also maintained in 1843/1844 that Malthus's notion that dictates the tendency of population to always '... multiply in excess of the available means of subsistence ... [as] the root of all misery and all vice.' as a '... vile, infamous theory, ... [a] hideous blasphemy against nature and mankind' (Engels 1844).

Yet for all his strong words, Engels did not completely rule out the notion that Malthus's ideas might have had at least some modicum of truth to them (Benton 1996). Others too were cynical about Malthus's ability to make such charges. Patten (1912) for instance observed Malthus's complete lack of awareness of economic influences on population's choices. Equally, David Grigg (1982) regarding Malthus's view that population increases are restrained by food supplies writes in his book *The Dynamics of Agricultural Change* that Malthus completely ignores technical advances in agricultural production and the idea of choices and voluntary birth control. By way of illustration, Grigg (1982) points to food increases throughout most of the nineteenth century, which continually grew faster than the population. Johnson too convincingly argues that major increases in population in the developed world in the mid-nineteenth century had come about more so through advances in knowledge and technologies that had successfully reduced mortality rather than increased fertility (Davis 1956; Johnson 2000).

Moreover, contrary to Malthus's view that food restricted population numbers, Boserup (1965), Johnson (2000) and even Woodruff (1909) among others suggested that it is in fact increased population that is pivotal in increasing or spurring growth within the economy and the agricultural industry in particular: a form of agricultural and population advancement through forced adaptation. Grigg (1982), on the other hand, while observing many flaws in Malthusian theory, fell short of agreeing with this notion, conceding instead that agricultural growth or change was more likely a result of multitude factors beyond just food and population. Nonetheless, the core precept still remains the subject of much conjecture, as Weisdorf offered in 2005, that this

> ... 'chicken-and-egg issue' remains unresolved; did human societies domesticate plants and animals as an adaptive response to population pressure or did domestication give rise to a larger population? (Weisdorf 2005, p. 566)

However, besides these vocal detectors perhaps Malthus's fiercest and most vociferous contemporary critic however, was writer and philosopher William Hazlit who in a series of open letters rebuffed not only much of Malthus's doctrine (especially in his treatment of the poor), but also his fundamental contribution to science in general. Hazlit, in describing Malthus's Essay on Population, offered that

> ... I have no hesitation in saying that his work is the most complete specimen of illogical, crude and contradictory reasoning ... Argument threatens argument, conclusion stands opposed to conclusion ... (Hazlit 1807, p. 15)

He further added that

> There is hardly a single statement in the whole work, in which he seems to have a distinct idea of his own meaning. The principle is neither new, nor does it prove anything new; least of all does it prove what he meant it to prove ... His whole theory is a contradiction; it is a nullity in the science of political philosophy. (Hazlit 1807, p. 16)

5.3.3.1 Lasting Legacy

Despite such continuous and fierce opposition, Malthus also enjoyed much praise and commendation. For all the criticisms of the Malthusian concept, it cannot be denied that simply by raising the profile of the debate, his work has undoubtedly contributed to the ongoing discourse of population growth and food supply. In the commemoration of the centenary of his death, Bonar and colleagues (1935) summed up the work of Malthus.

> Is he [Malthus] worthy of remembrance? ... Yes ... If only for the impulse he gave to the serious study of a branch of economics which before him, hardly a branch at all ... [Although] I do not say it was universally well received, but it excited universal attention. (Bonar et al. 1935, p. 222)

The economist John Maynard Keynes too, co-author of the commemoration and mindful of Malthus's indelible contribution to the perennial debate, summed up his ability for intellectualising the population problem:

> ... from being a caterpillar of a moral scientist to a chrysalis of an historian he could at last spread the wings of his thought and survey the world as an economist! (Bonar et al. 1935, p. 223)

At its core, the idea of natural resources versus population growth is a valid and pertinent point in the food security debate; yet the difficulties arise when Malthus attempts to further qualify his arguments with contradictory and unsupported opinion rather than empirical evidence. That said however, on balance, it cannot be denied that his work has undoubtedly positively contributed to the ongoing dialogue of population growth and food supply. Indeed, such provocative and lasting thoughts marshalled the beginning of a popular concept whose dominant discourse continues to preoccupy philosophers and policy makers alike (Prentice 2001; Sulistyowati 2002; Ross 2003). Not surprisingly too, just as often his theories perennially appear as an authoritative stamp used to promote, re-enforce or rebuke proffered hypotheses on population pressures past and present (Brentano 1910; Patten 1912; Flugel 1915; Wolfe 1928; Hiller 1930; Spengler 1949; Meade 1961; Ehrlich 1967; Galor and Weil 1999; Kögel and Prskawetz 2001; Trewavas 2002; Lagerlöf 2003).

5.3.4 Optimum Population and Sustainable Carrying Capacity

Running parallel to the Malthusian debate is the equally controversial theory of optimum population capacity. It is fundamentally similar to Malthus's population and natural resources theory, but it attempts to analyse the problem from the other side of the argument, that of a finite natural resource supply supporting a certain population size. This is not the same as a maximum number of people, instead it infers that the world's population is regulated by the sustainable 'carrying capacity' of the land on which it relies for sustenance (Roughgarden 1979; Daily et al. 1994). The idea proposes a fundamental limit to population numbers that the earth can support sustainably, and this is oftentimes but not always related to limiting factors of food and natural resources (Lidicker Jr 1962; Grigg 1982; Van Den Bergh and Rietveld 2004; Gilland 2006).

As with Malthus's theory, it is an age-old concept and one that too has had a long developmental history. It has often been quoted that Anton van Leeuwenhoek, an eminent microbiologist, in a letter to the Royal Society of London, made the first real calculation of optimum population capacity in 1679 (Leeuwenhoek 1686; Neill 1926; Egerton 1968; Van Den Bergh and Rietveld 2004; Gilland 2006). Leeuwenhoek's 'back of the envelope' calculation extrapolated a global number based on the density of the Netherlands at the time to arrive at a figure of 13,385 million (Leeuwenhoek 1686, pp. 14–16). However, in reality, this was an exercise more to do with its metaphoric inference on his ground-breaking spermatozoa studies than it was any real attempt at offering any viable or valuable contribution to the study of optimum capacity. However, despite sporadic interest throughout the seventeenth and eighteenth centuries, it was not until the late nineteenth and early twentieth century that the idea of optimum capacity resurfaced with any great fervour. Limiting factors were many and varied. Early thoughts from Ravenstein in 1891, for instance, concentrated on population growth and limiting factors of cultivatable land while Swedish economist Knut Wicksell pondered the limits of optimum population in respect of economic and technological resources (Felkin 1891; Gottlieb 1945; Uhr 1951; Spengler 1983; Ravenstein 1990). Yet others like Woodruff (1909) more simplistically proposed that

> The saturation point for population closely corresponds to the mean annual rainfall ... the more there is the more grass and grain ... (Woodruff 1909, p. 21)

These early salvoes were the opening rounds in a concept that paralleled the food population debate and one that was to further feature in the coming decades of the twentieth century.

References

Abramitzky, R. and F. Braggion (2009). Malthusian and Neo-Malthusian Theories: Malthus' Legacy. http://www.stanford.edu/~ranabr/Malthusian%20and%20Neo%20Malthusian1%20for%20webpage%20040731.pdf.

Agropolis Museum (2009). *History of Food & Agriculture: Beginning and Development of Agriculture.* Montpellier, France: Agropolis Museum.

Al-Hassani, S., E. Woodcock and R. Saoud (2007). *Muslim Heritage in Our World.* Manchester, UK: Foundation for Science Technology and Civilisation Publishing.

Bender, B. (1975). *Farming in Prehistory: From Hunter-Gatherer to Food Producer.* London: John Baker Ltd.

Benton, T., Ed. (1996). *The Greening of Marxism.* New York: The Guilford Press.

Binford, L. (1968). Post-Pleistocene Adaptations. In *New Perspectives in Archaeology.* R. Binford and L. Binford, eds. Chicago, IL: Aldine Publishing Company.

Biraben, J.-N. (1980). An Essay Concerning Mankind's Evolution. *Journal of Human Evolution* 9(8): 655–663.

Bonar, J., C. R. Fay and J. M. Keynes (1935). The Commemoration of Thomas Robert Malthus. *The Economic Journal* 45(178): 221–234.

Boserup, E. (1965). *The Conditions of Agricultural Growth: The Economics of Agrarian Change under Population Pressure*. London: G. Allen and Unwin.

Brentano, L. (1910). The Doctrine of Malthus and the Increase of Population during the Last Decades. *The Economic Journal* 20(79): 371–393.

Britannica (2009). Encyclopædia Britannica Online. 2009.

Brunt, L. (2003). Mechanical Innovation in the Industrial Revolution: The Case of Plough Design. *The Economic History Review* 56(3): 444–477.

Chalmers, C. (1852). *Notes, Thoughts, and Inquiries*. London: Princes Street, Soho, John Churchill.

Childe, G. (1936). *Man Makes Himself*. Oxford: Oxford University Press.

Cohen, J. E. (1995). *How Many People Can the Earth Support?* New York and London: W.W. Norton.

Crosby, A. W. (1972). *The Columbian Exchange: Biological and Cultural Consequences of 1492*. Westport, CT: Greenwood Publishing Co.

Daily, G. C., A. H. Ehrlich and P. R. Ehrlich (1994). Optimum Human Population Size. *Population and Environment* 15: 469–475.

Danhof, C. H. (1949). American Evaluations of European Agriculture. *The Journal of Economic History* 9 ((Supplement) The Tasks of Economic History): 61–71.

Davis, K. (1956). The Amazing Decline of Mortality in Underdeveloped Areas. *The American Economic Review* 46(2 Papers and Proceedings of the Sixty-eighth Annual Meeting of the American Economic Association): 305–318.

de Graffigny, H. (1898). *Gas and Petroleum Engines*. London: Whittaker and Co.

Durand, E. D. (1916). Some Problems of Population Growth. *Publications of the American Statistical Association* 15(114): 129–148.

Egerton, F. N. (1968). Leeuwenhoek as a Founder of Animal Demography. *Journal of the History of Biology* 1(1): 1–22.

Ehrlich, P. (1967). Paying the Piper: There Is No Longer Any Hope of Feeding the Population of the World. *New Scientist* 36: 652–655.

Elliott, A. G. (2008). *Gas and Petroleum Engines*. Charleston, SC: BiblioBazaar.

Engels, F. (1844). Umrisse Zu Einer Kritik Der Nationalokonomie, Trans. M. Milligan and D.J. Struik, Outlines of a Critique of Political Economy. In *Collected Works*. London: Lawrence and Wishart.

Farnworth, E. R. (2003). *Handbook of Fermented Functional Foods*. Boca Raton, FL: CRC Press.

Felkin, R. W. (1891). On Acclimatisation. *Scottish Geographical Journal* 7(12): 647–656.

Flugel, J. C. (1915). Ethics and the Struggle for Existence. *International Journal of Ethics* 25(4): 518–539.

Fouts, L. X. (1921). Jefferson the Inventor and His Relation to the Patent System. *Journal of the Patent and Trademark Office Society* 4(316).

Fox, P. F. (1993). *Cheese: Chemistry, Physics and Microbiology*. New York: Chapman and Hall.

Freen, R. H. (1996). Ancient Greek Philosophical Concerns with Population and Environment. *Population and Environment* 17(6): 447–458.

Frey, S. (1996). A Glossary of Agriculture, Environment, and Sustainable Development. Kansas State University. http://www.oznet.ksu.edu/library/MISC2/SB661.PDF (accessed 3 July 2009).

Furon, R. (1958). *Manuel De Prehistoire General, Geologie Prehistorique: Evolution De Phumanit, Areheologie, Prehistorique, Les Metaux Et La Protohistoire*. Paris: Payor.

Galor, O. and D. N. Weil (1999). From Malthusian Stagnation to Modern Growth. *The American Economic Review* 89(2 Papers and Proceedings of the One Hundred Eleventh Annual Meeting of the American Economic Association): 150–154.

Gardner, B. L. (2002). *American Agriculture in the Twentieth Century: How It Flourished and What It Cost*. Cambridge, MA: Harvard University Press.

Gilland, B. (2006). Population, Nutrition and Agriculture. *Population & Environment* 28(1): 1–16.

Gimenez, M. E. (2008). The Population Issue: Marx vs Malthus. http://www.colorado.edu/Sociology/gimenez/work/popissue.html.

Glick, T. (1977). Noria Pots in Spain. *Technology and Culture* 18(4): 644–650.

Gopher, A., S. Abbo and S. Lev-Yadun (2001). The 'When', the 'Where' and the 'Why' of the Neolithic Revolution in the Levant. *Documenta Praehistorica* 28: 49–61.

Gottlieb, M. (1945). The Theory of Optimum Population for a Closed Economy. *The Journal of Political Economy* 53(4): 289–316.

Greene, K. (1999). V. Gordon Childe and the Vocabulary of Revolutionary Change. (Archaeologist and Author). Kevin Greene. *Antiquity* 73(279): 97–109.

Grigg, D. (1982). *The Dynamics of Agricultural Change: The Historical Experience*. New York: St Martin's Press.

Guisepi, R. A. (2009). *World Civilizations: The Origins of Civilizations: The Agrarian Revolution and the Birth of Civilization*: Online http://history-world.org/neolithic1.htm

Haub, C. (2002). How Many People Have Ever Lived on Earth? *Population Reference Bureau* 30(8): 3–4.

Hayden, B. (1992). Models of Domestication: Transitions to Agriculture. In *Prehistory*. A. Gebauer and T. Price, eds. Madison, WI: Prehistory Press.

Hayden, B. (1995). A New Overview of Domestication. In *Last Hunters, First Farmers*. T. Gebauer and A. Gebauer, eds. Santa Fe, NM: School of American Research Press.

Hazlit, W. (1807). *A Reply to the Essay on Population, by the Rev. T. R. Malthus. In a Series of Letters to Which Are Added, Extracts from the Essay; with Notes*. Paternoster Row, London: Longman, Hurst, Rees and Orme.

Heldman, D. R. (2003). *Encyclopedia of Agricultural, Food, and Biological Engineering*. New York: Marcel Dekker.

Hiller, E. T. (1930). A Culture Theory of Population Trends. *The Journal of Political Economy* 38(5): 523–550.

Hills, R. L. (1989). *Power from Steam: A History of the Stationary Steam Engine*. Cambridge: Cambridge University Press.

Johnson, D. G. (1997). Agriculture and the Wealth of Nations. *American Economic Review* 87(2): 1–12.

Johnson, D. G. (2000). Population, Food, and Knowledge. *The American Economic Review* 90(1): 1–14.

Kauffman, K. D., Ed. (2003). *Advances in Agricultural Economic History*. Oxford: Elsevier Science.

Kögel, T. and A. Prskawetz (2001). Agricultural Productivity Growth and Escape from the Malthusian Trap. *Journal of Economic Growth* 6(4): 337–357.

Kuo-Chün, C. (1958). Organized Leadership and Agricultural Technology in Modern China. *Agricultural History* 32(1): 25–31.

Lagerlöf, N.-P. (2003). From Malthus to Modern Growth: Can Epidemics Explain the Three Regimes? *International Economic Review* 44(2): 755–777.

Leeuwenhoek, A. v. (1686). *Ontledingen En Ontdekkingen Van Levende Dierkens in De Teel-Deelen Van Verscheyde Dieren, Vogelen En Visschen; Van Het Hout Met Derselver Menigvuldige Vaaten; Van Hair, Vlees En Vis; Alsmede Van De Groote Menigte Der Dierkens in De Excrementen*. Leydon: Cornelis Boutesteyn.

Lidicker Jr, W. Z. (1962). Emigration as a Possible Mechanism Permitting the Regulation of Population Density Below Carrying Capacity. *The American Naturalist* 96(886): 29–33.

LL Cavalli-Sforza, P. Menozzi and A. Piazza (1993). Demic Expansions and Human Evolution. *Science* 259(5095): 639–646.

Malthus, T. R. (1798). *An Essay on the Principle of Population, as It Affects the Future Improvement of Society with Remarks on the Speculations of Mr. Godwin, M. Condorcet, and Other Writers*. London: Printed for J. Johnson, St. Paul's Church-Yard.

Malthus, T. R. (1803). *An Essay on the Principle of Population; or, a View of Its Past and Present Effects on Human Happiness; with an Enquiry into Our Prospects Respecting the Future Removal or Mitigation of the Evils Which It Occasions*. London: Printed for J. Johnson, St. Paul's Church-Yard.

Martin, J. H. (1991). Iron, Liebigs Law, and the Greenhouse. *Oceanography* 4(2): 52–55.

McEvedy, C. and R. Jones (1974). *Atlas of World Population History*. New York: Facts on File.

McMichael, P. (1995). *Food and Agrarian Orders in the World-Economy*. Oxford: Greenwood Publishing Group.

McNeill, J. R. (2003). Europe's Place in the Global History of Biological Exchange. *Landscape Research* 28(1): 33–39.

Meade, J. E. (1961). Mauritius: A Case Study in Malthusian Economics. *The Economic Journal* 71(283): 521–534.

Miller, R. (1980). Water Use in Syria and Palestine from the Neolithic to the Bronze Age. *World Archaeology* 11(3: Water Management): 331–341.

Morris, T. N. (1933). *Principles of Fruit Preservation Jam Making, Canning and Drying*. London: Chapman and Hall.

Neill, R. M. (1926). *Microscopy in the Service of Man*. London: Williams & Norgate.

Nuvolar, A. (2004). Collective Invention During the British Industrial Revolution: The Case of the Cornish Pumping Engine. *Cambridge Journal of Economics* 28(3): 347–363.

Ó Gráda, C. (2009). *Famine: A Short History*. Princeton: University Presses of California, Columbia and Princeton.

Ogburn, W. F. and D. Thomas (1922). Are Inventions Inevitable? A Note on Social Evolution. *Political Science Quarterly* 37(1): 83–98.

Olmstead, A. L. (1975). The Mechanization of Reaping and Mowing in American Agriculture, 1833–1870. *The Journal of Economic History* 35(2): 327–352.

Patten, S. N. (1912). Theories of Progress. *The American Economic Review* 2(3): 61–68.

Playfair, L., Ed. (1847). *Chemistry in Its Application to Agriculture and Physiology: By Justus Liebig*. Philadelphia: T.B. Peterson.

Powell, B. (1988). *Scottish Agricultural Implements*. UK: C.I. Thomas and Sons.

Prentice, A. M. (2001). Fires of Life: The Struggles of an Ancient Metabolism in a Modern World. *Nutrition Bulletin* 26(1): 13–27.

Price, T. and A. Gebauer (1995). Last Hunters-First Farmers. New Perspectives on the Prehistoric Transition to Agriculture. In *New Perspectives on the Transition to Agriculture*. T. Gebauer and A. Gebauer, eds. Santa Fe, NM: School of American Research Press.

Rasmussen, W. D. (1977). *Agriculture in the United States: A Documentary History*. Westport, Connecticut: Greenwood Press.

Ravenstein, E. G. (1990). Ravenstein on Global Carrying Capacity: Originally Published In: Proceedings of the Royal Geographical Society, New Monthly Series, Vol. 13, 1891. *Population and Development Review* 16(1): 153–162.

Rindos, D. (1987). *The Origins of Agriculture: An Evolutionary Perspective*. New York: Academic Press.

Rosen, A. M. (2007). *Civilizing Climate: Social Responses to Climate Change in the Ancient near East*. Plymouth, MA: Altamira Press.

Ross, E. B. (2003). Malthusianism, Capitalist Agriculture, and the Fate of Peasants in the Making of the Modern World Food System. *Review of Radical Political Economics 2003* 35: 437–461.

Roughgarden, J. (1979). *Theory of Population Genetics and Evolutionary Ecology: An Introduction*. New York: Macmillan.

Ruxin, J. N. (1996). *Hunger, Science, and Politics: Fao, Who, and Unicef Nutrition Policies, 1945–1978, Chapter II. the Backdrop of Un Nutrition Agencies, by Joshua Nalibow Ruxin*. London: University College London.

Salvaggio, J. E. (1992). Fauna, Flora, Fowl, and Fruit: Effects of the Columbian Exchange on the Allergic Response of New and Old World Inhabitants. 13(6): 335–344.

Sauer, C. (1952). *Agricultural Origins and Dispersals*. New York: American Geographical Society.

Spengler, J. (1983). Knut Wicksell, Father of the Optimum. *Atlantic Economic Journal* 11(4): 1–5.

Spengler, J. J. (1949). The World's Hunger: Malthus, 1948. *Proceedings of the Academy of Political Science* 23(2): 53–72.

Sulistyowati, C. (2002). Can Gmo Ensure Food Security Amid the Population Growth. Institute of Development Studies and Technological Assistance. *BIC News*. http://www.bic.searca.org/news/2002/oct/indo/28.html

Time (1936) After Breasted. *Time Magazine Online*. http://www.time.com/time/magazine/article/0,9171,755744,00.html.

Tomlinson, H. (1978). 'Not an Instrument of Punishment': Prison Diet in the Mid Nineteenth Century. *International Journal of Consumer Studies* 2(1): 15–26.

Trewavas, A. (2002). Malthus Foiled Again and Again. *Nature* 418: 668–670.

Tull, J. (1762). *Horse-Hoeing Husbandry or, an Essay on the Principles of Vegetation and Tillage* London: Printed for A. Millar.

Uhr, C. G. (1951). Knut Wicksell: A Centennial Evaluation. *The American Economic Review* 41(5): 829–860.

UN (1973). The Determinants and Consequences of Population Trends: New Summary of Findings on Interaction of Demographic, Economic and Social Factors, Vol. I, Population Studies, No. 50. Sales No. E.71.XIII.5. New York: United Nations.

UNDP (2006). Human Development Report 2006—Beyond Scarcity: Power, Poverty and the Global Water Crisis. New York: United Nations Development Programme.

UNEP (1996). Glossary of Environmental Terms. United Nations Environment Programme. http://www.nyo. unep.org/action/ap1.htm (accessed 10 August 2009).

UNPP (2009). *World Population Prospects: The 2008 Revision Population Database*: United Nations Population Fund.

UOR (2009). University of Reading. Agriculture, Policy and Development, History of Agriculture. University of Reading. http://www.ecifm rdg.ac.uk/history.htm (accessed 4 June 2008).

USCB (2008a). *Historical Estimates of World Population*. Washington, DC: US Census Bureau.

USCB (2008b). *Total Midyear Population for the World: 1950–2050*: US Census Bureau.

Van Den Bergh, J. C. J. M. and P. Rietveld (2004). Reconsidering the Limits to World Population: Meta-Analysis and Meta-Prediction. *Bioscience* 54(3): 195–204.

Vorzimmer, P. (1969). Darwin, Malthus, and the Theory of Natural Selection. *Journal of the History of Ideas* 30(4): 527–542.

Watson, A. (1983). *Agricultural Innovation in the Early Islamic World*. Cambridge: Cambridge University Press.

Watson, A. M. (1974). The Arab Agricultural Revolution and Its Diffusion, 700–1100. *The Journal of Economic History* 34(1 The Tasks of Economic History): 8–35.

Weisdorf, J. L. (2005). From Foraging to Farming: Explaining the Neolithic Revolution. *Journal of Economic Surveys* 19(4): 561–586.

Wolfe, A. B. (1928). The Population Problem since the World War: A Survey of Literature and Research. *The Journal of Political Economy* 36(5): 1928.

Woodruff, C. E. (1909). *The Expansion of the Races*. New York: Rebman Company.

Zhang, W. (2008). A Forecast Analysis on World Population and Urbanization Process. *Environment Development and Sustainability* 10(6): 717–730.

6

Governance, Philosophy, Politics and Economics

Strictly speaking this next section has more to do with philosophical enlightenment than any underlying political theory; however, it is included in this book to illustrate the progression of ideas that has ultimately culminated in current thoughts on global governance, specifically in the areas of peace and food security.

6.1 Intellectual Origins of Political Internationalist Discourse: Governance and Responsibility

While the first real organisation of global reach was the League of Nations in 1919 (Chapter 9), the emergence of supra-nationalism goes way back in history, yet despite its early beginnings global governance has been slow and beset with conflict and suffering. Some like Aksu (2008a) and Tittler and Jones (2004) feel it began with notions of a European 'perpetual peace', which developed out of the fourteenth century with the likes of French lawyer and political pamphleteer Pierre du Bois (Tittler and Jones 2004; Aksu 2008b). Others however suggest it began with Desiderius Erasmus, the sixteenth century humanist writer of the Renaissance whose castigation of noblemen for their 'unjust' and 'unnecessary' wars permeated much of his work; in particular the Querela Pacis (Complaint of Peace) and which led to calls of arbitration supervised by the Church (Erasmus 1917; Roosevelt 1999). Others still cite the treaties of the fifteenth century as the emerging impetus of supranational collective governance (Lesaffer 2004).

6.1.1 Perpetual Peace: Utilitarianism

Whatever date one chooses to ascribe to the beginnings of internationalism it was not really until the seventeenth/eighteenth century that the idea gained popular momentum, a period referred to as the 'Age of Enlightenment' (Louden 2007; Aksu 2008b). From among the many proponents of this cooperative social responsibility, several key figures stand apart from the rest and can collectively be credited with influencing future political ideology. Perhaps one of the first people to put forward a practical proposal for a universal and perpetual world peace was Emeric Crucé, a French monk whose insightful work proposed a federation of states governed by a permanent 'Council of Ambassadors' (Crucé 1909; Roosevelt 1999; Louden 2007). Shortly after this another man, described as one of first true altruistic humanitarians, was William Penn, prominent Quaker and founder of the American colony Pennsylvania (Mead 1912). Penn was horrified by the constant ravages of war throughout seventeenth century in continental Europe and wrote an essay in 1693 proposing the establishment of a European parliament, state or federal assembly whereby differences could be settled not through conflict but equitably, collectively and peacefully through legal, social and moral grounds (Penn 1693; Griffis 1901; White 1919; Salmon and Nicoll 1997; Aksu 2008b). This effort in marshalling a moral force to eliminate war and providing a collective and democratic means of establishing peace is not confined to these examples alone. Other eminent men worthy of note include renowned thinkers and writers of the eighteenth century such as Bellers, Saint-Pierre, Alberoni, Rousseau and others (see Table 6.1).

While several authors, notably Penn and Kant, casually refer to supra-nationalism in a European context, many consider the wider issues of governance as a whole without geographical or political boundaries (Mead 1912; Aksu 2008b). In doing so, according to Aksu (2008b), their collective thoughts and ideals transcended specificity and reach out to the global ethic of mankind, further elevating the

TABLE 6.1

Notable Pioneers, Edicts and Treaties Foreshadowing European Perpetual Peace and Global Governance

Author	Date	Writings, Edicts and Treaties
Pierre du Bois	1306	'De Recuperatione Terrae Sanctae'
Edward III of England and John II of France	1360	Treaty of Bretigny
Henry VII and Charles VIII of France	1492	Treaty of Boulogne
James IV of Scotland and Henry VII of England	1502	'The Treaty of Perpetual Peace'
Desiderius Erasmus	1521	'Querela Pacis' [Complaint of Peace]
Attributed to Henry IV (but actually, Maximilien de Béthune, Duc de Sully)	1596	'Le Grand Dessein' (The Grand Design) beginning with the Edict of Nantes
Emeric Crucé	1623	'Le Nouveau Cynée ou Discours d'Estat' (The New Cyneas, or a Discourse on the State)
Hugo Grotius	1625	'De Jure Belli Ac Pacis Libri Tres' (On the law of war and peace)
Tommaso Campanella	1623	'Civitas solis' (The City of the Sun)
William Penn	1693	'An Essay towards the Present and Future Peace of Europe'
John Bellers	1710	'Some Reasons for an European State'
Charles Irenee Castel, abbe de Saint-Pierre	1712 1713	'Memoire pour rendre le paix perpetuelle en Europe' 1712 and 'A Project for Settling an Everlasting Peace in Europe' 1713
James Francis Edward Stuart ('King James III')	1722	'Declaration for a Lasting Peace in Europe'
Cardinal Alberoni	1736	'Scheme of a "Perpetual Diet" for Establishing the Public Tranquillity'
Gottfried Wilhelm Leibniz	1715	Letter from Leibniz to Abbe de Saint-Pierre
Emeric Vattel	1741 1758	'Defense of the Leibnizean System' 1741 and 'The Law of Nations' 1758
Jean-Jacques Rousseau	1756	'Abstract' and 'Judgement' of the Abbe de Saint-Pierre's 'Project for Perpetual Peace'
Pierre Andre Gargaz	1779	'A Project of Universal and Perpetual Peace'
Charles Alexandre de Calonne	1796	'Considerations on the Most Effectual Means of Procuring A Solid and Permanent Peace'
Jeremy Bentham	1789	'A Plan for an Universal and Perpetual Peace'
Immanuel Kant	1795	'Perpetual Peace: a Philosophical Essay'

Sources: Crucé, *Le Nouveau Cynee Ou Discours D'estat Representant Les Occasions Et Moyens D'establir Une Paix General, Et La Liberte Du Commerce Par Tout Le Monde*, Allen, Lane and Scott, Philadelphia, PA, 1909; Erasmus, D., *Desiderius Erasmus, the Complaint of Peace [1521]*, The Open Court Publishing Company, London, Chicago, 1917; de Béthune, M. and Ogg, D., eds., *Grand Design of Henry IV. From the Memoirs of Maximilian De Bethune Duc De Sully (1559–1641)*, Sweet and Maxwell Ltd., London, 1921; Grotius, H., *On the Law of War and Peace*, The Clarenden Press, Oxford, 1925; Lee, S.J., *Aspects of European History, 1494–1789*, Methuen and Company Ltd., London, 1984; Hoffmann, S. and Fidler, D.P., *Rousseau on International Relations*, Clarendon Press, Oxford, 1991; Roosevelt, G., *Pacifism and Just War Theory in Europe from the 16th to the 20th Centuries*, Global Policy Forum, New York, 1999; Williams, D., *The Enlightenment*, Cambridge University Press, Cambridge, 1999; Riley, P., *Leibniz: Political Writings*, Cambridge University Press, Cambridge, 2001; Gale, T., *Tommaso Campanella*, Online: Encyclopedia of World Biography, 2004; Tittler, R. and Jones, N.L., eds., *A Companion to Tudor Britain*, Blackwell Publishing, Oxford, 2004; Beck, S., *History of Peace Volume 1: Guides to Peace and Justice from Ancient Sages to the Suffragettes*, World Peace Communications, Santa Barbara, CA, 2005; Louden, R.B., *The World We Want: How and Why the Ideals of the Enlightenment Still Elude Us*, Oxford University Press, New York 2007; Aksu, *Early Notions of Global Governance Selected Eighteenth-Century Proposals for 'Perpetual Peace' with Rousseau, Bentham, and Kant—Unabridged*, University of Wales Press, Wales 2008a, *Peace & Change: A Journal of Peace Research*, 33, 368–387, 2008b.

notion of 'perpetual peace' (Louden 2007; Aksu 2008a, 2008b). Like Penn and others before him, Kant espoused the universal applicability of utilitarianism and in doing so laid out the fundamental preconditions or key universal principles to a workable and agreeable global institution (Louden 2007; Aksu 2008b). Collectively such forward thinking set the tone of both philosophical and political ideology and essentially foreshadowed the notion of modern global governance and international relations. Table 6.1 illustrates some of the highlights of this movement.

6.1.2 Social and Political Interdependence

Thus, the eighteenth century's 'Age of Enlightenment' together with residual European tensions arising out of the French Revolution as well as the Napoleonic Wars of the late eighteenth and early nineteenth centuries were coupled with the dissolution of the Holy Roman Empire. Together all this combined to provide the impetus for a growing trend of multilateralism throughout the nineteenth century (Denemark et al. 2007). For the first time the realisation that nations could act cooperatively for the benefit of both national and international objectives encouraged international cooperation and interdependence (Hirsch 1995). This was achieved initially via diplomatic channels and treaties then by the turn of the nineteenth century via the growth of international conferences and river commissions. This was a prolific period with the bulk of the 7000 treaties of the last 400 years being signed in the 100 years or so between 1850 and the mid-1960s (Denemark et al. 2007). Such treaties, according to Denemark et al. (2007), have been responsible for the birth of some of the great international organisations of the nineteenth and twentieth centuries.

6.1.3 The Concert of Europe

This cooperation continued through into the latter part of the nineteenth century with public international bodies such as the telegraph, postal and railway unions. However, although initially narrow in their mandates, they did in fact pave the way for later successful international cooperation at both the economic and political levels (Hirsch 1995; Denemark et al. 2007). In the aftermath of the Napoleonic Wars, a secure and stable peace was sought. Out of the subsequent Congress of Vienna 1814–1815, the victors strove to re-draw boundaries to maintain order and re-establish a balance of power through the creation of the Concert of Europe; the first truly multilateral forum designed to enforce the decisions of the Congress (Moran 1995; Haftendorn et al. 2002; Denemark et al. 2007). After many successes and much criticism, conclude Haftendorn et al. (2002), the coalition was badly weakened by a mixture of renewed European revolutionary upheavals in 1848, industrial and colonial competition as well as shifting balances of power (Henig 2002). As a multilateral organisation, the Concert of Europe was effectively over by the time of the Crimean War (1854). However, down but not out it still continued to be a forum of international collaboration, albeit of lesser importance. Although the concert eventually ceased to be a potent organisation with the advent of World War I, it did serve as a blueprint for both the League of Nations and the United Nations (Moran 1995; Haftendorn et al. 2002; Henig 2002).

6.2 Western Economic Development

Not surprisingly, within the arena of food and agriculture, research shows that the rise of national economic wealth shaped by political ideology influenced not only the Western national diet but the international policy agenda as well. This rise of Western or industrialised economies was also an important factor in the economic development of developing countries in the twentieth century too. As such, the following predominantly explores this growth from the Western perspective. Without drawing too deeply on a full history of political and economic philosophy, this research benefits from a fundamental understanding of both: the rise of internationalism, described as the increase of political integration and cooperation among nations for the benefit of all; and development economics, a branch of economics which specifically deals with the development process of low-income countries. An overview of historic economic thought is set out.

TABLE 6.2

History of Economic Thought

Period	Notes
Classical Greek	Practitioners included Kung Fu Tsu (Confucius), Meng Tzu, Plato, Aristotle.
Medieval Scholastic	A period characterised by moral ideology rather than political interests.
Mercantilism	About the seventeenth century monarchical or nationalist warring nations focused on increasing wealth; it is a period that has been described as the birth of capitalism.
Modern	The modern period of economic thought can be subdivided into several stages or schools of thought:
	Classic political economics of the Adam Smith era whose book *The Wealth of Nations* in 1776 suggests that the ideal economy is a self-regulating market system;
	Marxism: Marxian economics infers the exploitation of labour by capital and holds that the true value of things is defined by the labour used in its production;
	Neo-classical economics formed from about 1870 espouses the marginal utility theory that both supply and demand are joint determinants of price affecting both the output and the distribution of income;
	Keynesian economics: Keynes' book *The General Theory of Employment, Interest and Money* in 1936 forwards the idea of a mixed economy where both the state and the private sector play an important role in the economy;
	Chicago School of Economics, known for its advocacy of free market and monetarist ideals. Milton Friedman offers that, if left to themselves, market economies are inherently stable and that depressions or downturns result sololey from government intervention.

Sources: Smith, A., *An Inquiry into the Nature and Causes of the Wealth of Nations*, Printed for Messrs. Whitestone, Chamberlaine, W. Watson, Potts, etc., Dublin, 1776; Deane, P., *The Evolution of Economic Ideas*, Cambridge University Press, Cambridge, 1978; Birchall, J., *Evolutionary Economics*. John Birchall, 2002; Gimenez, M. E., The Population Issue: Marx vs. Malthus, 2008.

Briefly, economic historians talk of four progressive epochs of economic thought (Table 6.2) (Deane 1978). Following the classical Greek period, the breakdown of governance and law and the waning of the Carolingian Empire around the ninth and tenth centuries, the medieval scholastic economic era emerged. This was characterised by feudalism whereby the rule of law, taxes and economic activity were governed by the feudal or manorial system, often with the aid of a parliament chaired by an authoritative assembly of ecclesiastics and tenants-in-chief (North and Thomas 1999; Bloch 2001). This began to be challenged around the thirteenth and fourteenth centuries but ultimately lasted until about the seventeenth century when the next stage in the economic progression when mercantilism witnessed the revolutionary change of the feudal or agrarian society in favour of modern capitalism (Hobsbawm 1954).

6.2.1 Feudalism and Mercantilism

In economics, this book's interest begins with the challenge to feudalism by mercantilist ideology. Following the beginning of the demise of feudalism in Europe around the thirteenth to fifteenth centuries, newly formed parliaments or governments began to trade a decentralised land-based economy for a centralised money-based economy. Up till this point development policy designed to improve the welfare of society as we think of it today simply did not exist. During this time, mercantilism took a foothold in a period of economic and social malaise. Trading increased and the privatisation of assets and wealth followed while the widespread pursuit of capital for capital's sake trickled down to the commoners. Later, political instability of the seventeenth century born out of peasant revolts, wars, rebellions and religious crusades led to what has been described as a century of war (Hobsbawm 1954; Deane 1978; Bloch 2001). Although largely European, this 'crisis' was not confined to the continent; indeed, it involved much of the world including China, India and the Americas (Parker and Smith 1997). As the cost of war increased so governments needed money to finance armies and standing navies (Deane 1978). This funding relied on taxes imposed by the central government, and there were only two ways in which this could be increased: by increasing the size of the tax base or through increases in production. As it turned out, a combination of the two was often the most fruitful (Carrasco and Berg 1999). Thus, as necessity provided the impetus for the promotion of economic development as a legitimate objective of state policy so the Westphalian Treaties of 1648 concentrated power in the hands of sovereign states. At this time, the transition from feudalism to capitalism was largely complete.

Building on these developments and partly driven by ambitious men in search of personal financial gain, the governments of the day further promoted this new economic paradigm of growth. In this model, the market and the entrepreneurial private citizenry was seen as the driving force behind increasing economic success while the government sought to facilitate much of this through legislation and intervention. This became known as mercantilism and as it progressed and developed as an idea, trade objectives sought to maintain a strong and self-sufficient economy. Furthermore, much of this was facilitated by government intervention policies that favoured the host countries. Simultaneously, central ideas of growing domestic involvement aimed to create jobs, promote industries and increase trade. Interestingly at this time, trade involved maximising exports whilst minimising imports; thus, a favourable trade balance was seen as central to wealth creation at the time. In the words of Thomas Mun,

> ... wee must ever observe this rule; to sell more to strangers yearly than wee consume of theirs in value. (Mun 1664)

Simultaneously, operating on an ever-widening trade base, merchants and bankers realised their exploits were becoming more and more reliant on national governments' economic policies. This interdependence of governments, producers and consumers marked the final split away from feudalism towards the beginning of modern economic philosophy entrenched in capitalism (Deane 1978). It was during this time that developmental economic philosophy really began.

6.2.2 Colonialism: Western Wealth Creation

Fuelling this capitalist growth further was the Colonialist period from the fifteenth century onwards which saw countries from Europe building colonies and empires on other continents. While this practice grew so, capitalism and the desire to create new markets for European goods increased the 'legitimate' exploitation of these colonies. As Western countries divided up much of the undeveloped world among themselves, domestic industries were supplanted by new capitalist modes of production. It was initially led by the Spanish and Portuguese exploration of the Americas, Africa, the Middle East, India and East Asia in the fifteenth century. This was quickly followed in the sixteenth and seventeenth centuries by France, England and Holland who fought and established their own empires in competition with each other (Birchall 2002). Ever-growing demand for raw materials led to the colonial realignment of land use where all too often indigenous populations were forcibly ejected from their lands and new industries and plantations replaced traditional activities. Accordingly, these colonies became suppliers of raw materials to the West and consumers of manufactured or finished goods with the net result of transferring much wealth out of these colonies into their host occupying countries.

6.2.3 Free Trade versus Protectionism

From the late eighteenth century, mercantilism was being challenged by the ideas of free trade. In Britain in particular there had been a long history of legislation familiarly coined the 'Corn Laws', which up to about the 1660s were too numerous and contradictory to be of any real policy value (Barnes 2006). However, after this period, several successive legislative acts (also coined the Corn Laws) were implemented and designed to protect Britain from competition against less expensive foreign imports. These Corn Laws exercised political and economic strategies through the liberal use of tariffs and protectionism (Barnes 2006; Vaidya 2006). Britain was not alone in the field of intervention either (Vaidya 2006). Previous to the American Revolution, individual colonies raised 'bounties' or taxes on the back of imports whether colonial or international (Hill 1893). This did not change under the new Union of 1776, instead the new federal government introduced one of the first substantive legislations which operated to protect trade and to raise revenues: the Tariff Act of 1789 (Hill 1893).

However, all this was now about to be challenged. Without doubt, two of the staunchest proponents of free trade were David Ricardo and Adam Smith. By 1776 Smith was openly criticising the concept of mercantilism and prevailing government interventionism promoting instead a different model: that of free trade. Smith was not against total abstinence of government intervention but rather a reduction to the bare minimum necessary for aspects like national security. In Smith's view, voluntary exchange was mutually beneficial, which must ultimately result in market equilibrium where the flow of goods one way is commensurate with an equal flow of payments or goods in kind the other way (Smith 1776). This view that trade was best achieved with the minimum interference from the government was also championed by Ricardo. Free trade, it was said, was underpinned by the laws of comparative advantage which also stated that the trade of goods and services was of mutual benefit to all parties. This, in their collective view, was the fundamental reason why some nations prospered economically over others.

About this time, some decolonisation saw America gain their independence from their European masters. However, at about this time industrialisation brought with it a new wave of colonisation with Great Britain, France, Holland and Belgium turning their attention to Africa, India and South East Asia. Furthermore, as time went by so new alternative theories of economic development emerged. Competing specifically with capitalism and the theories of Smith and Ricardo, Karl Marx in his book *Das Kapital* (1867) studied the capitalist system (Marx 1887). Fundamentally while he agreed with the notion that capitalism was a necessary evil to promote the free movement of goods and services, he did not see this as the ultimate end game. For him it was a stepping stone and one that he felt was plagued by its own inequities. Principally he felt the capitalist system relied on the exploitation of the working people and in its complicity the government was subservient to capitalist ideals. This, Marx felt created a class system consisting of the bourgeoisie (the capitalist middle class) and the proletariat (the working class). It was also a system that Marx believed would eventually be the architect of its own downfall. In this scenario, Marx postulates that as the capitalist system concentrates power and money in fewer hands so the bourgeoisie who fail to compete successfully are propelled into servitude of the economy and ultimately becoming proletariat like the rest. While such theories continue to provide much of the opposition to capitalism today contemporary economists like Alfred Marshall disagreed and saw capitalism as the benefactor of people destroying, not creating, a class-based society. While Marshall, like Smith before him, advocated limited state intervention so John Maynard Keynes on the other hand preferred more direct input. In attempting to find answers for the near collapse of the capitalist system in the 1929 depression era, Keynes proposed his circular theory of money. In this theory one person's spending goes towards another's earnings and in turn that person's spending goes towards another's earnings, *ad infinitum*. Keynes's solution then to the periodic market depressions was for governments to intervene by priming the markets to encourage spending and increase the flow of money. This in fact was the ideology held by many of the leaders who introduced the Bretton Woods institutions.

6.3 Social Awakenings: Philanthropy and Humanitarianism

Running parallel to these developments and building on the political and philosophical paradigms of the time, an emerging collective internationalism, both politically and institutionally, was beginning to take foothold in the collective conscience of societies. In this environment, international and political humanitarianism as well as philanthropy flourished and the first real tendrils of international peace took seed. These ideas were developed over time and were eventually brought to the public attention with repeated incidents of war and famine.

6.3.1 Famine: The Black Horseman of the Apocalypse

As with the term 'food security', 'famine' too shares similar problems of definition and conceptualisation (Watkins and Menken 1985). Loosely defined by Ó Gráda for example, famine 'represents the upper end of the continuum whose average is hunger' (Ó Gráda 2009). The term 'famine' can also be highly emotive and is one that Ó Gráda (2009) insists needs to be used with caution. By way of example,

Malthus's simple and stark view was for many years highly influential and set the tone for much debate that followed (Watkins and Menken 1985; Ó Gráda 2009). In Malthus's view:

> Famine seems to be the last, the most dreadful resource of nature. The power of population is so superior to the power in the Earth to produce subsistence … that premature death must in some shape or other visit the human race … But should [that] fail … gigantic inevitable famine stalks in the rear, and with one mighty blow levels the population with the food of the world. (Malthus 1798, pp. 139–140)

Bearing this in mind there are major difficulties with any discussion on famine; first, as suggested, a lack of universal agreement makes it difficult to establish what constitutes a famine; and second as with much of pre-history, famine records rely heavily on anecdotal evidence often tainted with hidden biases, evasions and omissions. Such oral evidence can as well suffer chronological confusion and complete omissions of the more distant past (Ó Gráda 2009). All in all the scale and dimension of the phenomenon and any credible mortality numbers from such famines are notoriously difficult to estimate with any degree of accuracy (Devereux 2000). Acute shortages of foodstuffs however have existed since ancient times, and while historical records are fragmented its devastating potential is without question.

The historical frequency of famines too for the above reasons is difficult to determine and once again depends to a large extent on the definition and time frame used (Watkins and Menken 1985; Speakman 2006; Ó Gráda 2009). Bearing these and other factors in mind, various estimates place the number of famines throughout history at anywhere between 400 to as many as 5000 (Walford 1879; Keys et al. 1950; Prentice 2005; Speakman 2006). Nonetheless, Ó Gráda (2009) stipulates that while the total number is almost impossible to know, 400 major famines based largely on the first real assimilation of historical figures by Cornelius Walford in 1879 is a number that is quite often quoted (Walford 1879). That said, the purpose of this book is not to put a definitive figure on the number of famines throughout history, instead the discussion solely aims to highlight the frequency, severity and contentious nature of the famine debate and to create context of the phenomenon within the food security arena.

So what causes famine? Besides Malthus's harsh analysis, famine over the years has variously been ascribed to the whims of Gods, the punishment of kings and acts of nature (Mehta 1929; Hutchinson 1998). In reality however, many of the most devastating famines on record have been linked to: drought; flood; harvest failure; infestation of pests; 'excessive milking of the agricultural cow' (Mehta 1929); from natural ecological events such as volcanic eruptions (Ó Gráda 2009); and more recently as a result of political, economic and social consequences (Hutchinson 1998). Indeed, the earliest recorded famines (resulting from prolonged droughts) are inscribed on Egyptian stone pillars (Egyptian Stelae) back in the third millennium BC (Prentice 2001; Ó Gráda 2009). Common characteristics of famines too were set down in the early Mesopotamia period, about AD 500, (the famines of Edessa) as being characterised by high food prices, spousal or child desertion, social or public intervention, substitute foods, migration and infectious disease. Many of these symptoms too, according to Ó Gráda and others, have changed little over the years to the present day (Devereux 2000; Ó Gráda 2009).

With regard to famines in particular and humanitarianism in general, before any real coordinated international effort was introduced, intervention was mainly operated at the local, national and regional levels. In British-administered India, for instance, the famines of 1837–1838 and 1866 (Orissa) formed the backdrop for the first thorough studies into the cause and possible relief of famine (Walker 2008). As a result, the Indian Famine Codes were introduced in the 1880s. These were established at the provincial level with the aim of creating linkages and channels of information that aided in the detection and preparedness of such emergencies (Mehta 1929; Watkins and Menken 1985; Maxwell and Frankenberger 1992; Walker 2008; Ó Gráda 2009). The charity minded would then open *sadavrats* or *annakshetras* (centres where free food was given out) and transport surpluses from one region to make up shortfalls in others (Mehta 1929). These codes also introduced a sophisticated monitoring system. Among indicators used to determine the likelihood of an emergency were stress detectors of market price, rainfall and agricultural production while at the social level stressors included credit availability, number of beggars and migration statistics, etc. The establishment of the code in fact was very successful and was used as a blueprint for British-administered Sudan in the 1920s as well as other food-stressed areas. Importantly, the Indian Famine Codes in particular also introduced the idea of culpability and accountability in the event of non-action as the following quote illustrates:

> The famine code, like the code of Justinian, will ever remain a monument signalising the establishment of the 'reign of law' in the midst of warring policies, and what was once but a pious wish has been made a detail of famine administrative routine: Every District Officer would be held personally responsible that no death occurred from starvation which could have been avoided by any exertion or arrangement on his part. (Mehta 1929)

In this way, hunger was seen to be avoidable and a collective social responsibility, a notion that was perhaps, ahead of its time.

6.3.2 Humanitarianism: Local and International Intervention

Continuing concern over such horrors prompted many individuals and groups to take things into their own hands. Nowadays though, humanitarianism is frequently seen or discussed in the modern perspective, and it is often viewed as contingent on nation-state interests. Its divorce from its historical context has resulted in the notion of humanitarianism being redefined in today's terms as altruistic and unproblematic (Paras 2007). In reality, argues Para (2007), humanitarian ideals and nation or state interests are and have always in fact been mutually intertwined. Within the context of this book and avoiding a larger philosophical debate, such political or ideological distinctions serve simply to illustrate that humanitarianism as will be seen later in this study is perhaps not always altruistic and free from political influence.

Apart from examples of humanitarianism from as far back as Greco-Roman times, Walker (2008) cites the middle of the nineteenth century, coinciding with the advent of globalisation, as a time that allowed those with philanthropic ideals to make a difference on a world stage (Walker 2008). However, it is worth remembering too that this period coincided with the full Age of Enlightenment, which no doubt served to prod people's moral conscience. Walker talks of four stages or era's of humanitarianism: containment (famine relief); compassion (moral outrage and war relief); change (the neutrality of relief and human right); and welfare (Walker 2008). He also credits Henry Dunant, founder of the Red Cross in 1863, as the founder of the modern humanitarian movement. However, to leave it there would be to miss out on much relevant work leading up to this point. In Germany in 1790 for instance, American born Benjamin Thompson, also known as Count Rumford, delivered a combined programme of teaching and feeding to the hungry vagrant children and poor adults of Munich (Rumford 1796). By establishing the Poor People's Institute, adults and to some extent children were required to work for their food and clothing. During non-working times, however the children were educated in reading, writing and arithmetic. Because of cost constraints, Count Rumford was also constantly looking to develop meals which would provide good nutrition at the lowest possible price. Rumford went on to replicate this success in England, Germany, Scotland, France and Switzerland and in his heyday he was feeding as many as 60,000 people daily from his soup kitchens in London (Gunderson 2003). On the back of Rumford's successes, other privately funded societies sprung up. One of the first organisations with the express purpose of school feeding was perhaps the Society for Feeding Needy School Children, which originated in Dresden in 1880. By this time, others too were being regularly organised, and it was not just about food either; the new humanitarian outlook sponsored by teachers and doctors included providing Vacation Colonies where the sick and infirm children from cities were given holidays in the country. From these humble beginnings, it was not long before the welfare of children in particular took on political dimensions with the proposed Leipzig government-backed legislation aiming to provide school meals in all cities of Germany. Although defeated, the bill effectively witnessed the beginning of expansion of school feeding by local societies (Brown 1962). In America too, local churches and faith-based agencies raised funds and provided skills through volunteers and training and by the late nineteenth century widespread school feeding programmes had been successfully introduced throughout several countries in Europe and the United States (Acheson 1986; Gunderson 2003; Riddell 2007).

6.3.3 International Peace

Building on these early humanitarian beginnings and the ideologies of the earlier mentioned perpetual peace and Concert of Europe movements was the growing notion that relief and arbitration can be brought to bear in the world of conflict. It began with the Red Cross and the vision of one man, Henry Dunant.

6.3.3.1 The Red Cross

After witnessing first hand the horrors of the French and Austrian conflict in Solfarino in 1863, business-man Henry Dunant organised relief for the injured and dying (Dunant 1986; Walker 2008; Red Cross 2009). Upon his return to Geneva, his writings led to the Geneva Society for Public Welfare commission-ing a five-man team (Dunant and four others) known as the 'Committee of the Five'. The committee was first renamed the International Committee for the Relief of the Wounded before later being transformed into the International Committee of the Red Cross (ICRC) (ibid). From these small beginnings, the next logical progression was the development of collaboration with an international remit (Walker 2008).

6.3.3.2 Geneva Conventions and the Permanent Court of Arbitration

As becomes clear later in this study, peace, conflict and ultimately food security are inextricably linked. To this end, both the Geneva Convention and the Hague Peace Conferences are briefly introduced. After the creation of the ICRC, further proposals by Henry Dunant and his team suggested the founding of an international agreement for the express purpose of recognising

> ... the status of medical services and of the wounded on the battlefield. (Red Cross 2009)

One year later the Geneva Convention was founded and formed the basis of

> ... modern humanitarian law governing the treatment of soldiers and civilians during conflict. (Red Cross 2009)

The Geneva Conventions are a set of four agreements (Table 6.3) comprising rules that apply in armed conflict that seek to protect people who are not or are no longer part of the hostilities. These include the sick or wounded combatants, prisoners of war, civilians and medical and religious personnel.

6.3.3.3 The 1899 International Peace Conference, Hague

A few decades after the Geneva Convention, the first peace conference was convened at The Hague. It was primarily called upon to alleviate the arms race between Russia, Germany and Britain. However, the agenda was expanded and changed and instead the conference re-emerged as a conference on peace and arbitration (Mahan 1899; Holls 1900; Griffis 1901; Higgins 1904; Hague 2005). This first confer-ence was held in 1899 and out of this came the Permanent Court of Arbitration (PCA), the oldest court of international resolution. Both the Geneva and the Hague conventions and protocols offer unique and distinctive approaches to conflict. The Geneva Convention occupies itself with the humanitarian treat-ment of civilians and combatants whilst The Hague Conventions deal with the use of weapons of war. From these early foundations, it can be seen that a growing movement of humanitarianism emerged; notable early dates this period are summarised in Table 6.4.

TABLE 6.3

The Geneva Conventions

First Geneva Convention, 1864	To better the conditions of the wounded and sick in armed conflict in the field.
Second Geneva Convention, 1906	To better the conditions wounded, sick and shipwrecked members of armed forces at sea.
Third Geneva Convention, 1929	Relates to the proper treatment of prisoners of war.
Fourth Geneva Convention, 1949	Relates to the protection of civilian persons in time of war.

TABLE 6.4

Early Dates in the Humanitarian Calendar

Date	Organisation, Agencies or Conventions
1790: Benjamin Thompson, a.k.a. Count Rumford	Count Rumford delivered teaching and feeding programmes to the poor and needy.
1880: Society for Feeding Needy School Children	Society for Feeding Needy School Children originated in Dresden.
1863: International Committee of the Red Cross (ICRC)	Founded by Henry Dunant who organised relief for the injured and dying (Walker 2008).
1864: The Geneva Convention	'… modern humanitarian law governing the treatment of soldiers and civilians during conflict.' (Red Cross 2009).
1899: The Hague Peace Conference—The Permanent Court of Arbitration (PCA)	An international organisation based in The Hague established by one of the acts of the first Hague Peace Conference (Hague 2005).

References

Acheson, E. D. (1986). Tenth Boyd Orr Memorial Lecture: Food Policy, Nutrition and Government. *Proceedings of the Nutrition Society* 45: 131–138.

Aksu, E. (2008a). *Early Notions of Global Governance Selected Eighteenth-Century Proposals for 'Perpetual Peace' with Rousseau, Bentham, and Kant—Unabridged*. Wales: University of Wales Press.

Aksu, E. (2008b). Perpetual Peace a Project by Europeans for Europeans? *Peace & Change: A Journal of Peace Research* 33(3): 368–387.

Barnes, D. G. (2006). *A History of English Corn Laws: From 1660–1846*. London: Routledge.

Beck, S. (2005). *History of Peace Volume 1: Guides to Peace and Justice from Ancient Sages to the Suffragettes*. Santa Barbara, CA: World Peace Communications.

Birchall, J. (2002). Evolutionary Economics. John Birchall. http://www.themeister.co.uk/economics/evolutionary_economics.htm (accessed 15 January 2010).

Bloch, M. (2001). *Feudal Society, Volume 1: The Growth and Ties of Dependence*. London: Routledge and Keegan Paul.

Brown, C. (2002). State and Nation in Nineteenth Century International Political Theory. In *International Relations in Political Thought : Texts from the Ancient Greeks to the First World War*. C. Brown, T. Nardin and N. Rengger, eds. Cambridge: Cambridge University Press.

Brown, S. C. (1962). *Count Rumford, Physicist Extraordinary*. New York: Doubleday.

Carrasco, E. R. and K. J. Berg (1999). *The E-Book on International Finance and Development: Part One —V. The 1980's: The Debt Crisis and the Lost Decade*: The University of Iowa Center for International Finance and Development (UICIFD).

Crucé, E. (1909). *Le Nouveau Cynee Ou Discours D'estat Representant Les Occasions Et Moyens D'establir Une Paix General, Et La Liberte Du Commerce Par Tout Le Monde*. Philadelphia, PA: Allen, Lane and Scott.

de Béthune, M. and D. Ogg, Eds. (1921). *Grand Design of Henry IV. From the Memoirs of Maximilian De Bethune Duc De Sully (1559–1641)*. London: Sweet and Maxwell Ltd.

Deane, P. (1978). *The Evolution of Economic Ideas*. Cambridge, MA: Cambridge University Press.

Denemark, R., M. Hoffman, L. Twist and H. Yonten (2007). Dominance and Diplomacy: Trends in the Utilization of Multilateral Agreements by Global Powers. ISA 48th Annual Convention. Chicago, IL: International Studies Association.

Devereux, S. (2000). Famine in the Twentieth Century. Brighton: Istitute for Development Studies.

Dunant, H. (1986). *A Memory of Solferino*. Geneva, Switzerland: International Committee of the Red Cross.

Erasmus, D. (1917). *Desiderius Erasmus, the Complaint of Peace [1521]*. London, Chicago: The Open Court Publishing Company.

Gale, T. (2004). *Tommaso Campanella*. Online: Encyclopedia of World Biography.

Gimenez, M. E. (2008) The Population Issue: Marx vs Malthus. http://www.colorado.edu/Sociology/gimenez/work/popissue.html.

Griffis, W. E. (1901). Reviews: The Peace Conference at the Hague, and Its Bearings on International Law and Policy by F. W. Holls. *The Annals of the American Academy of Political and Social Science* 17: 116–119.

Grotius, H. (1925). *On the Law of War and Peace*. Oxford: The Clarenden Press.

Gunderson, G. W. (2003). *The National School Lunch Program: Background and Development*. New York: Nova Science Publishers.

Haftendorn, H., R. O. Keohane and C. A. Wallander (2002). *Imperfect Unions: Security Institutions over Time and Space*. Oxford: Oxford University Press

Hague (2005). The First Hague Peace Conference. The Hague Appeal for Peace. http://www.haguepeace.org/index.php?action=history&subAction=conf&selection=when (accessed 15 February 2011).

Henig, R. B. (2002). *The Origins of the First World War*. London: Routledge.

Higgins, A. P. (1904). *The Hague Peace Conferences and Other International Conferences Concerning the Laws and Usages of War: Texts Conventions with Commentary*. London: Stevens and Sons Limited.

Hill, W. (1893). The First Stages of the Tariff Policy of the United States. *Publications of the American Economic Association* 8(6): 9–162.

Hirsch, M. (1995). *The Responsibility of International Organizations toward Third Parties: Some Basic Principles*. Dordrecht, Boston and London: Martinus Nijhoff.

Hobsbawm, E. J. (1954). The General Crisis of the European Economy in the 17th Century. *Past & Present* 5(1): 33–53.

Hoffmann, S. and D. P. Fidler, Eds. (1991). *Rousseau on International Relations*. Oxford: Clarendon Press.

Holls, F. W. (1900). *The Peace Conference at the Hague, and Its Bearings on International Law and Policy*. New York: Macmillan.

Hutchinson, C. F. (1998). Social Science and Remote Sensing in Famine Early Warning. In *People and Pixels: Linking Remote Sensing and Social Science*. D. Livereman, E. F. Moran, R. R. Rindfuss and P. C. Stern, eds. Washington, DC: National Academy Press.

Keys, A. J., J. Brojek, O. Henschel, O. Michelson and H. L. Taylor (1950). *The Biology of Human Starvation*. Minneapolis, MN: University of Minnesota Press.

Lee, S. J. (1984). *Aspects of European History, 1494–1789*. London: Methuen and Company Ltd.

Lesaffer, R. (2004). *Peace Treaties and International Law in European History: From the Late Middle Ages to World War One*. Cambridge: Cambridge University Press.

Louden, R. B. (2007). *The World We Want: How and Why the Ideals of the Enlightenment Still Elude Us*. New York: Oxford University Press.

Mahan, A. T. (1899). The Peace Conference and the Moral Aspect of War. *The North American Review* 169(515): 433–447.

Malthus, T. R. (1798). *An Essay on the Principle of Population, as It Affects the Future Improvement of Society with Remarks on the Speculations of Mr. Godwin, M. Condorcet, and Other Writers*. London: Printed for J. Johnson, in St. Paul's Church-Yard.

Marx, K. (1887). *Capital: A Critique of Political Economy—Volume I Book One: The Process of Production of Capital*. Moscow: Progress Publishers.

Maxwell, S. and T. Frankenberger (1992). *Household Food Security: Concepts, Indicators and Measurements: A Technical Review*. New York and Rome: UNICEF and IFAD.

Mead, L. A. (1912). *Swords and Ploughshares or the Supplantng of the System of War by the System of Law*. New York and London: G. P. Putnam's Sons.

Mehta, V. N. (1929). Famines and Standards of Living. *The Annals of the American Academy of Political and Social Science* 145: 82–89.

Moran, D. (1995). The Fog of Peace: The Military Dimensions of the Concert of Europe. US Army War College Annual Strategy Conference: Strategy Through the Lean Years: Learning from the Past and the Present, Carlisle Barracks, Pennsylvania.

Mun, T. (1664). *England's Treasure by Forraign Trade: Or the Balance of Our Forraign Trade Is the Rule of Our Treasure*. London: John Mun.

North, D. C. and R. P. Thomas (1999). *The Rise of the Western World: A New Economic History*. Cambridge: Cambridge University Press.

Ó Gráda, C. (2009). *Famine: A Short History*. Princeton: University Presses of California, Columbia and Princeton.

Paras, A. (2007). 'Once Upon a Time … ': Huguenots, Humanitarianism and International Society". ISA's 48th Annual Convention, Chicago, IL: International Studies Association.

Parker, G. and L. M. Smith (1997). *The General Crisis of the Seventeenth Century.* London: Routledge.

Penn, W. (1693). *An Essay Towards the Present and Future Peace of Europe by the Establishment of an European Dyet, Parliament, or Estates (1693): Beati Pacifici, Cedant Arma Togae.* London: Tace Sowle.

Prentice, A. M. (2001). Fires of Life: The Struggles of an Ancient Metabolism in a Modern World. *Nutrition Bulletin* 26(1). 13–27.

Prentice, A. M. (2005). Early Influences on Human Energy Regulation: Thrifty Genotypes and Thrifty Phenotypes. *Physiology & Behavior* 86(5): 640–645.

Red Cross (2009). The Beginning of the Red Cross Movement. International Committee of the Red Cross. http://www.redcross.org.uk/standard.asp?id=87373 (accessed 5 January 2010).

Riddell, R. C. (2007). *Does Foreign Aid Really Work?* Oxford: Oxford University Press.

Riley, P., Ed. (2001). *Leibniz: Political Writings.* Cambridge: Cambridge University Press.

Roosevelt, G., Ed. (1999). *Pacifism and Just War Theory in Europe from the 16th to the 20th Centuries.* New York: Global Policy Forum.

Rumford, B. C. (1796). *Essays Political, Economical and Philosophical, Volume 1.* London: Printed for T. Cadell Jun. and W. Davies (successors to Mr. Cadell) in the Strand.

Salmon, T. C. and W. Nicoll (1997). *Building European Union: A Documentary History and Analysis.* Manchester: Manchester University Press.

Smith, A. (1776). *An Inquiry into the Nature and Causes of the Wealth of Nations.* Dublin: Printed for Messrs. Whitestone, Chamberlaine, W.Watson, Potts, etc.

Speakman, J. R. (2006). Thrifty Genes for Obesity and the Metabolic Syndrome—Time to Call Off the Search? *Diabetes and Vascular Disease Research* 3(1): 7–11.

Tittler, R. and N. L. Jones, Eds. (2004). *A Companion to Tudor Britain.* Oxford: Blackwell Publishing.

Vaidya, A. K., Ed. (2006). *Globalization: Encyclopedia of Trade, Labor, and Politics.* Santa Barbara, CA: ABC-CLIO.

Walford, C. (1879). *The Famines of the World: Past and Present.* London: Edward Stanford.

Walker, P. (2008). The Origins, Development and Future of the International Humanitarian System: Containment, Compassion and Crusades. ISA's 49th Annual Convention, Bridging Multiple Divides, San Fransisco, CA, International Studies Association (ISA).

Watkins, S. C. and J. Menken (1985). Famines in Historical Perspective. *Population and Development Review* 11(4): 647–675.

White, T. R. (1919). The Amended Covenant of the League of Nations. *The Annals of the American Academy of Political and Social Science* 84: 177–193.

Williams, D. (1999). *The Enlightenment.* Cambridge: Cambridge University Press.

7

Science, Technology and Philosophy

It is ultimately in the understanding of biological life, its physiology, the various organic compounds, the genetics and the chemical reactions among much else that paved the way for the full realisation of the body's needs and processes. However, progress was stubbornly slow owing to ancient beliefs about all things life, health and spiritual. I say ancient but in reality although such beliefs emerged from ancient civilisations it took till the sixteenth and seventeenth centuries and the persistence of science and scientists for these ideas to be properly challenged. Furthermore, science at this time also opened up new areas of disciplines and found new and innovative ways of addressing many of the challenges of the day that ultimately improved many aspects of agriculture, food processing, safety and by extension good health. With these ideas in mind, this chapter explores some of these persistent beliefs and explores just how science smoothed the transition of the food security technologies.

7.1 Origins of Life: Creationism, Evolution, Holism or Reductionism?

Beliefs in the origins of life consist of a varied mix of faith, evolutionary biology, chemistry and physics. For millennia, the main prevailing theory of life's origins was concerned with the idea of spontaneous generation; the notion that life could arise from inanimate matter. The story spans over two millennia and is full of some of history's most notable and colourful characters and as such is deserving of more detailed analysis beyond that touched upon in this book. However, the deeper connotations of theology, philosophy and metaphysics aside, the origin of life is not a gratuitous foray into ancient belief systems, instead what it illustrates are strong contemporary convictions as they pertain to food and health. Such belief systems persisted for close to 2000 years, and some argue even stifled intellectual debate, so indoctrinated were the views of spontaneous generation, vitalism and mechanism. However, despite, or even in spite of this much succeeding science, thought and discovery were eventually undertaken in part to answer these very questions of life and in doing so led to major advances in the chemical and bio-sciences and a fuller comprehension of biotechnologies in general. Indeed, Louis Pasteur often argued that the sciences, specifically microbiology and medicine, could not properly progress until the idea of spontaneous generation was put to rest (Schlager 2002). Indeed the subsequent resolution of the debate around the seventeenth- and eighteenth centuries witnessed many such advances. These contributed directly to new health paradigms, food safety techniques and food preservation procedures, particularly in the form of sterilisation, disease prevention and storage techniques such as freezing, drying and canning, etc.

7.1.1 Spontaneous Generation and Vitalism

Many ancient civilisations regarded Gaian* or Mother Earth theories as fundamental to their socio-cultural belief systems (Kluckhohn 1959). 'Mother Earth' is replete with connotations of life-generating forces and many cultures thought of Earth as a nurturing entity able to spontaneously generate† life. Plato (429–347) too adopted the notion of living Earth but went further and suggested she was also endowed with

---

* Gaia is an ancient belief system that proposes Earth is a living entity which fosters and maintains natural equilibrium and which is also seen as generating and sustaining life.
† Spontaneous generation in general is variously known as archebiosis or abiogenesis. It is an obsolete theory nowadays but at the time the sudden or spontaneous generation of life that was commonly believed to be the origins of life up until the sixteenth and seventeenth centuries. The reproduction of life from the same progeny is referred to as homogenesis while heterogenesis (or xenogenesis) describes the birth of a living thing from that emerges from a different form.

intelligence and soul (Schlager 2002; Scofield 2004). Even the history of Western science and Greek philosophy encompassed notions of Gaia, the bountiful embodiment of Earth. Indeed, the first Western thinker to elucidate the theory was perhaps Thales of Miletus (on the Ionian coast of Asia Minor). A pre-Socratic Greek philosopher of the seventh century BC, Thales regarded Earth as animistic. Animism is the doctrine that all things, even inanimate objects such as rocks and phenomena such as thunder have spirits or souls. He also hypothesised that Earth contained one elemental substance water, and that water had the

> … potentiality to change to the myriad things of which the universe is made, the botanical, physiological, meteorological and geological states. (Feldman 1945)

Further elaborating his philosophies of life, Thales also signed up to the idea that life could spontaneously generate fully formed from inanimate objects. Others too, notably, Anaximander (611–547), Anaximenes (588–524), Xenophanes (576–480), Empedocles (495–435), Democritus (b. 450), Anaxagoras (500–428) and Plato (429–347), attempting to naturally explain things previously ascribed to the gods, adopted various incarnations of the concept of spontaneous generation (Osborn 1894; Wilkins 2004). However, perhaps the most sophisticated of all the Greek views on this subject came from Aristotle (384–323 BC) when he offered three main methods of life generation: sexual and asexual reproduction as well as spontaneous generation (Schlager 2002). In one of his many books, *The History of Animals* Aristotle writes

> So with animals, some spring from parent animals according to their kind, whilst others grow spontaneously and not from kindred stock; and of these instances of spontaneous generation some come from putrefying Earth or vegetable matter, as is the case with a number of insects, while others are spontaneously generated in the inside of animals out of the secretions of their several organs. (Aristotle 2007, book 5)

Building on such convictions, it was also believed, particularly by Aristotle, that every living organism was imbued with some sort of vital force or spirit, above and beyond or simply distinct from purely biochemical reactions. Vitalism, as it was known, theorises that all living things are alive as a result of this vital force, as opposed to a mechanistic view of life which results from simply appropriately arranged non-living matter. Thus, the notion of spontaneous generation and vitalism was furthered by some of the greatest thinkers of the time (Bechtel and Richardson 1998).

7.1.2 Vitalism versus Biological Mechanism

This preoccupation with the notion of spontaneous generation and the question that life might not be wholly explicable by the laws of physics and chemistry alone remained salient for almost 2000 years. Indeed, even continuing into the seventheenth century and beyond, the theory of life's spontaneous generation was still accepted dogma as evidenced by Flemish physician, chemist, and physicist Jan Bapista van Helmont in his alchemic recipe for mice:

> If you press a piece of underwear soiled with sweat together with some wheat in an open mouth jar, after about 21 days the odor changes and the ferment, coming out of the underwear and penetrating through the husks of the wheat, changes the wheat into mice. (Knott 1905)

However, while spontaneous generation was generally accepted, the manner in which it occurred still remained a great source of debate (Fox et al. 2007).

Meanwhile although biological mechanism had been touched upon by the ancient Greeks (Democritus and his atomist theory), it was not really brought forward as a major contender to rival vitalism until about the seventeenth century and the arrival of Thomas Hobbes (1588–1679), Pierre Gassendi (1592–1655), René Descartes (1596–1650), Robert Boyle (1627–1691), Robert Hooke (1635–1703) and many others (Macintosh 1983; Bechtel and Richardson 1998; SEP 2005). Mechanism is the theory that all natural phenomena can be simply explained by the laws of nature. Contrasting the views of vitalism, mechanism holds that phenomena or biological systems are solely determined by mechanical principles. That is to say, humans and animals are like 'automata' likened to the interaction of material components in a machine governed by the physical laws of the universe (Durbin 1988; Bechtel and Richardson 1998; Grene and Depew 2004). Indeed, in the opening lines of Hobbes's great work of political philosophy *The Leviathan* he famously exemplifies this very notion by likening the commonwealth or state to an artificial man imbued with influence beyond the biological. Humans, he said, comprise an artificial collective that may well interrelate but only at the behest of an outside or external agent (Berkowitz 2008). While this may sound like vitalism the distinction is more fundamental in that biological life can simply be explained exclusively as the product of a complex series of chemical and physical reactions in which the sum is greater than the parts.

It is more than simple semantics or philosophical posturing though. Mechanism became the ideological stance by which the new scientific approach distinguished itself from its Aristotelian parentage. In this way, it came to dominate the epistemological landscape of virtually all Western science from that point on (Durbin 1988). It was in this environment that the first real challenge to the ideas of spontaneous generation was made. Francesco Redi was an Italian physician and poet and disputing the widely held belief that maggots spontaneously arose from rotting meat, he set out to challenge this assumption in 1668. His postulated that maggots did not arise from out of thin air but were instead hatched from eggs laid by flies. To test his theory, he laid out meat in a several flasks; some open to the air, others sealed and a few others covered with gauze. As he had predicted, maggots did indeed appear but only in the open flasks in which the flies had access to the meat on which they could lay their eggs (Schlager 2002). However, while Redi's experiment was very illuminating, it did not end conventional wisdom as to the spontaneity of life. Moreover, even though he set out to challenge assumptions about maggots, he still believed himself in the general principle that other microscopic organisms could spontaneously arise, believing intently, as many others did into the nineteenth century, in the doctrine of vitalism.

However, one discovery more than any other took the debate onto the next level and that was with the invention of the microscope. Advances marked by Zaccharias Janssen, Galileo, Anthon Van Leeuwenhoek, Robert Hooke and others in microscopy stimulated a certain momentum or inertia in the seventeenth-century debate on the origins of life. The big surprising discovery made by these microscopes was that microscopic life was everywhere. These invisible entities or 'little animals' were called infusoria or animalcules. While for some this gave the justification that all life was complex and therefore beyond the realms of simple biological processes, others saw it as possible answers to many of the 'apparent' contradictions. At about this time, a well-known experiment by Louis Joblot in 1718 used an infusion of boiled hay in a closed vessel and showed that it did not in fact give rise to micro-organisms (Kudo 1946). John Needham, however, similarly tested broths in 1749 only this time using meat and claimed to conclusively show that life could indeed be generated spontaneously from non-living things. Aiming to refute his claim however, Lazzaro Spallanzani repeated Needham's experiment of twenty years before. This time though unlike Needham, Spallanzani boiled the broth and closed it off to the elements (Magner 2002). Spallanzani, opposed to spontaneous generation, observed that the flasks remained sterile as long as they remained sealed. However, the unsealed flasks rapidly took on a cloudy appearance with the growth of microbes and he quickly concluded that the broth must have been contaminated by exposure to air. Needham countered by interpreting the results differently, suggesting that while boiling might kill microbes, subsequent exposure to the air would see the broth being re-populated with a new vital force. In this way, Needham saw this as further sign that the atmosphere contained the 'vital' spark necessary for life.

This was definitely not the end of the argument and the debate between the creationists and the reductionists raged on with prominent figures championing both sides of the dispute. Even the influential philosopher Immanuel Kant jumped into the foray in 1790 in his Critique of Judgement claiming:

> It is absurd for men to hope that another Newton will arise in the future who shall make comprehensible by us the production of a blade of grass according to natural laws which no design has ordered. (Kant 1914)

Diehards however, still continued to cling on to their beliefs, even when Friedrich Wöhler accidentally created the organic compound urea in 1828 by heating silver cyanate and ammonium chloride. Up till this point, creating an organic compound from inorganic matter was considered implausible, as it was thought that the ubiquitous 'vital' life force was needed for all organic compounds. In this experiment,

Wöhler had inadvertently tied the organic–inorganic issue directly into the spontaneous generation debate. This fuelled more debate and people like Jacob Berzelius (1779–1848) argued in 1836 that chemistry could eventually replicate all of the reactions occurring inside living organisms and that there is no special or 'vital' force exclusively the property of living matter. Others however like Leibig, while believing in the ability to re-create organic compounds without the so called vital force was in theory possible, argued that for the same compounds to be involved in movement, digestion, metabolism or 'life' as it were, there would indeed be need for that supernatural spark or life force—vitalism.

While, micro-organisms themselves and the experiments of Redi, Joblot, Needham and Spallanzini failed to resolve the controversy it did mark the beginning of the end. It took the work of Louis Pasteur (1822–1885), Kant's Newton of the microbial world, to finally dispel the myth of the appearance of microbes. Pasteur, taking up the challenge of the French Academy of Science's reward offered to anyone who could shed further insight on the so-called question of spontaneous generation, designed a series of similar nutrient-rich experiments to those previously carried out. In challenging the previous work by Felix Pouchet, Pasteur's innovative breakthrough came with the specially made beakers with long twisted necks that, while open to ambient air, were found to be a barrier in which airborne microbes were unable to traverse. Based on his essay of 1861 'Mémoire sur les corpuscules organisés qui existent dans l'atmosphère', Pasteur won the prestigious award in 1862. Armed with this knowledge over the proceeding few years, Pasteur successfully perfected techniques of sterilisation which had tremendous practical use in the prevention of disease as well as food and beverage safety. In fact the Pastuer–Pouchet debate was finally resolved by physicist John Tyndall who revealed, through the use of light, the presence of tiny microscopic spores and particles. His work convinced much of the scientific community that abiogenesis (spontaneous generation) did not actually occur.

Further experiments in the same vein progressively undermined the idea of the spontaneity of life as people instead increasingly focused on new emergent evolutionary paradigms. Charles Darwin, undoubtedly the most vociferous of this theory, wrote on heterogeny that

> ... our ignorance is as profound on the origin of life as on the origin of force or matter ...
> (Darwin 1863, p. 554)

He further offered that

> ... the nature of life will not be seized on by assuming that Foraminifera are periodically generated from slime or ooze. (Darwin 1863, p. 554)

7.1.3 Darwinism, Complexity and the Pre-Biotic Soup

So, in light of the beginning of the end of theory of spontaneous generation, a vacuum was left that scientists, philosophers and theologians aimed to fill. The debate then took a fork with some hardy followers of conventional orthodoxy refusing to accept such findings while others began to explore the equally divisive evolutionary theory of Charles Darwin. The publication of Darwin's *On the Origin of Species by Means of Natural Selection* in 1859 suggested that life evolved over thousands of years rather than as acts of spontaneous generation. It was a controversial theory that generated huge public debate. Conflict centred on the interpretation of the theory, with some expounding that it supported the end of the spontaneous generation theory and others arguing the opposite. In this latter view, even though previous experiments and the new theory of evolution might have inferred the 'law of biogenesis' (suggesting that life can only come from previously existing life), the problem with this view however, was that according to the many believers of the old paradigm, Darwin's theory did not solve the puzzle of the first living organism. In their view, if all species had evolved from preceding life forms, the question is raised as to from where did the earliest forms of life themselves evolve? In Darwinism, it seemed to many, the unifying concept of nature appeared to bridge the gap between living and non-living matter (Schlager 2002). Over the next couple of generations, holding out against Cartesian mechanism, the last vestiges of modern vitalism were championed by several people, notable of whom was Henri Bergson (1874–1948). Bergson in 1907 hypothesised that an 'élan vital' (vital impetus or force) overcame the barrier of generating living bodies from inert matter (Bechtel and Richardson 1998). However, as science advanced to explain the workings of human anatomy and physiology in less supernatural terms, vitalism was finally consigned to the realms of the superfluous. Or has it? Nowadays, vitalism or one of its various incarnations are more and more serving as the basis of modern alternative therapies such as *chi, prana, chakras, yin & yang* and *reiki* among numerous others.

So what of the origin of life? Well, today the notion of spontaneous generation has all but been replaced with the chemical evolutionary hypothesis known as the 'primordial soup' theory. This theory was first espoused by the Russian biochemist A. I. Oparin in his 1924 booklet *The Origin of Life*. In the booklet, he played with the idea that over huge geologic timeframes, inorganic material might combine and organise itself into ever complex carbon or organic compounds. In this concentrated organic soup, it is opined that the first replicating organic molecules essential for life arose. This was also strengthened by similar ideas by the Scottish biologist J. B. S Haldane (Flores et al. 2000). Further strengthening the primordial soup theory are the proponents of complexity theory, most notable of whom are Roger Lewin (1992), Mitchell Waldrop (1992) and Stuart Kauffman (1993). Complexity theory stipulates that simple systems generate complex patterns. As in the chemical evolution theory, complexity theory asserts that evolution occurs through the emergence of new variables that naturally develop over time. That is to say, simple systems are adaptive in which many independent variables interact with each other allowing for the spontaneous self-organisation that is sometimes seen in complex systems. This, it is said, happens naturally without any overarching force or planning rather it is the result of organisms, variables or agents constantly adapting to each other. This demonstrative patterned behaviour is often likened to the stock market metaphor, which sees numerous individual units acting independently in response to market stimuli. The overall effect is the collective movement of price whether up or down (Lewin 1992; Waldrop 1992; Kauffman 1993; Inayatullah 1994).

With the origins of life largely consigned to scientific investigation, this opened the way for new understandings in biology, chemistry and the new science of genetics among others. Before we explore how these all came together, a little background in microbiology is necessary.

7.2 Biotechnology and the Bio-Sciences

People have been manipulating living organisms to address problems or to modify products for improving plants or animals for human needs for millennia (Peters 1993; AU/ ICE 1997). Often motivated by such diverse goals as curiosity, scientific advancement and necessity, biotechnology is intricately wrapped up in the history of the sciences, particularly the bio-sciences. In unravelling the biotechnological timeline, progress can be thought of as taking two different but interrelated pathways: lab-based scientific discovery or curiosity, as well as practical, real-world applications shaped by necessity and/or improvements. In some cases, the practical everyday necessities became the driving force that pushed real issues into the lab. Other times, the leading objective was vice versa whereby lab-based science for science sake found technologies and solutions which were then adapted to real-life situations. For ease of narrative, the history of biotechnology in respect of food and food security can be divided into two categories: microbiology (micro-organism research including zymotechnology) and genetic engineering (hybridisation, cross-breeding, genetic modification).

7.2.1 Zymotechnology

As long ago as 8000–10,000 years our ancestors were producing wine, beer and bread through fermentation. While the precise dates of fermented food technologies are lost to antiquity so too are its geographic origins, although the Indian subcontinent, Ancient Egypt and Babylon have been put forward as potential originators (albeit probably by accident) of the process. Whosoever came across the principles, next to dry curing, fermentation was used extensively and is the second oldest form of food preservation (Farnworth 2003). From these humble beginnings, zymotechnology, a term coined by Georg Ernst Stahl in 1697, is the study and application of such fermentation technology. Zymotechnology then, can ultimately be thought of as a forerunner to many of the later biotechnologies. At its very elemental level, fermentation involves micro-organisms such as yeasts, bacteria and moulds which when mixed with appropriate foods digest this food and produce two by-products: carbon dioxide gas and alcohol (Beck 1960). However, up till this point (seventeenth century), fermentation technologies were still largely domestic or small-scale affairs (Anderson 2005). Two things helped propel this into the large-scale industrial industry that it is today: the advent of the microscope and the realisation that manipulating the conditions of fermentation allowed for the improvement of both quality and yield (NHM 1989/90).

7.2.2 Micro-Organisms and Food Microbiology

The discovery of micro-organisms helped identify the processes involved in fermentation. This in turn was aided by the discovery of the microscope. Lenses for correcting sight had been around since 1268 (Roger Bacon); however, the microscope was credited as having been invented by one of two possible Dutch eyeglass makers: Hans Lippershey or the father and son team Hans and Zaccharias Janssen. Sometime between 1590 and 1608, it was found that by placing several lenses in a tube images could be greatly magnified. Hearing of these experiments, Galileo Galilei (1564–1642) started to experiment on his own. He described the principles of lenses and light rays in his book *Sidereus Nuncius*, The Sidereal (Starry) Messenger in 1610 and began improving on both the microscope and telescope (King 1955/2003; Galilei 1989). The rise of the popularity and improvements in magnification of the microscope in the seventeenth century allowed for greater insight into the microscopic world, and although Robert Hooke was one of the first to examine the living cells of plants and organisms described in his book *Micrographia* in 1665 (Hooke 2007), it was actually Anton Van Leeuwenhoek who was one of the first to observe micro-organisms using a microscope of his own design in the 1670s. This was a pivotal time and a whole new world of protozoa, moulds, yeasts and bacteria were discovered giving rise to collective names such as infusoria or animalcules (later termed protozoa) (Schlager 2002). While yeast cells were first seen by Leeuwenhoek in 1680, he did not realise that they were indeed living organisms that caused fermentation. In fact first person to theorise this was German chemist Georg Stahl in 1697 although it took Schwann Cagniard de la Tour and F. Kuetzing (1836–1838) to suggest that yeast was in fact a living organism and responsible for much fermentation. The scientific establishment of the time criticised and ridiculed these findings, the harshest critics came especially from the previous opponents of vitalism, Justus von Liebig, J.J. Berzelius and Friedrich Woehler, who saw it as giving credence to the concept of spontaneous generation. In their view, fermentation was considered primarily a chemical rather than a biological process. Much debate and further research ensued by the likes of Lavoisier and Guy-Lussac and others, but more persuasively by French chemist Louis Pasteur (1822–1895) who finally brought closure and conclusively proved through a series of classic investigations during the 1850–1860s that indeed yeast was a living organism (Shurtleff and Aoyagi 2007). This period culminated in a marked change in people's perception of the world around them, and food microbiology, in this respect, was finally put on a scientific footing.

Although fermentation was shown to be the result of the action of living micro-organisms, it did not explain the nature of the fermentation process itself. It was later discovered however that the biological process of fermentation induced the chemical breakdown of foods. In this respect, Leibig and his contemporaries had been partly right. The breakdown of foods, as in the digestive experiments of Reamur, Payen, Persoz and later Swann, was shown to be caused by 'ferments', so-called because it was seen that both digestion and fermentation were in some way organically allied. So it was in 1858 that Traube suggested such 'ferments' were due to chemical substances; later the terms 'enzym' (1881) and 'enzyme' (1890s) were coined by Kuehne and Roberts, respectively. The next step was taken by Eduard Buchner who ground up yeast and extracted a juice (which contained the enzymes) and which when mixed with sugar would ferment just as the living organism had done. This effectively unified the organic and inorganic camps who finally agreed that enzymes were responsible for the fermentation processes in both, which in turn were produced by micro-organisms (Shurtleff and Aoyagi 2007).

7.2.2.1 Germ Theory

Despite proposals by Girolamo Fracastoro (a.k.a. Hieronymus Fracastorius 1478–1553) in 1546, germ theory was not really accepted for another couple of hundred years. Even with the heralded arrival of the microscope and Leeuwenhoek's animules or animalcula, people failed to make the connection between bacteria and the transfer of disease (Leeuwenhoek 1677). The most widely accepted notion of infection at this time was miasma theory. This held that diseases were caused by the presence of bad air carrying poisonous vapours or decaying matter. The next significant step was when Ignaz Semmelweis (1847) discovered that doctors' failing to wash their hands before examining patients increased the risk of mortality, particularly in women after childbirth. Furthering the debate, John Snow, an English physician, traced the source of the London cholera outbreak in 1854 to drinking water. Once again however, the work of the prolific scientist Pasteur in the 1860s as well as Robert Koch in 1876 who, working with anthrax and cattle, finally established a causal link between microbes and disease. This not only helped disprove the idea of spontaneous generation but led directly to the rise of germ theory and a means of sterilisation as well as further potential life-saving treatments (Shurtleff and Aoyagi 2007). Collectively, these ideas led to Pasteur's 1880 general germ theory of infectious disease, which stipulated that each disease was caused by a specific micro-organism. Significantly at this time too, Koch also discovered the method for isolating micro-organisms in pure culture.

Collectively, fermentation and germ theory collaborated to provide the foundations of a whole new development paradigm. For the first time, people began to accept the fact that they shared their environment with multitudes of minute organisms that exerted an ongoing powerful influence on human life. This new world view, among other things, provided tremendous stimulus for new research on fermented foods, viruses and diseases and a whole host of other possibilities.

7.2.3 Early Genetic Engineering: Hybridisation and Cross-Breeding

Early genetics (up to the twentieth century) was concerned with the notion of manipulating traits in species of plants and animals through the use of non-molecular cross-breeding, inbreeding and selective breeding techniques to produce hybrids in a process called hybridisation (Bateson 1899). While both plants and animals were involved, it was perhaps the plant biology that lent itself to more productive and more progressive research on account of the relative short time spans between experimentation and results.

7.2.3.1 *Plants*

Plant science and understanding of plant genetics, engineering and hybridisation had a slow start. While plant science is often viewed as a relatively modern discipline, in reality its fundamental techniques go back to the beginnings of early civilisation when cultivated crops became vital for survival. Primitive farmers found that by selecting seeds from the best plants of the season they could maintain and strengthen desirable traits in successive harvests. However, despite these early beginnings, debate rather than experimentation dominated the landscape until perhaps the late sixteenth and early seventeenth centuries largely, as some have suggested, through the reluctance to accept that plants rely on sexual reproduction to propagate. The breakthrough came with English botanist Nehemiah Grew in 1682 who, addressing the Royal Society, offered his view (sharing that of Sir Thomas Millington's) as to the sexual nature of the plants. He suggested that the stamens (pollen-producing) and pistils were in fact the male and female equivalent *sex* organs (Bradley 1717; Zirkle 1932). Although it was well received it was not until Rudolf J. Camerarius (as credited by Julius Sachs in his *Text-book of Botany, Morphological and Physiological, 1875*) who in 1694, offered the first real experimental proof that pollen was indeed necessary for the development of seed and the propagation of plants (Bennett 1875; Sachs and Vines 1882, Harshberger 1894; Knuth 1906).

This period also coincided with the development of the microscope and once again the late seventeenth century proved to be a pivotal time in scientific progress. This upped the curiosity stakes, and it took the likes of Robert Hooke and like-minded others to further the understanding of plant science by putting it, as with much else of the day, on a scientific footing. In the interim period, the evolution of hybrid knowledge had taken a significant step forward when in 1588 Jacob Theodor von Bergzabern, a physician and herbalist from Heidelberg, published a 'lively' account of multi-coloured corn kernels from the New World in his book *Krauterbuch* (Zirkle 1934). This was the first of many similar descriptions around this period, yet it seems they were viewed more as delightful curiosities than for what they really were: natural hybridisation. Moreover, up till this point hybridisation in nature was considered a rarity and was also thought to be responsible for inferior quality plants (Wissemann 2007). One of the first known records of this hybridisation process that showed an inkling of an understanding of the concept came much later in a letter written by Cotton Mather to James Petiver (a fellow of the Royal Society) in 1716. In the letter, he talked enthusiastically about red and blue Indian corn cross-pollinating the usual yellow varieties which were

> … so infected … as to communicate the same Colour unto them … (Zirkle 1934)

At about this time too the first artificial hybrid was also recorded. Thomas Fairchild (1667–1729) is credited with having transferred the pollen of a sweet William (*Dianthus barbatus* L.) into the pistil of a carnation (*Dianthus caryophyllus* L.) to produce what became known as Fairchild's Mule (Bradley 1717). While today this is accepted practice and is considered pretty unremarkable, in the early eighteenth century this early form of genetic engineering incited as much of a storm, both from scientific and religious perspectives as genetically modified plants do today (Leapman 2000). Much progress was made over the ensuing decades with one leading light given credit for undertaking the first systematic study of plant hybridisation, Joseph G. Kolreuter (1733–1806) (Bradley 1717). Kolreuter, according to Sachs, truly expanded the knowledge of plant reproduction by experimenting with artificial hybrids and showing that pollen-transmitted traits were passed from parent to child.

Paralleling the advancing plant science in general was the growing knowledge of cell science itself. Shortly thereafter in 1833, the botanist Robert Brown discovered the cell structure known as the cell nucleus and a few years later in 1839 Johannes Purkinje described and named the complex, translucent semi-fluid substance that constituted the living matter within the cells themselves as protoplasm* (Wolf 2003). This coincided with the theory by Gerardus Johannes Mulder that the albuminous substance of all living tissue was made from carbon, hydrogen, nitrogen, oxygen, sulphur and phosphorus. He coined this substance 'protein' in 1838. It was at this time too that Theodor Schwann first came to regard the cell as the basic element of the structure of tissues. However, he mistakenly considered the formation of new cells to involve the aggregation of existing extracellular substances. In fact, it was not until around 1852 that Robert Remak and others shortly after Rudolf Virchow and Albert Kölliker demonstrated that such cells originated by the division of pre-existing cells. Further realisation led to the notion that as cells divided so too did the nucleus (Mazzarello 1999). Further work by Friedrich Miescher in 1869 aimed to chemically isolate and analyse these cell components. After eventually isolating the nucleus, Miescher managed to successfully extract a 'new' substance, he called 'nuclein'. This nucleic acid from the nuclei is more commonly known today as deoxyribonucleic acid (DNA). Within a few decades, both chromatin (later chromosomes) and mitochondria (organelles) were discovered, and Walther Flemming named the process of cell division as mitosis (Mazzarello 1999).

Meanwhile back to evolution and plant science. Other leading lights of the mid-nineteenth century were Lamarck, Weismann, De Vries, Nillson, Burbank and Hays. Collectively, they pushed the boundaries of hybridisation ever further by producing such things as higher more yield giving crops as well as more hardier disease-resistant plants (Buffum 1911; Shull 1911; Orton 1918). However, two people during this period are worth mentioning for their role in the evolution of hybrid science and modern genetic engineering; these are Charles Darwin and Gregor Mendel. Darwin's famous postulations on evolution and natural selection suggested that minute variation within genera as well as hybridised new lineages over time accounted for the sheer breadth of the variety of biota on Earth. Gradual evolution and hybridisation in Darwin's view were natural processes that lent credibility to the contemporary debate on the origins of life (Lotsy 1918). In his book *On the Origin of Species* he writes at great length on both of these processes and offers

> On the view that each species has been independently created, I can see no explanation of this great fact in the classification of all organic beings; but, to the best of my judgment, it is explained through inheritance and the complex action of natural selection, entailing extinction and divergence of character … (Darwin 1859, p. 129)

---

* Protoplasm is a term rarely used in modern biology anymore. *Protoplasm* was commonly thought of as the living content of a cell surrounded by a plasma (cell) membrane. The cell was viewed as a bag of protoplasm, a colloidal jelly-like chemical substance of infinite complexity with such fragility that it needed to be studied intact, as a whole.

Mendel too, while working with hereditary traits in peas, helped shape the new direction of thinking. He discovered that one inheritable trait would habitually be dominant over its recessive alternative. This suggested that there was a genetic basis for such traits which would allow prediction of such characteristics in their progeny (Mendel 1866). In the meantime in the same year as Mendel postulated his theory of heredity, Ernst Haeckel, in his book *Generelle Morphologie*, ascribed the property of inheritance of the cell's inheritable characteristics to the nucleus (Wolf 2003). These were remarkable breakthroughs although sadly, at the time, Mendel's work languished in the shadow of Darwin's more sensational previous publications. That was until 1900 when Hugo de Vries, Erich Von Tschermak and Carl Correns published research corroborating Mendel's theory of heredity. In what became known as Mendelian inheritance or Mendelian genetics, it offered an alternative to the existing blending genetic theory of the time. This work effectively re-discovered Mendel's theories and propelled them to the fore (Druery and Bateson 1901; Reinhardt and Ganzel 2003).

These experiments in plant science were not in the abstract either. Economic needs, agricultural necessity as well as agricultural reform helped bridge the gap between pure and applied science. Fuelling this was the United States Department of Agriculture's (USDA) policy of promoting agricultural diversity and the encouragement of new experimental hybrids with desirable traits. Indeed, by the end of the nineteenth century, USDA researchers, in the hope of reducing imports and increasing exports, had crossed more than 20,000 raisin grapes, 116 crosses of pear varieties as well as hundreds of hybrid pineapples, oranges and wheat (for yield and disease resistance) (Paul and Kimmelman 1988). This and the rediscovery of Mendel's work set the foundation of modern genetic understanding, which was taken up with gusto around the turn of the twentieth century. This period then was marked by great advances in all aspects of microbiology which underpinned much work in the fields of nutrition, disease and agricultural growth; the highlights are outlined in Table 7.1.

TABLE 7.1

Biotechnology Timeline

6000–4000 BC	Yeast was used to make beer by the Sumerians and the Babylonians made leavened bread with brewer's yeast about 6000 years ago.
5000 BC	Yoghurt was perhaps discovered by accident by the Mesopotamians 7000 years ago.
1000 BC	The Chinese use the fermented liquid extracted from soybean.
400 BC	Hippocrates (460–377 BC) speculated that the male semen contributed to a child's heredity. He also wondered whether there was a similar fluid in women since children receive traits from each parent.
320 BC	Aristotle (384–322 BC), rejecting Hippocrates theories, insisted that all inheritance comes from the father.
250 BC	The Greeks practice crop rotation to maximise soil fertility.
AD 100	Powdered chrysanthemum is used as an insecticide in China.
AD 600	Japanese seasoning of fermented rice, barley and/or soybean with salt and the kōjikin fungus.
AD 1000	Hindus observed that some diseases were passed down in the family and that children inherit all their parents' characteristics.
1300	In Mexico, the Aztecs harvested *Spirulina geitleri* algae from lakes as a food source.
1590–1608	Hans and Zaccharias Janssen or Hans Lippershey were credited with inventing the microscope.
1663	Plant cells are first described by Hooke.
1670s	Anton van Leeuwenhoek discovered microscopic protozoa.
1697	Zymotechnology first coined by Georg Ernst Stahl.
1724	Cross-fertilisation in corn was discovered.
1796	The world's first vaccination: Edward Jenner inoculated an eight-year-old boy with cow pox virus to protect against smallpox.
1766–1802	The word 'biology' first appears. Although often attributed to Karl Friedrich Burdach in 1800, it was in fact mentioned by Michael Christoph Hanov in 1766. Two others worth considering are Gottfried Reinhold Treviranus and Jean-Baptiste Lamarck, both in 1802.
1809/1810	Nicolas Appert devised a technique using heat to sterilise and can food winning a 12,000 franc prize (first offered in 1795) from Napoleon.

continued

TABLE 7.1 (Continued)

Biotechnology Timeline

1824	Henri Dutrochet discovered that tissue is composed of living cells
1830	Gerardus Johannes Mulder coined the albuminous substances of all living tissue—protein.
1831	Robert Brown discovered the cell nucleus in plants in 1831.
1855	The *Escherichia coli* bacterium was discovered and later became a major research, development and production tool for biotechnology.
1856	Karl Ludwig discovered a technique for keeping animal organs alive outside the body (in vitro) by pumping blood through them.
1859	Charles Darwin (1809–1882) elucidated his natural selection hypothesis in his book *On the Origin of Species.*
1862	The Organic Act established the US Department of Agriculture (USDA) was directed its commissioner 'to collect new and valuable seeds and plants … and to distribute them among agriculturalists'.
1863	Louis Pasteur invented the process of pasteurisation, heating wine sufficiently to inactivate microbes.
1865	Gregor Mendel presented his findings to the Natural Science Society and later published his paper titled 'Experiments in Plant Hybridisation' in 1866.
1869	Friedrich Miescher in 1869 chemically isolated a cell nucleus called 'nuclein', more commonly known today as deoxyribonucleic acid (DNA).
1878	Joseph Lister described the 'most probable number' (MPN) method for the isolation of pure cultures of bacteria. It was an important step in understanding infectious diseases.
1879	Flemming discovered chromatin, rod-like structures inside the cell nucleus that help with DNA.
1881/2	Robert Koch successfully introduces nutrient agar, a microbiological growth medium, which became a standard tool in microbiological experimentation.
1883	August Weismann coined the term 'germ-plasm', a concept of the physical basis of heredity that both the male and female parents contribute equally to the heredity of the offspring and that the chromosomes must be the bearers of heredity.
1883	A vaccine for rabies by Louis Pasteur in 1883, which was successfully tested in 1885.
1884	Robert Koch and Jacob Henle introduced the first method used to establish the etiology of a specified infectious disease. Known as Koch–Henle postulates, or 'Koch's postulates' for short.
1888	Waldeyer suggested a name for the threads within the nucleus of a cell. Building on Fleming and Hertwig's findings Heinrich Waldeyer named the threadlike structures chromosomes.
1897	Eduard Buchner demonstrated that fermentation can occur with an extract of yeast in the absence of intact yeast cells. This discovery of cell-free fermentation is a founding moment in biochemistry and enzymology.

Sources: Thompson, E.E., *Bull. Med. Libr. Assoc.*, 50, 236–242, 1962; Brock, T.D., *Robert Koch: A Life in Medicine and Bacteriology*, Science Tech Publishers, New York, 1988; Yuan, J., Viruses, HIV, Prions, and Related Topics: The Small Pox Story. *Human Virology at Stanford*, 1998; McLaughlin, P., *J. Hist. Biol.*, 35, 1–4, 2002; Farnworth, F. R , *Handbook of Fermented Functional Foods*, CRC Press, Boca Raton, FL, 2003; Smith, M.E., *The Aztecs*, Blackwell Publishing, Oxford, 2003; Anderson, R., *J. Brew. Hist.*, 121, 5–24, 2005; Alvarez, P.J.J and Illman, W.A., *Bioremediation and Natural Attenuation: Process Fundamentals and Mathematical Models*, John Wiley, New Jersey, 2006; Hooke, *Micrographia*, BiblioBazaar, 2007; Inglis, T.J.J., *J. Med. Microbiol.*, 56, 1419–1422, 2007; D'souza, J.I. and Killedar, S.G., *Biotechnology and Fermentation Process*, Nirali Prakashan, Pune, India; Britannica, Encyclopaedia Britannica Online, 2009; Tribby, D., In *The Sensory Evaluation of Dairy Products*, S. Clark, M. Costello, F.W. Bodyfelt and M. Drake, eds., AVI Van Nostrand Reinhold, New York, 2009; HoV, Library Treasures: Louis Pasteur Letter. *The History of Vaccines*. The College of Physicians of Philadelphia, 2010; Nobel Prize, Eduard Buchner: The Nobel Prize in Chemistry 1907, Nobel Foundation, 2011.

7.3 Health and Nutrition

Food and health and by extension food security have unravelled in their own ways at their own time. From time to time over the centuries though one discovery or epiphanous breakthrough in one discipline led to corresponding breakthroughs in others. The true confluence of health and nutrition probably came about as a result of advances in science, particularly chemistry and biology in the seventeenth and eighteenth centuries. However, even before this health and nutrition were ideologically and medically linked in the minds of some of the greatest thinkers of all time. In ancient times, two proponents of diet and health, Hippocrates and Aristotle have often been credited with early notions of health and diet's synergistic relationship (Nichols 1992). Fondly known as the father of medicine, the Greek philosopher Hippocrates (460–377 BC) closely linked food with health and disease and extolled the virtues of diet and exercise (Jones 1923). He based his medical knowledge on empirical observation and the study of the human body, and he scoffed at contemporary views that considered illness to be caused by disfavour of the gods or the work of spirits. Instead, Hippocrates firmly believed that illness had a physical and a rational explanation. He also considered the importance of a natural healing process of rest, fresh air and a good diet and also noted that some individuals were better able to cope with disease and illnesses than others. Subsequent medical and philosophical debate was further imbued through his teachings, which took a rational approach to health by paying attention to fact and experience. In these observations, he considered disease a natural phenomenon and held that the body must be treated as a whole.

Aristotle too, in his ancient Greek treatise on the art of persuasion *The Rhetoric* in 350 BC, expounds the virtues of deliberation. In food and health for instance Aristotle suggests that

> Things are productive in three ways; first in the way that the being healthy is productive of health; or as food is so of health, or as exercise is, because usually it does produce health. (Hobbes 1857, p. 40)

His seminal contemplations sought to determine not so much 'ends' but the 'means to ends' and in this endeavour he was, alongside Hippocrates perhaps among the first to link diet and health (Nichols 1992).

7.3.1 Humoralism: The Balance of Diet and Health

Returning to Hippocrates, his legacy was two-fold: the Hippocratic Oath and the Hippocratic Corpus. Of interest is the Corpus, which is a collection of 70 or so medical works that espoused his collected teachings. His teachings among other things refined and adapted the ancient idea of humors (or humours) and applied them to human physiology. Humoralism is essentially a theory that health and illness result from a balance of bodily liquids or humors. It was built on a variation of Empedocles's theory of the four roots: Earth, air, fire and water. Building on this, Hippocrates held that two basic themes, hot-cold and wet-dry, combined in four different ways to embody the elements of Earth. In this way, the four humors related to black bile (Earth, of the spleen), blood (fire, of the heart), yellow bile (air, of the gall bladder) and phlegm (water, of the lungs). In turn, Aristotle (384–323 BC), Theophrastus (371–287 BC), Galen in the second century AD, along with others, further developed the concept of the humors relating them to temperament and the four seasons (Table 7.2; Lettsom 1778; Magendie 1829; Adams 1849; Marketos 1997; Bockler 1998).

Hippocrates applied his theory of humoralism to medicine. The goal of these early medical practices was to keep the four humors in balance or 'eucrasia' through correct diet and lifestyle. Out of humoral balance or disharmony the body was considered to be in 'dyscrasia'. Therefore, by connecting human physiology so closely with the elements the teachings go on to intimately associate diet and the environment with a populations' health and characteristics. Accordingly, this set the stage for the legacy of the doctrine of the four 'humors', which became orthodox in Europe for the next 2000 years (Cohen 1955; Bockler 1998).

TABLE 7.2

Relationships of the Humors

Humor	Season	Element	Organ	Qualities	Temperament
Blood	Spring	Air	Heart	Hot and wet	Sanguine
Phlegm	Winter	Water	Brain	Cold and wet	Phlegmatic
Yellow bile	Summer	Fire	Liver	Hot and dry	Choleric
Black bile	Autumn	Earth	Spleen	Cold and dry	Melancholic

Sources: Magendie, F., *An Elementary Compendium of Physiology: For the Use of Students*, printed for John Carfrae and Son and Longman, Rees, Orme, Brown and Green, Edinburgh, 1829; Cohen, H., *Proc. Royal Soc. Med.*, 48, 155–160, 1955.

However, it was not till the turn of the seventeenth century, aided by the development of the microscope and the subsequent germ theory of disease, that the role of the four humors in Western medicine was eventually challenged. As the doctrine came under question so the science linking health and nutrition progressed. In fact, it was in the early studies on food intake and excreta by Santorio (Sanctorius) and later the body's blood circulation in 1628 by William Harvey that the convergence of food nutrition and health finally had its scientific genesis. This was further cemented by perhaps the first book on physiology in English in 1659 entitled, *The Natural History of Nutrition, Life and Voluntary Notion* written by Walter Charleton, later one of the Royal Society's founding members. Charleton's assertions of the role of food in the body was remarkably similar to those of today, that is to say partly for maintenance and partly for energy (Carpenter 1994).

7.3.2 Linking Diet and Health

Further linking health and nutrition was the story of scurvy. As early as 1600 Sir James Lankester, commander of the East India Company's first fleet, provided lemon juice daily for his sailors (Power 1928). However, it was not for almost another two centuries in (1796) that the issue of daily citrus juice became compulsory in the Royal Navy, virtually extinguishing scurvy in the Navy completely. By the late eighteenth century, many scholars had commented on how food was used in the body; however, the next step towards improved nutritional understanding came with the so-called chemical revolution in France. The goal of this period was to understand how the body worked physiologically and the role of food in that process. In this endeavour, no one did more to advance nutritional investigation at this point than Antoine Lavoisier (Nichols and Reeds 1991; Carpenter 1994, 2003; Weaver 2006).

7.3.2.1 *Lavoisier, Boussingault, Liebig and Others*

Lavoisier, the father of modern chemistry, had a particular interest in metabolism. He noted that carbon was the end product of oxidative activity and proposed the connection between oxygen absorbed into the body and carbon dioxide excreted. He started by placing guinea pigs into a calorimeter* and measured the amount of ice that melted (direct calorimetry) showing that this was quantitatively related to the amount of carbon dioxide given off by the animal (Passmore 1982). These respiratory exchanges were the first studies that eventually led to a full understanding of the metabolic process. In these experiments, working with Pierre-Simon Laplace, Lavoisier demonstrated the role of oxygen in animal and plant respiration as essentially a slow combustion of organic material. Further research estimated the heat output per unit of carbon dioxide, which yielded the same energy ratios for candle flames as for animals. To Lavoisier, the conclusion was clear: animals produced energy through a process of combustive reaction involving oxygen (Lavoisier 1780; Lavoisier and Laplace 1784; Lusk 1906; Underwood 1944; Nichols and Reeds 1991; Bensaude-Vincent 1996). Attention now turned to energy inputs and outputs of animals. Jean Baptiste Boussingault had already proposed from François Magendie's single food experiments with dogs that foods that did not contain nitrogen could not support life. Therefore, in conclusion it was determined that the nutritional value of a vegetable substance resided in the gluten and vegetable albumin (its nitrogenous compounds) (Carpenter 2003). Boussingault's experiments in the 1830s first with plants and then with animals resulted in two findings. First, only plants were able to synthesise atmospheric nitrogen (indirectly) during growth, and second by examining animal excreta, Boussingault established that the nitrogen content of the food animals ingested would indeed be sufficient to meet their needs, thus negating the need to obtain nitrogen directly from the atmosphere. Moreover, in light of such experiments, Jean-Baptiste Dumas, a colleague of Boussingault, deduced that animals were only capable of oxidising the materials that it obtained from its food (Dumas and Boussingault 1844).

   The leading German organic chemist of the time, Justus von Liebig, improved on Lavoisier's methods for organic analysis and in so doing helped create a new discipline of organic chemistry. In answer to Dumas's findings about oxidation however, Liebig wrote that Dumas must be wrong and that animals must be able to convert carbohydrates to fat requiring 'reduction' (later metabolism) rather than through

---

* Calorimeter is a device which measures the heat energy outputs of individuals either directly via direct heat measurement or indirectly through calculations of the respiratory gases.

oxidation alone (Carpenter 2003). Building on this and other research, Liebig improved on Dumas's work and further postulated that it was in fact proteid (protein), carbohydrates and fat that was metabolised or burned in the body to produce the energy the body needed (Lusk 1906). It was also known at the time that protein contained nitrogen, and Liebig further suggested, based on the law of the conservation of energy, that it might be possible to establish equilibrium between nitrogen intake from protein and nitrogen excretion from urine and faeces (Nichols and Reeds 1991).

7.3.2.2 Voit

This had implications beyond scientific research to that of practical nutritional value. Carl Voit, a student of Liebig, established beyond doubt in 1857 that such equilibrium could indeed be met, and in doing so he also determined that any imbalance in the nitrogen content of the excreta was mirrored by a corresponding loss or gain in body tissue (Lusk 1906). Further experiments in 1866 by Voit and Pettenkofer on carbon dioxide and nitrogen waste using respiration chambers suitable for a man also confirmed Liebig's theory that it was indeed not just protein but carbohydrate and fat that were broken down in the body and used as energy (Passmore 1982). Once again, it is Lavoisier who is recognised as having put forward the notion that body heat is a product of the oxidation of the body's ingested substances (Zeigler 1922). Later however, although the process of metabolisation was not fully understood, it was determined that energy released in this manner manifested itself as heat and as such it was proposed that ' ... heat may become a measure of the total activity of the body' (Lusk 1906, p. 31).

7.3.3 The First Dietary Studies

By now, elements of the many disparate strands of the sciences were converging on the new discipline of nutrition. Indeed much work was now coming together that would ultimately lend scientific credibility to a subject that was working its way to prominence in both the public and political spheres. Although many scientists at this time (the mid-nineteenth century) were working on dietary standards, the first formal action by any government to introduce public policy of dietary recommendation based on scientific principles was at the request of the British Privy Council in 1862/1863. These first dietary recommendations emerged out of the pioneering work of Edward Smith and his food intake surveys among low-income groups in Northern England (Acheson 1986; Carpenter 1991; McArdle et al. 1999). Smith's studies calculated a minimum basic daily requirement equal to 4300 grains of carbon and 200 grains of nitrogen, enough to avert starvation and disease (equivalent to about 2800–3000 calories; Tomlinson 1978; Oddy 1983; Harper 1985). He also recommended reduced rations (10%) for women in light of their smaller stature and reduced needs.

7.3.3.1 Frankland

Research continued, and after visiting Edward Frankland, Voit returned to Munich in 1860 with a Thompson calorimeter to further his research. Several years later in 1866, Frankland found a way to combust organic material by oxidising it in a calorimeter with a mixture of potassium chlorate and potassium nitrate and measuring the heat it produced. Using this method, he calculated the values of 29 foodstuffs in 'heat units' (the same value as calories). This effectively established the first direct measurement of heat from food energy and introduced the quantitative concept of food energy (Frankland 1866; McLoed 1905; Ensminger et al. 1993/4; Russell 1996; Carpenter 1998).

7.3.3.2 Voit

Simultaneously using the calorimeter, Voit and Pettenkofer's experiments on a fasting man determined how much food was burned in the body and whether there was an overall net loss or gain. From these studies, Voit prepared a table in 1866 in which he calculated the metabolic production of a man to be between 2.25 and 2.4 million g-calories or small calories (2250–2400 kcal equivalent). In doing so, Carl Voit was the first to establish the practice of defining the potential energy of food in caloric terms (Hargrove 2006).

7.3.3.3 Rubner

Further studies by Voit initiated work on the interchangeability of fats and carbohydrates in the diet. This idea was taken up by Rubner, a student of Voit, who by 1884 had calculated the per gram calorie conversion rates of the major food groups, establishing the standard energy equivalent factors for protein (4.1 kcal/g) fat (9.3 kcal/g) and carbohydrates (4.1 kcal/g). This formed the basis of Max Rubner's 1884 isodynamic law which stated that foodstuffs replaced each other: '... in accordance with their heat-producing value ...' (Rubner 1885; Lusk 1906; Mudry 1974; Nichols 1992; Ensminger et al. 1993/4; Hwalla and Koleilat 2004).

7.3.3.4 Atwater

Wilbur Olin Atwater, who worked with Rubner in Voit's lab, took this knowledge back to America and continued working on calorimetry. He built the first human calorimeter in 1892 at Wesleyan University in collaboration with Edward Bennett Rosa as well as a bomb calorimeter for measuring the energy values in foods themselves. The original machine was known as the Atwater-Rosa calorimeter. However, this was later refined by Francis Benedict who began working with Atwater shortly after this time, and the final contraption became known as the Atwater-Rosa-Benedict calorimeter (Chambers 1952). Between 1894 and 1900, Wilbur Atwater had also refined Rubner's earlier energy conversion factors allowing for the losses due to metabolism and digestion in the body. He calculated conversion factors for protein, fat and carbohydrate based on average calculations of different animal and plant foods of 5.65 kcal/g, 9.4 kcal/g and 4.15 kcal/g, respectively. Note here that these figures represented actual energy conversion factors of the food and not what the body absorbed. Atwater ended up creating two main conversion factor tables, the reason being the different quality of protein, fats and carbs in different foods resulted in substantial variability in specific energy values. As an example, it was found that the protein energy contained in potatoes was equal to only 2.78 kcal/g, whereas eggs were found to contain protein energy equivalent to 4.36 kcal/g. Such differences warranted a comprehensive table of specific group values which would fully reflect these different food conversion factors. The other table was general food conversion factors which applied to all foods irrespective of specific food composition. The above figures of 5.65, 9.4 and 4.15 kcal/g were recalibrated to represent actual metabolisable energy, that is what was actually absorbed by the body, the revised general figures equated to about 3.9 kcal/g for both protein and carbs and 8.99 kcal/g for fats. This is taken across both animal and vegetable diets and is applied to mixed diets where the calculation of specific foods might be unwarranted (Atwater and Woods 1896; FAO 2003). The importance of this work is still relevant today; indeed, many of the caloric composition tables in use today are based on the rounded Atwater values of 4 kcal/g for both protein and carbs and 9 kcal/g for fats (Nichols 1992; Ensminger et al. 1993/4).

7.3.3.5 Basal Metabolism

Although comparative experiments determining basal metabolic rates (at rest) as measured by oxygen absorption and subsequent carbon dioxide production were conducted as early as 1843 by Scharling, the term did not come into common usage until Adolf Magnus-Levy coined the expression 'Grundumsatz' or 'basal metabolism' in 1899 (Benedict and Emmes 1915; Henry 2005). It was a term that was used to describe the chemical changes in which the body's 'living substance' was produced or maintained. More specifically, the term reflects the process by which energy is transformed or liberated and made available to the body (Murlin 1910). In other words, food is ingested for fuel and essential nutrients in the form of protein, fat, carbohydrate, minerals and water. These are then broken down or metabolised for energy, to be used immediately or stored in the tissue as protein, fat or carbohydrate in the form of glycogen. Metabolised energy is needed for the many chemical reactions in the body, which fuels muscular activity allowing us to breathe, digest food and maintain a healthy physically active life (more in Appendix D).

Basal metabolic rate (BMR) accounts for between 45% and 70% of a person's energy expenditure (FAO/WHO/UNU 2004). To be accurately measured, basal metabolism has to be measured in the absence of absolute physical activity even the bodily functions of ingestion. In other words,

> The heat production of the individual [has to be measured] in a state of complete muscular repose 12–14 hours after the last meal, i.e., in the postabsorptive condition ... (Harris and Benedict 1919, p. 370)

Basal metabolism at this time was quantified using calorimetric techniques either directly by measuring the heat output or indirectly through the calculation of oxygen consumption and carbon dioxide excretion which was then converted to their energy equivalents (Murlin 1910; Scrimshaw et al. 1994). At the time Francis Gano Benedict was recognised as one of the world's foremost pioneers in the field of respiratory metabolism. Working on the calorimeter with Atwater in 1895 over the next 12 years, they conducted over 500 experiments to do with exercise, rest and diet using the Atwater-Rosa respiration calorimeter. Benedict was also involved in the development of metabolism studies based on age, sex, height and weight, which culminated in the Harris–Benedict equation of metabolism.

Thus, by the turn of the twentieth century, the practice of estimating energy requirements in caloric values was firmly established. There were two approaches in determining such requirements: the first (and less accurate) relied on measuring a person's food intake, ensuring the person was neither gaining nor losing weight and the second, more difficult but more accurate method involved taking account of the nitrogen balance, that is, measuring the nitrogen losses in faecal, urinary and respiratory excreta and providing sufficient energy to cover these losses plus extra for physical needs. Around this time though, it was the former empirical method that prevailed (WHO 1973; Schürch and Scrimshaw 1989). As a direct result of all this scientific effort, dietary studies were acquiring greater credibility. At the forefront, driving the new discipline ever forward were Voit and his contemporaries. Table 7.3 summarises some of the first scientific daily nutritional recommendations undertaken by scientists of the day.

TABLE 7.3

Earliest Recommended Daily Requirements

Date/Authority	Load (Default Man)	Protein	Fat	Carbohydrate	Calories
1862/3: Edward Smith					
Smith responding to a British Privy Council request studied the poor and malnourished in Northern England and calculated that 2800 kcal of energy per day is sufficient for a working man. These were the first dietary standards based in scientific study.					
Men					2800
Women					Less 10%
1865–1868: Lyon Playfair					
Playfair recommended an intake of 3000–3140 calories for a man of light work to 3750 calories for an hardworking man.					
Men	Light work				3029
	Medium work	119	51	531	3140–3525
	Hard work				3750
1866: Carl Voit					
After visiting Frankland in the United Kingdom, Voit returned to Germany and, while working with Pettenkofer, built a respiration chamber that could fit a man inside. He recommended of 2250–2400 calories for an average 70-kg working man.					
Men	Medium work				2400
1881: Carl Voit					
Building on his previous work, Voit suggests a man's daily intake undertaking moderate work at 3050–3055 calories and hardworking men's intake to be 3370 calories and hardworking soldiers' to be 3574 calories.					
Men	Medium work	118	56	500	3055
	Hard work	145	100	500	3574

continued

TABLE 7.3 (Continued)

Earliest Recommended Daily Requirements

Date/Authority	Load (Default Man)	Protein	Fat	Carbohydrate	Calories
1885: Rubner					
	Light work	125	40	377	2445
	Medium work	127	52	509	2868–3092
	Hard work	165	70	565	3362
1889: Atwater					
Atwater also recommends a woman of light work to partake of 2300 calories.					
Men	Light work	100			2700–3000
	Medium work	125	125	450	3400–3520
	Hard work	150			4150–4060
1893: Dr. W. Prausnitz					
Dr. W. Prausnitz offers that for German Hospital patients a diet of 2350 calories and 110 g of protein for men and 100 g of protein and 2100 calories for women.					
Men		110	50	350–400	2350
Women		100	50	300–350	2100
1893: Sir Benjamin Thompson (Count Rumford)					
Known as 'Count Rumford'; suggests between 3000 and 3500 calories for maximum and minimum working men's rations.					
Men	Minimum work	110	90	420	3000
	Maximum work	125	125	450	3500

Sources: Atwater, W.O., Public Health Papers and Reports. 27th Annual Meeting of the American Public Health Association, Minneapolis, The Berlin Printing Company, 1889; Thompson, B., *Plain Words About Food: The Rumford Kitchen Leaflets*, Rockwell and Churchill Press, Boston, MA, 1899; Paton et al., *A Study of the Diet of the Labouring Classes in Edinburgh : Carried out under the Auspices of the Town Council of the City of Edinburgh*, Otto Schulze, Edinburgh, 1901; Lusk, G., *The Elements of the Science of Nutrition*, W.B. Saunders, Philadelphia, London, 1906; Hutchinson, R., *Food and the Principles of Diatetics*, W. Wood and Co., New York, 1916; Harper, A.E., *Am. J. Clin. Nutr.*, 41, 140–148, 1985; Carpenter, K.J., *J. Nutr.*, 121, 1515–1521, 1991; Ensminger et al., *Foods & Nutrition Encyclopedia*, CRC Press, London, 1993/4; McArdle et al., *Sports and Exercise Nutrition*, Williams and Wilkins, Baltimore, MD, 1999; Hargrove, J.L., *J. Nutr.*, 136, 2957–2961, 2006.

Note: Workloads are predicated on those adopted by Graham Lusk in 1906. *Light*: Men engaged in occupations involving light muscular work such as writers, draughtsman, tailors, physicians. *Medium*: Ideal for labourers working 8–10 hours a day such as mechanics, joiners, garrisoned soldiers and farmers. *Hard*: Such as soldiers in the field, shoemakers and blacksmiths.

References

Acheson, E. D. (1986). Tenth Boyd Orr Memorial Lecture: Food Policy, Nutrition and Government. *Proceedings of the Nutrition Society* 45: 131–138.

Adams, F. (1849). *The Genuine Works of Hippocrates*. London: The Sydenham Society.

Alvarez, P. J. J. and W. A. Illman (2006). *Bioremediation and Natural Attenuation: Process Fundamentals and Mathematical Models*. Hoboken, NJ: John Wiley.

Anderson, R. (2005). The Transformation of Brewing: An Overview of Three Centuries of Science and Practice. *Journal of Brewing History* 121: 5–24.

Aristotle (2007). *The History of Animals*. South Australia: eBooks@Adelaide.

Atwater, W. O. (1889). Public Health Papers and Reports. 27th Annual Meeting of the American Public Health Association. Minneapolis, MN: The Berlin Printing Company.

Atwater, W. O. and C. D. Woods (1896). The Chemical Composition of American Food Materials. *US Office of Experiment Stations Bulletin* (28):1–47.

AU/ICE (1997). Ice Cases Studies—Intellectual Property Rights & Biotechnology: Indian Seed Conflicts. *Inventory of Conflict and Environment (ICE)* (41): http://www1.american.edu/ted/ice/grainwar.htm.

Bateson, W. (1899). Hybridisation and Cross-Breeding as a Method of Scientific Investigation. *Journal of the Royal Horticultural Society*(24): 59–66.

Bechtel, W. and R. C. Richardson (1998). *Vitalism*. London: Routledge.

Beck, C. W. (1960). Georg Ernst Stahl, 1660–1734. *Journal of Chemical Education* 37(10): 506–510.

Benedict, F. G. and L. E. Emmes (1915). A Comparison of the Basal Metabolism of Normal Men and Women. *The Journal of Biological Chemistry* 20(3): 253–262.

Bennett, A. W. (1875). Sir Thomas Millington and the Sexuality of Plants. *Nature* (13): 85–86.

Bensaude-Vincent, B. (1996). Between History and Memory: Centennial and Bicentennial Images of Lavoisier. *Isis* 87(3): 481–499.

Berkowitz, P. (2008). Leviathan Then and Now: The Latter-Day Importance of Hobbes's Masterpiece. *Policy Review* (151).

Bockler, D. (1998). Let's Play Doctor: Medical Rounds in Ancient Greece. *The American Biology Teacher* 60(2): 106–111.

Bradley, R. (1717). *New Improvements of Planting and Gardening, Both Philosophical and Practical, Explaining the Motion of Sap and Generation of Plants*. London: W. Mears.

Britannica (2009). Encyclopædia Britannica Online. 2009.

Brock, T. D. (1988). *Robert Koch: A Life in Medicine and Bacteriology*. New York: Science Tech Publishers.

Buffum, B. C. (1911). Effect of Environment on Plant Breeding. *The Journal of Heredity* 6(1): 63–72.

Carpenter, K. J. (1991). Biographical Article: Edward Smith (1819–1874). *Journal of Nutrition* 121(3): 1515–1521.

Carpenter, K. J. (1994). *Protein and Energy: A Study of Changing Ideas in Nutrition*. Cambridge: Cambridge University Press.

Carpenter, K. J. (1998). Early Ideas on the Nutritional Significance of Lipids. *The Journal of Nutrition* 128(2): 423S–426S.

Carpenter, K. J. (2003). A Short History of Nutritional Science: Part 1 (1785–1885). *Journal of Nutrition* 133(March): 638–645.

Chambers, W. H. (1952). Max Rubner. *Journal of Nutrition* 48 (Supplement): 3–12.

Cohen, H. (1955). The Evolution of the Concept of Disease. *Proceedings of the Royal Society of Medicine* 48(3): 155–160.

D'souza, J. I. and S. G. Killedar (2008). *Biotechnology and Fermentation Process*. Pune, India: Nirali Prakashan.

Darwin, C. (1859). *On the Origin of Species by Means of Natural Selection, or the Preservation of Favoured Races in the Struggle for Life*. London: John Murray.

Darwin, C. (1863). The Doctrine of Heterogeny and Modification of Species. *Athenaeum* 1852(25 April): 554–555.

Druery, C. T. and W. Bateson (1901). Gregor Mendel: Experiments in Plant Hybridization. *Journal of the Royal Horticultural Society* 26: 1–32.

Dumas, M. J. and M. J. B. Boussingault (1844). *The Chemical and Physiological Balance of Organic Nature*. London: H. Bailliere Publisher.

Durbin, P. T. (1988). *Dictionary of Concepts in the Philosophy of Science*. New York: Greenwood Press.

Ensminger, A. H., M. E. Ensminger, J. L. Konlande and J. R. K. Robson (1993/4). *Foods & Nutrition Encyclopedia*. London: CRC Press.

FAO (2003). Food Energy—Methods of Analysis and Conversion Factors. *FAO Food and Nutrition Paper 77*. Food and Agriculture Organization.

FAO/WHO/UNU (2004). Human Energy Requirements Report of a Joint FAO/WHO/UNU Expert Consultation: Rome, 17–24 October 2001. Rome: UNU/WHO/FAO.

Farnworth, E. R. (2003). *Handbook of Fermented Functional Foods*. Boca Raton, FL: CRC Press.

Feldman, A. (1945). Thoughts on Thales. *The Classical Journal* 41(1): 4–6.

Flores, J. C., T. C. Owen and F. Raulin (2000). First Steps in the Origin of Life in the Universe. Sixth Trieste Conference on Chemical Evolution. Trieste, Italy: Kluwer Academic Publishers.

Fox, J. G., S. Barthold, M. Daviso, C. Newcomer, F. Quimby and A. Smith (2007). *The Mouse in Biomedical Research*. Oxford: Elselvier.

Frankland, E. (1866). On the Origin of Muscular Power. *London, Edinburgh and Dublin Philosophical Magazine*, 4th series(31): 182–199.

Galilei, G. (1989). *Sidereus Nuncius, or, the Sidereal Messenger*. Chicago, IL: University of Chicago Press.

Grene, M. and D. J. Depew (2004). *The Philosophy of Biology: An Episodic History*. Cambridge: Cambridge University Press.

Hargrove, J. L. (2006). History of the Calorie in Nutrition. *Journal of Nutrition* 136: 2957–2961.

Harper, A. E. (1985). Origin of Recommended Dietary Allowances – an Historic Overview. *The American Journal of Clinical Nutrition* 41: 140–148.

Harris, J. A. and F. G. Benedict (1919). *A Biometric Study of Basal Metabolism in Man Carnegie Institute of Washington Publication No. 279*. Washington, DC: Washington Carnegie Institute.

Harshberger, J. W. (1894). James Logan, an Early Contributor to the Doctrine of Sex in Plants. *Botanical Gazette* 19(8): 307–312.

Henry, C. (2005). Basal Metabolic Rate Studies in Humans: Measurement and Development of New Equations. *Public Health Nutrition* 8(7A): 1133–1152.

Hobbes, T. (1857). *Aristotles: Treatise on Rhetoric*. London: Henry G. Bohn.

Hooke, R. (2007). *Micrographia*. BiblioBazaar.

HoV (2010). Library Treasures: Louis Pasteur Letter. *The History of Vaccines*. The College of Physicians of Philadelphia. http://www.historyofvaccines.org/content/blog/library-treasures-louis-pasteur-letter (accessed 21 February 2011).

Hutchinson, R. (1916). *Food and the Principles of Diatetics*. New York: W. Wood and Co.

Hwalla, N. and M. Koleilat (2004). Dietetic Practice: The Past, Present and Future. *Health Journal* 10(6): 716–730.

Inayatullah, S. (1994). Life, the Universe and Emergence. *Futures* 26(6): 683–696.

Inglis, T. J. J. (2007). Principia Ætiologica: Taking Causality Beyond Koch's Postulates. *Journal of Medical Microbiology*(56): 1419–1422.

Jones, W. H. S. (1923). *Hippocrates: The Sacred Disease*. London: Heinemann Ltd.

Kant, I. (1914). *Critique of Judgment*. London: Macmillan.

Kauffman, S. A. (1993). *The Origins of Order: Self-Organization and Selection in Evolution*. New York: Oxford University Press.

King, H. C. (1955/2003). *The History of the Telescope*. New York: Courier Dover Publications.

Kluckhohn, C. (1959). Recurrent Themes in Myths and Mythmaking. *Dædalus: the Journal of the American Academy of Arts and Sciences* 88(2): 268–279.

Knott, J. (1905). Part IV. Medical Miscellany: Jan Bapista Van Helmont. *Dublin Journal of Medical Science* 120(2): 132–148.

Knuth, P. (1906). *Handbook of Flower Pollination: Based Upon Hermann Muller's Work - the Fertilisation of Flowers by Insects*. Oxford: Clarendon Press.

Kudo, R. R. (1946). *Protozoology*. Springfield, IL: Charles C. Thomas.

Lavoisier, A. (1780). Expériences Sur La Respiration Des Animaux Et Sur Les Changements Qui Arrivent À L'air Par Leur Poumon. In *Histoire De L'académie Royale Des Sciences, Avec Les Mémoires De Mathématique Et De Physique Pour Le Meme Annee*. Académie royale des sciences. Paris: Académie royale des sciences.

Lavoisier, A. and P.-S. Laplace (1784). Mémoire Sur La Chaleur. In *Histoire De L'académie Royale Des Sciences, Avec Les Mémoires De Mathématique Et De Physique Pour Le Meme Annee*. Académie royale des sciences. Paris: Académie royale des sciences,.

Leapman, M. (2000). *The Ingenious Mr. Fairchild: The Forgotten Father of the Flower Garden*. London: Headline Book Publishing.

Leeuwenhoek, A. v. (1677). Observations, Communicated to the Publisher by Mr. Antony Van Leewenhoeck, in a Dutch Letter of the 9th of Octob. 1676. Here English'd: Concerning Little Animals by Him Observed in Rain-Well-Sea. And Snow Water; as Also in Water Wherein Pepper Had Lain Infused (Trans. Henry Oldenburg). *Philosophical Transactions of the Royal Society of London* 12: 821–831.

Lettsom, J. C. (1778). *History of the Origin of Medicine*. London: Printed by J. Phillips for E. and C. Dilly in the Poultry.

Lewin, R. (1992). *Complexity: Life at the Edge of Chaos*. New York: Macmillan.

Lotsy, J. P. (1918). Evolution by Means of Hybridization. *Molecular and General Genetics MGG* 20(1): 42–45.

Lusk, G. (1906). *The Elements of the Science of Nutrition*. Philadelphia and London: W. B Saunders.

Macintosh, J. J. (1983). Perception and Imagination in Descartes, Boyle and Hooke. *Canadian Journal of Philosophy* 13(3): 327–352.

Magendie, F. (1829). *An Elementary Compendium of Physiology: For the Use of Students*. Edinburgh: Printed for John Carfrae and Son and Longman, Rees, Orme, Brown and Green.

Magner, L. M. (2002). Microbiology, Virology, and Immunology. In *A History of the Life Sciences, Revised and Expanded*. Boca Raton, FL: CRC Press.

Marketos, S. G. (1997). History of Medicine. *Athens University Medical School*. http://asclepieion.mpl.uoa.gr/parko/marketos2.htm.

Mazzarello, P. (1999). A Unifying Concept: The History of Cell Theory. *Nature Cell Biology*. (1): E13–E15.

McArdle, W. D., F. I. Katch and V. L. Katch (1999). *Sports and Exercise Nutrition*. Baltimore, MD: Williams and Wilkins.

McLaughlin, P. (2002). Naming Biology. *Journal of the History of Biology* 35(1): 1–4.

McLoed, H. (1905). Edward Frankland—Born Jan 18th 1825, Died August 9th 1899. *Journal of the Chemical Society, Transactions* 87: 565–618.

Mendel, G. (1866). Versuche Über Plflanzen-Hybriden. *Verhandlungen des naturforschenden Ver-eines in Brünn* Bd. IV für das Jahr 1865(Abhand-lungen): 3–47.

Mudry, J. J. (1974). *Measured Meals: Nutrition in America*. Albany: University of New York.

Murlin, J. R. (1910). The Metabolism of Development: I Energy Metabolism in the Pregnant Dog. *American Journal of Physiology* 26: 134–155.

NHM (1989/90) Where Did Biotechnology Begin? *Access Excellence*. http://www.accessexcellence.org/RC/AB/BC/Where_Biotechnology_Begin.php.

Nichols, B. L. (1992). A History of Nutrition Research Reflected in the Usda Tables of Food Composition. 17th National Nutrient Databank Conference, Baltimore, Maryland, National Nutrient Databank.

Nichols, B. L. and P. J. Reeds (1991). History of Nutrition: History and Current Status of Research in Human Energy Metabolism. *The Journal of Nutrition*(121): 1889–1890.

Nobel Prize (2011). Eduard Buchner: The Nobel Prize in Chemistry 1907. Nobel Foundation. http://nobelprize.org/nobel_prizes/chemistry/laureates/1907/buchner-bio.html (accessed).

Oddy, D. J. (1983). Urban Famine in Nineteenth-Century Britain: The Effect of the Landcashire Cotton Famine on Working-Class Diet and Health. *The Economic History Review, New Series* 36(1): 68–86.

Orton, W. A. (1918). Breeding Disease Resistant Plants. *American Journal of Botany* V(6): 279–283.

Osborn, H. F. (1894). *From the Greeks to Darwin: An Outline of the Development of the Evolution Idea*. New York: Macmillan.

Passmore, R. (1982). Reflexions on Energy Balance. *Proceedings of the Nutrition Society* 41: 161–165.

Paton, D. N., J. C. Dunlop and E. Inglis (1901). *A Study of the Diet of the Labouring Classes in Edinburgh: Carried out under the Auspices of the Town Council of the City of Edinburgh*. Edinburgh: Otto Schulze.

Paul, D. B. and B. A. Kimmelman (1988). Mendel in America: Theory and Practice, 1900–1919. In *The American Development of Biology*. R. Rainger, K. R. Benson and J. Maienschein. Philadelphia, PA: University of Pennsylvania Press.

Peters, P. (1993). *Biotechnology: A Guide to Genetic Engineering*. Dubuque, IA: Wm. C. Brown Publishers, Inc.

Power, D. (1928). The Surgeons Mate by John Woodall. *British Journal of Surgery* 16(61): 1–5.

Reinhardt, C. and B. Ganzel (2003). A History of Farming. *Living History Farm*. http://www.livinghistoryfarm.org/farminginthe30s/crops_03.html.

Rubner, M. (1885). Calorimetrische Untersuchungen I and Ii. *Zeitschrift für Biologie* 21: 250–334 and 337–410.

Russell, C. A. (1996). *Edward Frankland: Chemistry, Controversy and Conspiracy in Victorian England*. Cambridge: Cambridge University Press.

Sachs, J. and S. H. Vines (1882). *Text-Book of Botany: Morphological and Physiological*. Oxford: Clarendon Press.

Schlager, N. (2002). *Science in Dispute*. Farmington Hills, MI: Thomson Gale.

Schürch, B. and Nevin S. Scrimshaw (1989). Activity, Energy Expenditure and Energy Requirements of Infants and Children. London: International Dietary Energy Consultancy Group.

Scofield, B. (2004). Gaia: The Living Earth-2,500 Years of Precedents in Natural Science and Philosophy. In *Scientists Debate Gaia: The Next Century*. S. Schneider, J. Mille, E. Cris and P. Boston, eds. Cambridge, MA: Massachusetts Institute of Technology.

Scrimshaw, N. S., J. C. Waterlow and B. Schürch (1994). Energy and Protein Requirements. *Proceedings of an Idecg Workshop*. London: International Dietary Energy Consultative Group.

SEP (2005). *Democtitus*. Stanford, CA: Stanford University.

Shull, G. H. (1911). Hybridisation Methods in Corn Breeding. *The Journal of Heredity* 6(1): 63–72.

Shurtleff, W. and A. Aoyagi (2007). *History of Soybeans and Soyfoods: 1100 B.C. To the 1980s*. Lafayette, CA: Soyinfo Center.

Smith, M. E. (2003). *The Aztecs*. Oxford: Blackwell Publishing.

Thompson, B. (1899). *Plain Words About Food: The Rumford Kitchen Leaflets*. Boston, MA: Rockwell and Churchill Press.

Thompson, E. E. (1962). Doctors, Doctrines, and Drugs in Ancient Times. *Bulletin of the Medical Library Association* 50(2): 236–242.

Tomlinson, H. (1978). 'Not an Instrument of Punishment': Prison Diet in the Mid-Nineteenth Century. *International Journal of Consumer Studies* 2(1): 15–26.

Tribby, D. (2009). Yoghurt: A Brief History. In *The Sensory Evaluation of Dairy Products*. S. Clark, M. Costello, F. W. Bodyfelt and M. Drake, eds. New York: AVI Van Nostrand Reinhold.

Underwood, E. A. (1944). Lavoisier and the History of Respiration. *Proceedings of the Royal Society of Medicine* 37(6): 247–262.

Waldrop, M. M. (1992). *Complexity: The Emerging Science at the Edge of Chaos and Order*. New York: Simon and Schuster.

Weaver, L. T. (2006). The Emergence of Our Modern Understanding of Infant Nutrition and Feeding 1750–1900. *Current Paediatrics* 16(5): 342–347.

WHO (1973). Energy and Protein Requirments: Report of a Joint FAO/WHO Ad Hoc Expert Committee - Technical Report Series No. 522. *WHO Technical Report Series No. 522 FAO Nutrition Meetings Reports Series No. 52*. Geneva: World Health Organization 24.

Wilkins, J. S. (2004). Spontaneous Generation and the Origin of Life. *Talk Origins*: http://www.talkorigins.org/faqs/abioprob/spontaneous-generation.html.

Wissemann, V. (2007). Plant Evolution by Means of Hybridization. *Systematics and Biodiversity* 5(3): 243–253.

Wolf, G. (2003) Friedrich Miescher: The Man Who Discovered DNA. *Miesceriana*. http://www.bizgraphic.ch/miescheriana/html/the_man_who_dicovered_dna.html.

Yuan, J. (1998) Viruses, Hiv, Prions, and Related Topics: The Small Pox Story. *Human Virology at Stanford*. http://virus.stanford.edu/pox/history.html.

Zeigler, M. R. (1922). The History of the Calorie in Nutrition.In *The Scientific Monthly Volume15*. J. M. Cattell. New York: American Association for the Advancement of Science.

Zirkle, C. (1932). Some Forgotten Records of Hybridization and Sex in Plants, 1716–1739. *Journal of Heredity* 23(11): 432–448.

Zirkle, C. (1934). More Records of Plant Hybridization before Koelreuter. *The Journal of Heredity* 25(1): 3–18.

Summary of Part II

The growth of agricultural civilisations and the domestication of crops and livestock brought with it new problems. By relying on harvested food rather than that gathered from the wild, societies had to consider, perhaps for the first time collective continuity or security of supply. Thus, the growing international dimension of food witnessed the introduction of new varieties, new diseases and an explosion of international trade. Concerns over the expansion of the global population focused attention on an apparent finite food supply. People like Malthus brought valid questions into the public domain that had up till that point perhaps been the preserve of intellectualists alone. Simultaneously, spurred on by both the industrial and agricultural revolutions, the intensification and mechanisation of agriculture by the eighteenth/ nineteenth centuries witnessed rapid increases in the productivity curve. Food, it now seemed, need not be the stumbling block of growth and prosperity of a growing world.

Separately food, health and nutrition were on their own historical trajectory. The relationship between nutritional status and health has been the object of inquiry for millennia. Initial intellectualism that conceived of a relationship between food and health in the time of Hippocrates took another full millennia and then some to satisfy this relationship in science. However, this did not prevent great leaps of intuitive understanding. Health and vitality in the formative years were wrapped up with spiritualism and mysticism with the ideas of spontaneous generation and the humoral system of health. In spite of this, Hippocrates and his followers touched upon some fundamental notions of disease, health and exercise that provided good solid foundations later in science. The humoral theory of medicine continued throughout the Middle Ages before being taken over by the New Sciences. However these practices are not to be consigned to history, what was once considered medical dogma is still widely used throughout Chinese and Eastern cultures to this day. Nevertheless, it was the preceding eighteenth century and the advance of technological innovation together with notable research from Black, Priestley, Cavendish, Liebig and in particular Lavoisier with his respiration-calorimeter studies that ultimately shepherded the modern ideas of physiology, metabolism and nutrition.

Meanwhile, growing internationalism brought with it a need for suitable and appropriate responsibility. Ideas based on the principles of perpetual peace were raised from many quarters in an attempt to halt a pervading sense of the growing frequency and senselessness of conflict. Casualties of war too came to the attention of a now sensitive collective social conscience and with it the beginnings of international humanitarianism. This ultimately led to many new instruments and institutions that aimed to redress the many injustices and from these humble beginnings a new global governance was born.

By this time, developed Western countries had already undergone several hundred years of expansionism through territorial acquisition and colonisation. While it is true to say many of these countries benefited at the time from planned economic management and infrastructural improvement, terms of trade were generally such that conquering governments increased their own domestic wealth on favourable terms. This was an era dominated by mercantilist orthodoxy where a positive net balance of trade was viewed as the prevailing economic paradigm. Adam Smith and his contemporaries questioned this approach, offering instead an economic model based on free trade rather than protectionism. Thus, a see-sawing of protectionist and free trade policies followed.

In the meantime, the advent of the microscope and advances in organic chemistry, agricultural innovation through hybridisation and genetic manipulation saw new varieties of crops more adapted to the environment and requiring less agricultural inputs arise. Further progress in biotechnology and the biosciences improved fermentation technologies as well as advances in food microbiology.

Lastly, it can be seen that the link between health and nutrition was finally postulated at the scientific level. In determining the energy requirements of individuals, experiments in the fields of chemistry, nutrition and physiology came together around the latter half of the nineteenth century. This allowed the first dietary studies based on scientific principles to be made. Health and nutrition too began to feature in the political landscape with many governments and institutions taking a healthy interest in the pro-active feeding of the vulnerable, in particular children.

Part II then, aimed to provide supporting material of the concept of food security. The topics covered included three broad areas of interest: the socio-cultural evolution; governance, philosophy, politics and economics; and finally science and technology. What is perhaps surprising in an initial overview of these chapters is the sheer breadth of disciplines in which today's food security concept touches upon. Of interest too is the fact that this is not just an academic exercise; instead it highlights a continuum of ideas that have criss-crossed over time building momentum before reaching a critical mass of confluence in food and health issues around the eighteenth and nineteenth centuries. Throw into the mix emerging internationalism as well as collective social awakening and the makings of the modern concept of food security is born.

Part III

History: Twentieth Century

What's the use you learning to do right,
when it's troublesome to do right
and ain't no trouble to do wrong,
and the wages is just the same?

Mark Twain
The Adventures of Huckleberry Finn

While the previous sections outlined some of the key underlying and fundamental dimensions that acted as a backdrop for the emergence of the food security concept, the twentieth century is a period that lays claim to the popularisation of the concept of food for all. With this in mind, this section aims to catalogue the evolution of the concept in chronological order from several perspectives: sociological, political, economic, and science and technology. For this purpose, the twentieth century is divided not by any conventional means but into periods that best advance the narrative. The purpose of this is to highlight and to evaluate the contemporary motivational changes that have shaped the construct of today's environment. Such analysis of the societal influences too helps to further emphasise and integrate the multi-dimensional nature of the phenomena and also helps with further evaluation of the issues in the proceeding sections. The story unfolds mainly from a Western perspective not by design but by circumstance as much of the prevailing contemporary thought on nutritional, political and economic issues stem from the developed world's view. Furthermore, it is important to understand too that just because events unfold in chronological proximity, that one necessarily influenced the other cannot be inferred. Moreover, as far as was prudent I have attempted to mimic the language of the day, with the use of such words, phrases and terminologies as hunger and malnutrition or protein-calorie-malnutrition, etc. as I believe it helps contextualise contemporary thought and understanding.

8

Twentieth Century: The Feeding of Nations—A New Global Enthusiasm

8.1 Background

Although the emerging science of nutrition at the time might have relegated hunger to that of a purely technical matter, experts' inference that such *mal*-nutrition might be avoided through proper planning essentially concreted hungers political trajectory. Instead and in large part due to a collective social awakening and in part thanks to a growing body of sensationalist journalism, hunger issues were now receiving a great deal of attention. This effectively helped fully politicise the notion of hunger and starvation and spurred interventionism at the government level in many countries (Aronson 1982; Vernon 2007). However, interventionism itself was not new, and it has already been shown that governments, particularly Western governments, had for a long time been involved in agricultural markets with the Corn Laws in the United Kingdom and elsewhere through the exercising of political and economic strategies of tariffs and protectionism (Barnes 2006; Vaidya 2006). Thus, although agricultural trade, hunger and humanitarianism descended onto the political arena, at that time adequate food was still largely seen as an issue of production and supply rather than of equitable access.

8.2 Governance, Politics and Socioeconomics

Stagnating yields, war, national policy, agricultural specialisation and increased nutritional understanding all colluded to change the landscape of food production, policy and intervention at around the turn of the century. Comparable to other industries too, the close of the previous century also coincided with the end of a long period of low farm incomes, especially in the United States (Gardner 2002). It was also a time of exponentially increasing population growth and despite previous advances in agriculture, stagnating crop yields began to draw attention to agricultures' potential limitations in providing for this growth. In terms of trade too, many focused on the negative aspects of global business both for the economy and by extension agriculture and food security (Michie and Smith 2000). Increased reliance on international trade, suggested some, threatened to foster over-dependence and expose importing nations to economic or trade shocks. Fundamentally, according to Michie and Smith (2000) and in keeping with the prevailing economic theories of the day, imports were seen as a drain on foreign earnings while growing exports were seen to increase a nation's overall wealth. An over-reliance on imports, it was thought, could hinder domestic growth and if allowed to accumulate could lead to ever-increasing 'trade gaps' or balance of payments disequilibrium (Michie and Smith 2000). Simultaneously at about this time increased calls for self-sufficiency in terms of agricultural production ushered in a period of much innovation. America, for instance, reliant on Britain up to this point for much of its seed, livestock and technology branched out with the aid of collective societies that successfully lobbied for the establishment of dedicated governmental agencies of agriculture. Moreover, with the aid of government intervention, US farmers also invested heavily in new land and equipment at this time. In Britain and Europe too, specialised agencies and agricultural acts of the late 1800s and early 1900s emphasised the importance of food production and industry protection through imports and export tariffs. This also led to the first real international organisation interested in agricultural issues. The International Institute of Agriculture

(IIA) which was founded by David Lubin, a Polish-American merchant, in 1905 acted as an international clearing house of information. The institute helped raise international awareness of agricultural data and increased the importance of global agriculture's profile (Britannica 2009).

One of the combined consequences of the earlier agricultural and industrial revolutions together with agricultural trade policies of the time was the widespread displacement of many long-established trading partners. This came about through the increased efficient use of resources and specialisation, which in turn ultimately witnessed many changes in the locations of traditional agricultural suppliers (Michie and Smith 2000). When grain production became unprofitable for European farmers for instance, the Australians and North Americans stepped in providing cheap and reliable substitutes. Many European farmers too responded by specialisation in dairying, cheese making and other products, which allowed them to compete globally (Clayton and Black 1943). In effect, the demand and specialisation for food and raw materials became pivotal in the realignment of world trade. Thus, the prevailing pre-occupation with food production and supply at the time led policy makers in the industrialised world to invest much effort in securing, through acts and legislation, adequate food provisions for itself and its allies or close trading partners.

This was also a period that saw governments become more involved not just in agricultural issues but those of the nations health too. Although involved in public food and health policy since the end of the previous century, it took the advent of the World War I for governments, particularly the British and American governments, to become truly engaged in the population's diet and health (Beardsworth and Keil 1997). Up till this point, the feeding of nations was considered largely the responsibility of market-driven forces of supply and demand and little regard was actually devoted to the quality of the diet itself. However, Professor Ernest Starling, University College, London, noted in 1919 that with the advent of the Great War, governments awoke to the monumental difficulties of properly nourishing the population in terms of both quantity and quality. It was also realised, according to Starling, that to leave this in the hands of the market in times of war was just too great a risk (Starling 1919a). His rationale focused on the fact that although most previous conflicts only affected small regional populations and communities, the constant threat of famine and pestilence had still endured. Scaling this up to the Great War, Starling inferred, when every able-bodied man was signed up and the majority of the rest of the population was involved in support industries, governments had to realise the enormously damaging potential of doing nothing (Starling 1919a).

Thus, the emergence of specialisation together with the onset of war provided both problems and opportunities in food production, supply, security and incomes. During the World War I too, as able farmers were diverted to war and machinery and fertilisers became scarce, so agricultural production fell in Europe (excluding Britain) by as much as a third in some areas (Pollen 1917; Pinchot 1918; Dewey 1989). Simultaneously however, the American agricultural industry thrived and as the war progressed so the United States was well placed to become the chief suppliers of food for the warring nations of Europe (Encyclopedia 2006). It was about this time too that the United States nationalised their grain trades and suspended trading in commodity futures and assumed complete control of grain futures in order that a steady supply of grain at fixed prices was available to the Allied Forces. Although strictly a temporary measure, this marked a significant shift in thinking away from *laissez-faire* free trade to considerable interventionism (Santos 2006).

8.2.1 Rationing and Food Control Measures

Back in Britain, as a means of mitigating the worst effects of perceived in security during the war, the Defence of the Realm Act (DORA) was introduced in 1914. It was initially introduced as an aid in the general war effort and which saw the government effectively taking control of key industry logistics; particularly in the supply of materials. DORA also came to play an important role in the area of the food supply too. This fitted with the plans of the new Asquith coalition government of 1915, which tentatively also favoured a more pro-active interventionist food policy in which more land was to be cultivated and price supports put in place. However, according to Dewey (1989), it was not until the appointment of Lloyd George as Prime Minister in 1916 that the full measures of interventionism came into force with a raft of policies to stimulate production, guarantee prices and if needed, compulsory powers to enforce cultivation (Clynes 1920; Dewey 1989).

Before the war, Britain's initial food situation was fairly stable. Britain would regularly grow about one-fifth of the wheat and one-half the country's total consumption of beef, mutton and bacon. This continued through the early stages of hostilities with Britain still able to import a further 13 million tons of food annually, relying heavily on the United States and Canada for as much as 65% of its total food import needs (Pollen 1917; Pinchot 1918). Even up to midway through the war, Britain continued to appear to have a 'laissez-faire' attitude to food policy. It was not suffering the same shortages as continental Europe and a steady import trade from the America's and West Indies among others continued to secure sufficient provisions (Starling 1919b; Dewey 1989). However in 1916, after a poor harvest, an increasingly successful U-boat campaign and reduced allied harvests, things changed. Food shortages began to disrupt daily lives. This was exacerbated by panic buying and black market profiteering and it became clear that DORA was failing (Clynes 1920). In response, the government formed the Ministry of Food in 1916/17 with the purpose of regulating the supply and consumption of food whilst also attempting to stimulate production. Up till this point (December 1916), the UK's food policy fell under the auspices of the Board of Agriculture, which in turn was a product of the Board of Agriculture Act of 1889. In the United States, similar controls were implemented under the Lever Food Control Act of 1917 whereby the Department of Agriculture retained control of all on-farm activities while market organisation and pricing became the preserve of the Food Administration (Clayton and Black 1943).

Britain and America were not alone either with many other countries in the continents of Europe and Africa also being hard hit in the war of attrition (Lysaght et al. 1917; Starling 1919a; Cary 1920; Bennett 1949). Other countries, such as Armenia, Belgium, Mount-Lebanon and even Germany suffered widespread persecution and famine. Whether from policy failure or war itself is difficult to determine yet during this period many people died from tuberculosis, rickets, influenza, dysentery, scurvy, keratomalacia and hunger. There were even reports of cannibalism and exhumed bodies being used for food (Lysaght et al. 1917; Starling 1919a; Cary 1920; Bennett 1949). As a result of these experiences, it has been noted that for the preservation of food and economic security, especially during wartime, food and agriculture came to be seen as an industry of significant strategic economic and social importance (Tanner 2004).

Back in Britain, even despite these realisations the assumption of the control of the country's food supply by the government was initially slow. Even before the Ministry of Food was formally set up in 1916, attempts were made to regulate the UK's food supplies. In this endeavour the Royal Sugar Commission was appointed in August 1914 followed by the Royal Commission on Wheat Supplies in October 1916. However, it was not until 1917 with the appointment of Lord Rhondda as Food Controller did nutritional scientific principles first find their proper place in the determination of policy in Britain. Even despite all this and regardless of increasing government intervention in food production and market mechanisms and the subsequent modifications to DORA, it eventually became necessary, towards the end of the war for the government, to introduce a voluntary rationing scheme, which was ultimately made compulsory in January 1918 (Starling 1919b; Howard 2002; National Archives 2011). In the United States too, the commitment to provide for export contracts and their own servicemen meant that rationing in the United States was not to be avoided either (Armour 1917). High-profile nutritionists and physiologists like Graham Lusk had conceded that while Americans had enjoyed a better standard of nutrition than Europe before the war, the time had come to advocate balance and moderation of diet. This included reduced meat intake for better health and economy as well as decreased consumption of wheat for the benefit of all in the inter-allied war effort (Lusk 1918). For an initial few years during the war at least, the US agricultural industry, being the main provider of staple foods to Europe, flourished and farm incomes grew (Black 1942). However, the end of the war saw a sudden decline in demand for American goods, which further placed downward pressures on prices and incomes at the time (Encyclopedia 2006).

In the midst of all this, two fledgling movements—the bio-dynamic and organic movements—owed their beginnings to the popular ideas of Rudolf Steiner.

8.2.2 Bio-Dynamics and Organic Farming

Organic farming was the norm in agriculture since early man stopped hunting and gathering and began growing his own food. This changed after about 1845 with the agricultural and chemical revolutions spearheaded by Leibig and others. This brought with it a whole new industrial approach to farming and crop cultivation. In fact, the prevailing approach advocated the intensification and industrialisation on an unprecedented scale with the use of synthetic or chemical pesticides, herbicides and fertilisers. However, by 1920, the common practice of synthetic fertiliser use led to some European farmers becoming disillusioned with what they saw as reduced seed vitality, declining animal health, pest as well as disease problems including to some, a loss of flavour. To help with the problem, a group of land owners approached Dr. Rudolf Steiner, an Austrian responsible for the Steiner schools movement, to ask for his help. A short while later, Steiner gave a series of eight lectures in which the fundamentals of a new agricultural method; bio-dynamics, were outlined. Steiner's particular viewpoint embraced a spiritual, almost ethereal attitude to agriculture fully embraced in a holistic approach to the cycle of life. Thus, by treating the Earth as a single organism, for Steiner the regeneration of life through agriculture was a simple process of harmonising the elements contained within, as the following quote from his fifth lecture reveals:

> Spiritual Science always tries to look into the effects of living things on a large scale … the wide circumference of Nature's workings—that is the talk of Spiritual Science. But we must first know how to penetrate into these wider workings of Nature. (Steiner 1958)

Thus, in the genesis of Steiner's bio-dynamic beginnings, a full picture of the complex, inter-related relationships at work in nature is elucidated with an almost religious zeal (Steiner 1958; Bradshaw 2003).

8.3 Science and Technology

Technology at the turn of the century was moving at a fast pace. Indeed, in these heady times one month's scientific breakthrough could well end up being another month's policy. Technological advances and improved infrastructure too brought on by the wide uptake of technology in steam, diesel and electric power together with refrigeration and other food preservation techniques allowed previously remote producers, suppliers and markets to trade more effectively. Indeed, advances in technological thinking was no more evident than in the field of biotechnology.

8.3.1 Biotechnology

At the beginning of the twentieth century, biotechnology began to close the gap between industry and agriculture. Industries were now employing the latest biotechnological methods on an industrial scale to aid in the provision of solutions to food shortages and other societal crises. Biotechnologies also paved the way for bacterially generated key industrial chemicals such as glycerol, acetone and butanol. This was particularly advanced in Germany where acetone produced by plants was used to make bombs (or paint solvents). Also in Britain and Germany major outbreaks of disease within overcrowded industrial cities led to the introduction of sewage purification systems based on microbial yeast (glycerol) activity (Bud 1993). Furthermore, pioneering research by Max Delbrück and colleagues realised the potential of surplus brewer's yeast (used in the brewing industry) as a feeding supplement for animals. This proved enormously beneficial for Germany too during the war as they managed to replace as much as half of their imported protein needs for animal feed with surplus brewer's yeast (Ugaldea and Castrillob 2002).

Importantly too, as mentioned earlier around 1900, three scientists, Hugo DeVries, Erich Von Tschermak and Carl Correns, independently rediscovered Mendel's seminal work and consequentially the modern notion of genetics was born. Other important work around this time included advances in the fields of chromosomes, fertilisation, genes and bacteria; see Table 8.1. Lastly, the term 'biotechnology' was finally coined by Hungarian agricultural engineer Károly (Karl) Ereky in 1917 in two publications: one 'Food Production and Agriculture' and the other 'Large Scale Development of Pig Fattening' (Rao 2008). At the time its meaning represented the lines of enquiry by which products could be produced from raw materials with the aid of living organisms (Fári and Kralovánszky 2006).

8.4 Health and Nutrition

During the nineteenth century and coinciding with the free movement of individuals through better railway links, communicable diseases spread widely across Europe. Initial international health efforts were slow and beset with poor international agreement although one of the first partially successful attempts was the International Sanitary Conference in Paris in 1851 (WHO 2010). The notion was borne out of the idea that

TABLE 8.1

Key Early Dates in Biotechnology

1900	The science of genetics was effectively born when Mendel's work was rediscovered by three scientists—Hugo DeVries, Erich Von Tschermak, and Carl Correns.
1901	Beijerinck identified free-living aerobic nitrogen fixers.
1902	The term 'immunology' first appears.
1902	Walter Stanborough Sutton suggested that Mendel's 'factors' were located on chromosomes and that these were paired and may even be the carriers of heredity. Sutton eventually gave Mendel's 'factors' the name we use today—'genes'.
1902	Archibald Garrod firmed up the connection between Mendelian heredity and the biochemical pathways of reproduction in the organisms.
1903	Independently, Walter Sutton and Theodor Boveri proposed that each egg or sperm cell contained one of each of the chromosome pairs, thus pairing Mendel's factors or genes with the fertilisation of the egg and ergo heredity.
1905	Edmund Wilson and Nellie Stevens proposed that separate X and Y chromosomes were responsible for determining sex.
1906	The term 'genetics' is introduced.
1907	The first in vivo culture of animal cells is reported.
1909	Genes are linked with hereditary disorders.
1909	Wilhelm Johannsen building on Suttons terminology defined the terms 'gene' to describe the carrier of heredity; 'genotype' to describe the genetic makeup of an organism and 'phenotype' to describe the actual organism.
1910	Thomas Hunt Morgan proved that indeed Sutton's ideas that genes were carried on chromosomes was correct.
1912	Lawrence Bragg saw that X-rays could be used to study the molecular structure of simple crystalline substances which eventually led to the development of X-ray crystallography and the ability to further explore the three-dimensional structures of nucleic acids and proteins.
1914	Bacteria are used to treat sewage for the first time in England.
1915	Phages, or bacterial viruses, are discovered.
1917	The term 'biotechnology' was coined by Hungarian agricultural engineer Károly (Karl) Ereky in 1917.
1918	The German army employed acetone produced by plants to make bombs.

Sources: Bud, R., *The Uses of Life: A Hstory of Biotechnology*, Cambridge University Press, Cambridge, 1993; Thackray, A., *Private Science: Biotechnology and the Rise of the Molecular Sciences*, University of Pennsylvania Press, Pennsylvania, 1998; Fári et al., History of the term biotechnology: K. Ereky & his contribution, Presentation at the Fourth Congress of Redbio, Goiânia, Brazil, 2001; Fári, M.G. and Kralovánszky, U.P., *Int. J. Hort. Sci.*, 12, 9–12, 2006; BioTech Institute, *Timeline of Biotechnolgy: 20th Century*. Biotechnology Institute, Washington, DC, 2010. http://www.biotechinstitute.org/what_is/timeline.html

would see nations protected from epidemics without overly hindering trade. Several European countries met to set up protocols and determine appropriate quarantine precautions against the spread of cholera. While in principle the sanitary code was agreed upon, it was never actually ratified. A few other sanitary conferences between 1851 and the close of the century, as well as a other isolated attempts at international cooperation offered only small successes however, collectively they did set important precedents for governance in the global health arena. Building on this, perhaps the first real discussion in favour of a permanent international agency to deal exclusively with questions of health was considered at the fourth International Sanitary Conference held in Vienna in 1874. Although despite the early recommendations, it was not until 1903 that such a body was firmly agreed upon and which ultimately led to the creation in Paris in 1907 of the Office International d'Hygiène Publique (OIHP). Among the many functions of the OIHP, they were to continue the international sanitary conferences and implement their conventions as well as conducting studies on epidemiological diseases. They were also particularly active in the attempt, through quarantine protocols to prevent the spread of plague and cholera. Thus through the organisation of 12 European countries the OIHP became the first real successful international health organisation (EoN 2009).

8.4.1 International Standards of Dietary Nutrition

Simultaneous to the increasing involvement of production and supply considerations by the British government, several events coincided to promote a new philosophy regarding the quality of the diet and in turn the nation's nutritional health. This was spurred on in part by the rejection, on the grounds of poor health, of more than one-third of British applicants to fight in the Boer War (1899–1902). The government was horrified at what in reality was widespread malnutrition and responded with a more proactive public policy including free school meals and infant welfare centre grants. Similarly, in 1917–1918, as a prelude to conscription 2.5 million Britain's were also deemed 'C3' or unfit for military service. Once again, the problems of nutrition shocked the government, and it helped further politicise public health issues. This promoted an increasing interventionist stance from the government through monitoring and regulating of suitable dietary standards (Starling 1918; Smith 2009). About this time too, other countries began to adopt their long-recognised duty to relieve famine, implement food safety measures and take responsibility for the surveillance and nutritional status of their populations, especially when it came to children (Acheson 1986).

At this pivotal time, governments around the world began looking to adopt nutritional guidelines based upon scientific principles. Fortuitously, in this endeavour, new technology and understanding was emerging in the field of food microbiology, health and nutrition, particularly in the area of vitamins and minerals (Beardsworth and Keil 1997). This came about as nutritionists' interests were being partly diverted by the increasing chemical and physiological understanding of micro-organisms (DuBois 1940; Passmore 1982). It began with the previous nineteenth century's nutritional research in which scientists had realised that healthy diets needed to include proteins, fats and carbohydrates (Carpenter 2004). However, simultaneous research into deficiency diseases such as scurvy, pellagra and beriberi also led to the realisation in 1901 that

> ... there occur in various natural foods substances which cannot be absent without serious injury to the peripheral nervous system ... (Grijns 1936, p. 38)

Thus, pioneering work led by notable scientists including Eijkman, Grijns, Hopkins, Funk, McCollum, Osborne and Mendel led to the discovery of 'vitamines' (as it was then called), nutrients and other essential minerals and trace elements. Collectively such advances provided a solid foundation on which the relatively new science of health and nutrition grew (Carpenter 2003). From this point, for the next two decades, many of the more common vitamins were discovered, the importance of which was subsequently confirmed as necessary for the vitality and general health of the public (Table 8.2) (Carpenter 2003). This also established this period as what come to be known as 'the golden age of nutrition' (Carpenter 2003; Ordovas and Mooser 2004; Go et al. 2005).

Meanwhile caloric studies around the early part of the century in Russia, Sweden, France, Italy, Japan and others, writes Chittenden, were all conducting research that appeared to back up the 3000 calorie per day allowance set by Voit two decades earlier (Chittenden 1906, 1907). However, there was still no widely held universal consensus, leading Chittenden to bemoan what he saw as a general lack of

TABLE 8.2

Timeline for the Discovery of the Vitamins up to 1944

Vitamin	Year of Discovery or Proposal
Thiamin	1901
Vitamin C	1907
Vitamin A	1915
Vitamin D	1919
Vitamin E	1922
Niacin	1926
Biotin	1926
Vitamin K	1929
Pantothenic acid	1931
Folate	1931
Riboflavin	1933
Vitamin B_6	1934

Source: Based on Carpenter, K.J., *J. Nutr.*, 133, 3023–3032, 2003.

scientific evidence and suggesting that one man's instinct, habit or craving for foods did not represent sufficient scientific rigour to support many prominent views of dietary needs of the time (Chittenden 1904, 1906, 1907; Carpenter et al. 1997). Another cautionary note at the time from Chittenden led him to question just how such dietary studies in general actually prescribed the real needs of the body beyond just throwing light upon existing dietary habits. He cites as supporting this argument his own prophetic studies, which showed that the body's physiological needs could in fact be met with much lower standards of diet than the increasingly generally accepted 3000 calories of the time (Chittenden 1904; Carpenter et al. 1997). Lusk too, a student of Voit and someone who did more to highlight the field of nutrition than almost any other individual around this time, like Chittenden also saw 3000 calories daily as a little excessive. Their respective recommendations can be seen in Table 8.3.

With regard to the advancement of metabolic studies, two men, Arthur Harris and Francis Benedict, teamed up at the Carnegie Institution of Washington to collaborate on the analysis of certain biometric physiological measurements to determine a formula for calculating predictive basal metabolic rates (BMR) without the need for calorimetry. The subsequent paper published in 1918 stands up after nearly 100 years and indeed Harris and Benedict's formula is still used in many areas of BMR analysis today. Harris and Benedict's extensive statistical treatment of the data for 136 men, 103 women and 94 newborn infants found a link between the surface area of the body, a person's height and their weight. The formula made by the use of the multiple regression equations offered BMR values equal to the following:

For men, BMR $= 66.4730 + 13.7516\,w + 5.0033\,s - 6.7550\,a$

For women, BMR $= 655.0955 + 9.5634\,w + 1.8496\,s - 4.6756\,a$,

where BMR is the total heat production per 24 hours, w is the weight in kilograms, s is the stature in centimetres and a is the age in years (Harris and Benedict 1918).

All this research was to prove its worth as shortly afterward amidst continuing concerns over malnourishment, the Royal Society (of Britain) founded the Physiological War Committee in 1914 to advise the current government on questions involving physiological principles of nutrition (Royal Society 1917). The Royal Society was asked by the president of the board of trade to draw up recommendations for minimum dietary requirements. In response, a subcommittee of the Royal Society's War Committee the Food Committee was set up in 1916 (Starling 1919b; Beardsworth and Keil 1997; Moore 2008). After some research the committee concluded that

> … [We] are convinced that the dietary requirements of a nation engaged on active work cannot 'be satisfactorily met on a less supply … than 100 grams protein, 100 grams fat, 500 grams carbohydrate, equal approximately to 3,400 calories per 'man' per day. (Royal Society 1917)

However for an average woman doing an average day's work, the society recommended a total of 2400 calories per woman per day (Table 8.4) (Starling 1919a).

TABLE 8.3

Early Lusk and Chittenden Daily Nutritional Values

Date/ Authority	Load (Default Man)	Protein	Fat	Carbohydrate	Calories
1904–07: Russell Chittenden					
Chittenden's studies concluded that equilibrium for 'the brain worker' could be maintained on much less than the often-quoted 3000 calories per day and proposed only half of the 118 grams proposed by Atwater. For himself being sedentary in nature, Chittenden believed 2000 calories was adequate (Chittenden 1904, 1906, 1907; Lusk 1906; JoN 1944)					
		>50			2500–2600
1906: Graham Lusk (Lusk 1906; Hutchinson 1916)					
					2488–2562

TABLE 8.4

Royal Society's Daily Recommended Food Allowance

Date/Authority	Load (Default Man)	Protein	Fat	Carbohydrate	Calories
1917–18: Royal Society (of London)					
The Royal Society believed that approximately 3000–3400 calories per man based on modified Atwater's figures were sufficient at about this time					
	100	100	500	3000–3400

Sources: Royal Society, *The Food Supply of the United Kingdom: A Report Drawn Up by a Committee of the Royal Society*, Royal Society, London, 1917; Starling, E.H., *The Oliver-Sharpey Lectures on the Feeding of Nations: A Study in Applied Physiology*, Longmans, Green and Co, London, 1919; DuBois, E.F., Biographical memoir of Graham Lusk, 1866–1932. In *National Academy of Sciences Biographical Memoirs*, Vol. XXI, 3d Memoir, Biographical Memoirs, National Academy of Sciences, Washington, DC, 1940.

TABLE 8.5

Inter-Allied Scientific Table of Daily Calorie Requirements

	Calories	
Group	**Gross (allowing for waste)**	**Net**
Male		
Sedentary	2750	2500
Heavy	3500–5500	3200–5000
Soldier		3700–3800
Average Male	3300–3400	3000
Female		
Laundress		3291
Typewriter	2100	1900
Average Female	2650	2400
Children		
0–6	1650	1500
6–10	2100	2310
10–13	2500	2750
14+	As average adults	

Source: Recreated from Starling, E.H., *The Feeding of Nations*, Longmans, Green and Co, London, 1919.

Thus, the impetus for national nutritional standards was founded, humanitarianism notwithstanding, in the early war efforts of both the Boer and the Great Wars (Beaton 1991). This took on further inertia when in 1917 it was decided to pool all allied nations' nutritional supplies and resources in order to achieve complete unity of action (Starling 1919a). To this end, the Inter-allied Conference at Versailles in November 1917 appointed an Inter-allied Scientific Food Commission whose task was to examine, inter alia, the nutritional needs of nations with a view to the planning of food shipments from North America to Europe (Starling 1919b; Beaton 1991; Iacobbo 2004).

The Inter-allied Scientific Food Commission met in Paris in 1918 and drawing heavily on the work of the Royal Society adopted many of its recommendations on food requirements. The first problem they perceived was how to calculate the nutritional requirements of apparent heterogeneous populations where individuals' energy needs vary according to '… age, sex, size, occupation and environment'. Aided by Chittendon and Lusk (US representatives of the Commission) (G.R.C. 1944) the solution it seemed was to simplify the problem into two component parts—that of an average person's basal metabolic rate (BMR) and second to add any additional requirements based on the type of work to be achieved. The commission calculated an average person's BMR to be 1687 calories per day. Factoring in an average days work and the results came in at just over 3000 calories (3136.8). However, this did not take into account the differing occupational requirements; Table 8.5 highlights the results of the commission's recommendations.

8.5 Period Summary

The turn of the century was a seminal time for food and nutrition security. These remarkable and humble beginnings could only have happened with the fortuitous confluence that emerged from two centuries of scientific research, political and economic thought as well as social enlightenment and public pressure. Accordingly many governments of the day were willingly or unwittingly fully drawn into the food security debate. On top of this, early internationalism in the form of the Office International d'Hygiène had set the bar for multilateral cooperation in aspects of health and sanitation. Further bilateral and multilateral governance of food issues was necessitated by the war, which effectively showed that in times of hardship and uncertainty the national planning of the food resource was vital for the nation's health and economy. Nutritional science too was advancing in great leaps, the discovery of vitamins and other nutrients as well as the of scientific application of nutritional principles firmly imbued this new science with kudos and credibility. Moreover, the first tentative steps in introducing a semi-international standard of dietary guidelines was furthered by the likes of the Royal Society and the Inter-allied Commissions. Indeed, much of today's solid nutritional groundings owes a great deal to this early period.

8.6 Key Dates, People, Acts, Reports and Surveys

Table 8.6 highlights some of the more important events as well as noteworthy individuals and key reports, etc. that occurred during this period.

TABLE 8.6

Key Dates: Turn of the Century

Date	People, Organisations, Agencies, Acts or Conventions
1905: Education (Provision of Meals) Act	In England the 1905 Education (Provision of Meals) Act was the result of the efforts of 365 private, charitable organisations to improve child nutrition.
1908: International Institute of Agriculture (IIA)	Agricultural reformer David Lubin's activities led to the inception (1905) of the International Institute of Agriculture to ameliorate conditions of rural life and act as a world clearinghouse for information on crops, prices, and trade.
1909: The Haber-Bosch Process	A process developed by Fritz Haber and Carl Bosch that directly synthesised ammonia from hydrogen gas and atmospheric nitrogen to be used in the manufacture of fertiliser. First used industrially in 1913 and then by the German government during the War.
1914: Royal Society's War Committee	The Royal Society's (of London) Physiological War Committee was initially formed in 1914 to outline a plan to blockade German food supplies. In 1916, a sub-committee, the Food (War) Committee was set up to advise the government on questions of food and nutrition.
1914: The UK Defence of the Realm Act (DORA)	The Defence of the Realm Act (DORA) was introduced in 1914 citing everything people were not allowed to do in time of war both in terms of security and to prevent food shortages. As WW1 progressed so DORA began to fail. Panic buying and a healthy black market distorted the equitable distribution of food so DORA was modified. However, this did not work either and with increasing malnutrition the authorities, through the Ministry of Food introduced rationing in 1918. The strategy worked and no one starved as a result.
1916: Ministry of Food	The Board of Agriculture, (later the Ministry of Agriculture, Fisheries and Food (MAFF)), was established under the Board of Agriculture Act 1889. It was preceded, however, by an earlier Board of Agriculture, founded by *Royal Charter* on 23 August 1793 as the Board or Society for the Encouragement of Agriculture and Internal Improvement, which lasted until it was dissolved in June 1822. Under the Board, the Ministry of Food was created to regulate the supply and consumption of food and to encourage food production.
1916: US Federal Farm Loan Act	The first massive intervention in US agriculture was the Federal Farm Loan System.

continued

TABLE 8.6 (Continued)

Key Dates: Turn of the Century

Date	People, Organisation, Agencies, Acts or Conventions
1917: The Lever Food Act	Officially known as 'An Act to Provide Further for the National Security and Defense by Encouraging the Production, Conserving the Supply, and Controlling the Distribution of Food Products and Fuel', the Lever Food Control Act of 1917 authorised the US regulation of price, production, transportation, and allocation of feeds, food, fuel and beverages for the remainder of World War I.
1917: Inter-allied Scientific Food Commission	Inter-allied Scientific Food Commission with members of Great Britain, the United States, Italy and France were tasked with examining the scientific aspect of the Inter-allied food problem and to propose any appropriate measures.

Sources: Baldwin, S.E., *Am. J. Int. Law*, 1, 808–829, 1907; Kates, C.S., *Ann. Am. Soc. Sci. Acad. Polit. Social Sci.*, 40, 110–116, 1912; Starling, E.H., *The Oliver-Sharpey Lectures on the Feeding of Nations: A Study in Applied Physiology*, Longmans, Green and Co, London, 1919; Gunderson, G.W., *The National School Lunch Program: Background and Development*. Ova Science Publishers, New York, 2003; Moore, K., *Notes Rec. Royal Soc.*, 62, 315–319, 2008; Smith, R., *Stud. Hist. Philos. Sci. C: Stud. Hist. Philos. Biol. Biomed. Sci.*, 40, 179–189, 2009.

References

Acheson, E. D. (1986). Tenth Boyd Orr Memorial Lecture: Food Policy, Nutrition and Government. *Proceedings of the Nutrition Society* 45: 131–138.

Armour, J. O. (1917). Food Shortage: An Appeal to Physicians. *The Journal of the American Association* LXVIII(18): 1339–1340.

Aronson, N. (1982). Nutrition as a Social Problem: A Case Study of Entrepreneurial Strategy in Science. *Social Problems* 29(5): 474–487.

Baldwin, S. E. (1907). The International Congresses and Conferences of the Last Century as Forces Working toward the Solidarity of the World. *The American Journal of International Law* 1(3): 808–829.

Barnes, D. G. (2006). *A History of English Corn Laws: From 1660–1846*. London: Routledge.

Beardsworth, A. and T. Keil (1997). *Sociology on the Menu: An Invitation to the Study of Food and Society*. London: Routledge.

Beaton, G. H. (1991). Human Nutrient Requirement Estimates: Derivation, Interpretation and Application in Evolutionary Perspective. *Food, Nutrition and Agriculture* (2/3): 3–15.

Bennett, M. K. (1949). Food and Agriculture in the Soviet Union, 1917–48. *The Journal of Political Economy* 57(3): 185–198.

BioTech Institute (2010). *Timeline of Biotechnolgy: 20th Century*. Washington, DC: Biotechnology Institute. http://www.biotechinstitute.org/what_is/timeline.html (accessed 12 January 2011).

Black, J. D. (1942). *Parity, Parity, Parity*. Cambridge, MA: The Harvard Committee on research in the social Sciences.

Bradshaw, J. (2003). A Brief History of Bio-Dynamics—An Australian Perspective. *Biodynamic Growing* 1: 4–6.

Britannica (2009). Encyclopædia Britannica Online.

Bud, R. (1993). *The Uses of Life: A Hstory of Biotechnology*. Cambridge: Cambridge University Press.

Carpenter, K. J. (2003). A Short History of Nutritional Science: Part 3 (1912–1944). *Journal of Nutrition* 133(Oct.): 3023–3032.

Carpenter, K. J. (2004) The Nobel Prize and the Discovery of Vitamins. *Nobel Prize*: http://nobelprize.org/nobel_prizes/medicine/articles/carpenter/index.html.

Carpenter, K. J., A. E. Harper and R. E. Olsondagger (1997). Experiments That Changed Nutritional Thinking. *The Journal of Nutrition* 127(5): 1017S–1053S.

Cary, R. L. (1920). Social and Industrial Conditions in the Germany of Today. *Annals of the American Academy of Political and Social Science* 92: 157–162.

Chittenden, R. H. (1904). *Physiological Economy in Nutrition: With Special Reference to the Minimal Proteid Requirement of the Healthy Man. An Experimental Study*. New York: Frederick A. Stokes Co.

Chittenden, R. H. (1906). A Discussion on Over-Nutrition and Under-Nutrition. *The British Medical Journal* 2(2391): 1100–1103.

Chittenden, R. H. (1907). *Physiological Economy in Nutrition, with Special Reference to the Minimal Proteid Requirement of the Healthy Man.* New York: F. A. Stokes Co.

Clayton, C. F. and J. D. Black (1943). The Food Situation in a World at War: Wartime Food Administration-U. S. A. *The ANNALS of the American Academy of Political and Social Science* 225: 96–105.

Clynes, J. R. (1920). Food Control in War and Peace. *The Economic Journal* 30(118): 147–155.

Dewey, P. E. (1989). *British Agriculture in the First World War.* London: Routledge.

DuBois, E. F. (1940). Biographical Memoir of Graham Lusk, 1866–1932. In *National Academy of Sciences Biographical Memoirs, Vol. XXI, 3d Memoir, Biographical Memoirs (National Academy of Sciences (US)).* Washington, D.C: National Academy of Sciences.

Encyclopedia (2006) Agriculture. *Funk & Wagnalls New Encyclopedia.* http://www.history.com/encyclopedia. do?articleId=200414.

EoN (2009). *Economic and Social Development: Regional Commissions.* New York: United Nations.

Fári, M. G., R. Bud and P. U. Kralovánszky (2001). *History of the Term Biotechnology: K. Ereky & His Contribution: Presentation at the Fourth Congress of Redbio* Goiânia, Brazil.

Fári, M. G. and U. P. Kralovánszky (2006). The Founding Father of Biotechnology: Károly (Karl) Ereky. *International Journal of Horticultural Science* 12(1): 9–12.

G.R.C. (1944). Russel Henry Chittenden: February 18 1856- December 26, 1943: An Appreciation. *The Journal of Nutrition* 28 (1): 2–6.

Gardner, B. L. (2002). *American Agriculture in the Twentieth Century: How It Flourished and What It Cost.* Cambridge, MA: Harvard University Press.

Go, V. L. W., C. T. H. Nguyen, D. M. Harris and W.-N. P. Lee (2005). Nutrient-Gene Interaction: Metabolic Genotype-Phenotype Relationship. *Journal of Nutrition* 135(Supplement: International Conference on Diet, Nutrition, and Cancer): 3016S–3020S.

Grijns, G. (1936). Prof. Dr. G. Grijns' Researches on Vitamins, 1900–1911, and His Thesis on the Physiology of the N. Opticus, Translated and Reedited by a Committee of Honour on Occasion of His 70th Birthday. *Journal of American Medical Association* 107(6): 453.

Gunderson, G. W. (2003). *The National School Lunch Program: Background and Development.* New York: Nova Science Publishers.

Harris, J. A. and F. G. Benedict (1918). A Biometric Study of Human Basal Metabolism. *Proceedings of the National Academy of Sciences* 4(12): 370–373.

Howard, M. E. (2002). *The First World War: A Very Short Introduction.* New York: Oxford University Press.

Hutchinson, R. (1916). *Food and the Principles of Diatetics.* New York: W. Wood and Co.

Iacobbo, M. (2004). *Vegetarian America: A History.* Westport, CT: Praeger Publishers.

JoN (1944). Russel Henry Chittenden: February 18, 1856–December 26, 1943. An Appreciation. *The Journal of Nutrition* 28(1): 2–6.

Kates, C. S. (1912). Origin and Growth of Rural Conferences. *The Annals of the American Social Science Academy of Political and Social Science* 40: 110–116.

Lusk, G. (1906). *The Elements of the Science of Nutrition.* Philadelphia and London: W B Saunders

Lusk, G. (1918). *Food in War Time.* Philadelphia and London: W.B. Saunders.

Lysaght, E. E., T. W. W. Bennett and R. Ball (1917). The Farmers and the Food Problem. *An Irish Quarterly Review* 6(21): 21–34.

Michie, J. and J. G. Smith, Eds. (2000). *Managing the Global Economy.* New York: Oxford University Press.

Moore, K. (2008). The Royal Society's War Committee on Engineering 1914–19. *Notes and Records of the Royal Society* 62(3): 315–319.

National Archives (2011). Britain 1906–1918: Defence of the Realm Act (Dora). British National Archives. http://www.nationalarchives.gov.uk/education/britain1906to1918/g5/background.htm (accessed 21 January 2011).

Ordovas, J. M. and V. Mooser (2004). Current Opinion in Lipidology. *Genetics and Molecular Biology* 15 (2): 101–108.

Passmore, R. (1982). Reflexions on Energy Balance. *Proceedings of the Nutrition Society* 41: 161–165.

Pinchot, G. (1918). Essentials to a Food Program for Next Year. *The Annals of the American Academy of Political and Social Science* 78: 156–163.

Pollen, A. (1917). The Food Problem of Great Britain; the Shipping Problem of the World. *The Annals of the American Academy of Political and Social Science* 74: 91–94.

Rao, C. K. (2008). Who Coined the Terms 'Biotechnology' and 'Genetic Engineering', and When? *Current Science* 95(11): 1512–1513.

Royal Society (1917). *The Food Supply of the United Kingdom; A Report Drawn up by a Committee of the Royal Society*. London: Royal Society.

Santos, J. (2006). Political Economy of Enterprise: Going against the Grain: Why Did Wheat Marketing in the United States and Canada Evolve So Differently? Business and Economic History Annual Conference, Toronto, Business History Conference.

Smith, R. (2009). The Emergence of Vitamins as Bio-Political Objects During World War I. *Studies in History and Philosophy of Science Part C: Studies in History and Philosophy of Biological and Biomedical Sciences* 40(3): 179–189.

Starling, E. H. (1918). The Significance of Fats in the Diet. *The British medical Journal* 2(3005): 105–107.

Starling, E. H. (1919a). *The Feeding of Nations*. London: Longmans, Green and Co.

Starling, E. H. (1919b). *The Oliver-Sharpey Lectures on the Feeding of Nations: A Study in Applied Physiology*. London: Longmans, Green and Co.

Steiner, R. (1958). *The Agriculture Course*. London: Bio-Dynamic Agricultural Association.

Tanner, J. (2004). *Incorporated Knowledge and the Making of the Consumer: Nutritional Science and Food Habits in the USA, Germany and Switzerland (1930s to 50s)*. Bielefeld, Germany: ZIF (Zentrum für Interdisziplinäre Forschung/Centre for Interdisciplinary Research).

Thackray, A. (1998). *Private Science: Biotechnology and the Rise of the Molecular Sciences*. Pennsylvania: University of Pennsylvania Press.

Ugaldea, U. O. and J. I. Castrillob (2002). Single Cell Proteins from Fungi and Yeasts. *Applied Mycology and Biotechnology* 2: 123–149.

Vaidya, A. K., Ed. (2006). *Globalization: Encyclopedia of Trade, Labor, and Politics*. Santa Barbara, CA: ABC-CLIO.

Vernon, J. (2007). *Hunger: A Modern History*. Cambridge, MA: Belknap Press.

WHO (2010). Website of the World Health Organization. http://www.who.int/ (accessed 2 April 2010).

9

The Inter-War Years: 1919–1939

9.1 Governance, Politics and Socioeconomics

In the arena of governance and politics, the cessation of hostilities proved to be a busy period in the fledgling food security concept. Post-war reconstruction as well as existing social and economic issues needed to be addressed on an unprecedented global scale. One of the first efforts in post-war global governance was the introduction of a new far-reaching institution aimed at promoting international collaboration on many fronts. Thus out of the embers of World War I grew the League of Nations.

9.1.1 The League of Nations 1919

Internationally it was felt that an organisation capable of promoting international peace and co-operation was needed (Hitchcock 1919; League of Nations 1919; Northedge 1986). It was conceived before the war ended, fleshed out under the Treaty of Versailles and formerly created at the 1919 Paris Peace Conference. It was the by far the most ambitious attempt to construct a peaceful global order to date. However, one major economic and political power, the United States failed to sign up. The Americans felt the creation of a global political power would restrict their freedom of action and more importantly subject the United States to control of what was seen as a superstate or *supra*-national power. Moreover, the United States felt it might be 'bound' to fight in unjust wars and on sides not of their choosing (White 1919). Yet even despite America's lack of involvement, the League worked surprisingly well. Several other institutions affiliated to the League were also created at around this time, they were: the International Labour Organization; the Permanent Court of International Justice; and the Health Organization.

The Covenant of the League of Nations was the League's mandated charter and it covered two overriding goals. The first was the establishment of a multilateral body in which the nations of the world could find refuge in discourse: essentially a place to settle their disagreements without recourse to conflict. In this endeavour it sought to secure or preserve peace through dialogue, arbitration and conciliation. In the second goal, the League aimed to foster international cooperation within economic and social fields (Hitchcock 1919; League of Nations 1919; Kohn 1924; Pipkin 1933). Although only a fledgling organisation it was the first real international forum in which issues of, among others, economic and social standing could be discussed and addressed. Its underlying concept, that of multilateral dialogue is one of the organisations founding principles and served as a blueprint for its successor organisation; the United Nations and its many imitators. Perhaps the League's greatest efforts were concentrated in the area of economic and social affairs and it was in this area that the League excelled; this is particularly noticeable in their work on health and nutrition (Section 9.2.2).

Meanwhile, in the field of Western agricultural policy, initial optimism associated with victory and the highly geared industrialisation of agriculture became subdued as surpluses and then depression crept in.

9.1.2 Surpluses and Depression

Increasing demand for agricultural products during the war continued into the post-war reconstruction period. Many government policies in the form of interventionism, price support and protectionism increased output and aimed to control costs. Indeed in America, despite legislation requiring middle men to sell all food products at cost plus a token pre-war profit margin, apparent profiteering from unscrupulous merchants created universal complaint from the American consumer. One commentator, arguing against excessive legislative measures, observed that

> The unalterable law of supply and demand will [ultimately] determine the price of food.
> (Crutchfield 1919)

However despite apparent anti-regulation sentiments Crutchfield, acknowledging the role of American industry and agriculture in a world of post-war reconstruction was actually in favour of the duality of regulated pricing and controlled demand. This he cited would help stabilise prices and supply at a time when the

> … the whole world is looking to the American market for an undue share of our supplies …
> (Crutchfield 1919)

Nevertheless, whether primarily through the 'inelasticity' of the agricultural sector or such Western government policies, growth in agricultural output towards the end of the war and for the next few years increased markedly (Holmes 1924; Ostrolenk 1930; Michie and Smith 2000). In Eastern Europe and much of the developing world however the story was different; poverty and depression continued unabated (Durand 1922; Taylor 1926). Indeed, agricultural shortfalls were common, and a growing trade gap and the resultant economic disparity between the developing and under-developed nations was also beginning to emerge.

   This was during a period (1919–1926) that saw over 6 million tons of food aid being shipped to Europe (Shaw 2007). While this was good news predominantly for American production, continued gains and resultant increases in yield and productivity from the industrialisation of agriculture however, eventually led to the 'West's' first real glut of agricultural surpluses (Ostrolenk 1930). Such overproduction was seen at the time as both a hindrance and an opportunity: a hindrance in the reduced prices and incomes for farmers and producers, and an opportunity in the potential for increasing existing and developing new export markets (Mayne 1947). Ultimately however, such surpluses ended up overheating an industry at a time when real prices were falling. On top of this falling demand from a near bankrupt Europe as well as increased competition coincided with the re-entry of military personnel into the civilian labour market. All these factors combined with changes in fiscal and monetary policies, particularly in America, brought about great unemployment and depressed purchasing power, eventually leading to a worldwide economic depression (see Section 9.1.4).

   In Russia at this time however, famine was lurking around the corner. For several years, after the Revolution in 1917 and the subsequent civil war, food crops especially wheat declined measurably. Further magnified by drought, massive crop failures caused widespread hunger and death. The initial disaster though was greatly exacerbated when Lenin refused to acknowledge the famine and sent no aid. It has been estimated since, that as many as between four and five million people died of famine during this period (Chamberlin 1934; Gantt 1936; Bennett 1949). Rapid recovery took place over the next few years before fully recovering by about 1927/28 to pre-war levels; however, after collectivisation of land, food stocks fluctuated somewhat and once again localised regional outbreaks of famine occurred in 1932–1933. Like Lenin before him, failures of Soviet central planning were compounded by Stalin's decision to withhold food, reserving it instead for the Red Army after fearing imminent conflict with Germany and Japan. This coupled with a priority of funding the war machinery through the exportation of much needed food supplies led to huge loss of life (Gantt 1936). Various estimates at the time ranged between 2 and 15 million deaths (Gantt 1936) but more recently, likely numbers ranging between 3 and 7 million have been quoted (Dyson and Ó Gráda 2002).

9.1.3 The Rise of American Economic Power

In the midst of all this and no doubt exacerbated by the consequences of the war and social upheaval was the disintegration of centuries old dynasties and empires such as the Habsburgs and the Romanovs. Under these conditions, these political and economic environment changed and the realignment of global politics quickly followed (Blainey 2000). Taking the place of long-established monarchies throughout

Europe were new emerging democratic, communist and fascist successor states. In Russia, widespread economic upheaval brought on by the war effort was instrumental in the 1917 Russian and subsequent Bolshevik revolutions. Political borders were redrawn and a number of smaller states emerged from the disbanded German, Austro-Hungarian and Ottoman empires. This as well as the cost of the war and the devastated internal infrastructures of many countries in turn created an economically weaker Europe while on the other hand economic opportunities presented by World War I allowed the United States to emerge as the economic superpower of the day.

In the social arena too changes were taking place with the emergence of a new socio-cultural ideology that embraced artistic, literary, philosophical, musical and cultural movements.

9.1.4 Declaration of the Rights of the Child

After the success of the recently formed Save the Children's fund in 1919 by British social reformer Eglantyne Jebb, she turned to another issue close to her heart—that of children's rights. She headed to Geneva in 1923 to the International Union armed with a simple well-defined ideologue asserting the rights of children. The resulting Declaration of the Rights of the Child, or the Declaration of Geneva as it is known, was well received and eventually adopted by the League of Nations a year later in the form of the World Child Welfare Charter (UNICEF 2009).

The initial 1923 document was groundbreaking by the standards of the day and included the following stipulations:

- Children must be provided the means for normal material and spiritual development.
- Children that are hungry must be fed, sick children must be nursed, backward children must be helped, delinquent children must be reclaimed and the orphan and the waif must be sheltered and succoured.
- Children must also be the first to receive relief in times of distress.
- Children must be protected against every form of exploitation.
- Children must be allowed to develop good moral upstanding vis-à-vis their fellow men.

Another area too was receiving attention. After waning interest in the optimum capacity of population theory, the end of the war and the prospects of continued food shortages also re-ignited public debate about population capacity and food (Ostrolenk 1930; Mukerjee 1933; Wolfe 1934).

9.1.5 The First World Population Conference, 1927

Global economic depression and agricultural uncertainty was prevalent. Once again the perennial debate of population overcrowding and the pressure on the world's natural resources was brought to the fore (Holmes 1924; Merritt-Hawkes 1928). The debate was held at Geneva in 1927 by the League of Nations and was the first real organised attempt to pool information and views with regard to the underlying science of the population problem (Carr-Saunders 1927). This first World Population Conference was also quite possibly the first real international forum to consider the question of, among other things, optimum population (De Gans 2002). One of the key scientists at this conference was Professor Raymond Pearl from Johns Hopkins University who, based on his previous important empirical study of the US population together with his work with rats, yeast cells and fruit flies, extrapolated to the human population the now-familiar growth curve resembling a flattened 'S' or 'logistic growth' curve as it became known (Pearl and Reed 1920; Pearl 1927; De Gans 2002). This growth, according to Pearl, was restricted in large part by the

> … actual and potential resources for the support of growth. (Pearl 1927, p. 22)

At the conference, Pearl further postulated that population growth was primarily the result of fertility mortality and migration and only secondarily resulted from economic and social factors (Pearl 1927;

De Gans 2002). His work also signalled a departure from previous prediction trends predicated on historic growth rates and asserted the idea of alternative future conditional predictions dependant on combinations of different alignments of inter alia, social and economic variables (Pearl 1927).

Despite being highly divisive and polarising it was considered by many, a great success. (Carr-Saunders 1927; Merritt-Hawkes 1928; Connelly 2006). Indeed it was the first time, according to birth control pioneer and sponsor of the conference Margaret Sanger (1927), that biologists and sociologists collaborated with the express purpose of utilising scientific analysis of social problems in the search for solutions to economic dilemmas (Sanger 1927). Nonetheless, despite this renewed momentum in population pressures, there was still no consensus over the limiting factors determining population ceilings. On this score Hiller, in 1930, explored several previous studies and found limiting checks on population growth ranging from temperature, light and nutrition to air and humidity (Hiller 1930). However, he quickly dismissed these studies as

> ... invalidated by methods used in their construction ... [that contain] ... logical inadequacies.
> (Hiller 1930, p. 524)

More interestingly, Hiller fully dismissed the notion of population prediction per se as untenable, suggesting instead that complex technological, social and cultural influences as determinants of population numbers by which

> ... any attempt to predict long time population cycles will probably go as far astray as did the forecasts of Malthus. (Hiller 1930, p. 550)

This underlying theme of population concern was also further reinforced in a book by Speier and Kähler titled *The War of Our Time* (1939) when Staudinger commented that

> ... interdependently social and economic conditions influence population development. ... A study of the world-wide implications of population trends gives us only further proof that it is necessary to build up an international system of social and economic co operation if we are to prevent the starvation of millions of human beings ... (Staudinger 1939)

It was about this time that the global economy was hit with yet another, this time more devastating, depression.

9.1.6 The Great Depression

With the war a not so distant memory the 1920s witnessed steady recovery in global output and trade in nearly all but the agricultural sectors (Kindleberger 1986; Michie and Smith 2000). The collapse in demand after the World War I for American agricultural products and the increasing mechanisation of agriculture together with a growing industrial sector all led to free-falling real farm incomes (Black 1942). In other sectors though, from about the mid-1920s till about 1929, the steady recovery became a boom; however this was short-lived, and due either in part to reparations and war debts (Keynesian) or to poor adherence to the gold standard (monetarism) the world collectively fell into the 'Great Depression' (Kindleberger 1986). This compounded an already difficult economic situation and had the effect of depressing agricultural markets even further and contributing to further widespread unemployment (Ostrolenk 1930). In dealing with this price-reducing overproduction, there was much activity on the policy front. In America an attempt to pass the McNary–Haugen Bill that would see surpluses dumped in foreign countries whilst simultaneously raising prices in the domestic market failed. Instead the government in America introduced the Federal Farm Board (FFB) in 1929. While the FFB had some successes, ultimately it failed to address the almost annual problem of 'over productivity'. In its wake, the Agricultural Adjustment Act (AAA) of 1933 aimed to reduce cultivated acreage and compensate farmers for non-production. Another method also employed by the AAA was an early environmental-based policy that combined reduction policies and environmental protection with the introduction of payments to farmers for shifting acreage of soil-depleting crops such as corn and wheat to soil-conserving plants such as grasses and legumes (Hardin 1943). More importantly for the farmer, the AAA, in an attempt to re-balance a flailing sector introduced policies of agricultural parity. This was an important weapon in the interventionists arsenal and through legislation it aimed to ensure Congress:

Establish and maintain such balance between the production and consumption of agricultural commodities . . . as will reestablish prices to farmers at a level that will give agricultural commodities a purchasing power with respect to articles that farmers buy ... (USDA 1985)

Such policies essentially aimed to equate farmers' incomes with costs of labour and material and to place them on par other industries' incomes (Tenny 1938). These assistance or intervention policies, it was assumed, would encourage and protect real farm incomes at a time when farms and farm workers were vulnerable but more importantly, it was meant to reflect the agricultural sector's importance within the national economy (Black 1942; Varnee 1955). The period 1910–1914 was considered the golden era of farming; as a result this was used as the base period for parity. Parity thus meant the maintenance of farm product prices by federal government support to ensure that farming remained an essential part of the US economy.

Meanwhile in the economic arena, the pre-1930s world trading system was based on 'relative' free trade backed by the gold standard. With competitive currency devaluations and strategic manoeuvring in trade policy, the standard finally collapsed in the 1930s. Some consider this failing of classical economic paradigms as paving the way for the new Keynesian revolution, focusing on national or domestic variables of fiscal and monetary policy (Goodrich 1947). Shortly after the collapse of the gold standard however, the spectre of agricultural surpluses once again raised its head.

9.1.7 Surpluses and Aid in the United States

Aiming to address the issue the United Kingdom together with the major wheat-exporting countries combined forces to address the need to control and dispose of the incredible wheat surpluses (more than double the total wheat exports of all countries in 1938–1939). Prevailing ideas at the time drew consensus on the need to reduce production altering trade barriers and other production, stock and quota controls to stabilise supplies and by extension prices (Thompson 1943). Simultaneously, the US Department of Agriculture's aim at this time was two-fold: as well as removing price depressing surpluses, where surpluses existed there were plans to make these available to school children and low-income families (Waugh 1943). The department commenced buying surpluses and making them available to state welfare agencies in the early 1930s. Initially such apparent altruism was a convenient measure of controlling farm prices at the gate, and no attempt was made to provide satisfactory nutritional diets. However, this was to change, and it was recognised that the two problems of farm incomes and food prices were the flip side of the same coin. With millions of people unable to buy the farmers' products, it was realised that farmers' incomes came to suffer while people simultaneously went hungry. As a result, it soon became clear that much larger welfare feeding programmes were needed, allowing both problems to be addressed at the same time. By 1938 and 1939, the food stamp programme was introduced that allowed people on relief to purchase orange stamps to the value of their normal affordable food expenditure and in return receive 50% in value of blue stamps that could be used to buy food commodities determined by the department to be surplus (Waugh 1943). Other initiatives too like the school lunch programme benefitted from this new government outlook on social welfare and collectively such distribution programmes thrived during a period when huge surpluses hung over the market and which led Waugh to famously comment that

> We were confronted with the striking dilemma of want and hunger in the midst of plenty. (Waugh 1943)

One of the consequences of this growing social awareness especially during the latter half of the World War I was a need for governments or institutions to act as a singular entity. Previous to this it was only a handful of aid agencies and a few individual government programmes that effectively operated with an international remit. However as the following sections show what followed was a new era of collaboration with objectives firmly sighted in the international arena (Belshaw 1947).

9.2 Health and Nutrition

In the area of health and nutrition, this was a very active period, kept busy by the most part initially through advances in scientific understanding and then subsequently through the wide-reaching work of the League of Nations.

9.2.1 Health and Nutrition Policy

In Britain, despite previous governmental health initiatives in 1831, 1848 and 1858, from 1871 the nation's policy of health was moved to the President of the Local Government Board. This meant that by the early part of the twentieth century public health services were provided for under the Poor Law through a system of voluntary hospitals, workhouses and hospitals. As mentioned this was distributed under the auspices of the local government board and was a relatively hit or miss affair that allowed some but not all, access to these services. Moreover, the services were uneven in quality and application. As a result the board, also responsible for sanitation and environmental health, was eventually reorganised in 1919 under the new Ministry of Health. Shortly afterwards local government functions were transferred to the Minister of Housing and Local Government, leaving the Health Ministry fully in charge of the nation's health.

It was about this time that continued promising research into vitamins and nutrients led researchers to view health itself in more nutritional terms. This was aided by the fact that science was beginning to show that infant growth was directly related to the food he or she consumed. Building on these influences, a report by the Minister of Health in Britain in 1921 outlined the need for a balanced portfolio of vitamins and minerals in addition to the main sources of energy, as essential in the continuing health of the population. It was noted too that milk, vegetables and fruit were particularly rich in these substances, and such 'protective foods' should be made more readily available. It was concerns like these in both England and the United States that ensured continued school feeding and national milk distribution programmes into the 1930s. It is also worth mentioning too that many such programmes were led by active organisations such as the Committee Against Malnutrition and the Children's Minimum Council (later the Children's Nutrition Council) that provided much of the impetus to improve national nutrition particularly when it came to children (BMJ 1938, 1954; Lloyd and Shore 1938; Acheson 1986).

These realisations also encouraged further investigation into nutrition resulting in a growing understanding of the roles of the various nutrients within the diet. As understanding progressed, the evidence for requirement needs began to shift in focus from the predominant observation-of-intake techniques of seemingly healthy groups to experimental studies whereby nutrient intakes were varied to determine minimum needs that denied any clinical sign of deficiencies (Beaton 1991). It was this period that the United Kingdom came to the fore becoming the epicentre of a global nutritional movement (Ruxin 1996).

9.2.2 Health Organization of the League of Nations

In the wake of the Great War (1918/19), influenza cut a swath of destruction across the globe killing conservatively between 15 and 20 million people. On top of this, a further 2 million or so cases of typhus put unprecedented burden on the Office International d'Hygiène Publique (OIHP). Separately, in early 1920, a plan was approved by the League of Nations for a permanent international health organisation, the Health Organization of the League of Nations (HOLN). Shortly after there was a proposal by the League to combine the duties of the League's Health Organization with those of the OIHP into a single international health organisation. However, as the United States was a member of the OIHP and not the League, any unification meant the United States becoming a member of the League. The United States had made it clear this was not in their interests and along with other reasons negotiations broke down, leaving two international health organisations working independently with the similar remits (WHO 2010). There were differences however, while the OIHP's main remit continued to be directed towards the supervision and improvement of international quarantine measures, the HOLN among other things collected and disseminated data on the status of epidemiological diseases of significance. Initially, the HOLN dealt almost exclusively with health problems in Europe and North America; however, this soon expanded with many of its surveys taking account of global nutritional standards. According to Ruxin (1996) too, HOLN was a slow starter with a limited nutritional programme. In this early incarnation, Ruxin comments that

> … prevalence of nutritional disorders in the developing world principally remained a scientific point of ignorance. (Ruxin 1996)

This was soon to change though and HOLN went on to work with many of the most important nutritionists of the time, shaping and moulding both policy and understanding. However, even before HOLN moved into full swing other events, and studies were underway that were to add to the changing face of nutritional knowledge and application, not least of which were the continuing advances in nutritional understanding. Heralding this new understanding were the dietary studies of Britain and America.

9.2.3 Dietary Energy Requirements

The debate over calorific requirements continued into the 1920s. Two countries were at the forefront: Britain and America.

9.2.3.1 Britain

While many studies showed disunity of consensus, there was in fact a growing convergence in the very general sense of dietary make-up. Most studies around this time too also continued to use calorimetry in their methodologies although one exception was Miss E. M. Bedale. Despite the early clinical use of basal metabolic rate (BMR), the first real study to use BMR to estimate human energy requirements and by extension food requirements was Bedale's. In her 1923 publication of school children's energy requirements, she examined different groupings of children's activities and calculated energy expenditure in relation to their calorie intakes using BMR. This early pioneering approach was largely ignored by the wider nutrition community, in large part it would seem owing to lack of credibility due to her gender (Bedale 1923; McNaughton and Cahn 1970; Henry 2005).

Meanwhile, in the general level of caloric requirements and as alluded to earlier, Chittenden's previous sceptical view with regard to the general over-prescriptive nutritional recommendations was much criticised and was '… neither quickly nor universally accepted' (Carpenter et al. 1997). Yet his misgivings continued to reverberate with early work by Henry Sherman during this period appeared to confirm Chittenden's doubts. In 1920 Sherman published results of a sort of meta-analysis of 109 previous researchers' experiments and found protein recommendations ranging from 21 to 65 g for an average 70 kg man with a general average indicating a requirement of 0.635 g of protein per kilogram of body weight. This equated to 44.4 g per day for an average 70 kg person and echoing Chittenden's earlier view, Sherman suggested that

> Many of the published experiments which were designed to test the amounts of protein required in normal nutrition are now seen to have given misleadingly high results … (Sherman 1920, p. 98)

9.2.3.2 Joint Memorandum of the BMA and MoH

Further discord came about with the British Medical Association's (1933) report which caused a political furore by citing 3400 calories as that needed for daily health and maintenance and by calculating the minimum cost of such a diet. The tabloids picked up on this and compared it with the previous Ministry of Health's (MoH) 3000 calories and implied that unemployment pay at the time would not even cover the cost of sufficient basic food for good health (BMJ 1933, 1934). The MoH was incensed and the report led to a bitter confrontation between the two bodies. Eventually it was felt that the gap had to be closed somehow and a conference between the Nutrition Advisory Committee of the Ministry of Health (MoH)

TABLE 9.1

BMA and MoH Agreed Daily Requirements

Group	Calories
Male	
Light work	2600 3000
Medium work	3000–3400
Heavy work	3400–4000
Female	
Active	2800–3000
Housewife	2600–2800
Children	
Boy 14–18	3000–3400
Girl 14–18	2800–3000
Children	
12–14	2800–3000
10–12	2300–2800
8–10	2000–2300
6–8	1700–2000
3–6	1400–1700
2–3	1100–1400
1–2	900–1200
All round adult average	3000

Source: Re-created from BMJ, *Br. Med. J.*, 1, 900–901, 1934.

and the Nutrition Committee of the British Medical Association (BMA) in 1934 aimed to address their differences. A tense meeting ensued but nonetheless, agreement was eventually reached (BMJ 1934; Smith 2003). Out of this controversy came a joint memorandum proposing a set of sliding caloric scales based on age sex and activity level and for purposes of aggregation, an average figure of 3000 calories per person was also agreed (Table 9.1) (BMJ 1934).

Ironically, despite such misgivings and although undernourishment was still prevalent in the United Kingdom, these inter-war years actually saw improvement in Britain's general dietary health (Beardsworth and Keil 1997).

9.2.3.3 America

Nutritional research in the United States at this time was furthered by the work of nutritionist Hazel Stiebling. Based on her work at the Sherman laboratory, Stiebling, working for the USDA, proposed the first dietary standards taking account of vitamin and mineral requirements (Harper 1985). Table 9.2 shows the original values, although she later collaborated with Esther Phipard, expanding the recommendations to include thiamine and riboflavin.

Importantly too, this marked an ideological transition from guidelines that aimed solely to prevent starvation and provide energy for work to an overall view of optimum provisions for health, nutrition and well-being (Harper 1985).

9.2.4 League of Nations Mixed Committee's Report

During all of this, a rumbling undercurrent was stirring. Shocked by the devastating impact of the Great Depression on people's health and living conditions throughout the world, the international community took decisive interest (Périssé 1981). Already by the 1930s, many European governments had begun improving the state of children's nutrition with the provision of free or subsidised milk and school meals; however, it was with the relentless work of the League of Nations that real international progress was

TABLE 9.2

USDA Recommended Daily Requirements by Hazel Stiebling

	Energy	Protein	Calcium (g)	Phosphorous (g)	Iron (g)	Vitamin A (Units)	Vitamin C (Units)
Child under 4	1200	45	1	1	0.006–0.009	3000	75
Boy 4–6, girl 4–7	1500	55	1	1	0.008–0.011	3000	80
Boy 7–8, girl 8–10	2100	65	1	1	0.011–0.015	3500	85
Boy 9–10, girl 11–13	2400	75	1	1.2	0.012–0.015	3500	90
Moderately active woman, boy 11–12, girl over 13	2500	75	0.88	1.2	0.013–0.015	4000	95
Very active woman, boy 13–15	3000	75	0.88	1.32	0.015	4000	100
Active boy 15+	3000–4000	75		1.32	0.015	4000	100
Moderately active man	3000	67	0.68	1.32	0.015	4000	100
Very active man	4500	67	0.68	1.32	0.015	4000	100

Source: Re-created from Stiebling, E.H., *Food Budgets for Nutrition and Production Programmes*, US Department of Agriculture, Washington, DC, 1933.

finally made (Carpenter 2007). It began with the realisation of the need for an agreed set of nutritional standards that could be used for comparison across regions and nations. In recognition, during the mid-1930s the Health Organization of the League (HOLN) initiated several international nutritional studies, the first of their kind. These studies covered a 10-year period and resulted in League's Mixed Committee's Final Report of 1936, the most up-to-date global study containing the 'first international table of calorie and protein requirements by age and sex' (League of Nations 1936a; Périssé 1981).

It had started in 1925 when the Health Organization of the League, the International Institute of Agriculture and the International Labour Organization together undertook to research nutrition in respect of agriculture, social and economical problems (League of Nations 1936a; Mayne 1947). By 1931 the League's Health Secretariat's Nutrition Division (based in Switzerland) recruited Wallace Aykroyd as the first nutritionist in what was perhaps the first ever role specifically tasked with international nutritional advice and responsibility (Carpenter 2007). In 1932 the League held two conferences of experts in Rome and Berlin to assess the fundamental principles of adequate diets as well as the role of nutrition in relation to the economic crisis (League of Nations 1936a; Périssé 1981; Carpenter et al. 1997) This was a prelude to the League's Expert Commission on Nutrition Conference, which was held in London in 1935 (Medical Science 1936).

9.2.4.1 Burnet and Aykroyd Report

Meanwhile, on the suggestion of the Yugoslav delegate, the Health Organization of the League of Nations had collected information on the food situation in a few representative countries. Overseen by the Director of the Organization, Dr. Frank G. Boudreau, it resulted in the Burnet-Aykroyd report 'Nutrition and Public Health'. The report drew on the work of John Boyd Orr of the UK and from Hazel K. Stiebling of the United States and showed that even in the midst of plenty or excessive food production the poor were still suffering from hunger. The report went on to address the significance of good nutrition in general human health and well-being and to illustrate practical measures for improving such nutrition. Of overriding significance in this report was the formal realisation that at the heart of the problem was under-consumption through low purchasing power or poverty (Burnet and Aykroyd 1935; League of Nations 1936a; Harper 1985). The report was finally submitted in 1935 at the sixteenth full assembly of the League of Nations. It was a pivotal work that not only recognised the previous ideas of John Boyd Orr (later to become the first Director-General of the FAO) but also built on the 10 years of health and nutrition work by scientists and the Health Organization of the League of Nations. It was also timely brought to an international forum at a time when important discoveries were being made in the field of diet and nutrition. The report itself centred around three main objectives: nutritional needs, the resources available to fulfil these needs and how best to utilise these resources for the benefit of the population in general (Burnet and Aykroyd 1935; League of Nations 1936a). By recognising that health had become a problem of social and economic significance the report proposed a marriage of health and agriculture, one in which the new role of science could have a major impact (Passmore 1980)

9.2.4.2 Mixed Commission of the League of Nations

The Burnet-Aykroyd report was discussed in detail by the League of Nations Assembly. One of many people enthused by the report was Stanley Bruce, the former Australian Prime Minister who, fortified by the report in an address to the League of Nations, echoed the suggestion of the merging of agriculture and health responsibilities (FAO 1970). As a result of the meeting several things in particular stood out. A technical committee was appointed to review the report's Physiological Basis of Nutrition and secondly a Mixed Committee was appointed and tasked with submitting a report on the whole wide-ranging issues of nutrition in relation to health and economics (Campbell 1938a). The Mixed Committee was chaired by Lord Astor who oversaw the nutrition work of the health organisation of the League, the International Institute of Agriculture and the International Labour Organization (Royal Statistical Society 1936). Lastly, the Secretary General of the League issued a forthright statement highlighting

1. The importance of a good diet for mothers and infants, school children, and other vulnerable groups
2. The need for examining the current level of nutritional understanding, its gathering and dissemination
3. The need for proper adequate and scientifically sound principles of dietary standards (Philpott 2009)

The mixed committee produced two reports: the interim and final reports. The interim or preliminary report titled 'Problem of Nutrition' consisted of four volumes: (1) An Interim Report of the Mixed Committee on the Problem of Nutrition; (2) Report on the physiological basis of nutrition; (3) Nutrition in various countries and (4) Statistics of food production, consumption and price (League of Nations 1936a; Royal Statistical Society 1936). In the report, recommendations to the League and governments in general suggested among other things the setting up of national nutritional committees to promote nutritional policies (Mcdougall 1940). This report was subject to further revision and ultimately led to the final report on the health and economic aspects of diet in 1937 entitled Final Report of the Mixed Commission of the League of Nations on the Relation of Nutrition to Health, Agriculture and Economic Policy (League of Nations 1937). The report noted that in the face of increased agricultural output,

> … markets were glutted and producers were faced with ruin; at the same time … extensive was the failure [of mankind] to satisfy normal nutritional requirements (League of Nations 1937)

It also continued that while

> … owners of food stocks were unable to find remunerative markets, some parts of the world were suffering from famine; in others, large sections of the population were suffering from serious malnutrition … [and] Hence the question naturally arose whether it was not the duty of the public authorities to assume the responsibilities inherent in a 'nutrition policy'. (League of Nations 1937)

The Final Report of the Mixed Committee on the Relations of Nutrition to Health, Agriculture and Economic Policy was a remarkable document and, at that point, the most authoritative examination of nutrition and its broader aspects of social, economic and agricultural issues ever undertaken. It became a best seller (among the League's publications) and was voted by the *New York Times* as the most important book of the year (Mcdougall 1940). The report as well as accompanying technical papers outlined the League's nutritional recommendations and although no scientific justification was offered in support of such guidelines, it did set forth a four-fold blueprint calling for general dietary standards including the first dietary recommendations for special groups, a call for nutritional surveillance and for global nutritional education (Périssé 1981; Harper 1985; Carpenter et al. 1997). Calorie recommendations for the general population were issued for an adult, male and female in a temperate climate not engaged in manual work giving an allowance of 2400 calories per day. On top of Table 9.3 shows the supplements for muscular activity, which are added to the basic requirement of 2400 calories per day. The goal of '… striving after optimum nutrition rather than minimum nutrition', along with Stiebling's efforts were now firmly cemented (League of Nations 1937).

While the report was perhaps more oriented towards the industrial nations, it did look at the developing world, although not to the extent that some would have liked. Among the many conclusions and recommendations, the report suggested was that in most countries of the world, serious malnutrition was occurring, but perhaps more importantly, was the recognition that there already existed the means and

TABLE 9.3

The League of Nations Nutritional Recommendations

Age	Coefficient	Calories	Protein (g/per kg)	Calcium (g)	Iron (mg)
Infants					
0–6 months		100 (per kg of body weight)			
6–12 months		90 (per kg of body weight)			
Children					
1–2	0.35	840	3.5	1.0	
2–3	0.42	1000	3.5	1.0	
3–5	0.5	1200	3.0	1.0	
5–7	0.6	1440	2.5	1.0	
7–9	0.7	1680	2.5	1.0	
9–11	0.8	1920	2.5	1.0	
11–12	0.9	2160	2.5	1.0	
12–15	1.0	2400	2.5	1.0	
15–17	1.0	2400	2.0	1.0	
17–21	1.0	2400	1.5	1.0	
Men	1.0	2400	1.0	0.75	10
Women	1.0	2400	1.0	0.75	10+
Pregnant (1–3 months)	1.0	2400	1.0	1.5	10+
Pregnant (4–9 months)	1.0	2400	1.5	1.5	10+
Nursing	1.25	3000	2.0	1.5	10+
		per hour of work			
Light work		Up to 75 calories per hour of work			
Moderate work		75–150 per hour of work			
Hard work		150–300 per hour of work			
Very hard work		300 calories and upwards per hour of work			

Sources: League of Nations, *The Problem of Nutrition: Volume I, Interim Report of the Mixed Committee on the Problem of Nutrition*, League of Nations, Geneva, 1936a; League of Nations, *The Problem of Nutrition: Volume II, Report on the Physiological Bases of Nutrition*, Technical Commission of the Health Committee of the League of Nations, Geneva, 1936b; League of Nations, *Nutrition: Final Report of the Mixed Commision of the League of Nations on the Relation of Nutrition to Health, Agriculture and Economic Policy*, League of Nations, Geneva, 1937.

wherewithal to address this issue. It also made clear that while ignorance was a big issue in bringing about such horrors, poverty undoubtedly played a central and important part. This further reinforced the move away from food being solely a provision of energy to seeing diet as a holistic enterprise for optimum health and well-being (League of Nations 1937; Hanekamp and Bast 2007). The report also repeatedly used the example of national free milk provision policies as an important and simple means for improving the health of populations, particularly for mother's and children.

It can be seen then that this period between the two wars witnessed great leaps in social responsibility and scientific understanding in the field of nutrition. Although in the early 1920s, there was still concern that protein standards continued to be predicated on opinion, by the end of the 1930s evidence-based nutritional research allowed answers to be found to the many questions of energy requirements and healthy lifestyles of the day (Hwalla and Koleilat 2004). Furthermore, despite a continuing lack of universal consensus nutritional science nevertheless flourished and a growing convergence began to emerge (Table 9.4). It was also acknowledged at the time and credited to the combined work of the ILO and the League in the mid-1930s that

> The provision of food adequate in quantity and quality will have a more profound effect upon national health than any other single reform. (McDougall 1943)

By 1938 the HOLN had already spread its message far and wide with nutrition committees springing up in 21 countries around the world. However even in spite of such progress there was still a lack of information on the extent or gravity of under-nutrition throughout the world. In response, the 1939 meeting

TABLE 9.4

Nutritional Recommended Requirements for Adults 1919–1939

Date/Authority	Load (Default Man)	Protein	Fat	Carbohydrate	Calories
1932: Ministry of Health's Advisory Committee on Nutrition (ACN)					
MoH offered a standard of 3000 calories as sufficient for daily requirements (Ministry of Health 1932; BMJ 1934).					
					3000
	67–75				
1933: British Medical Association (BMA)					
The BMA's report of 1933 caused a political furore by citing 3400 calories as that needed for health and maintenance and calculating the minimum cost of such a diet (BMJ 1933, 1934).					
					3400
1933: USDA					
Under the guidance of Hazel Stiebling the fist recommendations were made containing suggested vitamin and mineral allowances for calcium, phosphorous, iron and vitamins A and C (Stiebling 1933).					
					3000–4500
1934: Joint Ministry of Health and British Medical Association Memorandum					
The controversy and political fallout caused between the MoH and the BMA resulted in joint meetings to address the issues out of which came an agreed sliding scale of values (BMJ 1934; Smith 2003).					
	Men				2600–4000
	Women				2600–3000
1932: League of Nations Conference of Experts					
The League met to consider dietary standards in order that the state of nutrition could be compared across countries. The starting point was 2400 calories to which are added supplements depending on the activity level of the individual: light work adds 75 calories per hour of work, moderate work 75–150 calories per hour, hard work 150–300 calories per hour and very hard work up to 300 calories per hour of work (League of Nations 1936a, 1937; Hwalla and Koleilat 2004).					
		Not less than 1 g per kg			2400

of the HOLN Health Committee, the League called for extensive nutritional surveys. Unfortunately, the outbreak of World War II put these ideas on hold (Ruxin 1996).

9.2.5 Growing Multilateralism

This period however did lead to a growing awareness of food and nutrition in connection with health benefits that fostered many national and, despite Ruxin's (1996) suggestions to the contrary, international studies and articles encompassing the developing world (Firth 1934; Gilks 1935; Worthington 1936; Watson 1937; Auchter 1939). That said, in Ruxin's defence, the League's Nutrition report of 1935 did highlight a shortfall in studies in Asia and other developing countries of the time. Ultimately, though building on 10 years of studies and reports the 1935 meeting of the League of Nations was the first of its kind; it was a forum in which many ideas in the fields of food, nutrition and public health were brought together and synthesised into one 'lofty' multilateral global objective to expand profitable nutrition throughout the world (League of Nations 1937). It was also a period of realisation of the multi-dimensionality of the problems inherent in achieving food security, and in this ideologue it also recognised the need for a multilateral world food security mechanism or organisation that could bring together both food and agriculture under one remit (Campbell 1938b; McDougall 1943; FAO 1970; Shaw 2007).

9.2.6 Poverty, Not Supply

Up to this point the problem of adequate food had been considered one of production and supply, yet the Burnet-Aykroyd report concluded that despite many economists talk of surpluses and excessive food production it was in fact, poverty and inadequate or low purchasing power that led to much hunger and starvation around the world (Burnet and Aykroyd 1935). This idea of insufficient purchasing power was the central idea that formed the backbone of Amartya Sen's 1981 essay on entitlement and deprivation and one that contributed to his Nobel Prize on Welfare Economics in 1998. Yet it was an idea that had its genesis long before this and even the Burnett and Aykroyd report as the many other examples attest to in this book. Nonetheless, the mixed committee's 1937 report was perhaps the first internationally oriented scientific publication that categorically placed poverty in the frame

> Poverty and ignorance remain formidable obstacles to progress in the disparity between food prices and incomes and increases the difficulty experienced by the poorer sections of the community in obtaining an adequate supply of the proper foods. [As a result] millions of people in all parts of the globe are … suffering from inadequate … development or … malnutrition … That this situation can exist in a world in which agricultural resources are so abundant … remains an outstanding challenge to constructive statesmanship and international cooperation. (League of Nations 1937)

9.3 Science and Technology

9.3.1 Green Shoots of Revolution: Industrial Fertilisers

In the field of agrarian science little has had more impact on the growth of crop production and by extension population explosion than the economical industrial synthesis of ammonia (Smil 2001). Up to about the turn of the century, crop yields worldwide began to stagnate and in the United States at this time wheat yields were reaching only about 1.7% higher in 1909 than the decade earlier (Webb et al. 2008). This was problematical and growing speculation, crop failures and trade constraints were being readily translated into higher global food costs and prices. It was fast becoming acknowledged too that actual physiological limits were being reached given the condition of many of the soils.

Consequently, it was felt that any substantive increase would have to come from yield increases through agricultural technology rather than through expansion of cultivated areas. Unfortunately what was holding back further yield increases was the shortage of nitrogen for fertilisers.* In the first two decades of the 1900s, farmers in the United States and Europe lacked adequate access to nitrates, and despite nitrates being one of the top three commodities (by volume) globally shipped, too many farmers still lacked credit and purchasing power for its acquisition. However, moves were underway to synthesise artificial nitrogen compounds in the laboratory in the hope of cheaper more readily available stocks. Indeed, a process had already been developed in 1909 by Fritz Haber and Carl Bosch† that could directly synthesise ammonia from hydrogen gas and atmospheric nitrogen for use in the manufacture of fertiliser. It was first used industrially by BASF Limburgerhof agricultural research centre in 1913 and then by the German government. However, during the war, the German government shifted the conversion of ammonia away from fertiliser channelling it instead into the production of high explosives. Both chemists won the Nobel Prize for their work in chemistry: Fritz Haber in 1918 and Carl Bosch in 1931. This fixing of nitrogen (the Haber–Bosch process) was at first a closely guarded secret of the Germans until after the war when as part of the treaty of Versailles, BASF, the company responsible for its industrial scale manufacturing process was obliged to share the knowledge under license to a new purpose built plant in France. This opened up new possibilities for the agricultural sectors around the world as new plants sprung up to take advantage of this 'marvellous' new industrial technology (Smil 1999).

9.3.2 Plant Hybridisation

Selective breeding for millennia had helped produce improved strains of both crops and livestock. Before about 1910 for instance farmers selected the best ears of corn from current crops for use as seed for the next year, however it was also recognised that this type of selective breeding solely on the basis of appearance was of limited success. Three geneticists, George H. Shull, Edward M. East, and East's student Donald F. Jones, were to change all this when they separately began work on the genetics of corn. Independently they showed that by crossing breeds of corn their offspring were not only superior to their parents but in fact sometimes surpassed the original plants. Unfortunately, continued experiments with inbreeding varieties showed degeneration of the quality of hybrids after the first generation. As a result the only way to guarantee continued high yields was to return to the original parental combination. This would have consequences in which farmers could not be able to re-use that year's crop seeds for the following year and which meant that if farmers were to go that route they would now be tied to seed companies for their annual supply. However, these new hybrids had the potential to greatly improve the productivity of agriculture during the first part of the century, and foreseeing the benefits the farmers jumped on board.

   The big breakthrough in productivity growth came with the first commercially available hybrid corn. Advancement in this field can best be shared by two people, George S. Carter (1921) and Henry Wallace (1923). Commonly quoted as the first, Henry Wallace in 1913 became interested in the methods that Shull and East had pioneered and eventually, through help from the government of America, Wallace developed one of the first commercial hybrid corns in 1923 (Berlan and Lewontin 1986). Carter on the other hand, also building on the work of Shull, East and others pioneered the first double-crossed method of hybridisation in 1921 (Cook 1937). Thus by 1923 this new method of breeding became profitable

---

* Fertilisers: Every plant relies on several basic compounds to grow: nitrogen, phosphates as well as numerous other trace elements. While plants can only absorb nitrogen in the nitrate form, micro-organisms in the soil convert atmospheric nitrogen into nitrate. Soil pH (acidity or alkalinity) must also be adjusted to meet optimal conditions for each crop. The major development of the Green Revolution was the use of chemical fertilisers to adjust soil pH balances and achieve the right levels of all the important compounds such as nitrates and phosphates.
† Haber–Bosch Process: This is a method for chemically fixing nitrogen involving mixing nitrogen from the air with hydrogen (generated by heating natural gas), which was then burned to yield a nitrogen–hydrogen mixture. This is then compressed (10–80 MPa) and heated to between 200°C and 700°C in the presence of a metal oxide catalyst to create ammonia for fertiliser use.

TABLE 9.5

Biotechnology Key Dates: 1919–1939

1921	The first commercial hybrid corn.
1925	Congress cut the decades-old free Seed Distribution Program, which consumed more than 10% of the USDA's total budget in 1921.
1926	Hermann Muller discovered that X-rays induced genetic mutations, providing researchers a way to induce mutations.
	Henry Agard Wallace founded the Hi-Bred Company, a hybrid corn seed producer known today as Pioneer Hi-Bred International, Inc.
1928	Flemming discovered penicillin.
1934	Desmond Bernal showed that protein molecules could be studied using X-ray crystallography.
	Martin Schlesinger purified bacteriophage and found equal amounts of protein and DNA.
1938	The term 'molecular biology' was coined.

Sources: NHM, Where Did Biotechnology Begin? *Access Excellence*, 1989/90; BioTech Institute, Timeline of Biotechnolgy 20th Century, Biotechnology Institute, Washington, DC, 2010.

not only for the farmer but also for industry too (Morrison 1947; Sprague 1962; Smith et al. 2004). Simultaneous experiments in wheat also allowed, by the mid-1920s, widespread commercially available hybrids (APS 2010). The uptake of these new hybrids was immediate and expanded to such an extent that in America by the 1930s grain yields increased considerably, further adding to overproduction mentioned in the previous sections.

9.3.3 Biotechnology

Advances in microbiology and fermentation technology continued throughout the 1930s. It was discovered that micro-organisms could now be mutated through physical and chemical manipulation to become faster growing and tolerant of less oxygen. This new micro-organism hybridisation further allowed for the development of new technologies in food fermentation (Shurtleff and Aoyagi 2007). While understanding of the chemical functions stemming from the emergence of post-war molecular biology was still to come, it was nevertheless an exciting time. In microbiology, for instance the discovery by Alexander Fleming in 1928 of the antibiotic penicillin (derived from the mold Penicillium) was a seminal moment in the collective field of bioscience. Other such discoveries led to the subsequent development of large-scale fermentation production, particularly in the pharmaceutical industry. This spawned new interest and rapid increase in life sciences research, which in turn witnessed the production of new antibiotics as well as increasing the range of enzymes and vitamins (AU/ ICE 1997). Other notable moments in biotechnology are outlined in Table 9.5.

9.4 Period Commentary

Government or state intervention policies advocating inter alia agricultural price support, regulating economic stability, trade protectionism and rationing when necessary, together with bilateral and multilateral trade agreements had become standard political fodder in the previous inter-war years (Markham 2002; IAAE 2008; IGC 2009a). This period also shaped a new international political landscape concerned with the individual's economic and social welfare which took on a formal global remit with the incorporation of the League of Nations. The League's work on nutrition was a breakthrough in international cooperation, it effectively galvanised consensus and encouraged a growing momentum in the nature and relationship of food, nutrition and social welfare (League of Nations 1936a; FAO 1970; Passmore 1980; Shaw 2007). The 1935 conference and subsequent nutritional reports were also pivotal in recognising the multi-dimensionality of the food security problem. Importantly too at this time it was found that nutrition was no longer solely a physiological problem, but one of economic, agricultural, industrial and commercial concerns too (Burnet and Aykroyd 1935). The report also put forward a strong case for the need for education in nutrition, dietary diversity and health (Ravenel 1936). These advances ultimately encouraged the widespread acceptance of a new standard of food requirements, much higher than was merely needed to satisfy hunger. Effectively the confluence of the above strands cemented the multi-dimensionality and the interdependence of food, nutrition and politics. In fact, this period had finally witnessed the full and complete politicisation of hunger and malnutrition that had perhaps started at the turn of the century.

The humanitarian and social welfare sector was also promoted with the arrival of organisations and agencies like Save the Children, the World Child Welfare Charter and the US Commodity Credit Corporation (CCC) (UNICEF 2005; Save the Children 2009). Furthermore the role of development economics had also gained momentum. In particular, this development ideology coincided with the idea that countries had a moral and social obligation to develop their colonial administered states as well as the need to advance the lot of the poorer nation's economies in general (Garside 2002). At the forefront of this movement once again was perhaps the League of Nations. Throughout this period, the League's Health Organization was calling; often loudly and openly for policies of nutritional improvement to be spearheaded by economic improvement, nutritional education and increased food supplies, reflecting both camps' views who saw malnutrition as either a problem of ignorance or of economics. Moreover, the role of women in development was also noted; although often subservient, their role in care giving, employment and household nutrition was seen as crucial to future nutritional goals (League of Nations 1937). Importantly too, this period saw the dual notion's of farm parity price support and sustainability attain new status at policy level with the United State's Agricultural Adjustment Acts of 1933 and 1938. As well as introducing parity pricing the act importantly recognised the need for conservation of natural resources when it offered as rationale for new legislation in its pre-amble as

> ... for the purpose of conserving national resources, preventing the wasteful use of soil fertility, and of preserving maintaining, and building the farm and ranch land resources in the national public interest ... (Tenny 1938)

Science and technology too witnessed much progress as understanding of chemicals and micro-organisms was applied with great success to the study of food nutrition and health (Orr 1939; Ruxin 1996). Preliminary work in the previous century on protein and energy led to further work in the identification and classification of vitamins in 1911–1919. In turn this led to better understanding of the roles of vitamins and nutrients within a healthy diet (Carpenter 2003a, 2003b). By uncovering the mysteries of diet and health, it seemed, leading personalities in the field such as Funk, Mellanby, Orr, Platt, Aykroyd and others took on the mantle of caretakers of the 'national well-being' (Ruxin 1996). All in all, this period ultimately became known as the 'Golden Age of Nutrition' (Carpenter et al. 1997; Go et al. 2005). Remarkably then it can be seen that this period clearly shows the many disparate strands of the modern concerns of food security such as poverty, economic development, equality, environmentalism and health, all coming together in one overarching social construct. Indeed, while many modern observers maintain that the idea of food security emerged from the 1970s (Chapter 23), research indicates that it was in fact the generations leading up to and culminating in the inter-war years that can perhaps be described as the foundation from which the modern concept of food security truly emerged. At the heart of this convergence was the golden age of nutritional science and the incorporation of multilateral agencies with global social and economic remits. In this era it seems, the seeds of a multilateral world food security apparatus were sown and the notion of food as a right had finally come of age (Barona Vilar 2008).

9.5 Key Dates, People, Acts, Reports and Surveys

Table 9.6 highlights some of the more important events, noteworthy individuals and key reports, etc. that took place during this period.

TABLE 9.6

Key Dates: Inter-War Years 1919–1939

Date	People, Organisation, Agencies, Acts or Conventions
1918/19: Influenza Epidemic	The influenza pandemic of 1918–1919 killed somewhere between 20 and 40 million people worldwide. Estimates vary greatly.
1919: The International Labour Organization (ILO)	Initially established as an agency of the League of Nations in 1919 dealing with the reconstruction and the protection of labour unions. After the demise of the League, the ILO became a specialised agency of the UN in 1946 (ILO 2010).
1919: The International Federation of Red Cross and Red Crescent Societies (IFRC)	Expanded international activities of Red Cross were made to include assistance to emergencies outside of war (IFRC 2009).
1919: Save the Children	Founded to combat the intense civilian suffering caused by the continued a blockade against Germany and Austrian-Hungary (Save the Children 2009).
1919: League of Nations (LoN)	Creation of the first international multilateral organisation with the mandate for peace and cooperation (League of Nations 1919).
1920: Health Organization of the LON	
1920: Save the Children International Union (SCIU) (L'Union Internationale de Secours aux Enfants)	Founded in 1920 by Eglantyne Jebb and sister Dorothy Buxton, founders of Save the Children in the UK (UNICEF 2005).
1924: World Child Welfare Charter	The League adopted the SCIU's principles of the declaration of the rights of the child (UNICEF 2005).
1927: The (LoN) International Economic Conference (Geneva)	The League of Nations Assembly's first International Economic Conference in 1927. Its two objectives were to strengthen international trade laws (halting widespread tariff increases) and to deal with important commercial, industrial and agricultural problems (UNOG 2010).
1929: US Agricultural Marketing Act	The 1929 Act created the Federal Farm Board which directly intervened to influence price stabilisation, crop surpluses and export prices (Markham 2002).
1929: The International Association of Agricultural Economists (IAAE)	Established in 1929 by 11 countries to hold conferences to share knowledge and address challenges throughout the world (IAAE 2008).
1929: British Colonial Development Act	Colonial Development Act, providing agricultural and economic development in the colonies with the purpose of stimulating domestic investment whilst advancing British commerce and trade (Garside 2002).
1930: Wheat Meetings	1930–1931 saw 16 International meetings exclusively for international wheat importers and exporters, although no agreement was reached, the meetings led to the first wheat conference in 1931 (IAAE 2008; IGC 2009a).
1930–34: US Grain Stabilisation Corporation (GSC)	The FFB set up the Grain Stabilization Corporation to buy grain to shore up prices of key crops (Markham 2002).
1930: US Agriculture Marketing Act established the Foreign Agricultural Service Division (FASD)	Tasked with maintaining and expanding international export opportunities the FASD was superseded by the Foreign Agricultural Service in 1953 (National Archives 1995; Otto 1999; Swanson 2003).
1931: International Wheat Conferences	First and Second International Wheat Conferences of wheat importers and exporters (Rome, London), although no agreement was reached it led to the third conference and the first successful agreement among members (IGC 2009a).
1931: The International Relief Association (IRA)	Formed to assist the opponents of Hitler and the Nazis however was suspended in 1933 (Montgomery 2008).
1933: International Rescue Committee (IRC)	American branch of the IRA formed at the behest of Albert Einstein to continue the work of the IRA (IRC 2009).
1933: 1st International Wheat Agreement—Wheat Advisory Committee	Third International Wheat Conferences signed between the exporting and importing countries agrees the 1st International Wheat Agreement and the introduction of the Wheat Advisory Committee (which became the IWC in 1942). The agreement failed in the 1st year (League of Nations 1933; Meerhaeghe 1998; IGC 2009a, 2009b).

continued

TABLE 9.6 (Continued)

Key Dates: Inter-War Years 1919–1939

Date	People, Organisation, Agencies, Acts or Conventions
1933: US Agricultural Adjustment Act (AAA)	The United States intervened to control production with the AAA (PL. 73–10). The idea was to restrict agricultural production and ergo surpluses by paying farmers to reduce the cropped area effectively raising the value of crops (Fite 1962; NAL 2010).
1933: Kwashiorkor first coined	Meanwhile in the Gold Coast, a British nurse, Cecilly Williams, first described and coined the term 'kwashiorkor' in 1933 as the inadequate protein consumption an infant suffers often induced by poor breastfeeding. Kwashiorkor, or the sickness of weaning, as it became known to some is taken from the GA language of coastal Ghana.
1933: US Commodity Credit Corporation (CCC)	A vehicle for managing government food stocks and surpluses and to protect and stabilise farm income and prices. It formed the basis of the first food aid and became part of USDA in 1939 (FSA 2008).
1934: US Export & Import Bank	Was incorporated to promote US exports through the use of concessional loans.

References

Acheson, E. D. (1986). Tenth Boyd Orr Memorial Lecture: Food Policy, Nutrition and Government. *Proceedings of the Nutrition Society* 45: 131–138.

APS (2010). Population Genetics of Plant Pathogens: Interactions among Evolutionary Forces and the Genetic Structure of Pathogen Populations. The American Phytopathological Society (APS). http://www.apsnet.org/edcenter/advanced/topics/PopGenetics/Pages/InteractionsGeneticStructure.aspx (accessed 15 February 2011).

AU/ ICE (1997) Ice Cases Studies—Intellectual Property Rights & Biotechnology: Indian Seed Conflicts. *Inventory of Conflict and Environment (ICE)* (41). http://www1.american.edu/ted/ice/grainwar.htm.

Auchter, E. C. (1939). The Interrelation of Soils and Plant Animals. *Science* 89(2315): 421–426.

Barona Vilar, J. L. (2008). Nutrition and Health. The International Context During the Inter-War Crisis. *Social History of Medicine* 21(1): 87–105.

Beardsworth, A. and T. Keil (1997). *Sociology on the Menu: An Invitation to the Study of Food and Society*. London: Routledge.

Beaton, G. H. (1991). Human Nutrient Requirement Estimates: Derivation, Interpretation and Application in Evolutionary Perspective. *Food, Nutrition and Agriculture* (2/3): 3–15.

Bedale, E. M. (1923). Energy Expenditure and Food Requirements of Children at School. *Proceedings of the Royal Society of London. Series B, Containing Papers of a Biological Character* 94(662): 368–404.

Belshaw, H. (1947). The Food and Agriculture Organization of the United Nations. *International Organization* 1(2): 291–306.

Bennett, M. K. (1949). Food and Agriculture in the Soviet Union, 1917–48. *The Journal of Political Economy* 57(3): 185–198.

Berlan, J.-P. and R. C. Lewontin (1986). The Political Economy of Hybrid Corn. *Monthly Review, July–August*, 38: 35–47.

BioTech Institute (2010). Timeline of Biotechnolgy 20th Century. Washington, DC: Biotechnology Institute. http://www.biotechinstitute.org/what_is/timeline.html (accessed 12 January 2011).

Black, J. D. (1942). *Parity, Parity, Parity*. Cambridge, MA: The Harvard Committee On Research In The Social Sciences.

Blainey, G. (2000). *A Short History of the World*. Ringwood, Victoria: Viking.

BMJ (1933). British Medical Association. Report of Committee on Nutrition. *British Medical Journal*(Supplement): 1–16.

BMJ (1934). The Nutrition Question. *British Medical Journal* 1(3828): 900–901.

BMJ (1938). Malnutrition among School Children. *British Medical Journal* 2(4053): 585–586.

BMJ (1954). Correspondence: Women and Child Nutrition. *British Medical Journal* 1(4874): 1322.

Burnet, E. and W. R. Aykroyd (1935). Nutrition and Public Health. *Quarterly Bulletin of the Health Organisation* IV(2): 152.

Campbell, J. M. (1938a). The Nutrition Report. *Royal Institute of International Affairs* 17(2): 251–253.

Campbell, J. M. (1938b). The Nutrition Report. *International Affairs (Royal Institute of International Affairs 1931–1939)* 17(2): 251 253.

Carpenter, K. J. (2003a). A Short History of Nutritional Science: Part 1 (1785–1885). *Journal of Nutrition* 133(March): 638–645.

Carpenter, K. J. (2003b). A Short History of Nutritional Science: Part 2 (1885–1912). *Journal of Nutrition* 133(April): 975–984.

Carpenter, K. J. (2007). Biographical Article: The Work of Wallace Aykroyd: International Nutritionist and Author. *The Journal of Nutrition* 137: 873–878.

Carpenter, K. J., A. E. Harper and R. E. Olsondagger (1997). Experiments That Changed Nutritional Thinking. *The Journal of Nutrition* 127(5): 1017S–1053S.

Carr-Saunders, A. M. (1927). The Population Conference at Geneva. *The Economic Journal* 37(148): 670–672.

Chamberlin, W. H. (1934). The Ordeal of the Russian Peasantry. *Foreign Affairs* 12(3): 495–507.

Connelly, M. (2006). Seeing Beyond the State: The Population Control Movement and the Problem of Sovereignty. *Past & Present* 193: 197–233.

Cook, R. (1937). *Yearbook of Agriculture: A Chronology of Genetics*. Washington, DC: US Department of Agriculture.

Crutchfield, J. S. (1919). Food in the Reconstruction Period. *The Annals of the American Academy of Political and Social Science* 82: 7–10.

De Gans, H. A. (2002). Law or Speculation? A Debate on the Method of Forecasting Population Size in the 1920s. *Population (English Edition)* 57(1): 83–108.

Durand, E. D. (1922). Agriculture in Eastern Europe. *The Quarterly Journal of Economics* 36(2): 169–196.

Dyson, T. and C. Ó Gráda, Eds. (2002). *Famine Demography: Perspectives from the Past and Present*. UK: Oxford University Press.

FAO (1970). Report of the Conference of FAO: Annex D—Commemorative Address by Professor M. Cépède, Independent Chairman of the Fao Council. Rome: Food and Agriculture Organization.

Firth, R. (1934). The Sociological Study of Native Diet. *Africa: Journal of the International African Institute* 7(4): 401–414.

Fite, G. C. (1962). Farmer Opinion and the Agricultural Adjustment Act, 1933. *The Mississippi Valley Historical Review* 48(4): 656–673.

FSA (2008). About FSA: About the Commodity Credit Corporation. USDA Farm Service Agency. http://www.fsa.usda.gov/FSA/webapp?area=about&subject=landing&topic=sao-cc (accessed 1 January 2011).

Gantt, W. H. (1936). A Medical Review of Soviet Russia: Results of the First Five Year Plan. *British Medical Journal* 2(3939): 19–22.

Garside, W. R. (2002). *British Unemployment 1919–1939: A Study in Public Policy.* Cambridge: Cambridge University.

Gilks, J. L. (1935). The Relation of Economic Development to Public Health in Rural Africa. *Journal of the African Society* xxxiv(cxxxrv): 31–40.

Go, V. L. W., C. T. H. Nguyen, D. M. Harris and W.-N. P. Lee (2005). Nutrient-Gene Interaction: Metabolic Genotype-Phenotype Relationship. *Journal of Nutrition* 135(Supplement: International Conference on Diet, Nutrition, and Cancer): 3016S–3020S.

Goodrich, L. M. (1947). From League of Nations to United Nations. *International Organization* 1(1): 3 21.

Hanekamp, J. c. and A. Bast (2007). New Recommended Daily Allowances: Benchmarking Healthy European Micronutrient Regulations: Let Governments Take Care of Safety. *Journal of Environmental Liability* 4: 155–162.

Hardin, C. M. (1943). The Food Production Programs of the United States Department of Agriculture. *The Annals of the American Academy of Political and Social Science* 225: 191–200.

Harper, A. E. (1985). Origin of Recommended Dietary Allowances—An Historic Overview. *The American Journal of Clinical Nutrition* 41: 140–148.

Henry, C. (2005). Basal Metabolic Rate Studies in Humans: Measurement and Development of New Equations. *Public Health Nutrition* 8(7A): 1133–1152.

Hiller, E. T. (1930). A Culture Theory of Population Trends. *The Journal of Political Economy* 38(5): 523–550.

Hitchcock, G. M. (1919). In Defense of the League of Nations. *The Annals of the American Academy of Political and Social Science* 84: 201–207.

Holmes, O. L. (1924). The Economic Future of Our Agriculture *The Journal of Political Economy* 32(5): 505–525.

Hwalla, N. and M. Koleilat (2004). Dietetic Practice: The Past, Present and Future. *Health Journal* 10(6): 716–730.

IAAE (2008). About IAAE: History. International Association of Agricultural Economists. http://www.iaac-agecon.org/about/history.html (accessed 25 January 2010).

IFRC (2009). Red Cross and Red Crescent Movement: History. The International Federation of Red Cross and Red Crescent Societies. http://www.ifrc.org/who/history.asp?navid=03_09 (accessed 13 January 2010).

IGC (2009a). 60 Years of Successive Agreements: Before 1949: The Early Years. London: International Grains Council 4.

IGC (2009b). Grains Trade and Food Security Cooperation: The International Grains Agreement. The International Grains Council. http://www.igc.org.uk/en/aboutus/default.aspx#igc (accessed 21 February 2010).

ILO (2010). Ilo: Origins and History. International Labor Organization. http://www.ilo.org/global/About_the_ILO/Origins_and_history/lang--en/index.htm (accessed 12 March 2010).

IRC (2009). Who We Are. International Rescue Committee. http://www.theirc.org/about (accessed 12 January 2010).

Kindleberger, C. P. (1986). *The World in Depression, 1929–1939*. Berkely and Los Angeles, CA: University of California Press.

Kohn, G. F. (1924). The Organization and the Work of the League of Nations. *The Annals of the American Academy of Political and Social Science* 114: 5–77.

League of Nations (1919). *Covenant of the League of Nations, 28 April 1919*. League of Nations.

League of Nations (1933). *Final Act of the Conference of Wheat Exporting and Importing Countries, Held in London from August 21 to 25, 1933, with Appendices and Minutes of Final Meeting. Signed at London, August 25, 1933*: League of Nations. 3262.

League of Nations (1936a). The Problem of Nutrition: Volume I Interim Report of the Mixed Committee on the Problem of Nutrition. Geneva: League of Nations.

League of Nations (1936b). The Problem of Nutrition: Volume II Report on the Physiological Bases of Nutrition. Geneva: Technical Commission of the Health Committee of the League of Nations.

League of Nations (1937). Nutrition: Final Report of the Mixed Commision of the League of Nations on the Relation of Nutrition to Health, Agriculture and Economic Policy. Geneva: League of Nations.

Lloyd, D. J. and A. Shore (1938). *Chemistry of the Proteins*. London: J. and A. Churchill.

Markham, J. W. (2002). *A Financial History of the United States: From JP Morgan to the Institutional Investor (1900–1970)*. Armonk, NY, and London: M. E. Sharpe.

Mayne, J. B. (1947). FAO the History. *Review of Marketing and Agricultural Economics* 15(11): 418–426.

McDougall, F. L. (1940). *Food for a Hungry World*. Evanston, IL, September.

McDougall, F. L. (1943). The Food Situation in a World at War: International Aspects of Postwar Food and Agriculture. *The Annals of the American Academy of Political and Social Science* 225: 122–127.

McNaughton, J. W. and A. J. Cahn (1970). A Study of the Food Intake and Activity of a Group of Urban Adolescents. *British Journal of Nutrition* 24: 331–344.

Medical Science (1936). League of Nations and Nutrition. *Irish Journal of Medical Science* 11(1): 1.

Meerhaeghe, M. A. G. v. (1998). *International Economic Institutions*. Cambridge, MA: Kluwer Academic Publishers.

Merritt-Hawkes, O. A. (1928). The Population Conference at Geneva. *Journal of Heredity* 19: 313–315.

Michie, J. and J. G. Smith, Eds. (2000). *Managing the Global Economy*. New York: Oxford University Press.

Ministry of Health (1932). Advisory Committee on Nutrition Report to the Minister of Health on the Criticism and Improvement of Diets. London: HMSO, 1932. London: Ministry of Health.

Montgomery, B. (2008). E=Mc2, the Theory of Relativity, and IRC: Einstein's Great Ideas. International Rescue Committee. http://www.mynewsletterbuilder.com/tools/view_newsletter.php?newsletter_id=1409639110 (accessed 2 January 2010).

Morrison, G. (1947). Hybrid Corn: Science in Practice. *Economic Botany* 1(2): 5–19.

Mukerjee, R. (1933). The Criterion of Optimum Population. *The American Journal of Sociology* 38(5): 688–698.

NAL (2010). *Agricultural Adjustment Act of 1933: Definition*: USDA The National Agricultural Library. Agriculture Fact Book, USDA.

National Archives (1995). *Federal Records: Records of the Foreign Agricultural Service*. US National Archives.

NHM (1989/90) Where Did Biotechnology Begin? *Access Excellence*: http://www.accessexcellence.org/RC/AB/BC/Where_Biotechnology_Begin.php.

Northedge, F. S. (1986). *The League of Nations: Its Life and Times, 1920–1946*. Leicester: Leicester University Press.

Orr, J. B. (1939). Diet and Nutrition: Nutrition Problems, Dietary Requirements for Health. *The Canadian Medical Association Journal* 41(1): 78–80.

Ostrolenk, B. (1930). The Surplus Farm Lands. *The Annals of the American Academy of Political and Social Science*(148): 207–211.

Otto, J. S. (1999). *The Final Frontiers, 1880–1930: Settling the Southern Bottomlands*. Westport, CT: Greenwood Press.

Passmore, R. (1980). Wallcae Ruddel Aykroyd. *The British Journal of Nutrition* 43(1): 245–250.

Pearl, R. (1927). The Biology of Population Growth. In *Proceedings of the World Population Conference Held at the Salle Centrale, Geneva, August 29th to September 3rd, 1927*. M. Sanger. London: Edward Arnold.

Pearl, R. and L. J. Reed (1920). On the Rate of Growth of the Population of the United States since 1790 and Its Mathematical Representation. *Proceedings of the National Academy of Sciences of the United States of America (PNAS)* 6: 275–288.

Périssé, J. (1981). Joint FAO/WHO/UNU Expert Consultation on Energy and Protein Requirements: Past Work and Future Prospects at the International Level. Rome: Food and Agriculture Organization.

Philpott, J. (2009). How Healthy Are Government Dietary Guidelines? Part 1: Origin and Evolution of Dietary Guidelines. *The Nutritiona Practitioner* (Summer 2009): 15.

Pipkin, C. W. (1933). Relations with the League of Nations. *The Annals of the American Academy of Political and Social Science* 166: 124–134.

Ravenel, M. P. (1936). Nations Health: A Review of Selected Books of Interest to Public Health Worker. *American Journal of Public Health* 26(4): 432–438.

Royal Statistical Society (1936). Current Notes. *Journal of the Royal Statistical Society* 99(4): 830–835.

Ruxin, J. N. (1996). *Hunger, Science, and Politics: Fao, Who, and Unicef Nutrition Policies, 1945–1978, Chapter II the Backdrop of Un Nutrition Agencies, by Joshua Nalibow Ruxin*. London: University College London.

Sanger, M. (1927). *Proceedings of the World Population Conference, Geneva, August 29th–September 3, 1927*. London: Edward Arnold.

Save the Children (2009). History: Creating the Foundation. Save the Children. http://www.savethechildren.org/about/mission/our-history/ (accessed 8th Jan 2010).

Shaw, J. (2007). *World Food Security: A History since 1945*. UK: Palgrave Macmillan.

Sherman, H. C. (1920). Protein Requirement of Maintenance in Man and the Nutritive Efficiency of Bread Protein. *The Journal of Biological Chemistry* 41: 97–109.

Shurtleff, W. and A. Aoyagi (2007). *History of Soybeans and Soyfoods: 1100 B.C. To the 1980s*. California: Soyinfo Center, Lafayette.

Smil, V. (1999). Detonator of the Population Explosion. *Nature* 400(6743): 415–416.

Smil, V. (2001). *Enriching the Earth: Fritz Haber, Carl Bosch, and the Transformation of the World Food Production*. Massachusetts: Massachusetts Institute of Technology.

Smith, C. W., J. Betrán and E. C. A. Runge (2004), *Corn: Origin, History, Technology, and Production*. New Jersey: John Wiley.

Smith, D. F. (2003). Commentary: The Context and Outcome of Nutrition Campaigning in 1934. *International Journal of Epidemiology* 32: 500–502.

Sprague, G. F. (1962). Hybrid Corn.In *After a Hundred Years—Yearbook of Agriculture, 1962*. Washington, DC: US Department of Agriculture.

Staudinger, H. (1939). Problems of Population. In *War in Our Time*. H. Speier and A. Kähler, eds. New York: W.W. Norton & Co.

Stiebling, H. K. (1933). *Food Budgets for Nutrition and Production Programs*. Washington, DC: USDA.

Swanson, R. (2003). Foreign Agricultural Service Act of 1930. US: USDA Foreign Agricultural Service 8.

Taylor, A. E. (1926). World Food Resources. *Journal of Foreign Affairs* 5(1): 18–32.

Tenny, L. S. (1938). The Agricultural Adjustment Act of 1938: A Symposium. *The Journal of Land & Public Utility Economics* 14(2): 162–166.

Thompson, R. J. (1943). The United Nations Conference on Food and Agriculture. *Journal of the Royal Statistical Society* 106(3): 273–276.

UNICEF (2005). United Nations Childrens Fund. United Nations International Children's Emergency Fund (UNICEF). http://www.unicef.org/about/who/index_history.html (accessed 30 January 2010).

UNICEF (2009). The Convention on the Rights of the Child. United Nations International Children's Emergency Fund (UNICEF). http://www.unicef.org/rightsite/sowc/ (accessed 30 January 2011).

UNOG (2010). UNOG Registry, Records and Archives Unit, 1870- (Archive). UN Office at Geneva. http://biblio-archive.unog.ch/detail.aspx?ID=404 (accessed 23 February 2010).

USDA (1985). Possible Economic Consequences of Reverting to Permanent Legislation or Eliminating Price and Income Supports *Agricultural Economic Report* National Agricultural Library 99.

Varnee, D. B. (1955). A New Concept of Parity. *Journal of Dairy Science* 38(8): 935–939.

Watson, M. (1937). Malaria and Nutrition in Africa. *Journal of the Royal African Society* XXXVI(CXLV): 405–420.

Waugh, F. V. (1943). The Food Distribution Programs of the Agricultural Marketing Administration.*The Annals of the American Academy of Political and Social Science* 225: 169–176.

Webb, P., J. Gerald and D. R. Friedman (2008). More Food, but Not yet Enough: 20th Century Successes in Agriculture Growth and 21st Century Challenges. *Food Policy and Applied Nutrition Program*. Boston, Massachusetts: School of Nutrition Science and Policy, Tufts University.

White, T. R. (1919). The Amended Covenant of the League of Nations. *The ANNALS of the American Academy of Political and Social Science* 84: 177–193.

WHO (2010). Website of the World Health Organization. World Health Organization. http://www.who.int/ (accessed 2 April 2010).

Wolfe, A. B. (1934). On the Criterion of Optimum Population. *The American Journal of Sociology* 39(5): 585–599.

Worthington, E. B. (1936). On the Food and Nutrition of African Natives *Africa: Journal of the International African Institute* 9(2, Problems of African Native Diet): 150–165.

10

World War Two: 1939–1945

10.1 Governance, Politics and Socioeconomics

As busy as the previous inter-war years were for nutritional advancement so the World War II saw governance and multilateral cooperation on an as yet unprecedented scale. Commenting on free trade and the notion of comparative advantage, a staple of international economics at the time, Karl Brandt noted in the 1939 compilation 'War in Our Time' that nation's seemingly increasing over-reliance on foreign resources was borne out of political expediency rather than necessity (Brandt 1939). Indeed, in arguing against the over-simplistic and oft-parried phrase 'the have's and have-nots' in the economic sense, Brandt suggested that it was in fact this over-reliance and the neglect of a country's own resource base that was more likely at the heart of the widening economic gaps rather than any desire to get rich off the back of other countries. In rectifying this situation Brandt offered that in fact nations were historically generally very adaptable when it came to food and raw materials and instead of following an economic course that relied too heavily on imports, a rethink might be what was needed. Despite sentiments seemingly to the contrary too, Brandt was not against free trade as such not by a long shot instead he proposed that in times of supposed free trade when protectionist barriers or quotas rendered the playing field uneven, then a perhaps a fundamental shift of focus towards domestic reliance might be one answer. In this insightful look at resource allocation within countries in general, he touched on many of the natural resource issues that were to become prevalent in the 1990s. Although his focus was economics, trade and foodstuffs in particular, he did advocate thrift in consumption and sustainability in the sympathetic use of local natural resources. Furthermore, in his writings, there is more than a hint of the modern day food sovereignty movement in his advocacy of policy self-determination and food self-reliance (Brandt 1939).

On the institutional front, despite the Leagues' great leaps in economic and social cooperation, in particular with regards to the fields of health and nutrition, the inability of the institution to prevent the hostilities of World War II sounded the death knell for the first truly international organisation.

10.1.1 The End of the League

The League suffered many shortcomings which are beyond the remit of this study; however, one of the major limitations involved aggressor states and the League's inability to offer little more than moral outrage. The problems began perhaps in 1931 when Japan, in defiance of both the Council and the Assembly, waged war against China. Furthermore, in 1935 when Italy invaded Ethiopia, sanctions were limited and only half-heartedly supported whilst finally in 1936 the illegal German reoccupation of the Rhineland left the League looking impotent. Indeed, even as Hitler's armies marched into Poland in 1939, not a single member of the League called for a meeting of the Council or Assembly (Encyclopedia 2002). It was now clear for all to see that as the League had ultimately failed in its main duty to prevent the outbreak of World War II; it was effectively undermined as a competent organisation and was now largely seen as powerless and incapable (Shaw 2007).

Once again as war broke out so the spectre of food shortages raised its head and one of the first priorities for the Allied powers was the production and provision of foodstuff on an enormous scale. As one commentator acknowledged

> ... the food and nutrition problems of a nation at war are tremendously complex and involve a wide network of interrelated activities ... (Clayton and Black 1943, p. 105)

10.1.2 Food Provisioning, Rationing and Production

For those nations throughout Europe heavily reliant on imported foodstuffs, once more widespread rationing became an everyday occurrence (Bacon 1943; Volin 1943; Whipple 1943; Lloyd 1943; Richter 1943). Reduced reliance in imports and increased acreage under production helped secure measured success in many parts; however, not all faired the same. Much inequality and deficiencies still existed and stretched existing resources to the limits (ibid). To offset some of these shortages and to help in the war effort, one important measure of aid was brought in by the United States; the Lend-Lease Agreement. The agreement was passed in 1941 and allowed those countries the United States deemed suitable to buy goods and services on a loan basis on favourable terms (Clayton and Black 1943). Procurement was expedited by the Office of Lend-Lease Administration, which authorised transfer to the Lend-Lease countries of the goods and services concerned. During this time countries and regions fared quite differently.

10.1.2.1 Britain

While the war broke in Britain during a period of food surpluses, the UK Ministry of Food once again introduced rationing in 1940 amid fears of shortages brought about by the worries of a successful submarine and bomber campaign (Black 1943; Lloyd 1943). Food imports were slashed, and farmers increased acreage under production by as much as 50%. During this period, home production was micro-managed jointly by the Ministry of Food and the Agricultural Department. The Ministry of Food too became the sole purchaser of the country's food imports, and intervention in pricing aimed at keeping staples like bread and meat affordable became commonplace (Clayton and Black 1943). Thanks to the US lend-lease programme and Britain's wartime food planning and equitable distribution through rationing, Britain's general nutritional status was quite possibly better than was achieved in the United States at that time (Cassels and Hall 1943; Lloyd 1943).

10.1.2.2 Europe

In central and northern Europe too, before the war broke, European countries, particularly Austria, Czechoslovakia, Germany and Sweden, were largely self-sufficient to the tune of 80% production to consumption (on a caloric basis) (Richter 1943). Hostilities changed all this. By midway through the war, Belgium and Norway's population, reliant on imports of grain, feedstuff, fats and oils suffered badly from sub-optimum diets. This was further exacerbated by German takings of fish from Norway and meat from Belgium. Sweden and Switzerland on the other hand with continued access to imports fared better, while Denmark still produced much of its requirement of food, even after exports to Germany. So much so in fact that

> ... enough was left for the Danes to maintain the highest food allowances in all of rationed Europe. (Richter 1943)

In France, Spain and Italy however, rationing was widely introduced although peasant revolts and a thriving black market ensured it was the urbanised population that suffered most (Bacon 1943). Greece, with an agricultural workforce largely consisting of women and children, was less affected than other more highly developed regions (Whipple 1943). In Russia too, midway through the war, collectivised Russian agriculture, despite heavy mechanisation and every effort to increase the acreage under cultivation, was failing, and its population was increasingly under grave threat of mass hunger and starvation. So much so that substantial allied assistance was heavily anticipated (Volin 1943).

10.1.2.3 Africa and Asia

In Central Africa, the problem of supply was considered mainly one of education to bring about more systematic methods of food production (McDougall 1943). It was realised that to achieve this goal, extensive international action was required including considerable financial aid in some quarters. In Asia however (Japan proper, Manchuria, China, Philippine Islands, Netherlands Indies, British Malaya, French Indo-China, Thailand, Burma and India), the story was different. Self-sufficiency in rice and other principal foodstuffs were not affected drastically by the war, and the status quo continued (Ladejinsky 1943).

10.1.2.4 The Americas

The all-out wartime food programme came slowly to the United States. Amid pre-war abundance, people were used to thinking in terms of agricultural surpluses rather than shortfalls (Cassels and Hall 1943). However, all that changed as military and lend-lease requirements as well as shortages of labour, machinery and agricultural chemicals threatened the continuation of the status quo (Clayton and Black 1943). Previous agricultural abundance gave way to a rationing system based on need rather than ability to pay. Shortages came as a shock to the Americans, but they rallied to the challenge, increasing production and supplying the war effort. Helping in this effort was the inclusion of the Secretary of Agriculture into the Defense Council in May of 1940; this effectively put agriculture on a war footing. Several aims of the Advisory Commission to this council were: the maintaining of parity between industrial and farm prices; gearing up of production in certain areas; using regional agricultural surpluses in others; and finding employment for idle rural labour (Clayton and Black 1943). However, despite such ambitious aims, it soon became clear that these measures were not living up to expectation, and in 1941 President Roosevelt inaugurated the Office of Production Management responsible for production and purchasing.

The central and south Americas on the whole, despite some regional variations in central America, were largely self-sufficient in production and in many instances (in the temperate zone countries) producing large surpluses (Almonacid 1943). Thus, it was seen that rationing throughout much of the world during the war brought focus of the nutritional question of what was considered adequate nourishment as well as concentrating minds on the means of increasing food production on a massive scale (WHO 1958).

However, out of the ashes of World War II came one of the major breakthroughs in the multilateral fight against food security—the United Nations (UN).

10.1.3 The United Nations

With the end of the League of Nations in 1946 what was needed was a new international organisation with stronger executive powers responsible, among other things, for peace and cooperation once hostilities of the war were settled (UNOG 2009). To this end, the League's failure to prevent hostilities in 1939 did not destroy the belief in the concept of an international organisation for peace. On the contrary, determination to learn from mistakes and to build a new body more able to maintain international peace was envisaged. Indeed, even before the end of the League itself, this new multilateral body was in actual fact, already fully conceived. Setting the blueprint for the new 'United Nations' was the Inter-Allied Declaration (or the London Declaration) of June 1941. Meeting at St. James's Palace in London, representatives of the United Kingdom, Australia, Canada, New Zealand, the Union of South Africa as well as the governments-in-exile of Belgium, Czechoslovakia, France, Greece, Luxembourg, the Netherlands, Norway, Poland and Yugoslavia each pledged not to sign a separate peace document with Germany and declared that

> The only true basis of enduring peace is the willing cooperation of free peoples in a world in which, relieved of the menace of aggression, all may enjoy economic and social security … .
> (UN 2006)

A couple of months later, off the coast of Newfoundland, a declaration summarising the key points of agreement between President Roosevelt and Prime Minister Winston Churchill aboard the warship *USS Augusta* was issued. Although not a treaty as such but rather a document of expression, the Atlantic Charter of August 1941, as it became known, was further reinforced by Roosevelt in his State of the Union speech to Congress in 1941 when he elucidated his Four Freedoms concept. These freedoms, he insisted, were essential to the development of mankind throughout the world and consisted of the freedom of speech, the freedom of worship, the freedom from want and ultimately the freedom from fear

(Roosevelt 1941; UN 1945). Collectively these notions established a wider, more permanent ideology of general social security.

On 1 and 2 January 1942, 27 nations signed up to what became the United Nations Declaration. This effectively bound governments to the war effort and against any separate peace deal with the aggressor powers (UN 2006). By 1943, although declarations of interest had been signed, the basis for such a world organisation had yet to be properly defined. This was achieved at the Moscow Declaration of October 1943 where the Foreign Ministers of Great Britain, the United States and the Soviet Union along with an ambassador from China drew up the Four-Power Declaration which recognised the need to establish an international organisation for the 'maintenance of international peace and security'. Two months after this declaration, Roosevelt, Stalin and Churchill, meeting at Teheran, hammered out agreement on the scope and timing of operations against Germany.

While the principles were laid down, the structure of the proposed United Nations was tackled at Dumbarton Oaks (a private mansion in Washington, DC). By October 1944 negotiations identified four main bodies: a General Assembly (with provision for an Economic and Social Council); a Security Council of eleven members; an International Court of Justice; and the fourth, a Secretariat. In 1945 despite the sudden death of President Roosevelt, the United Nations Conference on International Organization was convened in San Francisco as scheduled where representatives of 50 countries met to establish the final text that would become the Charter of the United Nations. The United Nations was born.

One astute economist from Australia, and a member of the Economic Committee of the League of Nations, Frank McDougall, observed of the UN's responsibilities at the time:

> If the United Nations decide to make freedom from want of food the first step towards the attainment of the President's third freedom—freedom from want—this will require national action in every country and international action to assist countries lacking technical knowledge and financial means to secure improvements in food production. It has therefore been proposed that the more economically advanced United Nations should pledge themselves, first to institute policies designed to ensure that the foods required for diets adequate for health are within the purchasing power of their own citizens, and second to provide technical and financial aid to enable the less advanced nations to progress towards the accomplishment of this objective. (McDougall 1943)

This was indeed what the UN had planned for itself, although just how it fared in this endeavour is unraveled in the course of following sections.

10.1.3.1 UN Conference on Food and Agriculture: Creation of the Food and Agriculture Organization

Even before the proposals for the United Nations were formally agreed (in fact, pre-dating the San Francisco Conference by two years), President Roosevelt convened the first UN conference—the Conference on Food and Agriculture at Hot Springs, Virginia, in May and June 1943 (UN 1943). This came about in part due to the efforts of Frank McDougall who in 1942 wrote a report entitled the 'Draft memorandum on a United Nations Programme for Freedom from Want of Food'. This caught the attention of the president and eventually the two men met. McDougall was passionate and urged the president that the first problem that the United Nations should tackle after the war should be the problem of food. As a result the following year, Roosevelt called for a United Nations Conference on Food and Agriculture (FAO 1995). The conference reiterated the notion of the 1935 League of Nations commission for the need for universal multilateral action to reduce occurrences of hunger and malnutrition through the introduction of a permanent organisation responsible solely for the domains of food and agriculture (Parran 1943; Thompson 1943; Evang and McDougall 1944). Essentially, an organisation was needed to act as a central clearing house for information and one that could promote stability through international commodity agreements, maintain adequate reserves and when necessary, the orderly and equitable disposal of any surpluses (Shaw, 2007). This first UN Conference on Food and Agriculture led to the creation of an Interim Commission on Food and Agriculture. The commissions went on to draft a Constitution for the new permanent organisation to be known as the Food and Agriculture Organization (UN 1943; Phillips 1981).

The 1943 conference on food and agriculture was a decisive moment in the history of food security. It aspired to many goals and building on Roosevelt's third freedom, defined as its main objectives the goal of 'freedom from want of food suitable and adequate for the health and strength of all peoples' (Thompson 1943). Furthermore, the conference also declared that

> The first cause of hunger and malnutrition is poverty. It is useless to produce more food unless men and nations provide the markets to absorb it. There must be an expansion of the whole world economy to provide the purchasing power sufficient to maintain an adequate diet for all. (UN 1943)

The scope of this aim covered three main areas including improved national nutritional diets, increased production through worldwide agricultural expansion and better resources for the achievement of such aims. The conference also identified several important issues regarding food security for all:

- Freedom from want entailed a secure, adequate and suitable supply of food for every man, woman and child
- To encourage the growth of national nutrition organisations to gather and exchange information and experience
- It also recognised the interdependence of producers and consumers and that agricultural policy must be considered multilaterally through concerted and coordinated action by member governments
- The conference espoused the fundamental ideals of sustainability too in providing sufficient nutritious food whilst maintaining productivity of the lands and conserving land and water resources
- To encourage agricultural credit, cooperative movements, to ensure land tenure, education and research
- The recognition of vulnerable or special groups of populations such as the young and pregnant women and the need to place these groups at the top of food security priority
- Building on previous studies previously mentioned the connection was made clear between many prevalent diseases and dietary deficiencies.

The conference effectively galvanised much of the accepted wisdom of the day and sought to bring to bear the full might of a multilateral organisation in its solutions (Burnet and Aykroyd 1935; UN 1943; Carlson 1944; Evang and McDougall 1944; Schultz 1945; Phillips 1981). In this endeavour the conference was succinctly summed up in one of its resolutions that stipulated

> There has never been enough food for the health of all people. This is justified neither by ignorance nor by the harshness of nature. Production of food must be greatly expanded; we now have knowledge of the means by which this can be done. It requires imagination and firm will on the part of each government and people to make use of that knowledge. (UN 1943)

It was envisaged too that in order to achieve this, much work was needed beginning with cooperation of existing agencies; the introduction of national nutritional organisations; more agricultural investment and credit agencies; widespread cooperative movements; improved education and research; sustainability and conservation particularly in the fields of land and water resources; occupational adjustments; improved land tenure; international commodity agreements; more equitable distribution measures; humanitarian aid policies and ultimately monetary, financial and commercial mechanisms to secure expanding world economic growth and increased international security (McDougall 1943; Thompson 1943; Phillips 1981).

In particular, remarks McDougall, these measures would have to be applied across much of the developing world (McDougall 1943). For at the time, as much as 60% of the working population of the world was in agriculture:

> Therefore, on the basis of numbers, the social condition of the peasants, share croppers, and other farm workers is the outstanding social problem of the world. (McDougall 1943, p. 126)

The deliberations of the Conference also showed that farmers could not rely on adequate returns nor would consumers be in a position to purchase the food they needed, if progress was not engaged on the international stage to raise the general level of employment in all countries. This ultimately meant the general worldwide expansion of economic activity (UN 1943). In effect, the Conference advocated expanded production and increased consumption whilst touching on the need to achieve stability in commodity prices, availability and incomes. All this to be achieved through increased regulation of production and other instruments of development (Thompson 1943).

This forward thinking body that placed food and agriculture at the forefront of the first United Nations' conference was considered by then Surgeon General of the United States, Thomas Parran, to have been a rare insight by Roosevelt and one in which the global needs and expansion of a healthy and beneficial life were first addressed in such a forum (Parran 1943). As a result, in 1945 a new body of the United Nations, the Food and Agriculture Organization, was duly created and held its first meeting in Quebec that year. It was headed by the British nutritionist Sir John Boyd Orr.

10.1.3.2 Relief and Rehabilitation Administration (UNRRA)

As the war drew to a close, the need for broader international action and co-ordination beyond just theory, as some had charged the Allied Committee with, was presented to the UN. This resulted in the United Nations Relief and Rehabilitation Administration (UNRRA) of 1943 (Foreign Office 1943; House of Commons 1944; League of Nations 1946). This subsumed the recently formed Office of Foreign Relief and Rehabilitation (OFRRO), which was set up in 1942 by the US State Department (Williams 2005), and its remit was both 'broad and sweeping' (League of Nations 1946). Fundamentally, in the UNRRA's brief three-year tenure it aimed to administer

> ... relief of victims of war in any area under the control of any of the United Nations through the provision of food, fuel, clothing, shelter and other basic necessities, medical and other essential services. (League of Nations 1946, p. 92)

10.1.3.3 Bretton Woods Agreement

Another of the fledgling United Nations founding conferences was the UN Monetary and Financial Conference, more commonly known as the Bretton Woods Conference of 1944. The need for such a financial agreement was partly seen as a response to practices that contributed to the 1930s depression era whereby nations attempted to gain an economic competitive edge by raising barriers to foreign trade. Although some practices such as the devaluation of currencies to compete for export markets actually became counterproductive and saw world trade rapidly decline and living standards plummet. The aims of this new agreement were ostensibly the regulation of international monetary and financial order as well as the expedition of post-war reconstruction (Bordo and Eichengreen 1993; IMF 2010a). The agreement that came out of this conference came into effect in 1945 and created among others the International Monetary Fund (IMF) and the International Bank for Reconstruction and Development (IBRD), later collectively known as the World Bank Group (WB).

The International Monetary Fund (IMF): The IMF came into formal existence in 1945 and began operations in 1947. The breakdown in international monetary cooperation described in previous sections led IMF's founders to initiate an institution with the remit to oversee the international monetary system—a system of exchange rates and international capital and current payments that enabled the free flow of goods and services among countries. The new organisation would facilitate trade through exchange rate stability and aimed to achieve this using the Par-Value System (next chapter) as well as serving as the lender of last resort to countries.

The International Bank for Reconstruction and Development (IBRD) was originally created to finance post-war reconstruction while nowadays its mission has expanded to fight poverty and the extension of long-term investments in underdeveloped nations for the benefit of development and structural improvements. This is achieved by means of the groups' financing member states. Collectively the IMF and the IBRD are known today as the World Bank.

Elsewhere in the field of governance, people were not happy with the status quo of agriculture. For these people, the industrialisation and intensification of farming was abhorrent. As a result and in the midst of all this, the birth of the organic movement had begun to take root.

10.1.4 Organic Farming and Environmentalism

In 1924, Rudolf Steiner introduced the fundamentals of a new organic method of agriculture called bio-dynamics. Sharing its fundamental belief in natural agriculture, bio-dynamics was essentially the forerunner to the modern organic movement. From these humble beginnings, the world's first organic association was the Australian Organic Farming and Gardening Society (AOFGS), which began in Sydney in 1944 (Paull 2008). However, prior to this, one keen agriculturist and writer Lord Northbourne, born Walter Ernest Christopher James, tended his gardens at Northbourne Court and at Home Farm in the United Kingdom according to the biodynamic principles espoused by Steiner. By avoiding all chemical fertilisers, pesticides and herbicides as well as by composting and returning all organic wastes back to the soil, Lord Northbourne was perhaps one of the first true organic farmers in Britain. His importance within the organic movement cannot be underestimated as in 1938/1939 the first conference on organic farming was held at Northbourne Court. Furthermore, in 1940 he wrote an influential book *Look to the Land* in which he foretold of the fears that were to pre-occupy Rachel Carson and the public in the 1960s:

> In the long run, the results of attempting to substitute chemical farming for organic farming will very probably prove far more deleterious than has yet become clear. (Northbourne 1940, p. 61)

Moreover, in the same book, Northbourne has been credited as having been the first to coin the phrase 'organic farming' (James and Fitzgerald 2008). Shortly prior to this though, inspired by the works of Sir Albert Howard and Sir Robert McCarrison, Lady Eve Balfour started the Haughley Experiment on her farm in Suffolk. It began in 1939 with the intention of investigating the claims of the benefits of organic husbandry. Her approach was to manage three side-by-side parcels of land in which the full cycle of the food chains could be studied during many successive generations of plants and animals. By 1945, the need to administer the experiment and correlate the information led to a body, the Founders Committee, which later (1946) became the Soil Association (SA). A year or so later under financial strain, the experiment was taken over by the SA. Incidentally, the experiment lasted over 40 years, and soil analyses results of the organic soil showed comparatively higher moisture, organic C, and mineral N, P, K, and S content (Balfour 1975; Blakemore 2000).

10.2 Health and Nutrition

At the outbreak of World War II, nutritional health policies in the United Kingdom and the United States as well as other countries were becoming more and more bound up with national health directives (Beardsworth and Keil 1997). Improved scientific knowledge had allowed rations to be more precisely determined at the national level. Interest was also moving towards the nutritional needs of the disaggregated population, first at the level of population sub-groups, then to the household and to the individual (Beaton 1992). Although ostensibly the work of respective National Nutrition Committees was abruptly halted as war reared, some domestic nutritional studies continued (McDougall 1943). At the forefront of such studies was Britain, and armed with new knowledge and the rationing culture the government embarked on several nutritional education programmes, one of which the Food Leaders Scheme, aimed particularly at women. At the heart of such policies was the almost evangelical belief that nutrition was a hearty antidote to many social ills of the day (Ruxin 1996).

At about this time, both the US and British governments sought to establish the ideal nutritional requirements of soldiers in the field. Before the United States had entered the war, a report by the influential National Research Council (NRC) assured the government that the American people were in no immediate danger of experiencing a shortfall in protein supply. Although it did concede that were the country called upon to export its high-protein foods to the allied nations then a protein shortage might indeed just occur (ibid). Thus, building on the earlier pre-occupation with proteins in the diet the NRC fuelled this early obsession that later came to dominate much nutritional literature in the ensuing years. As the war drew to a close, this report along with McDougal's 1942 report and others placed the foreseen problem of post-war food supply high on the international agenda. However, even before this time others had not forgotten the lessons of World War I and were already thinking of the needs of post-war reconstruction.

10.2.1 1941 Inter-Allied Committee on Post-War Requirements

In 1940, the British Prime Minister in a House of Commons speech promised food and relief for the peoples of Europe at the end of the war. Shortly afterwards, under the initiative of the British Government, the Allied powers met in June 1941 at St. James's Palace to consider relief and rehabilitation after the cessation of hostilities (House of Commons 1944). At this meeting they passed a resolution stating that

> ... the only true basis of enduring peace is the willing co-operation of free peoples in a world, in which, relieved of the menace of aggression, all may enjoy economic and social security; and that it is their intention to work together, and with other free peoples, both in war and peace to this end. (House of Commons 1944)

This meeting, two months before the signing of the Atlantic Charter (previous section), reconfirmed the principles of mutual cooperation for the benefit of all, while a further meeting in September agreed a resolution to secure supplies of food, raw materials and other needs of the post-war liberated countries (ibid). Out of these meetings, the Inter-Allied Committee (IAC) was set up with responsibilities of gathering information with which to formulate policy (Foreign Office 1943; House of Commons 1944; League of Nations 1946). In their efforts, the Inter-Allied Committee greatly advanced the use of food balance sheets and used them extensively in their 1942/43 studies of post-war requirements (Foreign Office 1943; FAO 2001). The Committee was eventually transformed into the European Advisory Commission (EAC) (Williams 2005).

10.2.2 Recommended Dietary Allowances

By this time, it was the work by Mitchell, Stiebling and Roberts, working for the US Research Council of the National Academy of Sciences who, building on these collective advances, produced the first widely accepted Recommended Dietary Allowances (RDA) in 1941. It began with the Committee on Food and Nutrition which was set up at the request of the US Federal Government in 1940 under the National Research Council and was later renamed the Food and Nutrition Board in 1941 and established as a permanent body. The Board presented their recommended allowances to the Spring Meeting of

TABLE 10.1

Food and Nutrition Board's 1941 Recommended Dietary Allowances

Energy (kcal)	Protein (g)	Calcium (g)	Phosphorus (g)	Iron (mg)	Vitamin A (IU)	Vitamin B1 (IU2)	Vitamin C (mg)	Riboflavin (mg)	Nicotinic acid (mg)	Vitamin D (IU)
2775	66	0.91	—	12	4696	516	71	2.3	15.5	210

Source: NRC *Recommended Dietary Allowances: Report of the Food and Nutrition Board*, Reprint and Circular Series No. 115., National Research Council, Washington, DC, 6, 1943.

the American Institute of Nutrition in 1941 (Editorial 1943; Harper 1985; Rosenberg 1994; Aggett et al. 1997; McArdle et al. 1999; Harper 2003). However, although these allowances were not published by the National Academy until 1943, the table did appear in 1941 in the *Journal of the American Dietetic Association* (Table 10.1) (ADA 1941; NRC 1943; Harper 1985; Mayer 1986; Harper 2003).

10.3 Science and Technology

In science and technology, during the war years much emphasis was placed on the application of existing technologies for the war effort. However, there were also some advances in the areas of food production technologies. This was a time that mass production of food using automation really began to take off. Concentrated, frozen and dehydrated foods such as concentrated citrus juices produced in mass quantities for shipping overseas to the military were perfected as was the fortification of flour with vitamins and iron (1940). Aseptic processing and packaging, which involved the high-temperature, short-time sterilisation of a food and its container independently, then the filling of the container with the product in a sterile atmosphere, was also developed. This allowed the improvement of the quality of food through the retention of valuable nutrients as well as increased safety through pasteurisation and sterilisation processes (IFT 1999; Gardner 2002).

   Other events saw the large-scale production of penicillin as well as the introduction of the electron microscope, which was successfully used to identify and characterise a bacteriophage, a virus that infects bacteria (1942). Within a year separate studies by Salvador Luria and Max Delbruck who together performed the first quantitative study of mutation in bacteria and along with previous advances in this area witnessed the beginning of bacterial genetics as a distinct discipline. This led in the following year to the discovery that DNA was shown to be the material substance of the gene and responsible for generational transformations. This was counter to the accepted wisdom of the day, which thought that only proteins were intricate enough to express all of the complicated genetic combinations. It was also discovered too about this time that genes in fact could be transposed from one position on the chromosome to another (BioTech Institute 2010).

10.4 Period Commentary

World War II broke amid a period of surpluses in the West, at a time when agricultural policy was still predicated on restraining output without unduly depressing prices (Black 1943). However, it was not long before much needed allied imports from the East Indies, the Americas and Asia were failing to get past the German blockades and stocks became low. With this in mind, the lessons of the World War I were not forgotten, and domestic agricultural production was geared up and swelled to meet expected demand (Reid and Hatton 1941; Clayton and Black 1943; Hardin 1943). As a result, Western governments and agencies, through legislative acts and policies, colluded to secure the anticipated necessary agricultural supplies needed for themselves, the war effort and their allies. The importance of food at this time was also reflected in the United States when they incorporated it for the first time in the National Defence Council, while the US lend-lease act of 1941 also contributed not only vital war materials but also essential food supplies too (Clayton and Black 1943).

   By mid way through the war, people were turning their attention to the perceived problems of post-war reconstruction, food and nutrition when political and social instability was widely expected to be at its most challenging. Indeed, referring to such instability at the time Black noted in 1943 that after hostilities the suffering people must be fed and that food would be the first step in restoring responsible government in Europe otherwise, he contended that

> A democracy is weak unless it has the strong support of all its citizens. A government that lets its people suffer from hunger and malnutrition in the midst of plenty no longer is safe from the infiltration of perilous isms and from secret hostile groups. (Black 1943, p. 2)

Further echoing this view, Frank McDougall's relentless advocacy of improved nutrition through health and agriculture during the 1930s led to the Australian government's own advocacy of such policies in the League of Nations. Furthermore in achieving such harmony of nutrition McDougall recognised the importance of how this might fit or be achieved within the bigger picture of post-war global reconstruction. Later, writing in the *Annals of the American Academy of Political and Social Science*, McDougall voiced the general opinion that was to become (and continues to be) the mainstay of the economic paradigm of development

> There is general agreement that in order to secure their own economic prosperity and to enable the world to enjoy political security, the United States and the United Kingdom, acting in agreement with the other United Nations, must so adjust monetary and financial mechanisms and commercial policies as to secure an expanding world economy. (McDougall 1943)

Such views not only cemented the prevailing global development paradigm of the time, it also acted to illuminate, for those still sitting on the fence, the relationship between economic prosperity and the health and nutritional well-being of individuals and by extension, the nation's well-being.

Meanwhile, the end of the League of Nations, with its failure to prevent the outbreak of World War II did not mark the end of this line of ideology. Indeed, it was understood more fervently that the suffering and the many challenges to be faced, post-war, was a global problem that needed an equally strong global remit in the form of the United Nations. Although the term 'United Nations (UN), was first coined by Roosevelt in 1941 whilst referring to the Allies in their fight against the Axis powers, it was later used to signify the signatories of the declaration in Washington 1942 and their agreement with the principles of the Atlantic Charter. After several subsequent meetings to determine a blueprint for the organisation' the United Nations was ratified by its members in 1945 (Sweetser 1946; Shaw 2007; UN 2009a). Although the league was now abandoned, the UN kept much of its ideals and some of its structure (UN 1945; Sweetser 1946). Originally, the organisation's charter established six principal organs: the General Assembly; the Security Council; the Economic and Social Council; the Trusteeship Council; the International Court of Justice; and the Secretariat (UN 1945, 2009b). It also relied on specialised agencies, several of which were autonomous or semi-autonomous and responsible for particular areas or domains in which they specialised.

One such organisation, the FAO, came out of the first UN conference; the Conference on Food and Agriculture at Hot Springs in 1943 (Marrack 1948). This conference also led to the statement further backing the views of Black and McDougall and one that remains as salient today as it was then by acknowledging that

> ... the first cause of hunger and malnutrition is poverty ... (UN 1943)

It was realised by this time too that the when the war finally ended, the relief organisations of the United Nations would have to meet the needs of a 'large proportion if not the majority, of the urban population of continental Europe (McDougall 1943). Causes of hunger were researched too' and by the end of the war the correlation between poverty and hunger had been firmly established and enshrined in the first UN conference on food and agriculture. Also of note during the Hot Springs conference was the fledgling idea of sustainability when, in its official summary, it offered that

> The types of food most generally required to improve people's diets and health are in many cases those produced by methods of farming best calculated to maintain the productivity of the soil and to increase and make more stable the returns to agricultural producers. In short, better nutrition means better farming. (UN 1943)

This was an important time for nutrition too as great leaps were made in nutritional knowledge with what was perhaps the first widely accepted Recommended Dietary Allowances in 1941 by the National Academy of Sciences. Of note too at this time is the notion that some of the previous disparate strands of food security were beginning to come together, and this was no more elucidated than with the

Declaration of the first UN Conference on Food and Agriculture mentioned before but worth repeating here:

> It is useless to produce more food unless men and nations provide the markets to absorb it. There must be an expansion of the whole world economy to provide the purchasing power sufficient to maintain an adequate diet for all … (AIS 1946, p. 10)

Thus, by the end of the war, it can be seen that the food security ideology was beginning to take on much of the complexity recognised in today's concept and as such it was realised that to alleviate the burdening problem it meant tackling a whole host of social issues head on.

10.5 Key Dates, People, Acts, Reports and Surveys

Table 10.2 highlights some of the more important events, noteworthy individuals and key reports, etc. that helped shape this period of the 20th century.

TABLE 10.2

Key Ideas and Dates World War II: 1939–1945

Date	People, Organisation, Agencies, Acts or Conventions
1939: US Food Stamp Program	Food distribution programmes developed in the United States during a period of huge surpluses and massive unemployment. Later became known as the Supplemental Nutrition in Assistance (SNAP) (Waugh 1943; Dimitri et al. 2005; Landers 2007; National Archives 2009).
1941: The term 'genetic engineering' is first used	The term 'genetic engineering' is first used by Danish microbiologist A. Jost in a lecture on sexual reproduction in yeast at the technical institute in Lwow, Poland.
1941: Recommended Dietary Allowances (RDA)	Building on work by the LoN, Burnett and Ackroyd and many others in nutritional research, the Committee on Food and Nutrition set up under the US Research Council of the National Academy of Sciences produced the first Recommended Dietary Allowances in 1941, subsequently published in 1943. The committee was renamed the Food and Nutrition Board in 1941 (Harper 1985; Mayer 1986; Rosenberg 1994; Harper 2003).
1941: The Atlantic Charter	Winston Churchill and President Franklin Roosevelt met aboard the *USS Atlanta* and signed a declaration giving the first indication that the two powers would strive for the creation of a new world organisation once peace was restored after World War II (UN 1943; UNOG 2009).
1941: US Lend Lease Act	Contributed vital war materials and essential food supplies to Allied nations (Flicker 1967; Hoffmann and Maier 1984).
1942: International Wheat Council (IWC)	Supersedes the 1933 Wheat Advisory Committee with headquarters in Washington; in turn the international seat of IWC was established in London in 1949 (League of Nations 1933; IGC 2009a, 2009b).
1942: OXFAM	Started by a group in Oxford as the Oxford Committee for Famine Relief (OXFAM 2007).
1943: United Nations Relief and Rehabilitation Administration (UNRRA)	Began in 1943 to provide relief and healthcare work to areas liberated from the Axis powers after World War II. UNRRA provided billions of dollars of aid. It ceased operations in Europe in 1947 and in Asia in 1949. UNRRA was dissolved in 1946 and its functions transferred to the Interim Commission of WHO (WHO 2009).
1943: Catholic Relief Services (CRS)	Catholic Relief Services founded by the Catholic Bishops of the United States to aid World War II survivors in Europe (CRS 2009).
1944: Norman Borlaug and the Rockefeller Foundation	Norman Borlaug has often been referred to as the father of the green revolution; his work with the Rockefeller Foundation and further studies in the field of crop hybridisation in the 1950s and 1960s led to the Nobel Peace Prize in 1970 (Herdt 1998; USAID 2009).

continued

TABLE 10.2 (Continued)

Key Ideas and Dates World War II: 1939–1945

Date	People, Organisation, Agencies, Acts or Conventions
1944: UN Monetary and Financial Conference—The Bretton Woods Agreement	Commonly known as Bretton Woods conference, signatories sought to regulate international monetary and financial order after World War II. The agreements established the International Monetary Fund (IMF) and the International Bank for Reconstruction and Development (IBRD) (IMF 2010b).
1943: Plant Breeding as foreign aid	The Rockefeller Foundation collaborated with the Mexican government in the Mexican Agricultural Program, the first use of plant breeding as foreign aid.
1945: CARE	Cooperative for American Remittances to Europe later renamed Cooperative for Assistance and Relief Everywhere (CARE 2009).
1945: United Nations (UN)	International Organisation for the economic, judicial and social development (UN 1945, 2000).
1945: Food and Agriculture Organization (FAO)	A specialised United Nations agency responsible for the economic and social development of food and agriculture (Phillips 1981). FAO deals with all major aspects of agriculture, fisheries and forestry, through both its Regular and Field Programmes (FAO 1946; Phillips 1981; FAO 2010).
1945: World Bank Group (WB): International Bank for Reconstruction and Development (IBRD) and International Monetary Fund (IMF)	The IBRD was originally created to finance post-war reconstruction. Nowadays its mission has expanded to fight poverty, while the function of the IMF is to maintain orderly currency practices in international trade (IMF 2010b).
1945: UN Educational, Scientific and Cultural Organization (**UNESCO**)	UN Educational, Scientific and Cultural Organization (UNESCO) founded in 1945 is a specialised United Nations agency. The organisation serves as a clearing house of ideas and the dissemination and sharing of such information and knowledge (UNESCO 2009).

References

ADA (1941). Recommended Allowances for the Various Dietary Essentials. *Journal of American Dietetic Association* 17: 565–567.

Aggett, P. J., J. Bresson, F. Haschke, et al. (1997). Recommended Dietary Allowances (RDAs), Recommended Dietary Intakes (RDIs), Recommended Nutrient Intakes (RNIs), and Population Reference Intakes (PRIs) are not Recommended Intakes. *Journal of Pediatric Gastroenterology and Nutrition* 25(2): 236–224.

AIS (1946). *Pamphlet No. 4, PILLARS OF PEACE—Documents Pertaining to American Interest in Establishing a Lasting World Peace: January 1941–February 1946*. Pennsylvania: Book Department, Army Information School.

Almonacid, P. N. (1943). The Food Situation in a World at War: The Other Americas. *The ANNALS of the American Academy of Political and Social Science* 225: 93–95.

Bacon, L. (1943). The Food Situation in a World at War: Southwestern Europe. *The Annals of the American Academy of Political and Social Science* 225: 87–88.

Balfour, L. E. B. (1975). *The Living Soil and The Haughley Experiment*. London: Faber and Faber.

Beardsworth, A. and T. Keil (1997). *Sociology on the Menu: An Invitation to the Study of Food and Society*. London: Routledge.

Beaton, G. H. (1992). Human Nutrient Requirement Estimates: Derivation, Interpretation and Application in Evolutionary Perspective. *Food, Nutrition and Agriculture* (2/3): 3–15.

BioTech Institute (2010). Timeline of Biotechnolgy 20th Century. http://www.biotechinstitute.org/what_is/timeline.html (accessed 12 January 2011).

Black, J. D. (1943). Food: War and Postwar. *The Annals of the American Academy of Political and Social Science* 225: 1–5.

Blakemore, R. J. (2000). Ecology of Earthworms under the 'Haughley Experiment' of Organic and Conventional Management Regimes. *Biological Agriculture and Horticulture* 18: 141–159.

Bordo, M. D. and B. Eichengreen (1993). The Bretton Woods International Monetary System: A Historical Overview. In *A Retrospective on the Bretton Woods System: Lessons for International Monetary Reform*. M. D. Bordo and B. Eichengreen, eds. Chicago: University of Chicago Press.

Brandt, K. (1939). Foodstuffs and Raw Materials. In *War in Our Time*. H. Speier and A. Kahler, eds. New York: WW Norton.

Burnet, E. and W. R. Aykroyd (1935). Nutrition and Public Health. *Quarterly Bulletin of the Health Organisation* IV(2): 152.

CARE (2009). History of CARE International. http://www.careinternational.org.uk/57/history/history-of-care-international.html (accessed 12 January 2010).

Carlson, A. J. (1944). Symposium on Civilian Wartime Problems in Nutrition: From the Standpoint of the Physician: The Importance of Food in Wartime. *California and Western Medicine* 61(6): 281–285.

Cassels, J. M. and F. L. Hall (1943). Food Supplies for Our Civilian Population. *The Annals of the American Academy of Political and Social Science* 225: 106–115.

Clayton, C. F. and J. D. Black (1943). The Food Situation in a World at War: Wartime Food Administration-U. S. A. *The Annals of the American Academy of Political and Social Science* 225: 96–105.

CRS (2009). About Catholic Relief Services. http://crs.org/about/ (accessed 12 January 2010).

Dimitri, C., A. Effland and N. Conklin (2005). Economic Information Bulletin: Number 3: The 20th Century Transformation of US Agriculture and Farm Policy. United States Department of Agriculture, USDA.

Editorial (1943). The Marriage of Public Health and Agriculture (United Nations' Conference on Food and Agriculture). *American Journal of Public Health* 33(7): 847–848.

Encyclopedia (2002). Science of Everyday Things. *Encyclopedia.com*. http://www.encyclopedia.com (accessed 13 July 2009).

Evang, K. and F. L. McDougall (1944). The Hot Springs Conference. *Proceedings of the Nutrition Society* 2(3–4): 163–176.

FAO (1946). Report of the First Session of the Conference held at the City of Quebec, Canada, 16 October to 1 November, 1945. Washington, DC: FAO, 89.

FAO (1995). *Dimensions of Need: An Atlas of Food and Agriculture*. Rome: Food and Agriculture Organization.

FAO (2001). *Food Balance Sheets: A Handbook*. Rome: Food and Agriculture Organization.

FAO (2010). The Food and Agriculture Organization of the United Nations. http://www.fao.org/about/en/ (accessed 26 January 2011).

Ficker, H. (1967). Fifty Years of Foreign Loans and Foreign Aid by the United States, 1917–1967. Wasington, DC: Library of Congress, Legislative Reference Service, 40.

Foreign Office (1943). Report to Allied Governments by the Inter-Allied Committee on Post-War Requirements. UK Foreign Office, HMSO.

Gardner, B. L. (2002). *American Agriculture in the Twentieth Century: How It Flourished and What It Cost*. Cambridge, MA: Harvard University Press.

Hardin, C. M. (1943). The Food Production Programs of the United States Department of Agriculture. *The ANNALS of the American Academy of Political and Social Science* 225: 191–200.

Harper, A. E. (1985). Origin of Recommended Dietary Allowances—An Historic Overview. *The American Journal of Clinical Nutrition* 41: 140–148.

Harper, A. E. (2003). Symposium: Historically Important Contributions of Women in the Nutrition Society: Contributions of Women Scientists in the US to the Development of Recommended Dietary Allowances. *The American Society for Nutritional Sciences* 133: 3698–3702.

Herdt, R. W. (1998). *The Rockefeller Foundation: The Life and Work of Norman Borlaug, Nobel Laureate*. New York: The Rockefeller Foundation.

Hoffmann, S. and C. Maier, Eds. (1984). *The Marshall Plan: A Retrospective*. Boulder: Westview Press.

House of Commons (1944). House of Commons Debates, 25 January 1944: United Nations (Relief Administration). *House of Commons* 396: cc567–cc632.

IFT (1999). 20th Century Marks Acievements in Food Science. http://www.ift.org/cms/?pid=1000562.

IGC (2009a). *60 Years of Successive Agreements: Before 1949: The Early Years*. London: International Grains Council, 4.

IGC (2009b). Grains Trade and Food Security Cooperation: The International Grains Agreement. http://www.igc.org.uk/en/aboutus/default.aspx#igc.(accessed 21 February 2011).

IMF (2010a). Cooperation and reconstruction (1944–71). http://www.imf.org/external/about/histcoop.htm (accessed 16 January 2011).

IMF (2010b). International Monetary Fund Website, http://www.imf.org/external/ (accessed 4 January 2011).

James, C. and J. A. Fitzgerald, Eds. (2008). *Of the Land and the Spirit: The Essential Lord Northbourne on Ecology & Religion*. Bloomington, IN: World Wisdom Inc.

Ladejinsky, W. I. (1943). The Food Situation in a World at War: The Food Situation in Asia. *The Annals of the American Academy of Political and Social Science* 225: 91–93.

Landers, P. S. (2007). The Food Stamp Program: History, Nutrition Education, and Impact. *Journal of the American Dietetic Association* 107(11): 1945–1951.

League of Nations (1933). Final Act of the Conference of Wheat Exporting and Importing Countries, held in London from August 21 to 25, 1933, with Appendices and Minutes of Final Meeting. Signed at London, August 25, 1933. *League of Nations* 3262: 73–88.

League of Nations (1946). *Transit Department: Food, Famine and Relief 1940–1946*. Geneva, Switzerland: League of Nations.

Lloyd, E. M. H. (1943). The Food Situation in a World at War: In the United Kingdom. *The Annals of the American Academy of Political and Social Science* 225: 83–85.

Marrack, J. R. (1948). Hot Springs and FAO: Summary. *International Journal of Food Sciences and Nutrition* 2(3): 93–96.

Mayer, J. (1986). Social Responsibilities of Nutritionists. *Journal of Nutrition* 116: 714–717.

McArdle, W. D., F. I. Katch and V. L. Katch (1999). *Sports and Exercise Nutrition*. Baltimore, MD: Williams and Wilkins.

McDougall, F. L. (1943). The Food Situation in a World at War: International Aspects of Postwar Food and Agriculture. *The Annals of the American Academy of Political and Social Science* 225: 122–127.

National Archives (2009). Federal Register: Food and Nutrition Service. National Archives of the United States. *US National Archives* 74.

Northbourne, L. (1940). *Look to the Land*. London: J. M. Dent & Sons.

NRC (1943). Recommended Dietary Allowances: Report of the Food and Nutrition Board, Reprint and Circular Series No. 115. Washington, DC: National Research Council, 6.

OXFAM (2007). How did Oxfam start? http://www.oxfam.org.uk/coolplanet/kidsweb/oxfam/history.htm (accessed 8 March 2011).

Parran, T. (1943). A Blueprint for the Conquest of Hunger. *Public Health Reports* 58(24): 893–899.

Paull, J. (2008). The Lost History of Organic Farming in Australia. *Journal of Organic Systems* 3(2): 2–17.

Phillips, R. W. (1981). *FAO: Its Origins, Formation and Evolution 1945–1981*. Rome: Food and Agriculture Organisation.

Reid, R. and E. H. Hatton (1941). Price Control and National Defense. *Northwestern University Law Review* 36(255): 283–284.

Richter, J. H. (1943). The Food Situation in a World at War: Central and Northern Europe. *The Annals of the American Academy of Political and Social Science* 225: 85–87.

Roosevelt, F. D. (1941). XXII Annual message to the Congress, 1941. Department of State, United States Government Printing Office.

Rosenberg, I. H. (1994). Nutrient Requirements for Optimal Health: What Does That Mean? *Journal of Nutrition* 124(9 Suppl): 1777S–1779S.

Ruxin, J. N. (1996). Hunger, Science, and Politics: FAO, WHO, and Unicef Nutrition Policies, 1945–1978. In *The Backdrop of UN Nutrition Agencies*. J. N. Ruxin, ed. London: University College London.

Schultz, T. W., Ed. (1945). *Food for the World*. Chicago, IL: University of Chicago Press.

Shaw, J. (2007). *World Food Security: A History since 1945*. London: Palgrave Macmillan.

Sweetser, A. (1946). From the League to the United Nations. *The Annals of the American Academy of Political and Social Science* 246: 1–8.

Thompson, R. J. (1943). The United Nations Conference on Food and Agriculture. *Journal of the Royal Statistical Society* 106(3): 273–276.

UN (1943). *United Nations Conference on Food and Agriculture, May 18-June 3, 1943: Final Act and Section Reports/Department of State, United States of America. Washington, D.C.–US G.P.O.* United Nations Conference on Food and Agriculture, Washington, DC: United Nations Government Printing Office.

UN (1945). Charter of the United Nations, 24 October 1945, 1 UNTS XVI. U.N., United Nations.

UN (2000). U.N. History. http://www.un.org/aboutun/history.htm (accessed 11 March 2011).

UN (2006). The Declaration of St. James's Palace. http://www.un.org/aboutun/charter/history/ (accessed 13 March 2011).

UN (2009a). History of the United Nations. http://www.un.org/aboutun/unhistory/ (accessed 10 October 2010).

UN (2009b). Structure and Organization of the United Nations. http://www.un.org/en/aboutun/structure/index.shtml (accessed 11 December 2010).

UNESCO (2009). About UNESCO: What is it? What does it do? http://portal.unesco.org/en/ev.php-URL_ID=3328&URL_DO=DO_TOPIC&URL_SECTION=201.html (accessed 30 February 2011).

UNOG (2009). The end of the League of Nations. http://www.unog.ch/80256EE60057D930/(httpPages)/0207 6E77C9D0EF73C1256F32002F48B3?OpenDocument (accessed 12 November 2010).

USAID (2009). FrontLines: Borlaug, Father of Green Revolution, Dies. http://www.usaid.gov/press/frontlines/fl_oct09/p01_borlaug091002.html (accessed 5 March 2011).

Volin, L. (1943). The Food Situation in a World at War: The Russian Food Situation. *The Annals of the American Academy of Political and Social Science* 225: 89–91.

Waugh, F. V. (1943). The Food Distribution Programs of the Agricultural Marketing Administration. *The Annals of the American Academy of Political and Social Science* 225: 169–176.

Whipple, C. E. (1943). The Food Situation in a World at War: Southeastern Europe. *The Annals of the American Academy of Political and Social Science* 225: 88–89.

WHO (1958). Nutrition. In *The First Ten Years of the World Health Organization*. Geneva, Switzerland: WHO.

WHO (2009). Archives of the United Nations Relief and Rehabilitation Administration (UNRRA). http://www.who.int/archives/fonds_collections/bytitle/fonds_2/en/index.html (accessed 12 March 2010).

Williams, A. J. (2005). 'Reconstruction' before the Marshall Plan. *Review of International Studies* 31(3): 541–558.

11

The Post-War Years

11.1 Governance, Politics and Socioeconomics

Immediately after the war, agricultural production and economic systems were in upheaval. The cost of financing the war had taken its toll both economically and socially and building on the lessons of the World War I and to avoid general malaise governments and institutions acted quickly and decisively. The first orders of business were the financial and agricultural reconstruction of the warring nations.

11.1.1 The New Gold Standard

The war had ravaged the financial stability of the world and the regulation of international monetary, and financial order was given over to the UN's fledgling World Bank Group (the IMF and the IBRD). One of the initial outcomes of the Bretton Woods agreement was the introduction of a fixed monetary exchange rate system based on the system similar to the previous gold standard of 1875–1914. Countries had previously found the gold standard difficult to maintain as it meant keeping large gold reserves in the face of great currency volatility. The new proposed par-value* system differed from traditional gold standards in that countries fixed their exchange rates relative to the US dollar and in turn the United States aimed to fix the price of gold at $35 per ounce. Implicit in this system was that all currencies pegged to the dollar also had a fixed value, or were pegged as it were to gold. One of the overriding benefits of this system was that it negated the need to hold vast reserves of gold. The measure was initially successful in stabilising a jittery global economy; now nations could once again trade without fear of wild fluctuations in terms of trade brought about by unstable exchange rates.

On the social front the newly established United Nations had set up the Human Rights Commission in 1948 chaired by Roosevelt's widow, Eleanor.

11.1.2 United Nations Human Rights Commission (1948)

The declaration was a remarkable achievement of cooperation considering that the then 58 member states were all representative of different ideologies, political systems as well religious and cultural beliefs. After a long process of negotiation and deliberation (18 months), the declaration sought to ensure that the draft text was fully reflective of these diverse cultural customs. More than this though it also came at a time too when war-weary nations were looking for focus and direction both politically and socially. By incorporating a common set of social, legal religious and philosophical values, the stage was set for a unified vision of a more just and equitable world. The Universal Declaration of Human Rights (UDHR) was finally drafted on 10 December (UN 1948). It consisted of 30 articles ranging from the right to freedom and equal opportunity (Article 1) to the right of education (Article 26) to the freedom of movement (Article 13). It can be divided predominantly into two parts: Articles 3–21 deal with civil and political rights associated with freedom from torture, slavery and arbitrary arrest as well as the rights to free movement, free speech, privacy and a fair trial; Articles 22–27 on the other hand deal chiefly

---

* Par-value system (a.k.a. the Bretton Woods system) worked by pegging other nation's currencies at a fixed rate against the US dollar and the US dollar in turn was pegged against gold. Any adjustments were made solely to correct any fundamental disequilibrium in exchange rates and even then only with the agreement of the IMF. The system lasted until 1971, when the US government suspended the dollar's convertibility into gold, thus ending the era of a de-facto gold standard.

with economic, social and cultural rights including work, education, community and leisure (ibid). Of note however in the food security arena is Article 25 parts 1 and 2, which clearly articulates every man woman and child's right to food:

1. Everyone has the right to a standard of living adequate for the health and well-being of themselves and of their family, including food, clothing, housing and medical care and necessary social services, and the right to security in the event of unemployment, sickness, disability, widowhood, old age or other lack of livelihood in circumstances beyond his control.
2. Motherhood and childhood are entitled to special care and assistance. All children, whether born in or out of wedlock, shall enjoy the same social protection (UN 1948).

With regard to agriculture and the immediate shortages of the war era, once again rising productivity after the war and agricultural output driven by mechanisation and technological uptake led yet again to problems of surpluses in the West.

11.1.3 Aid and Trade

These surpluses were initially exacerbated by the needs of war and occurred even despite agriculture's decreasing share of total economic activity (Belshaw 1947; Dimitri et al. 2005a). However, enterprising legislation and genuine altruism in the short run, even if it served domestic policy as well, ensured that much surplus was used for the general good.

11.1.3.1 The Marshall Plan

At the outset of the Second Great War, as in the First, governments prompted protectionist policies of preferential and bilateral trade agreements that promoted domestic production at the expense of imports. Compounded by continuing price support policies, the accumulation of large surpluses of food was not really surprising. Indeed it was foreseen by one contemporary commentator, noting of the potentiality of post-war surpluses, that

> ... the [FAO] ... will soon be faced with the problem of disposing of supplies which cannot be marketed at profitable prices ... [if] the goal of freedom from want is to be brought within even approximate reach, the problem of expanding world purchasing power must be solved .... A progressive expansion of world purchasing power, associated with national policies which facilitate rather than impede necessary transfers of resources, provides the setting in which surpluses will be least depressing on price and most easy to handle ... (Belshaw 1947)

This was not the case across the board however; some countries were still food impoverished and suffering as a result of hostilities and still found it difficult to import necessary supplies due to a lack of sufficient foreign reserves.

In keeping with the renewed philosophy of global co-operation and aid the 1947/1948 US Foreign Assistance Act had enabled the European post-war economic recovery to properly begin. This Act in turn paved the way for Truman's Secretary of State, George C. Marshall's European Recovery Programme (ERP) or more commonly, the 'Marshall Plan' (Borchard 1947). Tasked with European reconstruction after the war, the goals were enormous and, not unlike the UN Relief and Rehabilitation Administration's (UNRRA) efforts before it, without political agenda (Williams 2005). Previous commendable efforts of the UNRRA stood as a blueprint for the plan and in some observers' minds under the leadership of Governor Lehman, the UNRRA constituted a 'vital link in the development of the whole concept of reconstruction' (Williams 2005, p. 550). Overall the Marshall Plan involved a 4-year, multi-billion dollar programme of financial, technical and food aid for the effective redevelopment of Europe. It came to an end in 1951 and despite criticisms has been hailed as one of the most remarkable achievements of humanitarianism ever and one that truly recognised the economic and social interdependence of nations (Borchard 1947; Crawford 1949; Scott 1951; Weir 1952; Hoffmann and Maier 1984; FAS 2004; USAID 2009a, 2009b).

TABLE 11.1

Agricultural Trade Development Assistance Act: PL 480

Title	Title Objectives	Administered by
Title 1	Trade and Development Assistance: providing for government-to-government sales of US agricultural commodities to developing countries on credit or grant terms	USDA
Title 2	Emergency & Development Assistance Program	USAID
Title 3	Food for Development	USAID
Title 5	Farmer to Farmer	USAID

Sources: Toma, P.A., *The Politics of Food for Peace*, University of Arizona Press, Tucson, 1967; Eulau, H., *Polit. Res. Q.*, 21, 530–532, 1968; Sullivan, R.R., *West. Polit. Q.*, 23, 762–768, 1970; USDA, History of the US Department of Agriculture USDA, US Department of Agriculture, 2009.

As the Marshall Plan slowed down in 1949/50, large stocks of grain continued to accrue in the United States, leading to storage expenses and in some instances to depressed producer prices. Overproduction in the absence of strong European aid exports, was once again threatening to became a serious problem. Domestically the United States increased government intervention, and while legislation like the passage of the US Agricultural Act of 1949 helped to alleviate the problem to some degree; more was definitely needed. By this time Eisenhower, trading on election promises, managed to push through the Agricultural Trade Development and Assistance Act in 1954. In a trade-off for supporting flexible price supports rather than traditional fixed 90% parity pricing, farmers received

> … the promise of increased export sales to needy, friendly governments abroad. (Reinhardt and Ganzel 2003)

The Act more commonly known as Public Law 480 (PL 480) contained four Titles (Table 11.1), which allowed for the direct donation of US agricultural commodities for emergency and non-emergency programmes alike.

11.1.3.2 Public Law 480 (1954) and Food-for-Peace (1961)

In its inception, PL 480 was heralded on the basis that persistent food surpluses should be tied more closely to the emerging nations of Asia, Latin America and Africa's economic development needs (Sullivan 1970). From the start, the debate was politically charged, and it has often been cited that a one-time domestic surplus response, was transformed into a US foreign policy programme to the detriment of its original 'altruistic' objectives (Toma 1967; Sullivan 1970, USAID 2009).

As this was unfolding, there followed much debate about the role of government intervention and the regulation of prices and productivity as a matter of principle (Dimitri et al. 2005a). Under the aegis of the Food and Agriculture Organization (FAO) in 1946/7, efforts by the Committee on Commodity Problems (CCP) had failed to introduce an International Commodity Clearing House (ICCH) to address a proposed World Food Reserve. This was due in part to a report that concluded that technical and financial difficulties rather than a physical world food shortage were in fact bigger obstacles in the relief of emergency famine. However, its failure has also been attributed by some to the unwillingness of the United States and United Kingdom to once again devolve such power to a multilateral organisation over which they had little or no control (UN 1945; FAO 1957b, 1958; GATT 1960). Some good however did come out of these meetings, as it was realised that if these agricultural surpluses were released onto world markets without due consideration, export trade, particularly of primary-producing countries, including those of some of the underdeveloped countries, could be severely damaged. This threat was so great that in 1954 the CCP established a Consultative Subcommittee on Surplus Disposal (CSD) to monitor such surpluses and consider their effects (ActionAid 2003; CCP 2005a).

As the debate rumbled between aggregate political interests and the collection of non-governmental humanitarian and voluntary agencies, one commentator, Eulau (1968), adopted a more philosophical approach. In his view, the original purpose of the PL 480 act was never really solely to do with human-itarianism; for him, he saw it primarily as a domestic response to American agricultural food surpluses and one in which almost 'incidentally' been fixed the needy. From this, he is not surprised that the subsequent and gradual transformation of the Act into an instrument of American foreign policy took place. Indeed, on this point he was very comfortable with the notion that aid could serve two masters. Also, in his view the separation of domestic and foreign policy on this issue was never really going to be that straightforward or that simple. Moreover, irrespective of its political ulterior motives, it was still humanitarianism, as Elau succinctly summed up

> If congressmen are willing to support a programme such as Food for Peace, their private motiva-tions or constituency interests, however selfish, do not defile, tarnish or befoul its humanitarian consequences. (Eulau 1968)

Following on from this in 1961, the PL 480 programme was expanded and renamed the Food for Peace programme by President Kennedy before being officially renamed the Food for Peace Act in 2008 (USAID 2009a). Yet despite continuing measures, US surpluses continued, and in 1956 with the advent of the Agricultural Act, also known as the soil bank programme, it was hoped that by taking farmland out of production, adjustments could be made between supply and demand for farmer's agricultural products. This measure was aimed at solving both the problem of overproduction and farmer incomes at the same time. In the Act, farmers were either paid to take cropland out of production or to grow more soil-conserving crops. However, farmers continued to produce more than was being consumed, so much so that the government had spent $1.5 billion ($11.45 billion in 2007 dollars) buying up surplus commod-ities in 1953 and 1954 alone. Moreover, this had little effect on farmers' incomes as by the mid-1950s the farmer's disposable incomes still stood at only 48% of non-farm families (Reinhardt and Ganzel 2003).

Simultaneously, these events marked the beginning of the donation of US farm surpluses to voluntary agencies (Sullivan 1970). The passage of the Agricultural Act authorised the Secretary of Agriculture to donate surplus farm products to voluntary agencies registered with the Department of State's Advisory Committee on Voluntary Foreign Aid. And so it was at this time that large amounts of US grain were distributed around the world. On the face of it, this served to satisfy America's major foreign policy aim; the stabilisation and recovery of Europe, while others in fact argued that it best suited US domestic pol-icy by opening up new markets for US grains. Although in truth it more likely suited both the United States and recipient countries at a time when choice was limited (Sullivan 1970; Webb et al. 2008). This was soon to change however, as the expansion of humanitarian activities over time disrupted the delicate government–NGO partnership balance. On the one hand was the government, its domestic, foreign and humanitarian policy objectives and on the other the voluntary organisations of various flavours such as NGOs like CARE International and Catholic Relief Services. Over time, it seemed the political-stroke-altruistic balance was becoming distorted with domestic agricultural interests gaining preference over foreign policy objectives. This resulted in foreign disposal programmes that were less related to the needs of the recipients and more favourable to pressing domestic American agricultural surpluses (Sullivan 1970). While at the time this was seen as problematic in some quarters, there were few alterna-tives. This was a story that was to play out in later years. For now, these and other legislative Acts of the time, as well as the previous Marshall plan, had collectively heralded a new era of US foreign policy that laid the foundations for a permanent US food aid programme linked to domestic interests.

Meanwhile on a tangential note, the goal of international cooperation and free trade came under the spotlight.

11.1.3.3 General Agreement on Tariffs and Trade

In the face of an enormous number of international barriers to trade that directly flew in the face of the underlying principles of free trade and global equitable cooperation, there were loud calls for change. In response the UN's Economic and Social Committee (ECOSOC) called for an International Trade Organization (ITO) to coordinate multilateral trade policies in 1946, by 1947 an international treaty, the General Agreement on Tariffs and Trade (GATT), was created and designed to operate under the ITO on its inception. As it was the ITO never came to fruition; however, the GATT treaty became an important instrument that was to prove its worth over the ensuing decades.

Elsewhere in the world the late 1950s witnessed the birth of the European Community and what was quite possibly the worst ever famine in human history, the Chinese Great Leap Forward Famine.

11.1.3.4 The European Economic Community

The establishment of the European Economic Community in 1957 was the culmination of several years of negotiations that began with the European Coal and Steel Community (ECSC) in 1952. However, after the failure of the proposed European Defence Community (EDC) in 1954, the idea temporarily waned. Despite this setback many still yearned for European unity even if it meant seceding a little national sovereignty. This was encouraging, and further negotiations at the Messina Conference in 1955 followed and a draft report on the creation of a European Common Market was created in 1956. Finally, a group met in Brussels and hammered out the various options and in March of 1957 in Rome the 'Treaty of Rome' was at last signed and entered into force on 1 January 1958.

The establishment of the EEC and the creation of the Common Market was predicated on two main objectives. The first was to transform trade and manufacture within the community by reducing inter-union barriers, and the second aimed to construct a politically unified Europe. These objectives were fleshed out through the creation of common policies. In the economic sphere, the treaty aimed to promote the harmonious development of European economic activity based on the principle of free competition and through the actions of the four freedoms of the free movement of people, services, goods and capital. In establishing a customs union, the EEC Treaty looked to abolish quotas and customs duties between member states at the same time introducing a common external tariff for non-members. The development of these common policies sought to improve the free movement of people, creating job opportunities and to raise standards of living among other social goals (Europa 2007, 2010).

11.1.4 The Great Leap Forward Famine (1958–1961)

This decade also witnessed the largest famine of the twentieth century and almost certainly of history in the 1958–1961 Chinese Great Leap Forward Famine. Up to this point in the century, China had already suffered several large famines; however, rather than being prepared, the political atmosphere of the day suppressed the extent of the unfolding horror until it was too late. At the heart of the problem was Chairman Mao Zedong's attempt to transform China from a traditional agricultural nation and place it on a more industrial footing. Peasants were told to abandon their smallholdings and work collectively in large production farms. Surplus workers were then instructed to produce steel in small foundries. Such collectivisation saw massive social and economic upheaval and together with unfavourable weather conditions, the overconsumption of available food and under-investment in agriculture, the scene was set for famine on an unprecedented scale. Compounding this delicate tinderbox was the Communist Party's preference for centralised control of information, which when coupled with the propensity to only report good news, the scale of the impending disaster was ultimately suppressed. Eventually though when the leadership did learn of the scale of the problem it did little by way of mitigation and further clamped down on the information flow. This tight control of news was so effective that it took nearly 20 years for the full extent of the atrocity to unfold. The total death toll has been placed at between 20 and 40 million lives and resulted in the single greatest peacetime demographic disaster of the twentieth century (Watson 1999; Devereux 2000; Ó Gráda 2009).

On a lighter note, fully 35 years after the League of Nations adopted the first Declaration of the Rights of the Child, the United Nations revisited the notion.

11.1.5 Declaration of the Rights of the Child (1959)

After a decade of heightened activity concerning child welfare and the full realisation of the importance of child nutrition and care, particularly of the under-5s, the UN General Assembly adopted the

Declaration of the Rights of the Child in 1959; a name given to a series of related proclamations recognising children's rights. Beginning with Eglantyne Jebb's (Save the Children International Union [SCIU]) 1923 declaration of children's rights, which was adopted by the League of Nations in 1924. The SCIU then merged with the International Union of Child Welfare in 1946. By then (1946), the UN International Children's Emergency Fund (UNICEF) was created to replace the UNRRA and continued to work for the benefit of children, and by 1948 the special rights of the child were once again articulated in Article 25(2) of the Universal Declaration of Human Rights. In 1959 however, the United Nations decided to formalise these rights in a special non-binding declaration that decreed that every child, no matter their or their family's race, colour, sex, religion, nationality, social origin or political views should be entitled to protection of rights vis-à-vis education, health, freedom, dignity and discrimination among others (UN 1959; UNICEF 2009).

11.2 Health and Nutrition

The post-war period was a time of new institutions and collaborations in the field of health and nutrition. Few were more active at this time than the United Nations stable of organisations.

11.2.1 Post-War Nutritional Reconstruction

At the outset of World War II much of the work of both the Office International d'Hygiène Publique (OIHP) and the League of Nations' (HOLN) health units were put on hold. However, fear of possible widespread post-war epidemic rallied the newly created UNRRA and in its first meeting (1943) it placed health among its primary responsibilities. Immediately after the war the main focus of the warring nations was on reconstruction and development, and the problems posed by this daunting task provided the first real test for international action. Leading the international effort were the United States (the Marshall Plan) and the newly formalised United Nations. One of the main questions that needed answering though was whether the international community was to actively rebuild economic and social infrastructure or simply to provide temporary relief through emergency supplies (Gillespie 2003). At first the allied forces in the form of the Allied Committee on Post War Reconstruction Requirements (the Leith Ross Committee) focused on planning the relief requirements of each nation. However, this was superseded by the Office of Foreign Relief and Rehabilitation Operations (OFRRO) set up by Roosevelt in 1942 and headed by Herbert Lehman. This body took a more proactive and activist role which the British-dominated Allied Committee lacked. OFRRO, it was envisaged, would not only ship supplies but would in fact plan and rebuild the fractured economies. In the midst of all this, the plight of the mother and child was also recognised as central in rebuilding national and international child health and welfare services (ibid). In 1943, the newly established UNRRA (also headed by Lehman) adopted the basic model of the OFRRO, providing much needed relief supplies. Milk became the staple of much of UNRRA's activities, as it provided a quick and easy solution in the face of hampered research of alternative nutritional methods. It was also easily administered through existing infant and school feeding programmes. Another of the big logistical issues at the time was the repatriation of war refugees. On top of the mammoth task of transferring foodstuffs, the responsibility of the refugees fell under the umbrella of the UNRRA. In late 1946/7 after much political debate both from its supporters and from its opponents, the UNRRA's tasks were delegated to the International Refugee Organization (IRO) (House of Commons 1946; USHMM 2010). The IRO was a temporary UN specialist agency (1946–52), which was later succeeded by the Office of the United Nations High Commissioner for Refugees (UNHCR) (Broughton 1997).

About this time, debate over the future of such international health policies became the subject of increasing infighting among new emerging rival organisations, particularly the FAO, World Health Organization (WHO) and UNICEF. The first on the scene was the FAO.

11.2.1.1 Food and Agriculture Organization

While the final meeting of the League of Nations did not take place until April 1946 it was all but abandoned as an effective policy-making body, and in this light the UN, for all practical purposes was regarded as a continuation of the organisation (Sweetser 1946; Goodrich 1947). One of the first bodies of the new UN was the FAO; proposed in 1943 and officially created in 1945. It was headed by nutritional scientist Sir John Boyd Orr* (Director General) who quickly appointed Dr. W. R. Aykroyd as Director of the FAO's Nutrition Division. Although explicitly the new organisation was dedicated to improving global agriculture, forestry and fisheries, Orr, Aykroyd and others saw their role as part of a broader more general welfare objective: "to end hunger and raise general standards of nutrition" (Gillespie 2003). The next on the scene was the WHO.

11.2.1.2 The World Health Organization

In its first meeting also in 1946, the UN Economic and Social Council called for a conference to consider the creation of a single, permanent specialised health organisation. In July of the same year, the conference adopted a mandate that would see the establishment of a World Health Organization to carry on the previous functions of the League and the OIHP (EoN 2009). In the meantime, an interim commission of the proposed WHO continued the work of UNRRA's health functions until the WHO's constitution was finally ratified in 1948. At its initial inception, there was little scope for the WHO to be involved in programmes such as maternal and child health. This was seen by some as a joint Anglo-American attempt to reign in the future organisation to create a mere 'hygiene' institution rather than a much needed universal public health organisation. However, fierce opposition prevailed and the final draft of the new institution, approved by the UN ECOSOC, included the objective of maternal and child health welfare (WHO 1946).

Then came UNICEF.

11.2.1.3 The UN International Children's Emergency Fund

Previous to the last meeting of UNRRA in 1946, US Army film makers made a short 19-minute documentary, *Seeds of Destiny* which contained harrowing images of begging and foraging children. The film highlighted the plight of post-war children, many of whom were barely surviving in hospitals and orphanages. The film was shown at the last meeting of UNRRA's governing council, which then went on to propose to the United Nations that its considerable surplus funds be used to continue the relief effort for children (UNICEF 2005). Later that year a new child welfare organisation, the UN International Children's Emergency Fund, was established to provide emergency relief assistance to the children of post-war Europe. The eager and practical, proposed new head of the organisation, Maurice Pate, conditionally accepted the job only if the mandate included all needy children, no matter their creed or colour, and more importantly the political persuasion of their governments. It was a controversial point that aimed to place children above politics (ibid). It was also a point that was eloquently summed up in a lecture given in honour of UNICEF receiving the Nobel Peace Prize in 1965, by the then chairman of its executive board, Zena Harman, who said that

> Child suffering could not be distinguished by virtue of its cause or origin. Children in desperate
> need anywhere and everywhere required help and attention. (Nobel Prize 2010)

In the beginning, UNICEF was to be a temporary affair with an original mandate that recognised the importance of child health as primary. The basic needs of the child it considered were protection

---

* This period also saw the Scottish nutritionist John Boyd Orr become the first Director General of the newly created FAO. His legacy is that he was perhaps one of the most influential people working in nutrition at the time. His groundbreaking 1936 book, *Food Health and Income*, had become one of the most important works on nutrition before World War II and was largely influential in bringing together nutritional science and policy (Ruxin 1996). He consistently argued that the poor were largely malnourished while the wealthy were overwhelmingly healthy. In this argument, the poor helped propel the nutritional debate, straddling not only science and education but towards economics and governance. As a practical solution to poverty, Orr constantly advocated the improved socio-economic status of individuals.

against disease; sufficient and adequate food; clean safe water; shelter and clothing; and a suitable environment in which the child can grow emotionally and socially. It had already been established that schools provided an excellent means by which the monitoring and distribution of health care to nutritionally deficient children could be achieved. As a result, in the beginning, and by the end of World War II, UNICEF was primarily focused on supplementary feeding programmes of school aged children.

In summary, the three organisations' mandates involved much overlap. At the very outset, FAO recognised the need for inter-collaborative effort from similar like-minded organisations both within and without the UN umbrella. In its first conference, the FAO encouraged cooperation with the International Children's Emergency Fund (later UNICEF), the ILO and the soon to be formed WHO. The WHO, already undertaking the work of the UNRRA and the FAO, initially cooperated in the hope of attracting the approximately $35 million of surplus funds of the soon to be wound-up UNNRA. Unfortunately for both organisations, the surplus went to the new children's organisation UNICEF and although UNICEF's mandate required it to seek the technical advice of WHO and FAO in its programmes, its financial dominance gave it a governing role in post-war international institutional health policies (Gillespie 2003). In the governance vacuum that followed, the WHO struggled for acceptance and having taken over the work of the UNRRA without its associated funds found it difficult to develop its own agenda. Meanwhile in the same year (1946) the FAO had reaffirmed its commitment to increasing world trade and overcoming nutritional shortfalls. To achieve this, increased agricultural productivity combined with liberal international trade and national nutritional targets would all be collectively targeted through expanded economic and physical infrastructures, as well as through the revival of League of Nations national nutrition committees (Aykroyd 1948).

By December 1950 however, with the easing of the European crisis, discussions ensued over the future of UNICEF. Being a temporary organisation, its initial remit was coming to an end. War-torn Europe was beginning to get back on somewhat of a level footing, and it was thought any continuing work with children could be absorbed by other UN organisations. Vociferous noises were made by the WHO for its immediate abolition, while, on the other hand, the FAO's Nutrition Division, despite some reservations, was accepting of the continuation of the organisation. Even Eleanor Roosevelt argued that the group had completed its remit. In the end, it was Pakistan's UN delegate Ahmed Shah Bokhari's plea that children in developing countries were in continuous need of the kind of work carried out by UNICEF that swayed the argument. In his view, the ravages of war-weary European children could be equated with the normal plight of children in developing countries everywhere. The Fund was rolled over and given another 3 years; although at the end of this period, in 1953, it was finally decided to continue the work of UNICEF indefinitely. At this time the name was changed to the UN Children's Fund, although the acronym 'UNICEF' was retained.

By now, the three agencies began to fall into their respective roles. The WHO concentrated on nutritional issues in respect of disease prevention and the maintenance of health, while the FAO focused on the nutritional aspects of food production and consumption. Although having said that, the boundaries were difficult to define with any great clarity and the nutritional work of both WHO and FAO became inextricably linked (Ruxin 1996). In its new role, UNICEF transposed the hitherto successful school-based nutrition programme into its new remit. This involved the United States and Northern Europe providing surplus dried milk powder supplemented with vitamins A and D and other foods, and while governments paid for much of the transportation and distribution costs, both UNICEF and FAO provided staff to supervise distribution (UNICEF 2005; Nobel Prize 2010). It was not long before the reliance by UNICEF on imported milk feed was increasingly seen as unsatisfactory and in collaboration with the FAO, a move was made to replace imported milk with a locally produced equivalent. However, this too came to be seen as problematic, and after much debate, it was concluded that an alternative source of protein based on locally produced foods should take precedence. This was also further reinforced by Martha Eliot, deputy director of the WHO (in 1949), who warned of the dangers of creating reliance on relief, especially milk-based relief that unnaturally stimulated demand and redefined local preferences in a way that could never be maintained by local production. In this endeavour, Eliot attempted to persuade UNICEF to realign its considerable resources away from the sole aim of emergency feeding to assist nations in developmental programmes of infrastructure (Gillespie 2003).

In the midst of all this, the FAO conducted the First World Food Survey, which highlighted that between half and two-thirds of the world's population were undernourished.

11.2.1.4 FAO First World Food Survey (1946)

The first world food survey involved 70 countries and was aided by many nutritionally knowledgeable governments and people around the globe. It was also the first large-scale project by the FAO. The rationale given was that while it was widely held that global hunger and malnutrition was prevalent, concrete proof was needed; this was summed up at the time:

> ... it is well known that there is much starvation and malnutrition in the world [yet] vague knowledge that this situation exists is not enough; facts and figures are needed if the nations are to attempt to do away with famine and malnutrition. (FAO 1995)

However, given the breakdown of statistical gathering during the war, the approach relied on some estimation and imprecise data. Nonetheless by estimating a pre-war baseline of calorie availability and comparing it with postulated minimum nutritional standards, it was possible to make some solid determinations. The report served to provide concrete scientific evidence of sorts, confirming long-held beliefs that widespread hunger and malnourishment did indeed exist throughout the world; with more than half the world's population subsiding on less than 2250 calories per day (FAO 1995).

After determining the pattern of food intake, the next step was to set long-range nutritional targets based on perceived needs to 1960; based on a projected 25% population increase (FAO 1946; McMillan 1946; Belshaw 1947). Based on 'realistic' rather than 'ideal' targets, the goal was to achieve intermediate targets for each country grouping. Countries with pre-war low-calorie intakes, for instance, were designated as needing between 2550 and 2650 calories per person per day compared to an average intake 3000+ calories for the 12 pre-war countries with the highest calorie consumption. However, even with a moderate projection of 25% population increase, such targets would involve large increases in general world production of all food groups. More specifically, this meant increases of 21% for wheat, 80% pulses, 34% fats and a massive 163% increase in world fruit and vegetable production. If, on the other hand, population targets were estimated at 35% increase above then present numbers, then a massive 90% overall global production increase would be required (FAO 1946; McMillan 1946).

By 1948 the FAO had fully recognised the importance of food, energy and nutrition as central to its remit (Weisell 1995).

11.2.1.5 International Nutritional Requirements (1949)

In determining whether there were sufficient food supplies to meet a population's nutritional needs, it was important to understand more fully the human energy and nutrient requirements (FAO 2002). Thus, carrying on where the League left off, the Food and Nutrition Division of the FAO in 1949 convened an expert committee (Committee on Calorie Requirements) to undertake the first of several nutrition-related requirement reviews (Aykroyd 1956; Périssé 1981; FAO 2003). In 1948, building on the 1947 report of the FAO Calorie Conversion Factors and Food Composition Tables, the Committee on Calorie Requirements produced the first real Food Composition Tables for international use (Chatfield 1949; FAO 1950; Aykroyd 1956). These preliminary tables and the subsequent 1949 revisions (reported in 1950) were calculated in large part using Atwater's food energy conversion factors of the turn of the century (Wu Leung et al. 1968). The first tables expressed the calorie values and the protein, fat and carbohydrate content of many common foods (Chatfield 1949). These were subsequently revised and expanded first in 1954 then in 1955 by Merrill and Watt to incorporate important vitamins and minerals (Chatfield 1954; WHO 1955; Aykroyd 1956; Wu Leung et al. 1968).

These recommendations came at a time when the FAO had, in collaboration with the WHO, noted that

> Food regulations in different countries are often conflicting and contradictory. [While] Legislation ... often varies widely from country to country. New legislation [is] not based on

scientific knowledge [and] is often introduced … [with] little account … taken of nutritional principles … . (WHO 1950, p. 24)

Indeed, as this quote illustrates, the primary purpose of such expert meetings was, and continues to be, to advise the director generals of the FAO, WHO and other interested parties on the scientific issues of energy and nutritional requirements in order that appropriate policy can be formulated (Weisell 1995).

Also by 1949, not only did the FAO's Calorie Requirements Committee introduce the first international food composition tables, it set the first recommended specific average requirements for a 'reference' man (65 kg) and woman (55 kg) using a factorial approach.* The factorial approach was a systematic approach that entailed establishing a baseline for energy needs for a person of particular gender and of a certain age. Recommendations could then be adjusted depending on four variables: physical activity, body mass, age and climate to reach the final energy requirement estimate for the individual (FAO 1950; Passmore 1964; FAO/WHO 1973; Weisell 1995). Based on the factorial approach, the male and female reference adults—both 25 years of age and living in a temperate zone with an average annual temperature of 10°C—were calculated to need 3200 and 2300 kilocalories per day, respectively (FAO/WHO 1973; WHO 1973). This method effectively allowed nutritional requirements to be assessed on a national basis and contributed to the notion of international comparability (FAO/WHO 1973, 1985).

The 1950 report also importantly emphasised several points, most notable was that energy requirement figures were in fact average requirements of the group which they represented and should not be taken to mean actual individual requirements (FAO 1950; FAO/WHO 1985). Second, it was deemed appropriate that energy expenditure calculations, where possible, should be used in determining energy requirements marking an ideological departure from many previous studies (ibid).

By this time, food balance sheets had come into their own.

11.2.1.6 Food Balance Sheets

Food balance sheets' usefulness in analysing and understanding a country's food situations was well understood by the FAO, so much so that in 1948 the FAO recommended that governments develop their own food balance sheets to be published regularly. By 1949, the FAO published the *Handbook for the Preparation of Food Balance Sheets*, which is still published on a regular basis today (FAO 2001).

11.2.2 Protein Energy Requirements

While much work had been carried out with regard to energy requirements, individual's needs in the proportions of the main energy-giving nutrients was also being further researched (FAO 2002). By the early 1950s, it was becoming increasingly recognised that sufficient energy did not necessarily equate to adequate nutrition. Of concern was the idea that people were not getting enough protein and that protein malnutrition could evolve to replace energy malnutrition as the most devastating form of malnutrition (Allen 2003). It was first realised through the work of Nevin Scrimshaw and others that protein deficiency in infants first developed when breast milk was no longer the single or sole source of food. Thus, cheap alternatives to milk protein were sought in an effort to slow down and prevent kwashiorkor, as it was known in weaning infants.

11.2.2.1 Kwashiorkor and PEM

During the 1950s, children's nutritional welfare was still high on the agenda (WHO 1950, 1951, 1953, 1955). The First Joint FAO/WHO Expert Committee on Nutrition noted that high child and infant mortality rates, particularly in the tropical regions, were blamed on wide spread disorders such as kwashiorkor

---

* Factorial approach to nutritional estimation is a breakdown of the required solution into its component parts. Traditionally, the factorial approach was used in protein calculations to determine nitrogen losses in urine, faeces and sweat, etc. These were then added to estimations needed for tissue deposition. Later, however, the factorial approach was used as a term in reference to calculating calorie requirements mathematically and systematically, i.e. BMR x PAL.

(protein deficiency) (WHO 1950, 1953). Added to this, it was considered that in some countries it was pre-school children that suffered more severely than any other group:

> In many parts of the world, children after weaning are usually fed on a diet consisting mainly of cereal preparations and obtain little or no milk … The etiology of these syndromes is not at present clear, but there is evidence that they result from serious dietary deficiencies during the earliest years. The ill-effects … [of which] may be reflected in adult life. (WHO 1950, p. 16)

The FAO/WHO's report was further strengthened by the Brock and Autret report published a year or so later, which further detailed the extent of kwashiorkor in Africa at the time (Brock and Autret 1952).

Although the term kwashiorkor was coined by Cecilly Williams in the early 1930s, it had not up till this point been adopted as a catchall phrase for protein malnutrition. Aiding this usage was the specific protein malnutrition meeting of the FAO/WHO expert committee in Gambia in 1952, which temporarily used the term to better distinguish from other malnutrition syndromes caused by such factors as vitamin deficiencies. From this point on, kwashiorkor became synonymous with protein malnutrition.

In response to the perceived problem of protein deficiency, the joint committee advocated nutrition education for both professionals and lay people alike and subsequently launched a global investigation to assess the extent of the problem (WHO 1951, 1958; Weisell 1995). Complementing these strategies, further work in the area of vitamin and mineral content composition, as well as requirements to support the FAO's 1949 energy and calorie requirement tables was instigated (WHO 1951). What followed was the Second World Food Survey.

11.2.3 FAO Second World Food Survey (1952)

The Second World Food Survey provided global estimates of caloric and protein distribution and concluded that the situation was much worse than the first survey of 1946. The FAO estimated per capita caloric availability from acreage, yields and production. After adjusting for losses, seed and feed, total world output was calculated and based on a daily standard of 2600 calories/capita. From this it was shown that over two-thirds of the world's population was indeed undernourished. This led to the belief that kwashiorkor was one of, if not the most serious and widespread, nutritional disorder in the world at the time (Goldsmith 1955; Aykroyd 1956).

This report along with other work in this field led the FAO and WHO to consider protein the single most important nutrient deficiency of the time. With much to be discussed, the difficult problem of protein requirements became the focal point of a conference sponsored jointly by FAO, WHO and the Josiah Macy Jr. Foundation in June 1955. Following the conference, the WHO established the Protein Advisory Group (PAG), a group of clinical nutritionists drawn from WHO's Expert Advisory Panel on Nutrition at the end of 1955 (WHO 1958). Initially, the PAG's role was that of adviser to the FAO, the WHO and UNICEF. It aimed to advise the organisations on high-protein weaning foods and recommended that protein be the foundation of most nutritional programmes from then on. Swelled by top nutritionists from developing countries, the PAG ultimately became very influential with its main focus on research and education. The PAG was eventually rolled into a tripartite (FAO/UNICEF/WHO) programme in 1961 (Aykroyd 1956; PAG 1967; WHO 1972; Carpenter 1994; Weisell 1995).

Shortly after this in October 1955, an FAO expert committee met to consider the question of protein requirements (FAO 1955). As with energy requirements, the time had come to formally address the notion of adequate recommendations for protein. While many such allowances had been made by several previous studies, in particular the League of Nations and others, the FAO utilised the notion of a 'reference protein' and armed with up to date studies attempted to establish requirements of protein with a 'high nutritive value'. The results for adults were calculated to be 0.35 g/kg body weight for adults compared with the League's 1 g/kg in its 1936 report. As far as children and adolescents were concerned, limited data meant that results were predominantly estimated at 0.7–0.8 g/kg. Although tentative, the report stimulated further research, which was taken up by the 1963 FAO/WHO Expert Group (FAO/WHO 1973; WHO 1973; Allen 2000).

TABLE 11.2

Christensen's Energy Expenditure Rates at Differing
Work Levels

Work Rate	kcal/min
Very light	Less than 2.5
Light	2.5–4.9
Moderate	5.0–7.4
Heavy	7.5–9.9
Very heavy	More than 10.0

Source: Christensen, *Physiological Valuation of Work in the Nykroppa Iron Works.* In Ergonomics Society Symposium on Fatigue, W.F. Floyd and A.T. Welford, eds., Lewis, London, 93–108, 1953.

11.2.4 Work Rates

By 1956 the second FAO Calorie Committee reconfirmed the usefulness of the concept of the male and female 'reference person', although some minor revisions were made to the allowances. The original reference levels were endorsed and depending on physical activity,

> … a range of energy expenditure between 2400 and 4000 kcal/day for men and between 1600 and 3000 for women would appear to include most human beings. (FAO 1957a, p. 16)

The second committee also re-acknowledged the difficulty in assessing the different physical activity or industrial work rates by which the reference person was modified. For such purposes the committee referred to the Christensen classification of work (Table 11.2). Passmore (1964), in his review however, bemoaned the multitude of differing classifications in use at the time and offered that it would be more scientifically productive if a single agreed scale were used (Passmore 1964).

Of particular significance too was the notion that the FAO's report marked a departure from previous practices by Voit, Atwater, Lusk, NRC and others who traditionally based energy requirements on food intake. Instead the FAO 1957 report advocated the use of energy expenditure to calculate energy requirements although it was not until much later that the FAO actually made the switch due to technological and methodological challenges (FAO 1957a; Henry 2005).

11.2.5 Under-5s

A major shift too in the focus of several agencies resulted from the 1958 Behar, Ascoli and Scrimshaw study, when its findings revealed that significant cause of death (nearly 20%) of kwashiorkor cases occurred in pre-school children. Furthermore, a startling analysis of the age distribution among child kwashiorkor deaths overall, highlighted that 58% of these occurred in children under the age of 5 (Behar et al. 1958). The implications were enormous and UNICEF quickly realigned its policies to make children under the age of 5 their priority programme recipients. Further focal shifts concentrated on improving the nutrition of pregnant and nursing mothers as well as the period between post-weaning and pre-school ages.

11.2.6 Education and Applied Nutrition Projects

Since the early days of humanitarianism, field workers in developing countries had cited ignorance as a major factor that caused or exacerbated malnutrition in children. By the mid-1950s as evidenced at the fourth meeting of the Joint FAO/WHO Expert Committee on Nutrition in 1954, there was enough impetus for the FAO to set out broad guidelines for nutrition education programmes and training. This was reiterated several times by the FAO over the ensuing years and was seen as a favourable way to empower individuals in important issues of health, care and nutrition. The seed was sown institutionally and UNICEF took up the mantle too, and in 1954 UNICEF built on these ideas and by 1957 began to consider alternative methods of channelling high-protein foods to young children through education and empowerment. The answer came in the form of applied nutrition projects (ANPs). These were in situ programmes designed to foster nutrition education, training as well as schemes for improving production of local protective foods.

11.2.7 Relationship between Malnutrition and Infection

People had long speculated of the link between health and infection. By the mid to late 1950s, two successive Joint FAO/WHO Expert Committees on Nutrition expressed concern over a lack of knowledge concerning the relationship between nutrition and infection. It finally took the work of Scrimshaw, Taylor and Gordon however to shed light on this growing link. In 1959, in a paper in the *American Journal of the Medical Sciences*, Scrimshaw and colleagues documented, for the first time the scientific correlation between malnutrition and diarrhea. Through the extensive and cyclical interactions between malnutrition and infection, the authors suggested that malnutrition resulted in increased susceptibility to infection; moreover, that infection reciprocated this synergistic relationship in turn causing further deterioration of nutritional status. These interactions too, it seemed, were amplified when the combined effects of malnutrition and infection were present simultaneously. As a result, the study's findings suggested that it was not simply a case of intervening to break the cycle, instead a dual assault on nutrition and infection was needed if an optimal outcome was hoped for (Scrimshaw et al. 1959; Keusch 2003).

11.3 Science and Technology

Technological developments in agriculture have been particularly influential in driving change in the farm sector. Following World War II, technological developments occurred at an extraordinarily rapid pace with advances in mechanisation and increasing availability of chemical inputs. This in turn led to ever-increasing economies of scale that perpetuated a seemingly endless cycle of growth (Dimitri et al. 2005b). In this setting the first tendrils of the green revolution that was to come to full fruition in the 1950/1960s began with Norman Borlaug, an American scientist researching new disease-resistant high-yielding crop varieties of wheat in Mexico in the 1940s. Coupled with the dissemination of the Haber-Bosch process of fertiliser manufacturing, made available after the war, as well as the growth in mechanisation and increased irrigation. Industrialised world was about to witness a bonanza of agricultural bounty of unprecedented proportions.

Meanwhile by 1946, in the field of genetics Edward Tatum and Joshua Lederberg showed that bacteria could sometimes exchange genetic material. Independently in the same year Max Delbruck and Alfred Day Hershey discovered that in a process of genetic recombination, the genetic material from different viruses could be combined to form a new type of virus. At this time too, advances in the field of biology saw the isolation and growth of animal cells in vitro. This period also saw for the first time in 1950 the artificial insemination of livestock using frozen semen: this was a tremendous breakthrough and one that satisfied a long-held dream of farmers. Also pushing the boundaries was electron microscopy, which by 1952 was able to show the tiny well-formed anatomical structures or organelles that existed inside every living *eukaryotic* cell (Appendix D). However, perhaps the *pièce de résistance* during this period was reached in 1953 with the discovery by James Watson and Francis Crick of the molecular structure of DNA (Watson and Crick 1953a). In their proposal, the structure of DNA molecules was double-stranded, helical, complementary and anti-parallel. This was a major breakthrough in the science of genetics as up till this point there was still no proven vehicle for the transfer of genetic material from adult to offspring. Indeed up till Oswald T. Avery's work in 1944, it was still believed that cell proteins in fact carried the stuff of heredity, although just how duplication might have occurred prior to cell division was still only guesswork. After Avery's work with bacteria, he correctly postulated that it was in fact the nucleic acid, the DNA of the cells and not the proteins that actually contained the genetic information of life. Building on these experiments, Alfred Hershey and Martha Chase and their kitchen blender experiments in 1952 furthered the theory, paving the way for the Watson and Cricks discovery. What Watson and Crick's discovery had done was to open the door to the possibility that the structure of DNA was so complex, that it was indeed capable of carrying the coded messages of genes and thus capable of replication of this vital store of information prior to cell division (Watson and Crick 1953b). This buoyed belief in the DNA gene transfer theory and ultimately the principles of Mendelian inheritance. Such a discovery had enormous implications, not least was that by understanding how the genetic information was copied, opened up untold possibilities in finding cures for human diseases. What perhaps Watson and Crick did not realise was the impact such research was also to have on the way we looked at the food we eat. In one respect, it opens up a whole Aladdin's cave of possibilities whilst on the other it threatens a Pandora's box of ills (Sections 14.3, 16.1 and 18.2.1).

Lastly in the area of science, in 1958 the National Seed Storage Laboratory (NSSI) in Fort Collins Colorado was opened becoming the first long-term seed storage facility in the world.

11.4 Period Commentary

Much happened in this period to shape the food security arena in both policy, institutional and agricultural terms (Argeñal 2007). Immediately after the war, the main world focus was on reconstruction and in the physical, economical, political and social global landscapes. On the political and economic front, old colonial empires disintegrated and new political powers emerged to take their place and, in vogue at the time, was the propensity to favour industry as the economic growth engine of choice. In achieving this, a rural exodus began in earnest and cities sprung up everywhere heralding a new optimism, a new modernity (FAO 1995). In the social arena, the health and nutrition of peoples became a monumental goal after hostilities had ceased, and few were more aware of this fact than Frank McDougall. Indeed in summarising much of the League of Nations and the International Labor Organization's (ILO) work, McDougall, a member of the Economic Committee of the League recognised the enormous task and necessity of meeting the nutritional needs of much of continental Europe after the war (McDougall 1943; Richter 1943). He remarked that

> The provision of food adequate in quantity and quality will have a more profound effect upon national health than any other single reform. (McDougall 1943)

In his summary, he also understood the need for the adoption of sound standards of international nutritional diets and their subsequent favourable effects on world agriculture and trade (ibid). He was also without doubt too in his recognition for the need for international cooperation and coordination of effort and assistance (McDougall 1943). This was somewhat echoed in the first UN conference on food in 1943 when it was acknowledged that poverty was the first cause of hunger and that to alleviate this dilemma there would have to be an expansion of the whole world economy (UN 1943).

In a rare moment of clarity and in an effort to aspire to such lofty goals, governments around the world pooled their collective optimism and in a exceptional time of multilateral cooperation, achieved critical mass with the incorporation of the United Nations. Although it was the second real coordinated effort to introduce a singular world body responsible for, among other things, economic growth and peace, it learned from the many lessons of the League of Nations. However, while the synergistic relationship between hunger and its many causes had been established at a global institutional level, initial collective multilateral solutions were hampered by stakeholder self-interests and a general reluctance to devolve so much perceived power to an institutional body (Shaw 2007). One of the early ideas to combat food shortages was the establishment of a World Food Board (WFB); it was put forth by Boyd Orr, then Director General of the newly formed FAO in 1946. However, after initial enthusiasm the United States and United Kingdom flagged in their support and the idea was eventually shelved. These and other similarly unsuccessful ideas including the International Commodity Clearing House (ICCH) largely failed because as Shaw (2007) suggests, the United Kingdom and United States were once again unwilling to devolve such power to a multilateral organisation over which they had little or no control (Mayne 1947; Shaw 2007).

Despite these setbacks, the post-war years were a productive and groundbreaking time for hunger and nutritional governance. The plight of war as well and the suffering of many of its most vulnerable citizens continued to cause public and institutional anxiety. By the middle of the 1940s, the United Nations and its newly formed specialised agencies—the FAO, WHO and UNICEF—were uniquely positioned to act. The new agencies were quick to take up the humanitarian mantle and collectively the new organisations dominated the international arena. These organisations in turn were backed by the technical and financial cooperation of many nations; however, some were better placed than others to help. Most notable and by far the most dominant of these was the United States. Both unilaterally and through these organisations, the United States provided a great deal in the way of financial and technical assistance through the Marshall plan, the Lend Lease Act and others. Other countries too helped channel funds through the newly created World Bank, providing opportunity and access for reconstruction (Goodrich 1947).

Thus, as post-war reconstruction slowed, so these institutions and voluntary organisations closed out operations in Western Europe and simultaneously opened up theatres of work in the less developed countries of Asia, Latin America and Africa. UNICEF continued its famous milk feeding programmes only to find them of negligible value. Giving way to the realisation of the inability of milk alone to stem protein malnutrition, the FAO, WHO and UNICEF looked for alternative high-quality protein foods. Also by 1950, the FAO had embarked on the Expanded Technical Assistance Programme, effectively broadening the Nutrition Division's reach to more and more underdeveloped countries.

The combined efforts of the three leading organisations featured heavily in periodicals of the time, leading to the frequent tag of 'super organisations' that sought to end to humankind's misery and hunger (Ruxin 1996). However, things were not quite so rosy. Overlapping mandates, political infighting and being charged with placing ideology above realism, the institutions were in danger of going by the way of the League of Nations and becoming impotent. That said, the relations between FAO and WHO were better than those between the FAO and UNICEF, although they were still somewhat fractious, leading both to affirm a 'Gentlemen's Agreement' in 1959 stipulating their individual as well as mutual areas of nutritional responsibility (ibid). Alliances too between the various governments, institutions and non-governmental organisations (NGO) were also tested. This was particularly noticeable in the movement of US surplus foods where many have accused the government and indeed the aid agencies of supporting an American foreign policy agenda of finding markets for agricultural surpluses.

In the 1950s, after initial difficulties, the growth and collaboration between the three main specialised organisations in nutritional matters took on greater importance (WHO 1958). The FAO's nutritional interests focused on the relation to food of production, distribution and consumption; the WHO's emphasis was on health and the prevention of disease; while UNICEF concentrated efforts on children's health and well-being (WHO 1950, 1958).

In the nutritional sphere, the problems of poverty and hunger were viewed differently depending on which side of the economic divide one lived. While policy adjustments in wealthier countries were seen as sufficient in many cases to raise standards of nutrition, in the lesser developed nations, more vigorous policies would be needed in order to raise consumption levels to meet the basic nutritional needs (Belshaw 1947). During this period, several elemental shifts resulted from a period of extensive experience and research. Overridingly, it had already been noted that causes of hunger and malnutrition were many and interconnected; what changed was the fundamental way in which nutrition came to be understood and, by extension, tackled. Changes in nutritional thinking acknowledged the importance of economic factors as well as social and cultural customs in the diet. Moreover, emphasis was also placed on the quality of the diet rather than just the quantity, taking into account not just calories but vitamins and minerals too (WHO 1953). Furthermore in this area, the initial focus of children's aid in general was found to be misguided, it was now understood that it was in fact the under-5s that were more susceptible to disorders of malnutrition than those of school-age children. This, coupled with the limited successes of milk and other aid programmes, highlighted the limitations of then collective current ideology: the notion that they could feed all children. This led to the strong conviction of change best summarised by the proverb

> Feed a man a fish and he eats for a day; teach a man to fish and he eats for a lifetime.

Two other breakthroughs also propelled changes in the way nutrition came to be understood. First, the role of protein deficiency and kwashiorkor (its clinical manifestation) highlighted the role of a balanced diet and marked the beginning of an obsession that was to last several decades. It also prompted the WHO in 1958 to state that

> Kwashiorkor is the now the main nutritional disease with which the organisation is concerned. (WHO 1958)

Second was the discovery of the connection between malnutrition and infection. The implications were enormous and meant that malnutrition could no longer be tackled by feeding alone. Collectively, these realisations marked a shift in development ideology which meant that new ways were needed to promote sustainable nutritional changes in mothers and children. One of the outcomes was the applied nutrition programmes (APN), which started in 1957. These sought to combine education with aid and practical assistance and went on to become a major policy of the 1960s.

Fuelling much of the aid at the time were agricultural surpluses, and as was becoming a regular cyclical occurrence, so the end of World War II once again witnessed agricultural policies aimed primarily at increasing domestic productivity and protecting prices. Despite some attempts such as the US Agricultural Act of 1956, otherwise known as the soil bank programme, which sought to take land out of crop production, surpluses persisted. As a result, large amounts of grain were ultimately distributed around the world primarily in the name of goodwill. But not everyone was convinced. Loud overt calls of foul play were made, particularly of the American camp, of supporting domestic and foreign policy objectives above humanitarian needs. This was seen by many at the time and since, as a deceitful way to facilitate the opening of new markets for US exports (Webb et al. 2008).

It was not all bad however; one overriding benefit of the rapid adoption of mechanical and chemical technologies of the time helped improve the 'struggle against natural difficulties' (Daniel 1944, p. 55). For while agriculture used to depend largely on geographical location, weather, soil and labour conditions, etc., improvements in knowledge and farming techniques as well as the uptake of technology, allowed farming to adapt more readily to harder climes and more labour challenges, in turn allowing the farmer now rather than the geography to dictate cropping patterns.

Lastly, when it came to BMR and its increasing use, Quenouille and colleagues examined the numerous Body Mass Index (BMR) studies that had taken place over the years producing the first large-scale study of world literature on the subject. Such was its impact that this publication eventually led to the Quenouille standards of BMR. This research was also important for its place as the forerunner to the modern Schofield equations (FAO/WHO/UNU equations) that are used today (Quenouille et al. 1951; Henry 2005).

11.5 Key Dates, People, Acts, Reports and Surveys

Summing up the previous period, Table 11.3 highlights some of the more important events, individuals and key reports, etc. that helped promote the growth of the food security concept.

TABLE 11.3

Key Dates: The Post-War Years

Date	People, Organisation, Agencies, Acts or Conventions
1946: Administrative Committee on Coordination (ACC)	The Administrative Committee on Coordination (ACC) was established in 1946 at the behest of ECOSOC. Its main functions were to ensure the coordination of UN programmes and to supervise implementation of agreements between the UN and the specialised agencies as well as to eliminate duplication. A number of sub-committees evolved with a focus on coordination and cooperation of work programmes throughout the UN system. The ACC became the CEB in 2001 (Mezzalama et al. 1999; CEB 2010; IANWGE 2010).
1946: Nutrition Division of FAO	Introduced to raise levels of nutrition throughout the world and any programmes predicated on increasing food supplies. The division was aided by the Standing Advisory Committee on Nutrition which met three times 1946, 1947 and 1948, which was itself replaced by the Joint FAO/WHO Expert Committee on Nutrition who continued to advise the division until the joint committees' end in 1954 (WHO 1950; Aykroyd 1956).

TABLE 11.3 (Continued)

Key Dates: The Post-war Years

Date	People, Organisation, Agencies, Acts or Conventions
1946: International Trade Organization (ITO)	The UN Economic and Social Committee 1946 called for a conference to draft a charter for an International Trade Organization (ITO), a specialised agency to compliment the Bretton Woods institutions. The aim was to create the ITO at the Havana UN Conference on Trade and Employment in 1947. The charter was completed in 1948, which set out the basic rules for international trade. The ITO Charter, however, was never approved, instead GATT stood in its place (ITO 1948; WTO 2010).
1946: International Emergency Food Council (IEFC)	The International Emergency Food Council (IEFC) was organised by FAO to deal with the world food crisis after World War II. It existed independently until 1948, when FAO absorbed and renamed it as the International Emergency Food Committee (IEFC) (FAO 1947; NAL 2010).
1946: The UN International Children's Emergency Fund (UNICEF)	Established in December 1946 to provide emergency relief assistance to the children of post-war Europe and China although changed its emphasis to give long-term benefit to children everywhere, particularly those in developing countries in 1953. Because of this, the name was changed to the UN Children's Fund, but the acronym 'UNICEF' was retained (UNICEF 2005).
1946: The Division for the Advancement of Women	The Division for the Advancement of Women was established in 1946 as the Section on the Status of Women, Human Rights Division, Department of Social Affairs (DAW 2009).
1947: UK's Agriculture Act	Consolidated and extended state support for using predominantly deficiency payments guaranteeing a certain market price while consumers continued to pay market prices, thus farmers were protected irrespective of market forces. This was a policy predicated on stability and efficiency (Bikistow 1998; OPSI 2007).
1947: General Agreement on Tariffs and Trade (GATT)	The GATT was a provisional agreement intended as a stopgap while the ITO was being drafted. However with the breakdown of the ITO, these provisional agreements lasted from 1947 to 1994 when the WTO took up the mantel. The GATT's role was to established procedures to substantially reduced tariffs between member nations and thus facilitate better world trade. There were eight rounds of trade talks spanning more than 30 years (WTO 2010).
1947: The Economic Commission for Europe (ECE)	One of five regional commissions of ECOSOC, with headquarters in Geneva established to help mobilise action for the economic reconstruction of post-war Europe and to increase European economic activity both within and outside the region (UNECE 2010).
1947: Economic and Social Commission for Asia and the Pacific (ESCAP)	One of five regional commissions of ECOSOC, headquarters in Bangkok, and a Pacific Operations Centre based in Port Vila, Vanuatu. Established in 1947. It serves a region that contains more than half the world's population (ESCAP 2010).
1946: UN Population Commission	The Population Commission was established by the Economic and Social Council as a subsidiary body to study and advise the General Council on the size, structure and changes in the world population (POPIN 1994).
1946: Meals for Millions	One man, Clifford E. Clinton had the vision to end world hunger; he formed Meals for Millions in 1946, which joined forces in 1960 with Kennedy's Freedom from Hunger Foundation (Freedom from Hunger 2010).
1947: Interim Coordinating Committee for International Commodity Arrangements (ICCICA)	ECOSOC recommended the establishment of Interim Coordinating Committee for International Commodity Arrangements (ICCICA) to help with trade and tariff reductions (GATT 1952; Khan 1982).
1947/1948: US Foreign Assistance Act a.k.a. Marshall Plan	Secretary of State Marshall's European Recovery Programme (ERR) brought about by the US Foreign Assistance Act of 1947 (a.k.a. 'Marshall Plan') involved a 4-year, multi-billion dollar programme of financial technical and food aid for the redevelopment of Europe. It came to an end in 1951.
1947: GATT 1st Round Geneva	In 1948 the agreement signed by 23 countries—45,000 tariff reductions were made affecting over $10 billion worth of trade, comprising 20% of the total global market at the time (Morrison 1986; WTO 2010).

continued

TABLE 11.3 (Continued)

Key Dates: The Post-War Years

Date	People, Organisation, Agencies, Acts or Conventions
1948: FAO Standing Advisory Committee	The Standing Advisory Committee. 'the problem of assessing the calorie and nutrient requirements of human beings, with the greatest possible degree of accuracy, is of basic importance to FAO' (Weisell 2002).
1948: Economic Commission for Latin America and the Caribbean (ECLAC)	One of five regional commissions of ECOSOC with headquarters in Santiago, Chile. Established in 1948 ECLAC integrated Latin American and Caribbean economies into the global economy (UNIHP 2009).
1948: The World Health Organization (WHO)	The WHO is a specialised agency of the UN that coordinates and disseminates information on international public health. Based in Geneva, Switzerland, the agency continued the mandate and legacy of its League of Nations predecessor, the Health Organization (WHO 1998, 2010).
1948: Universal Declaration of Human Rights	The Declaration adopted by the United Nations General Assembly arises from the experiences of World War II and represents the first global appearance of rights to which all human beings should be entitled (UN 1948).
1949: Committee on Commodity Problems (CCP)	In 1949 FAO set up the Committee on Commodity Problems (CCP). As a standing committee of the FAO council, its responsibilities were to continually review international commodity problems affecting production, trade, distribution, consumption and related economic matters (CCP 2005b).
1949: International Wheat Council (IWC)	The International seat of the Wheat Council (IWC) was established in London (IGC 2009b).
1949: GATT 2nd Round Annecy	Around 5000 total tariff reductions were achieved at these talks (Morrison 1986; WTO 2010).
1950: The UN High Commissioner for Refugees (UNHCR)	The UN High Commissioner for Refugees is responsible for the protection of refugees. It draws on the expertise of other UN organisations such as food production (FAO), health measures (WHO), education (UNESCO) and child welfare (UNICEF). It also cooperates closely with the World Food Programme in providing basic supplies to refugees (UNHCR 2009).
1950: World Vision	World Vision founded in 1950 by the Reverend Bob Pierce. First focusing on orphans and other children in need South Korea, then expanding throughout Asia then than 90 countries today (World Vision 2009).
1951: GATT 3rd round Torquay	8700 tariff concessions were made at the Torquay round of GATT (WTO 2010).
1953: Foreign Agricultural Service (FAS) of the US Department of Agriculture (USDA)	Partnering with USAID, the FAS concentrates predominantly on the agribusiness aspect of humanitarian aid (National Archives 1995; Otto 1999; FAS 2004).
1954: US Agricultural Trade Development and Assistance Act (Food for Peace)	The US federal law that established Food for Peace, the primary US overseas food assistance programme (FAS 2004).
1954: US Agency for International Development (USAID)	USAID is the US government's lead agency for humanitarian aid assistance in developing countries and the top provider of food aid in the world. Initially created as a way to help the US agriculture industry (USAID 2009c, 2010).
1954: The FAO Consultative Sub-Committee on Surplus Disposal (CSSD)	Established in 1954 as a subcommittee of the CCP, the CSSD's role was designed to monitor the disposal of agricultural surpluses used as food aid (ActionAid 2003). Its aim was to use food aid in a way that avoided the displacement of commercial imports and safeguarding export interests whilst at the same time seeking to avoid aid damaging the long-term food self-reliance of the recipient country (ActionAid 2003; CCP 2005a).
1955–1956: GATT 4th Round Geneva	$2.5 billion in tariffs were eliminated or reduced (WTO 2010).
1957: The European Economic Community (EEC)	The European Economic Community (EEC) (renamed the European Community (EC) in 1993) was established by the Treaty of Rome in 1957 (Europa 2007, 2010).

TABLE 11.3 (Continued)

Key Dates: The Post-War Years

Date	People, Organisation, Agencies, Acts or Conventions
1957: The Joint Research Centre (JRC)	In 1957, Treaty of Rome established both the European Economic Community (EEC) and the European Atomic Energy Community (Euratom). The Joint Research Centre was established under the Euratom treaty. The JRC has, however, expanded to include other fields important to policy making including life sciences, energy, security and consumer protection offering research-based policy support (JRC 2009).
1957: UK's Agriculture Act	As with 1947 Act the UK's 1957 Agriculture Act set out long-term assurances of price stability (UOR 2009).
1958: Economic Commission for Africa (ECA)	One of five regional commissions of ECOSOC with headquarters in Addis Ababa, Ethiopia. The ECE was the first intergovernmental organisation in Africa. Its covers countries whose economic and social conditions differ widely and where many countries and dependent territories are among the poorest in the world (UNECE 2000; EoN 2009).
1959: UN Declaration of the Rights of the Child	In 1959 the UN General Assembly adopted the Declaration of the Rights of the Child offering special protection, opportunities and facilities to develop physically, mentally, morally, spiritually and socially in conditions of freedom and dignity (UN 1959).

References

ActionAid (2003). Food Aid: An Actionaid Briefing Paper. ActionAid.

ADA (1941). Recommended Allowances for the Various Dietary Essentials. *Journal of American Dietetic Association* 17: 565–567.

Aggett, P. J., J. Bresson, F. Haschke, O. Hernell, B. Koletzko, H. N. Lafeber, K. F. Michaelsen, J. Micheli, A. Ormisson, J. Rey, J. S. d. Sousa and L. Weaver (1997). Recommended Dietary Allowances (RDAs), Recommended Dietary Intakes (RDIs), Recommended Nutrient Intakes (RNIs), and Population Reference Intakes (PRIs) Are Not Recommended Intakes. *Journal of Pediatric Gastroenterology and Nutrition* 25(2): 236–224.

AIS (1946). *Pamphlet No. 4, Pillars of Peace—Documents Pertaining to American Interest in Establishing a Lasting World Peace: January 1941–February 1946.* Pennsylvania: Book Department, Army Information School.

Allen, L. H. (2000). Ending Hidden Hunger: The History of Micronutrient Deficiency Control. *Background Analysis for the World Bank–UNICEF Nutrition Assessment Project,* Washington, DC: World Bank, 111–130.

Allen, L. H. (2003). Interventions for Micronutrient Deficiency Control in Developing Countries: Past, Present and Future. *The Jornal of Nutrition* 133(Supplement: Animal Source Foods to Improve Micronutrient Nutrition in Developing Countries): 3875S–3878S.

Almonacid, P. N. (1943). The Food Situation in a World at War: The Other Americas. *The Annals of the American Academy of Political and Social Science* 225: 93–95.

Argeñal (2007). History of Food Security in International Development: Background Paper Developed for Christian Children's Fund Honduras. Honduras: Christian Children's Fund.

Aykroyd, W. R. (1948). Objectives of the Nutrition Division. Rome: FAO Archives.

Aykroyd, W. R. (1956). Fao and Nutrition. *Proceedings of the Nutrition Society* 15(1): 4–13.

Bacon, L. (1943). The Food Situation in a World at War: Southwestern Europe *The Annals of the American Academy of Political and Social Science* 225: 87–88.

Balfour, L. E. B. (1975). *The Living Soil and the Haughley Experiment.* London: Faber and Faber.

Beardsworth, A. and T. Keil (1997). *Sociology on the Menu: An Invitation to the Study of Food and Society.* London: Routledge.

Beaton, G. H. (1992). Human Nutrient Requirement Estimates: Derivation, Interpretation and Application in Evolutionary Perspective. *Food, Nutrition and Agriculture* 1(2/3): 3–15.

Behar, M., W. Ascoli and N. S. Scrimshaw (1958). An Investigation into the Causes of Death in Children in Four Rural Communities in Guatemala. *Bulletin of the World Health Organization* 19: 1093–1102.

Belshaw, H. (1947). The Food and Agriculture Organization of the United Nations *International Organization* 1(2): 291–306.

Bikistow, G. (1998). Measuring Regional Variation in Farm Support: Wales and the UK, 1947–72, *Agricultural history review* 46(1): 81–98

BioTech Institute (2010). Timeline of Biotechnolgy 20th Century. Biotechnology Institute. http://www.biotechinstitute.org/what_is/timeline.html (accessed 12 January 2011).

Black, J. D. (1943). Food: War and Postwar. *The Annals of the American Academy of Political and Social Science* 225: 1–5.

Blakemore, R. J. (2000). Ecology of Earthworms under the 'Haughley Experiment' of Organic and Conventional Management Regimes. *Biological Agriculture and Horticulture* 18: 141–159.

Borchard, E. (1947). Intervention—the Truman Doctrine and the Marshall Plan. *The American Journal of International Law* 41(4): 885–888.

Bordo, M. D. and B. Eichengreen (1993). The Bretton Woods International Monetary System: An Historical Overview. In *A Retrospective on the Bretton Woods System: Lessons for International Monetary Reform.* Chicago, IL: University of Chicago Press.

Brock, J. F. and M. Autret (1952). Kwashiorkor in Africa. *Bulletin of the World Health Organization* 5(1): 1–71.

Broughton, L. (1997). Lester B. Pearson's Role in the Formation of the Food and Agriculture Organization (FAO) and in Other United Nations Activities. Canada: United Nations Agency in Canada.

Burnet, E. and W. R. Aykroyd (1935). Nutrition and Public Health. *Quarterly Bulletin of the Health Organization* IV(2): 152.

CARE (2009). History of Care International. Cooperative for Assistance and Relief Everywhere. http://www.careinternational.org.uk/57/history/history-of-care-international.html (accessed 12 January 2010).

Carlson, A. J. (1944). Symposium on Civilian Wartime Problems in Nutrition: From the Standpoint of the Physician: The Importance of Food in Wartime. *California and Western Medicine* 61(6): 281–285.

Carpenter, K. J. (1994). *Protein and Energy: A Study of Changing Ideas in Nutrition.* Cambridge: Cambridge University Press.

Cassels, J. M. and F. L. Hall (1943). Food Supplies for Our Civilian Population. *The Annals of the American Academy of Political and Social Science* 225: 106–115.

CCP (2005a). Consultative Subcommittee on Surplus Disposal: Fortieth Report to the CCP. Committee on Commodity Problems: Sixty-fifth Session Rome, Italy, 11–13 April 2005, Rome: Food and Agriculture Organization.

CCP (2005b). A Historical Background on Food Aid and Key Milestones. Committee on Commodity Problems: Sixty-fifth Session, Rome: Food and Agriculture Organization.

CEB (2010). The Chief Executives Board (CEB): Overview. The Chief Executives Board. http://www.unsystemceb.org/reference/ceb (accessed 23 March 2010).

Chatfield, C. (1949). Food Composition Tables for International Use: Fao Nutritional Studies No. 3. Washington, DC: Nutrition Division, FAO UN 56.

Chatfield, C. (1954). Food Composition Tables Minerals and Vitamins for International Use: FAO Nutritional Studies No. 11. Washington, DC: Nutrition Division, FAO UN 56.

Clayton, C. F. and J. D. Black (1943). The Food Situation in a World at War: Wartime Food Administration—U.S.A. *The Annals of the American Academy of Political and Social Science* 225: 96–105.

Crawford, D. M. (1949). United States Foreign Assistance Legislation, 1947–1948. *The Yale Law Journal* 58(6): 871–922.

CRS (2009). About Catholic Relief Services. Catholic Relief Services. http://crs.org/about/ (accessed 12 January 2010).

Daniel, A. (1944). Regional Differences of Productivity in European Agriculture. *Review of Economic Studies* 12(1): 50–70.

DAW (2009). Brief History. UN Division for the Advancement of Women. http://www.un.org/womenwatch/daw/daw/history.html (accessed 23 February 2010).

Devereux, S. (2000). Famine in the Twentieth Century. Brighton: Institute for Development Studies.

Dimitri, C., A. Effland and N. Conklin (2005a). *Electronic Information Bulletin: Number 3: The 20th Century Transformation of US Agriculture and Farm Policy.* http://www.ers.usda.gov/publications/eib3/eib3.htm.

Dimitri, C., A. Effland and N. Conklin (2005b). *Economic Information Bulletin: Number 3: The 20th Century Transformation of US Agriculture and Farm Policy*: USDA.

Editorial (1943). The Marriage of Public Health and Agriculture (United Nations' Conference on Food and Agriculture). *American Journal of Public Health* 33(7): 847–848.

Encyclopedia (2002). Science of Everyday Things. *Encyclopedia.com*. Encyclopedia.com. http://www.encyclopedia.com (accessed 13 July 2009).

EoN (2009). *Economic and Social Development: Regional Commissions* New York. The United Nations.

ESCAP (2010). United Nations Economic and Social Commission for Asia and the Pacific: General Description. UN ESCAP. http://www.unescap.org/about/index.asp (accessed 28 March 2010).

Eulau, H. (1968). Book Reviews: The Politics of Food for Peace. By Peter A. Toma. *Political Research Quarterly* 21: 530–532.

Europa (2007). *Treaty Establishing the European Economic Community, EEC Treaty*. European Union Publications Office.

Europa (2010). Website of the European Union. *The EU at a Glance*. Europa. http://europa.eu/index_en.htm (accessed 21 June 2010).

Evang, K. and F. L. McDougall (1944). The Hot Springs Conference. *Proceedings of the Nutrition Society* 2(3–4): 163–176.

FAO (1946). World Food Survey. Washington, DC: Food and Agriculture Organization.

FAO (1947). Report of the First Session of Council of FAO 4–11 November 1947. Washington, DC: Food and Agriculture Organization 35.

FAO (1950). Calorie Requirements. Report of the Committee on Calorie Requirements 1949: FAO Nutritional Studies No. 5. Washington, DC: Food and Agriculture Organization.

FAO (1955). Protein Requirements. Report of the F.A.O. Committee. *FAO. Nutritional Studies No. 16*. Rome: Food and Agriculture Organization.

FAO (1957a). Calorie Requirements. Report of the Second Committee on Calorie Requirements: FAO Nutritional Studies No. 15. Rome: Food and Agriculture Organization.

FAO (1957b). Report of the Conference of FAO. *Conference Reports*. Rome: Food and Agriculture Organization.

FAO (1958). *National Food Reserve Policies in Underdeveloped Countries*. Rome: Food and Agriculture Organization.

FAO (1995). Dimensions of Need: An Atlas of Food and Agriculture. Rome: Food and Agriculture Organization.

FAO (2001). *Food Balance Sheets: A Handbook*. Rome: Food and Agriculture Organization.

FAO (2002). Food, Nutrition and Agriculture. Rome: FAO Food and Nutrition Division.

FAO (2003). Food Energy—Methods of Analysis and Conversion Factors. *FAO Food And Nutrition Paper 77*. Rome: Food and Agriculture Organization.

FAO/WHO (1973). FAO/WHO Ad Hoc Committee of Experts on Energy and Protein: Requirements and Recommended Intakes. Rome: Food and Agriculture Organization.

FAO/WHO (1985). Energy and Protein Requirements: WHO Technical Report Series, No. 724. Geneva, Switzerland: FAO/WHO/UNU.

FAS (2004). Timeline of US Agricultural Trade and Development. USDA Foreign Agricultural Service. http://www.fas.usda.gov/info/50th/timelines.htm (accessed 12 January 2010).

Ficker, H. (1967). Fifty Years of Foreign Loans and Foreign Aid by the United States, 1917–1967. Washington, DC: Library of Congress, Legislative Reference Service 40.

Foreign Office (1943). *Report to Allied Governments by the Inter-Allied Committee on Post-War Requirements*. Washington, DC: HMSO.

Freedom from Hunger (2010). Website of the Freedom from Hunger. Freedom From Hunger http://www.freedomfromhunger.org/ (accessed 15 June 2010).

Gardner, B. L. (2002). *American Agriculture in the Twentieth Century: How It Flourished and What It Cost*. Cambridge, MA: Harvard University Press.

GATT (1952). *The History Structure and Functions of the ICCICA*. Washington, DC: World Trade Organization.

GATT (1960). General Agreement on Tariffs and Trade: Seventeenth Session; Disposal of Commodity Surpluses. New York: UN General Agreement on Tarrifs and Trade.

Gillespie, J. A. (2003). International Organizations and the Problem of Child Health, 1945–1960. *Dynamis* 23: 115–142.

Goldsmith, G. A. (1955). The Kwashiorkor Syndrome. *The American Journal of Clinical Nutrition* 3(4): 337–338.

Goodrich, L. M. (1947). From League of Nations to United Nations. *International Organization* 1(1): 3–21.

Hardin, C. M. (1943). The Food Production Programs of the United States Department of Agriculture. *The Annals of the American Academy of Political and Social Science* 225: 191–200.

Harper, A. E. (1985). Origin of Recommended Dietary Allowances—an Historic Overview. *The American Journal of Clinical Nutrition* 41: 140–148.

Harper, A. E. (2003). Symposium: Historically Important Contributions of Women in the Nutrition Society: Contributions of Women Scientists in the US to the Development of Recommended Dietary Allowances. *The American Society for Nutritional Sciences* 133: 3698–3702.

Henry, C. (2005). Basal Metabolic Rate Studies in Humans: Measurement and Development of New Equations. *Public Health Nutrition* 8(7A): 1133–1152.

Herdt, R. W. (1998). *The Rockefeller Foundation: The Life and Work of Norman Borlaug, Nobel Laureate*. New York: The Rockefeller Foundation.

Hoffmann, S. and C. Maier, Eds. (1984). *The Marshall Plan: A Retrospective*. Boulder, CO: Westview Press.

House of Commons (1944). *House of Commons Debates, 25 January 1944: United Nations (Relief Administration)*. 396.

House of Commons (1946). *House of Commons Debates, 25 January 1944: UNRRA Winding Up*. 430.

IANWGE (2010). Un Interagency Collaboration: History of the Collaboration on Women and Gender Equality. UN Inter Agency Network on Women and Gender Equality. http://www.un.org/womenwatch/ianwge/uninteagcoll.htm (accessed 12 February 2011).

IFT (1999). 20th Century Marks Achievements in Food Science. The Institute of Food Technologists. http://www.ift.org/cms/?pid=1000562 (accessed 11 January 2011).

IGC (2009a). *60 Years of Successive Agreements: Before 1949: The Early Years*. London: International Grains Council 4.

IGC (2009b). Grains Trade and Food Security Cooperation: The International Grains Agreement. The International Grains Council. http://www.igc.org.uk/en/aboutus/default.aspx#igc (accessed 21 February 2010).

ITO (1948). ITO Press Release: Speech to Be Delivered to the Chief Delegate of China at the Closing Session of the Havana Conference. UN Conference on Trade and the Employment, Havana, International Trade Organization.

James, C. and J. A. Fitzgerald, Eds. (2008). *Of the Land and the Spirit: The Essential Lord Northbourne on Ecology & Religion*. Bloomington, IN: World Wisdom Inc.

JRC (2009). European Commission Joint Research Centre: History. European Commission Joint Research Centre (accessed 22 June 2010).

Keusch, G. T. (2003). The History of Nutrition: Malnutrition, Infection and Immunity. *Journal of Nutrition* 133: 336S–340S.

Khan, K. (1982). *The Law and Organisation of International Commodity Agreements*. Hague: Martinus Nijhoff.

Ladejinsky, W. I. (1943). The Food Situation in a World at War: The Food Situation in Asia. *The Annals of the American Academy of Political and Social Science* 225: 91–93.

Landers, P. S. (2007). The Food Stamp Program: History, Nutrition Education, and Impact *Journal of the American Dietetic Association* 107(11): 1945–1951.

League of Nations (1933). *Final Act of the Conference of Wheat Exporting and Importing Countries, Held in London from August 21 to 25, 1933, with Appendices and Minutes of Final Meeting. Signed at London, August 25, 1933*: League of Nations. 3262.

League of Nations (1946). Transit Department: Food, Famine and Relief 1940–1946. Geneva, Switzerland: League of Nations.

Lloyd, E. M. H. (1943). The Food Situation in a World at War: In the United Kingdom. *The Annals of the American Academy of Political and Social Science* 225: 83–85.

Marrack, J. R. (1948). Hot Springs and Fao: Summary. *International Journal of Food Sciences and Nutrition* 2(3): 93–96.

Mayer, J. (1986). Social Responsibilities of Nutritionists. *Journal of Nutrition* 116: 714–717.

Mayne, J. B. (1947). FAO the History. *Review of Marketing and Agricultural Economics* 15(11): 419–426.

McArdle, W. D., F. I. Katch and V. L. Katch (1999). *Sports and Exercise Nutrition*. Baltimore, MD: Williams and Wilkins.

McDougall, F. L. (1943). The Food Situation in a World at War: International Aspects of Postwar Food and Agriculture. *The Annals of the American Academy of Political and Social Science* 225: 122–127.

McMillan, R. B. (1946). "World Food Survey"—A Report from F.A.O. *Review of Marketing and Agricultural Economics* 14(10): 372–378.

Mezzalama, F., K. I. Othman and L.-D. Ouedraogo (1999). Review of the Administrative Committee on Coordination and Its Machinery. Geneva: UN Joint Inspection Unit 31.

Morrison, A. V. (1986). Gatt's Seven Rounds of Trade Talks Span More Than Thirty Years—General Agreement on Tariffs and Trade. Swanton USA: Business America 8–10.

NAL (2010) *Consumers Guide*. USDA The National Agricultural Library. Agriculture Fact Book, USDA.

National Archives (1995). *Federal Records: Records of the Foreign Agricultural Service*: US National Archives.

National Archives (2009). *Federal Register: Food and Nutrition Service*: US National Archives. 74.

Nobel Prize (2010). UNICEF: Achievement and Challenge. Nobel Foundation. http://nobelprize.org/nobel_prizes/peace/laureates/1965/unicef-lecture.html (accessed 15 March 2011).

Northbourne, L. (1940). *Look to the Land*. London: J. M. Dent & Sons.

NRC (1943). Recommended Dietary Allowances: Report of the Food and Nutrition Board, Reprint and Circular Series No. 115. Washington, DC: National Research Council 6.

Ó Gráda, C. (2009). *Famine: A Short History*. Princeton: University Presses of California, Columbia and Princeton.

OPSI (2007). *Agriculture Act 1947*: HMSO: Office of Public Sector Information.

Otto, J. S. (1999). *The Final Frontiers, 1880–1930: Settling the Southern Bottomlands*. Westport, CT: Greenwood Press.

OXFAM (2007). How Did Oxfam Start? OXFAM. http://www.oxfam.org.uk/coolplanet/kidsweb/oxfam/history.htm (accessed 8 January 2010).

PAG (1967). PAG Bulletin Number 7: II. Articles—Development of Supplementary Food Mixture (CSM) for Children. *Protein Advisory Group News Bulletin*. Geneva, Switzerland: World Health Organization 24.

Parran, T. (1943). A Blueprint for the Conquest of Hunger. *Public Health Reports* 58(24): 893–899.

Passmore, R. (1964). An Assessment of the Report of the Second Committee on Calorie Requirements (FAO, 1957). Rome: Food and Agriculture Organization.

Paull, J. (2008). The Lost History of Organic Farming in Australia. *Journal of Organic Systems* 3(2): 2–17.

Périssé, J. (1981). Joint FAO/WHO/UNU Expert Consultation on Energy and Protein Requirements: Past Work and Future Prospects at the International Level. Rome: Food and Agriculture Organization.

Phillips, R. W. (1981). FAO: Its Origins, Formation and Evolution 1945–1981. Rome: Food and Agriculture Organization.

POPIN (1994). Background Document on the Population Programme of the UN. Geneva, Switzerland: United Nations Population Fund.

Quenouille, M., A. Boyne, W. Fisher and I. Leitch (1951). Statistical Studies of Recorded Energy Expenditure of Man. Basal Metabolism Related to Sex, Stature, Age, Climate, and Race. *Technical Communication No. 17*. Aberdeen: Commonwealth Agricultural Bureau.

Reid, R. and E. H. Hatton (1941). Price Control and National Defense. *Northwestern University Law Review* 36(255): 283–284.

Reinhardt, C. and B. Ganzel (2003) A History of Farming. *Living History Farm*: http://www.livinghistoryfarm.org/farmingin the30s/crops_03.html.

Richter, J. H. (1943). The Food Situation in a World at War: Central and Northern Europe. *The ANNALS of the American Academy of Political and Social Science* 225: 85–87.

Roosevelt, F. D. (1941). *Xxii Annual Message to the Congress, 1941*: United States Government Printing Office.

Rosenberg, I. H. (1994). Nutrient Requirements for Optimal Health: What Does That Mean? *Journal of Nutrition* 124(9 Suppl): 1777S–1779S.

Ruxin, J. N. (1996). *Hunger, Science, and Politics: FAO, WHO, and UNICEF Nutrition Policies, 1945–1978, Chapter II the Backdrop of UN Nutrition Agencies, by Joshua Nalibow Ruxin*. London: University College London.

Schultz, T. W., Ed. (1945). *Food for the World*. Chicago: University of Chicago Press.

Scott, S. L. (1951). The Military Aid Program. *The Annals of the American Academy of Political and Social Science* 278(7/8): 47–55.

Scrimshaw, N. S., C. E. Taylor and J. E. Gordon (1959). Interactions of Nutrition and Infection. *American Journal of Medical Science* 237(3): 367–403.

Shaw, J. (2007). *World Food Security: A History since 1945*. UK: Palgrave Macmillan.

Sullivan, R. R. (1970). Politics of Altruism—Introduction to Food-for-Peace Partnership between United States Government and Voluntary Relief Agencies. *Western Political Quarterly* 23(4): 762–768.

Sweetser, A. (1946). From the League to the United Nations. *The Annals of the American Academy of Political and Social Science* 246: 1–8.

Thompson, R. J. (1943). The United Nations Conference on Food and Agriculture. *Journal of the Royal Statistical Society* 106(3): 273–276.

Toma, P. A. (1967). *The Politics of Food for Peace*. Tucson: University of Arizona Press.

UN (1943). United Nations Conference on Food and Agriculture, May 18-June 3, 1943: Final Act and Section Reports/Department of State, United States of America. Washington, DC: US G.P.O. United Nations Conference on Food and Agriculture, Washington DC: United Nations Government Printing Office.

UN (1945). *Charter of the United Nations, 24 October 1945, 1 Unts XVI*: United Nations.

UN (1948). *Part a of General Assembly Resolution 217 (III). International Bill of Human Rights: Universal Declaration of Human Rights* United Nations. Resolution 217.

UN (1959). *Declaration of the Rights of the Child*: United Nations. A/RES/1386 (XIV).

UNECE (2000). Overview of the Economic Commission for Africa. Economic Commission for Africa. http://www.uneca.org/adfiii/riefforts/hist.htm (accessed 23 June 2010).

UNECE (2010). Economic Commission for Europe: Inception. *Economic Commission for Europe*. United Nations. http://www.unece.org/oes/history/history.htm (accessed 21 March 2010).

UNESCO (2009). About UNESCO: What Is It? What Does It Do? United Nations Educational, Scientific and Cultural Organization. http://portal.unesco.org/en/ev.php-URL_ID=3328&URL_DO=DO_TOPIC&URL_SECTION=201.html (accessed 30 February 2011).

UNHCR (2009). About Us: Basic Facts. Office of the United Nations High Commissioner for Refugees. http://www.unhcr.org.au/basicfacts.shtml (accessed 13 April 2010).

UNICEF (2005). United Nations Children's Fund. United Nations International Children's Emergency Fund (UNICEF). http://www.unicef.org/about/who/index_history.html (accessed 30 January 2010).

UNICEF (2009). The Convention on the Rights of the Child. United Nations International Children's Emergency Fund (UNICEF). http://www.unicef.org/rightsite/sowc/ (accessed 30 January 2011).

UNIHP (2009). UN Regional Contributions: Latin America and the Caribbean. *Briefing Note No.19*. New York: UN Intellectual History Project.

UNOG (2009). The End of the League of Nations. United Nations Office at Geneva, Switzerland. http://www.unog.ch/80256EE60057D930/(httpPages)/02076E77C9D0EF73C1256F32002F48B3?OpenDocument (accessed 12 November 2009).

UOR (2009). University of Reading: Agriculture, Policy and Development, History of Agriculture. University of Reading. http://www.ecifm.rdg.ac.uk/history.htm (accessed 4 June 2008).

USAID (2009a). Food for Peace. United States Agency for International Development. http://www.usaid.gov/our_work/humanitarian_assistance/ffp/history.html (accessed 4 January 2010).

USAID (2009b). The Marshall Plan. United States Agency for International Development. http://www.usaid.gov/multimedia/video/marshall/timeline.html (accessed 12 January 2010).

USAID (2009c). United States Agency for International Development: History. United States Agency for International Development. http://www.usaid.gov/about_usaid/usaidhist.html (accessed 12 May 2010).

USAID (2010). Our Work: A Better Future for All. United States Agency for International Development. http://www.usaid.gov/our_work/ (accessed 12 January 2010).

USDA (2009). History of the US Department of Agriculture (USDA). US Department of Agriculture. http://riley.nal.usda.gov/nal_display/index.php?info_center=8&tax_level=2&tax_subject=3&level3_id=0&level4_id=0&level5_id=0&topic_id=1033&&placement_default=0 (accessed 12 February 2010).

USHMM (2010). US Holocaust Memorial Museum: Politics—UNRRA. US Holocaust Memorial Museum http://www.ushmm.org/museum/exhibit/online/dp/politic4.htm (accessed 21 February 2010).

Volin, L. (1943). The Food Situation in a World at War: The Russian Food Situation. *The Annals of the American Academy of Political and Social Science* 225: 89–91.

Watson, F. (1999). One Hundred Years of Famine—a Pause for Reflection. *Field Exchange*(8): 20–23.

Watson, J. D. and F. H. Crick (1953a). Genetic Implications of the Structure of Deoxyribonucleic Acid. *Nature* 171: 964–967.

Watson, J. D. and F. H. Crick (1953b). A Structure for Deoxyribose Nucleic Acid. *Nature* 171: 737–738.

Waugh, F. V. (1943). The Food Distribution Programs of the Agricultural Marketing Administration *The Annals of the American Academy of Political and Social Science* 225: 169–176.

Webb, P., J. Gerald and D. R. Friedman (2008). More Food, but Not yet Enough: 20th Century Successes in Agriculture Growth and 21st Century Challenges. *Food Policy and Applied Nutrition Program*. Boston, Massachusetts: School of Nutrition Science and Policy, Tufts University.

Weir, C. (1952). Dollar Exports in the Marshall Plan Period *Journal of International Affairs (Royal Institute of International Affairs 1944-)* 28(1): 9–14.

Weisell, R. (2002). Measurement and Assessment of Food Deprivation and Undernutrition: Part Iv - Summary of the Draft Findings of the Joint FAO/WHO/UNU Expert Consultation on Human Energy Requirements. International Scientific Symposium, Rome: FAO.

Weisell, R. C. (1995). Food, Nutrition and Agriculture: FAO Celebrates 50 Years—Expert Advice on Energy and Nutrient Requirements - an Fao Tradition. Rome: Food and Agriculture Organization.

Whipple, C. E. (1943). The Food Situation in a World at War: Southeastern Europe *The Annals of the American Academy of Political and Social Science* 225: 88–89.

White, T. R. (1919). The Amended Covenant of the League of Nations. *The Annals of the American Academy of Political and Social Science* 84: 177–193.

WHO (1946). Official Records of the World Health Organization No1 Minutes of the Technical Prepatory Committee for the International Health Conference. Paris: WHO 80.

WHO (1950). Joint FAO/WHO Expert Committee on Nutrition: Report on the First Session 1949.Technical Report Series No. 16. *World Health Organization Technical Report Series No. 16*. Geneva, Switzerland: World Health Organization 24.

WHO (1951). Joint FAO/WHO Expert Committee on Nutrition: Report on the Second Session 1951. Technical Report Series No. 44. *World Health Organization Technical Report Series No. 44*. Geneva, Switzerland: WHO 24.

WHO (1953). Joint FAO/WHO Expert Committee on Nutrition: Third Report 1953. Technical Report Series No. 72. *World Health Organization Technical Report Series No. 72*. Geneva, Switzerland: World Health Organization 24.

WHO (1955). Joint Fao/Who Expert Committee on Nutrition: Fourth Report 1953. Technical Report Series No. 97. *World Health Organization Technical Report Series No. 97*. Geneva, Switzerland: WHO 24.

WHO (1958). The First Ten Years of the World Health Organization: Ch 22 Nutrition. Geneva, Switzerland: WHO.

WHO (1972). Nutrition: A Review of the Who Programme. Ii. *World Health Organization Chronicle* 26(5): 195–206.

WHO (1973). Energy and Protein Requirments: Report of a Joint FAO/WHO Ad Hoc Expert Committee— Technical Report Series No. 522. *WHO Technical Report Series No. 522 FAO Nutrition Meetings Reports Series No. 52*. Geneva, Switzerland: WHO 24.

WHO (1998). Fifty Years of the World Health Organization in the Western Pacific Region, 1948–1998:Report of the Regional Director to the Regional Committee for the Western Pacific. Forty-Ninth Session. Manila, Philippines: World Health Organization, Regional Office for the Western Pacific.

WHO (2009). Archives of the United Nations Relief and Rehabilitation Administration (UNRRA). World Health Organization. http://www.who.int/archives/fonds_collections/bytitle/fonds_2/en/index.html (accessed 12 March 2010).

WHO (2010). Website of the World Health Organization (WHO) http://www.who.int/ (accessed 2 April 2010).

Williams, A. J. (2005). 'Reconstruction' before the Marshall Plan. *Review of International Studies* 31(3): 541–558.

World Vision (2009). Website of World Vision. http://www.worldvision.org.uk/ (accessed 23 May 2010).

WTO (2010). Understanding the WTO: Basics—the Gatt Years: From Havana to Marrakesh. World Trade Organization. http://www.wto.org/english/thewto_c/whatis_e/tif_e/fact4_e.htm (accessed 30 March 2010).

Wu Leung, W.-T., F. Busson and C. Jardin (1968). Food Composition Table for Use in Africa. Maryland, Rome: US Department of Health/FAO.

12

The Development Decade: 1960s

12.1 Governance, Politics and Socioeconomics

The UN proclaimed the 1960s the first decade of development and it became a time of hope for ending hunger in the Third World. There was also growing optimism vis-à-vis agricultural productivity and agrarian reform; however, growing concern over the problems of malnutrition and the economic and social development prospects of poorer countries still weighed heavy (SOFA 2000). This period also saw the UN Organisations—the WHO, FAO and UNICEF—joined by new international players, policies and initiatives and numerous conferences that many believed would unite them in a common cause. The decade began with two new UN's enterprises: the Freedom from Hunger Campaign and the World Food Programme. This was quickly followed by the first World Food Conference, the creation of the Common Agricultural Policy and the arrival of the US Department of Agriculture onto the world food analysis stage. Lastly, the birth and growth of modern environmentalism began with Rachel Carson's attack on what was seen as questionable agricultural practices.

12.1.1 The Freedom from Hunger Campaign

The brainchild of FAO's then Director General B. R. Sen, the Freedom From Hunger Campaign (FFHC) was initiated on a temporary 5-year mandate and aimed to draw unprecedented attention to the plight of hunger and malnutrition in developing countries. Beginning in 1960, it was launched at a time when it was acknowledged that progress against global poverty had indeed been slow; moreover, continued population pressures threatened to worsen the situation. The Freedom from Hunger Campaign was to be a focus of cooperation for international organisations, governments and various NGOs alike. It achieved this objective in measurable success and in March of 1963 it culminated in the 'World Freedom from Hunger Week' campaign. During this week global campaigns were launched by the FAO and the public to attract worldwide attention to the injustice of hunger and to the perennial concerns of population growth outstripping agricultural production. A year later, at the request of President Kennedy, the American Freedom from Hunger Foundation was founded to mobilise support of the FAO's FFHC. This foundation joined forces with Clifford E. Clinton's Meals for Millions organisation and both were later combined (1979) to form the group Freedom from Hunger (Freedom from Hunger 2010). Building on this early momentum, three months later, 1300 participants from over 100 countries converged on Washington to discuss the issues at the UN's 1963 World Food Congress (UN 1963). As with the FFHC, the Congress drew attention to the plight of the hungry and emphasised the greater role of the developing countries in helping themselves (SOFA 2000).

Meanwhile in 1960, the FAO had commissioned a study on the use of global food surpluses as aid. In a meeting in Rome the following year, convened to discuss these findings among other things, the director of the US Food for Peace Programme, George McGovern, proposed a 3-year multilateral food aid trial programme. The result was the World Food Programme (WFP) (ODI 2000; CCP 2005; Mousseau 2005; Rucker 2007; Walker 2008).

12.1.2 World Food Programme

The World Food Programme (WFP) was established as a frontline food arm of the UN, jointly sponsored by both the FAO and the UN. At its inception, it immediately became responsible for humanitarian aid in the context of world hunger. Although it was scheduled to start operations in 1963, it did in fact start in 1962. This marked the beginning of global multilateral food aid organisation that aimed to combat hunger, promote economic and social development and provide relief assistance in emergencies (WFP 2008). Working closely with the FAO and other organs of the UN as well as various NGOs, the WFP sought to fulfil their objectives through three main programme allocations comprising emergencies, feeding programmes and projects. The improvement of quality of life and nutrition through the use of surpluses as well as other more direct financial instruments was considered so important that in 1965 the work of the temporary body, through parallel resolutions of both the FAO and the UN General Assembly, established the WFP on a continuing basis.

In the meantime with the stalling of UNICEF's milk supplement programmes in 1959, many of its established projects were to a large extent subsumed by the WFP and others. UNICEF had to adapt and turned to the relatively new nutrition programmes such as the applied nutrition projects (ANP). However, programmes were slow and somewhat generic and as a result progress was lacking and studies of their effectiveness soon earmarked a radical shift away from an individual projects approach to a country-specific approach. Indeed, these early UNICEF studies revealed that too many centrally planned projects lacked country-specific insight needed to identify and implement the most effective programmes. It was realised too that by decentralising decision-making, UNICEF could effectively give local staff, in part-nership with national governments, the autonomy to decide which projects would be best suited to which regions, social situations or circumstances (Ruxin 1996).

Separately, in the United Kingdom, these early years also witnessed the creation of the newly created European Economic Union (EEC) Common Agricultural Policy (CAP). This was to become a behemoth over the years introducing wide ranging policies with equally wide ranging consequences.

12.1.3 EU Common Agricultural Policy

In 1962 Article 39 of the Treaty of Rome saw the birth of the Common Agricultural Policy* (CAP). Justification for its implementation was based on the idea that the agricultural economy, in times of development and prosperity, would not reap the same benefits as other industries. Specifically it was argued that as incomes increased so people would spend proportionately less of their money on food and this would have negative consequences for the agricultural sector. In order to redress the balance, the EEC put in place incentives and introduced policies to ensure member countries' goods were stable and profitable in the domestic market whilst being competitive in the international arena. At the same time the CAP was to provide a good means of income for its farmers (Subramanian 2003; EU 2009).

Meanwhile, UNICEF was once again raising the profile of children's nutritional issues.

12.1.4 The Bellagio Conference

In 1964, with the support of the Rockefeller Foundation, UNICEF sponsored the international conference on children in development planning at the Villa Serbelloni in Bellagio, Italy. The conference marked a departure from UNICEF's normal policies of assistance and in its aspirations aimed to encourage recipient governments to accept greater responsibility in the role of nutrition as an explicit and integral part of develop-ment planning. The prevailing belief was that no real progress could be made on the health status of children if these issues did not factor at the national planning level. Both FAO and WHO advisors suggested that to achieve this objective, national nutrition councils should be established that implement country-wide devel-opment projects so as to place nutrition directly and firmly in the realm of national planning. These councils,

---

* At the heart of the CAP is a price support system that aims to set a target price for a commodity. In maintaining this the EU impose a levy on cheaper imports and also intervene to buy commodities at a predetermined level to maintain the stability of the internal market.

it was suggested, would also act as the main focal or liaison points for the UN agencies. However, as Ruxin (1996) pointed out, this required a hitherto unseen level of commitment on the part of governments on policies of national nutrition. Such a concerted cooperative effort sought to integrate the policies of various ministries of health, education and agriculture into overarching nutrition strategy subverting the previous bureaucratic treadmill. Wearily, as national initiatives began, it was clear that humanitarian interests alone would not suffice in the much needed the support and funding of the new nutritional aspirations. What was needed was the insight that nutritional improvement equalled economic growth. This was not long to follow.

So far the UN, collective organisations were at the vanguard of collecting data for international analysis. While the United States Agricultural Service (USDA) was, up to this point, very active in the global arena, its role, albeit influential, was one of supply and policy advocation; this was about to change.

12.1.5 USDA Joins the Club

The year was 1962 and marked the USDA's first incursion into global food security analysis arena with their 'World Food Budget, 1962 and 1966'. The first report, according to one critic could have been better prepared and suggested that

> … on the basis of a number of hastily-prepared balance sheets, drew up a most depressing 'geography of hunger. (Poleman 1972)

The report also claimed that most of the developing countries' (containing 1.9 billion people) diets were nutritionally inadequate and compounding this, it was suggested, was that the problem was unlikely to be resolved anytime soon while the population continued to expand so rapidly (USDA 1961; Poleman 1972). Moreover, it was offered that 'diet-deficits' used in this and subsequent reports by the USDA were an arbitrary construct, leaving the same critic-stroke-cynic to once again bring up that oft-quoted observation that

> … it is difficult not to conclude that promotion of the notion of hunger in the developing world was good politics for the USDA, which was faced with increasingly bothersome surpluses. (Poleman 1972)

In the interim period, the domestic front also needed attention and by the early 1960s, the USDA once again placed the perennial problem of agricultural overproduction and farmer incomes at the forefront of policy. Continuing the practice that started in the late 1940s, and which continued into the 1950s with legislations like the soil bank programmes and others, the American government, however, was losing the battle. By the end of the previous decade, it seemed the soil bank programme had achieved little more than taking 28 million acres of marginal farmland out of production whilst at the same time paying farmers subsidies that were reinvested in mechanisation which in turn resulted in yet more surpluses. Furthermore, it was also seen that federal farm programme costs were soaring. The USDA once again, under the administrations of Presidents Kennedy and Johnson, made overproduction a primary objective of farm policy. As well as taking more land out of production, measures were introduced to expand export markets for agricultural products. It was a time that saw farmer incomes in relation to non-farmer's increase from about 50% to 75% (Reinhardt and Ganzel 2003).

Ultimately, the development decade culminated in the high point for the UN with the creation of the International Bill of Rights as protection against oppression and discrimination.

12.1.6 International Bill of Human Rights

The International Bill of Human Rights is the informal and collective name given to the three instruments: one General Assembly Resolution and two international treaties. These are the 1948 Universal Declaration of Human Rights (UDHR), the 1966 International Covenant on Economic, Social and Cultural Rights (ICESCR) and the 1966 International Covenant on Civil and Political Rights (ICCPR). On acceptance of the Declaration of Human Rights in 1948, the Commission on Human Rights set upon the task of translating these principles into international treaties. Because of the breadth of the task, the UN decided to draft two international agreements representing civil and political rights in one and economic, social and cultural rights in another. For the next two decades, members debated the treaties until in 1966 consensus was finally reached. Although adopted in 1966 the two international covenants were only ratified and entered into force in 1976. This effectively solidified many of the pre-existing provisions of the Universal Declaration (1948) and collectively ushered in a new era of 'rights' for the individual (UN 1948, 1966a, 1966b; Perry 1998). In both international covenants, it is recognised that 'human rights derive from the inherent dignity of human beings' and by extension people are self-determined in their political and religious views. This principle of non-discrimination was now binding to those that ratified the treaties, and the onus was now on signatories to ensure the 'equal right of men and women to the enjoyment of all human rights' (UN 2011).

More specifically, the two international treaties added further dimensions idealised in the 1948 UDHR. The ICESCR, on the one hand, recognised the right of people's self-determination and their rights to: social security (Article 9); their right to family life and the protection of children (Article 10); rights to an adequate standard of living, including food, clothing and housing (Article 11); good health (Article 12); a good standard of education (Articles 13 and 14); and their full and free participation in cultural life (Article 15) (UN 1966b). On the other hand, the ICCPR secured individuals rights to personal security and judicial fairness (Article 9); a person's political rights (Article 1); and the right of non-discrimination (Article 4) (UN 1966a).

12.1.7 Urbanisation

At this juncture of the decade, concerns were beginning to be voiced about a new phenomenon; rural–urban migration. It was a problem that was raised with the early nutritional programmes, as planners expressed concerns at the concentration of rural projects at the expense of a rapidly changing demographic. This was also noted by McLaren who also saw this shift as potentially aggravating the situation with regard to marasmus while Jeliffe too was certain that prevailing rural disorders of diarrhea, malnutrition and respiratory infection would be transposed onto the urban environment. However, despite these early musings, it was not until the late 1968 with a report into nutritional problems of urban and peri-urban areas by the joint FAO/WHO Inter-Secretariat that the problem reached the heights of UN's policy staging area (Ruxin 1996).

In the meantime, with all the collective gains in nutritional governance, there was a slow burning environmental movement which having begun with the opposition to the industrialisation of British agriculture during the late 1930s, prospered in the immediate post-war years. Building on the previous ideas of organic farming by Lord Northbourne, the environmental movement also took immeasurable impetus from the tireless work of Rachel Carson.

12.1.8 Environmentalism

It began with the serialisation of an as yet unpublished work by Rachel Carson in the *New Yorker*. By the time the book of the serialisation 'Silent Spring' came out in 1962, the well-known writer had become an equally well-known social critic. The book catalogued, through case studies, the harmful effects that chemicals and synthetic pesticides had on the environment and publically challenged ecologically destructive practices. The book generated a lot of concern and controversy in the public's eye, not least because Carson eloquently pointed out, that the many of the potential long-term effects of such chemical use may have untold consequences for both the environment as well as for humans. Such controversy and public concern eventually led to the creation of environmental legislation and agencies to regulate the use of these chemicals. Ironically, attempts to discredit Carson ultimately endowed her book with more publicity and roped in many more supporters. One reviewer, US Supreme Court Justice William O. Douglas, opined it was 'the most important chronicle of this century for the human race', while another, Loren Eisely, claimed it was a 'devastatingly, heavily documented, relentless attack upon human carelessness, greed and irresponsibility'. Carson's accusations were so horrifying that President Kennedy ordered his Science Advisory Committee to investigate. The report of the investigation, which came out in 1963, ultimately vindicated Miss Carson. Indeed, in an article by Bruce Frisch in *Science Digest*, he noted that the President's Committee found that there were indeed many unknown long-range risks both to man and animal alike. True global environmentalism then can be said to have started as one woman's cautionary tale and which has ultimately turned into a movement of global proportions, that is, one that was to play a leading role in future policy at all levels of government (Carson 1962; Frisch 1964; Baker 2003).

12.2 Health and Nutrition

On the health and nutrition front, there was much change afoot. The established primacy of protein in dietary nutritional disorders was being questioned as was the ideology of the applied nutrition programmes. Also the Third World food survey continued to highlight the extent of the global situation, while infection was finally linked with malnutrition. It was also the period that an extraordinary White House report was published, articulating the problems facing the global abolition of hunger and malnutrition.

12.2.1 Nutrition Programmes

APNs were seen as a way to educate people in nutritional, agricultural and health issues and to show people how to produce foods rich in all the essential protein and vitamins. With the support of the FAO, UNICEF's first nutrition projects initially sought to encourage agricultural production and educate people about such practices. A typical project would have involved a baseline nutritional survey to determine the extent and need of change, followed by education and training using on-the-ground field staff. These in turn would pass on this knowledge to the population who in turn would put in place production projects to increase the availability of high-protein foods through school, community and home gardens. Later however, the projects became more individualised in a regional or group context and were tied more closely to economic aspects of the food challenges such as purchasing power, currency exchanges and markets for products, etc.

Progress once again was slow and the projects came under fire. An independent audit of WHO, FAO and UNICEF programmes in nutrition training and education by Platt and two other consultants in 1962 highlighted these many concerns. Interestingly and perhaps prophetically the consultants, while acknowledging that ignorance and poverty were major stressors of nutritional problems, argued that the problem of ignorance was not confined to the recipients alone; rather politicians and other policy makers ignorant of good nutritional practice were just as culpable. Importantly too, there was concern that the so-called specialists administering such programmes in the field were not particularly knowledgeable or experienced either (Ruxin 1996). Indeed, many well-intentioned national programmes failed, according to Platt, on these grounds and the report went on to suggest that it was up to nutritionists to inform national leaders about appropriate diets. These views were echoed by others including Scrimshaw and Moises Behar of the Instituto of Nutrition of Central America and Panama (INCAP). Scrimshaw believed that the problem or unsustainability of ANPs was that they were a reactionary and unrealistic approach to malnutrition. Behar, on the other hand, equally convinced and parroting the views of early UNICEF studies, thought the problem stemmed from the notion that such policies were pushed onto individual's from people hidden away in their ivory towers, with little or no regard for local cultural traditions or issues. Whatever the fault, the programmes themselves eventually withered towards, the end of the 1960s as funding dried up and political will failed to re-ignite the required impetus.

In the separate area of protein research, it was time to question the respective roles of kwashiorkor and marasmus.

12.2.2 Protein and the Growing Kwashiorkor–Marasmus Debate

By the beginning of the 1960s, the Protein Advisory Group (PAG) was already an influential body and yet while the group stated that it was interested in caloric intake as well as protein, the reality was that protein always took prominence much to the chagrin of many nutritionists (Mclaren 1966). The cracks too were starting to appear in the notion of the primacy of kwashiorkor in protein deficiency as evidenced by the Sixth Joint FAO/WHO Expert Committee on Nutrition in 1962 when it expressed concerns that perhaps the focus on protein was becoming excessive. In the report, the committee suggested that not just kwashiorkor but rather all aspects of the problem of protein-calorie-deficiency disease would indeed benefit from wider focus (FAO 1962). In strengthening this view, the Expert Committee talked of protein-calorie-malnutrition and also designated the term 'protein-calorie deficiency diseases' to include both kwashiorkor and marasmus and all its sequelae. Further de-emphasising kwashiorkor's primacy was a study in southern India supported by the WHO, which at that time had found that marasmus actually featured twice as much as kwashiorkor. Even Scrimshaw, who had previously spent much effort and research on the problem of kwashiorkor, began to shift focus towards marasmus and marasmic-kwashiorkor (Scrimshaw 2010). However despite this shift in thinking, the focus on protein was still strong. In fact building on this and the PAG's momentum, the FAO Expert Group met again in 1963 to re-examine protein requirements and despite a plethora of studies, intake levels for children still remained divisive. In the end, the committee adopted the factorial approach to calculating protein requirements. This involved the summation of core BMR values plus an allowance for daily individual variability. However, it was later concluded by the 1971 expert group that the 1963 protein requirements had perhaps been incorrectly assessed due to the overestimation of nitrogen losses (Passmore 1964; FAO/WHO 1965, 1973; WHO 1973).

By this time, the Third World Food Survey was published highlighting once again the extent of the global malaise.

12.2.3 The Third World Food Survey

It had been felt by many FAO statisticians that the previous World Food Survey of 1952, which stated that nearly two-thirds of the population were malnourished, had vastly exaggerated the problem. In spite of this, the Third World Food Survey published in 1963, noted that although better than the pre- and post-war periods, the global nutritional situation was still in fact dire. Taking advantage of new methodologies in sampling, analysis and interpretation, the Third World Food Survey placed the global undernourished figure at between 10% and 15% and up to 50% considered to be suffering from general hunger and malnutrition (FAO 1963; Gillin 2006). Bearing in mind too the population increase over the intervening period, it was still a large and unacceptable proportion. Once again, it was acknowledged that rapid expansion of food production was needed if future populations were to be fed. In this race against population growth, it was projected that the food supply would need to increase by 50% to 1975.

Meanwhile by 1964 the FAO had changed tack and refocused supplementary food's emphasis away from high protein to a more appropriately balanced protein–calorie ratio. Further, in developing countries, the challenge of producing low-cost protein-rich concentrates made from oilseeds and fish were gradually seeing encouraging returns.

12.2.4 Malnutrition, Infection and Mental Development

Although Scrimshaw and colleagues had highlighted the connection between malnutrition and infection, nearly a decade earlier, the idea was slow to take root. Some, like Burgess, head of WHO's Nutrition Unit at the time, were eager to grab the bull by the horns; however, it was not really acted upon until the WHO convened an expert committee on nutrition in 1965, which Scrimshaw himself chaired. In cementing the relationship between nutrition and infection in the minds of their peers, the committee suggested that high mortality rates in children (1–4 years) could very well be attributable to the, as yet still undefined, relationship between malnutrition and infectious disease (WHO 1965). Following this the need for further studies on the topic as well as calls for increased cooperation between food programmes and ministries of health, were called for, in order that health components could be provided for in every project (Ruxin 1996). And so it was, 2 years later, in the seventh joint FAO/WHO report, the connection was slowly becoming accepted

> That the interrelationship of malnutrition and infection constitutes a major public health problem is slowly becoming apparent. (FAO/WHO 1967)

Leading this convergence was diarrhea, which often accompanied kwashiorkor and marasmus together with the realisation that in a good number of cases, this signified an underlying problem of nutrition. This stimulated many studies, which culminated in the 1968 WHO monograph on *Interactions of Nutrition and Infection*. in which Scrimshaw partook and according to him provided

> ... extensive evidence for the role of infections in precipitating clinical malnutrition and for the impact of malnutrition on morbidity and mortality from infection. (Scrimshaw 2003)

It was noted too that kwashiorkor was possibly precipitated by a child who, if suffering from malnutrition and infection, also simultaneously suffered a bout of diarrhea. With these findings, years of speculation were vindicated as the notion of a synergistic relationship between malnutrition and infection became increasingly accepted.

The consequences of malnutrition on children did not stop at the physical body either. As the decade drew to a close, studies were revealing another potentially more harrowing symptom of malnutrition: that of the child's cognitive learning ability. Once again Scrimshaw was at the forefront of this knowledge dissemination and writing in the *American Journal of Nutrition*, he stated that

> Evidence is already available to suggest that malnutrition during the first few years of life does have an adverse effect on subsequent learning and behavior. (Scrimshaw 1967)

Scrimshaw went on to review many studies and suggested that the probability of mental retardation due to insufficient nutrition was in all likelihood very high. Thus, the possibility of a link between retardation and malnutrition was seen by Scrimshaw as having such far-reaching consequences that he urged more studies on the subject and for economic planners and governments to

> ... increase the investment in programmes for reducing the synergistic impact of malnutrition and infection on the preschool children of developing countries ... (Scrimshaw 1967)

Meanwhile even the Science Advisory Committee (SAC) to the White House became involved in the nutritional debate when, at the behest of President Johnson, the committee published a three-volume tome called the *The World Food Problem* in 1967.

12.2.5 White House

In its preamble, the report commented on the gravity of the problem declaring that

> ... the problem of hunger has lingered on and the shadow of starvation and impending famine has grown ever darker. Hunger's unceasing anguish drains hope, crushes aspirations, and obstructs the generation of programmes of self help. The threat of starvation sets man against man and citizen against government, leading to civil strife and political unrest. (SAC 1967, p. IV)

Perceptively the voluminous White House report also elucidated many of the difficulties and challenges faced by the hunger issues of the time and openly wondered whether oversimplification of the subject, compassion, fatigue and improper comprehension contributed to the prevailing lack of progress. It also acknowledged that just because we might have some of the answers, this did not always directly translate into appropriate or successful policies and programmes in the developing world. In conclusion, among its many and varied recommendations, the report espoused the need for more adaptive research and called for increased governmental aid on every level (SAC 1967).

12.3 Science and Technology

On the scientific front, the 1960s also witnessed another surge of agricultural knowledge dissemination and scientific advances that led to great returns in both productivity and yield. New protein sources and advances in genetic engineering led to scientific leaps and also public fears.

12.3.1 Green Revolution

The so-called green revolution of the 1960s and 1970s was in reality an extension of the green revolution of the 1940s (Swaminathan 1990; Borlaug 2000; Webb et al. 2008). Adoption of green revolution technologies at this time, especially in wheat and rice, were facilitated by conducive government policies, wide availability of inorganic fertilisers as well as advances in irrigation facilities. Not only were crop yields improving from better soil nutrient technology and pest control measures but also from successful plant hybridisation (Khush 1999). Dwarf varieties of grain and rice (later known as green revolution grains) were bred to increase yields compared to their biomass. Such dwarf varieties led to improvements in the harvest index (grain to straw ratio) and saw pre-revolution indices of 0.3 (30% grain and 70% straw) increase to post-revolution harvest indices of 0.5 (ibid). Leading this onward march was Mexico in 1962, in which the first strains of high-yield wheat varieties were planted on a commercial basis and which led, a few years later, to Mexico changing its fortunes and finally becoming a net exporter of wheat.

Meanwhile the miracle of science saw new disciplines emerge, grow and converge in the ever-growing field of biotechnology.

12.3.2 Single Cell Proteins

The search for low-cost high-protein rich foods was underway long before the UN and aid agencies came knocking at the doors of multinational companies. The world hunger problem and the agencies did however expedite a growing enthusiasm for alternative, if not unconventional sources of protein. Since ancient times, people had harvested microbial filamentous blue green algae (Spirulina) from lakes, dried it in the sun and used it as food. The focus in the 1950s/1960s, however, centred on microbial sources such as yeast, algae, fungi and bacteria. Yeast's importance as an animal feed supplement was already widely known and indeed the German government in the World War I had already successfully used yeast to replace approximately half of the needed imported animal protein. Also during World War II, Germany had already successfully integrated yeast-based foods like sausages and soups into the nation's diet (Najafpour 2007).

The idea was simple; yeast contained between 45% and 55% protein, and if a way could be found to produce this on a cheap commercially viable, industrial scale, then the potential was limitless. Yeast can grow and feed on a number of foods (substrates) and the search for low-cost substrates focused on by-products such as cheese, whey, molasses, ethanol, starch and even, surprisingly, hydrocarbon substrates such as oil and natural gas (Ugaldea and Castrillob 2002). Bacteria too, growing on unwanted waste, was also an option; however, sticking with yeast, the 1960s was an exciting time. Researchers at the oil producer British Petroleum (BP) headed by Alfred Champagnat at BP's Lavera Oil Refinery in France 1962 developed a technology for producing single-cell protein by feeding yeast on waxy n-paraffins (linear saturated hydrocarbons). Toprina, as the yeast protein was called, was cleared after nutritional and toxicological tests and by 1967 was being produced on an industrial scale to furnish the agricultural industry with poultry and cattle feed (Najafpour 2007). In separate but related experiments, BP used methanol from methane conversion (natural gas) and marketed it as 'Pruteen', which was used as a 'milk substitute in calf feeding'. By now, several plants around the world had taken up the idea and were collectively producing around a quarter of a million tons of food yeast. These new forms of protein were described by Carol L. Wilson at Massachusetts Institute of Technology (MIT) in 1966 as single cell proteins* (SCP) as the connotations of microbial protein just was not appetising (Ward 1977).

Meanwhile, new legislation protecting new breeds or varieties of plants was negotiated by the International Union for the Protection of New Varieties of Plants (UPOV). The FAO established new crop research centres, and scientists were awarded higher status in their research. At the forefront of this mini biotechnological revolution, the 1960s witnessed great leaps in genetic understanding,

---

* Single cell proteins (SCP) are the dried microbial cells of micro-organisms such as algae, bacteria, actinomycetes and fungi. Feeding off materials such as biomass, hydrocarbons, manufacturing food processing wastes as well as human and animal excreta and are used as a substitute for protein-rich foods, in human and animal feeds.

particularly in the field of the nature and structure of DNA. Crick and Watson were awarded the Nobel Prize, the genetic code was cracked and mouse and human cells were successfully combined in hybridisation.

Given the nature of such progress and experimentation, it was probably not surprising then that the public reacted with fear and scepticism. This was further propagated by books such as *The Biological Time Bomb* by Gordon Rattray Taylor who offered up a platter of science and extrapolation and along the way informed the public of eugenics and potential lethal doomsday bugs among other things (Taylor 1968). In response to such concerns, scientists, industry, and governments increasingly aimed to link this new technology to the more positive, practical functions that bio-technologies promised. One such figure, Joshua Lederberg, stressed the importance of microbial genetic research in the curing of human diseases, etc.

12.4 Period Commentary

The 1960s burst forth with a flurry of institutional and policy activity as well as new ideas on nutrition. The UN had proclaimed the 1960s as the first decade of development and it became a time of hope for ending hunger in the Third World. Existing multilateral bodies were joined by new international institutions such as the International Rice Research Institute (IRRI), the Overseas Development Institute (ODI), the UN Development Programme (UNDP) and the Organization for Economic Development (OECD), which sprang up with global remits around food and development agendas. At the same time, new programmes and campaigns like the International Freedom from Hunger Campaign (FFHC) and the American Freedom from Hunger Foundation (FFHF) as well as the World Food Programme (WFP) and the EEC's Common Agricultural Policy (CAP) all aimed to educate and feed people and foster wider political and social integration. It was also a period that mobilised government and non-government support (Argeñal 2007; Riddell 2007). Alliances changed, relationships were re-evaluated and re-affirmed, while new initiatives like the joint FAO and the International Bank for Reconstruction and Development (IBRD) arose to further support agricultural development.

All the while there was an increasing propensity for industrialised countries to engage in bilateral and multilateral aid effort to administer the growing overseas development assistance (ODA) programme. Underpinning this move was the inception of a host of new multilateral instruments that were created (Riddell 2007). Many however, including Mikesell (1968), cautioned against the notion that 'aid' per se would be of automatic benefit and instead suggested that a beneficial and supportive institutional environment would be a better precondition to a more permanent solution (Mikesell 1968).

The period's political landscape ultimately marked a subtle shift in global governance away from the proprietary domains of the FAO, WHO and UNICEF to include the affected countries themselves. This inclusive need, highlighted by many reports and surveys of the time, sought to not only share responsibility with national governments, but also reflected the growing understanding that the individual had a role to play too in his or her own nutritional status too. This tangible de-centralisation, both implicitly and explicitly acknowledged that in-country projects could well deliver measurable improvement in results, but only with the full support of domestic governments and agencies alike. This idea was further extended, albeit tangentially, when Marcel Autret (FAO) asserted the need for a multi-disciplined approach, which would see nutritionists working closely with educators, agronomists, technologists as well as economists (Johnston and Greaves 1969).

Urbanisation became a growing issue at this time too. With mass migrations from the rural countryside, many nutritionists and planners believed they imported with them wider challenges of social importance. Challenges that, included among others, the growth of industrialisation; a more integrated social infrastructure requiring concomitant health and nutrition policies; shelter; and water needs. Emphasis too was beginning to fall on science and the food technologists in particular to come up with improved pest-resistant, high-yielding hardier grains. At the forefront of this 'green revolution', the 1960s witnessed the dissemination of an agricultural revolution that began in the 1940s. In this movement, few men did more to advance this cause than the 'father of the green revolution' and Nobel Laureate Norman Borlaug (Herdt 1998; USAID 2009).

Meanwhile the FAO, WHO and UNICEF had long tried to woo commercial interests into produc-ing healthy infant foods for years. The difficulty was two fold: how do you persuade companies to put nutritionally adequate foods over profit, and secondly how do you convince governments that nutritional investment in its people's would ultimately reflect in the economic health and wealth of the nation? This proved especially difficult as, at the time some US firms were taking advantage of markets in developing countries to promote nutritionally inadequate products. In elucidating the problem of nutritional invest-ment not having a place in a country's national balance of accounts, Maurice Pate, UNICEF's Executive Director, had stumbled on one of the overriding difficulties of selling the idea of improved nutrition (Ruxin 1996).

Trade issues were also prominent during the 1960s as the continuing GATT talks sought to further decrease the extent of international trade barriers. Up till now, many tens of thousands of such barriers were re-organised, re-negotiated or simply withdrawn and this continued into the Dillon and Kennedy rounds; the Kennedy round also incidentally included for the first time the non-tariff barriers to trade as well.

During the 1950s and by the early 1960s, the scientific community recognised the importance of the link between energy and protein (Weisell 1995; FAO 2002). It was also known that diets in most develop-ing countries suffered high protein deficiency 'both in absolute terms and relative to calories' (Scrimshaw 1996). This marked a turning point where increasingly it was believed that protein malnutrition was one of the main causes of malnutrition worldwide, affecting about 25% of the global population. This 'pro-tein-gap' became a focal point, leading nutritionists to focus on the development of high-protein dietary supplements, particularly aimed at pre-school children. With advances in biotechnologies, in particular in the area of single cell proteins, as well as traditional approaches such as oil and fish concentrates, it was hoped then that the new impetus would transform the nutritional landscape and close the so-called protein gap (Allen 2003). Sadly, it was not to be and in a report to the UN on protein needs, Scrimshaw succinctly made it clear that there were no real short-cuts or quick-fixes to solving the problem. However, out of this, the concept of a protein crisis nevertheless came to dominate the agenda's of the UN's body of agencies as well as nutritionists around the world. Many too were now questioning the primary role of kwashiorkor as a catchall for protein disorders. Part of the problem, it seemed, lied in the lack of charac-teristic features of marasmus (protein disorder's other main protagonist) that eluded a precise definition. Moreover, so much attention was paid to kwashiorkor that research into marasmus became secondary. This was about to change as by the mid-1960s it was becoming increasingly recognised by nutritionists that marasmus was indeed the more prevalent of the protein-calorie-malnutrition (PCM) disorders. One doctor working in the field in Beirut, Dr. Donald Mclaren, noted that

> Marasmus is more widespread, has a more complex aetiology, and more difficult to diagnose, to treat, or to prevent, and has a poorer prognosis. (Mclaren 1966)

McLaren also noted in his studies that kwashiorkor tended typically to be a disease of the 'pre-school' child aged 1–4 years, whereas marasmus, chiefly affected infants in the first years of life. The focus on kwashiorkor at the time was for McLaren a continued source of annoyance; however, further frustration for the doctor came with the tendency to treat all of the disorders of PCM as one single disease. This, in McLaren's view, was not surprising although he felt would be of serious consequence as the many dis-orders along the PCM spectrum required different treatments and offered as many different prognosis. In conclusion McLaren controversially and prophetically stated in his prognosis of the problem that the treatment of nutritional disorders would not necessarily be best achieved through the use of nutritional measures. In his opinion

> The Whole Society is the Patient. (Mclaren 1966)

In recognition of the changing climate, the UN adopted of the term protein-calorie-malnutrition (PCM) to represent the spectrum of diseases to incorporate both marasmus and kwashiorkor. However, at this time too, while few doubted the importance of protein in the diet, cracks were now beginning to appear in the panacea of protein. The obsession and incessant quest for all things protein led opposing camps on a collision course. The two camps consisted of those who that felt protein was paramount and at the heart of the debate, and those who, while accepting of the importance of protein, felt that the current emphasis was too great and not warranted. Out of this period then, it was beginning to be realised that neither protein nor calorie intake should be considered in isolation (Weisell 1995; FAO 2002).

In other developments, nutritionally associated mortality and cognitive learning entered the debate. On firming up the link between nutrition, mortality and infection, Behar, Scrimshaw and colleagues at INCAP conducted a study on mortality and the roles of malnutrition and infection in children between one and four. Their studies showed kwashiorkor to be responsible for approximately two-fifths of deaths between them while diarrhea and infection accounted for the remainder. Such findings reinforced the call by Platt and others for a more integral approach to the problem through policies that targeted several aspects of childhood health to include nutrition, clean water and sanitation. Furthermore, questions of nutrition and cognitive learning came up with studies ominously linking the two. To quote Ruxin, from his exceptionally encyclopedic and insightful nutritional treatise

> … knowledge of the effects of malnutrition on learning and behaviour suggested that the intel-
> lectual potential of hundreds of millions of children was at stake and provided further thrust to
> nutritional undertakings. (Ruxin 1996)

Calorie intake too came under scrutiny. Passmore (1964), an FAO/WHO consultant, in a detailed analysis of the second FAO report on calorie requirements, noted the difficulty in accurately calculating and grading physical work (the factor by which the reference adult was adjusted to arrive at final energy requirements). He recognised too the methodological difficulty in achieving uniformity of results while many in the industry continued to use wide-ranging scales for grading physical activity. Furthermore, in his extensive review, Passmore suggested too that perhaps men and women may be leading more sedentary lives than had previously been offered. If this were the case he questioned, then FAO estimates might have overestimated calorie requirements (Passmore 1964).

Despite all this effort however, progress was unfortunately slow. This was almost certainly compounded by the resurgent political infighting within and across the halls of the UN agencies—the FAO, WHO and UNICEF. Whereas the crux of division between these organisations in the 1950s was in respective areas of finance and responsibility, the 1960s differences centred on '… the quality and control of technical advice'. It had already been a point of contention in the 1950s but the notion of technical advice being given from centralised, less experienced staff became increasingly difficult for field workers to accept in the 1960s. Resentment and negotiation eventually led to the realisation of the veracity of the complaint, which in turn helped push through the notion of de-centralisation.

Outside of this internal malady, the general failure of the tumultuous efforts of combined agencies, treaties and interventions to produce tangible results led some to adopt an increasingly pessimistic rhetoric. Critics from both inside and outside of the central organisations were many as the general nature of nutrition policies increasingly came under fire. As with Scrimshaw and Behar's earlier attacks, it was not long before other critics jumped on the bandwagon lashing out at organisations like the WHO and FAO. Donald McLaren, a professor at the American University of Beirut, while agreeing for the need to expand the food supply, firmly believed that the crux of the problem once again is the feet of ignorance. At first glance though this did not seem to be in opposition to the UN goals, as the UN had widely advanced the elimination of the ignorance of infant care and feeding practices as central to ending hunger. McLaren however favoured a more proactive approach that would see hunger education at all levels of family and community take prominence and in this his McLaren felt the UN efforts were lacking.

Ultimately, the 1960s was also a time where philosophical differences in nutritional approaches began to emerge highlighting many divergent approaches to understanding and treatment in both an institutional and clinical sense. It turned out to be a time of reflection and re-appraisal. UNICEF's role had once again changed while the WHO became more proactively involved in nutritional issues, relative to previous years. Moreover, the WHO, in a rare openly reflective note, found itself questioning whether international nutritional surveys had indeed adequately plotted the seriousness and prevalence of the global problem of malnutrition (WHO 1965). Externally too, the humanitarian sector's voluntary agencies became more entwined with these multilateral agencies as combined goals sought to streamline efforts and magnify results. Lastly, more and more in the 1960s the rhetoric of hunger and malnutrition took on the language of economics and development, as it became widely accepted that malnutrition could not be tackled in isolation, but rather through the expansion of economic and social development, thus finally echoing calls that were made many decades earlier.

12.5 Key Dates, People, Acts, Reports and Surveys

Table 12.1 highlights some of the events and noteworthy reports, etc. that helped shape the 1960s food security landscape.

TABLE 12.1

Key Dates: The Development Decade

Date	People, Organisation, Agencies, Acts or Conventions
1960: International Rice Research Institute (IRRI)	The Philippines together with the Ford and Rockefeller Foundations established the IRRI, an autonomous, non-profit rice research and education organisation (IRRI 2007).
1960: Overseas Development Institute (ODI)	ODI, a leading independent think tank on issues of international development and humanitarianism based in Britain was created (ODI 2010).
1960: International Freedom from Hunger Campaign (FFHC)	In 1960, the UN Food and Agriculture Organization (FAO) launched a 5-year campaign of International Freedom from Hunger to mobilise non-governmental support (Osmanczyk 2002; FAO 2010).
1960: American Freedom from Hunger Foundation (FFHF)	Founded to mobilise Americans to get involved in the FAO's 5-year global Freedom from Hunger Campaign. In 1979, the FFHF and the Meal for Millions organisations merged to form the Freedom from Hunger organisation (Freedom from Hunger 2010).
1960–1962: GATT 'Dillon' 5th Round Geneva	Reduced over $4.9 billion in trade tariffs (WTO 2010).
1961: Organization for Economic Development (OECD)	OECD works with democratic governments and the global market economy to, among other things support sustainable economic growth (OECD 2010).
1962/3: World Food Programme (WFP)	The WFP was proposed in 1961 by George McGovern, director of the US Food for Peace Programme, scheduled to start in 1963, it in fact started in 1962. The WFP's governing body, the Intergovernmental Committee (IGC), responsible for the supervision over the WFP, was changed to Committee on Food Aid Policies and Programmes (CFA) in 1975. In 1965 resolutions of the FAO and the UN General Assembly established WFP on a continuing basis for as long as feasible and desirable (WFP 2008).
1962: The European Common Agricultural Policy (CAP)	The EEC first established the Common Market directly in response to over a decade of severe food shortages during and after World War II. Based on the principles initially set out at the Stresa Conference (1958) the Common Agricultural Policy (CAP) was created in 1962. The main decision-making body for CAP was the Agricultural Council. The body was predominantly geared towards increasing agricultural productivity and food self-sufficiency. Through a combination of farm price supports and food import protectionism. It was expanded in 1973 to include the United Kingdom, Denmark and Ireland (Europa 2010a, 2010b).
1962: FAO/WHO Codex Alimentarius Commission	The FAO/WHO Codex Alimentarius Commission was set up to develop international food standards and guidelines (WHO/FAO 2006; FAO 2010).
1963: The United Nations Research Institute for Social Development (UNRISD)	An autonomous UN agency conducting research on contemporary problems of development. Through its multi-disciplinary research, UNRISD promotes discussion and contribution to key social development issues (UNRISD 2010).
1964–1967: GATT "Kennedy" 6th Round	Concerned with the usual multilateral tariff cuts, a new GATT Anti-Dumping Agreement dealing with problems of Developing Countries was also reached. Some of the GATT negotiation rules were also more clearly defined. The Kennedy Round was also the first time that negotiations looked at non-tariff barriers to trade as well (WTO 2010).

TABLE 12.1 (Continued)

Key Dates: The Development Decade

Date	People, Organisation, Agencies, Acts or Conventions
1964: UN Conference on Trade and Development (UNCTAD)	UN Conference on Trade and Development met in Geneva in 1964 and recommended the establishment of a permanent UN body to deal with trade in relation to development. The General Assembly, noting the importance of international trade as an instrument of economic development established UNCTAD as one of its permanent organs in 1964 (UNCTAD 2002).
1965: The United Nations Development Fund (UNDP)	Based on the merging of the United Nations Expanded Programme of Technical Assistance (1949) and the UN Special Fund (1958) the UNDP was established in 1965. The UNDP works on a number of development issues and helps developing countries attract and use aid effectively (Winderl 2006; UNDP 2010).
1966: The Institute of Development Studies (IDS)	The IDS is a leading organisation of international development focusing on research, teaching and communications. Its goal is to understand the world and to apply academic skills to real-world challenges (IDS 2010).
1966: International Maize and Wheat Improvement Center (Centro Internacional de Mejoramiento de Maíz y Trigo, (CIMMYT))	Born out of the Rockefeller, Ford and Mexico collaboration the work led to the founding of the Office of Special Studies to ensure food security in Mexico and beyond. An agreement was signed with Mexico's agricultural minister in 1963 which developed the CIMMYT. The CIMMYT developed into a collaboration of international researchers (Hesser 2006; Byerlee and Dubin 2008; Ortiz et al. 2008).
1966: UN Covenant on Economic, Social and Cultural Rights and: The International Covenant on Civil and Political Rights	These rights were adopted by the UN General Assembly in 1966. Both recognise that human rights are inherent in the dignity of human beings. Article 1 of each Covenant affirms that all peoples are free to determine their own political status and have the right to pursue their economic, social and cultural development (UN 1966a, 1966b).
1967: The Food Aid Convention (FAC)	The FAC was agreed as part of the International Grains Agreement. Intended as a safety net, its primary objective is to ensure a minimum availability of food aid to meet emergency requirements in developing countries. It is run by the London-based International Grains Council (IGC 2009b).
1968: International Grains Arrangement (IGA)	International Grains Arrangement, 1968 (IGA) consisting of two separate instruments: the Wheat Trade Convention, 1967 (WTC) and the Food Aid Convention, 1967 (FAC). The WTC consults on prices, rights and obligations, while the FAC provided food aid to developing Countries (Meerhaeghe 1998; IGC 2009a).
1969: Indicative world Plan for Agriculture (IWP)	Detailed report on the present and future problems of agriculture (Fassbender 1970; Goldsmith et al. 1972).
1968: The First International Conference on Human Rights	The first International Conference on Human Rights was held in Teheran, Iran.
1969: United Nations Population Fund (UNFPA)	The Population Commission recommended to ECOSOC an expanded population programme and subsequently in 1969 the Fund was put under the administration of UNDP (POPIN 1994).

References

Allen, L. H. (2003). Interventions for Micronutrient Deficiency Control in Developing Countries: Past, Present and Future. *The Journal of Nutrition* 133(Supplement: Animal Source Foods to Improve Micronutrient Nutrition in Developing Countries): 3875S–3878S.

Argeñal (2007). History of Food Security in International Development: Background Paper Developed for Christian Children's Fund Honduras. Honduras: Christian Children's Fund.

Baker, R. (2003). *Rachel Carson's Silent Spring and the Beginning of the Environmental Movement in the United States.* Bloomington, IN: Indiana University.

Borlaug, N. E. (2000). The Green Revolution Revisited and the Road Ahead. 30th Anniversary Lecture. Oslo: The Norwegian Nobel Institute.

Byerlee, D. and H. J. Dubin (2008). Crop Improvement in the Cgiar as a Global Success Story of Open Access and International Collaboration. Microbial Commons International Conference. Ghent, Belgium. Microbial Commons.

Carson, R. (1962). *Silent Spring*. New York: Houghton Mifflin.

CCP (2005). A Historical Background on Food Aid and Key Milestones. Committee on Commodity Problems: Sixty-fifth Session, Rome, Food and Agriculture Organization.

EU (2009). The Common Agricultural Policy Explained. Brussels: European Union 11.

Europa (2010a). Agriculture: Meeting the Needs of Farmers and Consumers. *The EU at a Glance*. Europa. http://europa.eu/pol/agr/index_en.htm (accessed 18 June 2010).

Europa (2010b). European Union History: 1960–1969 the 'Swinging Sixties'—A Period of Economic Growth. *The EU at a Glance*. Europa. http://europa.eu/abc/history/1960–1969/index_en.htm (accessed 22 June 2010).

FAO (1962). Joint FAO/WHO Expert Committee on Nutrition. Sixth Report. World Health Organization *Technical Report Series No. 245*. Rome: Food and Agriculture Organization/World Health Organization, Geneva 68.

FAO (1963). Third World Food Survey, FFHC Basic Study No. 11. Rome: Food and Agriculture Organization 102.

FAO (2002). Food, Nutrition and Agriculture. Rome: FAO Food and Nutrition Division.

FAO (2010). The Food and Agriculture Organization of the United Nations. UN Food and Agriculture Organization. http://www.fao.org/about/en/ (accessed 26 January 2010).

FAO/WHO (1965). Joint FAO/WHO Expert Committee on Protein Requirements—Nutrition Studies No. 37. Rome: FAO/WHO.

FAO/WHO (1967). Joint FAO/WHO Expert Committee on Nutrition Seventh Report *FAO Nutrition Meetings Report Series no. 42/ WHO Technical Report Series No. 377*. Rome: FAO/WHO 68.

FAO/WHO (1973). FAO/WHO Ad Hoc Committee of Experts on Energy and Protein: Requirements and Recommended Intakes. Rome: Food and Agriculture Organization.

Fasbender, K. (1970). The FAO Indicative World Plan *Journal of Intereconomics* 5(6): 190–192.

Freedom from Hunger (2010). Website of the Freedom from Hunger. Freedom From Hunger. http://www.freedomfromhunger.org/ (accessed 15 June 2010).

Frisch, B. H. (1964). Was Rachel Carson Right? *Science Digest* 39–45.

Gillin, E. (2006). Fao: The Statistics Division History. FAO Statistics Division. http://www.fao.org/economic/ess/the-statistics-division-history/en/ (accessed 12 January 2011).

Goldsmith, E., R. Allen, M. Allaby, J. Davoll and S. Lawrence (1972). Introduction: The Need for Change. *The Ecologist* 2(1): 2–7.

Herdt, R. W. (1998). The Rockefeller Foundation: The Life and Work of Norman Borlaug, Nobel Laureate. New York: The Rockefeller Foundation.

Hesser, L. F. (2006). *The Man Who Fed the World: Nobel Peace Prize Laureate Norman Borlaug and His Battle to End World Hunger*. Dallas, TX: Durban House Publishing.

IDS (2010). About the Institute of Development Studies. IDS. http://www.ids.ac.uk/go/about-ids (accessed 21 June 2010).

IGC (2009a). 60 Years of Successive Agreements: Before 1949: The Early Years. London: International Grains Council 4.

IGC (2009b). Grains Trade and Food Security Cooperation: The International Grains Agreement. The International Grains Council. http://www.igc.org.uk/en/aboutus/default.aspx#igc (accessed 21 February 2010).

IRRI (2007). What Is IRRI? International Rice Research Institute. http://www.irri.org/about/about.asp (accessed 21 June 2010).

Johnston, B. F. and J. P. Greaves (1969). Manual on Food and Nutrition Policy. *Nutritional Studies no. 22*. Rome: FAO 95.

Khush, G. S. (1999). Green Revolution: Preparing for the 21st Century. *Genome* 42(4): 646–655.

Mclaren, D. S. (1966). A Fresh Look at Protein-Calorie Malnutrition. *The Lancet* 288(7461): 485–488.

Meerhaeghe, M. A. G. v. (1998). *International Economic Institutions*. Massachusetts: Kluwer Academic Publishers.

Mikesell, R. F. (1968). *The Economics of Foreign Aid*. Chicago, IL: Aldine Publishing Company.

Mousseau, F. (2005). *Food Aid or Food Sovereignty? Ending World Hunger in Our Time*. Oakland, CA: Oakland Institute.

Najafpour, G. D. (2007). *Biochemical Engineering and Biotechnology*. Netherlands/Oxford: Elevier.

ODI (2000). Briefing Paper. Reforming Food Aid: Time to Grasp the Nettle. London: Overseas Development Institute.

ODI (2010). About ODI: Our Mission, People and Organisation. Overseas Development Institute. http://www.odi.org.uk/about/default.asp (accessed 23 July 2010).

OECD (2010). About the Organisation for Economic Development. Organisation For Economic Development. http://www.oecd.org/home/0,3305,en_2649_201185_1_1_1_1_1,00.html (accessed 30 March 2010).

Ortiz, R., H.-J. Braun, J. Crossa, J. H. Crouch, G. Davenport, J. Dixon, S. Dreisigacker, E. Duveiller, Z. He, J. Huerta, A. K. Joshi, M. Kishii, P. Kosina, Y. Manes, M. Mezzalama, A. Morgounov, J. Murakami, J. Nicol, G. O. Ferrara, J. I. Ortiz-Monasterio, T. S. Payne, R. J. Peña, M. P. Reynolds, K. D. Sayre, R. C. Sharma, R. P. Singh, J. Wang, M. Warburton, H. Wu and M. Iwanaga (2008). Wheat Genetic Resources Enhancement by the International Maize and Wheat Improvement Center (CIMMYT). *Genetic Resources and Crop Evolution* 55(7): 1095–1140.

Osmanczyk, E. J. (2002). *Encyclopedia of the United Nations and International Agreements G-M*. London: Taylor & Francis.

Passmore, R. (1964). An Assessment of the Report of the Second Committee on Calorie Requirements (FAO, 1957). Rome: Food and Agriculture Organization.

Perry, M. J. (1998). *The Idea of Human Rights: Four Inquiries*. New York: Oxford University Press.

Poleman, T. T. (1972). Food, Population, and Employment: Ceylon's Crisis in Global Perspective. *Staff Paper*. New York Cornell University: Department of Agricultural Economics 31.

POPIN (1994). Background Document on the Population Programme of the UN. Geneva, Switzerland: United Nations Population Fund.

Reinhardt, C. and B. Ganzel (2003) A History of Farming. *Living History Farm*: http://www.livinghistoryfarm.org/farminginthe30s/crops_03.html.

Riddell, R. C. (2007). *Does Foreign Aid Really Work?* Oxford: Oxford University Press.

Rucker, A. (2007) Key Players in Food Aid. *eJournal USA*. http://www.america.gov/st/health-english/2008/June/20080616003149xjyrrep0.4639551.html.

Ruxin, J. N. (1996). *Hunger, Science, and Politics: FAO, WHO, and UNICEF Nutrition Policies, 1945–1978, Chapter II the Backdrop of Un Nutrition Agencies, by Joshua Nalibow Ruxin*. London: University College London.

SAC (1967). The World Food Problem: A Report of the President's Science Advisory Committee, Volume Lll Report of the Panel on the World Food Supply. Washington, DC, The White House: The President's Science Advisory Committee (PSAC).

Scrimshaw, N. (2003). Historical Concepts of Interactions, Synergism and Antagonism between Nutrition and Infection. *Journal of Nutrition* 133(316S–321S): 367–403.

Scrimshaw, N. S. (1967). Malnutrition, Learning and Behavior, *The American Journal of Clinical Nutrition* 20(5): 493–502.

Scrimshaw, N. S. (1996). Human Protein Requirements: A Brief Update. *Food and Nutrition Bulletin* 17(3): 185–190.

Scrimshaw, N. S. (2010). History and Early Development of INCAP. *Journal of Nutrition* 140(2): 394–396.

SOFA (2000). The State of Food and Agriculture 2000. Rome: Food and Agriculture Organization.

Subramanian, S. (2003). An Analysis of the Relationship between the Agricultural Policies of the European Union and the United States, and Their Implications on the World Food Economy *Journals of the Stanford Course: Engr 297b*(Controlling World Trade): 32.

Swaminathan, M. S. (1990). Sir John Crawford Memorial Lecture: Changing Nature of the Food Security Challenge: Implications for Agricultural Research and Policy. Washington, DC: Consultative Group on International Agricultural Research.

Taylor, G. R. (1968). *The Biological Time Bomb*. New York: The World Publishing Co. .

Ugaldea, U. O. and J. I. Castrillob (2002). Single Cell Proteins from Fungi and Yeasts. *Applied Mycology and Biotechnology* 2: 123–149.

UN (1948). *Universal Declaration of Human Rights*. United Nations. A/RES/217 A (III).

UN (1963). *Operational Files of the Secretary-General: U Thant: Speeches, Messages, Statements, and Addresses—Not Issued as Press Releases*. New York: United Nations.

UN (1966a). *International Covenant on Civil and Political Rights and [First] Optional Protocol*: United Nations. A/RES/2200 A (XXI).

UN (1966b). *International Covenant on Economic, Social and Cultural Rights*: United Nations. A/RES/2200 A (XXI).

UN (2011). Universal Declaration of Human Rights. United Nations. http://www.un.org/rights/HRToday/declar.htm (accessed 23 March 2011).

UNCTAD (2002). About the United Nations Conference on Trade and Development. http://www.unctad.org/Templates/Page.asp?intItemID=1530&lang=1 (accessed 12 February 2010).

UNDP (2010). United Nations Development Fund: Frequently Asked Questions. UNDP. http://www.undp.org (accessed 24 June 2010).

UNRISD (2010). What Is the United Nations Research Institute for Social Development. The United Nations Research Institute for Social Development. http://www.unrisd.org/80256B3C005BF3C2/SectionHomepages/FBF509B36533B0FD80256B4A00636A37?OpenDocument (accessed 21 January 2010).

USAID (2009). Frontlines: Borlaug, Father of Green Revolution, Dies. US Agency for International Development. http://www.usaid.gov/press/frontlines/fl_oct09/p01_borlaug091002.html (accessed 5 March 2010).

USDA (1961). World Food Budget, 1962 and 1966. *Foreign Agricultural Economic Report 4*. Washington, DC: United States Department of Agriculture, Economic Research Service (ERS).

Walker, P. (2008). The Origins, Development and Future of the International Humanitarian System: Containment, Compassion and Crusades. ISA's 49th Annual Convention, Bridging Multiple Divides, San Francisco, CA, International Studies Association (ISA).

Ward, W. A. (1977). Calculating Import and Export Parity Prices: Training Material of the Economic Development Institute, CN-3. Washington DC: World Bank.

Webb, P., J. Gerald and D. R. Friedman (2008). More Food, but Not yet Enough: 20th Century Successes in Agriculture Growth and 21st Century Challenges. *Food Policy and Applied Nutrition Program*. Boston, MA: School of Nutrition Science and Policy, Tufts University.

Weisell, R. C. (1995). Food, Nutrition and Agriculture: FAO Celebrates 50 Years—Expert Advice on Energy and Nutrient Requirements—An FAO Tradition. Rome: Food and Agriculture Organization.

WFP (2008). The World Food Programme: The First 45 Years. World Food Programme. World Food Programme. http://www.wfp.org/aboutwfp/history/index.asp?section=1&sub_section=2 (accessed 25 April 2010).

WHO (1965). *Who Activities in Nutrition, 1948–1964*. Geneva, Switzerland: World Health Organization.

WHO (1973). Energy and Protein Requirments: Report of a Joint FAO/WHO Ad Hoc Expert Committee—Technical Report Series No. 522. *WHO Technical Report Series No. 522 FAO Nutrition Meetings Reports Series No. 52*. Geneva, Switzerland: World Health Organization 24.

WHO/FAO (2006). Understanding the Codex Alimentarius. Rome: Joint FAO/WHO Food Standards Programme 40.

Winderl, T. (2006). UNDP for Beginners: A Beginners Guide to the United Nations Development Fund. Copenhagen, Denmark: UNDP JPO Service Centre 25.

WTO (2010). Understanding the WTO: Basics—The GATT Years: From Havana to Marrakesh. World Trade Organization. http://www.wto.org/english/thewto_e/whatis_e/tif_e/fact4_e.htm (accessed 30 March 2010).

13

Famine, Oil and the Food Crisis: 1970s

13.1 Governance, Politics and Socioeconomics

In the political atmosphere of the 1970s, great social unrest followed the oil and food crisis while developing countries took on more debt in the name of development. By 1974, a seminal World Food Conference ushered in a new era of ideological expansion, opening the way for new programmes and initiatives. Meanwhile, the obsession with protein waned and the hitherto powerful Protein Advisory Group was all but disbanded. The wider acceptance of tackling nutritional issues which had become increasingly wrapped up in policies of economic and development in the last decade now took on dimensions of primary health care. To further aid efforts, the need was also acknowledged for a system of nutritional surveillance overseeing food and health.

13.1.1 Economic Uncertainty

Very early in the decade, debt and a growing destabilised proxy gold standard weighed heavily. Up to this point, the Bretton Woods agreement had kept currency exchange rates at a dollar-gold parity that so far successfully stabilised currencies and national balance of payments. By 1971 however, local wage rises and growing transnational capital movements led to large US deficits, leading in turn to an over-valued and unmanageable dollar. As a result, the United States decided to uncouple the dollar from the gold standard, foreshadowing a new era of floating exchange rates and unpredictable world trade. This effectively released nations to re-orient and re-negotiate 'their own competitive position in the world economy' (McMichael 1994, p. 2). By extension this freed up the global capital market, which allowed for the exponential increase in world trade and globalisation. McMichael (1994) also suggested that as well as this de-coupling, spatial changes in sovereignty and boundaries which emerged after the dissolution of the Cold War Bloc allowed new markets and trading partners to be established. This, in turn, he thought, further destabilised international trade. At this time too agricultural trade increased in importance as a global commodity. One particular effect of this global, rather than national competitiveness was that the agro-food industry underwent considerable re-structuring that saw increased specialisation with increased contract farming and a resultant redundancy in some national policies (McMichael 1994).

By now, the growing sustainable movement that had already been tentatively associated with agricultural production had permeated into the public and political conscience.

13.1.1.1 Sustainable Development

Industrialisation of countries and the ravages of fossil fuels as well as soil degradation and deforestation were rife and the fear was that if left unchallenged, developing countries' sprint for progress would have dire consequences. It was felt that everyone, developed and developing countries alike, had a responsibility to future generations, and in 1972, Stockholm hosted the first United Nations Conference on the Human Environment (UNCHE). While the conference recognised that the industrial divide that separated the developed and developing countries would need to be bridged, it urged countries to do so in a manner that was respectful of the environment (Earth Summit 2002). Moreover, it was held that the problems of the environment were of a global nature and that solutions would require extensive international cooperation. One of the outcomes of the conference was the creation of the United Nations

Environmental Programme (UNEP) to aid and encourage UN agencies to integrate environmental measures into their programmes.

At about this time, the world witnessed what has been popularly described as a 'food crisis' on a global scale.

13.1.1.2 World Food Crisis

Precipitating this event was a number of concurrent factors of policy, economy and agricultural production that when combined collectively saw what was in reality the first truly global agricultural crisis. Beginning with the high grain stocks of grain-exporting countries during the late 1960s, governments, through policy, sought the rapid reduction to more realistic levels as well as the parallel slowdown in general production capacity. This was achieved in some measure; however, on the back of this, poor grain harvests, particularly in the Soviet Union, saw world food production dip for the first time since World War II. Simultaneously, and starting in 1972, crude oil barrel prices had increased 10-fold over the following 2 years. This resulted in a five-fold increase in fertiliser prices, further adding to the problem of shortages. Collectively, lower grain yields from less fertiliser use and growing demand from wealthier countries, in particular the huge imports of USSR and China in 1972, coupled with other factors such as the exchange rate volatility from the recent 'break' from the gold standard, saw stocks dwindle and grain prices double. This placed enormous pressure further threatening the food security of food-importing nations. During the next 3 years however, the situation barely improved as world food supplies remained low and perilously dependent on each successive year's production (Maxwell and Frankenberger 1992; Mitchell et al. 1997; Gerster-Bentaya and Maunder 2008; UOR 2009).

Against this backdrop, the 1960s and 1970s also saw the emergence of new independent states. Released from the grip of colonialism, these new states joined the growing ranks of developing nations vying for economic development and independence (Section 22.5.3; Goodrich 1947).

13.1.1.3 Third-World Debt

Early World Bank lending to these countries was predicated on a 'shared' economic development ideology with 'project lending' loans to nations favouring infrastructure, industry and agriculture (Carrasco and Berg 1999b). However, growth and investment was uneven and in some places non-existent. In turn, more people moved out of the poorer un-invested agricultural sectors and migrated to the industrialised big cities. This further impoverished rural agriculture to the extent that a new approach was endorsed by the World Bank (Carrasco and Berg 1999b). In the meantime, the problems of international trade and unfair trading practices seemed to antagonise the development of the poorer and emerging countries and their ability to trade freely on the open market (CFS 2008a). As a result, the reliance of developing countries on foreign debt and food imports grew while their self-sufficiency decreased (FAO 1996).

13.1.1.4 Debt and Economic Reforms

In an effort to redress this balance, the World Bank and others maintained that economic growth did not need to be at the expense of the quality of life. Running parallel with project lending, the World Bank proposed a new type of loan based on growth with equity. It aimed to facilitate social welfare programmes and the ultimate eradication of absolute poverty. Loans helped to develop programmes improving health, nutrition, family planning, education and the under-employed sectors of the community. In this way, social welfare, security of food and development became interlinked with economic growth and a person's ability to meet their basic needs and achieve quality of life (Munasinghe 1998; Carrasco and Berg 1999b; Slusser 2006). However, the debt burden was to grow as huge wealth created by the Oil Producing Exporting Countries (OPEC) from increased oil prices in the 1970s was deposited in the US and European banks (Eichengreen and Lindert 1989; Carrasco and Berg 1999a). This helped fuel cheap lending to developing countries (formerly Third World) at a time of growing energy prices and higher import costs, due in part to a strong dollar (Lipson 1981). This can be illustrated in the 18-fold increase in interest payments on the capital alone paid by the 12 largest non-oil-producing least developed countries

(LDCs) between 1970 and 1980 from $1.1 billion (6% of export earnings) to 18.4 billion (14% export earnings) (Lipson 1981; Munasinghe 1998; Slusser 2006).

As the decade matured, the ongoing food crisis led to much international negotiation and to the much heralded World Food Conference of 1974 (FAO 2003).

13.1.2 World Food Conference

The World Food Conference (WFC) was held in Rome by the UN under the auspices of the FAO and, after examining global problems of food production and consumption, the conference adopted a Universal Declaration on the Eradication of Hunger and Malnutrition declaring that

> Time is short. Urgent and sustained action is vital. The Conference, therefore, calls upon all peoples expressing their will as individuals, and through their Governments and non-governmental organisations, to work together to bring about the end of the age-old scourge of hunger. (UN 1974)

Moreover, it was also firmly established that

> Every man, woman and child has the inalienable right to be free from hunger and malnutrition in order to develop fully and maintain their physical and mental faculties. (UN 1974)

One important affirmation emerging from this conference too was that the causes of famine and food insecurity were not so much failures in food production but rather structural problems relating to social deprivation of poverty and inequality. It was also re-affirmed that the majority of the developing world's poor populations were concentrated in rural areas. It seemed that echoes from near history and the work of the League of Nations were finally being heard.

Other outcomes of the conference were numerous, and several bodies, funds and initiatives were created to combat the unacceptable global situation. The conference also requested that FAO, WHO and UNICEF unite in their nutritional research efforts while other outcomes of the Universal Declaration included

- The recognition of the role of society and its sufficient resources, technology and ability to achieve this goal;
- The fundamental responsibility of governments to cooperate in increased agricultural production whilst ensuring a more efficient and equitable distribution of food among peoples especially in the LDCs. There was also a need to commit to reducing incidents of chronic malnutrition and diseases among the lower income and other vulnerable groups;
- Governments should initiate appropriate policies of food and nutrition into overall socioeconomic and agricultural development plans and remove obstacles in the forms of inappropriate agrarian, tax, credit and investment policies;
- The recognition of the key role of women in the rural economy in particular, agricultural production and to ensure appropriate education and financial opportunities are made available to women 'on equal terms with men';
- The introduction of sustainable and fair use of marine and inland water resources as sources of food and economic prosperity;
- To provide technical and financial assistance on reasonable terms to developing countries and promote equitable transfer and dissemination of food production research and technology;
- To conserve natural resources and preserve the environment;
- Countries should recognise the interrelation between global trade the world food problem and understand the effects of domestic farm support programmes as well as tariff and non-tariff barriers on food-exporting developing countries;
- To ensure adequate world reserves of basic foodstuffs through the participation in Global Information and Early Warning Systems, ensuring sufficient stocks or funds to meet international emergency food requirements proposed by the International Undertaking on World Food Security, and to help stimulate rural development and employment projects (UN 1974).

It was also determined that for better understanding, monitoring and response to the problems of food security, early warning systems would be needed

> ...to monitor the food and nutrition conditions of the disadvantaged groups of the population at risk, and to provide a method of rapid and permanent assessment of all factors which influence food consumption patterns and nutritional status. (UN 1974)

The development of this food security early warning system was the joint responsibility of FAO, WHO and UNICEF. The WHO was designated lead agency and in 1975 a Joint Expert Committee met to develop the idea further. It was understood that in order to succeed, what were first needed were individual national systems of surveillance that could be used to compare cross-border data. This meant a system of indicators and methodologies that would be universally agreed upon. However, this was going to be difficult for it was also understood that while many elements of nutritional surveillance systems had existed in many developing countries for centuries, the challenge was going to be in the coordination of data gathering and universal methodologies for better analysis and cross-border comparisons (Maxwell and Frankenberger 1992; Argeñal 2007).

Other bodies also born out of the conference were to come to play important roles in the modern development of the food security concept. One such body was the International Fund for Agricultural Development (IFAD). The IFAD began operations in 1977 with the purpose of financing agricultural development projects primarily aimed at food production that improved national food self-sufficiency (IFAD 2009). While another body, the Committee on World Food Security (CFS) was incorporated to review and oversee developments in food security that included food production as well as physical and economic access to food (CFS 2009). The conference also called for the creation of 36-member ministerial-level World Food Council (WFC) reporting annually to the Economic and Social Council to review major problems and policy issues affecting world food security in general. The WFC was also expected to coordinate all UN agency policies related to nutrition, food trade, food aid and other related matters and it did this through the promotion of an inter-secretariat consultative process for the four Rome-based food organisations: the FAO, IFAD, WFC and WFP (UN 1975). The WFC's initial approach in addressing world food problems was to encourage developing countries to adopt national food strategies that would help them evaluate their present food situation in terms of current supply as well as future potential increases, processing, transport, distribution, marketing and their preparedness in meeting any food emergencies (FAO 2003; WFC 1981).

13.1.3 Food Security Is Born?

According to many, this was a time when people and institutions first attempted to properly define the concept of security of food or 'food security', as it had then become known. Early definitions focused on supply and absolute food availability, that is to say, aggregate supplies and stability at the national and global levels (FAO 2003; Devereux et al. 2004). One of the first definitions came out of the World Food Conference when it declared food security to involve

> Availability at all times of adequate world food supplies of basic foodstuffs to sustain a steady expansion of food consumption and to offset fluctuations in production and prices. (UN 1974; WFC 1975; FAO 2006)

By now other institutions were getting in on the nutrition act.

13.1.4 The World Bank

For over two decades, both institutionally and programmatically, the field of international nutrition was dominated by the UN tripartite: the WHO, FAO and UNICEF. Although agencies like USDA, the ILO and the UNDP had joined the foray, the arrival of the World Bank on the scene was little short of landmark. The rationale for its entry onto the world nutritional stage is provided by Ruxin (1996) who provides valuable insight into the motivation for the World Bank in this situation. Ruxin's research shows that the Bank's enthusiasm was bound up in two aspects. First, the Bank held less than favourable confidence in the work of the WHO, FAO and UNICEF programmes, feeling that their joint work to date had been too small and modest to have been of any great value, ill-targeted and dominated by medical and nutritionist experts rather than policy makers. Second, and, worse still, efforts to date were seen as small-minded in scale and breadth. As a result, officials at the Bank felt they alone had the potential to make inroads in 'nutrition in a manner that FAO, WHO, and UNICEF had been unable to do' (Ruxin 1996). At first, the animosity was mutual and cooperation was slow but in 1974 in a magnanimous display of unity, the World Bank, FAO, WHO and UNICEF held a meeting in Rome to discuss future joint nutritional work. The inclusion of the World Bank, irrespective of who had initiated the collaboration, was a minor coup in the fight against hunger. The advantages were immediately obvious, while the hitherto UN tripartite organisations between them held enormous knowledge and expertise, the World Bank's experience in the political arena was unmatched. Moreover, used to dealing with top level government officials as well as planners and economists, the World Bank, with its political savvy, it was hoped, could facilitate the introduction of nutritional components in many of its own projects. The Bank ultimately began its first nutrition projects in Brazil and Indonesia in 1977 with micronutrient supplementation (Heaver 2006).

13.1.5 The UN: A System in Change

Partly as a result of the recent roller-coaster ride in nutritional thinking, officials were left somewhat reluctant to once again become over-reliant on the expert views of nutritionists and experts. Indeed the negative repercussions of the protein gap fiasco alone had major consequences for the image of nutritional research. More and more people began to feel that research, in this area, was 'academic, irrelevant and a waste of time' (Waterlow and Payne 1975, p. 117), preferring instead to utilise what was already known in practical solutions. Such feelings further dwindled the influence of these specialists in favour of planners and policy makers.

These views coupled with the new bodies and initiatives emerging out of the food conference left some feeling confused about who was responsible for what. What was needed was improved rationalisation and co-ordination among the UN agencies. Thus by 1975, the FAO, WHO and UNICEF, looking for more integration and fluidity in their policies, adopted structural changes to best assist in their organisational undertakings. The Protein Advisory Group (PAG) too came under assault. Its previous work concentrating on protein almost to the exclusion of all else left it vulnerable to criticism with many favouring its decommissioning. Indeed as part of its role, the newly created UN Advisory Committee on Coordination (ACC) was responsible for co-ordination of UN work related in this field and as it worked tirelessly to eliminate duplication and overlap. Under these circumstances it was not surprising that the PAG fell on its radar. In its report, the ACC recommended the formation of a sub committee on nutrition which would act as a clearing house for the relevant agencies working in the field of nutrition, and under this new proposed arrangement, it was suggested that the PAG would be succeeded by a nutrition advisory panel. As a result, the PAG ceased to operate in 1977 and was replaced by an Advisory Group on Nutrition (AGN) which would work both independently and under the guidance of a Sub-Committee on Nutrition (SCN). The SCN, in turn, would report to the ACC and be accountable for reviewing the overall direction and coherence of current global nutritional problems of the world.

13.2 Health and Nutrition

13.2.1 Nutrition

In light of the close relationship between protein and energy, the next (third) FAO/WHO Joint Expert Committee meeting was convened in 1971 to consider simultaneously, both energy and protein requirements in dietary matters and to examine further any interrelationship between the two (FAO/WHO 1973; WHO 1973a; Périssé 1981; Weisell 1995; FAO 2002). While complimenting long-serving recommendations of the 1957 Calorie Committees report, a cautionary note by Dr. DeMaeyer, however, in the 1971 meeting's background documents, observed that the suggested energy recommendations of the 1957 report were perhaps illusory. In his introductory remarks, Dr. DeMaeyer referred to the subject of physical activity and the continuing difficulty in factoring in accurate and reliable calculations for different rates of work and their suitability for extrapolating to the general (FAO/WHO 1973). He continues:

> … there has always been some ambiguity as to whether the international figures for recommended intakes should constitute ideal objectives or should be practical enough to enable each government to implement them without setting impossible goals. (FAO/WHO 1973)

However, in further attempting to establish comparable standards, the committee proposed four activity levels for adults—light activity, moderately active, very active and exceptionally active—whilst for children a moderate standard of activity was assumed (WHO 1973a; Périssé 1981). For protein requirements the report looked at the methodology of energy calculations and noting previous reports' guidelines, the committee recommended further standardisation in respect of methodology. This once again touched upon the equilibrium versus expenditure methods of calculating energy requirements and referred to the fact that while the majority of the report's dietary recommendations were based on measurements of food intake, the committee recognised that measurements based on energy expenditure would be of future importance (WHO 1985). This reinforced observations made at the first two FAO committee meetings of 1950/56. Ultimately, the final 1973 report continued to favour the 'reference' adult first introduced by the 1950 committee, although a few adjustments were made to the relative recommendations of both the reference adults and the children's requirements (FAO/WHO 1973; Beaton et al. 1979; Périssé 1981).

However, not long after this report, in response to concerns over, inter alia, perceived low levels of protein allocation recommended by the 1971 committee, an informal expert gathering was convened in 1975 and then in 1978 by the WHO/FAO to provide guidance on some matters not considered adequately addressed (Beaton et al. 1979; Scrimshaw 1996; Weisell 2002). Several areas under review including, whether it was suitable and realistic to use requirements taken from 'healthy well nourished individuals' as the basis for developing countries (WHO 1985). Other areas of interest included low protein allocations, insufficient attention on the requirements of women and older children, and information regarding the ability of local diets to meet protein needs (WHO 1985). While these issues came to light, they were not addressed until 1981 (see Chapter 14).

Meanwhile, despite the progress made in energy and protein requirements including maternal and child nutrition feeding programmes, little progress continued to be made in the area of micronutrient deficiencies (Allen 2000). By the early 1970s, many countries were already aware that deficiencies of iron, vitamin A and iodine were more widespread problems than protein energy malnutrition. As a result of this new holistic approach to nutrition, the WHO convened the first Expert Group on Trace Elements in 1973 to discuss these issues and make recommendations (WHO 1973b). Other accumulated doubts about this time, in particular that perhaps proteins' role in malnutrition had been overstated, was fast becoming mainstream, and the idea of a worldwide protein gap was largely untenable. In fact, it was concluded in 1977 that inadequate intake of energy (rather than protein) from insufficient food intake did indeed rank above all other types of malnutrition. As a result, in 1976, WHO adopted the term protein-energy-malnutrition (PEM), which like the protein-calorie-malnutrition (PCM) of previous years aimed to reflect a more balanced approach to nutrition (Allen 2000).

13.2.2 Nutrition, Infection and Primary Health Care

With the passing of the protein crisis and the popular recourse to a balanced approach, the relationship between nutrition and infection once again resurfaced. An important study by Leonardo Mata, then the chief of the INCAP Division of Environmental Biology, and colleagues investigated child health in the Guatemalan village of Santa Maria Cauque. In this study, Mata sought to finally clarify the relationship between malnutrition and infection. The study started in 1963 and went on for nearly 10 years in which time extensive data was gathered with regard to health and nutrition. Findings overwhelmingly illustrated that infection indeed played a major role in the onset of malnutrition (Mata et al. 1971, 1976,

1991). This was indeed remarkable proof for the need to treat both malnutrition and infection on an equal footing and among Mata's many recommendations included

- Education in nutrition and weaning practices;
- Increasing household water to reduce diarrheal infection;
- Increasing the number of beds per household to reduce onset of respiratory disease;
- Immunisation against such diseases as measles, whooping cough, tuberculosis and tetanus.

Moreover, Mata also reaffirmed the importance of improved health services, agrarian reform, family planning and most importantly wages. These sentiments were further echoed by Latham, writing in *Science magazine,* at the time, when he suggested that more efficiency in the fight against nutrition might be achieved if the two problems were tackled together (Latham 1975). These ideas were also taken up by the WHO, which in 1975, adopted a resolution calling for increased focus on primary health care services in the developing countries.

In an adjunct, by the time of the Alma Ata conference on health care in 1978, nutritional issues in respect of breastfeeding supplements were already hotting up. Over the preceding years, a decline in breast feeding was blamed, in the popular press and by some nutritionists, on Nestlé's and other food industry practices. By 1974, the charity 'War on Want' had taken things a step further by publishing a pamphlet called 'The Baby Killer', which was later amended to 'Nestlé Kills Babies' and released in Switzerland. What the industry was doing, it alleged, was promoting their products on the back of suggestions that mother's milk was inferior to their formulaic substitutes. Although the court case was lost by the charity, it was a moral victory in that a subsequent boycott of Nestlé's products altered views and practices in this area (Moorhead 2007).

13.2.2.1 Alma Ata Conference on Primary Health Care

Such accumulated widespread dissatisfaction with the status quo of the world's health, in particular world health services, about this time led to the joint WHO/UNICEF sponsored Primary Health Care (PHC) conference in Alma Ata in 1978 (PHC 1978). The conference immediately aimed to define and promote the notion of primary health care as an integral part of health improvement. This involved promoting many of the recommendations of Leonardo Mata's previous studies such as immunisation against infectious disease, family planning and provision of clean water as well as adequate sanitation. In conclusion, the PHC conference outlined the first international declaration underlining the importance of primary health care and in it strongly affirmed that health was a state of

> … complete physical, mental and social wellbeing, and not merely the absence of disease or infirmity, is a fundamental human right … (PHC 1978)

The declaration continued:

> The existing gross inequality in the health status of the people particularly between developed and developing countries as well as within countries is politically, socially and economically unacceptable and is, therefore, of common concern to all countries. (PHC 1978)

Although not part of the declaration itself, nutrition was actually seen as an important part of PHC ideology. Although curiously at the time, other research indicated that tackling infant infection through the use of surveillance and early treatment appeared to be more productive than the nutritional approach (Kiclmann et al. 1978). Nevertheless, as Ruxin (1996) pointed out, the cost–benefit analysis showed that little was lost by leaving out the nutritional component, so it made sense to carry on treating both together (Ruxin 1996). As a result, the international community quickly adopted the PHC's recommendations with its focus on a top–down medical service approach embracing principles of: social justice, affordability, self-reliance, decentralisation, community involvement and inter-sectoral collaboration (PHC 1978; Lee 1980; Allen 2000).

13.3 Science and Technology

In the field of science, the growing influence of biotechnologies was beginning to have a noticeable impact on the economy in general. Single cell proteins had became quite popular by the 1970s, and much hope was invested in the this technology as an important contributor to the challenges of food production. However, public resistance and costs refocused the industry towards the seemingly more lucrative and less finicky animal feed market.

13.3.1 Recombinant DNA: The Arrival of Genetic Engineering

In the meantime, great strides were being made in the field of genetics. Spurred on by Watson and Crick's discovery of the structure of DNA, genetic research witnessed a great leap with the 1973 discovery by Cohen and Boyer of a recombinant DNA (rDNA)* technique (Section 18.2.1). This technique, known as gene splicing, saw DNA segments (genes) being cut from one bacterium and transferred into the DNA of another. The practice of transferring these genes into new organisms gave rise to the term 'transgenic'; thus, the field of genetic modification or genetic engineering was born. Cohen and Boyer's discovery along with Stanford's subsequent patenting of the results were, among other things, to have enormous consequences in the field of food technology (Styhre 2009). Once again though, public fears tempered enthusiasm. In July 1974, a group of 11 eminent biologists from the National Research Council and headed by Paul Berg wrote in *Science* magazine of their concerns which they felt were theoretically inherent in such research. In their view,

> There is serious concern that some of these artificial recombinant DNA molecules could prove biologically hazardous ... recombinant DNA molecules might be more easily disseminated to bacterial populations in humans and other species, and thus possibly increase the incidence of cancer or other diseases. (Berg et al. 1974)

This unprecedented action by a group of American scientists led to a voluntary moratorium during which time could be found, among other things, to convene a conference to discuss the ramifications of this line of research. The conference was held at the Asilomar Conference Center on California's Monterey peninsula in 1975 and its conclusion, among strong opposition, was that research should proceed only under strict guidelines (Nobel Prize 2010). And so it continued, and undoubtedly the major discovery of the time was the use of rDNA in the microbial production of synthetic human insulin by Genotech, the newly found biotechnology company of Boyers (Styhre 2009). Other milestones of this period are summarised in the Table 13.1.

13.4 Period Commentary

The 1970s was a time of great change, both socially and ideologically. It was a time when food sustainability had increasingly come to be seen in the context of the wider environment. It was also an era that saw a ratcheting up of public sentiment affecting policy; the power of the people, as it were, was now a force to be considered. Meanwhile the food crisis of the early 1970s marked a dramatic turning point from the previous era of surpluses and abundance to one of uncertainty, characterised by unstable food supplies and high prices on the world market. This was followed by a period of rapid wealth creation as the OPEC cashed in on rising oil prices. This further created a conducive environment of lending, and there were no shortage of takers in the Third World. Unfortunately though, this money increased the fluidity of the financial markets, helping it to overheat. On top of this, in many developing countries, by the early 1970s, the world food situation was characterised by extreme food shortages brought on by a near doubling of grain prices. Couple this with the slow progress in the fight against hunger and malnutrition and the growing institutional and political concern, the 1974 World Food Conference (WFC) was called (Maxwell and Frankenberger 1992; Mitchell et al. 1997; Weingärtner 2004; UOR 2009).

---

* Recombinant DNA is a form of DNA that does not exist naturally, but rather, has been created artificially. Strands of DNA, or genes, are taken from one organism and implanted into another. By combining these strands scientists are able to create new strands of DNA.

TABLE 13.1

Genetic Engineering Milestones of the 1970s

1972	The first successful DNA cloning experiments were created using a restriction enzyme and ligase to form the first recombinant DNA molecule by Paul Berg in California.
1973	Scientists Stanley Cohen, Annie Chang and Herbert Boyer for the first time successfully transferred DNA from one life form into another creating the first recombinant DNA organism.
1974	In a letter to *Science* magazine, Paul Berg and others called for the National Institutes of Health (NIH) to enact a moratorium on certain DNA techniques until questions of safety could be addressed.
1975	A moratorium on recombinant DNA experiments was agreed at an international meeting at Asilomar, California.
1976	Boyer and Swanson founded Genentech, Inc. to develop and market products based on recombinant DNA technology.
1976	The NIH released the first guidelines for recombinant DNA experimentation.
1977	Bills to regulate recombinant DNA research was introduced in the United States although none of the bills were passed. Walter Gilbert and Allan Maxam at Harvard University devised a method for sequencing DNA using chemicals rather than enzymes. Genentech, Inc. produces the first human protein manufactured in a bacteria: somatostatin, a human growth hormone-releasing inhibitory factor.
1978	Genentech, Inc. were successful in laboratory production of human insulin using recombinant DNA technology.
1977	Genetically engineered bacteria are used to synthesise human growth protein.
1978	Scientists Hutchinson and Edgell show it is possible to introduce specific mutations at specific sites in a DNA molecule.

Sources: Reprinted from Berg et al., *Science New Series* 185, 303, 1974; Bud, R., *The Uses of Life: A History of Biotechnology*, Cambridge University Press, Cambridge, 1993; Peters, P., *Biotechnology: A Guide to Genetic Engineering*, Wm. C. Brown Publishers, Dubuque, IA, 1993; Dobson, A., *Environ. Values* 4, 227–239, 1995; *Biochemical Engineering and Biotechnology*, Najafpour, G.D., Copyright 2007, with permission from Elsevier; Nature, *Nat. Rev. Genet.*, 8(Supplement), 2007; Styhre, A., *Ephemera Theory Polit. Org.*, 9, 26–43, 2009; BioTech Institute, Timeline of Biotechnology 20th Century, Biotechnology Institute, 2010.

Faced with the current food climate, the WFC turned its attention from nutritional issues to those of scarcity of supply. Once again, as after World War II, food, and not nutrition, was the main concern of policy makers. The conference focused on increasing global food production and advocated self-sufficiency whilst proposing international systems of grain reserves to offset fluctuations in times of short supply (Devereux et al. 2004; UOR 2009). It also recognised, in the adoption of the WFC Universal Declaration, that vulnerable and lower income groups needed extra attention and urged stakeholders to integrate food and nutrition policies into over-arching socio-economic and agricultural development plans to better aid such groups (WFC 1975; FAO 2003). Furthermore the conference went on to highlight the importance of women in rural agriculture and the need for sustainability in production and development practices (WFC 1975). On the policy front, the conference highlighted the need for favourable international policy cooperation, both economic and agricultural, which further endorsed the reductionist approach to trade barriers. This approach no longer questioned whether national planning had to incorporate nutrition but rather how could it best be prioritised and incorporated. All in all, the conference was a landmark in the fight against hunger and malnutrition, and it finally seemed that many of the disparate strands of food security were being pulled together in some form of coordinated understanding.

Several bodies came out of the conference responsible for nutrition. Inter-agency cooperation and food security and just as important was the concept of a coordinated global surveillance system. As a result, throughout much of the 1970s, national early warning systems were developed in the form of food balance sheets, nutrition surveys as well as rainfall and price data that collectively helped forecast the state of nutrition and agriculture and to forewarn of any local food shortages (Ismail 1991).

With regard to nutrition, the first years of the decade's interest in protein, an issue that had dominated substantial resources and expertise in the previous decades, reached its peak before beginning what was quite a rapid fall from grace. A different view, although not a new one, emerged as the leading paradigm to replace the overemphasis on protein and that was both protein and calories tended to occur together in the cases of deficiency. Indeed, in an attempt to gain a better insight into their interrelationship, an ad hoc FAO/WHO Expert Committee met in 1971 was the first meeting to consider energy and protein requirements together (Payne 1971). Separately, all round, it seemed that increased protein and energy intake by children was the solution. On the back of this, nutrition rehabilitation centres and applied nutrition programmes were offered up as tried and tested strategies. However, not everyone favoured this approach and what followed was an almost ideological split between two nutritionist camps. This split perhaps had its genesis in the previous decade when nutritionists and planners were of two groups: those that favoured the simple supply-side solutions, while the other looked towards those countries in need to take the initiative in terms of national development policy. Although ideologically split however, reality dictated both camps had to compromise: while improvements in national social and developmental programmes filtered down through the population, more immediate pressing needs would need to be addressed. This situation in effect would require the combined efforts of both camps.

Ironically though, at a time when earlier views of examining the whole PCM/PEM spectrum was now becoming mainstream, the UN, in response to the lingering notion of a protein gap, began calling for more intensified protein programmes. Despite this early misdirection, the UN system came back on track and after the World Food Conference it aimed to rationalise its agencies and their efforts. In this endeavour, the UN was not above self-reproach and so it was that in 1975 the ACC was convened to ask some very straightforward and direct questions of its nutritional agencies. One question stood out from the rest. Why, with the full force of moral righteousness behind them, had the respective agencies not come up with a reasonable strategy for dealing with the situation? The answer, when it came, was not that surprising. It was asserted that although for a while, protein had been the focal point at the detriment of other more fruitful avenues, more importantly it was conceded that efforts had been hindered by the pure complexity of the situation and the solutions it required (Ruxin 1996). The solution to this dilemma was more inter-agency co-operation and joint programming initiatives, when combined, ensured the continuing frontline role of UN as governor of first resort.

It was also about this time too that a new player came onto the scene: the World Bank. Armed with a broad and scathing view of the WHO, FAO and UNICEF's work to date, the Bank hoped to inject a sense of purpose with upscaled programmes and with acceptance of many international governments. By now, however, nutrition was taking another turn. With Leonado Mata's outstanding research concerning nutrition, health planners began to reevaluate the role of nutrition in policy. Once again, after endless years of hearing that in developing countries, hunger and malnutrition were the greatest problems, research now re-focused attention on health and disease prevention as the biggest challenge. In the new primary health care initiative, it was interesting to see Leonado Mata's recommendations figure prominently in the embryonic view of this new line of thinking. Thus, as the decade drew to a close, a new health perspective ensured the continuing complexity of food security into the next decade.

13.5 Key Dates, People, Acts, Reports and Surveys

The 1970s decade was a time of reflection and great hope, Table 13.2, introduces the highlights of much progress in this period.

TABLE 13.2

Key Dates: Famine, Oil and the Food Crisis

Date	People, Organisation, Agencies, Acts or Conventions
1970: Committee on Natural Resources (CNS)	The importance of natural resources and a nations, sovereignty over their natural resources was emphasised in 1970 when the Economic and Social Council established the Committee on Natural Resources (Osmanczyk 2003).
1971: Office of the United Nations Disaster Relief Coordinator (UNDRO)	In response to many nations becoming independent in the 1960s and the uncoordinated assistance during times of hardship, the UN established the Office of the UN Disaster Relief Coordinator (UNDRO), headquartered in Geneva. Its main role is that of catalyst and coordinator of donors of aid and services (Osmanczyk 2003).

TABLE 13.2 (Continued)

Key Dates: Famine, Oil and the Food Crisis

Date	People, Organisation, Agencies, Acts or Conventions
1971: Doctors Without Borders/ Médecins Sans Frontières (MSF)	An international humanitarian organisation created in 1971 by doctors and journalists to relieve suffering around the world (MSF 2010).
1971: Earthwatch Institute (EI)	An international environmental charity formed in 1971 founded because of the need for new scientific research funding models and promoting the public's understanding of science of environmentalism (Earthwatch 2009).
1971: The Consultative Group on International Agricultural Research (CGIAR)	The pioneering work of the Ford-Rockefeller Foundation led initially to four research centres: the CIAT (Colombia, 1967); CIMMYT (Mexico, 1966); IITA (Nigeria, 1967); and IRRI (Philippines, 1960). In 1970 however, these four together with the FAO, UNDP and the World Bank proposed a worldwide network of agricultural research centers under a permanent secretariat. In 1971, the Consultative Group on International Agricultural Research was established. By 2006, there were a coalition of 15 International Research Centers (CGIAR 2008).
1972: The 1st UN Conference on the Human Environment	Held in Stockholm 1972, the conference was 'the UN's first major conference on international environmental issues'. Also known as the First Earth Summit, it focused on the enhancement and preservation of the environment. On recommendations of the conference, the UN created the Environment Programme (UNEP) to monitor changes and to encourage and coordinate sound environmental practices (Jackson 2007).
1972: Bread for the World	Group of Catholics and Protestants came together to address the causes of hunger. The organisation today has a global remit (BFW 2009).
1972–81: The end of the Bretton Woods System	By early 1960s, the US dollar was seen as overvalued. Under the Bretton Woods agreement however, the dollar was fixed against gold and so in 1971 President Nixon temporarily suspended the dollar-gold convertibility. Although an attempt was made to re-introduce the fixed exchange rate system, major global currencies were beginning to 'float' against each other by 1973 sounding the end of the Bretton Woods agreement and the beginning of a floating exchange rate system (IMF 2010).
1972: ActionAid	Founded in 1972 to fight worldwide poverty (ActionAid 2010).
1972: World Resources Institute (WRI)	Seeing the urgent need for a credible research-based environmental, resource, population and development organisation, the founders envisioned an institute that would carry out policy research and analysis on a global scale (WRI 2010).
1972: The UN Environment Programme (UNEP)	Born out of the first Earth Summit (First UN Conference on the Human Environment) of 1972, UNEP advocates education and promotion for the fair and just use as well as sustainable global development practices within the environment (UNEP 2010).
1973: Economic and Social Commission for Western Asia (ESCWA)	One of five regional commissions of ECOSOC. A regional economic commission for the Middle East, was first proposed in 1947–1948 including Arab nations and Israel. Instead in 1963 for 11 years, the UN Economic and Social Office in Beirut (UNESOB) was set up and in 1972, Lebanon revived the idea of a regional commission. In August 1973 the ESCWA was established to supersede UNESOB (EoN 2009a; UN ESCWA 2009; UN 2010).
1973–1979: GATT "Tokyo Round" 7th Round	A sweeping attempt to extend and improve the system of tariffs reductions and introducing a detailed set of 'codes' or rules dealing with non-trade barriers and voluntary export restrictions. This round produced the antidumping and subsidies agreement (Morrison 1986; WTO 2010).

continued

TABLE 13.2 (Continued)

Key Dates: Famine, Oil and the Food Crisis

Date	People, Organisation, Agencies, Acts or Conventions
1973: Centre for Research on the Epidemiology of Disasters (CRED)	CRED is a non-profit institution founded by epidemiologist Professor Michel F. Lechat focusing on training and research in the areas of relief, rehabilitation and development. In 1980, the Centre has been recognised as a World Health Organization (WHO) Collaborating Centre (CRED 2009).
1973: University of United Nations (UNU)	UNU is an educational facility to promote knowledge and dissemination through the strengthening of individual and institutional capacities (UNU 2009).
1974: World Food Conference (WFC)	The first World Food Conference was held in Rome in 1974 in response to the worsening global food situation in the early 1970s together with a general lack of progress in the world fight against hunger. The conference promoted a universal declaration for the eradication of hunger and malnutrition within a decade. To help achieve this the conference called for the creation of ministerial-level World Food Council (WFC) to review problems and policy issues. The Conference also recommended the reconstitution of the WFP's governing body as well as the incorporation, the International Fund for Agricultural Development (IFAD) and the FAO Committee on World Food Security (CFS) (UN 1975).
1974: The Worldwatch Institute (TWI)	The Worldwatch Institute is an independent research organisation focusing on climate and energy, food and agriculture and the green economy (Worldwatch 2008).
1974: World Food Council (WFC)	The World Food Conference established the World Food Council to review problems affecting the world food and situation and to use its influence on governments and UN bodies and agencies alike. Eventually, the WFC recommended the creation of an inter-secretariat consultative body promoting cooperation between the four Rome-based food organisations (FAO, IFAD, WFC and WFP). However, in 1996, rationalising duplicative efforts, the WFC was absorbed by the FAO and World Food Programme (WFP) (CCP 2005; EoN 2009b).
1975: Committee on Food Aid Policies and Programmes (CFA)	The World Food Conference of 1974 focused lots of attention on the issues of food aid, as a result, the World Food Council's governing body the Intergovernmental Committee's (IGC) remit was broadened to include the more general problems of food aid and related policies. The new governing body was named the Committee on Food Aid Policies and Programmes (CFA) (Phillips 1981).
1974: AGRIS	AGRIS is the agricultural sciences information system created by the FAO to promote information exchange and to synthesise world literature in agriculture (AGRIS 2010).
1974: The International Fund for Agricultural Development (IFAD)	A specialised agency of the UN, IFAD aims at financing agricultural development projects through low-interest loans and grants in the developing countries (Rucker 2007; IFAD 2009).
1974: The 1st World Population Conference (UNFPA)	The First World Population Conference was held in Bucharest, 1974 and it adopted a World Population Plan of Action (WPPA) stressing the relationship between population and economic and social development (UNFPA 2004).
1974: Committee on World Food Security (CFS)	Out the 1974 World Food Conference, the World Food Council established the FAO Committee on World Food Security. It was designed as a forum for review and follow-up of policies of world food security; food production, nutrition and access to food. It also promoted the use of food aid to support economic development and food security in vulnerable countries (CFS 2008b; IAAH 2008; CFS 2011).
1975: The International Food Policy Research Institute (IFPRI)	IFPRI research's sustainable solutions towards ending hunger and poverty. It is one of the 15 research centres under the umbrella of the Consultative Group on International Agricultural Research (CGIAR) (IFPRI 2010).

TABLE 13.2 (Continued)

Key Dates: Famine, Oil and the Food Crisis

Date	People, Organisation, Agencies, Acts or Conventions
1975: Institute for Food and Development Policy (IFDP) AKA Food First	The IFDP analyses and looks for solutions to the root causes of hunger, poverty and environmental degradation (FoodFirst 2009).
1975: FAO Global Information and Early Warning System (GIEWS)	FAO's Global Information and Early Warning System (GIEWS) meets the needs of donors in keeping country-specific assessments of food problems under continuous review and issuing reports on the global food situation (ODI 1997).
1976: FAO's Technical Cooperation Programme (TCO)	TCO supports the FAO by affording greater flexibility in responding swiftly to urgent situations. It addresses specific problems in the agriculture, fisheries and forestry sectors through immediate and tangible results by swiftly mobilising the technical expertise of the entire organisation (TCP 2009).
1976: International Emergency Food Reserve (IEFR)	The World Food Conference in 1974 led to the establishment in 1976 of the International Emergency Food Reserve (IEFR) requiring member countries to pledge food target donations to be used in quick response to emergencies. Targets were made up of food in kind and cash payments for food and overheads (ODI 1997; EoN 2009c; TCP 2009).
1977: The International Fund for Agricultural Development (IFAD)	IFAD established out of the 1974 WFC is a financial institution and specialised agency of the UN primarily concerned with financing agricultural development projects for food production in the developing countries (IFAD 2009).
1977: The Standing Committee on Nutrition (SCN)	The Protein Advisory Group was incorporated into the UN's Administrative Coordination Committee's Subcommittee on Nutrition (ACC/SCN) in 1977 (Allen 2000). The Standing Sub-Committee on Nutrition (SCN) is a forum for UN agencies, NGOs/CSOs and partners to coordinate and cooperate on nutrition related issues (SCN 2009).
1979: Action Against Hunger (ACF)	Action Against Hunger is an agency born out of the need for a different kind of philanthropy, which was fighting the frontline as well as interested in tackling the problem from the other side. This involved a new brand of humanitarian politics that tries to influence political actors and institutions and not just mitigate the worst emergencies on the ground (ACF 2010).
1979: International Year of the Child	UNICEF creates the International Year of the Child (UNICEF 2005).

References

ACF (2010). Action against Hunger International: Who We Are. Action Contre La Faim (ACF). http://www.actionagainsthunger.org/ (accessed 23 March 2010).

ActionAid (2010). Act!Onaid. ActionAid. http://www.actionaid.org.uk/100041/our_history.html (accessed 23 January 2011).

AGRIS (2010). About FAO's Agris. Food and Agriculture Organization. http://www.fao.org/agris/ (accessed 21 June 2010).

Allen, L. H. (2000). Ending Hidden Hunger: The History of Micronutrient Deficiency Control. *Background Analysis for the World Bank-UNICEF Nutrition Assessment Project*. Washington, DC: World Bank, pp. 111–130.

Argeñal (2007). History of Food Security in International Development: Background Paper Developed for Christian Children's Fund Honduras. Honduras: Christian Children's Fund.

Beaton, G. H., D. H. Calloway and J. Waterlow (1979). Protein and Energy Requirements: A Joint FAO/WHO Memorandum. *Bulletin of the World Health Organization* 57(1): 65–69.

Berg, P., D. Baltimore, H. W. Boyer, S. N. Cohen, R. W. Davis, D. S. Hogness, D. Nathans, R. Roblin, J. D. Watson, S. Weissman and N. D. Zinder (1974). Potential Biohazards of Recombinant DNA Molecules. *Science, New Series* 185(4148): 303.

BFW (2009). Bread for the World: Our History. Bread For the World. http://www.bread.org/about-us/our-history.html (accessed 30 January 2010).

BioTech Institute (2010). Timeline of Biotechnology 20th Century, Biotechnology Institute. http://www.biote-chinstitute.org/what_is/timeline.html (accessed 12 January 2011).

Bud, R. (1993). *The Uses of Life: A History of Biotechnology*. Cambridge: Cambridge University Press.

Carrasco, E. R. and K. J. Berg (1999a). *The E-Book on International Finance and Development: Part One— V. The 1980's: The Debt Crisis and the Lost Decade*. The University of Iowa Center for International Finance and Development (UICIFD).

Carrasco, E. R. and K. J. Berg (1999b). *The E-Book on International Finance and Development: Part One LV—The 1960s and 1970s: The World Bank Attacks Poverty; Developing Countries Attack the IMF*. The University of Iowa Center for International Finance and Development (UICIFD).

CCP (2005). A Historical Background on Food Aid and Key Milestones. Committee on Commodity Problems: Sixty-fifth Session. Rome: Food Agriculture Organization.

CFS (2008a). Agenda Item II: Assessment of the World Food Security and Nutrition Situation. Committee On World Food Security: Thirty-fourth Session. Rome: Food and Agriculture Organization.

CFS (2008b). Report of the CFS: Statement by the Deputy Director-General. Committee On World Food Security: Thirty-fourth Session. Rome: Food and Agriculture Organization.

CFS (2009). Report of the Thirty-Fifth Session of the Committee on World Food Security (CFS). Rome: Committee on World Food Security.

CFS (2011). FAO Committee on World Food Security (CFS). Committee on World Food Security. http://www.fao.org/economic/cfs09/cfs-home/en/ (accessed 15 January 2011).

CGIAR (2008). Consultative Group on International Agricultural Research: Who We Are: History of the CGIAR (accessed 23 June 2010).

CRED (2009). The Centre for Research on the Epidemiology of Disasters: History. The Centre for Research on the Epidemiology of Disasters. http://www.cred.be/history (accessed 12 June 2010).

Devereux, S., B. Baulch, K. Hussein, J. Shoham, H. Sida and D. Wilcock (2004). Improving the Analysis of Food Insecurity: Food Insecurity Measurement, Livelihoods Approaches and Policy: Applications in Fivims. Rome: FAO-Food Insecurity and Vulnerability Information and Mapping Systems (FIVIMS) 52.

Dobson, A. (1995). Biocentrism and Genetic Engineering. *Environmental Values* 4(3): 227–239.

Earth Summit (2002). Decision Making: Briefing Sheet. United Nations. http://www.earthsummit2002.org/es/life/Decision-making.pdf (accessed 20 July 2009).

Earthwatch (2009). Earthwatch Our History. Earthwatch Institute. http://www.earthwatch.org/europe/aboutus/history/ (accessed 21 June 2010).

Eichengreen, B. and P. H. Lindert (1989). *The International Debt Crisis in Historical Perspective*. Cambridge, MA: MIT Press.

EoN (2009a). *Economic and Social Development: Regional Commissions* New York: United Nations.

EoN (2009b). *Economic and Social Development: World Food Council*. New York: United Nations.

EoN (2009c). *The Encyclopedia of the Nations*. New York: United Nations.

FAO (1996). World Food Summit—Technical Background Documents 1–5: Volume 1: 3. Socio-Political and Economic Environment for Food Security. World Food Summit. Rome: Food and Agriculture Organization.

FAO (2002). Food, Nutrition and Agriculture. Rome: FAO Food and Nutrition Division.

FAO (2003). Trade Reforms and Food Security: Conceptualizing the Linkages. Rome: Food and Agriculture of the United Nations.

FAO (2006). Policy Brief: Food Security. Rome: Food and Agriculture Organization 4.

FAO/WHO (1973). FAO/WHO Ad Hoc Committee of Experts on Energy and Protein: Requirements and Recommended Intakes. Rome: Food and Agriculture Organization.

FoodFirst (2009). About the Institute for Food and Development Policy. Food First. http://www.foodfirst.org/ (accessed 24 March 2010).

Gerster-Bentaya, M. and N. Maunder (2008). *Food Security Concepts and Frameworks: What Is Food Security?* Rome: EC-FAO.

Goodrich, L. M. (1947). From League of Nations to United Nations. *International Organization* 1(1): 3–21.

Heaver, R. (2006). Good Work—But Not Enough of It: A Review of the World Bank's Experience in Nutrition. Washington, DC: World Bank: Health, Nutrition and Population Department 89.

IAAH (2008) Interview with Mr. Kostas Stamoulis, Secretary General of the Committee on World Food Security. 2 http://www.iaahp.net/fileadmin/templates/iaah/pdf/Interview_Stamoulis.pdf.

IFAD (2009). About the International Fund for Agricultural Development. The International Fund for Agricultural Development http://www.ifad.org/governance/index.htm (accessed 24 February 2010).

IFPRI (2010). About the International Food Policy Research Institute. IFPRI. http://www.ifpri.org/ourwork/about (accessed 13 February 2010).

IMF (2010) The End of the Bretton Woods System (1972–81). http://www.imf.org/external/np/exr/contacts/contacts.aspx (accessed 23 July 2010).

Ismail, S. J. (1991). Nutritional Surveillance: Experiences from Developing Countries. *Proceedings of the Nutrition Society* 50(03): 673–679.

Jackson, P. (2007). From Stockholm to Kyoto: A Brief History of Climate Change. *United Nations Chronicle* XLIV(2) http://www.un.org:80/wcm/content/site/chronicle/lang/en/home/archive/issues2007/pid/4819 (accessed 22 May 2011).

Kielmann, A. A., C. E. Taylor and R. L. Parker (1978). The Narangwal Nutrition Study: A Summary Review. *The American Journal of Clinical Nutrition* 31: 2040–2052.

Latham, M. C. (1975). Nutrition and Infection in National Development. *Science* 188(4188): 561–565.

Lee, P. R. (1980). Primary Health Care, Report of the International Conference on Primary Health Care. *Health Services Research. Summer* 15(2): 178–184.

Lipson, C. (1981). The International Organization of Third World Debt. *Journal of International Organization* 35(4 (Autumn)): 603–631.

Mata, L. J., R. A. Kronmal, B. Garcia, W. Butler, J. J. Urrutia and S. Murillo (1976). Breast-Feeding, Weaning and the Diarrhoeal Syndrome in a Guatemalan Indian Village. Acute Diarrhoea in Childhood. *Ciba Foundation Symposium* 42: 311–338.

Mata, L. J., J. J. Urrutia, C. Albertazzi, O. Pellecer and E. Arellano (1991). Influence of Recurrent Infections on Nutrition and Growth of Children in Guatemala. *Nutrition Reviews* 49(9): 269–272.

Mata, L. J., J. J. Urrutia and A. Lechtig (1971). Infection and Nutrition of Children of a Low Socioeconomic Rural Community. *American Journal of Nutrition* 24(2): 249–259.

Maxwell, S. and T. Frankenberger (1992). Household Food Security: Concepts, Indicators and Measurements: A Technical Review. New York and Rome: UNICEF and IFAD.

McMichael, P. (1994). *The Global Restructuring of Agro-Food Systems*. Ithaca, NY: Cornell University Press.

Mitchell, D. O., M. D. Ingco and R. C. Duncan (1997). *The World Food Outlook*. Cambridge: Cambridge University Press.

Moorhead, J. (2007). *Riots and Hunger Feared as Demand for Grain Sends Food Costs Soaring*. London: Guardian.

Morrison, A. V. (1986). GATT's Seven Rounds of Trade Talks Span More Than Thirty Years—General Agreement on Tariffs and Trade. Swanton, OH: Business America 8–10.

MSF (2010). Doctors without Borders: History and Principles. Médecins Sans Frontières. http://doctorswithoutborders.org/aboutus/ (accessed 21 June 2010).

Munasinghe, M. (1998). Special Topic I: Structural Adjustment Policies and the Environment: Introduction. *Environment and Development Economics* 4: 9–18.

Najafpour, G. D. (2007). *Biochemical Engineering and Biotechnology*. The Netherlands/Oxford: Elsevier.

Nature (2007). Milestones in DNA Technologies. *Nature Reviews Genetics*. 8(Supplement).

Nobel Prize (2010). Paul Berg: Asilomar and Recombinant DNA. Nobel Foundation. http://nobelprize.org/nobel_prizes/chemistry/laureates/1980/berg-article.html (accessed 7 March 2011).

ODI (1997). Global Hunger and Food Security after the World Food Summit. London: Overseas Development Institute.

Osmanczyk, E. J. (2003). *Encyclopedia of the United Nations and International Agreements N-S*. New York: Routledge.

Payne, P. R. (1971). FAO/WHO Ad Hoc Committee of Experts on Energy and Protein: Requirements and Recommended Intakes: Working Paper—The Definition of Requirements and Recommendations for Protein with Notes on the Magnitude and Degree of Uncertainty of the Factorial Components. Rome: Food and Agriculture Organization.

Périssé, J. (1981). Joint FAO/WHO/UNU Expert Consultation on Energy and Protein Requirements: Past Work and Future Prospects at the International Level. Rome: Food and Agriculture Organization.

Peters, P. (1993). *Biotechnology: A Guide to Genetic Engineering*. Dubuque, IA: Wm. C. Brown Publishers, Inc.

PHC (1978). Declaration of Alma-Ata. International Conference on Primary Health Care (PHC), Alma-Ata (Almaty), USSR.

Phillips, R. W. (1981). *FAO: Its Origins, Formation and Evolution 1945–1981*. Rome: Food and Agriculture Organization.

Rucker, A. (2007) Key Players in Food Aid. *eJournal USA*: http://www.america.gov/st/health-english/2008/Jun
    e/20080616003149xjyrrep0.4639551.html.

Ruxin, J. N. (1996). *Hunger, Science and Politics; Fao, Who, and UNICEF Nutrition Policies, 1945–1978, Chapter
    II the Backdrop of Un Nutrition Agencies, by Joshua Nalibow Ruxin*. London: University College London.

SCN (2009). The United Nations Standing Committee on Nutrition (Scn): Who We Are. World Health
    Organization. http://www.unscn.org/en/home/who-we-are.php (accessed 20 July 2010).

Scrimshaw, N. S. (1996). Human Protein Requirements: A Brief Update. *Food and Nutrition Bulletin* 17(3):
    185–190.

Slusser, S. (2006). *The World Bank, Structural Adjustment Programs and Developing Countries: A Review
    Using Resource Dependency Theory*. Montreal: Paper presented at the annual meeting of the American
    Sociological Association, Montreal Convention Center, Montreal, Quebec, Canada, August 10, 2006.

Styhre, A. (2009). The Production of Informational Objects in Innovation Work: Pharmaceutical Reason and
    the Individuation of Illnesses. *Ephemera: Theory & Politics in Organization* 9(1): 26–43.

TCP (2009). The Technical Cooperation Programme. Food and Agriculture Organization. http://www.fao.org/
    tc/tcp/ (accessed 15 June 2010).

UN (1974). Report of the World Food Conference, 5–16 November 1974. World Food Conference. Rome,
    United Nations.

UN (1975). Report of the World Food Conference 1974. New York: United Nations.

UN (1981). *Report of the World Food Council*: United Nations. A/RES/36/185.

UN (2010). World Summit on the Information Society: Regional Dimensions. New York: United Nations 4.

UN ESCWA (2009). Un Economic and Social Commission for Western Asia at a Glance. UN ESCWA. http://
    www.escwa.un.org/about/main.asp (accessed 10 June 2010).

UNEP (2010). United Nations Environment Program. UNEP. http://www.unep.org/Documents.Multilingual/
    Default.asp?DocumentID=43&ArticleID=3301&l=en (accessed 25 October 2010).

UNFPA (2004). State of World Population 2004: The Cairo Consensus at Ten: Population, Reproductive Health
    and the Global Effort to End Poverty. New York: UN Population Fund.

UNICEF (2005). United Nations Childrens Fund. United Nations International Children's Emergency Fund
    (UNICEF). http://www.unicef.org/about/who/index_history.html (accessed 30 January 2010).

UNU (2009). United Nations University: Towards Sustainable Solutions for Global Problems. United Nations
    University. http://unu.edu/about/ (accessed 30 July 2010).

UOR (2009). University of Reading: Agriculture, Policy and Development, History of Agriculture. University
    of Reading. http://www.ecifm.rdg.ac.uk/history.htm (accessed 4 June 2008).

Waterlow, C. and P. R. Payne (1975). The Protein Gap. *Nature* 258: 113–117.

Weingärtner, L. (2004). Food and Nutrition Security, Assessment Instruments and Intervention Strategies:
    Background Paper No. I. *Food Security Information for Decision Making*. Rome: EC-FAO Food Security
    Information for Action Programme.

Weisell, R. (2002). Measurement and Assessment of Food Deprivation and Undernutrition: Part IV—Summary
    of the Draft Findings of the Joint FAO/WHO/UNU Expert Consultation on Human Energy Requirements.
    International Scientific Symposium. Rome: FAO.

Weisell, R. C. (1995). Food, Nutrition and Agriculture: FAO Celebrates 50 Years—Expert Advice on Energy
    and Nutrient Requirements—An FAO Tradition. Rome: Food and Agriculture Organization.

WFC (1975). Report of the World Food Conference, Rome 5–16 November 1974. New York: United Nations.

WHO (1973a). Energy and Protein Requirments: Report of a Joint Fao/Who Ad Hoc Expert Committee—
    Technical Report Series No. 522. *WHO Technical Report Series No. 522 FAO Nutrition Meetings Reports
    Series No. 52*. Geneva: World Health Organization 24.

WHO (1973b). Trace Elements in Human Nutrition: Report of a WHO Expert Committee. *WHO Technical
    Report Series No. 532*. Geneva: World Health Organization.

WHO (1985). Energy and Protein Requirements: World Health Organization Technical Report Series 724.
    World Health Organization Technical Report Series 724. Geneva: World Health Organization.

Worldwatch (2008). About Worldwatch. Worldwatch Institute. http://www.worldwatch.org/About#about
    (accessed 23 June 2010).

WRI (2010). A Brief History of World Resources Institute. WRI. http://www.wri.org/about/wri-history
    (accessed 12 October 2010).

WTO (2010). Understanding the WTO: Basics—The GATT Years: From Havana to Marrakesh. World Trade
    Organization. http://www.wto.org/english/thewto_e/whatis_e/tif_e/fact4_e.htm (accessed 30 March 2010).

14

The Lost Development Decade: 1980s

14.1 Governance, Politics and Socioeconomics

The early debt crisis of the 1980s gave many, pause for thought; it was clearly having a detrimental effect on borrowing countries' domestic economies. The answer at the time was structural adjustment policies, that effectively were extra loans conditionally dependant on tight domestic fiscal and monetary changes. This followed a neoliberalist ideology that hoped to free the developing world from stagnation and promote development and growth. This decade also witnessed what popularly became known as food mountains: hordes of surplus food stored at great expense. There was also continuing talks about trade barrier reductions while new insights into the notion of food security were aired. Moreover, and finally at this time, the role of women in agricultural issues was beginning to be realised.

Early on in the decade, humanitarian issues seemed to be high on the agenda (Nash and Humphrey 1987). The first-generation humanitarian agencies of the previous decades invariably operated to mitigate emergencies on the 'frontline'. The second-generation agencies of the 1960s and 1970s, however, rose above the perceived shortcomings of this firefighting approach and aspired to 'humanitarian' politics in the hope of influencing political actors and outcomes more directly. This organisational shift effectively revolutionised how humanitarian agencies interacted with other stakeholders and governments alike (ACF 2010). This time also saw the UN agencies beginning to foster new relationships with a growing number of non-governmental organisations (NGOs). In the meantime, by the early 1980s, Prince El Hassan bin Talal of Jordan, who had already been, up to this point, very active with a long history of work in the humanitarian field, was becoming acutely aware of the vast array of humanitarian agencies, organisations and people working sometimes together and sometimes at odds with each other. Against this background, at the 1981 United Nations General Assembly (GA), Prince El Hassan bin Talal proposed the establishment of the New International Humanitarian Order. The idea was adopted, and the following year, the GA asked the former UN High Commissioner for Refugees, Saddrudin Aga Khan, to prepare a study on Human Rights and Mass Exoduses, which was submitted the same year. As a result, the UN proposed the

> … establishment, outside the UN framework, of an Independent Commission on International Humanitarian Issues composed of leading personalities in the humanitarian field or having wide experience of government or world affairs. (Majlis El Hassan 2009)

This was heeded and the Independent Commission on International Humanitarian Issues (ICIHI) was set up in 1983 co-chaired by Prince El Hassan bin Talal and Prince Sadruddin Aga Khan. The commission, during its mandate between 1983 and 1987, had a far reaching remit, producing a number of reports touching upon a range of humanitarian issues such as famine, deforestation, desertification, children, indigenous peoples, refugees and conflict that sought to progress research and aid in wider acceptance of such issues (Majlis El Hassan 2009). The commission was eventually succeeded in 1988 by the Independent Bureau for Humanitarian Issues (IBHI), which aimed to continue the commission's good work.

On the economic front, the burden of debt was at breaking point and not surprisingly the spiralling debt burdens of least developed countries (LDC) in the 1960/1970s led to the debt crisis of the 1980s.

14.1.1 Debt Crisis/Relief and Structural Adjustment Policies

See also Section 22.5.3. This grew out of a combination of factors. First cheap lending from increased OPEC oil revenues from the 1970s came to an abrupt end with the global recession in the early 1980s (Carrasco and Berg 1999). On top of this, the subsequent drop in borrowers' exports, together with increased interest rates and depleted foreign exchange reserves, created a situation of heavy debt burden in the LDCs (Eichengreen and Lindert 1989, Carrasco and Berg 1999; Slusser 2006). As a result in 1982, Mexico declared that it could no longer pay its foreign debt. Not long after this, similar announcements from other countries such as Brazil, Venezuela, Argentina and Chile sent shockwaves through the global financial sector. Rather than face a catastrophic failing of the sector, the International Monetary Fund (IMF) and the World Bank agreed to provide new loans whilst restructuring the existing ones (Sachs and Williamson 1986; Eichengreen and Lindert 1989; Carrasco and Berg 1999).

In return for this new arrangement, debtor countries agreed to stabilisation and neoliberal structural adjustment programmes or policies (SAP) designed to firstly correct domestic economic problems that precipitated the crisis while simultaneously aspiring to better match the borrowing country's resources with its economic activities and ensuring greater economic liberalisation (Munasinghe 1998; Mousseau 2005; Slusser 2006). In some cases, SAPs were simple. Recipient countries would turn their agriculture sector into cash crops for export to earn foreign exchange to help pay for food imports and to help pay off debts. For others, it meant structural economic reforms such as privatisation of state-owned entities and further deregulation of the economy. The strategy ultimately failed in part because the proposed financing did not materialise. But more importantly, where it did, this new lending simply added to debtor countries' already burgeoning debt whilst restraining domestic opportunities. The problems threatened to spiral and in light of this, many began to call for plans to provide debtor countries with debt relief rather than debt restructuring. Not surprisingly, many indebted countries ended up suspended debt payments and failed to comply with or adopt the IMF adjustment programmes. Subsequently, it was realised that a significant proportion of this debt would not in fact be repaid and eventually prompted some of the large creditor banks to write-off some of these as losses (Munasinghe 1998; Carrasco and Berg 1999; Slusser 2006).

Worse was to come, as Africa prepared for hunger and famine on a mass scale.

14.1.2 The 1984–1985 Famines in Africa

The 1984–1985 famine of Ethiopia (present-day Eritrea and Ethiopia) was brought on initially by drought and civil war. As the famine in the northern part of the country spread to the southern parts, major efforts by the international community to provide relief supplies were hampered and beset by problems of security and insurgency. The Ethiopian government was also complicit by withholding food shipments to rebel areas. As well as the resultant economic collapse and over one million famine-related deaths, the famine saw further millions became homeless and destitute. Such resurgent crisis showed the international community that large-scale food insecurities occurred even where there were sufficient food supplies. Problems of conflict, policy and transparency as well as security of aid had to be addressed (Watts 1991; Maxwell and Frankenberger 1992; Devereux 2000; Ó Gráda 2009).

By now the scourge of surpluses in the West returned highlighting the widening gap between the have's and have-not's.

14.1.3 Food Mountains

By the 1980s, the low stocks that had contributed to the global food crisis of the 1970s had recovered at the same time as global recession was lowering demand for grains (Mitchell et al. 1997). In Europe, this was fuelled in large part by the Treaty of Rome and the EU's Common Agricultural Policies (CAP) of increased productivity, which simultaneously boosted food production and provided a reasonable standard of living for farmers and farm workers. Such price support mechanisms intended to boost productivity, and farm incomes were also used to stabilise markets and assure availability of supply at reasonable prices (UOR 2009). However, increased production brought about by these CAP subsidies coupled with lower demand meant that food commodities began to accumulate at an alarming rate. This resulted in the first food mountains in the EU. This was quickly followed by widespread public indignation of the idea that millions of tonnes of food were piling high, unused and incurring storage costs, while millions were starving in the world. Not only this, the CAP itself was becoming extremely expensive to run and amid calls for CAP reform, measures were taken in an attempt to alleviate both problems. These included subsidising exports and the voluntary 1988 set-aside measures, which became compulsory in 1992. These measures ultimately helped to reduce food mountains to more manageable levels (Howarth 2000; Dobbs and Pretty 2001; Wakeman 2003).

The phenomenon was not limited to Europe and many developed countries too were suffering the same fate (FAO 1982). Even India, was at this time becoming prone to such mountainous overproductions (Chapman and Baker 2001; Ray 2004). In America, however, while production increased, it was partially offset by the different approach to protectionism than the United Kingdom. The first main difference was that US farmers were paid based on their production rather than their incomes and second and more importantly the United States used previous years' agricultural data to forecast production and surpluses and paid farmers to take land out of production (Subramanian 2003).

14.1.4 The Changing Concept of Food Security

14.1.4.1 Entitlement Theory

Up to this point, in the early 1980s, the prevailing emphasis on agricultural supply-side policies to address global hunger was not properly challenged until the notable Nobel Prize winning welfare economist Amartya Sen published his book, *Poverty and Famines* in 1981 (Sen 1981b). In the book, he challenged many prevailing assumptions of the day and suggested that even though a country may produce ample food for all, it was still feasible that many thousands might starve due to poverty and inequitable access (Sen 1981a, 1981b). Poverty as a cause of hunger itself was not a new idea and was indeed a target of UN anti-poverty programmes. Sen, however, was a product of his generation, for while many understood the direct links between poverty and hunger, contemporary solutions still concentrated on education and supply-side policies. What Sen effectively does with his analysis in the book is to not only once again bring to the centrestage a problem that had been elucidated in numerous preceding decades, but also insightfully highlight a twist on the poverty aspect concerning access. Sen's central tenet posits that a person's access or *in*access to the food they need was just as important as the availability of the food itself. He also charges societal behavioral responses and coping strategies in response to food stresses as primacy in acquiring food security (Maxwell and Frankenberger 1992; CFS 2007). Sen further suggests that potentially inadequate social systems of distribution, likely exacerbate the problems of the poor and the hungry and by questioning traditional assumptions of hunger and famine, as being solely an artefact of production or supply failure, he introduces his seminal theory of 'entitlement'.

In Sen's entitlement theory, he talks of the rights an individual has over the food they need, and suggests that such entitlements can be established by one of several ways:

Directly: Through on farm or local garden production by growing sufficient food for themselves;

Exchange: This refers to bartering food for trade or using wages to purchase their requirements; or

Transfer: The receipt of aid, gifts or inheritance whether food or finance.

Thus, by extension, says Sen, hunger and famine are not just about aggregate supply or poverty alone but in fact arise due, in large part, to a variety of socioeconomic variables that result in entitlement failure or a lack of 'access' to basic foods (FAO 1970, 2003b; Maxwell and Frankenberger 1992; Sen 1997; CFS 2007). This view was to drastically change the way food security was viewed. Building on this, while the 1974 World Food Conference's original concerns were with the volume and stability of food supplies, in 1983, the FAO expanded the ideas of Sen and others to incorporate the concept of access into their definition of food security:

> … ensuring that all people at all times have both physical and economic access to the basic food that they need. (FAO 1983)

Consequently access now became an important dimension in the fight against hunger. It was a way of determining people's ability to convert their various entitlements such as financial, land or social into food. This was further incorporated into a new report (1986) by the World Bank, who also amended their views to state that

> … access of all people at all times to enough food for an active, healthy life … (World Bank 1986)

Also a few years later in 1986, a World Bank report was to provide new innovative views on the dynamics of food insecurity: those of chronic* and transitory† dimensions.

14.1.4.2 Chronic and Temporal

This came about as it was determined that there was both long-term and short-term aspects to food security. It was postulated that if individuals or households were persistently unable to meet the food needs over a long period characterised by continuous, temporary ups and downs, then it was to be considered a long-term problem: chronic food insecurity. On the other hand, it was suggested that a short-term problem, whether inherently chronic or not, may afflict any person or household. This might be as a result of seasonal scarcities, illness or unemployment or crop failure, among others. This situation was determined to be temporary or transitory food insecurity (World Bank 1986).

In the meantime, as important as Sen's entitlement theory of access was, as the decade progressed, it was soon to be challenged, leading to a new integrated livelihoods perspective.

14.1.4.3 Livelihoods Approach to Food Security

Towards the end of the decade, a study by de Waal revealed behaviour that seemed to complicate Sen's clear-cut entitlement analysis. De Waal's 1989 analysis contradicted Sen's supposition of a passive individual who, faced with hunger, sold off his assets to feed himself. Sen's view, it seemed, completely ignored the presence of choice in an individual's actions and instead, De Waal's observed that people sometimes intentionally suffered hunger rather than lose their assets effectively reducing

> … consumption in order to preserve essential assets such as seed, farm tools and plough oxen …
> (Waal 1991, p. 33)

De Waal also proposed that people, especially those subjected to frequent crisis, took into account short and longer term goals when faced with hunger versus assets decisions. In conclusion then, De Waal noted that food was not always the first priority of people experiencing famine, but one objective out of many (Waal 1989; Devereux et al. 2004; Argeñal 2007). This paradigm shift became known as the livelihoods theory. One response to these revelations was the realisations that rather than wait for victims to migrate to famine camps as a last resort, the better international response would be to intervene earlier in an attempt to prevent loss of livelihood assets and livelihoods in the first place (Shoham and Lopriore

---

* Chronic food insecurity is constant and results from inadequate food intake over a longer period of time largely driven by endemic poverty whereby people are unable to buy it or to produce it for themselves (ibid). (World Bank 1986).

† Temporal can have two meaning; the first is of time while the second is concerned with the temporaroy. In this sense, temporal (or transitory) food insecurity results from a temporary decrease in food intake due to, among other things, price changes; production failures; a loss of income; war; flooding; drought; crop failure and pest infestations (World Bank 1986).

2007). It was noted too that by properly identifying and categorising vulnerable groups with common characteristics, a more proactive and pre-emptive approach to re-establishing food security in times of crisis could be achieved (Devereux et al. 2004).

Simultaneously, food security was once again being expanded and numerous definitions and concepts surrounding the issues were offered at an incredible rate.

14.1.4.4 Food Security: An Expanding Notion

Initially, the Committee of Food Security (CFS) formulated an early tripartite concept of food security encompassing availability, access and stability while the Organisation for Economic Cooperation and Development (OECD) adopted a slightly different approach to include availability, access and utilisation (FAO 2003b). This was similarly echoed and summed up by Maxwell and Frankenberger in 1992 in an exceptionally detailed study suggesting that the ideas forming around the notion of food security could be considered within four separate but related concepts. These were (1) availability or sufficient calories, (2) entitlement access, (3) security, instability or vulnerability of supplies to shocks and (4) the World Bank's time dimensions—chronic and transitory (Maxwell and Frankenberger 1992).

Lastly, there was some good news with the arrival of the US Food Security Act of 1985. The Act was seen as a forward-looking instrument designed to integrate policies and expand agricultural production through ecologically sound technologies, giving access to better quality diets for the rural and urban poor. The Act was well received with one contemporary observer enthusiastically suggesting that it was

> … a fine example of an integrated approach to production, conservation and consumption. It would be advisable for every country to adopt similar legislation which can provide the legal framework essential both for sustainable advances in biological productivity and for eliminating chronic hunger. (Swaminathan 1990)

The 1980s was also the decade that finally heralded the recognition of women's crucial role in agriculture as well as their vital function on important social issues.

14.1.4.5 Women in Agriculture

For as long as the UN has been operating, there have always been elements within the secretariat verbalising issues concerning women; these unfortunately had little significant impact on agricultural policy. This was about to change however as pro-active NGOs vied to raise the profile of women through pressure and persuasion. By this time, the previous decade had already introduced the idea that developing countries themselves were important architects of their own future and building on this, many new programmes had been started hailing a new direction of 'participation' programmes. Indeed, such programmes were an antidote to staid government-led solutions and embraced a participatory relationship with people and communities on the ground. Unfortunately, however, for a long time, these were seen merely as marginal exercises by national governments and UN agencies alike. The NGOs, however, as is part of their genetic make-up, kept pushing and in time it was shown that women indeed were not merely just hard and often primary agricultural workers, but they were also productive in other ways too. The important role of women, it seemed, was finally becoming fully recognised. By now, certain key publications were highlighting the role of women too. One of these was by Gail Omvedt, a scholar and activist. Gail's influential book *We Will Smash This Prison* is an account of women's movements in India and their long tradition of fighting oppression. Another was a study published by the UN Research Institute for Social Development (UNRISD) in which Gail further elaborated the role of women in India, particularly when it came to agricultural movements and greater land rights (Omvedt 1979, 1986; Devi 1995). Yet another study at this time that apparently captured the attention of officials and planners was Solon Barraclough's (Director of UNRISD) analysis of the Nicaraguan food system. Lastly though and perhaps more influential was a study by UNRISD itself in which it sought answers to questions of poverty and malnutrition. In its execution, the study was an exemplary analysis of the roles everyone played at each stage of the food chain, and while individually these studies were highly informative and very significant, collectively, these publications along with continued pressure from NGOs witnessed for the first time change in the nature of UN projects as they 'became increasingly gender sensitive (Wadlow 2007).

14.2 Health and Nutrition

In the arena of health, much debate was taking place regarding nutritional issues and, recommendations in particular. In the background however, while micronutrients were not ignored, in many respects, they appeared to be of secondary importance to protein and calorie considerations. This led to the emergence of a silent or hidden problem of micronutrient malnutrition. However, even before this at the beginning of the decade, as mentioned in the previous chapter, there was considerable concern following the FAO/WHO 1973 report over the level of recommended protein intake levels. Following this, the 1975 and 1978 ad hoc meetings of experts, armed with the results of recent studies, indeed found protein recommendations of the 1973 report to be on the low side. Out of this, the 1978 group concluded that a further full-scale expert consultation was needed; this happened in 1981 when the new estimate put recommendations back on track (Scrimshaw 1996). Importantly too at this meeting the idea of a reference man or woman was now considered too restrictive and did not adequately reflect the wide range of body sizes or individual patterns of physical activity of the general populace. At the same time, building on previous recommendations, there was now sufficient data to use basal metabolic rate (BMR) as the underpinning methodology in calculations in all but the under tens. In light of this, a factorial approach using age, BMR and physical activity levels (PAL) was introduced outlining a possible range of intakes dependant on age, size and activity. This new methodological research also accounted for the 4-year delay in publication (Durnin 1981; Schofield et al. 1985; FAO 2002). The 1981 meeting also paved the way for new methodological techniques in determining total energy expenditure, in particular the doubly labeled water technique (DLW).

Seven years later, there was sufficient movement in research that suggestions arose for another expert meeting of the FAO/WHO. However, the UN agencies were slow to act, and instead the International Dietary Energy Consultancy Group (IDECG), in which FAO was a participant, met to discuss progress. By now too, expert committees were being drafted to address requirements for vitamin A, iron, vitamin B_{12} and folate (FAO/WHO 1988). In America at this time too, the National Research Council's (NRC) Food and Nutrition Board (FNB) was continuing to provide guidelines on nutrient intakes. Continuing their customary 5- to 10-yearly review of nutritional information, work began in 1980 in the 10th edition of their flagship recommended dietary allowances (RDAs) and by 1985 the first draft had been prepared. Although after much deliberation and consultation, the 10th edition was not finally published until 1989, an edition incidentally which is still current as of 2011.

By now, it seemed the end of the decade was bringing nutritional science and policy together as never before. However, things were not all they seemed.

14.2.1 Hidden Hunger

Hidden or silent hunger are terms that have invariably been used to describe both situations where intake of food is not so sufficiently acute as to permit the detection of clinical symptoms or secondly to depict a lack of micronutrients in the diet, leading to nutritional deficiency symptoms (Kruif 1926; Breckenridge 1942; Macy and Williams 1945; McGovern 1969; Allen 2000; Scrimshaw 2003). As previously mentioned, the 1970s had already identified that iron, vitamin A and iodine deficiencies were more widespread problems than PEM; however, the problem did not garner enough attention and PCM still remained nutritional policies' prime target. By now, the term 'hidden hunger' fell firmly on the side of micronutrient disorders to represent the silent suffering and neglect of millions of people around the world. It was not until the mid-1980s however that micronutrients received the attention they needed from the international nutrition community, with growing interest in iodine and vitamin A. Iodised salt had been used for a long time to combat the diseases of cretinism and goitre although in many areas this was not kept up. As iodine deficiency disorders (IDD) began to return, renewed interest sharpened the focus on the problem. Meanwhile, further interest in micronutrient disorders at the time was likely propelled in large part by the work of Sommer and colleagues in Aceh, Indonesia. The huge study, covering some 25,000 children, concluded that a massive 34% reduction in mortality figures could be ascribed to the supplementation of vitamin A alone (Sommer et al. 1986; UNICEF 2004).

This was a wake-up call for the international community and ultimately led to governments to call for a World Summit for Children in 1990. The summit, held at the offices of the UN, asserted the importance of children in their role as future ambassadors of humankind. The conference also took place in a short time after the Convention on the Rights of the Child had been adopted (1989) in which world leaders enthusiastically and jointly avowed to

> … always put the best interests of children first—in good times and bad, whether in peace or in war, in prosperity or economic distress. (UNICEF 1990)

The major outcome of the conference was the World Declaration on the Survival, Protection and Development of Children, in which three of the 27 goals related to tackling the problem of iodine, vitamin A and iron deficiencies by the year 2000 (Allen 2000).

14.3 Science and Technology

14.3.1 Single Cell Protein's Partial Success

With regard to the promise of single cell proteins (SCP) of the previous decade, events in the 1990s took the wind out of the sails of SCP as a mainstream source of protein. Marked improvements in both crop production and plant breeding saw increased agricultural output beyond expectations. This, coupled with the lowering of trade barriers as a result of the successful General Agreement on Tariffs and Trade (GATT) negotiations signed in 1994, led to a boom in global market for goods and services, particularly in agriculture (the first time agricultural products had been included in the talks). With this de-regulation, the market price of plant origin protein decreased and industrial SCP processes were beginning to be discontinued. Despite this however, there has been one example of SCP success, Quorn™. During the late 1960s and 1970s, Rank Hovis McDougall's (RHM) research team was looking into using their waste products, starch, as a substrate to grow SCPs. After much research, they isolated the fungus mould, *Fusarium venenatum* as the best candidate for their objectives. After joining up with ICI, they had access to one of their decommissioned SCP vats and after further extensive testing, Quorn™ was given the all-clear signal in 1985 and launched as a meat-free mycoprotein to the public in 1994 (Ugaldea and Castrillob 2002).

As with SCPs, biotechnology was at the forefront of research, and it was not long before that the new biotech industry saw the introduction of the first genetically modified organisms for the improvement of plants.

14.3.2 Genetic Engineering

At this time, a major development in the field of biotechnology was made that expedited the process of modern genetic engineering (Table 14.1). This was the polymerase chain reaction (PCR) method of genetic amplification. Developed in 1983 by Kary Mullis, PCR bypassed the need to use bacteria for amplification and allowed Mullis to produce millions of copies of a DNA sequence in a few hours. This cell-free method of DNA cloning, built on the previous decades' advances and was invaluable in expediting rDNA technologies involving the insertion of DNA material into other organisms.

It was not long before the potential of genetically modified organisms was fully realised by the private sector and the first transgenic plants to benefit from new technology of genetic engineering (transgentics, transgenic, genetic modified organism [GMOs]) were the tobacco plants. It was found in 1983 that by transgenically introducing *Agrobacterium tumefaciens* vectors in tobacco plants strong plant cell growth was shown to result (An 1985; Vlasák and Ondřej 1992). On the basis of this, companies are now moving at tremendous pace placing much resources into researching potential commercial benefits. This was highlighted by Swaminathan who tells us that in America alone between 1987 and 1990 the US Department of Agriculture issued nearly 100 permits for field testing of genetically altered crops. Under

TABLE 14.1

Genetic Engineering Milestones of the 1980s

1980	The US Supreme Court ruled that genetically altered life forms can be patented. This allowed the Exxon oil company to patent an oil-eating microorganism.
1981	Scientists at Ohio University produced the first transgenic animals by transferring genes from animals into mice while the first gene-synthesising machines are developed.
1982	Humulin, Genentech's human insulin drug, is produced for the treatment of diabetes. It is the first biotech drug to be approved by the Food and Drug Administration.
1983	The polymerase chain reaction (PCR) is invented by Kary Mullis and others at Cetus Corporation in Berkeley, California. It uses heat and enzymes to make unlimited copies of genes and gene fragments and becomes a major tool in biotech research and product development worldwide.
1984	The DNA fingerprinting technique is developed by Alec Jeffreys to identify individuals, and the first genetically engineered vaccine is developed.
1986	The first field tests of genetically engineered plants (tobacco) are conducted. Also the first biotech-derived interferon drugs for the treatment of cancer are produced.
1987	Advanced Genetic Sciences' Frostban, the first authorised outdoor test of a genetically altered bacterium to inhibit frost formation on crop plants, is field tested on strawberry and potato plants. Calgene, Inc., receives a patent for the tomato polygalacturonase DNA sequence, which is used to extend the shelf life of the fruit.
1988	Geneticists Philip Leder and Timothy Stewart were awarded the first patent for a genetically altered animal, a mouse that was very susceptible to breast cancer.
1989	The Human Genome Project is a massive effort to map and sequence the human genetic code as well as the genomes of other species and is coordinated by the US Department of Energy and the National Institutes of Health.
	Scientists developed a recombinant vaccine to protect against the deadly rinderpest virus, which had destroyed millions of cattle in developing countries.

Sources: Reprinted from Bud, R., *The Uses of Life: A History of Biotechnology*, Cambridge University Press, Cambridge, 1993; Peters, P., *Biotechnology: A Guide to Genetic Engineering*, Wm. C. Brown Publishers, Dubuque, IA, 1993; Dobson, A., *Environ. Values*, 4, 227–239, 1995; *Biochemical Engineering and Biotechnology*, Najafpour, G.D., Copyright 2007, with permission from Elsevier; Nature, *Nat. Rev. Genet.*, 8(Supplement), 2007; Styhre, A., *Ephemera Theory Polit. Org.*, 9, 26–43, 2009; BioTech Institute, Timeline of Biotechnolgy 20th Century. Biotechnology Institute, 2010.

testing at that time were new or improved varieties of alfalfa, corn, cotton, cantaloupe melons, cucumbers, potatoes, rice, soybeans, squash, tobacco, tomatoes and walnuts (Swaminathan 1990).

14.4 Period Commentary

The 1980s witnessed remarkable movements in ideology and brought about through technological advances, nutritional thinking and social responses to crises. In this effort, the NGOs have often been the link or the driving force behind raising grassroots issues to levels of national awareness, and this was no less true of the food mountain situation that saw protectionist policies result in ridiculous surplus volumes of food at the same time as crises were unfolding in Africa. For some, like Professor David Harvey, the problems of food mountains were clear-cut; these, in his view, could be avoided by addressing fundamental issues of agricultural market policies that hindered or led to unfair terms of trade as well as artificially by undervaluing the developing world's agricultural market. In his view the solution was simple:

> The cure, then, would seem to be simple: eliminate the market support policies of the industrialised world, and stop under pricing food in the developing world. (Harvey 1988)

In other areas, productivity was noted by Johnson who, quoting from a study by US Department of Agriculture, showed that labour saving by the industrial nations by the 1980s had come leaps and bounds showing that by 1980 only 1%–2% of the labour used in 1800 was now needed to produce one ton of wheat or corn (Johnson 1997).

On a more sombre note, in 1980, in a disturbing twist of political aid, President Carter in direct response to the USSR's invasion of Afghanistan withheld phosphate and grain sales to the USSR. Also by 1985, the IMF began to implement structural adjustment policies to help combat a growing debt crisis. Meanwhile, the major developments of the decade with regard to food security were the fundamental paradigmatic shifts that saw two important theories emerge into an expanded notion of development. This emphasis on supply-side policies was now being challenged by a combination of factors including Sen's entitlement theory; a general lowering of living standards in the Third World as well as the famine crisis in Africa (Argeñal 2007). Sen's work was particularly influential at this time. In his book, *Poverty and Famines* he challenged many prevailing assumptions of the day and suggested that even though a country may produce ample food for all, it was still feasible that many thousands might still starve because of poverty and inequitable access to food (Sen 1981b). This was groundbreaking and although, as with much else in the food security arena, the entitlement idea had been around for a long time, it was not until Sen's work that the idea was articulated to a receptive audience. His work had a marked effect on prevailing theory and practice of the day and his entitlement theory shifted the emphasis away from aggregate supply towards access, both at individual and household levels. This was important because 'access', being entrenched in the social, political and economic spheres, effectively shifted the debate from being the sole preserve of agricultural production issues 'to societal causes of famine' (Blaikie et al. 1994).

At first, while some national governments still tended to focus on production and the need to become self-reliant, it became clear that Sen's work suggested that adequate supplies of food at the national or international level alone did not guarantee food security (Sen 1981a, 1981b). Food access became increasingly recognised as a key determinant of food security and in recognition of this, the FAO's Committee on World Food Security broadened its framework to adopt the notion of a tripartite concept: availability, access and stability (FAO 2003b; Argeñal 2007). Similarly, at the time, the OECD also suggested that food security had three dimensions although these varied slightly as availability, access and utilisation (FAO 2003b). As a result, policy makers began to explore individual and household food security rather than food security from the national perspective. At the same time, new emergent thinking began to favour a change in developmental economics leaning further towards the idea of global free trade (Warnock 1997).

A second major shift and one that came as quite a revelation was de Waal's notion that food was not always the first priority of people experiencing hunger (Waal 1989). Adding an unexpected twist on Sen's entitlement theory of a few years earlier, de Waal's book highlighted the plight of the African famines of 1984–1985. In this disaster people were found to deliberately go without food in order to protect their assets. It would seem that people consciously assessed their risks and options, and food was not always, it seemed, the top priority. Up till now, it had been assumed that the last suffering of people would be starvation, that is, they would exhaust all other assets to feed themselves until there was nothing left. De Waal's eloquent and brutal logic defied all accepted reason. This essentially widened the concept of food security to embrace not just access but the choices made by people at household and individual levels. This second paradigmatic shift came to be reflected in all levels of multilateral policy making and became known as the livelihoods-based approach (DeRose et al. 1998; FAO 2003a; Heidhues et al. 2004; FAO 2006; Argeñal 2007). Although de Waal's work might be seen to challenge that of Sen's, the truth is that Sen's seminal work effectively unlocked the door to a new paradigm while the livelihoods perspective on the other hand simply pushed the door open.

By now, the both the FAO and the World Bank broadened their emphasis on food security definitions to take account of these changes in thinking, that is, both now considered food security as needing to pay attention to both the supply and demand-side variables (Devereux et al. 2004; Shoham and Lopriore 2007). Also by 1986 the World Bank's influential report 'Poverty and Hunger' had also elucidated the distinction between chronic and transitory food insecurity (World Bank 1986; FAO 2003b). Thus, it can be seen that a variety of socioeconomic variables were now considered to be involved in people's access or livelihoods entitlements. This led to observations that people's behavioral responses and coping strategies in response to such stresses were an important aspect to food security itself. In response, international organisations, NGOs and the like began to incorporate socioeconomic indicators in their monitoring systems (Maxwell and Frankenberger 1992).

On the nutrition front, the 1985 FAO/WHO/UNU report on energy and protein requirements dropped the notion of a reference person and focused instead upon the consideration of recommended requirements based on body size and composition and for different levels of activities. In the same report, judgments were also made about desirable body mass indices (BMI). This was also a time that saw the 10th edition of the United States's recommended dietary allowances (RDAs) detailing recommended levels of macronutrients as well as for most vitamins and minerals too. On the institutional policy front, change was also on the horizon with the WHO and UNICEF reinventing ANPs and marketing them as Joint Nutrition Support Programmes (JNSPs) (Latham 1997). All the while however, a silent emergency was unfolding in the guise of hidden hunger that saw millions suffering from micronutrient deficiencies, many of which could be prevented quickly and cheaply (Allen 2000).

Lastly, in this decade, the formation of a new field of genetic engineering was being fostered by research into transgenic or genetically modified organisms (GMO). Numerous early attempts to affect plant characteristics via genetic modification took place in the 1960s and 1970s, however, at that time methods of molecular analysis were limited, and as such, experimental proof or verification of results was restricted to coarse DNA density testing and phenotype expressions. Modern contemporary integration verification techniques, most notably the PCR procedures, allowed for quicker and more productive experiments and ultimately successes. Thus, the first real proof of the worth of transgenic plants took place in this decade (Galun and Breiman 1997). However, this new field, with all its benefits and concerns, promised to bring biotechnology to the forefront of science in society, ensuring future collision of industry, public and government interests.

14.5 Key Dates, People, Acts, Reports and Surveys

The 1980s built on the challenges of previous decades. Much research and policy intervention was beginning to emerge that would shape food security issues in the coming years. Table 14.2 highlights some of these more notable events.

TABLE 14.2

Key Ideas: The Lost Development Decade

Date	People, Organisation, Agencies, Acts or Conventions
1980: Southern Africa Development Community (SADC)	Originally a loose alliance of nine states in southern Africa known as the Southern African Development Coordination Conference (SADCC) whose main aim was to lessen economic dependence on South Africa. Its role now is inter alia to achieve economic development and growth and alleviate poverty (SADC 2010).
1981: EuronAid	EuronAid—a European network of non-governmental organisations (NGOs) founded as a non-profit association in 1981 using huge food surpluses as an instrument of its development policy. Becoming in the process, an important channel for European food aid. EuronAid ceased operations in 2007 (EuronAid 2008).
1981: Amartya Sen, *Poverty and Famines*	Economist Amartya Sen's 1981 book *Poverty and Famines* set the stage for a shift away from the one-dimensional paradigm of production and supply to one encompassing 'entitlement' theory, the idea that poverty and access to food have as much to do with hunger and starvation as the amount of food available on the market. He won the Nobel prize in 1998 for his contributions to welfare economics (Nobel Prize 2010).
1981: World Food Day	World Food Day (WFD) was established by the FAO at the Organisation's Conference in November 1979. The date of the FAO's anniversary 16 October was chosen to be observed as World Food Day (Argeñal 2007).
1982: Mercy Corps	The Save the Refugees Fund was founded by Dan O'Neill in 1979 as a task force for Cambodian refugees. Teaming up with Ellsworth ('Ells') Culver in 1982, the two founded Mercy Corps with its focus shifting from just relief assistance to focusing on long-term solutions to poverty and hunger (MercyCorps 2009).
1984: European Lomé III Convention	Starting in 1975, the Lomé Convention sets out the principles and objectives in respect of aid, trade and politics of cooperation between the European Union and the ACP countries: African, Caribbean and Pacific Group of States. In 1984 the Lomé III Convention shifted its attention from industrial development to development and self-reliance based on self-sufficiency and food security (EC 2009).

TABLE 14.2 (Continued)

Key Ideas: The Lost Development Decade

Date	People, Organisation, Agencies, Acts or Conventions
1985: USAID Famine Early Warning System Network (FEWS NET)	USAID-funded FEWS Network (FEWS NET) is a collaboration of stakeholders providing timely vulnerability and early warning information on emerging food security issues (USAID 2010).
1985: Global Resource Information Database (GRID)	UNEP, Division of Early Warning and Assessment, Global Resource Information Database (DEWA/GRID), forms UNEP's global network of environmental information centres, which provides access to environmental information and data useful for stakeholders or decision makers (UN ISDR 2006; UNEP GRID 2008).
1986: AGROSTAT	AGROSTAT, FAO's comprehensive statistical database covering world agricultural information, becomes operational in 1986 changing its name in the mid-1990s to FAOSTAT (Wu 2001; FAO 2010).
1986–1993: GATT Uruguay Round 8th Round	The most ambitious round to date expanding into previously guarded areas such as services, capital, intellectual property and agriculture. Many considered the previous agricultural exemption to be an obstacle, and the Agreement on Agriculture of the Uruguay Round was the most substantial trade liberalisation agreement in agricultural products in the trade rounds' history (WTO 2010).
1988: The Monitoring Agricultural Resources Unit Mission (MARS)	Part of the Joint Research Centre (JRC) applying space technology in the acquisition of timely information on crop areas and yields (EC JRC 2010). Today, the project is made up of four actions: GeoCAP (ex-MARS PAC), AGRI4CAST 1992 (ex-MARS STAT), Food Security (foodsec 2001) and CID (Community Image Data portal).
1988: Intergovernmental Panel on Climate Change (IPCC)	The IPCC was established by UNEP and WMO in 1988 to assess existing knowledge of climate change science, possible impacts and possible responses. The preparation of the Assessment Reports on Climate Change is main remit of the IPCC (Agrawala 2004; UNEP 2010).

References

ACF (2010). Action against Hunger International: Who We Are. Action Contre La Faim (ACF). http://www.actionagainsthunger.org/ (accessed 23 March 2010).

Agrawala, S. (2004). Context and Early Origins of the Intergovernmental Panel on Climate Change. *Journal of Climatic Change* 39(4): 1998.

Allen, L. H. (2000). Ending Hidden Hunger: The History of Micronutrient Deficiency Control. *Background analysis for the World Bank-UNICEF Nutrition Assessment Project*. Washington, DC: World Bank, pp. 111–130.

An, G. (1985). High Efficiency Transformation of Cultured Tobacco Cells. *The Journal of Plant Physiology* 79(2): 568–570.

Argeñal (2007). History of Food Security in International Development: Background Paper Developed for Christian Children's Fund Honduras. Honduras: Christian Children's Fund.

BioTech Institute (2010). Timeline of Biotechnolgy 20th Century. Biotechnology Institute. http://www.biotechinstitute.org/what_is/timeline.html (accessed 12 January 2011).

Blaikie, P., T. Cannon, I. Davis and B. Wisner (1994). *At Risk: Natural Hazards, People's Vulnerability and Disasters*. London: Routledge.

Breckenridge, M. (1942). The Preschool Child's Nutrition. *The American Journal of Nursing* 42(5): 533–537.

Bud, R. (1993). *The Uses of Life: A History of Biotechnology*. Cambridge: Cambridge University Press.

Carrasco, E. R. and K. J. Berg (1999). *The E-Book on International Finance and Development: Part One— V. The 1980's: The Debt Crisis and the Lost Decade*. The University of Iowa Center for International Finance and Development (UICIFD).

CFS (2007). Assessment of the World Food Security Situation. Committee On World Food Security: Thirty-third Session. Rome: Food and Agriculture Organization.

Chapman, G. and K. M. Baker (2001). *The Changing Geography of Asia*. London: Routledge.

DeRose, L., E. Messer and S. Millman (1998). Case Study: The Importance of Non-Market Entitlements. In *Who's Hungry? And How Do We Know? Food Shortage, Poverty, and Deprivation.* Derose, L., E. Messer, and S. Millman, eds., Tokyo, New York and Paris: United Nations University Press.

Devereux, S. (2000). *Famine in the Twentieth Century.* Brighton: Institute for Development Studies.

Devereux, S., B. Baulch, K. Hussein, J. Shoham, H. Sida and D. Wilcock (2004). Improving the Analysis of Food Insecurity: Food Insecurity Measurement, Livelihoods Approaches and Policy: Applications in FIVIMS. Rome: FAO-Food Insecurity and Vulnerability Information and Mapping Systems (FIVIMS) 52.

Devi, S. M. (1995). *Women's Movements in Kerala: Challenges and Prospects.* Kerala, India: Mahatma Gandhi Universtity. Doctoral thesis.

Dobbs, T. and J. Pretty (2001). Economics Research Report: Future Directions for Joint Agricultural-Environmental Policies: Implications of the United Kingdom Experience for Europe and the United States. Madison, SD: South Dakota State University.

Dobson, A. (1995). Biocentrism and Genetic Engineering. *Environmental Values* 4(3): 227–239.

Durnin, J. V. G. A. (1981). Expert Consultation on Energy and Protein Requirements: Basal Metabolic Rate in Man. *Working paper submitted to the Joint FAO/WHO/UNU.* Rome: Food and Agriculture Organization.

EC (2009). The Lomé Convention. European Commission. http://ec.europa.eu/development/geographical/cotonou/lomegen/lomeitoiv_en.cfm (accessed 6 June 2010).

EC JRC (2010). European Commission Joint Research Centre—MARS: About Us. European Commission. http://mars.jrc.it/mars/About-us (accessed 12 March 2010).

Eichengreen, B. and P. H. Lindert (1989). *The International Debt Crisis in Historical Perspective.* Cambridge, MA: MIT Press.

EuronAid (2008). Euronaid: Code of Conduct. EEC. http://www.euronaid.net/index.html?id=020200000005 (accessed 12 January 2010).

FAO (1970). Report of the Conference of FAO: Annex D—Commemorative Address by Professor M. Cépède, Independent Chairman of the Fao Council. Rome: Food and Agriculture Organization.

FAO (1982). Report of the Council of FAO: Eighty-Second Session. Rome: Food and Agriculture Organization.

FAO (1983). World Food Security: A Reappraisal of the Concepts and Approaches. *Director General's Report.* Rome: FAO.

FAO (2002). Food, Nutrition and Agriculture. Rome: FAO Food and Nutrition Division.

FAO (2003a). Trade Reforms and Food Security: Conceptualising the Linkages: Food Security: Concepts and Measurement. Rome: Food and Agricultural Organization of the UN.

FAO (2003b). Trade Reforms and Food Security: Conceptualizing the Linkages. Rome: Food and Agriculture of the United Nations.

FAO (2006). Policy Brief: Food Security. Rome: Food and Agriculture Organization 4.

FAO (2010). The Food and Agriculture Organization of the United Nations. UN Food and Agriculture Organization. http://www.fao.org/about/en/ (accessed 26 January 2010).

FAO/WHO (1988). Requirements of Vitamin A, Iron, Folate and Vitamin B12. *Report of a Joint FAO/WHO Expert Consultation. Food Nutrition Service No 23.* Rome: FAO.

Galun, E. and A. Breiman (1997). *Transgenic Plants.* Covent Garden, London: Imperial College Press.

Harvey, D. R. (1988). *Food Mountains and Famines: The Economics of Agricultural Policies.* Newcastle upon Tyne: University of Newcastle upon Tyne.

Heidhues, F., A. Atsain, H. Nyangito, M. Padilla, G. Ghersi and J. Le Vallée (2004). Development Strategies and Food and Nutrition Security in Africa: An Assessment. 2020 Discussion Paper No. 38. Washington, DC: International Food Policy Research Institute.

Howarth, R. (2000). The Cap: History and Attempts at Reform. *Economic Affairs* 20(2): 4–10.

Johnson, D. G. (1997). Agriculture and the Wealth of Nations. *American Economic Review* 87(2): 1–12.

Kruif, P. d. (1926). *Microbe Hunters.* New York: Harcourt, Brace, Jovanovich.

Latham, M. C. (1997). Human Nutrition in the Developing World. *Food and Nutrition Series No. 29.* Rome: Food and Agriculture Organization.

Macy, I. G. and H. H. Williams (1945). *Hidden Hunger.* Lancaster: Jaques Cattell Press.

Majlis El Hassan (2009). Organizations: International: Independent Bureau for Humanitarian Issues. His Royal Highness Prince El Hassan bin Talal. http://www.elhassan.org/Public/English.aspx?Site_ID=1&Page_ID=788&RM=125 (accessed 12 March 2011).

Maxwell, S. and T. Frankenberger (1992). *Household Food Security: Concepts, Indicators and Measurements: A Technical Review*. New York and Rome: UNICEF and IFAD.

McGovern, S. (1969). Are We Well Fed?—The Search for the Answer. *Nutrition Today* 4(1): 10–11.

MercyCorps (2009). Who We Are: Mercy Corps Timeline: Our History. Mercy Corps. http://www.mercycorps.org/10638 (accessed 25 March 2010).

Mitchell, D. O., M. D. Ingco and R. C. Duncan (1997). *The World Food Outlook*. Cambridge: Cambridge University Press.

Mousseau, F. (2005). *Food Aid or Food Sovereignty? Ending World Hunger in Our Time*. Oakland, CA: Oakland Institute.

Munasinghe, M. (1998). Special Topic I: Structural Adjustment Policies and the Environment: Introduction. *Environment and Development Economics* 4: 9–18.

Najafpour, G. D. (2007). *Biochemical Engineering and Biotechnology*. The Netherlands/Oxford: Elsevier.

Nash, A. E. and J. P. Humphrey (1987). Proceedings of a Conference Held in Montreal: Causes of Refugee Problems and the International Response. Human Rights and the Protection of Refugees under International Law, Montreal, Canadian Human Rights Foundation and the Institute for Research on Public Policy.

Nature (2007). Milestones in DNA Tecnologies. *Nature Reviews Genetics*. 8(Supplement).

Nobel Prize (2010). Amartya Sen: The Sveriges Riksbank Prize in Economic Sciences in Memory of Alfred Nobel 1998. Nobel Foundation. http://nobelprize.org/nobel_prizes/economics/laureates/1998/ (accessed 15 June 2011).

Ó Gráda, C. (2009). *Famine: A Short History*. Princeton, NJ: University Presses of California, Columbia and Princeton.

Omvedt, G. (1979). *We Will Smash This Prison! Indian Women in Struggle*. London: Zed Press.

Omvedt, G. (1986). Women in Popular Movements: India and Thailand during the Decade of Women. Geneva: UN Research Institute for Social Development (UNRISD).

Peters, P. (1993). *Biotechnology: A Guide to Genetic Engineering*. Dubuque, IA: Wm. C. Brown Publishers, Inc.

Ray, S. (2004). Roles of Agriculture Project: Socio-Economic Analysis and Policy Implications of the Roles of Agriculture National Synthesis Report India. *Roles of Agriculture Project (ROA)*. Rome: FAO.

Sachs, J. and J. Williamson (1986). Managing the LDC Debt Crisis. The Brookings Papers on Economic Activity. Washington, DC: The Brookings Institution.

SADC (2010). Southern Africa Development Community. Southern Africa Development Community. http://www.sadc.int/index/browse/page/52 (accessed 23 June 2010).

Schofield, W. N., C. Schofield and W. P. T. James (1985). Basal Metabolic Rate—Review and Prediction, Together with an Annotated Bibliography of Source Material. *Human Nutrition Clinical Nutrition* 39C Suppl 1: 5–96.

Scrimshaw, N. (2003). Historical Concepts of Interactions, Synergism and Antagonism between Nutrition and Infection. *Journal of Nutrition* 133(1): 316S–321S.

Scrimshaw, N. S. (1996). Human Protein Requirements: A Brief Update. *Food and Nutrition Bulletin* 17(3).

Sen, A. (1981a). Ingredients of Famine Analysis: Availability and Entitlements. *Quarterly Journal of Economics* Vol. XCVI(3): 433–464.

Sen, A. (1981b). *Poverty and Famines: An Essay on Entitlement and Deprivation*. Oxford: Clarendon Press.

Sen, A. (1997). Lectures Given at Wfp/Unu Seminar—Rome, 31 May 1997: Entitlement Perspective of Hunger. Ending the Inheritance of Hunger. Rome: WFP/UNU.

Shoham, J. and C. Lopriore (2007). *Livelihoods Assessment and Analysis: Introduction to Livelihoods*. Rome: EC FAO.

Slusser, S. (2006). *The World Bank, Structural Adjustment Programs and Developing Countries: A Review Using Resource Dependency Theory*. Montreal: Paper presented at the annual meeting of the American Sociological Association, Montreal Convention Center, Montreal, Quebec, Canada, August 10, 2006.

Sommer, A., E. Djunaedi, A. A. Loeden, I. Tarwotjo, K. P. West, R. Tilden and L. Mele (1986). Impact of Vitamin A Supplementation on Childhood Mortality. The Aceh Study Group. *The Lancet* 327(8491): 1169–1173.

Styhre, A. (2009). The Production of Informational Objects in Innovation Work: Pharmaceutical Reason and the Individuation of Illnesses. *Ephemera: Theory & Politics in Organization* 9(1): 26–43.

Subramanian, S. (2003). An Analysis of the Relationship between the Agricultural Policies of the European Union and the United States, and Their Implications on the World Food Economy. *Journals of the Stanford Cannon Engr 297b(Controlling World Trade)*: 32.

Swaminathan, M. S. (1990). Sir John Crawford Memorial Lecture: Changing Nature of the Food Security Challenge: Implications for Agricultural Research and Policy. Washington: Consultative Group on International Agricultural Research.

Ugaldea, U. O. and J. I. Castrillob (2002). Single Cell Proteins from Fungi and Yeasts. *Applied Mycology and Biotechnology* 2: 123–149.

UN ISDR (2006). ISDR Secretariat Biennial Work Plan 2006–2007. Geneva, Switzerland: United Nations. International Strategy for Disaster Reduction 20.

UNEP (2010). The Intergovernmental Panel on Climate Change (IPCC). United Nations Environment Programme. http://www.unep.org/documents.multilingual/default.asp?DocumentID=43&ArticleID=20 6&l=en (accessed 23 June 2010).

UNEP GRID (2008). The UNEP, Division of Early Warning and Assessment, Global Resource Information Database: About Us. UNEP (accessed 27 April 2010).

UNICEF (1990). World Declaration on the Survival, Protection and Development of Children. Geneva: UNICEF.

UNICEF (2004). Eliminating Iodine Deficiency Disorders through Universal Salt Iodization in Central Eastern Europe, Commonwealth Independent States and the Baltic States (CEE/CIS/BS). Geneva, Switzerland: UNICEF 18.

UOR (2009). University of Reading: Agriculture, Policy and Development, History of Agriculture. University of Reading. http://www.ecifm.rdg.ac.uk/history.htm (accessed 4 June 2008).

USAID (2010). What Is the Famine Early Warning Systems Network (FEWS Net)? USAID. http://www.fews. net/ml/en/info/Pages/default.aspx?l=en (accessed 30 June 2010).

Vlasák, J. and M. Ondřej (1992). Construction and Use of Agrobacterium Tumefaciens Binary Vectors with A. Tumefaciens C58 T-DNA Genes. *Folia Microbiologica* 37(3): 227–230.

Waal, A. D. (1989). *Famine That Kills: Darfur, Sudan, 1984–1985*. Oxford: Clarendon Press.

Waal, A. D. (1991). *Evil Days: Thirty Years of War and Famine in Ethiopia*. New York: Human Rights Watch.

Wadlow, R. (2007) World Food Day: No Day Off for Women. *Toward Freedom* (October): http://www.toward-freedom.com/women/1139-world-food-day-no-day-off-for-women.

Wakeman, R. (2003). *Themes in Modern European History since 1945*. London: Routledge.

Warnock, K. (1997). Third World: Feast or Famine? Food Security in the New Millennium (Commentary). *Journal of Race and Class* 38(3): 63–72.

Watts, M. (1991). Entitlements or Empowerment? Famine and Starvation in Africa. *Review of African Political Economy* 51(The Struggle for Resources in Africa): 9–26.

World Bank (1986). Poverty and Hunger: Issues and Options for Food Security in Developing Countries. Washington, DC: World Bank Group.

WTO (2010). Understanding the WTO: Basics—The GATT Years: From Havana to Marrakesh. World Trade Organization. http://www.wto.org/english/thewto_e/whatis_e/tif_e/fact4_e.htm (accessed 30 March 2010).

Wu, J. M. (2001). Pangaea Central as the Coming Global Access to Legal, Scientific and Technical Information through the Resource Networks of Intergovernmental Organisations. *International Journal of Special Libraries* 35(2): 94–112.

15

The Era of the Conference: 1990s

The decade of the 1990s was littered with conferences and their repeated call to arms, new initiatives and declarations. It was a period that witnessed trade negotiations finally take account of agricultural products while the spectre of overpopulation once again raised its head. It was a time too of growing awareness of the fragility of equilibrium with regard to Earth and her natural resources. Also international focus was brought to bear on the human development perspective of development and growth and all the while people attempted to pin down the notion of food security only to see it being redefined time and time again. Consensus, it would seem, would be a long way off.

15.1 Governance, Politics and Socioeconomics

The Uruguay round of General Agreement on Trade and Tariffs (GATT) negotiations came to an end in 1994 fully integrating agricultural trade issues, essentially opening up the global agricultural market. At the same time, GATT itself was being updated; in its place the 75 existing members along with the European Commission (EC) became the founding members of the World Trade Organization (WTO), which was mandated to further supervise the liberalisation of international trade (Warnock 1997). These negotiations came at a time when the costs of price supports and protectionist policies were soaring. The common agricultural policy (CAP), for instance, in the early 1990s saw the storage costs of surpluses 'food mountains' cost around 20% of the CAP's annual budget (which incidentally topped €30 billion at the time), while a further 28% was spent on export subsidies. However, amid harsh criticisms of the inefficient use of resources, reforms were in the offing. It began in 1992 in recognition of growing surpluses and continued into the decade and by 1997 the EC presented a roadmap of future policy direction to include wider reforms of the CAP. The proposals' roots lay in the growing liberalisation of world trade as well as the challenges of the European Union's (EU) eastward enlargement. The roadmap took the form of an action plan, Agenda 2000, which was agreed in 1999 at the Berlin European Council. This agreement, in view of EU enlargement, sought to put in motion a plan of action between 2000 and 2006 that would provide a new financial framework to strengthen policies and to continue the agricultural reform, that had begun in 1988 and 1992. The CAP reforms also aimed to reduce supports and ultimately place EU agricultural products on a more competitive footing in the world markets. At the same time, the reforms also aimed to simplify and decentralise the application of legislation and take greater account of environmental considerations (EU 2010).

In America too, agricultural reform was on the agenda with the passing of the Federal Agricultural Improvement and Reform Act (FAIR) in 1996. FAIR essentially replaced many of the existing farm price-support programmes with a system of payments designed to encourage farmers to rely less on the government for support, but rather the global market (UOR 2009).

15.1.1 Growing Social Responsibilities

Separately, all the effort that was focused on world trade and economic growth and development, it seemed, was losing sight of the notion that it was people who in fact were 'the real wealth of a nation' (Cleveland and Douglas 2008).

15.1.1.1 Human Development

In answer to this growing realisation, the United Nations Development Programme (UNDP) introduced the Human Development Report (HDR). The rationale was simple, while the economic expansion of an economy was seen as necessary in the social and economic growth of a nation; this expansion did not necessarily translate into the social and economic growth of the individual. That is to say people were not necessarily better off in terms of health, education, freedom or meaningful work and leisure time. Thus, the notion of an HDR effectively put people back at the centre of the equation. Human development at its simplest was thought to involve the process of enlarging people's choice. In achieving measurable standards, the first report focused on three elements: longevity, knowledge and living standards. These were translated into variables corresponding to each: longevity—life expectancy at birth; knowledge—literacy figures; and command over resources (needed for a good standard of living)—per capita income; these were then statistically combined to produce an index ranging from 0 to 1 to form the human development index (HPI). The first report in 1990 became fascinating reading as for the first time the notion of national growth was no longer restricted to measures of gross national product (GDP). Indeed, the first HDR report showed that the various countries of the world, compared to GDP, were in fact much less unequal than had previously been thought. Another startling revelation too was the fact that many countries were seen to achieve great gains in some development indices, while only earning moderate national income and yet others seemingly found it difficult to translate good economic growth into social developments of education and improved life expectancy (UNDP 1990). In short, it seemed money was not everything and the implications of this go way beyond a book of this kind, although suffice to say a radical rethink on social values vis-à-vis the Holy Grail of development; the economic paradigm, might warrant some revision.

Over time however, the report introduced four composite indices altogether: the human development index; the gender-related development index; the gender empowerment measure; and the human poverty index. The reports collectively sought to promote the idea of a wider notion of human development and reflected the growing human rights perspective of food security, poverty, health and nutrition (Clay 2002). In inextricably linking human rights and food security, the report, according to the UN Development Programme (1994), paved the way for a potential future World Social Charter where 'the right to food is as sacrosanct as the right to vote' (UNDP 1994).

Meanwhile as development issues continued, the notion of sustainability once again came to the fore. Concerns of global climate change, deforestation and the general regard, or dis-regard, for the sustainability of the natural resource base pooled people's attention.

15.1.1.2 *Earth Summit: Environmentalism*

The result was the 1992 UN Conference on Environment and Development (UNCED) also known as the Earth Summit (ES), and it was the first in a series of global conferences held during this period that concentrated on development through sustainability (Earth Summit 2002). The conference sought ways to halt the deterioration of Earth's natural resource base and to lessen humanity's impact on the environment by promoting, among other things, a sustainable development programme (Agenda 21) to be overseen by the Commission on Sustainable Development (CSD) (UN 1993a, 1997b; Earth Summit 2002; Jackson 2007). Agenda 21 was a document that eschewed collective global responsibility through acting locally to effect worldwide improvements. In the area of food security, attention was drawn to sustainable agricultural development and the potentially worsening degeneration of the natural resource base that included land, marine and biodiversity. The conference itself was a monumental effort to address the concerns of a growing environmental movement and with its Agenda 21 document, it effectively, morally and officially bound food security to future policies of sustainability and environmental protectionism (UN 1992). The major outcomes of Agenda 21 were as follows:

15.1.1.2.1 *Social and Economic Dimensions*

- To ensure cooperation in accelerating sustainable development in the developing countries whilst attempting to combat poverty and effect change in consumption patterns. Also to protect and promote good health through sustainable human settlement development.

15.1.1.2.2 Conservation and Management of Resources for Development

- To protect the atmosphere and the natural resource base through an integrated approach to land resource management; sustainable agriculture and rural development; the management of toxic and hazardous wastes. At the same time, it was proposed to combat deforestation, drought and desertification; manage ecosystems and conserve biological diversity whilst also promoting the protection and management of freshwater resources.

15.1.1.2.3 Strengthening the Role of Major Groups

- This involved strengthening the role of indigenous people and their communities and the inclusion of women and children in sustainable and equitable development. Further, it was recognised that the need to strengthen the role of farmers, NGOs and business as well as industry in roles of sustainable development with the aid of good practice and research.

15.1.1.3 Means of Implementation

This was all to be achieved through the promotion of public awareness and with international institutional legal arrangements. Moreover, the promotion of environmentally sound technologies was seen in the light of creating better international cooperation and capacity-building in developing countries (UN 1992).

Another important outcome of the Rio Earth Summit was the growth in the NGO powerbase. NGOs vary in numbers, budgets, accountability and mandate, yet collectively they created enough pressure at the summit to influence decision making and ensure governments adopted agreements controlling sustainable development issues, notable of which were the greenhouse gas emissions agreements.

The conference also impacted on the recurring theme of Earth's overpopulation and carrying capacity.

15.1.2 Optimum Capacity Resurfaces

From the 1950s onwards, much interest has been generated within and across various disciplines regarding the maximum number of people that Earth could support. This resulted in a wide variety of estimates of upper limits to sustainable numbers over this time (OPT 2009). However, with so many limiting factors and differing estimates, the need for some form of consensus was recognised by the 1990s, and it was Joel E. Cohen, head of Rockefeller University's Laboratory on Populations, who undertook the first important collective study of estimates. His work explored 65 previous studies from 1679 and 1994 and found figures ranging from one billion to over 1000 billion (Cohen 1995; ER 1996). In his findings, he explained such wildly differing estimates were based on the complex nature of the debate and the many differing assumptions about how people chose to live. Cohen also suggested that any consideration of numbers would need to consider the question from several angles, that of population dynamics, economics, the natural environment and culture (including politics). He also suggested that any number would need to factor in many questions including at what average level of well-being, for how long, what type of diet, health infrastructure and with what technology? Ultimately Cohen offered that the numbers Earth might sustainably support will be determined

> ... not only by natural constraints but also by human choices. The choices we and our children have made and will make about everything from food and the environment to liberty, styles of life, and other dearly held values will in turn influence which natural constraints will matter. (Cohen 1995)

Importance was further attached to the notion of sustainable populations in the earlier Earth summit in Rio de Janeiro (UN 1992). In chapter five of the comprehensive plan of action, 'Agenda 21', the UN calls for the relationships between demographic trends and environmental change to be analysed and further calls for

> An assessment should also be made of national population carrying capacity in the context of satisfaction of human needs and sustainable development, and special attention should be given to critical resources, such as water and land, and environmental factors, such as ecosystem health and biodiversity. (UN 1992)

In the meantime, food security was gathering momentum and in 1996, the FAO hosted the World Food Summit to discuss global hunger.

15.1.3 The World Food Summit

At the time of the conference, the world food security situation was characterised by a tight 'demand-supply balance in world cereals markets' in which reduced stock holdings, higher prices and diminishing food aid flows persisted (FAO 1996b). It was also recognised as a result that the hardships endured by the developing countries warranted a further conference that would once again bring together heads of states and nations in an effort to alleviate the continued suffering. As well as renewing the commitment at the 'highest political level', its objective was to raise awareness within the media as well as the public at large but especially among key public and private sector decision makers (ibid). The summit built on much empirical evidence over the previous two decades and together with research by Sen and others essentially established the complexity of the food security concept as it is known today. The notion that food security could be dealt with in isolation had long since been discounted yet the full multi-dimensionality of the concept was only just now being fully realised. In short, the summit formally established that above all, poverty eradication was fundamentally essential to improving food security. The summit also fully emphasised that a rights-based approach to food security was needed, one that included engaging not just in resource or economic issues, but also rights violations (Food Summit 1996; UN 1997a; Argeñal 2007).

In conclusion, the summit adopted both the Rome Declaration and the Plan of Action, two agreements that outlined the summit's commitments and objectives as well as an actionable plan to achieve these goals. One of the summit's popular stated aims was to halve the number of undernourished people by 2015 from their 3-year-rolling-average number from 1990 to 1992. The (estimated) number of undernourished people in developing countries in 1990–1992 was 824 million (FAO 2003b).

15.1.3.1 The Rome Declaration

Convinced that the multi-dimensionality of food security required concerted international efforts, the summit members agreed to the stated commitments of the seven-point 1996 Rome Declaration which sought the following:

- To ensure a conducive political, social and economic environment best designed to aid in the eradication of poverty and for the promotion of durable peace
- To implement policies to reduce poverty and to improve physical and economic access to sufficient, nutritional and safe food and its effective use
- To combat pests, drought and desertification and pursue sustainable agriculture, food, fisheries, forestry and rural development policies essential for reliable food supplies at individual, household, national, regional and global levels
- To ensure that overall trade policies including food and agricultural trade foster food security by implementing fair market-oriented world trading systems
- To be fully prepared for natural and human-made emergencies and to ensure emergency food requirements are used in ways that encourage recovery and rehabilitation through future oriented development mechanisms
- To promote the best use of public and private investment in the search for solutions to sustainable food and agriculture systems as well as rural development
- To follow up and monitor the Plan of Action in cooperation with the international community (FAO 1996a; UN 1997a)

15.1.3.2 Plan of Action

In achieving the declaration's goals, the summit drew up a plan of action in which both national governments and the international community needed to recognise and address the multi-dimensional causes of food insecurity. Furthermore, the international community was asked to review existing and adopt new, more appropriate national policies, programmes and strategies in achieving the stated goals. This included legal mechanisms to advance land reform and to protect property, water and other rights. The plan also included provisions that required nations to increase investment in agriculture and rural development and support sustainable food security and rural development at the same time as integrating population concerns into strategies of development. Emergency response systems at regional, national and international levels were also addressed as was the need to better monitor progress of food security. Implicit in this aim was the requisite improvement in the analysis and dissemination of information at all levels. Fundamentally, the conference also recognised the importance of the new liberal trade-reducing negotiations of the GATT Uruguay Round and recommended countries adopt the spirit of the legislation to generate opportunities for economic growth and trade expansion of the developing countries. Women were also mentioned as vulnerable groups and the rural sectors' economic and social development. In the field of environmentalism, the plan was also forthright in recognising that natural resources used for food, agriculture, fisheries and forestry were threatened by desertification, deforestation, overfishing and inefficient use of water as well as loss of biodiversity and climate change. To address this problem, the plan recommended that countries increase food production to meet the needs of population growth and the increased demand from rising standards of living whilst ensuring sustainability of natural resources and environmental protection. Lastly but perhaps most importantly the declaration asked nations to ensure a peaceful and stable environment as a fundamental pre-requisite for sustainable food security (Food Summit 1996).

Meanwhile the Rio Earth Summit provided fuel and momentum to the growing band of environmental accountants.

15.1.4 Environmental Accounting

15.1.4.1 Footprints and Ecological Accounting

The emergence of environmental accounting became fairly mainstream in the 1990s as more and more people, governments and organisations wanted to quantitatively take stock of natural resources as well as any social, physical or economic impacts on the environment. It began in the 1970s and underpinning these concerns was the simple epiphany that all renewable resources came from Earth and that as yet there was no tally or accountability as to what was being taken, replenished or destroyed. As a result, articles, books and academic papers on ecological accounting began to appear with regularity (Krebs 1972; Hannon 1973, 1985; Stearns and Montag 1974). Many people typified this novel approach through the idea of quantifying the inputs and outputs of ecosystems. From this point, the concept of ecological accounting took several routes. One of the more familiar derived concepts was the ecological footprint made popular by Mathis Wackernagel and William Rees in 1990 and properly elucidated in their book in 1996 (Rees 1992; Wackernagel and Rees 1996). The ecological footprint is defined as

> A measure of how much biologically productive land and water an individual, population or activity requires to produce all the resources it consumes and to absorb the waste it generates using prevailing technology and resource management practices ... (GFN 2010)

Taking the footprint of a general person as an example, their use of resources is measured in terms of global hectares per person per year, and to calculate this, the demands placed on Earth are divided by the biological capacity of Earth to support that demand. It was a good start and the concept's growing popularity helped found the Global Footprint Network in 2003. Other similar measures springboarding from such concepts were also the water and carbon footprints, which collectively combined to form a powerful arsenal in the field of ecological accounting.

Separately, there was a growing need to incorporate such methods into the macro-economic System of National Accounts (SNA). Even as far back as 1976, proposals for a system of national environmental accounting had been made (Peskin 1976). However, movement on this front did not occur until the early 1990s.

15.1.4.2 Greening the National Accounts

Also referred to as green, ecological or resource accounting, environmental accounting is seen by some as an aid to national environmental policy making and crucially not, as in the previous environmental accounting methods, simply to collate data about the environment. Nor is this system an attempt to value environmental goods or services or analyse the social cost benefit of projects affecting the environment (Hecht 1997). Instead, this system works at the national or regional perspective and is essentially a modified SNA, which aspires to place the environment on equal terms with a country's financial and economic accounts. In doing so, it aims to assess the sustainability of a nation's development vis-à-vis all sectors of the economy and the environment. However, while the value of such measures were clearly understood and accepted at the disaggregated level, there was little or no global consensus on the cost or even of their use in the national accounts. In the absence of such international consensus, the UN stepped in. Following the 1992 Rio conference on the environment, the United Nations Statistics Division (UNSD) produced a handbook of national accounting entitled *Integrated Environmental and Economic Accounting* (1993), which introduced the System of integrated Environmental and Economic Accounting (SEEA). This handbook pioneered by the UN Environment Programme (UNEP) and the World Bank aimed to integrate environment accounting as a satellite system to complement existing systems of national accounts rather than attempting to modify existing SNAs themselves (UN 1993b, 2000; Bartelmus 2001). Reflecting however, as it did, the new and experimental nature of some of the suggested methodologies, the report was issued in interim form. The system was tested in several countries and due to lack of data and the controversiality of certain valuations, only parts of the SEEA were actually compiled. As a result, revisions were made through numerous meetings over several years, which resulted in updated handbooks in 2000 and 2003. Essentially though the system seeks to track environmental resource use, including resource depletion and environmental degradation over a specific time period. This is achieved through measures of environmental capacity and the ability of the environment to perform its environmental functions at the same time as providing for the economy, which is ultimately represented by a monetary value. This complicated statistical analysis includes taking account of the natural resource stocks and flows, that is, the mineral and energy resources as well as the biological, land and water resources and ecosystems. It is a mammoth task with many such systems having been rolled out over the countries; however, as to its efficacy as a policy tool, only time can judge its worth (SEEA 2003).

15.1.4.3 Life Cycle Analysis/Assessment

Lastly one other practice that began to flourish around this time was the idea of life cycle analysis or assessment (LCA) of the environment. LCA facilitates a holistic environmental view by quantifying environmental burdens over the entire life cycle of a product from raw material acquisition through to manufacturing or processing, consumption and ultimately to its disposal. Although relatively young, the concept does differ greatly in translation from theory to practice with many different methods being proposed and, as with the footprinting concepts, is open to broad criticism. That said, LCA, as a tool, might just turn out to be one of the most beneficial and promising methods for assessing the overall environmental impact of products and services (Joshi 1999).

The 1990s was also a profitable time for the development of the food security definition.

15.1.5 Food Security Definitions

Continually propelled by new insights, the notion of food security seemingly defied definition and was almost being adapted on a year-by-year basis. The 1990s was a particularly prevalent time for this. In 1990, for instance, Swaminathan (1990) told how he pleaded to have the FAO's food security definition replaced with one involving nutrition security and all that that would entail such as a balanced diet and safe drinking water. However, this was not to be. Separately though in 1990, Molly Anderson of the Life

Sciences Research Office (LSRO) of the Federation of American Societies of Experimental Biology offered a definition of food security as

> ... access by all people at all times to enough food for an active, healthy life and includes at a minimum: a) the ready availability of nutritionally adequate and safe foods, and b) the assured ability to acquire acceptable foods in socially acceptable ways (e.g., without resorting to emergency food supplies, scavenging, stealing, and other coping strategies). Food insecurity exists whenever [a] or [b] is limited or uncertain. (Andersen 1990; USDA 2009)

On the other hand, the United States used several definitions at this time. The USAID denoted food security, in 1992, to exist

> ... when all people at all times have both physical and economic access to sufficient food to meet their dietary needs for a productive and healthy life. (USAID 1992)

This was used for several resource programming purposes, including the programming of Development Assistance (DA), Development Fund for Africa (DFA) and Economic Support Fund (ESF), and for Titles II and III of the PL 480 programmes. However, with regard to Titles I and IIII of PL 480, an older version of food security was to remain

> Access by all people at all times to sufficient food and nutrition for a healthy and productive life. (USAID 1992)

By 1996, however, the UN, in the form of the Rome Declaration of the World Food Summit, had adopted a still more complex definition

> Food security exists when all people, at all times, have physical and economic access to sufficient, safe and nutritious food to meet their dietary needs and food preferences for an active and healthy life. (FAO 1996a, 2003b)

However, like Swaminathan in 1990, Frankenberger and colleagues preferred nutrition security over food security and in their definition suggesting that

> ... an individual is nutritionally secure when he or she has secure access to a nutritionally adequate diet and the food consumed is biologically utilized such that adequate performance is maintained in growth, resisting or recovering from disease, pregnancy, lactation, and physical work. (Frankenberger et al. 1997)

In other events around this time, the subject of micronutrients and the problems of hidden hunger pushed PEM to the side and by the close of the decade, a more balanced holistic view of nutrition was beginning to emerge.

15.2 Science and Technology

15.2.1 Genetic Engineering Comes of Age

With all the advances made in genetics in the previous decade, the 1990s finally bore fruit for the fledgling new commercial technology (highlights can be seen in Table 15.1). In 1994, the first commercial transgenic product was Monsanto's recombinant version of the natural cow growth hormone Bovine Somatotropin (BST). The product was successful, and rBST, also known as bovine growth hormone (BgH), helped boost milk production by a whopping 25%. Another product introduced, this time by the US company Calgene (later acquired by Monsanto), was a genetically modified tomato, which slowed

TABLE 15.1

Genetic Engineering Milestones of the 1990s

1990	Calgene, Inc., conducted the first successful field trials of genetically engineered cotton plants to withstand use of the herbicide Bromoxynil. GenPharm International, Inc., created the first transgenic dairy cow able to produce human milk proteins for infant formula.
1993	The FDA declares that genetically engineered foods are not inherently dangerous and subsequently do not require special regulation. George Washington University researchers successfully clone human embryos provoking protests from ethicists, politicians and critics around the world.
1994	The first genetically engineered food product, Calgene's Flavr Savr tomato, was approved by the US FDA.
1995	A European research team identified a genetic defect which appears to underlie the most common cause of deafness.
1996	The UK government announces that 10 people may have become infected with the BSE agent through exposure to beef.
1997	Scottish scientists clone Dolly, the sheep, using DNA from adult sheep cells.
1998	Scientists at Japan's Kinki University clone eight identical calves using cells taken from a single adult cow.

Sources: Nature, *Nat. Rev. Genet.*, 8(Supplement), 2007; Huckett, B.I., *Sugar Cane Int.*, 27, 3–5, 2009; BioTech Institute, Timeline of Biotechnolgy 20th Century. Biotechnology Institute, 2010; Dennis et al., eds., *Encyclopedia of Biotechnology in Agriculture and Food*, CRC Press, Boca Raton, FL, 2010.

the ripening process, extending its shelf life. This new tomato was called the Flavr-Savr™ and became the first commercially grown product to be granted a license for human consumption (Hoban 1995; Schneider and Schneider 2010). Two years later, Monsanto transgenically introduced the soil-dwelling bacterium *Bacillus thuringiensis* (Bt) into corn and by expressing the genes responsible for pest deterrent. In doing so Monsanto had successfully produced the first Bt corn that required less pesticide during growth. However, perhaps their most famous genetically engineered product of the time were the corn and soya plants, which Monsanto altered to be resistant to glyphosate. Roundup, or Roundup Ready, was Monsanto's tradename for glyphosate. It is a herbicide that was usually sprayed around plants and was particularly good at killing a broad range of weeds. By introducing the genes into soya and corn, farmers could now effectively save time and money by spraying the whole field indiscriminately (Roberts 2008).

Enthusiasm though was tempered by a public fear that was growing with every new patent. Protests around the world raised public awareness of such foods and hampered their acceptance. Letters and articles began to appear regularly in newspapers worldwide (O'Neil 1992), and as the backlash grew, one leading group of enterprising German protesters decided to lead their protest with a banner containing a very peculiar menu of genetically altered foods:

Starters
> Smoked trout fillets with the gene for human growth hormone
> Tomato salad with flounder-fish gene

Main course
> Grilled chicken with bovine growth hormone gene
> Baked potato with scorpion gene

Dessert
> Melon with virus gene

This menu was not the result of their overactive imaginations either; it seems all these foods had indeed already been tested and developed at that time by genetic engineers (Blythman 1993). Such was the feeling among the general public that it led one observer to wryly offer:

> If they want to sell us Frankenfood, perhaps it's time to gather the villagers, light some torches and head to the castle. (Lewin 1992)

Indeed, this new label 'Frankenfoods' was first coined by Paul Lewin; yet it was a turn of phrase that was to be repeatedly seen in the tabloids time and time again over the ensuing years. By 1998, this backlash had reached the European Union (EU), and in response to fears and public demand, several EU countries held an unofficial moratorium on the import and approval of new varieties of GM crops (EC 2000).

15.3 Health and Nutrition

15.3.1 Conferences, Conferences, Conferences

During the 1980s, both the WHO and FAO had reviewed requirements concerning protein, energy and several vitamins and minerals including vitamin A and iron, although no specific meeting was convened to discuss nutrients in particular. By now, sufficient new research indicated that besides preventing deficiency diseases, several vitamins and minerals actually played vital roles in mitigating or preventing diet-related chronic diseases. Previous to this finding, the last specific FAO/WHO expert meeting on trace elements took place in 1974 (WHO 1974). It was time to meet again and in 1990 a Joint FAO/WHO/ IAEA Expert Consultation on Trace Elements in Human Nutrition was held in Geneva. Of note was the large number of nutrients that came under discussion—19 in total including iodine, zinc, selenium, copper, molybdenum, chromium, manganese, silicon, nickel, boron, vanadium, fluoride, arsenic, aluminium, lithium, tin, lead, cadmium and mercury. The consultation looked at the nutrients specifically along a continuum from deficiency to toxicity. Moreover, the effects of the interaction of nutrients among themselves and in relation to the diet were also considered. Ultimately recommendations for safe levels of intake were given for many of these nutrients (Weisell 1991).

Other meetings at the time included the World Summit for Children in 1990, which adopted a Declaration on the Survival, Protection and Development of Children as well as a Plan of Action for implementing it. This involved inter alia targets of reducing child mortality rates by a third by 2000 as well as the reduction in the number of children malnourished by one-half (UNICEF 1990). The next year, a follow-up conference in Montreal—Ending Hidden Hunger: A Policy Conference on Micronutrient Malnutrition—aimed to pursue the goals of the previous World's Summit for Children by looking at ways to succeed in the micronutrient agenda. As an outcome of this conference, a joint WHO–UNICEF initiative to eliminate micronutrient disorders was initiated. This aimed to increase the public and institutional visibility of the problem of 'hidden hunger' and pledged the virtual elimination of iodine deficiency disorders (IDD) by the year 2000. Another goal of this meeting was to encourage donors to make ending vitamin A and iron deficiencies their next major public health priority with a view to finally ending these micronutrient disorders altogether. In achieving this, it was hoped the three major nutrient deficiencies would be treated in a holistic way rather than as three separate and perhaps competing goals (IDD 1991).

Buoyed with hope, the FAO/WHO convened The International Conference on Nutrition (ICN) in 1992 in which representatives yet again reaffirmed their position on nutritional issues with a World Declaration on Nutrition, stating that

> Hunger and malnutrition are unacceptable in a world that has both the knowledge and the resources to end this human catastrophe … . We recognize that globally there is enough food for all and … pledge to act in solidarity to ensure that freedom from hunger becomes a reality. (FAO/WHO 1992)

More meetings and subsequent reports concerning recommendations of nutrition and trace elements followed in relative quick succession:

- Fats and oils in human nutrition (FAO/WHO 1994);
- Energy and protein requirements (IDECG 1994);
- Trace elements in human nutrition and health (WHO/FAO/IAEA 1996);
- Preparation and use of food-based dietary guidelines (WHO/FAO 1996);
- Carbohydrates in human nutrition (FAO/WHO 1998);
- Human vitamin and mineral requirements (FAO/WHO 1998).

These meetings collectively updated a lot of the nutritional literature and helped provide guidelines on nutrient intakes for infants, children, adults as well as pregnant and lactating women. The ultimate scope of these consultations and their subsequent recommendations combined to form a solid foundation of all human nutrition at the time (FAO 1991, 1994, 1998; IDECG 1994; FAO/WHO 2001; WHO/FAO 2002).

As an aside, up till the 1970s, most nutrition reports tended to ignore issues of distribution of intake among individuals and instead applied general average recommendations applying to groups or populations. Although however, it was understood and widely recognised that people were actually misusing these recommendations and attempting to apply these as accepted daily intakes for the individual What was needed was an actual per person daily recommendation of a group of people that would satisfy practically all members of the same group (Beaton 1992). The problem was the notion of just how could one recommendation be applicable to whole populations (even assuming heterogeneity) then to groups and finally to the individual. The answer clearly was that it could not. This meant a conceptual shift in the way people looked at nutrition to take account of the variability of distribution among groups or populations. This issue was tackled first conceptually in the 1940s and then later by Lorstad in 1971 who applied statistical application to the problem of intake distributions. This involved calculating just how much above the average intake would be acceptable before the prevalence of inadequate intakes would be considered acceptably low. After much consultation, the concept was gradually being incorporated into the series of UN reports on nutritional needs such as the FAO/WHO (1973), FAO/WHO/UNU (1985) and so on.

By 1994, the Energy and Protein Requirements report of the IDECG had reviewed much previous data and determined on the strength of new data to adopt the factorial method (multiples of BMR), already used for adults in the 1985 report as the basis for calculations of children's total energy expenditure.

15.4 Period Commentary

By the late 1980s, the idea of a global free market had already emerged; however, it was not until the 1990s and the full realisation of the Uruguay trade negotiations that agricultural price support and protectionist policies fully gave way to the new orthodoxy of internationalism (McMichael 1994). Support polices and protected trading blocs began to loosen their grip on the long held notions of a two-tier economy. It would seem that free trade culminating from many decades of structural reform was finally in everyone's grasp, although it still had its detractors. Many people were sceptical of the benefits of liberalisation and critics of free trade argued that it tended to favour the rich industrialised countries and large transnational corporations in particular. This in turn, it has been said, often created dependency and impoverishment for poorer countries, an equal inequality so to speak. Furthermore where profits became the driving force, further neo-liberalisation seemed to engender short-termism to the detriment of the 'stewardship of natural resources and sustainable production methods' (Warnock 1997). Meanwhile in the social stakes, the 1990s also witnessed a mini coup in the form of the human development report. The idea behind the report was that at the end of the day economic development was really created by the people for the people. Somewhere along the way, it seemed we had lost sight of the real aims of growth and were effectively chasing growth for growth's sake. The report reprioritised individuals and sought to reflect this in the human development index (HDI).

Environmentalism in the 1990s too grew to become an irreversible force and its spotlight on the world stage continued to ensure this would remain the case for years to come. It was an opportunity to firmly establish environmental issues at the top echelons of global governance. By regarding the holistic nature of the environment and its natural resources food production policy, politics and even the national accounts would be changed forever. It was a time too that raised awareness of accounting for the environment in a measurable way. This need to account for the environment in an integrated manner with a country's system of national accounts (SNA) arose because it became acutely obvious that new scarcities of natural resources threatened the sustainable productivity of many economies. More than this though, public pressure and political prudence converged on the realisation that the quality of the environment and its natural resources were more than just ideological goals. The practicality of not accounting for such measures had huge private and social costs for the environment itself as well as for the future of humanity. The new environmentalism had come to energise organisations, NGOs and the public alike.

Meanwhile, the 1992 Earth Summit came at a serendipitous time for food security issues and building on the ideas of food sustainability first presented at the inaugural FAO conference in 1943, the Rio Earth Summit publicly voiced an increasing backlash against the environmental effects of laissez-faire agricultural policies and mismanagement. This was a problem that had already been recognised 2 years before. Swaminathan, in his 1990 Sir John Crawford Memorial Lecture, had already talked about the increased damage inflicted on the 'ecological foundations of sustainable agriculture' and suggested that we should all take a leaf out of the *US Food Security Act of 1985* book, which promoted an 'an integrated approach to production, conservation and consumption'. He further offered that by doing so great leaps could be made in eliminating chronic hunger by providing better access to quality diets (Swaminathan 1990). However, the conference provided more than a wake-up call for food practices vis-à-vis the environment, more importantly became a platform from which to announce sweeping social changes in responsibility. Backed by an increasing NGO powerbase, the conference in no short measure sought to hold every government, state and individual to account for the sake of future generations. The status quo, as it were, was being challenged in a way like never before, and it brought home some pretty harsh messages.

Shortly afterwards, in many regions of the world, food production was becoming tainted with the legacy of continual unsustainable or inadequate programmes and policies. Perhaps spurred on in part by the recent environmental conference, it was acutely felt that Earth's natural resource base was given a raw deal in terms of inefficiency and wastage. Moreover, faltering institutional direction, as well as inappropriate technologies and insufficient rural infrastructures, were all colluding to the detriment of the environment. These views did not, surprisingly lead to the 1992 Earth Summit, although they were similarly expressed, but rather to the 1996 World Food Summit (Food Summit 1996). However, while also drawing attention to the complexities of sustainability and appropriate policy reform, little progress had actually been made and widespread undernutrition continued throughout the world. Compounding this were concerns that agricultural capacity might struggle to meet future needs. Amid such apprehension, the 1996 World Food Summit was called and once again as in the 1974 WFC before it, the conference recognised the fundamental necessity for international cooperation and development in the fields of production, supply and world trade (UN 1997a). Importantly in this conference though, from the vast gathering of heads of state perhaps the summit's greatest achievement was the formal adoption of the right to adequate food. While this had already been voiced in the UN Convention on Economic, Social and Cultural Rights of 1976, the summit sought ways to revive and implement such grand notions (Hartmannshenn 2004; FAO 2006). This marked a milestone of achievement and paved the way towards the realisation of a rights-based approach to food security. A further important idea coming out of the summit was the shift away from the objective, often indicator-led targets to a more softer, subjective approach looking at quality of food, choices and options. This was partly enshrined by the earlier human development-led change in perspectives where the right to food was seen as sacrosanct as the right to vote (UNDP 1994).

By the mid-1990s then, food security was understood to be of considerable concern at both the individual and global levels, and definitions were broadened to incorporate ideas of nutritional balance and food safety among other things. The concept of food security was rapidly losing its simplicity in the eyes of many; instead it was beginning to reflect more a way of life than a goal in and of itself (FAO 2003a). Not for the first time the concept was formally recognised as a complex interplay of multidimensional variables, yet the 1996 WFS definition, as with many previous and subsequent definitions, came to illustrate the full absorption of the food security concept within the wider development issues of environmental sustainability, development economics, political welfarism, poverty, conflict, policy, etc. (FAO 1996a).

By the end of the decade, data needs became such that monitoring malnutrition, previously carried out periodically over perhaps 10 years, was now required on an annual basis. The FAO responded with the introduction of the annual publication the 'State of Food Insecurity' in 1999.

Lastly, in this period biotechnologies were coming along in leaps and bounds with several genetically modified food-related products coming into the market. This promised to be the birth of a new era of new improved food products, yet public fears and growing distrust of the scientific community clouded this new dawn with cries of Frankenfoods and other equally colourful neologisms.

15.5 Key Dates, People, Acts, Reports and Surveys

The 1990s was a productive time, replete with many conferences and biotechnological research. Table 15.2 notes some of the highlights that shaped this decade.

TABLE 15.2

Key Dates: The Era of the Conference

Date	People, Organisation, Agencies, Acts or Conventions
1990: UN World Summit for Children	The largest gathering of world leaders at this point in history assembled at the World Summit for Children. The World Summit adopted a Declaration on the Survival, Protection and Development of Children and a Plan of Action for implementing the Declaration in the 1990s.
1990: Human Development Report (HDR)	The Human Development Report (HDR) first launched in 1990 with the goal of putting people back at the centre of the development process. This went further than previous simple economic measures of social development to include a variety of social goals and welfare pertaining to quality of life.
1992: Department of Humanitarian Affairs (DHA)	In 1991, the General Assembly wanted to strengthen its response to both complex emergencies and natural disasters whilst improving the overall effectiveness of their humanitarian operations and created the DHA in 1992. Also created was the Emergency Relief Coordinator (ERC) combining and coordinating with the UN Disaster Relief Coordinator (UNDRO). The same resolution that created the DHA also created the Inter-Agency, the Standing Committee (IASC), the Consolidated Appeals Process (CAP) and the Central Emergency Revolving Fund (CERF) as coordinating bodies and tools of the ERC. Following the reorganisation in 1998, the DHA was incorporated into the OCHA (OCHA 2011).
1992: United Nations Conference on Environment and Development (UNCED): The Earth Summit (ES)	The UN Conference (UNCED), known as the Earth Summit, was a historic meeting. Two decades after the first global environment conference, the Earth Summit sought ways to halt the destruction of natural resources and reverse the pollution of the planet. From this meeting, an 800-page document called Agenda 21, a blueprint for action in achieving sustainable development, was created (UN 1997b).
1992: Commission on Sustainable Development (CSD)	As a result of the UN Conference on Environment and Development (UNCED) of 1992, the council established the CSD to review progress following UNCED to aid with policy guidance and activities in achieving sustainable development as well as in building partnerships (UN 1993a).
1992: The Center for Global Food Issues	The Center for Global Food Issues conducts research into agriculture and environmental concerns surrounding food production. The center assess policies, understanding and heightens awareness of the environmental impacts of various food and farming systems and policies (CGFI 2010).
1992: European Commission Humanitarian Aid (ECHO)	The EC Humanitarian Aid department (ECHO) established in 1992 helps provide emergency assistance to victims of natural human-made disasters outside the EU (EC 1996).
1993: Second World conference on Human Rights	The World Conference on Human Rights was held by the UN. It was the first human rights conference held since the collapse of the Cold War. The main result of the conference was the Vienna Declaration and Programme of Action to strengthen human rights work around the world. (Lawson and Bertucci 1996).
1993: Global Policy Forum (GPF)	Founded by an international group of citizens, the GPF through networks and coalitions aim to create a more equitable and sustainable global society (GPF 2009).
1994: FAO's The Special Programme for Food Security (SPFS)	This Programme launched to boost food production through national and regional programmes for food security (FAO SPFS 2009).
1994: Food Security Analysis Unit (FSAU)	The FSAU seeks to provide nutrition and livelihood security for the Somali people. It began in 1994 funded by USAID/OFDA and managed by WFP (FSAU 2010).
1995: International Grains Council (IGC)	The International Grains Agreement of 1995 linked two conventions: the Grains Trade Convention (GTC) and the Food Aid Convention (FAC). In the 1995 GTC, the International Wheat Council (IWC) became the International Grains Council (IGC), giving full recognition of all coarse grains rather than just wheat (IGC 2009).
1995: Oxfam International	Founded in 1995, Oxfam International is a combination of 13 organisations supporting humanitarian programmes throughout the world (Oxfam 2008).

TABLE 15.2 (Continued)

Key Dates: The Era of the Conference

Date	People, Organisation, Agencies, Acts or Conventions
1995: The World Trade Organization (WTO)	In 1993 GATT was updated. The 75 existing GATT members and the EC became the founding members of the WTO taking over from GATT to supervise the liberalisation of international trade. Its 150 members account for over 97% of world trade (Wahlberg 2008; WTO 2010).
1996: World Food Summit (WFS)	It was realised that progress of the 1974 World Food Conference would need to be accelerated. The 1996 World Food Summit renewed global commitment to eliminate hunger and malnutrition as well as raising awareness among decision makers, the media and with the public in general. It also set political, conceptual and technical frameworks in order to achieve these goals. The summit culminated in the Rome Declaration on World Food Security and the World Food Summit Plan of Action. The Summit pledged to cut the number of hungry to about 400 million by 2015 (UN 1997a).
1996: ReliefWeb	Motivated by the notion that timely and reliable information during crises is critical to improving and maximising responses during emergencies. ReliefWeb also operate on the premise that transparency and accountability among the humanitarian relief agencies, NGOs and the like is paramount to the success of their work (ReliefWeb 2009).
1997: The UN Development Group (UNDG)	The UN Development Group (UNDG) coordinates and harmonises the 32 UN funds, programmes, agencies, departments and offices that independently or co-dependently work for social and economic development (UNDG 2009).
1998: The Office for the Coordination of Humanitarian Affairs (OCHA)	In 1998 the DHA was reorganised, becoming the OCHA. Its remit was extended to include the coordination of policy development and humanitarian responses of UN agencies, funds and NGOs. It works primarily through the Inter-Agency Standing Committee (IASC) chaired by the ERC (OCHA 2011).
1998–2008: Food and Nutrition Technical Assistance (FANTA)	The Food and Nutrition Technical Assistance (FANTA) project was a 10-year project to improve knowledge, policy, strategy, programme development, implementation and monitoring within the field of food and nutrition. The project, funded by USAID, was renewed in 2008 for a further 5 years (Copeland et al. 2002; POPLINE 2010).
1999–2004: Food Insecurity and Vulnerability Information and Mapping (FIVIMS)	FIVIMS helps countries understand, through cross-sectoral analysis and evidence-based analysis, the underlying causes and the formulation and implementation of policies looking at food insecurity. This is further aided by the KIDS Database. In 2004/2005, linked international databases created an information exchange network developing a common international database: the Key Indicators Database System (KIDS). KIDS effectively allowed the dissemination and analysis of FIVIMS data (CFS 1998; FAO/GIEWS 2006; FIVMS 2008).

References

Andersen, S. A. (1990). Core Indicators of Nutritional State for Difficult to Sample Populations. *The Journal of Nutrition* 120 (Suppl 11): 1559–1600.

Argeñal (2007). History of Food Security in International Development: Background Paper Developed for Christian Children's Fund Honduras. Honduras: Christian Children's Fund.

Bartelmus, P. (2001). Accounting for Sustainability: Greening the National Accounts In *Our Fragile World, Forerunner to the Encyclopedia of Life Support Systems, Vol. II.* M. K. Tolba. Oxford: Eolss Publishers.

Beaton, G. H. (1992). Human Nutrient Requirement Estimates: Derivation, Interpretation and Application in Evolutionary Perspective. *Food, Nutrition and Agriculture*(2/3): 3–15.

BioTech Institute (2010). Timeline of Biotechnolgy 20th Century. Biotechnology Institute. http://www.biotechinstitute.org/what_is/timeline.html (accessed 12 January 2011).

Blythman, J. (1993). *Food & Drink: Do We Have to Eat Our Genes?: We Will Soon Be Offered Genetically Engineered Food. Joanna Blythman Does Not Think We Should Accept It.* London: The Independent on Sunday, 16 October 1993.

CFS (1998). Guidelines for National Food Insecurity and Vulnerability Information and Mapping Systems (Fivims): Background and Principles. Committee on World Food Security: Twenty-fourth Session. Rome. Committee on World Food Security.

CGFI (2010). About the Center for Global Food Issues. Hudson Institute. http://www.cgfi.org/about/ (accessed 18 June 2010).

Clay, E. (2002). Food Security: Concepts and Measurement: Paper for FAO Expert Consultation on Trade and Food Security: Conceptualising the Linkages. Rome: Food and Agriculture Organization.

Cleveland, C. J. and G. Douglas (2008). *Human Development Index*. Washington, DC: Environmental Information Coalition/National Council for Science and the Environment.

Cohen, J. E. (1995). *How Many People Can the Earth Support?* New York and London: W.W. Norton.

Copeland, R., T. Frankenberger and E. Kennedy (2002). Food and Nutrition Technical Assistance Project Assessment. Washington, DC: The United States Agency for International Development.

Dennis, R. H., B. W. Matthew and G. H. Dallas, Eds. (2010). *Encyclopedia of Biotechnology in Agriculture and Food*. Boca Raton, FL: CRC Press, Taylor & Francis Group.

Earth Summit (2002). Decision Making: Briefing Sheet. United Nations. http://www.earthsummit2002.org/es/life/Decision-making.pdf (accessed 20 July 2009).

EC (1996). *Council Regulation (EC) No 1257/96 of 20 June 1996 Concerning Humanitarian Aid*: EU. 1257/96.

EC (2000). Commission Takes Initiative to Restore Confidence in GMO Approval Process. Press release. Brussels: European Commission.

ER (1996) A Conversation with Joel Cohen. *Environmental Review Newsletter* 3 (1). http://www.environmentalreview.org/archives/vol03/cohen.html.

EU (2010). Cap Reform: A Policy for the Future. Council of the European Union. http://ec.europa.eu/agriculture/publi/fact/policy/index_en.htm (accessed 20 March 201).

FAO (1991). Food, Nutrition and Agriculture—2/3—Nutrient Requirements. Rome: Food and Agriculture Organization.

FAO (1994). Fats and Oils in Human Nutrition: Report of a Joint Expert Consultation. *Food and Nutrition Paper No. 57*. Rome: FAO.

FAO (1996a). Rome Declaration on World Food Security and World Food Summit Plan of Action. Rome: Food and Agriculture Organization.

FAO (1996b). World Food Summit—Technical Background Documents 1–5: Volume 1: 3. Socio-Political and Economic Environment for Food Security. World Food Summit. Rome: Food and Agriculture Organization.

FAO (1998). Carbohydrates in Human Nutrition. *FAO Food and Nutrition Paper—66*. Rome: Food and Agriculture Organization.

FAO (2003a). Trade Reforms and Food Security: Conceptualising the Linkages: Food Security: Concepts and Measurement. Rome: Food and Agricultural Organization.

FAO (2003b). Trade Reforms and Food Security: Conceptualizing the Linkages. Rome: Food and Agriculture Organization.

FAO (2006). Policy Brief: Food Security. Rome: Food and Agriculture Organization 4.

FAO SPFS (2009). Special Programme for Food Security. FOAAD and Agriculture Organization. http://www.fao.org/spfs/about_spfs/mission_spfs/en/ (accessed 20 January 2010).

FAO/GIEWS (2006). The Use of Geospatial Data for Food Insecurity and Agricultural Drought Monitoring and Assessment by the FAO GIEWS and Asia FIVIMS. Seventeenth United Nations Regional Cartographic Conference for Asia and the Pacific, Bangkok, United Nations.

FAO/WHO (1992). Final Report of the Conference: International Conference on Nutrition (ICN). Rome: Food and Agriculture Organization 67.

FAO/WHO (2001). Report of the Joint FAO/WHO Expert Consultation on Human Vitamin and Mineral Requirements. Bankok, Thailand: FAO/WHO 303.

FIVMS (2008). The FIVIMS Initiative: Food Insecurity and Vulnerability Information and Mapping Systems. http://www.fivims.org/index.php?option=com_content&task=blogcategory&id=20&Itemid=37 (accessed 6 June 2009).

Food Summit (1996). Rome Declaration on World Food Security World Food Summit. Rome, Food and Agriculture Organization.

Frankenberger, T., L. Frankel, S. Ross, B. Marshall, C. Cardenas, D. Clark, A. Goddard, H. Kevin, M. Middleberg, D. O'Brien, C. Perez, R. Rand and J. Zielinski (1997). Household Livelihood Security: A Unifying Conceptual Framework for Care Programs: 1995. The USAID Workshop on

Performance Measurement for Food Security. Arlington, VA: United States Agency for International Development.

FSAU (2010). Food Security Analysis Unit for Somalia: FAQ. FSAU. http://www.fsausomali.org/index.php@id=32.html (accessed 8 May 2010).

GFN (2010). Global Footprint Network. Global Footprint Network. http://www.footprintnetwork.org/en/index.php/GFN/ (accessed 12 March 2011).

GPF (2009). Global Policy Forum: About Us. Global Policy Forum. http://www.globalpolicy.org/about-gpf-mm/introduction.html (accessed 21 May 2010).

Hannon, B. (1973). The Structure of Ecosystems. *Journal of Theoretical Biology* 41(3): 535–546.

Hannon, B. (1985). Linear Dynamic Ecosystems *Journal of Theoretical Biology* 116(1): 89–98.

Hartmannshenn, T. (2004). Food Security: Guidelines for the Promotion and Execution of Food Security Projects by German Agro Action. Bonn, Germany: German Agro Action.

Hecht, J. E. (1997). Environmental Accounting: What's It All About? Washington, DC: the International Union for Conservation of Nature (IUCN), Green Accounting Initiative 8.

Hoban, T. J. (1995). The Construction of Food Biotechnology as a Social Issue. In *Eating Agendas: Food and Nutrition and Social Problems*. D. Maurer and J. Sobal, eds. New York: Walter de Gruyter Inc.

Huckett, B. I. (2009). A Compact History of Genetic Transformation and Its Influence on Crop Development *Sugar Cane International* 27(1): 3–5.

IDD (1991). Ending Hidden Hunger: The Montreal Micronutrient Conference. *IDD Newsletter* 7(4): 1–10.

IDECG (1994). Energy and Protein Requirements, Proceedings of an IDECG Workshop, November 1994. *Supplement of the European Journal of Clinical Nutrition*: 198.

IGC (2009). Grains Trade and Food Security Cooperation: The International Grains Agreement. The International Grains Council. http://www.igc.org.uk/en/aboutus/default.aspx#igc (accessed 21 February 2010).

Jackson, P. (2007). From Stockholm to Kyoto: A Brief History of Climate Change. *United Nations Chronicle* XLIV(2)   http://www.un.org:80/wcm/content/site/chronicle/lang/en/home/archive/issues2007/pid/4819 (accessed 22 May 2011).

Joshi, S. (1999). Product Environmental Life-Cycle Assessment Using Input-Output Techniques. *Journal of Industrial Ecology* 3(2–3): 95–120.

Krebs, C. J. (1972). *The Experimental Analysis of Distribution and Abundance*. New York: Harper and Row.

Lawson, E. H. and M. L. Bertucci (1996). *Encyclopedia of Human Rights*. London: Routledge.

Lewin, R. (1992). *Complexity: Life at the Edge of Chaos*. New York: Macmillan.

McMichael, P. (1994). *The Global Restructuring of Agro-Food Systems*. Ithaca, NY: Cornell University Press.

Nature (2007). Milestones in DNA Tecnologies. *Nature Reviews Genetics*. 8(Supplement).

O'Neil, M. (1992). Geneticists' Latest Discovery: Public Fear of 'Frankenfood'. *New York Times*, June 28, 1992.

OCHA (2011). Website of the UN Office for the Coordination of Humanitarian Affairs. United Nations. http://ochaonline.un.org/AboutOCHA/ (accessed 23 July 2011).

OPT (2009) Earth Heading for 5 Billion Overpopulation? *Optimum Population Trust*. http://www.optimum-population.org/releases/opt.release16Mar09.htm

Oxfam (2008). Oxfam International. OI. http://www.oxfam.org.uk/resources/policy/conflict_disasters/downloads/oi_hum_policy_humanitarianism.pdf (accessed 12 May 2010).

Peskin, H. M. (1976). A National Accounting Framework for Environmental Assets. *Journal of Environmental Economics and Management* 2(4): 255–262.

POPLINE (2010). Food and Nutrition Technical Assistance Project Assessment. *POPulation information onLINE*. Baltimore, MD: Johns Hopkins/USAID.

Rees, W. E. (1992). Ecological Footprints and Appropriated Carrying Capacity: What Urban Economics Leaves Out. *Environment and Urbanisation* 4(2): 121–130.

ReliefWeb (2009). About the ReliefWeb Project: Transparency in Partnerships Fuels ReliefWeb Success. ReliefWeb. http://www.reliefweb.int/help/whatwedo.html (accessed 12 March 2001).

Roberts, P. (2008). *The End of Food*. New York: Houghton Mifflin Harcourt.

Schneider, K. R. and R. G. Schneider (2010). Genetically Modified Food: What Are Gm Foods? Florida: The Institute of Food and Agricultural Sciences (IFAS), University of Florida.

SEEA (2003). *Handbook of National Accounting Integrated Environmental and Economic Accounting* (*SEEA*). New York: UN Statistics Division: Environmental-Economic Accounts Section.

Stearns, F. and T. Montag (1974). *The Urban Ecosystem: A Holistic Approach*. Stroudsburg, PA: Dowden, Hutchinson and Ross.

Swaminathan, M. S. (1990). Sir John Crawford Memorial Lecture: Changing Nature of the Food Security Challenge: Implications for Agricultural Research and Policy. Washington, DC: Consultative Group on International Agricultural Research.

UN (1992). Agenda 21. United Nations Conference on Environment and Development (UNCED). Rio de Janeiro: United Nations.

UN (1993a). *Institutional Arrangements to Follow up the United Nations Conference on Environment and Development*: United Nations. A/RES/47/191.

UN (1993b). Integrated Environmental and Economic Accounting: Interim Version; Handbook of National Accounting. New York: United Nations. Department of Economic and Social Information and Policy Analysis 182.

UN (1997a). Integrated and Coordinated Follow-up and Implementation of the Major International United Nations Conferences and Summits: Outcome of the World Food Summit. New York: United Nations.

UN (1997b). United Nations Conference on Environment and Development. United Nations. http://www. un.org/geninfo/bp/enviro.html (accessed 20 June 2010).

UN (2000). *Handbook of National Accounting Integrated Environmental and Economic Accounting an Operational Manual*. New York: United Nations, 260.

UNDG (2009). The UN Development Group. United Nations. http://www.undg.org/index.cfm?P=2 (accessed 9 March 2010).

UNDP (1990). Human Development Report 1990. New York: United Nations Development Programme.

UNDP (1994). Human Development Report 1994: New Dimensions of Human Security. New York: United Nations Development Programme.

UNICEF (1990). World Declaration on the Survival, Protection and Development of Children. Geneva, Switzerland: UNICEF.

UOR (2009). University of Reading: Agriculture, Policy and Development, History of Agriculture. University of Reading. http://www.ecifm.rdg.ac.uk/history.htm (accessed 4 June 2008).

USAID (1992). USAID Policy Determination: Definition of Food Security. Washington, DC: United States Agency for International Development.

USDA (2009). Food Security in the United States: Measuring Household Food Security. United States Department of Agriculture. http://www.ers.usda.gov/Briefing/FoodSecurity/measurement.htm (accessed 15 October 2009).

Wackernagel, M. and W. E. Rees (1996). *Our Ecological Footprint: Reducing Human Impact on the Earth*. Gabriola Island, B.C.: New Society Publishers.

Wahlberg, K. (2008) Food Aid for the Hungry? http://www.globalpolicy.org/socecon/hunger/relief/2008/01wahlberg.htm (accessed 12 January 2011).

Warnock, K. (1997). Third World: Feast or Famine? Food Security in the New Millennium (Commentary). *Journal of Race and Class* 38(3): 63–72.

Weisell, R. C. (1991). Trace Elements in Human Nutrition. Salient Issues from the Joint FAO/WHO/IAEA Expert Consultation. *Food, Nutrition and Agriculture* 2/3: 25–29.

WHO (1974). Handbook on Human Nutritional Requirements. *WHO Monograph Series, No. 61*. Geneva, Switzerland: WHO.

WHO/FAO (2002). *Vitamin and Mineral Requirements*. Rome: WHO/FAO.

WTO (2010). Understanding the WTO: Basics—The GATT Years: From Havana to Marrakesh. World Trade Organisation. http://www.wto.org/english/thewto_e/whatis_e/tif_e/fact4_e.htm (accessed 30 March 2010).

16

The Twenty-First Century: Ideological Convergence?

The twenty-first century began with the seminal Millennium Summit. The Summit aimed to specify targets and goals for action in many areas of social concern and the Millennium Development Goals (MDGs), as they were known set the tone for much discussion and debate that followed. Growing social responsibility to the citizens of Earth and to the natural environment played a large part in harnessing consensus for change. Gone were the days when the environment could be pillaged with impunity, as new ecological accounting paradigms attempted to promote accountability. Furthermore, this was another decade in which genetically modified organisms (GMOs) once again came under fire as nations battled with legislation. Lastly, a financial crisis following quickly on the tail of a partially human-made food price bubble gave governments a pause for thought. It seemed, for some at least, the very foundations of capitalist economics was being called into question.

16.1 Governance, Politics and Socioeconomics

16.1.1 Millennium Summit

In terms of renewed enthusiasm in social and welfare politics, the new decade dawned with the advent of the Millennium Summit. The summit was the largest gathering of world leaders up to that time and in its sights were large, broad-sweeping social objectives, reflecting many pressing issues of the day. Among consideration too was the continuing role of the UN in the coming new millennium, the direction it should take and it what form. In conclusion the summit closed with delegates ratifying the United Nations Millennium Declaration which emphasised that

> ... only through broad and sustained efforts to create a shared future, based upon our common humanity in all its diversity, can globalization be made fully inclusive and equitable. (UN 2000c)

This was to be achieved through the acceptance of a set of fundamental ideological values which included freedom, equality, solidarity, tolerance, respect for nature and shared responsibility. Furthermore, world leaders agreed to a set of eight measurable, time-bound goals for ending poverty, universal education, gender equality, child health, maternal health, combating HIV/AIDS, environmental sustainability and global partnership. Collectively these goals are known as the Millennium Development Goals. These eight goals break down into a further 21 quantifiable targets, measured by 60 indicators (Appendix H), (UN 2000c, 2006; UNDP/MDG 2010).

By now the institutional definition of food security was also once again the focus of revision and in 2001 the Committee on Food Security sought to include a social component of the concept:

> Food security [is] a situation that exists when all people, at all times, have physical, social and economic access to sufficient, safe and nutritious food that meets their dietary needs and food preferences for an active and healthy life. (FAO 2001)

This new emphasis on safe and nutritious foods as well as the social aspects of access and preference brought into sharp focus a widening concept that became increasingly interpretable to fit specific needs.

Even the FAO in 2003 accepted that while the international community had generally been accepting of the 'increasingly broad statements of common goals and implied responsibilities' it recognised that the practical response had in fact been narrower, focusing instead on simpler objectives (FAO 2003a).

Meanwhile, the FAO convened another World Food Summit 5 years after the original to examine inter alia the progress made in eliminating hunger.

16.1.2 World Food Summit: Five Years Later and Its Parallel Conference—The Forum for Food Sovereignty

The World Food Summit was convened in 2002 amidst disappointing data that showed an average annual reduction of undernourished figures of 6 million. This was well below the yearly target of 22 million needed to achieve the World Food Summit goal of halving the number by 2015 (to 400 million). With only 32 of the 99 developing countries at the time showing a decrease in the number of the undernourished, delegates considered ways to accelerate the efforts and yet again vowed to renew their commitment and resolved to meet the agreed 1996 target by 2015 (FAO 2002b). In parallel with the summit, a separate conference, the NGO/CSO Forum for Food Sovereignty* attended by non-governmental, civil society and farmers' organisations, was scathing in its condemnation of action to date. The participants had expressed concern at the 1996 World Food Summit that their action plan would fail and feeling vindicated, they sought to understand why such failures had indeed occurred. For the parallel conference, it seemed fairly straightforward as only one aspect of the plan they suggested had been implemented: that of trade liberalisation. This, however, in the parallel conference's view, was fundamentally flawed as it was contested that such policies were pushed onto the developing countries, while the developed countries maintained high support policies themselves. The NGOs and CSOs also suggested that the WFS+5 was watering down commitments of the 'proposed Code of Conduct on the Right to Food to a set of voluntary guidelines'. Further complaints were also directed at the promotion of genetically modified (GM) foods as a panacea by certain governments and transnational corporations (NGO/CSO Forum 2002). In their view, despite there being more than sufficient resources and political will to tackle the problem, the causes of widespread hunger lay in the 'international trade-led hegemonic economic model'. This, it was claimed, deprived countries and communities of control over their own economic and natural resources and for remedying the situation, the NGO/CSO Forum proposed food sovereignty as a developmental tool over many prevailing alternatives. However, it was acknowledged that for this to happen, it was further suggested that a new 'human sustainable development paradigm' was needed based on an equitable human rights approach (NGO/CSO Forum 2002).

Sentiments of the above meeting were also echoed a year later in 2003 when several UN agencies were concerned that trade distortions were still a large factor impacting on food security matters despite movement in the right direction. Reform of agricultural policies, especially in the European Community which began in the 1990s, continued into the twenty-first Century. In the mid-term review of the CAP's Agenda 2000 in 2003, for instance, considerable changes to the original agreement were made. The main thrust of the change centred around the decoupling of subsidies with production. This freed up farmers to respond more freely to market forces of demand and supply. Furthermore, many existing direct payment subsidy schemes (11 in total) were also simplified and were to be replaced instead with a single payment scheme (SPS) (UOR 2009). Although for many, such movement was perhaps not enough, it did show a commitment to increasing liberalisation and increasing inclusive globalisation (CFS 2003; COAG 2003).

As in the past, it seemed the twenty-first century was no different from many others in that the spectre of population growth and overcrowding was never far from concern. Once again the notion of carrying capacity populated the minds of many; although one particular study is worthy of note.

---

* Food sovereignty is the right of countries to define their own agricultural, pastoral, fisheries and food policies according to their own ecological, social, economical and cultural ideologies. It promotes the right to food through small and medium-sized agricultural enterprises that ultimately respect culture and diversity. Decentralisation and democratisation are key principles of this movement (NGO/CSO Forum 2002).

16.1.3 The Earth to Support Nearly 8 Billion

It was in 2004 that perhaps the most important study on the carrying capacity of the Earth to date was undertaken. The study by Van Den Bergh and Rietveld (2004) analysed 69 historic studies, many of which were previously explored by Cohen in 1995, in which were offered ceiling population estimates ranged from 0.5 billion to 1021 billion. One important difference in this study, however, compared to Cohen's, was the use of meta-analysis in its methodology. This proved fruitful as the benefits of such meta-analysis over primary research is particularly useful when looking at a range of similar studies. Rather than being restricted by narrow methods and concentrating strictly on outputs, meta-analysis allowed Van Den Bergh and Rietveld to look beyond the output of the previous research and into the mechanics of the studies. By examining the methodology and operational structures, researchers Van Den Bergh and Rietveld effectively used the opportunity to ask some fundamental questions about comparative differences and similarities within each of the studies that could throw some light on any consensus that might emerge (Van Den Bergh and Rietveld 2004). In this respect they were successful and their analysis found sustainable population limitations, including among other things water, energy, carbon, natural resource limitations, photosynthetic capacity and land (ibid). Their studies also showed that the two most often quoted limitations stated by others were spatial extrapolation (land) and limited resources (food), which collectively accounted for 71 out of the 94 collected estimates. This, they cautioned however, suggested that while these two factors did indeed receive the most attention it was just as likely that the understudied and more stringent or inflexible resources such as water, forest and non-renewable products would be of primary importance in the future (Van Den Bergh and Rietveld 2004). Finally, after discarding many studies that were either erroneous, speculative or insufficiently rigorous, the study concluded a population best estimate carrying capacity based on a median value of 7.7 billion and even this number, they cautioned, might not even be properly sustainable (ibid).

By the middle of the decade, attention once again turned to the millennium promises, which now seemed to be hanging under a cloud.

16.1.4 2005 World Summit

The 2005 World Summit was a follow-up meeting to review progress since the UN Millennium Summit. There were a number of concerns leading up to the summit and perhaps most notable were those of the United States. In amendments totalling 750, the US ambassador to the UN sought to, among others, eliminate any new pledges of foreign aid to impoverished nations; lose any provisions that called for action to halt existing or future climate change; and more importantly the notion to strike any mention of the Millennium Development Goals from the upcoming meeting (Shah 2005). These restrictions were seen by some as retaliation to what the United States perceived as an increasing power base of the UN and an increasingly subservient United States. With all this turmoil and political undercurrent undermining the summit, it is not surprising then that the outcome document was considered weak and ineffectual and largely undemocratic. Weaknesses included concerns that it was particularly vague in certain areas; did not tackle the issue of subsidies in respect of liberalisation of trade; poor countries were still subjected to often unfair, economic policies prescribed by the richer countries and institutions such as the IMF and World Bank; what little commitment to foreign aid was seen as unfair; and lastly any urgency regarding the MDGs was noticeably lacking. The above aside however, the outcome document was hailed as a winner among some, with a few of the more positive strengths relating to advances in the human rights front. In particular the improved ability of the international community to protect vulnerable populations with proposals emanating from the right to protect including the Peace Building Commission and a Human Rights Council. Also despite the US position, attendees did in fact reiterate the need to achieve the MDGs and furthermore restated the need for member states to continue to provide the UN with adequate resources as it too was in need (Shah 2005; UN 2005).

By 2007, it seemed that agricultural surpluses, predominantly in the European Union (EU), had resurfaced. Statistics highlighted that wine lakes of over 260 million bottles and mountains of more than 13 million tonnes of cereal, rice, sugar and milk products were being stored (Open Europe 2007). In the meantime, from an overall perspective, global investment in agriculture was by no means equal with investment of developing countries' governments falling off between 1979 to 2007 by one-third in Africa and by two-thirds in Asia and Latin America. Moreover, aid to the agricultural sector declined significantly from 18% of total assistance to just 4.6% over the same period.

At this time, the difficult GMO situation that had started in the 1990s was now beginning to show progress, although reluctantly.

16.1.5 The GMO Backlash

Since the 'de facto' EU ban on GM foods in the last decade, the United States in 2003, along with several other countries, officially complained to the World Trade Organization (WTO) of the EU's violation of international trade agreements. At first, EU officials questioned the action; however, by June it seemed the EU Parliament had diffused a potentially difficult political situation by ratifying a UN biosafety protocol, regulating international trade in genetically modified foods. However, the protocol still allowed countries the provision to ban imports if they felt there was not enough scientific evidence regarding the safety of the products. By July of the same year, two further European laws effectively ended the controversial moratorium. The first required labelling of food containing more than 0.9% of GMOs, applied to both human and animal feed (although this did not take into account animals fed with GM crops). The second made it mandatory to label any food contaminated by non-authorised GMOs of over 0.5% for a period of 3 years, after which all non-authorised GMO-contaminated food would be banned. These laws sought to ensure the traceability of GMO products from sowing to final product, albeit with the exception of imported GM-fed livestock. Although these laws effectively ended the embargo in 2004, the WTO ruling found in 2006 that the ban had violated international trade agreements. Although the ruling did not question the right of EU countries to future bans if there was sufficient evidence to warrant such measures. This left both sides claiming victory and several environmental groups claiming it achieved little or no real change. Furthermore resistance was, and is still exercised by farmers and consumers alike (Section 18.2.1) (EC 2003; Reinhardt and Ganzel 2003; FOE 2006).

16.1.6 Food Crisis and Financial Chaos

16.1.6.1 Global Food Price Rises and the Food Riots

It was about this time (2006–2008) that a massive surge in worldwide food prices began. Triggered by several factors, the resultant estimated prices between 2007 and 2008 rose significantly, with rice recording an increase by as much as 150%, corn by 75% and wheat by 35%. It was the result of a combination of factors and began with a drought, the worst for a century which hit Australian farmers and halved their wheat production in 2007. Underpinning this was the increasing trend of economic growth in such countries as China, India, Brazil and Russia and with this growth, individuals began eating more meat-based foods. Considering that it takes approximately 6.5 kg of wheat to produce 1 kg of beef (off-the-bone), the picture begins to come together. Furthermore, oil prices spiked and once at $60 a barrel, biofuels* then became more competitive. The United States, for instance, in a bid to reduce reliability on oil-exporting countries, increased conversion of their corn and other crops into ethanol and bio-diesel; however rather than import these crops, the United States attempted to be self-sufficient and diverted approximately a third of its corn crop for biofuel production (Appendix J.6). These factors collectively ended up pushing the price of wheat and other cereals up considerably. On top of this, a relatively new phenomenon of excessive food commodity speculation was understood to substantially contribute to the rising prices of both food and fuel. On this last point Frederick, Kaufman, contributing editor of *Harper's Magazine*, wrote an interesting article entitled 'How Wall Street Starved Millions and Got Away With It' in 2010. In this article he succinctly summed up the speculative role of large banks in artificially inflating food commodity prices of unconscionable (Phillips 2008; Vallely 2009; Kaufman 2010). However, as food prices spiked between 2006 and 2008 so

---

* Biofuels are made from corn, sugarcane/beet and cassava among others. Crops are converted into ethanol (more in Appendix J.6) as a greener alternative to fossil-based fuels. The industry is huge, led in many instances by Brazil, which incidentally has had the longest-running commercial biofuel programme (1970s). In 2007/08, the CFS estimated that a whopping 4.7% of annual cereal production was used in biofuel production (CFS 2008).

global unease was felt. In low-income countries, high import bills created domestic inflationary pressures, although disruption was not confined to this group or region alone. Soaring prices were leading to political instability with governments forced to control the prices of some basic staples. In Russia, companies were forced to freeze milk and bread prices while in Argentina people boycotted tomatoes when they became more expensive than meat. In Italy too, a one-day boycott of pasta was organised in protest at rising prices while in other areas food riots occurred, or nearly occurred in India, Yemen, Mexico, Burkina Faso and others, as market-led shortages and food price inflation soared (Vidal 2007).

Meanwhile, it was not long before that a global financial crisis filtered through the system to affect ordinary people, exacerbating conditions of poverty and hunger.

16.1.6.2 Financial Crisis

The financial crisis of 2007/08 began with institutions like the high-street banks who, chasing increased revenue, moved into investment banking spheres and began trading risk. Meanwhile, investment banks, not content with trading risk, also moved into the home loan market. Book-held risks were increasing and ultimately, cleverly created and vastly complex financial instruments effectively suffered a crisis of confidence, and trust in the whole system began to wane. Finally, the collapse of the US sub-prime mortgage market set in a chain reaction on a global scale with the result that the previous extended period of economic growth suddenly ceased and began to reverse trend. The housing market collapsed in many industrialised economies creating ripples throughout the world. Lending slowed and in some cases all but dried up and as lenders, fearing their risk exposure, began calling in loans. As confidence further tumbled, financial runs on many institutions was swift and brutal. Many institutions lacking sufficient deposits or reserves quickly went to the wall while others turned to governments for bail-out. The new capital injection, however, did little to assuage the frontal assault and as the crisis slowly began to subside so banks and financial institutions' lending policies became strict and counter-productive. This tightened the liquidity of the financial markets and led to widespread global disorder that had immediate and devastating impacts on food security (Rudd 2009).

The financial crisis had a profound effect on poverty and hunger in many developing countries by restricting their access to the financing they required to purchase their food requirements. Moreover, challenges of climate change and bioenergy became more relevant in light of the financial troubles and as a result, the UN called for a series of consultations and expert meetings on climate change and bioenergy in preparation for a High-Level Conference on World Food Security and the Challenges of Climate Change and Bioenergy in 2008. Before this conference, however, a meeting chaired by UN Secretary-General Ban Ki-moon and other UN agency heads strategised ways to address the global food crisis. The result was the High-Level Task Force (HLTF) on the Global Food Security Crisis. The HLTF was chaired by Ban and included the heads of the World Bank, the IMF, the World Food Programme, the FAO, the IFAD and the WTO, who collectively developed an action plan for discussion at the forthcoming Conference.

16.1.7 High-Level Conference on World Food Security: The Challenges of Climate Change and Bioenergy

The conference combined two strands affecting food security: climate and biofuels. It was a timely meeting engaging concerns over the recent food price rises, its causes, consequences and solutions as well as continuing climate concerns and fears of a runaway biofuel industry. In addressing the concerns of biofuels, it was noted that although there was no consensus over biofuels' effects on the global prices of feed and food crops, hope was expressed in the ability of second generation biofuels that would concentrate on bio-mass rather than food crops. It was also pressed home that whatever the future for biofuels, 'bioenergy should be developed in a socially, environmentally responsible way' (Campbell-Platt 2008). One major question that arose was whether in the existing climate it was more prudent to concentrate on increased food production rather than biofuel crops.

With regards to climate change, it was understood at the very beginning that the effects of climate change on the pattern and distribution of global agricultural production could be one of the biggest challenges that the world faced. In conclusion, the conference recognised the fragility of the world's food systems, in particular their vulnerability to shocks and the need to address questions of how to increase the resilience of food production in the face of challenges posed by climate change. In this endeavour, several suggestions were forthcoming. These included the importance of biodiversity in sustaining future production; as well as the recommendation that financial mechanisms, investment, technological development, transfer and dissemination should all combine to support climate change adaptation and mitigation programmes. Consideration too was also given to expected continuing high prices with action recommended by the conference to protect the world's most vulnerable countries and populations (Campbell-Platt 2008). Despite these commitments however, the situation was still dire a year later and the new State of Food Insecurity report announced over 1 billion undernourished, sounding alarm bells around governmental and institutional halls alike. It seemed that the situation was rapidly deteriorating and in 2009 the Third World Food Summit was held.

16.1.8 The Third World Summit on Food Security

Once again heads of state met in order to mitigate the worst effects of the financial crises on world food security. On top of this problem, other issues on the agenda included the implementation of the reform of global governance within the food security field, rural development and trade considerations. Interestingly, Jacques Diouf took the absence of several key leaders as indication of the lack of interest in the hunger problem. However, perhaps of greater note was the fact that out of this meeting one important view not expressed before at such a high-level meeting was the loss of confidence in the global market system as a means of ensuring food security. Venezuela put this down to fundamental problems of the capitalist system whereas others downplayed the crisis, suggesting that it was in fact just a glitch, or market jitters in response to food shortages. Whatever the reason, these jitters firmly placed food self-sufficiency back on the agenda, which was welcomed and opposed by an equal number of people (Christiaensen 2009; IISD 2009).

The end of the decade once again bought reflection and a renewed promise of achieving the MDGs by 2015.

16.1.9 2010 UN Summit on MDGs

Ten years after the groundbreaking Millennium Summit and the introduction of the MDGs, a summit was convened to explore how well these goals were being implemented. Pre-summit reports outlined a mixed bag of successes, failures and progress, which were then discussed at the meeting and in many parallel events in New York. With only 5 years to go, many felt that the goals were slipping further out of reach; however, in an one-side event titled 'Raiding the public till', general consensus was reached regarding the resources for achieving the MDGs by 2015. It was suggested that such resources would not indeed be out of reach

> … if governments were willing to reverse the process of socializing losses and privatizing gains—a process that characterizes the current model of globalisation. (Caliari 2010)

That said, it was also postulated that progress would only be made if civil society and trade unions provided counterbalances to the financial sector while others insisted the best way forward was to tap into the forces of economic growth through entrepreneurship and trade.

Ultimately, the outcome document of the conference called 'Keeping the Promise—United to Achieve the Millennium Development Goals' outlined an Action Plan for the achievement of the goals to 2015. It was acknowledged too that the MDGs were interconnected and mutually reinforcing and that in the absence of a single panacea, member countries were encouraged to design and implement specific development strategies tailored to their own situations. President Obama also emphasised a shift in development strategy. In his address to the UN, like his predecessor Bush, he aimed to shift the focus of measuring success of any of the MDG goals and development, in general, away from the long-held measures of financial and resource inputs to a more result-oriented output. This effectively, in theory at least, takes the politics out of target achievements, as targets generally speak for themselves. Lastly, the conference reiterated its goal of investing in countries committed to their own development and was marked by calls for action, pledges of support and recommitment to the previously established MDGs (UNDP 2011).

16.2 Health and Nutrition

By 2000, the FAO and WHO decided once again to convene an expert meeting to consider progress in the field of nutritional science. This time, they were joined by the United Nations University (UNU) and together met at FAO headquarters in 2001. Past research focus and findings were assessed and new, thoroughly tested methodologies such as the stable isotope technique or doubly labeled water method (DLW), a form of indirect calorimetry, were discussed. The consultation also recognised the technology gap that held developing countries back (FAO 2002a). The report of the meeting was published in 2004 and out of this report, several developments emerged. After many years of research and methodological tweaking, it was finally affirmed that the calculation of energy requirements for all ages should be based on measurements of energy expenditure. Also new energy requirement values for infants, children and adolescents were given and improved recommendations for physical activity levels (PAL) required to maintain fitness and health were made (FAO 2002a). While the previous 1985 expert consultation relied on a standard single PAL value, light, moderate and heavy load, the 2001 consultation recognised that it was not prudent to base daily energy calculations on the basis of a single presumed work load. Therefore calculations of PAL were now made based the energy cost of activities as a multiple of BMR per minute known as the physical activity ratio (PAR). Table 16.1 is a an example of various activities as used in the report.

Also, while measurements of total energy expenditure (TEE) by the DLW method was considered a gold standard approach, consideration was made in respect of the lack of sufficient worldwide data therefore recommendations at this meeting were based on the factorial approach using BMR and PAL to calculate TEE. An example of a moderate activity level factorial analysis can be seen in Table 16.2.

On a tangent, the oft-misused term 'hunger' came to the attention of the USDA's Economic Research Service (ERS) about this time. While they acknowledged that hunger would be of great benefit as an indicator to planners and policy makers, it was felt that hunger meant different things to different people and fell anywhere along a spectrum of mild food insecurity to prolonged clinical undernutrition. For the ERS, the problem was clear; hunger was an emotive term that lacked adequate definition in the food security arena and more importantly was sufficiently vague to defy precise meaning in statistical terms (USDA 2009). Another term that popped up around this time built on the previous almost three decades' concern over primary health care issues and was termed 'health security'. Health security came by as a concept out of the Health Extension Programme/Package (HEP) initiatives, which comprised both curative and preventive health services in a community-based delivery system aimed at creating a healthy living environment. Thus the main objective of HEP was to improve access through community-based health services with a strong focus on sustained preventive health actions and increased health awareness. Under these conditions, health security was only achieved when

> ... individuals or households have the capacity to identify, prevent, and manage significant risks to their health, supported by optimal health technologies, capable institutions ... and appropriate public policies. (Teweldebrhan 2005; Behailu et al. 2010)

Meanwhile the World Bank too was coming under fire for its less than stellar performance to date.

TABLE 16.1

Physical Activity Levels of the 2001 FAO/WHO/UNU Energy Consultation Report

PAL Categories Associated with Habitual Physical Activity	
Category	PAL value
Sedentary or light activity lifestyle	1.40–1.69
Active or moderately active life style	1.70–1.99
Vigorous or vigorously active life style	2.00–2.39[a]

Source: Adapted from Human Energy Requirements Report of a Joint FAO/WHO/UNU Expert Consultation: Rome, 17–24 October 2001, FAO/WHO/UNU, Rome, 2004. With permission.

[a] PAL value of 2.40 is difficult to maintain over prolonged periods.

TABLE 16.2

Example of a Factorial Calculation Energy Expenditure for a Population Group Engaged in a Moderately Active Lifestyle

Main Daily Activities	Time Allocation Hours	Energy Cost[a] PAR	Time × Energy Cost	Mean PAL[b] Multiple of 24-hour BMR
Sedentary or light activity lifestyle				
Sleeping	8	1	8.0	
Personal care (dressing, showering)	1	2.3	2.3	
Eating	1	1.5	1.5	
Cooking	1	2.1	2.1	
Sitting (office work, selling produce, tending shop)	8	1.5	12.0	
General household work	1	2.8	2.8	
Driving car to/from work	1	2.0	2.0	
Walking at varying paces without a load	1	3.2	3.2	
Light leisure activities (watching TV, chatting)	2	1.4	2.8	
Total	**24**		**36.7**	**36.7/24 = 1.53**
Active or moderately active lifestyle				
Sleeping	8	1	8.0	
Personal care (dressing, showering)	1	2.3	2.3	
Eating	1	1.5	1.5	
Standing, carrying light loads (waiting on tables, arranging merchandise)[c]	8	2.2	17.6	
Commuting to/from work on the bus	1	1.2	1.2	
Walking at varying paces without a load	1	3.2	3.2	
Low intensity aerobic exercise	1	4.2	4.2	
Light leisure activities (watching TV, chatting)	3	1.4	4.2	
Total	**24**		**42.2**	**42.2/24 = 1.76**
Vigorous or vigorously active lifestyle				
Sleeping	8	1	8.0	
Personal care (dressing, bathing)	1	2.3	2.3	
Eating	1	1.4	1.4	
Cooking	1	2.1	2.1	
Non-mechanised agricultural work (planting, weeding, gathering)	6	4.1	24.6	
Collecting water/wood	1	4.4	4.4	
Non-mechanised domestic chores (sweeping, washing clothes and dishes by hand)	1	2.3	2.3	
Walking at varying paces without a load	1	3.2	3.2	
Miscellaneous light leisure activities	4	1.4	5.6	
Total	**24**		**53.9**	**53.9/24 = 2.25**

Source: From Human Energy Requirements Report of a Joint FAO/WHO/UNU Expert Consultation: Rome, 17–24 October 2001, FAO/WHO/UNU, Rome, 2004.

Notes: Multiples of BMR in the above table are based on MJ (mega joules) per day; therefore to arrive at the caloric value based on 1 kJ = 0.239, equals MJ × 239.

[a] Energy costs of activities, expressed as multiples of basal metabolic rate, or PAR, are based on Annex 5 of the 1985 report (WHO, 1985).

[b] PAL = physical activity level, or energy requirement expressed as a multiple of 24-hour BMR.

[c] Composite of the energy cost of standing, walking slowly and serving meals or carrying a light load.

16.2.1 The World Bank

With the World Bank's entry into the field of nutrition in the early 1970s, it did so on the back of scathing criticism of WHO, FAO and UNICEF programmes and policies. The main charge was that nutritional issues were such that only large-scale interventionism would suffice and that the World Bank was uniquely positioned to offer such assistance. However, in a review of the Bank's progress in nutrition,

Richard Heaver, consultant to the Health, Nutrition, and Population Family (HNP) of the World Bank's Human Development Network, ironically summed up its efforts to date:

> ... though the Bank has developed the economic justification for large-scale investment in nutrition, and has the experience needed to scale up, it has failed to do so. (Heaver 2006)

Furthermore, Heaver noted that the Bank spent a paltry 2.5% of its human development lending budget on nutrition. This was in stark contrast to the initial noises made by the organisation in the early 1970s. This was an inconsistency that also defied the World Bank's own research, which identified nutrition as one of the best economic investments that could be made in a country's development programme. The review also challenged the disparate manner in which the Bank's different departments working on nutrition operated in a vacuum of coherence. In response, greater direction and improved information dissemination among groups within the Bank, and for the public at large was recommended. Heaver also suggested that if the Bank wanted to place more importance on nutrition, a branch-wide re-evaluation of practices within and without, the specific nutritional departments would need to be undertaken (Heaver 2006).

16.3 Period Commentary

This decade began where the 1990s left off: conference after meeting, and meeting after discussion. Much hope too was invested in the new millennium and once again global governance took centre stage. The Millennium Summit was a benchmark, a sea change in policy outlook, and for the second time in less than 5 years, the international community set real and tangible targets. The echo of past failures was drowned out by the fanfare of the Millennium Development Goals. The sombre rhetoric of hopelessness was replaced with a new optimism on the horizon. Laudable targets too sought to halve the prevalence of abject poverty as well as the proportion of people suffering from hunger. Goals were also set towards achieving in the reduction of child and maternal mortality as well as disease and infections like HIV/AIDS and malaria. Furthermore, other targets, aimed at promoting better environmental stewardship, were put in place as were commitments to promote gender equality and universal primary education. All this too was targeted to be achieved by 2015 and under the righteous banner of moral indignation: enough was enough, a new millennium, a new dawn.

As commendable as these aims were, the real coup came with the explicit recognition of the interdependence of the goals themselves. For instance, it was well understood by that time that poverty was a root cause of chronic hunger and conversely, hunger itself was an important protagonist of poverty (MDG 1). However, that this mutually reinforcing synergistic relationship also spawned other equally deleterious interactions was also widely accepted. In this way, hunger had been shown to adversely affect health and susceptibility to disease (MDG 6) while disease and infection undermined nutritional uptake. Further, such infection and disease was known to be promoted through poor sanitation and dirty water (MDG 7). It was also acknowledged that good inroads into mitigating some of the above involved the promotion of gender equality (MDG 3), which could be considered part of a larger universal primary education (MDG 2) programme. Progress too was also predicated on making inroads into the global trade and financial sectors (MDG 8). Thus the full interplay of food security was now mirrored in the MDGs. In summary, here we had in one groundbreaking meeting the motive, means and understanding to finally tackle the problem of hunger and malnutrition, a situation that had unerringly defied all previous efforts to resolve.

Early in the decade too, the food security definition had another makeover. The Committee on World Food Security's (CFS) broadened concept embraced the idea of food safety and nutritional balance; not only this, the refined concept also introduced an element of social consideration leading to a definition that was becoming as cumbersome as it was unwieldy.

Then came the 1996 Summit +5. Reality had once again kicked in and amid somewhat muted enthusiasm, targets were reiterated, goals were re-affirmed and objectives and priorities restated. The reality was that little progress had in fact been made and with still over 840 million people undernourished, the target of halving the number of undernourished was looking difficult if not impossible to achieve. An interesting caveat here is about the two separate goals: the 1996 goal to halve the *number* of undernourished and the MDG goal of halving the *proportion*. This research has yet to uncover any real good reason as to why the UN would attempt to consider two different targets like this. The only conclusion one can reasonably come to is that the second MDG target is a more achievable target, whether through cynicism or realism is yet to be determined.

Not long afterwards, a food price bubble began and rippled around the world. A convergence of events had colluded to rise prices many-fold over a short period and for a while protests and civil unrest threatened to escalate. The governments had to step in and in a series of high-level meetings sought to defuse the situation. Rising food prices too had a depressing effect on food aid as such aid is usually determined by budget; the same budget buys less food effectively reducing food aid at a time when it was most needed. On the positive side though, higher prices did encourage domestic growth offsetting some of the higher import costs while increased domestic growth too led to surpluses, which could be exported, earning extra valuable foreign currency. However, weighing up the overall advantages between the consumers against gains to farmers is difficult to determine who were the winners and losers. The Overseas Development Institute (ODI) suggests on this score that the overall economy actually suffers. They observe that as consumer spending on food increases, so reduced spending on other goods and services slows down overall national economic growth (Phillips 2008).

However, worse was to come and amid some measured government intervention successes, another storm was brewing. This time a global financial crisis threatened to reshape the global economic landscape in ways that were difficult to imagine. Odious debts wrapped up in the sub-prime housing market were unravelling, setting in motion a chain reaction that saw many casualties of new and centuries-old institutions crumble under their own stupidity. In its wake many picked over the bones of an economic model in crisis; capitalism, it seemed, came to be questioned at the highest echelons of the United Nations. However, although classical economic theory prevailed, the harrowing aftermath witnessed the hungry and malnourished rise to just over a billion people worldwide by 2009. Set against the minimal gains in lower malnutrition numbers over the preceding years, even decades, it was a devastating blow. Simultaneously, meeting after meeting had been organised to deal with general progress over targets and commitments. Now though these meetings took on a sense of urgency. Many meetings were held including the Sixteenth Session of the Commission on Sustainable Development (2008), ECOSOC's Special meeting on the Global Food Crisis (2008), High-Level Conference on World Food Security, the Challenges of Climate Change and Bioenergy (2008), G8 Summit (2008), High Level Meeting on Food Security for All (2009), G8 Summit (2009), High-Level Expert Forum: How to Feed the World in 2050 (2009) and Thirty-Fifth Session of the Committee for World Food Security (2009), and eventually, the Third World Summit on Food Security in 2009 and the 2010 Summit on MDGs.

Meanwhile a growing number of government agencies, organisations and communities were adopting the environmental accounting and ecological footprint models as indicators of sustainable resource use. Using one particular methodology, things were not looking good; the latest 2010 report of the Ecological Footprint Atlas highlighted that the per person demand on the Earth's resources in 2007 was equivalent to 2.7 hectares and the Earth's biological capacity was in the region of 1.78 hectares. According to their methodology, this represented a 34% overshoot; in other words, what was consumed in 2007 by the population would take the Earth just over one-and-half years to replenish naturally.

It can be seen then that the challenges of feeding an extra 2.3 billion people by 2050 whilst tackling poverty and hunger are not to be underestimated. Through the efficient use of sustainable resources adapted to climate change, rising water shortages, and competition over land and agriculture will have to find an additional 1 billion tonnes of cereals to meet the anticipated 50–70% increased food needs to feed the expected 9.2 billion people by 2050 (FAO 2009a).

16.4 Key Dates, People, Acts, Reports and Surveys

The new millennium was a profitable time for food security analysis. Table 16.3 places some of these highlights in perspective.

TABLE 16.3

Key Dates: The Twenty-First Century

Date	People, Organisation, Agencies, Acts or Conventions
2000: International Strategy for Disaster Reduction (ISDR)	In response to the rise of natural disasters worldwide, the UN adopted the UN/ISDR to build upon the work of the International Decade for Natural Disaster Reduction 1990–99 (UN 2000b).
2000: The International Food Relief Partnership Act	A US initiative to expand international food-aid programmes by grants to non-profit organisations to stockpile food; authorising the USAID to procure and store commodities overseas in anticipation of future emergencies (CBO 2000).
2000: EC-FAO Food Security Information for Action	The Information for Action Programme provides countries undergoing crisis or economic transformation the assistance to formulate more effective policies with the aid of interalia early warning systems, vulnerability mapping, nutritional surveys, statistical databases, needs assessments and policy analysis and formulation.
2000: UN Millennium Summit	Held at UN headquarters in New York the summit examined the challenges facing humanity in the twenty-first century. At the meeting, world leaders ratified the UN Millennium Declaration striving to 'free our fellow men, women and children from the abject and dehumanizing conditions of extreme poverty'. The Declaration's eight goals were outlined as a way to achieve these goals from which the Millennium Development Goals were developed (UN 2000a).
2000: The UN's Millennium Development Goals (MDG)	The Millennium Declaration set out a framework outlining a response to the challenges expressed at the Millennium Summit. Extensive consultation of these challenges established the Millennium Development Goals (MDGs), a set of quantifiable time-bound (2015) targets addressing such challenges (UN 2000c; UNDG 2009).
2001: Chief Executives Board (CEB) for Coordination	In 2001, as a result of a review 2 years earlier, the Administrative Committee on Coordination (ACC) became the United Nations System Chief Executives Board (CEB) for Coordination (Mezzalama et al. 1999; CEB 2010).
2001: The Earth Policy Institute (EPI)	The Earth Policy Institute (EPI) was founded by former president of the Worldwatch Institute, Lester Brown towards sustainable environmental policy solutions (EPI 2010).
2001: Global Environmental Change and Food Systems (GECAFS)	GECAFS launched a 10-year programme in 2001 of interdisciplinary research focused on understanding the links between food security and environmental change (GECAFS 2010).
2001: EuropeAid	EuropeAid's goal is to combine, to the extent possible, the EU Commission's external aid into one centralised body. As a result, the EU, through its bilateral and multilateral combined donations effectively became the world's biggest aid donor (ODI 2000; Mousseau 2005; EuropeAid 2009).
2002: World Summit on Sustainable Development (WSSD)—Earth Summit 2002	Known as the Johannesburg Summit, the meeting aimed to continue and build on the efforts of the 1992 Earth Summit and to adopt concrete steps and identify quantifiable targets for implementing Agenda 21 (UN DESA 2006).
2002: World Food Summit +5	The World Food Summit: Five Years Later renewed calls for the acceleration of effort to reduce world hunger. Up to this point it was 'the largest-ever global gathering of leaders to address hunger and food security' (FAO 2002c).
2002/4: Water Footprint Network (WFN)	A form of Environmental Accounting popularised by Arjen Hoekstra that uses the idea of virtual water or the water footprint to calculate global water use (WFN 2009).
2003: The Global Footprint Network (GFN)	GFN is a think tank looking at global environmental issues based on the concept of Ecological or Environmental Accounting (GFN 2009).
2003: International Alliance Against Hunger (IAAH)	Established on World Food Day 2003, the IAAH is designed to facilitate action and political will through partnerships of government and civil society to fight hunger and poverty (FAO 2003a; IAAH 2009).

continued

TABLE 16.3 (Continued)

Key Dates: The Twenty-First Century

Date	People, Organisation, Agencies, Acts or Conventions
2004: UNU Institute for the Environment & Human Security (UNU-EHS)	UNU-EHS is the UN University's Institute for the Environment & Human Security (UNU-EHS). It focuses on environmental hazards, degradation and vulnerability and societies' coping capacities (UNU-EHS 2005).
2004: Millennium Challenge Corporation (MCC)	MCC is an independent US foreign aid agency offering Compact and Threshold Grants to poor countries in their fight against poverty and economic development (MCC 2010).
2004: The Integrated Food Security Phase Classification (ICP)	Calls for improved analysis, greater comparability, increased rigor and greater transparency of evidence in food security analysis in Somalia led to the creation of the IPC. It has been developed as an add on to existing frameworks that helps classify the severity of food security situations (IPC 2007; FSAU 2010).
2005: The Central Emergency Response Fund (CERF)	Many shortcomings exist in the UN emergency relief model; funding, slow delivery and uneven distribution of resources. Responding to these problems the UN in 2005, set up CERF, the Central Emergency Response Fund replacing the older Central Emergency Revolving Fund. Its aims were to speed up the process and reliability of humanitarian aid (CERF 2007).
2005: UN World Summit	A summit discussing *inter-alia* progress on the Millennium Development Goals and the world's re-commitment to them (UN 2005).
2008–2013: Food and Nutrition Technical Assistance II Project (FANTA-2)	The second programme of FANTA, FANTA-2 extends the work of the original project of improving nutrition, policies, strategies and programmes of food security (FANTA2 2010).
2008: The Special Unit on Commodities	An autonomous unit on commodities under the aegis of UNCTAD it helps developing countries to respond to the challenges and of commodity markets (UNCTAD 2009).
2009: FAO World Summit on Food Security	The worsening of the global food security situation and fear that high food prices would push the undernourished over one billion led to an emergency food summit. The summit re-enforced the idea that poor countries need economic development and policy tools to boost agricultural production and recognised too the persistent lack of international coherence in global governance (FAO 2009b, 2009c).
2010: UN Summit on MDG's	United Nations Secretary-General Ban Ki-moon hosted the UN High-level Plenary Meeting of the MDG summit with the aim of accelerating progress towards all the MDGs by 2015.
2010: FAO Working Group on Animal Genetic Resources for Food and Agriculture	FAO Meeting to assess the genetic diversity and importance of livestock at the regional and global levels (FAO 2010).
2011: FAO & German Federal Ministry of Food, Agriculture and Consumer Protection IMCF Conference	Consultation on improving the dietary intakes and nutritional status of infants and young children through improved food security and complementary feeding in young children in Asia studying the effectiveness of approaches which combine food security and nutrition interventions.

References

Behailu, S., G. Redaie, D. Mamo, D. Dimtse and P. Newborne (2010). Promoting Sanitation and Hygiene to Rural Households in Snnpr, Ethiopia Experiences of Health Extension Workers and Community Health Promoters. Ethiopia: ODI/ Research-inspired Policy and Practice Learning in Ethiopia and the Nile Region (RIPPLE), 35pp.

Caliari, A. (2010). MDGs Still Achievable, but Role of Public Sector Is Key. Cooperation Internationale pour le Developpement et la Solidarite (CIDSE). http://www.cidse.org/Area_of_work/MDGs/?id=1731 (accessed 21 February 2011).

Campbell-Platt, G. (2008). Report on High-Level Conference on World Food Security: The Challenges of Climate Change and Bioenergy. High-Level Conference on World Food Security. Rome: FAO.

CBO (2000). *H.R. 5224: International Food Relief Partnership Act of 2000* US Congress.

CEB (2010). The Chief Executives Board (CEB): Overview. The Chief Executives Board. http://www.unsystemceb.org/reference/ceb (accessed 23 March 2010).

CERF (2007). Central Emergency Response Fund: What Is the CERF? United Nations. http://ochaonline. un.org/cerf/ContactUs/tabid/1830/language/en-US/Default.aspx (accessed 12 June 2010).

CFS (2003). Report of the 29th Session of the Committee on World Food Security. Rome: Committee on Food Security.

CFS (2008). Agenda Item II: Assessment of the World Food Security and Nutrition Situation. Committee On World Food Security: Thirty-fourth Session, Rome, Food and Agriculture Organisation.

Christiaensen, L. (2009) 2009 World Summit on Food Security. *Our World*. http://ourworld.unu.edu/en/2009-world-summit-on-food-security/(accessed 30 January 2011).

COAG (2003). Report of the 17th Session of the Committee on Agriculture. Rome: Committee on Agriculture.

EC (2003). Press Release: European Commission Regrets Us Decision to File WTO Case on GMO's as Unnecessary. Brussels: European Commission.

EPI (2010). The Earth Policy Institute: About. Earth Policy Institute. http://www.earth-policy.org//indcx.php?/about_epi/ (accessed 21 May 2010).

EuropeAid (2009). About EuropeAid. European Commission. http://ec.europa.eu/europeaid/who/about/index_en.htm (accessed 13 February 2010).

FANTA2 (2010). About Food and Nutrition Technical Assistance II Project (FANTA-2). US Agency for International Development. http://www.fantaproject.org/ (accessed 15 July 2010).

FAO (2001). The State of Food Insecurity in the World 2001. Rome: Food and Agriculture Organisation.

FAO (2002a). Food, Nutrition and Agriculture. Rome: FAO Food and Nutrition Division.

FAO (2002b). Report of the World Food Summit: Five Years Later. World Food Summit: Five Years Later. Rome: Food and Agriculture Organisation.

FAO (2002c) World Food Summit: Five Years Later Reaffirms Pledge to Reduce Hunger. *Summit News*: http://www.fao.org/worldfoodsummit/english/newsroom/news/8580-en.html.

FAO (2003a) FAO Urges International Alliance against Hunger on World Food Day. *FAO Newsroom*: http://www.deptan.go.id/kln/wfd/dg_wfd.htm.

FAO (2003b). Trade Reforms and Food Security: Conceptualising the Linkages: Food Security: Concepts and Measurement Rome: Food and Agricultural Organisation.

FAO (2009a). 2050: A Third More Mouths to Feed - Food Production Will Have to Increase by 70 Percent—FAO Convenes High-Level Expert Forum. *Media Centre*. FAO. http://www.fao.org/news/story/0/item/35571/icode/en/ (accessed 12 February 2011).

FAO (2009b). Declaration of the World Summit on Food Security. World Summit on Food Security, Rome: Food and Agriculture Organisation.

FAO (2009c). Secretariat Contribution to Defining the Objectives and Possible Decisions of the World Summit on Food Security on 16, 17 and 18 November 2009. Rome: Food and Agriculture Organisation.

FAO (2010). Status and Trends of Animal Genetic Resources—2010. *Intergovernmental Technical Working Group on Animal Genetic Resources for Food and Agriculture*. Rome: FAO 89.

FOE (2006). Looking Behind the Us Spin: Wto Ruling Does Not Prevent Countries from Restricting or Banning Gmos. *Briefing Paper*. Amsterdam: Friends of the Earth International.

FSAU (2010). Food Security Analysis Unit for Somalia: FAQ, FSAU, http://www.fsausomali.org/indox.php@id=32.html (accessed 8 May 2010).

GECAFS (2010). Global Environmental Change and Food Systems. GECAFS. http://www.gecafs.org/contact.html (accessed 23 June 2010).

GFN (2009). Global Footprint Network: At a Glance GFN. http://www.footprintnetwork.org/en/index.php/GFN/page/at_a_glance/ (accessed 20 June 2010).

Heaver, R. (2006). Good Work—But Not Enough of It: A Review of the World Bank's Experience in Nutrition. Washington, DC: World Bank: Health, Nutrition and Population Department 89

IAAH (2009). The International Alliance against Hunger: History and Vision. IAAH. http://www.iaahp.net/about-iaah/history-and-vision/en/?no_cache=1 (accessed 6 June 2010).

IISD (2009). Summary of the World Summit on Food Security 16–18 November 2009. *World Summit on Food Security Bulletin* 150(7) http://www.iisd.ca/ymb/food/wsfs2009/html/ymbvol150num7e.html (accessed 21 January 2011).

IPC (2007). The Integrated Food Security and Humanitarian Phase Classification System: Technical Manual Version 1. Rome/Somalia: FAO/ FSUA 47.

Kaufman, F. (2010). *The Food Bubble: How Wall Street Starved Millions and Got Away with It*. New York: The Harper's Magazine Foundation.

MCC (2010). The Millennium Challenge Corporation. MCC. http://www.mcc.gov/mcc/about/index.shtml (accessed 23 March 2010).

Mezzalama, F., K. I. Othman, and L. D. Ouedraogo (1999). Review of the Administrative Committee on Coordination and Its Machinery. Geneva: UN Joint Inspection Unit 31.

Mousseau, F. (2005). *Food Aid or Food Sovereignty? Ending World Hunger in Our Time*. Oakland, CA: Oakland Institute.

NGO/CSO Forum (2002). An Action Plan of the NGO/CSO Forum for Food Sovereignty. 2002 Rome NGO/CSO Forum for Food Sovereignty, Rome, International NGO/CSO Planning Committee.

ODI (2000). Reforming Food Aid: Time to Grasp the Nettle. *Briefing Paper*. London: Overseas Development Institute.

Open Europe (2007). Scandal of EU Waste; Anger over Food Mountains. Open Europe. http://www.openeurope.org.uk/media-centre/article.aspx?newsid=1759 (accessed January 23 2011).

Phillips, L. (2008). Senators Hear Testimony on Legislation to Reduce Food and Energy Prices by Curbing Excessive Speculation in the Commodity Markets. Washington, DC: Senate Committee on Homeland Security and Governmental Affairs.

Reinhardt, C. and B. Ganzel (2003) A History of Farming. *Living History Farm*: http://www.livinghistoryfarm.org/farminginthe30s/crops_03.html.

Rudd, K. (2009). The Global Financial Crisis. *The Monthly*(Feb): 20–29.

Shah, A. (2005) Draft Outcome Document Leads to Concerns of Weak Text. *Global Issues*: http://www.globalissues.org/article/559/united-nations-world-summit-2005#DraftOutcomeDocumentLeadstoConcernsofWeakText.

Teweldebrhan, H. A. (2005). Household Health Extension Package Challenges and Implementation Issues. Washington, DC: Alive & Thrive, 33pp.

UN (2000a). Millennium Summit: The UN Millennium Declaration. The Millennium Assembly of the United Nations. New York: United Nations.

UN (2000b). United Nations International Strategy for Disaster Reduction Secretariat (UNISDR). United Nations. http://www.unisdr.org/ (accessed 21 January 2010).

UN (2000c). *United Nations Millennium Declaration*: UN. A/RES/55/2.

UN (2005). 2005 World Summit Outcome. *Fact Sheet*. New York: United Nations Department of Public Information.

UN (2006). The UN Millennium Project. United Nations/UNDP. http://www.unmillenniumproject.org/ (accessed 12 February 2011).

UN DESA (2006). What Is Johannesburg Summit 2002? United Nations Department of Economic and Social Affairs Division for Sustainable Development. http://www.un.org/jsummit/html/basic_info/basicinfo.html (accessed 23 July 2010).

UNCTAD (2009). The Special Unit on Commodities. United Nations Conference on Trade and Development. http://www.unctad.info/en/Special-Unit-on-Commodities/About-the-Special-Unit-on-Commodities/Who-we-are/ (accessed 21 June 2010).

UNDG (2009). The Millennium Declaration and the MDGs. United nations. http://www.undg.org/index.cfm?P=70 (accessed 19 June 2010).

UNDP (2011). UNDP Millennium Development Goals. United Nations Development Programme. http://www.undp.org/mdg/summit.shtml (accessed 12 January 2011).

UNDP/MDG (2010). Millennium Development Goals (MDGs). United Nations Development Programme. http://www.undp.org/mdg/ (accessed 24 June 2010).

UNU-EHS (2005). Human Security in a Changing World. Bonn, Germany: UNU Institute for Environment and Human Security 33.

UOR (2009). University of Reading: Agriculture, Policy and Development, History of Agriculture. University of Reading. http://www.ecifm.rdg.ac.uk/history.htm (accessed 4 June 2008).

USDA (2009). Food Security in the United States: Measuring Household Food Security. United States Department of Agriculture. http://www.ers.usda.gov/Briefing/FoodSecurity/measurement.htm (accessed 15 October 2009).

Vallely, P. (2009). *The Big Question: Why Is So Much of the World Still Hungry, and What Can We Do About It?* London: The Independent.

Van Den Bergh, J. C. J. M. and P. Rietveld (2004). Reconsidering the Limits to World Population: Meta-Analysis and Meta-Prediction. *Bioscience* 54(3): 195–204.

Vidal, J. (2007). *Global Food Crisis Looms as Climate Change and Fuel Shortages Bite: Soaring Crop Prices and Demand for Biofuels Raise Fears of Political Instability*. London: Guardian News and Media Limited. 3 November 2007.

WFN (2009). Water Footprint Network: Why a Water Footprint Network? University of Twente. http://www.waterfootprint.org/?page=files/home (accessed 13 July 2010).

Summary of Part III

While Part I aimed to give a snapshot or cursory examination of the current food security concept, a modern understanding would not have been complete without delving into a little history; this was achieved in Parts II and III. From this and as should be becoming clear by now, modern issues of food security are increasingly being linked to wider economic and societal goals. The objective of Part III was to outline these contemporary motivational forces as they unfolded to better understand the evolution of the food security concept as a whole. Such a review then involved donning many diverse disciplinary cap including those of social historian, food scientist and technologist as well as those of governance, politics and philosophical ideology. A cautionary note must also be made too as it becomes somewhat easy to apportion motive and value judgments to historic events with the benefit of hindsight. With this in mind, I have been careful not to assume anything, rather I have sympathetically and hopefully objectively looked at tangible links between events and circumstances in a way that best tells the story of food security. Furthermore, in order to give an overview of the twenty-first century from the perspective of food security, it has not been possible to outline all of the events, conferences or ideological viewpoints that took place within the limited scope of this book. What was offered in lieu of this was a broad spectrum overview and as with such endeavours there are sure to be elements that might have been under-utilised or omitted entirely; these perhaps can be corrected in future updates.

In execution, I have attempted to illuminate the complex interplay of science, nutrition and politics and within this framework, it can be seen that the aetiology of the modern food security concept in fact had its genesis in philosophy and events much older than has previously been ascribed. This highlights the fact that we ignore the past at our peril. Lastly, I have forgone a more thorough summary of the twenty-first century in favour of these generalisations as a more comprehensive summary is inextricably wrapped up in the following sections. Instead, each period commentary serves as a de facto summary of each section.

Part IV

A Sectoral Analysis: Food Security and...

Hunger is insolent, and will be fed.

Homer
The Odyssey

Discussions and considerations of food security so far in this book have generally followed a six-pronged format agriculture, science, socio-cultural, health, environment and governance. While this does not necessarily reflect the true characterisation of the elements contained within each heading, it has been used as a convenient means of categorisation. The difficulty here is that many of these elements may actually belong to multiple categories possibly falling under several spheres of influence; in this way, cross-referencing is used to try to bring all of the threads of food security together. In following this convention the following chapter then provides summary knowledge and insights so far gained in the unfolding research of this book. However, far from being just a summary, this section also aims to synthesise many of the disparate elements discussed in previous chapters into an overarching appreciation of the breadth and complexity of the concept in its most expansive form. It should be noted too that this inclusivity is not necessarily aimed at being the definitive word on the subject, rather, it highlights the sheer breadth of consideration that needs to be taken into account when contemplating issues of food security. The boundaries, it could be argued are arbitrary however, I leave it up to the reader to determine at what point food security stops being an issue of food and becomes an issue of society.

17

Agriculture, Forestry and Fisheries

Most people would not deny that agriculture is linked to food security although beyond the obvious production issues, other links are not always overtly apparent nor are they always direct or linear.

17.1 Global Structure

In 2050 Earth's population is projected to reach just over 9 billion. Between now and then, agriculture will not only be called upon to provide the food, feed and fibre to meet present population needs, but it will have to increase productivity year-on-year by 0.35% if we are to provide the extra 70% required to meet this increased demand. Achieving these rising demands however is only winning half the battle, continuing challenges in respect of whether these demands can be met without seriously damaging the environment or excessively raising food prices into the bargain remains to be seen (GHI 2010). This comes at a time too when the agricultural landscape continues on its downward trajectory. In Western Europe and the United States, for instance, in the past two decades, these countries have seen their share of world agricultural production decrease considerably to around 1%–2% of gross domestic product (GDP). While change is slower in other quarters, with the African agricultural sector, for example, although declining, it still currently stands at between 20% and 40% of overall GDP whilst continuing to employ about 60% of the labour force.

So what does this mean for the global agricultural sector as a whole? Overall there are currently an estimated 525 million farms worldwide employing nearly 1 billion people, which is equivalent to about 35% of the global working population. Moreover, an estimated 2.6 billion men, women and children in total rely directly on these agricultural production systems for their livelihoods (GHI 2010). Looking at this more closely, the global agricultural landscape can be thought of as running on two tracks: the global track, largely driven be demand and catered for on an industrial scale; and the local track characterised by subsistence or small-scale market farming pushed along largely by necessity.

17.1.1 Global Track

Since the Uruguay rounds of General Agreement on Trade and Tariffs (GATT), global agricultural trade has become a big business increasing over the last 15 years by more than threefold. Placed in perspective though, this increase still represents only a small percentage of overall food produced at the global level. However, in this increasingly global system, a complex food chain has come to be dominated by large multinational corporations who continue to consolidate their power and leverage over consumers and agricultural producers alike. Paralleling this, industrialised farming has grown and concentrated its focus on both the domestic and international mass markets. This is becoming increasingly reflected in the propensity for large scale intensification of farming and the concomitant growth in overall farm sizes, particularly in the industrialised world (Table 17.1).

For decades, it seems growth in agricultural production has been made largely through the development of science and new technologies. This expansion of commercial-industrial relations places much strain on many small-scale or smallholder farmers in developing countries. In turn, smallholders find themselves increasingly in direct competition with capital intensive and more often highly subsidised production systems that can produce vastly increased volumes of commodities, and which are often sold more cheaply on the open market (IAASTD 2010). Furthermore, such systems are favoured by the growing strict grading and standardisation of commodity requirements of the supermarkets and wholesalers. All of which incidentally coincides with the steady decline of local food production systems.

TABLE 17.1

Farm Size by Region

Region	Average Farm Size (ha)
Africa	1.6
Asia	1.6
Latin America and Caribbean	67.0
Western Europe	27.0
North America	121.0

Source: UNCTAD report, *Technology and Innovation for Sustainable Agriculture,* Commission on Science and Technology for Development, UNCTAD, Geneva, Switzerland, 2010.

Global distribution of farms under 2 hectares

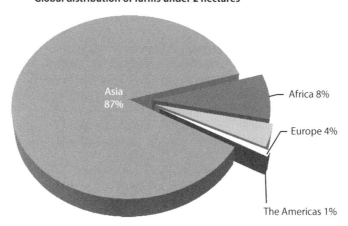

FIGURE 17.1 Regional distribution of small-scale farms. (Compiled from the IAASTD Global Report: Agriculture at a Crossroads, 2009).

17.1.2 Local Track

All in all, of the 525 million farms worldwide, between 400 and 450 million smallholder farms of two hectares or less, collectively provide a means of living (directly or indirectly) for up one-third of the global population. These farms also occupy about 60% of worldwide arable land and between them provide the bulk of many developing countries' food (Figure 17.1) (Hartmannshenn 2004). In another perspective, these smallholders effectively contribute about half the world's total food supply (Båge 2008; IAASTD 2009). It is not only the smallholders either that are productive on this small scale; among this agricultural cohort are herders or pastoralists; forest dwellers; hunters and gatherers fisher folk; gardeners and the landless and rural workers. These groups tend to be dominant in developing countries yet when such livelihoods constitute a large proportion of a country's farming activities then this predominantly subsistence-oriented approach becomes extremely vulnerable; particularly from direct drivers of change such as climate change, pests or disease, among others.

It would seem that these small farmers in developing countries have not really benefitted from the increased global traffic of agricultural trade either. International trade protectionism and the high-capital entry barriers of some industrialised farming sectors preclude any democratic involvement. This is unlikely to change either unless sufficient trade reforms in the form of reducing protectionist barriers such as subsidies and import tariffs are introduced to promote a true level-playing field. This is also unlikely to happen any time soon, or not at least until the negotiating position of the developing countries can be properly strengthened. Agriculture then can be seen as an essential element of economic growth as well as of food security of many developing countries. Crucially too however, there are many competing agricultural as well as economic development paradigms with as many advocates in each camp that push and pull at each other effectively hindering any real progressive consensus (Sections 17.6 and 22.4). Such dogmatic posturing adopted by many opposing camps however is unhelpful at best and at worst damaging to any structural solutions.

17.2 Production

(See also Science & Technology – Chapter 18.)

With the expected increase in demand from the growth of the population and the changing patterns of food consumption (Section 19.7), the challenge of meeting this demand lies with the extent that current agriculture can continue to increase production. This can only happen in one of two ways: by increasing the land under production or by increasing yields. Of course, there are also the off-farm food production enterprises but essentially these are a miniscule part of aggregate supply. Policy, technological advances and investment in land and land practices then will ultimately determine the extent to which both land and yields will provide. This is going to be particularly challenging, as yield gaps begin to close or optimum yields are reached and as well, on top of this, the extent to which available land and marginal lands can be brought into successful agriculture remains to be seen. On the policy front too, increasing production relies on a conducive environment that supports and incentivises its farmers, not just locally or in the domestic sphere but internationally too. In this regard, one aspect to watch with interest will be the impact of policy reforms in the developed world, in particular those that aim not just to reduce protectionist import or export quotas but those that also aim to reduce direct support to farmers (DFID 2004). Other policy changes might also include the direction of agricultural development strategies whether market-oriented versus subsistence farming or smallholdings as opposed to large farms. Thoughts also arise as to the degree to which countries pursue self-sufficiency. Whichever policy shifts are considered, there are consequences and some of which are not always clear-cut. Improving smallholder farms for instance (generally seen as a good idea) increases rural employment opportunities and incomes and while this increases the productivity and purchasing power of the individual, on the national scale, increased production generally leads to lower prices. Once again this is good news for the individual; however, a familiar response to such trends sees farmers raising production rather than decreasing it to compensate for lower prices. This has the effect of further depressing prices and encouraging swings where long periods of depressed prices are followed by swings in the other direction (Zhang et al. 2007). Indeed, there are ways to compensate for such swings, and it becomes imperative that suitable balance or trade-offs can be found that maximises the potential of as many stakeholders as possible.

All these considerations of course are predicated on continuing forms of sustainable agriculture. This in turn requires thought and consideration to the many farming practices that currently exist worldwide.

17.2.1 Modern Agricultural Practices

Agriculture encompasses a wide variety of specialties and techniques. In the 1960s Boserup, looking specifically at less developed countries, listed five different agricultural systems dependent on the length of fallow between periods of cultivation: slash and burn (forest-fallow; 15–20 years of fallow); bush-fallow (6–10 years); short-fallow (1–2 years); annual cropping (a few months); and multi-cropping (no fallow) (Boserup 1965). There are other systems in use too and while some of these practices have developed over time, some remain in use and unchanged throughout many parts of the world. Cropping systems vary among farms and regions and are wholly dependent on available resources and constraints such as geography and climate, government policy, economic, social and political pressures or resources as well as the philosophy and culture of the farmer. It is essential to understand here that different practices have different impacts vis-à-vis soil management, sustainability and ultimately food security.

17.2.1.1 Slash and Burn

Shifting cultivation (slash and burn) is a system in which land is cultivated temporarily and then abandoned or left to fallow for periods of time. If forest or woodland is used, it usually involves the burning of the trees and shrubland, which in turn releases nutrients into the soil. This newly nutrient-rich soil is then cultivated until such time when cropping is no longer productive at which time it is left to fallow for a period. The land is then effectively reclaimed by wild, natural vegetation until such time, many years later in some cases the land becomes fertile and can once again be farmed. This particular farming practice requires large amounts of land where at any one time the majority of the land is left to fallow and only the minority is in current use. That said, land left to fallow can still be used for firewood, timber and thatching, etc. Important considerations in the sustainability of this practice are concerned with whether there is a net nutrient loss (or degradation) over time. The skill in this system also understands how long each period of cultivation and fallow can continue without adversely affecting the soil as there is no optimum relationship between the two; it depends on many factors including time, type and intensity of cultivation, etc. Further consideration pertains to other variables such as the availability of such lands, the cultural acceptability of this type of practice and the proper management of the sometimes long fallow periods (Brush 1987).

17.2.1.2 Annual, Monocultivation and Polycultivation

Annual cultivation is the next few phases up the ladder and with it comes the beginning of agricultural intensification. This practice takes away the fallow periods (or drastically reduces them) between cropping seasons and as a result requires increased soil management and attention to the replenishment of nutrients (Boserup 1965). This usually entails greater nutrient and pest control inputs, which also has consequences in terms of financial cost sustainability issues as well as degradation (Section 21.2) and biodiversity (Section 21.5.1). Initially, while industrial advances or simply changes in agricultural paradigms led to the practice of monocultures and polycultures (multiple cropping and intercropping), and although both are in use, monocropping has become the dominant form of agriculture these days.

Monoculture: It is the practice of growing a single crop over a wide area. It is widely used in modern agriculture and is characterised by its large harvests with minimal labour (UOR 2009). It is worth mentioning too that although monoculture receives a lot of bad press, in and of itself the practice is not inherently environmentally 'unsound', rather the problems of monoculture reflect the poor management of this practice.

Polyculture: The practice of planting several crops sequentially in any one year is known as multiple cropping, while intercropping is when several crops are grown at the same time in the same area. Polycultures benefit from year-round production and diversity in the ecosystem as well as resulting in less leaching of the nutrient base.

17.2.1.3 Pastoralism or Pastoral Farming (Animal Husbandry)

These livestock production systems are often defined based on their source of feed whether grassland, mixed or landless. Ruminant livestock relies on plant material (grassland) found in shrubland, rangeland, pastures and marginal agricultural lands and is prevalent in areas where the climate or soil excludes the use of other forms of agriculture. Pastoralism then faces challenges of degradation and overgrazing and is the major source of food (milk and meat) and income, as well as a source of employment for this particular group, and this form of farming is highly vulnerable to different man-made and natural risks. In particular, being reliant on grass rangelands and ultimately the weather, another threat to this type of farming also includes the challenge of recurrent drought. Challenges faced by pastoral communities on the other hand might include the nearby expansion of sedentary agriculture or agricultural expansion projects that create land and resource competition. Another consideration in this type of farming is the conversion of traditional pastoral rangelands to wildlife parks and sanctuaries. This is particularly prevalent in Africa where hunting, wildlife parks and game reserves among others are further reducing pastoralists' available land.

One important consideration of which farming system is chosen over an other is partly determined by the climatic environment. Tropical environs allow for any or all the above practices while subtropical and arid environments are limited to such factors as sunshine and/or rainfall.

17.2.2 Agroforestry

(See also Deforestation [Section 21.2.3].)

Forests provide both employment and products either through wood and wood products or commodities as in agroforestry. Agroforestry or forest farming is the practice of combining trees and shrubs with crops and/or livestock. These can either be integrated in a spatial sense or over time (temporal) and usually concentrates on specialty high-value crops. By mixing forestry and agriculture, the biodiversity and ecological systems are wide and varied as well as robust (if properly managed), and the long-term wood harvest can be supplemented by relative short-term agroforestry. The added benefit of forest farming allows for valuable water retention and shade under the forest canopy while further protection is offered against wind and rain erosion. In general then, the successful practice of agroforestry relies on knowledgeable and careful selection of trees and crops which are needed to optimise production and to minimise any negative competitive effects. Several categories of crops are particularly suited to this type of farming. These include mushrooms, nuts, vegetables such as radish and beetroot, honey, herbs, fruits such as blueberries, elderberries, blackberries, raspberries edible flowers and sap products like maple syrup, birch sap wine.

17.2.3 Fisheries and Aquaculture

Fish is highly nutritious and contributes significantly to the animal proteins component of people's diets worldwide. In fact, it has been suggested that globally as much as 15%–20% of all animal proteins are derived from aquatic animals and singularly provides the main source of animal protein for nearly 1 billion people worldwide. Overwhelmingly much of the world's oceans, rivers and lakes are fished by small-scale artisanal fishers contributing vital food sources to the overall dietary make-up of many poor localised communities. Not surprising then that in 1996, fish and fisheries accounted for the income of an estimated 30 million people worldwide, 95% of which were in the developing countries (FAO 2011). Of note is the growing trend for these small fisheries to supply the international market and thus diverting traditional foods away from the local community in favour of incomes and alternative food sources. This came about largely as a result of growing demand in the industrialised nations and advances in processing, packing and transporting perishable goods over the last two to three decades. While this decreases the available fish for local consumption, there are those that argue that the increased processing and fisheries employment opportunities create more jobs and income. Moreover, it is argued too that by selling premium fish products on the international market, cheaper imported food substitutes effectively create a net inflow of foreign earnings. Others argue against this.

Traditional marine fishing is also receiving competition from aquaculture (or aquafarming); this is the cultivation of aquatic organisms whether fresh- or saltwater under controlled or farm conditions. According to the FAO, aquaculture is the fastest growing food-producing sector, now accounting for nearly 50% of the world's fish food (FAO 2011). Systems like these as well as more efficient processing that entails less spoilage and wastage together with a reduction of discards at sea, have the potential to considerably increase the productivity of the sector. However, poor fishery resource management has been seen to have a net negative impact on the aquatic environment and resources, creating in turn enormous pressures for the sustainability of the sector and resultant threats to long-term food security (Kurien 2004). How this develops is unclear although investment and current direction of thought seem to favour the growing aquaculture sector.

17.2.4 Intensification, Extensification, Concentration and Specialisation

By the mid-twentieth century, fears over population growth spurred a dramatic acceleration in global food production. Building on new technologies over this period and beyond with such things as agrochemicals, mechanisation and plant breeding, 'progress' has seen the intensification, extensification, concentration and specialisation of agriculture. The intensification of agriculture employs large amounts of capital relative to the land area cultivated whilst also involving large quantities of fertiliser, insecticides, fungicides and herbicides to improve crop yields and manage pests. Further, intensification also sees farming becoming increasingly mechanised with capital-intensive high-efficiency machinery for planting, irrigation, cultivating and harvesting. This kind of agriculture is often seen where population densities are high and/or where land values are also high; such situations also usually occur near primary markets. Extensive agriculture on the other hand relies on the natural fertility of the soil and availability of water. Unlike the large inputs of intensive agriculture, extensive practices relying on the natural productivity of the soil usually provide lower yields. This in turn means that large tracts of land are needed and results in this sort of agriculture taking hold in areas of low population density where land is plentiful and cheap in relation to labour and capital. It also means this sort of practice is oftentimes far from primary markets (Britannica 2009; UOR 2009).

This is just a snapshot of the many flavours of farming; whether working the land, with livestock, farming the seas or the forests, each has its benefits and each too faces many criticisms, pitfalls and challenges. Yet all are important and one method in one area might not be practical or desirable in another. The challenge in food security terms are also different with each practice; however, with regard to all of the food production methods, sustainability is of overriding importance. These are discussed in detail in Chapter 21 but now suffice it to say that land, soil and water management as well as overall good stewardship has a major impact in future on productivity, biodiversity, degradation and ultimately food security.

17.3 Sustainable Agriculture

In the past few decades, there has been concerted efforts towards implementing sustainability within agriculture. From a modern perspective too, sustainability, as has been shown, has become wholly entrenched in the idea of providing food security on a continuous basis. Due in large part to the green revolution, more food is being grown today than at any other time in human history, yet critics argue that modern agricultural practices are not sustainable. Many reasons have been cited for this situation, ranging from; the intensive application of fossil-based technologies (fertilisers, mechanisation, transport); to the reduction of biodiversity associated with modern farming techniques; to general environmental degradation (land, climate, etc.) (GFS 2011). Many such issues are global problems and addressing sustainability in general, and food security in particular is a challenge that will require a multinational, collaborative effort (GFS 2011). Sustainable agriculture is about producing food in a responsible, environmentally friendly manner. It is also about the efficient use of non-renewable and renewable resources in a manner that looks favourably on the ethos and inclusivity of the global allotment. While current inertia propels these 'apparent' modern concerns, it is worth remembering that sustainability is not a new concept either. Lack of diversity and the fear that 'Earth's becoming man-heavy' led Mehta in 1929 to quote from the expressive language of the Sanskritists when he pronounced this little nugget cautioning against short-termism:

> ... the excessive milking of the agricultural cow so that nothing is left for the nourishment of the calf. (Mehta 1929, p. 84)

With regard to fossil fuel inputs, there are strong prevailing backlashes from both the sustainability and environmental camps who both would like to see less usage overall. This touches on the whole portfolio of fossil inputs in the agricultural sector from the operation of farm machinery, transport, storage, processing and, of course, energy-intensive fertiliser use. An overarching approach with regards to overall chemical and energy inputs would see a more favourable balanced assortment of natural fertilisers, alternative fuels and restraint. Pest control measures too benefit from a similar ethos. The development of chemical pesticides and herbicides such as organochlorine and organophosphate compounds allowed for initial greater improvements in pest control and ultimately better crop yields. However, in response to the problems of pesticide abuse, overuse, misuse and public perception, the concept of integrated pest management (IPM) systems using alternative and natural controls arose and indeed feature more dominantly today. In terms of practical and sustainable use, IPM has at its disposal a range of practices in its arsenal including biological controls, such as biocontrol/predatory pest usage as well as mechanical or cultural controls like tillage, crop rotation, culling, cover cropping, intercropping, composting, avoidance and resistance. IPM aims to utilise as many of these that are feasible whilst minimising or leaving chemical pesticides and herbicides as a last resort.

Lastly in terms of agricultural paradigms and the perceived sustainability of alternative farming ideologies, these are looked at in more detail in Section 22.4.2.

17.3.1 Crop Losses and Food Wastage

One major under-researched area of sustainability touches on general food losses associated with farming practices as well as the worst profligacy of industry, retailer and consumer behaviour. According to a recent UNEP GRID-Arendal report, there has been very little effort focused on salvaging waste, especially knowing that

> Reducing such losses is likely to be among the most sustainable alternatives for increasing food availability. (Nellemann 2009)

Food losses and wastage however, is difficult to estimate with any degree of certainty; this is due in part to the fact that there is no universal method for measuring losses. Losses, or wastage, vary greatly by crop, country and climatic region (World Resources 1999). These can also be considered in respect of the different stages of food production and utilisation; these might be pre-harvesting, harvesting, handling/processing, storage, transport, retail and consumer. Placed in these terms, the potential for losses are aggregated and can be huge.

It has been variously calculated that food lost in the field between planting and harvest time can range from 20% to 40% depending on crop and wastage type (pest, pathogen, etc.) (Nellemann 2009). On the post-harvest side, estimates also vary. In dealing with fruit, for instance, the FAO contends that anywhere between 25% to as much as 50% can be lost to inefficient storage, processing or transport while up to 37% of post-harvested rice in the developing world can be lost before it is consumed (FAO 1989; World Resources 1999). When it comes to grains too, perhaps as much as 25% is also said to be lost during post-harvest handling, storage and infestation (Latham 1997). Fish does not escape our attention either and in terms of volume, the losses can be staggering with discarded produce accounting for about 23%–30% of landed catch; a whopping 30 million tonnes, which if we factor in spoilage and loss this can quickly reach as much as 40%. Worse yet, mortality figures of discarded fish at sea is not known although in some species it has been estimated to be as high 70%–80% (Nellemann 2009).

Furthermore, when it comes to a fickle public (or perhaps retailer) due to factors like the demand for perfect produce and the low financial value attached to food throughout the affluent world, losses from the retail sector as well as domestic wastage has been said to be unacceptably high. Indeed, some estimates put wastage from home use as high as 10% or in America, 25%–50% of the total economic value (depending on who you read) (Latham 1997; Nellemann 2009; WRI 2010). In one aspect of the wastage issue, it is not all bad for when it comes to wastage recovery, particularly at the industrial level, energy recovered from agricultural wastes is fast becoming increasingly economically feasible. Investments in this technology would enhance sustainability of existing systems and innovations in new management systems in expanding green energy (Nellemann 2009).

From this brief foray, it can be seen that losses due to pest, infestation, storage, processing, transport and wastage represents a major avenue of lost opportunity. The potential, in both increased availability and sustainability, is vast, so much so in fact that the currently discarded fish at sea alone could sustain an increase of more than 50% in fish supplies alone (Nellemann 2009). Moreover, the unexploited potential use of food waste as alternative sources of animal feed or for agricultural products is also considerable (ibid).

Another interesting concept when it comes to sustainability and improved production is the notion of *permaculture*.

17.3.2 Permaculture

Permaculture is an holistic approach to farming that mimics the relationships found in nature. It embodies a collection of moral and ethical values that include the education of the populace, the planning of population numbers and an environmentally friendly approach to utilising Earth's resources, which ensures only the lightest impact of agricultural systems. There are several principles laid out in achieving these goals including the following:

- To observe and interact with nature for better more efficient use of resources
- To catch and store the abundant natural energy provided by the sun, wind and water
- To encourage self-reliance rather than solely relying on global food systems, whether through marginal land use, urban gardening such as window boxes or perhaps replacing ornamental gardens with dual purpose multi-functioning aesthetic and edible or medicinal plants
- To apply self-regulation reducing inappropriate actions and behaviours
- To use and value renewable resources and services and also to produce no waste by connecting input and output elements so that they meet each other's needs

All in all, within the concept of sustainable food production, permaculture aims to be fruitful by integrating within the ecosystem rather than outside of it (Permaculture Ass. 2011).

17.4 Land Grabbing

One of the difficulties faced by poorer economies is whether to sell or lease domestic agricultural land to foreign entities. On the one hand, it provides much needed income, yet on the other, the fears are that it will lead to land conflict, evictions and increased land prices as well as water and other natural resource competition (RTFN 2010). Multinationals, governments and investment funds are buying up vast tracts of land to either ensure food security in their own countries or simply for profit. This is perhaps not surprising really when you consider that land in Africa for instance can be leased for as little as $1 a year in the cheapest areas. Paradoxically, The government of Ethiopia, in the face of some of the worst food insecurity in the world, is offering at up to 3 million hectares of its most fertile land to rich countries and while some might see this as sheer madness others offer that it brings foreign investment to the land where domestic measures have failed (Vidal 2010). The practice too is not confined to Africa either; in Romania the practice is widespread with around 12% or about one million hectares of its national stock of arable land having been sold to foreign interests (The Diplomat 2011). In other areas too like Saudi Arabia, for instance, the water stresses are encouraging the government to reduce domestic cereal production by at least 12% a year to conserve its valuable water resources whilst at the same time providing huge subsidised lending to companies interested in buying and cultivating land overseas. It is not only food pressures either that are claiming such lands. China, for instance, cultivates 2.8 million hectares of the Democratic Republic of Congo's land to produce palm oil for their biofuel industry while European biofuel companies have acquired large swathes of land, about 3.9 million hectares in Africa for their commercial purposes too. All in all the practice is widespread and involves numerous countries and organisations with mindboggling amounts of land, yet it is still something of an unknown quantity when it comes to food security (von Braun and Meinzen-Dick 2009). However, it is a practice that is gaining more and more attention on the global platform. So quick has been the phenomenon and so large the scale that the practice is receiving some measure of criticism amid fears of disruption to local food security, the lack of transparency or the perceived unfair contractual terms (ibid). While the full extent of land grabbing is not fully known, the FAO has suggested that over the last few years as much as 20 million hectares in Africa alone might have been acquired by foreign interests (RTFN 2009, 2010). Of course, as with everything else, there are benefits and disadvantages; on-the plus side, such land acquisitions might be seen as a welcome opportunity for much needed agricultural and rural investment. This, suggest proponents, will create on- and off-farm jobs for the local communities and facilitate yield-increasing practices and technologies. This, as it has been posited, can only benefit global price stability by ensuring increased supplies, more stability and more cost-effective output. Also of benefit are the trickle-down rural development projects such as new schools or health clinics that sometimes accompany such deals. However, on the flip side, there is concern in some quarters regarding whether the terms of such contracts benefit the investors more than the sellers and whether the sellers terms benefit local communities or their own interests. Furthermore, tied in with these concerns is the tangible fear that such practices will have an adverse effect on the local environment as well as on people's livelihoods (von Braun and Meinzen-Dick 2009). Ultimately whether such measures promote foreign food security at the expense of local populations or whether the advantages of big business, efficiency and investment aids in the aggregate global solution of food insecurity is one that few can properly foresee. As such this is one to watch with caution.

17.5 Food Systems

(See also Corporate Control: Transnational Corporations [Section 22.5.8].)

With regard to the supply chain, the recent past has witnessed many changes at every stage from growing, harvesting, processing, packaging, transporting, marketing, consumption as well as the disposal of food and food-related items. In each of these levels, the food chain has evolved through technological advances, globalisation and politicisation. Interestingly, in the more recent past every aspect of the food supply chain is becoming increasingly accountable at the social level. This accountability however, while a good step in the right direction, is predominantly limited to moral and ethical pressures alone and as has been seen so often, such influences can be limited. That said, this social accountability marks the culmination of much pressure both public and institutional with regard to sustainability, natural resource allocation and increased sense of fair trade (SOFI 2005; Ericksen 2008a). Global trends in the area of production over the past few decades have been identified by Erickson and others as the intensification of agriculture accompanied by a corresponding trend to larger farm sizes and the increasing fragmentation of marginalised smallholders. There has also been a large move towards 'value-added' foods in the processing sectors and a concentration of corporate businesses up and down the supply chain. Interestingly, this move (in many levels of the food chain) sees corporate concentration moving both horizontally (in the traditional model) as well as vertically, further enhancing control over the entire food system. Some of these changes were highlighted by Polly Ericksen in a recent paper contrasting food systems and their effects on societal outcomes and, building on Maxwell and Slater's work, draws attention to some fundamental shifts within the food system as a whole (Table 17.2).

In the same paper, Ericksen highlights both the independent and symbiotic relationship between food systems and environmental factors. By offering the notion that food systems and by extension food security affect environmental outcomes and that food security in turn is influenced by environmental considerations, this explicitly closes circle that sees environmentalism becoming inextricably

TABLE 17.2

Features of 'Traditional' and 'Modern' Food Systems

Food System Feature	Traditional Food Systems	Modern Food Systems
Principal employment in food sector	In food production	In food processing, packaging and retail
Supply chain	Short, local	Long with many food miles and nodes
Food production system	Diverse, varied productivity	Few crops predominate; intensive, high inputs
Typical farm	Family-based, small to moderate	Industrial, large
Typical food consumed	Basic staples	Processed food with a brand name; more animal products
Purchased food bought from	Small, local shop or market	Large supermarket chain
Nutritional concern	Under-nutrition	Chronic dietary diseases
Main source of national food shocks	Poor rains, production shocks	International price and trade problems
Main source of household food shocks	Poor rains, production shocks	Income shocks leading to food poverty
Major environmental concerns	Soil degradation, land clearing	Nutrient loading, chemical runoff, water demand, greenhouse gas emissions
Influential scale	Local to national	National to global

Source: Ericksen, P.J., *Glob. Environ. Change.* 18(1), 234–245, 2008a.

linked to the security of food (Ericksen 2008b). Furthermore, with the ongoing environmental concerns of water availability/pollution, energy use, land degradation and biodiversity, among others, this ensures the continuation of a food security concept that increasingly becomes difficult to separate from wider societal issues. In sum, the modern food chain systems are unrecognisable from those of just two decades ago and while back then good governance was spread evenly from one end the chain to the other, subsequent integration and concentration, on the governance front, is proving challenging at best. In response, this previously socially unregulated aspect of modern food culture is attracting increasing calls of public accountability in terms of fair play, good governance and open transparency.

17.6 Agricultural Development Paradigms

(See also Section 22.4, Development Paradigms.)

It has been shown that agricultural paradigms have shifted, sometimes quite radically over various time and spatial boundaries. At times, the agricultural sector has been treated as little more than a global allotment in which food was drawn out at will with little or no regard to its developmental potential in the wider economy. At others, it was noted that by investing in agriculture through rural education and industry, more labour could be quickly and readily drawn out of the sector into so-called more effective or prosperous industries. Further shifts in thinking eventually recognised the fact that poverty was predominantly a rural problem and one that could be effectively addressed through social and economic investment. Finally, it seems after a long and circular route, agricultural development paradigms have brought us back to the current concept that indeed the agricultural sector can have a valuable role to play in the overall strategy of both the economic and social development of a country or region. In this scenario, suggests McCalla (2007), the agricultural sector

> … becomes much more complex, more interdependent with the rest of the economy and charged with meeting multiple goals. [Where] Increasing food production is no longer a goal in itself. (McCalla 2007, p. 16)

Also when it comes to agricultural development paradigms, later discussion highlights the fact that there is far from widespread agreement on just how agriculture fits in with the overall development agenda. One such alternative view to the current paradigm is food sovereignty (discussed in Section 22.4.3.1). Briefly, food sovereignty is an alternative development paradigm that is in part a backlash against modern farming practices but more importantly it aims to wrestle control of the food security of millions of people away from the transnational corporations and globalised governance and places responsibility back in the hands of national interests. At the heart of this movement is the millions of livelihoods supported by small-scale farming, peasantry and herding that its proponents see as equally important in the overall global food security debate.

A cautionary note too, according to McCalla, suggests that despite progress, this is a debate that is not yet over and agriculture's role in future development strategies is perhaps one to watch.

17.7 Discussion

One of the overriding considerations in agriculture today is the sustainability of global food sources in relationship to many of Earth's natural resources. The amount of land used for farming for example, fluctuates globally for many reasons including both human and climactic factors, which might include anything from irrigation, deforestation, desertification, land reclamation, landfill and urban sprawl, among others. Yet understanding the needs of the land, the soil typology or its many nutrient needs and constraints, aids in good management practices and leads to increased productivity whilst maintaining sustainability. Indeed, implicit in good soil practice is the need to ensure replenishment of moisture and nutrients (through sustainable fertilisation, natural or mineral) and to minimise such losses through leaching and runoff. Good practices related to soil management include

- Maintaining/improving soil structure and carbon build-up of organic matter through the use of biowaste, crop rotation, manure application and conservation tillage practices among others;
- Reducing erosion by wind and water and to create a conducive habitat for soil biota through the practice of hedging, ditching and maintaining soil cover;
- Avoiding excessive inputs and runoff from fertilisers and other agro-chemicals and applying in amounts and timing that is appropriate to agricultural, environmental and human health considerations (FAO 2007).

In terms of water, in particular where rainfall is variable or insufficient for agricultural needs, which happens to some degree in most regions of the world, farmers turn to irrigation. This places huge pressure on governments to build large dams and mega-canal projects in an effort to reduce reliability on an otherwise intermittent resource. Although used in agriculture for thousands of years, the green revolution further improved and introduced more efficient irrigation systems. However, despite improvements and while worldwide agriculture represents 70% of total freshwater usage, in many instances, water reservoirs including groundwater and aquifers are being drained faster than they can be replenished (UNESCO 2009). Moreover, on top of this existing usage, the UN predicts that by 2025 irrigation demands will increase anywhere between 50% and 100%, making water resources an incredibly valuable natural resource and one which increasingly impacts on global political decision making (Section 21.3) (FAO 2007; GFS 2011).

When it comes to aquaculture, freshwater and marine fishery industries, it can be seen that between them they currently supply about 10% of the world's annual fish and fish products (in caloric terms) for human consumption (Nellemann 2009). This equates to about 110–130 million tonnes of seafood, of which 70 million tonnes are consumed directly by humans, 30 million tonnes are processed into fishmeal, while the remaining 30 million tonnes are discarded. Over time, global fishery landings have steadily declined since the 1980s through overfishing, degradation and climate change. However, with the 23% increase needed in landings to support the 56% growth in aquaculture production that is required to maintain per capita fish consumption at current levels to 2050, any collapse in marine ecosystems would have a major detrimental effect on production and prices. In fact the sad truth is, there are limits and the previous growth is unlikely to be sustained from natural sources; therefore, more reliance might be seen in the aquaculture sector where farmed species present more flexible opportunities (ibid).

Coming to industrial scale livestock farming, it can be seen that this practice uses high inputs of energy and increasingly larger proportions of cereal crop production. Having said that though much nonintensive livestock rearing does indeed take place on extensive grass or rangelands or other marginal lands not suited to arable production with little or no commercial inputs. However, the increasing use of cereal feed crops like maize and soybean as livestock feed is a widespread practice that can and does displace food crops (GEO 2000). In fact in 2008 alone, approximately 35%–40% of all cereal produced in that year was fed to livestock and this is not unique to that year either as similar figures are found year on year. In view of this, it has been said that if meat consumption increases as expected then livestock cereal usage could reach as much as 45%–50% by 2050. Moreover, the continuing expansion of land for livestock rearing is one of several key factors exacerbating the current rate of deforestation. In this sense, future food security would benefit, according to Nellemann (2009) and Swaminathan (1990), from new and novel uses of technology that would see developments like food or crop wastage, such as cellulose, being used as an alternative livestock feed. This may indeed come to pass with the new second generation biofuel technologies which aim to chemically and enzymatically degrade cellulose into glucose. If this succeeds in a commercially cost-effect manner, waste biomass has the potential to replace or at least supplement cereals as a feed source for both ruminants and monogastric animals. As a result, this could potentially release huge future cereal stocks for human consumption (Nellemann 2009).

Lastly, agriculture is a relatively stable industry and this is reflected in its unremarkable growth potential. For decades, modern agricultural practices have focused on increasing production through the introduction and development of new technologies such as monocultures and intensive farming. As a result, enormous yield gains have been achieved at ever decreasing costs yet such progress has not lived up to its developmental ideals. Apart from the notable gains in yields, modern farming has cost the environment dearly. Moreover, the gains have not been equitable among regions of the world, and it has not solved the structural problems of the developing countries. In sum, it can be said that while much liter ature and institutional goals attest to the redefining of agriculture's relationship with the environment, little in fact seems to be changing on the ground. Agriculture as an industry too has a crisis of identity. With many favouring large-scale aggregated farming techniques as the answer to solving food security, others vehemently deny this as the best way forward citing livelihoods, poverty, environmental and cultural concerns as primacy over the continuation of this approach. The debate over what to farm and how to farm it is a complicated one that clearly has no single solution. The status quo is clearly unsustainable and what is also abundantly clear is that there is a great deal of room here for the improvement in farming practices, of waste management and ideological divergence. Commonground needs to be found that satisfies as many as possible, from the organics and the food sovereign proponents to the capitalists and the internationalists; from fossil to renewables; multinationals to individuals; from economic to social development and of course many more. Yet just by elucidating these few simple areas of conflict, the scope of future challenges of agriculture and sustainable food security becomes immediately obvious.

References

Båge, L. (2008). *G8 Summit Special Report: Supporting Smallholders Is Crucial to Food Security. Financial Times*, July 7.

Boserup, E. (1965). *The Conditions of Agricultural Growth: The Economics of Agrarian Change under Population Pressure*. London: G. Allen and Unwin.

Britannica (2009). Encyclopaedia Britannica Online. 2009.

Brush, S. B. (1987). *Comparative Farming Systems*. New York: Guilford Press.

DFID (2004). Agriculture, Hunger and Food Security. London: Department for International Development (DFID).

Ericksen, P. J. (2008a). Conceptualizing Food Systems for Global Environmental Change Research *Global Environmental Change* 18(1): 234–245.

Ericksen, P. J. (2008b). What Is the Vulnerability of a Food System to Global Environmental Change? *Ecology and Society* 13(2): 1.

FAO (1989). Prevention of Post-Harvest Losses, Fruits Vegetables and Root Crops. A Training Maual. Rome: Food and Agriculture Organization.

FAO (2007). Good Agricultural Practices. *Good Agricultural Practices*. U.N. Food and Agriculture Organization. http://www.fao.org/prods/GAP/home/ (accessed 30 January 2011).

FAO (2011). Fisheries and Food Security. Food and Agriculture Organization. http://www.fao.org/FOCUS/E/fisheries/intro.htm (accessed 5 January 2011).

GEO (2000). *Global Environment Outlook*. Malta: United Nations Environment Programme.

GFS (2011). Global Issues. Global Food Security Resource Centre. http://www.foodsecurity.ac.uk/issue/global.html (accessed 12 March 2011).

GHI (2010). Gap Report: The Global Harvest Initiative: Measuring Global Agricultural Productivity Washington, DC: The Global Harvest Initiative.

Hartmannshenn, T. (2004). Food Security: Guidelines for the Promotion and Execution of Food Security Projects by German Agro Action. Bonn: German Agro Action.

IAASTD (2009). Iaastd Global Report: Agriculture at a Crossroads. *IAASTD Synthesis Report*. Johannesburg, South Africa: International Assessment of Agricultural Knowledge, Science and Technology for Development (IAASTD) 6.

IAASTD (2010). *Agriculture and Development: A Summary of the International Assessment on Agricultural Science and Technology for Development*. Johannesburg, South Africa: GreenFacts.

Kurien, J. (2004). Responsible Fish Trade. Towards Understanding the Relationship between International Fish Trade and Food Security. Rome: FAO and the Ministry of Foreign Affairs of Norway (MFA).

Latham, M. C. (1997). Human Nutrition in the Developing World. *Food and Nutrition Series—No. 29.* Rome: Food and Agriculture Organization.

McCalla, A. F. (2007). FAO in the Changing Global Landscape. *UCD. ARE Working Papers, vol. Paper 07–006.* Dublin: University College Dublin.

Mehta, V. N. (1929). Famines and Standards of Living. *The ANNALS of the American Academy of Political and Social Science* 145: 82–89.

Nellemann, C. (2009). The Environmental Food Crisis: The Environment's Role in Averting Future Food Crises. Geneva, Switzerland: United Nations Environment Programme (UNEP) GRID-Arendal.

Permaculture Ass. (2011). Website of the Permaculture Organisation. BCM Permaculture Association. http://www.permaculture.org.uk/ (accessed 12 February 2011).

RTFN (2009). Who Controls the Governance of the World Food System? *Right to Food and Nutrition Watch.* Germany: Brot für die Welt (Bread for the world).

RTFN (2010). Land Grabbing and Nutrition: Challenges for Global Governance. *Right to Food and Nutrition Watch.* Germany: Brot für die Welt (Bread for the world) 90.

Salter, R. M. (1950). Technical Progress in Agriculture. *Journal of Farm Economics* 32(3): 478–485.

SOFI (2005). State of Food Insecurity. Rome: Food and Agricultural Organization.

Swaminathan, M. S. (1990). Sir John Crawford Memorial Lecture: Changing Nature of the Food Security Challenge: Implications for Agricultural Research and Policy. Washington, DC: Consultative Group on International Agricultural Research.

The Diplomat (2011). Foreign Farmland Ownership Rises over Ten Per Cent. *The Diplomat: Bucharest* 7(2): 4.

UNESCO (2009). About UNESCO: What Is It? What Does It Do? United Nations Educational, Scientific and Cultural Organization. http://portal.unesco.org/en/cv.php-URL_ID=3328&URL_DO=DO_TOPIC&URL_SECTION=201.html (accessed 30 February 2010).

UOR (2009). University of Reading: Agriculture, Policy and Development, History of Agriculture. University of Reading. http://www.ecifm.rdg.ac.uk/history.htm (accessed 4 June 2008).

Vidal, J. (2010). *The Guardian Profile: Amartya Sen, Food for Thought.* London: Guardian News and Media Limited.

von Braun, J. and R. Meinzen-Dick (2009). Land Grabbing" by Foreign Investors in Developing Countries: Risks and Opportunities. Washington, DC: The International Food Policy Research Institute, 9pp.

World Resources (1999). Disappearing Food: How Big Are Postharvest Losses? US: EarthTrends.

WRI (2010). A Brief History of World Resources Institute. WRI. http://www.wri.org/about/wri-history (accessed 12 October 2010).

Zhang, X., M. Rockmore and J. Chamberlin (2007). A Typology for Vulnerability and Agriculture in Sub-Saharan Africa. Washington, DC: IFPRI.

18

Science and Technology

With regards to the security of food, science and technological advances over the last century in particular, has set the foundation upon which much progress has been built. This is especially evident in the remarkable progress of the biotechnology sector.

18.1 The Green Revolution

Typical references to the green revolution focus on the 1960s as a time when technological advances were becoming widely disseminated and taken up by many farmers around the world. However, the foundations of the green revolution were laid much earlier than this. Yields had already started to increase around the beginning of the industrial and chemical revolutions of the eighteenth century with new technologies and understanding in such inventions as hybridisation. That said, these increases were short-lived and after about 1850 or so, yields once again began to stagnate. From this point, as was so often the case in the past, the major means of increasing the world's food supply came at the expense of expanding the land base (Johnson 1997; Borlaug 2000; Johnson 2000; McCalla and Revoredo 2001; Webb et al. 2008). All that changed from about the 1940s onwards (Gardner 2002). This was a period characterised by scientific and technical progress in selective genetic breeding, hybrid crop varieties and the industrialisation and growing mechanisation of agriculture (Brewster 1945; Barton and Cooper 1948; Borlaug 2000; UOR 2009). However, the full impact of the looming new revolution was shortly arrested for while much of the science behind high-yielding agricultural production was known by the 1930s: three developments slowed the dissemination of these advances. First the great economic depression of the 1930s; second the unshared proprietary knowledge of Haber-Bosch ammonium nitrate (fertiliser) process; and third the advent of World War II (Borlaug 2000). At the end of the war, however, the allies needed to maximise food production and quickly. This led to multilateral technical-agricultural assistance programmes led by the United States and FAO as well as the Cooperative Mexican Government-Rockefeller Foundation agricultural programme of 1943. Such alliances collectively helped disseminate information throughout Europe Asia and Latin America. This knowledge, pioneered in no small measure by Nobel Laureate Norman Borlaug, saw new improved hybrid crop varieties better resistant to disease and pests as well as the new widely available synthetic fertilisers bringing better crops and increased yields per hectare (Sharma and Gill 1983; Perkins 1990; Herdt 1998; Borlaug 2000; Troyer 2004; Wu and Butz 2004; USAID 2009). By the 1960s the green revolution was effectively spreading technologies that, while already existed, had not been widely adopted outside of the industrialised nations up until this time. Shortly afterwards in 1968, this new movement received its familiar title in an address to the Society for International Development by then Administrator of the Agency for International Development (USAID), William S. Gaud, who said that

> These and other developments in the field of agriculture contain the makings of a new revolution. It is not a violet Red Revolution like that of the Soviets, nor is it a White Revolution like that of the Shah of Iran. I call it the Green Revolution. (Gaud 1968)

By the 1980s, the so-called green revolution technologies had become eagerly adopted, with the exception of sub-Saharan Africa, in most developing countries. The results were tangible with increased fertiliser use and more irrigation alone being responsible for over 70% of the crop yield increase: which incidentally was a staggering 78% between 1961 and 1999. Compare this to the 15% cropland increase and intensification of agriculture by 7% and the resultant gains have led to the growing realisation that agricultural research and development could decisively play a major role in ending future global food shortages (Nellemann 2009).

18.1.1 Criticisms of the Green Revolution

Nowadays, much of the food consumed in industrialised nations directly result from the green revolution crops. Despite such an accomplishment, there are many criticisms levelled at this revolution; two of the most prominently talked about are the heavy use of fossil fuel inputs and the environmental consequences of such technologies. In the past, traditional manure through the innovative use of fish and bone-meal, seaweed, dried blood and even sewage have all been used in the pursuit of higher yields. However, productivity gains witnessed throughout the green revolution relied to a large extent on the better understanding, production and use of improved fertilisers (Muir 2010). Of the three major nutrients needed for good plant growth—nitrogen, phosphorus and potassium—it is quite often the lack of nitrogen in the right form that was the limiting factor in a plant's optimum growth (Muir 2010). This is because plants cannot take advantage of the abundant supply of nitrogen in the atmosphere directly, instead they utilise it in its other forms; nitrates (NO_3) and nitrites (NO_2). Essentially, this is accomplished in two ways. First, through the biological fixation of nitrogen whereby the natural process of microbial and bacterial decay of the organic[*] matter in the soil is broken down into biologically preferable forms of nitrogen such as nitrate, ammonia, or amino acids. Secondly, artificially through the direct addition of inorganic nitrogen fertiliser to the soil.

These inorganic or mineral fertilisers, as they are often called, derive from both natural and synthetic sources. In recent years, close to 97% of nitrogen fertilisers is now manufactured from synthetically produced ammonia. The problem arises in the huge supplies of energy needed for such processes. This can be seen in the year-on-year figures that see the fertiliser industry consuming, on average nearly 5% of global natural gas supplies; which in itself equates to virtually 1.6% of total global energy production (McLaughlin et al. 2000; GEO 2007). This does not take into account other fossil fuels needed in the process either; put in perspective, to produce one ton of chemically fixed nitrogen in the form of anhydrous ammonia, it requires 30,000–40,000 cubic feet of natural gas as feedstock and a further 7 barrels or 1 ton of oil for energy to drive the process. Also for every one ton of fertiliser produced in this way, approximately 1.8 tonnes of CO_2 are produced. All in all, whichever way we look at it, this is quite an energy budget (Sundquist and Broecker 1985; McLaughlin et al. 2000; GEO 2007).

Also of concern to the critics of the green revolution technologies are the pesticides. While new improved green revolution fertilisers promote growth, pesticides are used to control the loss of valuable plants and crops. The need for some form of pesticide control cannot be underestimated, as each year approximately 30%–40% of global food production is lost to insects, birds, mammals, bacteria, viruses, fungi and weeds. Of particular note is the fact that insects alone account for 14% of such losses. Pesticides, a term incorporating fungicides, insecticides and herbicides, refer to any chemical, physical or biological agent that effectively destroys or controls such pests. They have had a chequered history and the original first generation compounds such as arsenic and cyanide were soon found to be highly toxic. After this and despite the initial hype and excitement of synthetic organic compounds of the second generation (like dichlorodiphenyltrichloroethane [DDT]), many of these too were also found to be quite toxic (Muir 2010). While modern day pesticides are fairly rigorously tested, concerns still remain as to the persistence of these chemicals in the soil as well as in the food and their potential toxicity to humans or the environment through excessive or mis-use.

The second major charge, related to the first, is that of the environmental consequences or degradation associated with the revolution. Improper use of modern day fertilisers, pesticides, agricultural persistent organic pollutants (POPs) and other technologies have been found to contribute to the long-range degradation of the soil. This degradation has been attributed to poor farming practices introduced as a result of the revolution, nitrate accumulation and heavy fertiliser runoff as well as from salinisation, eutrophication, chemical 'burning' and the loss of important trace elements and microbial species (discussed more in Chapter 21), (Eifert et al. 2002). Regarding this last point, there is also considerable concern regarding the green revolution and biodiversity. The widespread uptake of green revolution crop hybrids

---

[*] Organic compounds are carbon-containing substances that make up the various parts of the living organism's cells which work to carry out the chemical reactions that enable it to grow and maintain as well as reproduce itself.

has resulted in up to 90% reduction, in some cases, of varieties being grown. Such dependence on fewer crops bring with it an increased vulnerability in terms of fragility of the crops to unknown influences, the loss of dietary diversity and associated nutritional concerns as well as the potential impairment of future varieties (Section 21.5.1) (Darmawan et al. 2006).

Finally, on the socio-dynamic front, the green revolution witnessed major structural changes in the approach to farming, which encouraged larger, industrial type farms at the expense of their smallholder counterparts. Further, this revolution and its associated industrialisation witnessed increasing agricultural inputs, particularly energy to compensate for decreasing soil fertility. Thus, according to Pimentel et al. (1998), a cost-benefit trade-off between crop productivity and the health of associated ecosystems was needed in order to ensure that the gains of the green revolution were not achieved at the expense tomorrow's habitat. This was a message that began to take on more and more potency as the growing world witnessed a steady decline in the quality of the environment. Not surprisingly this view quickly became widespread and soon began to affect policies of energy, environmental pollution and agriculture, etc.

18.1.2 Evergreen Revolution

Effectively, the results of the green revolution have been remarkable by any measure of success. It no doubt played the largest part in the quadrupling of agricultural output in the United States alone between 1930 and 2000 while on the global front it can be seen that the 92% increase in global food production was achieved as a direct result of the green revolution as opposed to just 8% growth from the expansion of cultivated lands. Moreover, despite the continuing numbers of hungry and malnourishment in the world, more people, to date, through the green revolution, have achieved better nutrition at a cheaper cost than that at any previous time in history. Combined, these advances enabled the 'apparent' better management of land and resources, which greatly enhanced productivity through increased yields and which led to further changes in economies, population levels and the distribution of equity. Thus, the spectre of the Malthusian prophecy was kept in check, albeit temporarily (Brown 1981; Dyer 2006; Freedom 21 2008).

However, sustainability and the environmental questions of intensive agricultural productivity cannot be ignored, and with the yield gap slowing in many parts of the world, there are more recent calls for another more sustainable 'evergreen' revolution. And indeed, increased research and development spending, favourable policies and a new political and social setting is emerging to promote these new post-green or evergreen technologies. The idea is to sympathetically maintain equilibrium between the need to accelerate agricultural productivity and maintain a sustainable natural resource base, whilst simultaneously taking care of the environment. Much of the new focus aims at increasing the overall efficiency of inputs and effectively controlling pest and disease through the integrated use of complementary methods. Moreover, farming practices are continually being improved through the implementation of conservation tillage, drip irrigation, integrated as well as new multiple cropping practices among others (GHI 2010). The end result, it is hoped, would be the adoption of a new economic and social paradigm that fits with the modern ideology of sustainable economic growth and social development.

18.2 Biotechnology

Biotechnologies are profoundly important to food security. Discovering that juices could be fermented into wine, milk into cheese or crops into beer as well as the realisation that animals or crops could have desirable or undesirable traits bred in or out has been around for millennia. However, the latter half of the nineteenth and early twentieth centuries was an exciting time for the industry and the goal of food security. The coming together of the bio-sciences—microbiology, biochemistry, zymotechnology, bacteriology and geneticology—all colluded to propel scientific enquiry to new heights of respectability. As Friedrich Wöhler helped dispel the long-held *vital force* theory (vitalism) of life in 1828 and as French chemist Anselme Payen discovered the first enzyme, diastase (amylase), so Pasteur was to later promulgate his germ theory of infectious disease in the 1860s and 1870s. Indeed, the continuing preoccupation with this life force notion shaped much contemporary scientific and religious discourse and in the endeavour to provide proof, science and understanding accomplished a great deal, which ultimately led to the new field of biochemistry. At this time while it was already understood that enzymes could start or accelerate the breakdown of compounds outside of the cell, advancing research suggested that cells produced catalysts that performed many important chemical reactions of oxidation, fermentation, respiration as well as synthesis inside the cell. Such a mechanistic approach brought new understanding of human physiology and the subsequent study of nutrition on the cellular and intracellular levels. From this early period, nutritional research followed two main paths: first, the determination of the energy values of foods and second the pursuit of the optimal balance of the 'nutritional trinity' of protein, carbohydrate and fat (Swazey and Reeds 1978).

Such breakthroughs helped pioneer greater understanding of metabolic processes, which eventually led to the understanding and promotion of recommended dietary requirements. Elsewhere, biotechnology blossomed as discovery after discovery brought with it new techniques like chromatography, X-ray diffraction, spectroscopy and electron microscopy. In the areas of food manufacturing and processing also, microbiological and biochemical solutions helped pave the way for improved food safety and the long-term preservation of foods. Such scientific investigation also improved the possibilities of single cell proteins as a viable source of alternative proteins and the conversion of food crops into biofuels. But perhaps most importantly, biotechnologies led to the discovery of DNA and its subsequent manipulation—genetic engineering (Bechtel and Richardson 1998).

18.2.1 Genetic Engineering

While the green revolution popularised the use of conventional hybridisation, the advent of genetic engineering promised to exponentially propagate new breeds of plants and animals with beneficial traits that would see the green revolution pale by comparison. However, from the start, there were problems in the perception of modern genetic engineering in general which threatened and continues to threaten potential progress. There is much confusion and mistrust in this field, and with this in mind, it is worth considering a balanced view of the benefits and concerns of the debate. At its most basic, modern genetic engineering (modification) and traditional hybridisation (or cross-breeding) are often confused and transposed. Both methods aim to produce new crops and livestock exhibiting desirable benefits such as better yields; an increased tolerance to environmental pressures; and protection against disease, pests, insecticides and herbicides; however, they both go about it quite differently. Classical plant breeding involves the cross-breeding of related plant varieties at a non-genetic level. Such traditional hybridisation is commonplace and indeed happens naturally in the environment (GMES 2007). Moreover, such hybrids are often very fertile, producing progeny able to continue the lineage. Animals, on the other hand, do not cross-breed quite so easily and are quite often infertile. Genetic engineering (or genetically modified organisms (GMs or GMOs)) of plants and animals, on the other hand, is the practice of altering these organisms' characteristics at the genetic level. The reason this method is chosen over traditional cross-breeding or non-GM hybridisation is often one of time as genetic engineering is generally a much quicker process. Within the arena of genetic modification, then there are two distinct types or processes. The first method introduces gene varieties from the same variety or species and is known as non-transgenic modification. The second, transgenic genetic modification (or recombinant technique [rDNA]), involves introducing non-native genes from one organism of plant or animal into another plant or animal organism. These days too, this also includes transgenically crossing biological groups such as animals into plants, plants into bacteria and so on (Schneider and Schneider 2010). Although both methods are considered GMOs, it is this latter biotechnology, the recombinant DNA technology, that is often referred to as GMOs or GM products.

For many, the term has different connotations; some think of it as developing new types of hybrid animals or chimera with dire consequences while others see limitless potential in such things as therapeutic drug applications. Others still see the benefits of growing more nutritious, long-lasting and naturally pest-resistant crops. In all groups, however, there are those that are genuinely and seriously concerned about the unknown effects of GMOs.

18.2.1.1 Extent of GMOs in Production

GM crops are classified into one of three categories depending on traits: these include crops that benefit from enhanced input traits like herbicide, insect or environmental tolerance; those with value-added output traits such as longer shelf-life or nutrient enhancement; and finally those non-food and fibre industries such as pharmaceuticals and bio-based fuels. To the extent that GMOs, or transgenic crops in this respect, are currently used throughout the world; it can be seen that in 2010 GMOs were grown in 29 countries on six continents on over 148 million hectares of land. This represents a whopping 87-fold increase in the 15 years from 1996 to 2010. This figure represents close to 10% of the world's total 1.5 billion hectares (3.7 million acres) of cropland in use, making

> … biotech crops the fastest adopted crop technology in the history of modern agriculture. (James 2010)

Figures from the previous year show just to what extent GM crops are now being grown around the world. In 2009 by far the four most widely grown crops were the following:

- Soybean: 90 million hectares grown globally, of which 77% were genetically modified.
- Cotton: 33 million hectares grown globally, 49% of which were genetically modified.
- Maize: 158 million hectares grown globally, 26% of which were genetically modified.
- Canola: 31 million hectares grown globally, 21% of which were genetically modified.

Out of the top three producers in 2010, the United States, Brazil and Argentina accounted for over 100 of the 148 million hectares planted globally (see Figure 18.1) (James 2010).

While the above represents the major GM crops grown globally today, this list is not exhaustive. In the first 10 years of commercial GMOs, the USDA's Animal and Plant Health Inspection Service alone approved over 10,000 applications for GMO field testing (Fernandez-Cornejo et al. 2006). That said, this figure belies the actual number of food crops on the shelves today, as 5000 of these alone were to do with corn (maize). Examples of products that have been engineered include delayed-ripening tomatoes, virus-resistant squash, Colorado potato beetle-resistant potato, asbromoxynil-tolerant cotton and lyphosate-tolerant soybean, among many others. Moreover, approximately 90% of hard cheeses made today using chymosin (the primary component of rennet) are now commercially produced using genetically modified micro-organisms rather than the traditional mammalian stomach mucosa. In fact, traceability is one of the major difficulties and public sticking points of this method of production (next section).

The debate about and genetic engineering in general and GMO foods in particular is long and complex and is beyond the scope of this book; however, some discussion on the pros and cons is warranted in the name of balanced critique.

18.2.1.2 Benefits

From about the 1800s, when the world population reached about a billion people, land expansion was a fairly easy way of increasing agricultural productivity. However, with the increasing numbers, available land competed with non-farm uses such as urbanisation, industry and recreational needs and by the mid-twentieth century, marginal lands as well as lands cleared from deforestation were routinely being converted for agricultural production. Land expansion though is a finite option and although the green revolution had taken the edge off the need to rapidly expand land use further, these advances were beginning to slow down. Yield gaps across the world were closing and a new avenue of research, the genetic modification of organisms all but promised to take over where the green revolution left off. Furthermore, with the current need to increase food production by 70% over the next 40 years or so, sustainably and on approximately the same area of arable land, this might best be achieved, according to the International Service for the Acquisition of Agri-biotech Applications (ISAAA) by

> … integrating the best of conventional crop technology (adapted germplasm) and the best of crop biotechnology applications including novel traits. (James 2010)

So far, higher productivity through selective breeding and plant genetics are responsible for high-yielding varieties of some of the world's most important crops including wheat, rice, maize and sorghum.

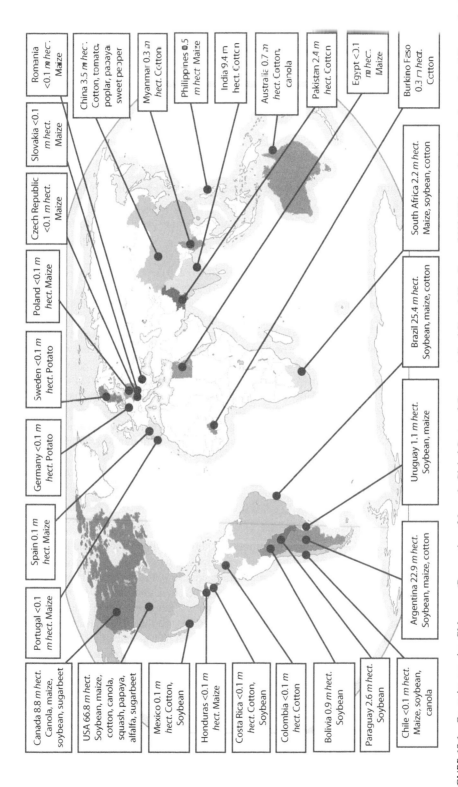

FIGURE 18.1 Countries growing GM crops. (Re-creation of James's *Global Status of Commercialized Biotech/Gm Crops: 2010*. With permission from International Service for the Acquisition of Agri-Biotech Applications. ISAAA.)

Through selective breeding, these new crop varieties allocate more of their photosynthetic energy to their grain relative to their vegetative matter in a ratio known as the 'harvest index' (HI). The higher the harvest index the more of a plant's biomass is devoted to edible seeds and kernels. By producing dwarf varieties of the originals with a higher HI, for instance, this also results in less energy use, more efficient use of the soils nutrients thus allowing more grain to be grown per acre. Present grain to vegetative increases of 20% take the grain to vegetative mass ratio to 50–55%, which, according to Muir (2010), is close to the physiological maximum of 60%. Other beneficial traits of plant genetics allow crops to be less sensitive to day-length and more responsive to fertiliser, pesticide and water inputs, ensuring more efficient and stable growth. The GM industry has also argued that GM foods will reduce production costs through the reduction of such chemical uses as well as a concomitant reduction in mechanical inputs. There are nutrition implications too since bioengineering could possibly create more foods with enhanced nutritious content such as that of the Golden Rice strand, which is bred with beta-carotene, a source of vitamin A and iron. Furthermore, seeds with enhanced plant enzymes effective under ambient conditions have been found to be a cheap and convenient method of storage and shipping. Genetic engineering science too can use transgenic plants as recombinant protein bioreactors as in single cell proteins which could provide cheap alternatives for large-scale protein users such as animal feed stocks, etc. In another benefit, large-scale GM biomass biotechnologies could very well be used in the second generation biofuel industry relieving direct pressure on food crops such as maize and sugar beet. Moreover, on the non-food front, food security might be aided in the fight against disease as plants are also capable of synthesising almost any kind of antibody from the smallest antigen-binding fragments to full length or even multimeric antibodies.

18.2.1.3 Concerns

Tempering the numerous proposed benefits of genetic engineering are many issues of concern from both public and professional perspectives that need to be addressed; this is particularly so as the full extent of what is achievable becomes known. Initially as genetic engineering crossed over into transgenic, recombinant DNA technologies public perception began to change. Initial caution and healthy debate was at once replaced with fear and mistrust at what was viewed as aberrant 'Frankenfoods' associated with unnatural technologies. More recently, concerns have taken on a plethora of many forms, a half-dozen or so, which are looked at here in brief.

18.2.1.3.1 Allergy

A big concern with GM foods is allergic reactions. Certain plant and animal allergenic proteins collectively found in wheat, milk, eggs, fish, tree nuts, peanuts, soybeans and shellfish have been found to cause over 90% of known food allergies. The fear that transgenic practices might insert one of these antigens into the DNA of an otherwise harmless organism is very real. Especially for those who might unknowingly consume such products. As a result, many regulations are put in place to prevent this with authorities requiring either evidence of compliance, or in lieu of this, a label on the foods concerned warning of the potential risks (Schneider and Schneider 2010).

18.2.1.3.2 New Viral and Bacterial Pathogens

In transgenetic plants or animals, the horizontal gene transfer process often uses viral and bacterial pathogens as well as other parasitic genetic material such as *plasmids and transposons* as vectors to transmit genes from one organism to another. While these vectors are generally rendered harmless before use, fears have been raised regarding the possibility that bacterial or viral genes might recombine to form new viral and bacterial pathogens. In turn, these may end up becoming potentially dangerous to plants, animals and humans. Further concerns include the potential to trigger cancers while some have also suggested there is evidence that horizontal gene transfers might also be spreading drug and antibiotic resistance among pathogens further complicating attempts to treat disease (Ho 2000; Cummins and Ho 2006).

18.2.1.3.3 Toxicity

Regarding toxicity, there is scant evidence of any peer-reviewed studies showing either the innocuous or harmful effects of GM foods. Episodes like the 37 deaths and the further 5000 permanently disabled or afflicted Americans that resulted from the genetically engineered brand of L-tryptophan still reverberate. Indeed, genuine fears that mycotoxins or other toxic artefacts remain present in some foods is palpable. Furthermore, contrary to biotech claims, some have suggested that farmers planting GM crops were using just as many toxic pesticides and herbicides as conventional farmers. This is because the herbicide- and pesticide-resistant crops mean just that; farmers can now spray as much as they need without risking the harvest. Lastly, there is also concern from the consumer's perspective regarding residual toxins that then subsequently enter the food chain (Pusztai 2002).

18.2.1.3.4 Resistance

Some have pointed out that in the quest to make crops that are pesticide or herbicide-resistant, there is a strong likelihood that existing pests will eventually become resistant to these measures or perhaps that new stronger pests or weeds will emerge to replace those killed off. This creates fears of superweeds or pests immune to such measures, and could result in the 'ramping' up of measures to prevent this and the escalation or viscous cycle of the immunity-resistance-immunity syndrome.

18.2.1.3.5 Bio-Diversity

There is also concern that with the growing use of improved plants and crops fewer crop varieties are being sown and the propensity for fewer, larger farms growing just a handful of these crops effectively encourages the simplification of cropping systems or cultivar biodiversity. Ecosystems too rely on biodiversity characterised by interactions between different species. As a result, fears reside with the notion that herbicide-resistant GM plants interact in a negative way, reducing not only the intended weeds but other proximal wild plants too.

18.2.1.3.6 Genetic Pollution

Just as hybridisation is a naturally occurring phenomenon so the current GM debate raises concerns regarding propagation. This has consequences and concerns for those who want to ensure genetic purity as bees and others potentially pollinate nearby cultivated non-GM plants with GM pollen. Once any artificially induced characteristics find their way into the natural gene pool, there is no undoing them and concerns that this might happen not just in cultivated plants but in wild species too fuels people's worries. Furthermore, such gene flows can occur not just through cross-pollination but also with persistent DNA in soils or composts in the form of decaying plant residues. Similarly, depending on the unbroken passage of DNA through the gastrointestinal tract, the use of manure from GM-feed fed animals could also be responsible for the transfer of novel genes into the soil. From here, just how long such DNA resides in the soil before it is degraded then depends on the soil structure and composition (Dale et al. 2002; Cummins and Ho 2006).

18.2.1.3.7 Stability and Pleiotrophy

It has been posited that current methods of gene transfer are not precise. While 'trait genes' can be manipulated with relative ease, control over the location or the number of copies inserted is more difficult. Location of genetic material in this sense is important as location of trait genes can determine whether the gene is expressed, silenced or affects other gene traits that may result in undesirable interactions. This also calls into question the simple model of gene expression, that ultimately one gene produces one effect. In reality, a single gene perhaps identified as controlling a single desirable trait may in fact affect multiple traits in a number of ways. This is known as pleiotrophy and the idea that one gene has the potential to disrupt non-targeted segments of the host DNA, giving rise to unintended traits exponentially multiplies the uncertainty surrounding transgenetic crops. This further compounds concerns of the potential for instability in transgenic plants in such characteristics as the plant's health, its fertility, production of toxins and allergens and the possible reduction in yields (Schahczenski and Adam 2006).

18.2.1.3.8 Proprietary Ownership

While the patenting of genetic modifications attracts large investment in agricultural research, proprietary ownership in the form of patented biologically based intellectual property rights (IPR) systems tends to drive up costs to the consumer. Unlike green revolution technologies, which arose from research funded by philanthropic foundations, and which became readily available to those who would derive benefit from them, there is a fear that the proprietary nature of GM study as well as inherent costs of research and development are going to put genetically modified benefits out of reach of those that can ill-afford such technologies. Fears too that traditional seed-saving practices of the majority of farmers are discouraged in the GM world continue to prevail. In this charge, the GM crop manufacturers have been accused of producing 'terminator' seeds and using Genetic Use Restriction Technologies (GURTs) or 'traitor' technologies, as some would call it. The seed manufacturers have spent small fortunes in developing their products and it has been suggested that in order to protect their investment, they have modified their GM plant characteristics so that they only produce infertile seed. Farmers then would have to buy new (patented) seeds every new cropping season or failing that, chemicals that, effectively 'switch-off' this sterility, further ensuring continued custom (Warwick 2000).

18.2.1.3.9 Regulatory Uncertainty

Finally, this brings into question the regulatory framework that surrounds GM products. Regulatory uncertainty questions whether the ground rules underpinning research of transgenic material is to be the same everywhere and whether it would be rigorously conducted and adhered too. There are already concerns that developing countries lack the capacity to establish any meaningful regulatory frameworks in this regard. However, around the world many countries are taking a cautious approach with various governments adopting regulation requiring labelling as in: Australia in 2000, the European Union in 1998, Japan in 2000, Russia in 2000, Hong Kong in 2000, South Korea in 2001 as well Taiwan and others. One interesting aside to this is now being questioned. That is the acceptability of GM contamination. In the EU, for instance, legislation (regulation [EC] No 1830/2003) was passed in 2003 concerning labelling of GM foods and required that GM content of foods with less than 0.9% of EU-authorised GMOs and less than 0.5% of unauthorised GMOs does not need to be recorded on the labels. This, it is said, is necessary to account for the 'adventitious' or 'technically unavoidable' presence of GMOs in the food; that is to say, GMOs are so widespread today that many manufactured foods inadvertently or otherwise use GM derivative products such as oil and flour (Ferrante and Simpson 2001). Lastly concerns have been brought forward regarding the US use of regulation in that the main regulating agency is the USDA's Animal and Plant Health Inspection Service (APHIS), which uses risk management tools that allow 'an acceptable level of possible collateral damage' rather than conducting studies of safety (Schahczenski and Adam 2006). In this way it is argued, if we cannot control for the 'adventitious' spread of GM's in the food chain, how on earth then, can we protect the environment?

18.3 Discussion

As the traditional expansion of the food supply through the acquisition of marginal lands, intensive agriculture, increased productivity, higher yielding commodities reach their limits, the potential to increase supplies slows and a shortfall threatens. At this point, new technologies such as GM or other as yet untested technologies, like in vitro meat growth, might ultimately play a greater role. As this happens, the potential benefits of biotechnologies, in particular GMOs within the food security arena, cannot be underestimated. Technological innovations in farming practices as well as manufacturing processes have allowed food crops to become fuel and allowed oil, fungi and bacteria to become single cell proteins and at the same time it has witnessed genetics cross the horizontal species barrier. Over the past two decades, genetic engineering has stimulated hopes in many fields of therapeutic drugs, for treating genetic disorders and for sustainable food production. Sustainable solutions to increasing food crop efficiency and productivity requiring less agricultural inputs promises to create a huge dent in the much needed agricultural gains over the next 40 years or so. However, any modification of life carries with it significant and complex ethical and safety issues and implications. There has been much discourse on the various divergent and convergent moral and ethical arguments of the principles of genetic engineering and the resultant commodification of such 'biocapital'. Yet there is much posturing on both sides that does little to resolve the public-scientific or scientific-scientific impasse. Many of the claims and counterclaims of the genetic engineered crops and livestock are hard to prove or disprove and the few studies available, are interpreted by both camps in different ways; often in ways that support their own views. This claim/counterclaim is an unproductive and infertile debate. GM science for-the-table is relatively young and if the general public are to reap the benefits of this new science what is clearly needed are independent long-term studies addressing their concerns in an honest open and transparent manner. That said, this would have to be done on a case by case basis, which would not only slow market entry for numerous products but increase costs considerably. It is interesting to note too that with such an uphill public perception problem alternative methods such as marker-assisted selection (MAS) are now being researched. This new approach drops transgenic GMOs in favour of manipulating an organism's existing genes. Just how the debate eventually pans out is yet to be decided, however, it will be one to watch with interest.

In other areas of technological advances, immediate solutions to several issues would see second generation biofuels taking the sting out of the food-fuel competition and hopefully at the same time disengaging the oil-food price convergence. The use of food grains as livestock feed might also be reconsidered in favour of alternative feed technologies such as single cell proteins or, after further testing, and if more viable, the use of GM crops. Another promising area of research is that of biomimetrics or biomimicry. This looks to nature for inspiration in solving many human problems. This innovation inspired by nature has been used for decades and one of the basic advantages is highlighted by the fact that nature does not produce waste. More than these closed-loop systems though, biomimicry helps find solutions to the way we farm foods, these might be promoted in practices like urban permaculture or multiple integrated cropping systems or polycultures.

All these aspects taken into consideration and in the fight against hunger and malnutrition, food production on a sustainable footing is paramount. Although caution and circumspection must inevitable play a part; in fact a large part, in this overall endeavour we must not be afraid of the new. Dismissing such technologies out of fear and without recourse to science or the full consideration of the benefits and costs is an archaic practice best left in the middle ages with the likes of spontaneous generation and humoralism. Another interesting challenge, however, is yet to be played out and that is the changing public-private landscape of agricultural research brought about by the bio-molecular revolution. With the ever increasing privatisation of our biological stock, it will indeed be one to follow and one too that may have untold consequences for future food security.

References

Barton, G. T. and M. R. Cooper (1948). Relation of Agricultural Production to Inputs. *The Review of Economics and Statistics* 30(2): 117–126.

Bechtel, W. and R. C. Richardson (1998). *Vitalism*. London: Routledge.

Borlaug, N. E. (2000). The Green Revolution Revisited and the Road Ahead. *30th Anniversary. Lecture*. Oslo: The Norwegian Nobel Institute.

Brewster, J. M. (1945). Farm Technological Advance and Total Population Growth. *Journal of Farm Economics* 27(3): 509–525.

Brown, L. R. (1981). World Population Growth, Soil Erosion, and Food Security. *Science* 214: 995– 1002.

Cummins, J. and M.-W. Ho (2006). Gm Food Animals Coming *ISIS Reports*. London: The Institute of Science in Society.

Dale, P. J., B. Clarke and E. M. G. Fontes (2002). Potential for the Environmental Impact of Transgenic Crops. *Nature Biotechnology* 20: 567–574.

Darmawan, k. kyuma, A. Saleh, H. Subagjo, T. Masunaga and T. Wakatsuki (2006). Effect of Green Revolution Technology from 1970 to 2003 on Sawah Soil Properties in Java, Indonesia: I. Carbon and Nitrogen Distribution under Different Land Management and Soil Types. *Soil Science & Plant Nutrition* 52(5): 634–644.

Dyer, G. (2006). How Long Can the World Feed Itself? *Energy Bulletin*: http://www.energybulletin.net/node/21736.

Eifert, B., C. Galvez, N. Kabir, A. Kaza, J. Moore and C. Pham (2002). The World Grain Economy to 2050: A Dynamic General Equilibrium, Two Sector Approach to Long-Term World-Level Macroeconomic Forecasting. *University Avenue Undergraduate Journal of Economics* (Online).

Fernandez-Cornejo, J., M. Caswell, L. Mitchell, E. Golan and F. Kuchler (2006). The First Decade of Genetically Engineered Crops in the United States. *Economic Information Bulletin Number 11*. Washington, DC: United States Departmentof Agriculture (USDA).

Ferrante, E. and D. Simpson (2001). Biological & Biomedical Sciences: A Review of the Progression of Transgenic Plants Used to Produce Plantibodies for Human Usage. *Journal of Young Investigators* 4(1) http://www.jyi.org/volumes/volume4/issue1/articles/ferrante.html (accessed 20 May 2011).

Freedom 21 (2008) Freedom21 Alternative to the U.N.'S Agenda 21 Program for Sustainable Development: V Part 2 Meeting Essential Human Needs. *Freedom 21*: http://www.freedom21.org/alternative/chapter5b.html.

Gardner, B. L. (2002). *American Agriculture in the Twentieth Century: How It Flourished and What It Cost*. Cambridge, MA: Harvard University Press.

Gaud, W. S. (1968). The Green Revolution: Accomplishments and Apprehensions. The Society for International Development, Shorehan Hotel, Washington, DC: US Agency for International Development.

GEO (2007) Planet's Tougher Problems Persist, UN Report Warns. *Global Environment Outlook 4*: http://www.unep.org/geo/geo4/media/media_briefs/Media_Briefs_GEO-4%20Global.pdf.

GHI (2010). Gap Report: The Global Harvest Initiative: Measuring Global Agricultural Productivity Washington, DC: The Global Harvest Initiative.

GMES (2007). Global Monitoring of Environment and Security. European Space Agency ESA/EU. http://www.gmfs.info/ (accessed 12 March 2011).

Herdt, R. W. (1998). The Rockefeller Foundation: The Life and Work of Norman Borlaug, Nobel Laureate. New York: The Rockefeller Foundation.

Ho, M.-W. (2000). Horizontal Gene Transfer—the Hidden Hazards of Genetic Engineering. *ISIS Reports*. London: The Institute of Science in Society

James, C. (2010). Global Status of Commercialized Biotech/GM Crops: 2010. Philippines: The International Service for the Acquisition of Agri-biotech Applications (ISAAA).

Johnson, D. G. (1997). Agriculture and the Wealth Of. Nations. *American Economic Review* 87(2): 1–12.

Johnson, D. G. (2000). Population, Food, and Knowledge. *The American Economic Review* 90(1): 1–14.

McCalla, A. F. and C. L. Revoredo (2001). *Prospects for Global Food Security: A Critical Appraisal of Past Projections and Predictions*. Washington, DC: International Food Policy Research Institute.

McLaughlin, A., G. J. W. Hiba and D. J. King (2000). Comparison of Energy Inputs for Inorganic Fertilizer and Manure Based Corn Production. *Canadian Agricultural Engineering* 42(1): 1–14.

Muir, P. (2010). *Bi301: Human Impacts on Ecosystems*. Oregon State University.

Nellemann, C. (2009). The Environmental Food Crisis: The Environment's Role in Averting Future Food Crises. Geneva: United Nations Environment Programme (UNEP) GRID-Arendal.

Perkins, J. H. (1990). The Rockefeller Foundation and the Green Revolution, 1941–1956. *Agriculture and Human Values* 7(3/4): 6–18.

Pimentel, D., M. Pimentel and M. Karpenstein-Machan (1998). Energy Use in Agriculture: An Overview. *Agricultural Engineering International: CIGR Electronic Journal*, www.agen.tamu.edu/cigr

Pusztai, A. (2002). Can Science Give Us the Tools for Recognizing Possible Health Risks of Gm Food? *Journal of Nutrition and Health* 16(2): 73–84.

Schahczenski, J. and K. Adam (2006). Transgenic Crops. Butte, Montana: National Center for Appropriate Technology (NCAT): National Sustainable Agriculture Information Service.

Schneider, K. R. and R. G. Schneider (2010). Genetically Modified Food: What Are GM Foods? Florida: The Institute of Food and Agricultural Sciences (IFAS), University of Florida.

Sharma, H. C. and B. S. Gill (1983). Current Status of Wide Hybridization in Wheat *Euphytica* 32(1): 17–31.

Sundquist, E. and W. Broecker (1985). *The Carbon Cycle and Atmospheric CO_2: Natural Variations Archean to Present*. Washington, DC: American Geophysical Union.

Swazey, J. P. and K. Reeds (1978). *Today's Medicine, Tomorrow's Science Essays on Paths of Discovery in the Biomedical Sciences*. US: US Department of Health, Education, and Welfare.

Troyer, A. F. (2004). Breeding Widely Adapted, Popular Maize Hybrids. *Euphytica* 92(1–2): 163–174.

UOR (2009). University of Reading: Agriculture, Policy and Development, History of Agriculture. University of Reading. http://www.ecifm.rdg.ac.uk/history.htm (accessed 4 June 2008).

USAID (2009). Frontlines: Borlaug, Father of Green Revolution, Dies. US Agency for International Development. http://www.usaid.gov/press/frontlines/fl_oct09/p01_borlaug091002.html (accessed 5 March 2010).

Warwick, H. (2000). Syngenta: Switching Off Farmers' Rights? London: ActionAid, Genewatch UK: Swedish Society for Nature Conservation, Berne Declaration, 26pp.

Webb, P., J. Gerald and D. R. Friedman (2008). More Food, but Not yet Enough: 20th Century Successes in Agriculture Growth and 21st Century Challenges. *Food Policy and Applied Nutrition Program.* Boston, MA: School of Nutrition Science and Policy, Tufts University.

Wu, F. and W. P. Butz (2004). *The Future of Genetically Modified Crops: Lessons from the Green Revolution.* US: RAND Corporation.

19

Socio-Cultural

19.1 Population Growth and the Earth's Carrying Capacity

One of the overriding pressures in achieving food security is the increasing demand for food from a growing population. Already by the middle of the twentieth century population growth was seen as an issue vis-à-vis the food supply; however, what was becoming more apparent by this time was the emerging awareness of the detrimental relationship population increase was having on other natural resources (Ruxin 1996). Unfortunately in the early years though few reports outside of the Food and Agriculture Organization (FAO), the US Department of Agriculture (USDA) and the World Bank were conducting globally representative surveys, and as a result many people relied heavily on the data and statistics contained within these publications. The reading was grim and by the 1950s and 1960s a plethora of reports proclaiming global Malthusian catastrophes were being regularly published. Hunger and malnutrition became buzzwords, almost clichés in the popular press too, with far too many suggesting that the world was on the brink of mass starvation. This was not confined to observers either; many academic authors as well, it seemed were convinced of the inability of humanity to feed itself and it was these growing fears that led ultimately to a White House Report in 1967 to look into the situation. This report looked at the problem of food and reflected on the situation at length, and weighing up the evidence eloquently suggested that the world's food problem's 'size and significance tend to be obscured by rhetorical overkill' (SAC 1967). Just how such a pessimistic outlook came about was not surprising according to several commentators, chief among them Thomas Poleman. The answer, it was suggested, linked back to the few truly global reports that were available and these, according to Pullman, were biased towards exaggeration (Poleman 1972). Despite this well-balanced and thorough report however, it was fundamentally flawed in that it too relied on a plethora of existing data. What was only beginning to emerge at this time was the sheer pervasiveness of the phenomenon of hunger and malnutrition. By the mid to late 1960s though, through increased research and information dissemination, the true extent of the problem was coming to light. Reports, like the 1967 White House report, while informative, only served to give critics further deniability. Others saw hunger and malnutrition as inherent in an ever-increasing populace. So what were the implications in terms of sustainability within these growing populations?

Population sustainability, as many have argued, revolves around the Malthusian concept of population versus food supply. Such contemporary insightfulness, it has been said, marshalled the beginning of a popular concept whose dominant discourse has as many advocates as opponents, and one which continued to preoccupy philosophers and policy makers since (Prentice 2001; Sulistyowati 2002; Ross 2003). Indeed ever since Anton Van Leeuwenhoek began estimating the carrying capacity of Earth in the seventeenth century the subject of the Earth's ability to sustain a population has been a contentious and hotly debated area of research. The notion that the world's natural resources are finite and can only support a limited number of people is not in question; what was well debated however was the perceived limiting factor within this equation (Roughgarden 1979). Despite rigorous and healthy debate though, as the idea trickled down through the centuries, it was still largely based on conjecture and subjectivity. Even up until the first half of the twentieth century, and despite advances in the ideas behind the notion, such growing impetus and general acceptance for the subject, suggested Gottlieb (1945), had still not gained academic credence. As a concept, he bemoaned the idea as being considered unscientific and of 'speculative construction' (Gottlieb 1945, p. 289). Yet, despite or even in spite of this, much interest was generated within and across various disciplines from the 1950s onwards over the years, resulting in a wide variety of estimates of upper limits to sustainable numbers (OPT 2009).

Indeed various studies have been put forward over the years, each based on the perceived limiting factor of that particular study. Many see agricultural land as the limiting factor; on this the FAO, for instance, calculates that the Earth can support nearly 9 billion on current and future potentially available agricultural land while others maintain that available freshwater might eventually surface as the most potent restrictive force (AEZ 2000). Ultimately two studies stand apart from the rest in the goal of quantifying a ceiling or upper limit to such population numbers: those of Cohen (1995) and Van Den Bergh and Rietveld (2004). Median values of these two studies covering research from 1679 to the present range from 12 billion in Cohen's study and 7.7 billion in Van Den Bergh's. However, with so many other wildly varying figures based on equally divergent limiting factors such as food, water, photosynthesis, carbon cycle, natural ecological and geophysical constraints, and with no semblance of convergence in estimates, it is clear the debate is set to continue (Constantino and Constantino 1988; Murai 1994; AEZ 2000; Gilland 2006; Cao et al. 2007; Ferguson 2008; OPT 2009).

With regard to the Malthusian prophecy and the food supply however, the truth of the debate is relatively simple to answer, at least from the historical perspective. Figure 19.1 highlights the true nature of attempts over the years to feed our growing populations and the answer to the charge of limited food supplies seems not to be the arithmetic growth as proposed by Malthus but rather a sigmoid growth curve similar to that of population growth.

In fact, the true food supply has more or less kept pace with population growth albeit not quite as smooth as the graph illustrates. Nonetheless, that said, the challenge comes with the divergence of the two lines. If, as has been predicted, population growth tails off by 2050/2060, then there is a good chance that current and future technologies may in fact prevail and continue to provide adequate food for the population. Although one to watch for the future will be just how the continuing capacity of the agricultural and technological sectors are able to maintain this almost parallel growth. In fact, sustainability in food terms rests with just how close these parallels can be maintained with divergence, ultimately determining the sustainability of population numbers.

Lastly the continuing debate over the carrying capacity of Earth emphasises too just how far apart professionals are in their views on this issue, and without agreement on such matters, any consensus on figures appears to be a long way off. Finally, whatever the limiting factor turns out to be, it is ultimately the collective choices we and our governments make concerning the distribution of material well-being; the diet and diversity of our peoples; cultural issues; available technologies; natural resource limitations; and fashions, tastes and values that ultimately will determine the confines of our existence (ER 1996).

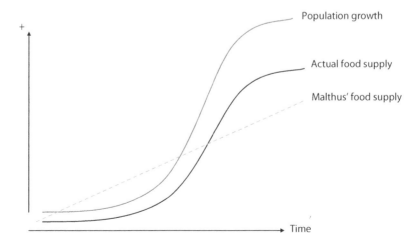

FIGURE 19.1   The true food–population relationship.

So what of population trends? Nowadays the present global population of 6.7 billion is growing by about 80 million people per year and is set to increase to an overall population of 7.94–8.33 billion by 2030 and to peter off at between 9.2 and 9.5 billion by 2050/2060 (ESA 2007; GEO 2007; Zhang 2008). The demographic split, however, will see less developed regions absorbing more of this growth from about 5.4 billion in 2007 to an estimated 7.9 billion in 2050, compared with more developed regions, whose population is expected to remain static at around 1.2 billion (ESA 2007). For some this seemingly two-track affair is of considerable concern and they see this growth as almost certain to aggravate food insecurity; for others however, like the FAO, believe that this populations growth's significance is somewhat overstated (FAO 2011). For the FAO the reason is simple, and they cite present-day global food production as an example of just how efficient and innovative we can be. Whether this remains the case is one of the future challenges that face both agriculture and biotechnology.

19.1.1 Urbanisation: The Rural–Urban Dynamic

Another major force altering the food equation is the continuing rural–urban shift. This has only been a recent historic cultural dynamic with Johnson and Ballioch suggesting that between 1300 and 1800, the figures remained largely unchanged. During this period, Europe (excluding Russia), India, China and much of the developed world was perhaps at about 10%–12% urbanisation (cities with 5000+ populations). By the end of the nineteenth century however, the figure had changed and in Europe the urban population had reached approximately 37.9%. Incidentally by 1900 there were over 16 cities with inhabitants over 1 million, up from 1 (Beijing) in 1800. From this point urbanisation continued to increase throughout the world, going from approximately 67%/33% rural/urban split in 1962 to reach a milestone 50%/50% in 2008 (Johnson 1997, 2000). This is not the end either, for the urban ratio is set to increase over the next two decades with urbanisation of the planet expected to reach approximately 4.72 to 5 billion by 2030. This equates to about 57.8% of projected population numbers. Most of this too will occur in Asia and sub-Saharan Africa where urban populations are expected to double in that time. At the same time, the rural population is expected to remain static or fall slightly accounting for between 3.12 and 3.41 billion over the same period (FAO 1995b; von Braun 1995a; Cohen 2006; UNFPA 2007; USDA/ERS 2008a; Zhang 2008).

It has been suggested that increasing poverty, hunger and malnutrition were some of the principal accelerants of this migration in developing countries. Exacerbating these drivers too it seems, was rising agricultural productivity which has ultimately led to concomitant decline in demand for rural agricultural labour. Yet while this might be true to some extent, the growing attraction of urbanisation is also at work here, and its drivers are little more complicated. On the surface, as well as the push of decreased agricultural labour demand, the pull of urbanisation has a lot to do with apparent higher standards of living, better social mobility, improved access to education as well as higher wages, diversity of employment opportunities and inherent choice associated with urban living. Importantly too urban dwellers typically have better accessibility to things like health care facilities, clean water and sanitation which in theory ensures better overall food security. All in all, these become very powerful pulls (Food Summit 1996; Ericksen 2008; USDA/ERS 2008a).

Urbanisation itself has a double impact on the food economy; more is discussed later but in brief, higher wages often mean more food choice and dietary diversity, which often leads to more meat and dairy consumption. Another impact is that of timing and recovery in the rural sector, and looks at whether drawing potential agricultural workers out of the rural landscape gives adequate time for the local agricultural economy to develop sufficient alternative strategies to cope with the shortfall. If the answer is negative, then these drivers too could see further populations being pushed into urban habitats (von Braun 1995b; King and Elliott 1996; Ravallion et al. 2007).

While the push and pull of urbanisation is a complicated matter, the perceived benefits are strongly identified as improvements all around. However, the reality is a little different and a little more complicated, for while in absolute terms rural poverty outstrips urban poverty, the urban poor as a group, are rising faster than in rural areas, perhaps by as much as 30%. Moreover, with 90% of the world's urban slums being located in the developing countries, there are concerns that food security in such area is set to worsen (USDA/ERS 2008a).

19.2 Poverty

One of the few areas of consensus regarding hunger issues is that poverty is all too often at the heart of the food security problem. This belief was built on many notions and reports over the years including the League of Nations, the UN and numerous others but more popularly Amartya Sen's entitlement theories. During this time, poverty vis-à-vis food security has had many different philosophical and practical focal points. Early on, poverty was largely defined in economic terms through measures of income or gross national product (GNP) per capita. By the 1970s however, the concept of poverty became a relative one, one of basic needs and one which evolved over time to included access to goods as well as to utilities such as education and health services. From about the 1980s, this was further elaborated by bestowing poverty with a more general and expanded social interpretation of well-being connected to political, moral and cultural values. By now the multi-dimensionality of poverty was largely acknowledged, and the realisation that the poor were not a homogenous group and were affected in completely different ways, opened the door to its full multi-contextual dimensions encompassing 'entitlements' of, economic, social, cultural and political value. As a result of these developments, the FAO and others employ the definition of poverty from the Organization of Economic Cooperation and Development (OECD) when it suggests that

> Poverty encompasses different dimensions of deprivation that relate to human capabilities including consumption and food security, health, education, rights, voice, security, dignity and decent work. (Gerster-Bentaya and Maunder 2008b)

In this way, poverty turns out not only to be an economic, but a philosophical construct too, and it is within this multi-dimensionality that poverty is more often than not analysed these days.

All things aside however, it is the economic construct of poverty that we are concerned with at present. In this regard, it has been shown that globally, there are around 1.2 billion people living on less than $1-a-day and a further 2.1 billion living on less than $2-a-day (World Bank 2009). The global poor are victims of multitude causation as disparate as they are regional. For some, this might be from a breakdown of entitlements through loss of employment, illness or disease while for others poverty might be perpetuated in the absence of adequate and secure property rights. Not surprisingly either, overpopulation has also come under fire as causative in issues of poverty, as has a lack of opportunities from a shortage of knowledge and skills gained through lack of formal education and training. Yet others still blame poverty on the legacy of colonialism and the unequal global trading structure as perpetuating a sort of modern neo-colonialism. Environmental issues do not escape blame either, or rather environmental degradation to be more precise, which for those living directly off the land can be financially devastating. It has also been said that bad governance and a lack of growth potential from poorer developing countries due to such factors as crushing external debt also has a major role to play too (Section 22.5.3). Or finally perhaps and more structurally, poverty is also seen as the result or symptom of the society-wide breakdown of social services and institutions, health and welfare policies. Lastly on the note of perpetuating poverty, the UN CFS assert that hunger itself, by compromising an individual's productive potential acts as its own barrier to escaping poverty (CFS 2007).

The glaring truth however in the continuing cycle of poverty, is that it no doubt results from any one, or a mixture of all of the above and perhaps more. What is known for sure too is the strong correlation that exists between poverty, health and undernutrition. Of particular note here is the perpetual 'vicious cycle' of poverty and hunger. Although the idea of a mutually reinforcing cycle of poverty and hunger is not new, what is of interest is the realisation that perhaps raising incomes alone would not necessarily be sufficient to end the poverty–hunger trap (CFS 2007). This once again taps into the wider social issues of hunger mentioned earlier to include a variety of non-income dimensions that collectively contrive to reduce an individual's ability to support their own needs (Louria 2000; Arcand 2001; Albala et al. 2002; UN 2002/3; Gerster-Bentaya and Maunder 2008a,b). Also of interest in this debate is the demographic make-up of the poor; the World Bank, for instance, tells us that over three-quarters of the poor in developing countries are the rural poor living on less than $1-a-day with the remaining 25% living in shantytowns or urban peripheries. These people are particularly vulnerable, especially when considering that in developing countries the poor can spend up to 80% of their income on food compared to just 10%–20% in the developing countries (USDA/ERS 2008a; WFP 2011). While alarming, it would also seem from analysis of the literature that the demographic status quo is set to change; predictions and estimates suggest that as a result of growing urbanisation many of the world's future poor will in fact come to reside in urban slums or shanty towns (von Braun 1995a; SOFI 2005).

Thus, from the previous reviews and this quick summary it can be seen that from whichever way we view the problem of poverty, the causes are many, complex and multifaceted. Addressing the issues of poverty remain the single biggest gift that we can give to the food insecure. Just how such change can be brought about is the subject of much speculation in terms of development paradigms, fiscal and monetary policies as well as adequate financial and institutional governance.

19.3 Rights and Human Capital Development

19.3.1 Human Development

During the latter half of the twentieth century, there has been considerable progress in the socioeconomic development in the lives of a large number of people throughout many countries. Having said that, there are still many more who have been lagging behind and by the late 1980s and early 1990s people began to look at the various aspects of development. The difficulty at first was apart from the obvious economic development, as measured by GNP or GDP, how were other concepts of human development, social capital or human development, for instance, to be defined let alone measured? Many different approaches were bandied about but finally by the 1990s the UN Development Programme (UNDP) brought out a much heralded forward-looking concept, the Human Development Report (HDR). This report defined human development in terms of longevity, knowledge and living standards and it became a benchmark for other measures of social well-being. Almost immediately it was noted with interest that the vastly divergent economic growth patterns did not always correspond to measures of social development; that is to say, some countries achieved better social development with less economic progress. This gave rise to much debate over the concept of economic growth for its own sake and helped rebalance the perspective that human development (or capital) was essentially the central goal of progress. More importantly it was starting to be realised that not only was human capital a major end goal but it was in fact a major input previously unaccounted for in neo-classical economic models (Hasan 2001). This new paradigm explicitly related economic productivity with the health, education and general well-being of the workforce—its human capital. In this way by employing healthy, educated and skilled workers, labour could then become more continuously productive, allowing firms to efficiently capitalise on both economic and social capital.

The idea of social capital or human development means different things to different people but central to all descriptions of the concept is the notion of a collection of social goals bound up in a 'rights' based ideology. It was not surprising then that with this new development paradigm that food security became inextricably linked with the 'rights' and 'development' of individuals on the grounds that humans needed adequate nutrition to fully develop their potential (MacAuslan 2009). In this way, human social development is concerned with the 'whole' person of which food security is but one component part. It could then be said that goals of food security, social development and economic growth for both the individual and the country became co-joined objectives (Clay 2002; FAO 2003b). This holistic view to development has gained much currency of late, especially since the failure of one or more of these objectives has repercussions on the others. Although one difficulty in this model is that, however well intentioned this might be, it exponentially complicates an already difficult concept of food security introducing yet more challenges in defining standards, cross-border equality and subjectivity.

Interestingly and despite these inherent complexities, human social development taps into the human rights dimension, most notably championed by the League and United Nations. Although for the sake of completeness, it is to be noted that the ideas of human rights have been around since man first sought council. Indeed many incarnations and attempts over the years to elucidate these rights has been a worldwide endeavour taking in the writings of the first king of Babylon (the Codes of Hammurabi, 1792–1750

TABLE 19.1

Selected Incarnations of Law and Human Rights

559–530 BC	The Persian Cyrus Cylinder
AD 697	The Irish Cáin Adomnái
1215	The English Magna Carta
1222	The Hungarian Golden Bull
1689	The English Bill of Rights
1776	The US Declaration of Independence and the Virginia Declaration of Rights
1789	The French Declaration of the Rights of Man and of The Citizen
1791	The US Bill of Rights

Sources: Hurlbut, E.P., *Essays on Human Rights and Their Political Guaranties*, Maclachlan, Stewart and Co., Edinburgh, 1847; Ishay, M.R., *History of Human Rights: From Ancient Times to the Globalization Era*, University of California Press, Berkeley and Los Angeles, 2008.

BC), the Hindu Gentoo Codes (Halhed 1776) plus numerous others through to the United Nations Declaration of Human Rights (1948), although it would not be fair to omit the numerous seminal landmarks in between such as those found in Table 19.1.

While such strides in human rights has a long and noble history, the idea of the human right to food has been equally well reflected in both moral, philosophical and ideological terms over a similar period.

19.3.2 Right to Food

The morality of human rights and the right to food is not a new concept. In fact although the FAO would have us believe, it is a concept first recognised by the Declaration of Human Rights in 1948, in reality, it is a notion that has scatterings throughout history, and one which has been indeed been evoked on numerous occasions (Mettrick 1929; UN 1948; FAO 2006). Even the ancient Babylonian slaves, despite doing 'service without pay', had the inalienable right to food (Johns 1904). Another example of this social construct saw the Rt. Rev. Thomas Sherlock, Bishop of London, building on the English philosopher John Locke's ideas (1714) when he suggested that

> There is not, I presume, a stronger natural right, than the right to food and raiment; this is founded in the common necessity of nature; and 'tis not to be thought that God sent men into the world merely to starve, without giving them a right to use in common so much of it as their necessities require. (Sherlock 1718, p. 25)

Further expanding on this notion, and it is worth quoting at length, was Edmund Burke's comments of the role of government and governance in his book reflecting on the French Revolution in 1790 when he wrote that

> Government is not made in virtue of natural rights, which may and do exist in much greater clearness, and in a much greater degree of abstract perfection: but their abstract perfection is their practical defect. By having a right to everything they want everything. Government is a contrivance of human wisdom to provide human wants. Men have a right that these wants should be provided for by this wisdom ... What is the use of discussing a man's abstract right to food or to medicine? The question is upon the method of procuring and administering them. In that deliberation I shall always advise to call in the aid of the farmer and the physician, rather than the professor of metaphysics. (Burke 1790, p. 88/89)

Interestingly here, Burke's separation from the philosophical viewpoint of the moral right to food and the adaptation of the practicalities of acquiring rightful nourishment marked an important departure at that time from contemporary doctrine. In a separate perspective, extrapolating from the individual to the community, the influential nineteenth century secularist Charles Cockbill Cattell also furthered the notion by stating unequivocally that

> Every industrious community has the right to food, clothing, shelter, and such social arrangements as will enable it to enjoy an average share of life. (Cattell 1874, p. 10)

Lastly and more recently though, food as a right was further given drive with the writings of Boyd Orr. In espousing the virtues of the 'new' science of nutrition in 1939 he indicated that

> It is the right of every citizen … to enjoy the benefits … of nutrition so that the health of every one … will be up to the level we now know is possible. (Orr 1939, p. 80; 1940)

With the impetus of moral momentum provided by the above and more, it is perhaps however a feather in the cap of the United Nations that the Human Rights Commission chaired by Eleanor Roosevelt drafted the Universal Declaration of Human Rights in 1948. It was the first time in history that an international body formally recognised food as a right that had been so often lacking in previous international mandates. Despite this early recognition however, it was only recently that the right to adequate food was officially enshrined in legislation (World Food Summit, 1996) and more recent still until the ethical and the human rights dimensions of the concept came into sharp focus (FAO 2006). The reason for this was that although the right to food is now judicially enshrined in over 40 countries worldwide, the fact that hunger was rising even before the 2007/2009 food and economic crises erupted, highlighted the fact that present solutions were, at best, inadequate (FAO 2006; SOFI 2009). After all

> It is a cruel mockery to tell someone they have the right to food when there is nobody with the duty to provide them with food. That is the risk with the rights rhetoric. (O'Niel 2002)

It has further been charged that although guidelines have been offered on how best to anchor a rights-based perspective in the fight against food insecurity; a fundamental lack of coherence between policies makes it difficult to co-ordinate any real progress, this is especially so considering that frameworks of the right to food are now based on voluntary codes (Hartmannshenn 2004; UN 2004).

The twentieth century was also a productive time for other rights too, in particular the rights of women and children.

19.4 Women and Children

The specific roles of women and children deserve special attention in any discussion on food security. While the role of children is perhaps more easily understood, the role of women in food security issues requires a little more elucidation. Women and children's issues are somewhat interlinked. It is more often than not that children receive their care and nutritional needs through the mother. In this respect, a malnourished mother is likely to give birth to an underweight baby susceptible to malnutrition, disease and ultimately premature death (Blössner and Onis 2005). With this in mind, beginning with the many voluntary non-governmental charities as well as key individuals like Count Rumford, the welfare of the child slowly began to take centre stage by the 1920s. Building on these early pioneering efforts, child welfare issues such as school feeding programmes and other dimensions of economic and sexual exploitation were finally elevated to the international collective conscience in the Geneva Declaration of the Rights of the Child of 1924. The need for special safeguards in respect of women and children was also noted in the Universal Declaration of Human Rights of 1948. However, as time progressed, public pressure, common sense and a growing international feminist movement in the 1970s helped place the gender question firmly on the international agenda with the UN's first world conference on women, held in Mexico City. Shortly afterwards, the first rights specifically enshrined for women, the International Bill of Rights for Women or more properly, the Convention on the Elimination of All Forms of Discrimination against Women (CEDAW) was adopted in 1979 (UN 2011). While the convention was the first human rights treaty to affirm the rights of women, it also recognised many of the underlying dynamics of culture and tradition as being influential forces that perpetuated gender roles in the society. Shortly after this in 1989, the United Nations finally drafted the Convention on the Rights of the Child (1989), which was subsequently enacted in 1990.

With specific regard to women and food security, their involvement is complex and often understated or under-acknowledged. Women's role's traverse the whole spectrum of the food security debate from primary caregivers, to agricultural labour to major bread winners and household custodians of nutritional security. With this in mind, it is worth looking at some of these in more detail.

19.4.1 Women in Agriculture

According to the International Labour Organization (ILO) in 2008, there were about 3 billion people of both sexes employed around the world. Of this figure, about 40.4% or 1.2 billion were women. Figure 19.2 highlights the individual proportions of total gender workforce by region and in each industry. It can be seen from this graph that there are only a relatively few women globally (18% of total women employed) working in the industrial sector compared with the nearly 27% of men. Moreover, continuing on the global perspective, it can be seen that while agriculture accounts for about 35% by far the biggest sector employing women is the service sector at about 46% (ILO 2009). However, a closer inspection of the regional disparities makes it clear that Asia and the Pacific as well as Africa continue to employ the majority of the female workforce in agriculture with sub-Saharan Africa and South Asia, specifically employing 64% and 65%, respectively. Incredibly, of all the women employed in the developed economies including the EU, a whopping 84% are employed in the service sector.

Taken together these figures are representative of each gender's total employment and considering the male global working population in 2008 was in the region of 1.8 billion and women at 1.2 billion, such isolated figures described above can be somewhat confusing. Figure 19.3 combines both genders and projects these figure as a total proportion of the 3 billion or so globally employed. From Figure 19.2 , it can be seen that globally the share of employees in agriculture in 2008 stood at 35.4% for women compared to 32.2% for men. This equates to approximately 424 million female and 579 million male global agricultural workers or combined about one-third of the total world labour market.

It can also be seen that huge reliance is still placed on the agricultural sector in both Africa, in particular sub-Saharan Africa and Asia and the Pacific. Having said that, later sections in the chapter look at some of these changing trends and highlights just how quickly agriculture is declining around the world.

In the past, the role of women in agriculture did not receive broad-based recognition. However, current trends in globalisation and more acknowledgment of the important roles of women; not just the concerning local food production but also in their valuable contribution in both industry and the service sectors, are redefining the relationships of women in economic and rural development paradigms (Hall 1998). Although sadly while some progress has been made, it has been and continues to be slow. Women continue to be plagued by low incomes, and limited access to, among other things, health care, education, credit and equitable access to land and natural resources. Fuelled by increasing competition for cheap, flexible labour as well as social and military conflict, job insecurity perpetuates difficult working

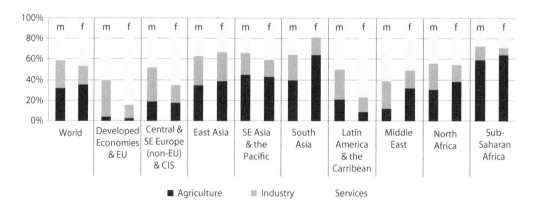

FIGURE 19.2 Respective proportions of total gender employment, 2008 preliminary estimates. (Compiled from the ILO Global Employment Trends for Women, International Labor Organization, Geneva, Switzerland, 78, 2009.)

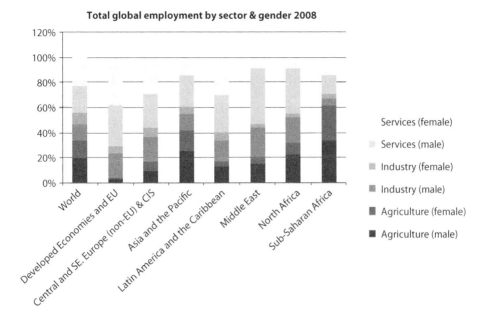

FIGURE 19.3 Total world employment by sector and by gender, 2008 preliminary estimates. (Compiled from the ILO Global Employment Trends for Women, International Labor Organization, Geneva, Switzerland, 78, 2009.)

conditions for women. This is all happening at a time when more risk of natural disasters, exacerbated by environmental change, is compounded by decreasing government support.

It is not just in agriculture either that women are active. The role of women in other areas of society range from caregiver to provider and educator. These roles however come with their own set of challenges.

19.4.2 Women: Roles and Gender Parity

Women play many vitally important roles in their own lives and in the lives of their families; yet they face many challenges when it comes to responsibility and gender biases. In this respect, particularly in regards to mothers, the responsibility of achieving balanced health and for alleviating food insecurity often falls more proportionally on their shoulders, more so in fact than any other member of the society.

Women are also the primary caregivers of children and in many instances of the household too. In particular, the mother is the crucial link in the protection and growth of the child from uterus to adulthood, providing both support and nourishment at all stages of his or her development. It has already been shown that children suffering chronic malnutrition fail to properly grow to their full potential both mentally and physically; yet this responsibility generally falls on the mother who, through pregnancy, breastfeeding and nurturing, provides a solid nutritional foundation on which a child's whole life depends. This is achieved often without help from other household members or institutional and governmental support. Moreover, women also tend to be the arbiters safeguarding family health, which includes not just the children but also the sick, the handicapped and the elderly, and either provide care themselves or are responsible for regular health visits to clinics and doctors. Women are also the family nutritional educators relaying information about the right foods to eat as well as the importance of good hygienic practices (SD 2001; Freedom from Hunger 2010; Save the Children 2010).

It does not stop there either; in many instances women also are responsible for choosing which foods the household eats; in this function, they also process and store food and find ways of feeding the family when stores run low (SD 2001). Over and above this, there are endless household chores that women in developing (and developed) countries continue to be responsible for. This brings to the fore questions of time constraints. Studies in Indian and Nepalese villages have shown, for instance, that in total, counting the work, household responsibilities and domestic chores women in these regions worked on average

16 hours a day compared to the average 8–9 hours of the men. Such extreme demands have consequences for the health and well-being of the women themselves but also of her immediate family. Time has to be found to prepare and cook food and in times like this households might rely on one cooked meal a day and left over food that might or might not be properly stored. In this regard, any measures that help to reduce this time-load of women would have enormous benefits. Things such as convenient access to wood for fuel, water or alternative child-care facilities coupled with readily accessible health care would contribute considerably in not only improving their quality of life but also that of their families (IFAD 2009).

Such notions bring up issues of gender equality. The relative status of women compared to men has on many occasions been voiced as an important consideration in the fight against food insecurity (Grebmer et al. 2009). A large body of evidence tells us that the greater the control of women over the family income often translates into improved level of food acquisition. However, the continual subjugation of women is perpetuated by many cultural biases where women's status continues to be seen as inferior to that of men. While gender parity might be a long way off, resources can still be provided which alleviate much of the pressure of time with such things as have already been mentioned as well as anything that raises women's status or enables them to cope more effectively. Any help in this direction will, in turn, see proportionally increased household food security (Hyder et al. 2005; IFAD 2009).

Despite all the above, there has been progress. With so much global momentum surrounding the issue of gender parity, strong movement is being seen on women's education in developing countries while local community employment opportunities for women, particularly by numerous multinationals, are providing much needed progress. Moreover, while businesses in the developed world have long understood the power and influence of the female consumer, businesses in the developing world are only just tuning in to the female dollar. On the political front too, several countries, predominantly developed ones, have introduced legislation aiming to close this gender gap by ensuring minimum requirements for women's membership in all sectors of work and social activity (Hausmann et al. 2010).

An interesting concept in measurement of this gender disparity is undertaken by the World Economic Forum in their annual Global Gender Gap Report. It is an index based on equitable patterns of global education, political empowerment and economic participation as well as health and survival of women. By combining these attributes into a single numerical indices, measures of gender gaps can be calculated where a score of one equals parity. From such metrics, it can be seen from the 2010 report that predominantly high income or developed countries occupied 8 of the top 10 slots reflecting the best gender-performing criteria: Iceland, Norway, Finland, Sweden, New Zealand, Ireland, Denmark and Switzerland. Surprisingly though, among the low- to middle-income (developing) countries, it seems that economic well-being is not always a pre-requisite for gender parity and looking at the tables several seemed to be doing well, of which the most notable are Lesotho (8), Philippines (9), Sri Lanka (16) and Mozambique (22). Regionally speaking, though the Nordic countries continue to occupy the top slots, in the Americas, the United States, Trinidad and Tobago, Cuba, Costa Rica and Argentina all show strong performance. Looking at the Middle East and North Africa too, Israel and the United Arab Emirates hold the top two positions while in Asia and the Pacific, New Zealand, the Philippines, Sri Lanka as well as Australia are all top performers. Lastly, it can also be seen that in sub-Saharan Africa, Lesotho, South Africa and Namibia have all taken great strides in closing the gender gap (Table 19.2).

Overall, the report, with a global coverage of 90%, tells us that as of 2010, great progress has been made in closing the gender gap. In particular, gaps have been reduced in respective health outcomes between men and women to the tune of 96% and on educational attainment by almost 93%. Noticeably lacking however is progress in the areas of economic participation and political empowerment with women having only achieved 59% and only 18% overall global parity, respectively. So while it can be seen progress is being made, there is still a long way to go both regionally as well as economically and politically.

19.5 Education and Employment

When it comes to food security vis-à-vis education and employment, a strong case has already been made regarding their combined importance. Education is at the heart of many attempts to improve human development. Although much of this is aimed at school children, an important barrier to adult

TABLE 19.2

Gender Gap Rankings, 2010

Country	Overall		Economic Participation and Opportunity		Educational Attainment		Health and Survival		Political Empowerment	
	Rank	Score	Rank	Score	Rank	Score	Rank	Score	Rank	Score
Top 5										
Iceland	1	0.8496	18	0.7540	1	1.0000	96	0.9696	1	0.6748
Norway	2	0.8404	3	0.8306	1	1.0000	91	0.9697	3	0.5614
Finland	3	0.8260	16	0.7566	28	0.9993	1	0.9796	2	0.5686
Sweden	4	0.8024	11	0.7695	41	0.9964	80	0.9729	4	0.4706
New Zealand	5	0.7808	9	0.7743	1	1.0000	91	0.9697	8	0.3792
Bottom 5										
Saudi Arabia	129	0.5713	132	0.3351	92	0.9739	53	0.9762	131	0.0000
Côte d'Ivoire	130	0.5691	106	0.5390	130	0.6923	1	0.9796	104	0.0655
Mali	131	0.5680	113	0.5137	131	0.6794	55	0.9761	81	0.1026
Pakistan	132	0.5465	133	0.3059	127	0.7698	122	0.9557	52	0.1545
Chad	133	0.5330	77	0.6265	134	0.5091	110	0.9612	122	0.0352
Yemen	134	0.4603	134	0.1951	132	0.6567	81	0.9727	130	0.0165

Source: Hausmann, Tyson et al., *The Global Gender Gap Report*, World Economic Forum, Geneva, 334, 2010. With Permission.

education is relatively high illiteracy rate. Despite progress over the last 20 years resulting in an improvement of 8% in global terms (6% men, 10% women), global illiteracy still affected over 796 million adults (15 years plus) in 2008, with about two-thirds of this number (64%) being women. Furthermore, two regions Southern Asia with 412 million and sub-Saharan Africa with 176 million accounted for nearly three-quarters (74%) of all adults worldwide unable to read and write. Also, once again gender disparity was at work in southern Asia where 73% of men compared to just 51% of women could properly read and write resulting in a gender parity index of 0.70. As a consequence illiteracy combined with general low levels of education can badly hinder the economic development of individuals and their country. This has an important bearing on food security as parental education has been found to be an influential factor in their own and their children's nutritional status, so much so in fact that children of illiterate parents are consistently seen to score more poorly on nutritional status indices (Mukudi 2003).

The rationale for this is fairly widely accepted; the better educated a person is the more reasoned choices can be made regarding nutrition and health as well as within many other aspects of economic and social inclusion. In this way, education empowers the poor in their fight for a better standard of living through improved opportunities of gainful employment prospects or perhaps the capacity to use the resources he or she owns more rationally. Improved education also means a better social skill set, which might also help with food security and better access to resources and community safety nets in times of need. Finally, it has been suggested that educated people tend to be better informed and are more likely to choose more valuable objectives in life. In this endeavour, education can provide a further psychological boost, making people generally more ambitious and self-confident (Hausmann et al. 2010; UIS 2010). In sum, investment in education can be seen to deliver many tangible and beneficial returns not only to the individual but also to the nation state. Given these considerations, it should come as no surprise then that a study in America of a low-income, multi-ethnic population found a close relationship between education and decreased food insecurity (Dollahite et al. 2003; Rosegrant and Cline 2003; CFS 2007).

With regard to employment and food security, a continually expanding developing world population poses many challenges. As Figure 19.4 highlights the trends in the various sectors is pretty largely self-evident. Of interest is that over the last 10 years in every region in the world without exception, agricultural employment numbers have been on a marked downward trend. The 10-year trends also show the rapid rises in almost all regions of industrial and service sectors. Thus the challenges posed by the increasing reliance on non-farm employment for developing countries is going to be considerable.

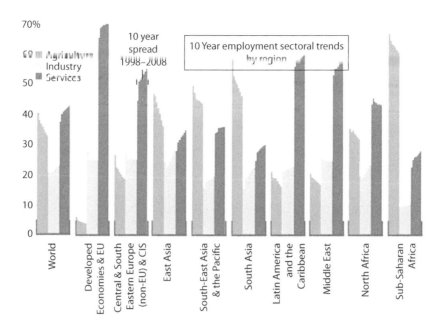

FIGURE 19.4 Global percentage of people employed by sector. (Compiled from ILO (2009), Global Employment Trends for Women, International Labor Organization, Geneva, Switzerland, 78, 2009.)

With global unemployment placed at around 6.5% (211 million people) for 2009, employment policies and programmes aimed at simultaneously addressing the problems of poverty and economic growth must be sustainable. Whether through the promotion of enterprise or the creation of national complimentary industries, employment strategies need to be found that are as socially inclusive as they are varied. Important too is the need to bring into the fold the informal, unregulated sector in order that it does not create poverty traps or promote social exclusionism. Furthermore, by diversifying and expanding the concentration of its industry and service sector economies, a nation effectively increases its ability to trade on a global setting. These and numerous other recommendations by others would collectively create an economic setting in which human self-regulation and choice can foster improved food security. A small caveat however suggests that although without doubt, such economic growth and diversification will bring with it the associated benefits of industrialisation such as improved infrastructure and better financial flows. This can only be achieved however, in the view of the author, through a more sympathetic pre-requisite development-oriented strategy of social progress in the vein of those already outlined.

19.6 Emergencies and Crises

19.6.1 Disasters

Collectively natural disasters, accidents, civil strife and armed conflicts are a major component of the food security problem. For simplification, such emergency situations can be thought of as belonging to two categories: natural and human-made. Natural disasters might include earthquakes, volcanoes, storms, floods, extreme temperatures, drought, wildfires, epidemics or insect infestations. On the other hand, complex or human-made disasters can include conflict and civil unrest, poor government and inadequate economic policies. As can be seen from Figure 19.5, the number of reported annual technological as well as natural disasters that either: killed 10 people or more; affected 100 or more people; involved the declaration of a state of emergency; or resulted in international assistance being requested by the country in question has markedly increased since the 1900s.

More worrying is the growth of natural disasters since the mid-to late 1970s. Fuelling this trend, as can be seen from the figure, are the frequent spikes in many extreme weather events such as floods

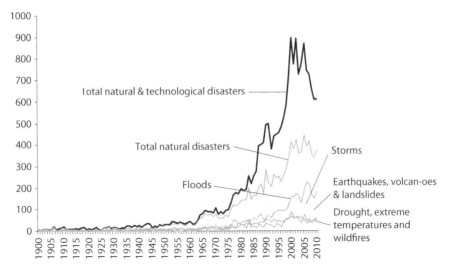

FIGURE 19.5 Number of reported disasters by type 1900–2010. (Courtesy of EM-DAT: The OFDA/CRED International Disaster Database (www.emdat.be), Université Catholique de Louvain, Brussels, Belgium.)

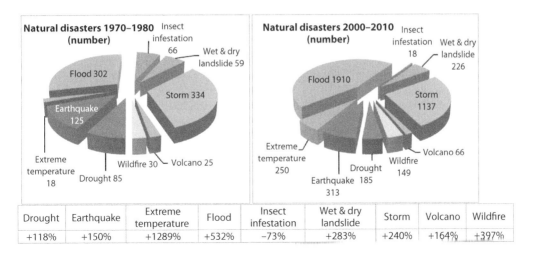

Drought	Earthquake	Extreme temperature	Flood	Insect infestation	Wet & dry landslide	Storm	Volcano	Wildfire
+118%	+150%	+1289%	+532%	−73%	+283%	+240%	+164%	+397%

FIGURE 19.6 Global natural disaster changes 1970/1980 to 2000/2010. (Courtesy of EM-DAT: The OFDA/CRED International Disaster Database (www.emdat.be), Université Catholique de Louvain, Brussels, Belgium.)

and to a lesser extent storms. In fact year on year the frequency of these two disaster types have come to dominate the natural emergency landscape. Looking at this period in a little more detail, Figure 19.6 illustrates the considerable increase in 10-year rolling averages of natural disasters in the 30-year period 1970/1980 to 2000/2010. It can be seen that while droughts and earthquakes have more then doubled (+118% and +150% respectively) in this period, it can also be shown on closer inspection that floods have increased by over 6-fold, wet and dry landslides by a near 3.5-fold and lastly extreme temperature events by nearly 14-fold.

Regionally speaking, the trends were fairly even across the board regarding storms and floods, while major upward trends were seen in incidents of extreme temperatures in Europe (65-fold). This is striking in its comparison to the overall global increase of nearly +1289% or 14-fold. Over the same period, it also seems North America saw more than its fair share of wet and dry landslides with an 18-fold overall increase. What this undoubtedly tells us is that global climate change over the last 30 years in

TABLE 19.3

Top 10 Disasters by Numbers Killed in Last 100 Years

Death Toll	Event	Location
1–2.5 million	1931 China floods	China
500,000	1970 Bhola cyclone	East Pakistan (now Bangladesh)
316,000+	2010 Haiti earthquake	Haiti
242,419	1976 Tangshan earthquake	China
234,000	1920 Haiyuan earthquake	China
230,210	2004 Indian Ocean tsunami	Indonesia
142,000	1923 Great Kanto earthquake	Japan
138,000+	2008 Cyclone Nargis	Myanmar
138,000	1991 Bangladesh cyclone	Bangladesh
120,000	1948 Ashgabat earthquake	Turkmen SSR, (now Turkmenistan)

particular is changing rather rapidly. With more incidents of extreme weather patterns and associated geological fallout such as floods, landslides and the like, vulnerability assessment and risk management programmes are an important addition in the arsenal of food security tools (FAO 2002, 2003a; UNEP/ GRID Arendal 2005; Vos et al. 2009; OFDA/CRED 2010).

One of the sad realities of an increasing number of worldwide disasters is that they occur so frequently these days that many people are rarely tuned into the effects of such events unless they are on a scale that stands out of the ordinary. People remember the Indian Ocean earthquake and tsunami of 2004, the 2010 Haiti earthquake or perhaps the Japanese earthquake of 2011, yet many are quickly forgotten. How many remember the Chinese drought and dust storms that affected 10 provinces causing widespread disruption, or the Javan eruption of Mount Merapi that saw 350,000 people evacuated or maybe the series of storms that wreaked havoc through much of the Australian state of Victoria? The reality is many of us suffer from compassion fatigue, and this has a debilitating effect on many issues of humanitarianism, not least the food security aspect. Large-and small-scale disasters affect many populations on a regular basis and in 2010 alone, 373 natural disasters were responsible for the deaths of over 296,800 people and affecting a further 208 million others. This collectively cost close to $110 billion (UN ISDR 2006). Figures like this are alarming yet many small-scale disasters go relatively unnoticed or are quickly forgotten by today's standards. Just as a reminder, Table 19.3 highlights the top 10 disasters by number killed in last 100 years.

And then there are the conflicts and associated challenges of civil strife.

19.6.2 War/Conflict and Social Displacement

Conflict and civil unrest as well as economic disaster are two major components of man-made emergencies and of these, war undoubtedly takes the most toll in mortality and displacement figures. In fact, estimates put the total mortality figures of all wars during the twentieth century at approximately 123 million. This breaks down to about 37 million military deaths, 27 million civilian casualties of war and a further 58 million attributable to war-related democide and famines (White 2010). Conflicts are started for numerous reasons ranging from territorial, economic or political ideology although one aspect of conflict trends is worth mentioning in that since 1946, interstate (conflict between states), has tailed off considerably as can be seen in Figure 19.7. Many have apportioned this trend to the effects of the United Nations and the central forum in which state disputes can be addressed without recourse to war. From this picture, it can be seen that most wars then tend to be internal conflicts and occur mostly in the developing regions of the world where poor economic resources prevail. This is especially so in Africa and Asia, which between them accounted for the largest number of internal conflicts in the past decade or so (Hartmannshenn 2004).

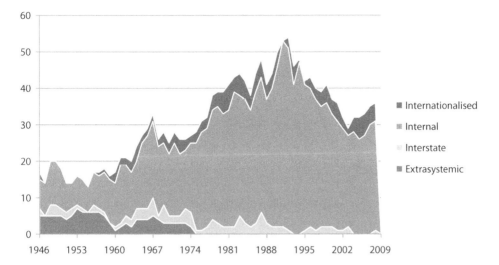

FIGURE 19.7 Number of conflicts by type 1946–2009. (Re-created from the UCDP/PRIO Armed Conflict Dataset; http://www.nsd.uib.no/macrodataguide/set.html?id=55&sub=1, 2009.) *Notes:* Extra-systemic conflict is between a state and a non-state group outside of its own. Interstate conflict occurs between two or more states. Internal conflict occurs between the government of a state and one or more internal. opposition group(s) without the intervention of other states. Internationalised internal conflict occurs between the government of a state and internal opposition with intervention from other states.

Whatever the reason for such conflicts, they are financially costly and just as importantly tie up valuable government resources whilst diminishing a country's ability to regulate its own food security. Furthermore, conflicts are a major impediment of development objectives and a point that is increasingly being remarked upon in the international community. What is more, while conflict can and does affect anyone whether rich or poor, the problems are particularly exacerbated when countries can ill-afford or are ill-prepared for any of the associated physical, financial and social costs. In this situation, lines of transportation are affected, markets become dysfunctional or collapse completely while crops might be abandoned, pillaged and plagued with landmines. Such tense situations can also act as a deterrent, discouraging further investment in the land for fear of future displacement. All the while, health service infrastructure too can come under enormous pressure and resultant disease and epidemics can prevail.

On top of this, another aspect of war that must be addressed if food insecurity is to be alleviated are the enormous number of refugees (FAO 2002).

19.6.2.1 Refugees and Displaced Populations

Sometimes the only solution open to individuals in a conflict zone is to flee the area. Sometimes, this means going from one region to another within the same borders to become the internally displaced persons (IDP), or to cross nearby borders at which time they become refugees. Every year, many millions of people have to abandon their homes as a result of conflict or political unrest and natural disaster. The main challenges in these situations are that those forcibly removed, or otherwise are at once cut-off from their livelihoods, and more often than not have little more assets to their name other than what they can carry. This places enormous burdens on their food and livelihood security and as a result situations of mass exodus often require external emergency assistance and very often this becomes the only buffer keeping these people from starvation (DeRose et al. 1998).

When it comes to numbers, the different conventions see the different categories of refugees IDPs being counted differently and this accounts for differences in official figures offered by the likes of. The United States Committee for Refugees and Immigrants (USCRI), The Office of the United Nations High Commissioner for Refugees (UNHCR) and the Geneva-based Internal Displacement Monitoring

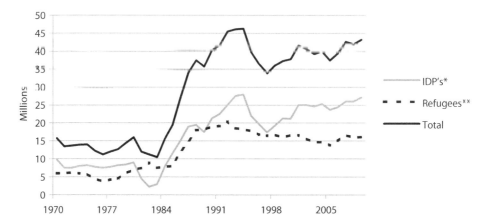

FIGURE 19.8 Refugee and IDP trends, 1970–2009. (From UNICEF, The Convention on the Rights of the Child. United Nations International Children's Emergency Fund (UNICEF), 2009; IDMC, Global IDP Estimates (1990–2009). The Internal Displacement Monitoring Centre (IDMC), 2011; INSCR, *The Integrated Network for Societal Conflict Research (INSCR)*. Center for Systemic Peace, 2011.) *Notes*: IDP's* include returned refugees returned IDPs, stateless and others of concern but not those displaced from natural disasters; Refugees** include pending asylum-seeker cases.

Centre (IDMC) among others. Figure 19.8 amalgamates these figures and displays the total refugees and IDPs for all countries, published as of 2010. Of considerable note is the sharp rise in all forcibly displaced people from about the middle and towards the end of the 1980s. While the number of conflicts coinciding with this rise also increased, this does not account for the magnitude of the trend. The Centre for Systemic Peace (CSP) suggests this spike to be a result of the potential consequences of four converging factors: the increased prevalence of armed conflicts in the poorer countries; the protractedness of such conflicts; the blurring of distinction between combatants and non-combatants; and lastly the general expansion of humanitarian capabilities in providing assistance to those affected. Combined, these factors could indeed account for the sudden increase in those forcibly displaced and seeking assistance.

The total figures are alarming and have continued to hover around 38–43 million over the last decade or so. Of these figures in 2009, the total 43.3 million consisted of 27.1 million internally displaced persons and 15.2 million refugees (10.4 million under UNHCR's responsibility while 4.8 million under UNRWA's). Once again the developing countries bore the brunt of the refugee problem, hosting four-fifths of the world's total refugees (UNHCR 2009; INSCR 2011).

Whether through conflict, natural disaster or being forcibly driven away from their lands, many people face imminent hunger issues and all too often there is very little stopping these crises becoming emergencies of famine.

19.6.3 Famine

Historically, there tends to be a great disparity in the agreed numbers of famines over the centuries; one study in 1930 uncovered over 1800 famines between 108 BC and AD 1911 in China alone. Ancel Keys and others, on the other hand, are commonly quoted as citing approximately 400 famines; although many now concede this might only in fact be scratching the surface. The point is, the number is almost impossible to determine with any degree of consensus so as a general rule 400 or so famines is becoming an accepted best guess (Watkins and Menken 1985; Prentice 2001; Ó Gráda 2009). Interestingly in Britain, from about AD 1600, famine had all but disappeared with Europe following suit by around 1850 (Walker 2008). Despite this however Devereux, citing geographer William Dando, suggests that such decreases were largely due to advances in industrialised countries and that in general, at the global level, famine figures have in fact been on the increase with 2, 10 and 25 million deaths reported in the seventeenth, eighteenth and nineteenth centuries respectively (Devereux 2000). Furthermore globally it has been said that a massive 70 million people plus have died as a result of famines in the twentieth century alone.

Also whereas many famines of the past have been linked to poor harvests and the vagaries of the weather, the variety and diversification of crops which started, and continued with the Columbian exchange, in fact did much to offset this aspect of vulnerability. In short, it seems that while the effects of famine have changed little in millennia, the causes themselves have shifted over time. It now seems that perhaps the one major distinguishing feature of twentieth-century famines is that they are nowadays more likely linked to conflict and civil strife or other human-made determinants (ODI 1997; Devereux 2000; Ó Gráda 2009). Table 19.4 highlights this trend in which it can clearly be seen that the worst famines of the twentieth century were often exacerbated either by conflict, poor government policy or a combination of both. In addition, it has also been suggested that the pervasive, all-embracing totalitarian state (prevalent in Africa) greatly increases the human cost in such circumstances through the exacerbation of poor governance or ill-conceived political policy (ibid). That said though, even this is slowly changing through a combination of early warning systems, government and public intervention and food aid, which collectively help to mitigate the worst effects of modern day famines (Johnson 2000; Ó Gráda 2009).

From the table, it can be seen that Asia, China and India between them accounted for the largest of all famines in the twentieth century, yet in Africa where they continually persist, deaths from famine are decreasing and number in the tens or hundreds of thousands rather than the millions. One of the major reasons that has been suggested for such decreases is the break in causality between crop failures and famine, through, among other things, swift humanitarian responses (Devereux 2000; Ó Gráda 2009). Another interesting finding of the above research also suggests that while the twentieth century might have indeed witnessed fewer famines than in the past, it has in fact contained some of the worst famines ever recorded in history. With this in mind, it is not surprising, suggests one researcher, that famine and all its emotive connotations have attracted much attention and indignation from concerned academics, the public and other critical observers alike (Ó Gráda 2009).

19.6.4 Coping Strategies and Priorities

For a long time, it was widely thought that given sufficient food and entitlement resources food security would naturally follow. However, this idea wrongly assumed that food security was the only priority for the hungry. While this might at first seem counterintuitive, it is now becoming widely accepted that food security is only one of a number of priorities for individuals, within households and at the national and global levels. Furthermore, it is also now understood that at times, food security is not always the uppermost priority in these choices. People and households will trade priorities in a juggling act dependent on the perceived needs of the time, although there appears to be little consideration of this within the food security debate. This is problematic when, faced with finite resources, all the goodwill in the world cannot force people to prioritise one goal over another. This is perhaps one of the great dilemma's of development objectives for while we as humanists are horrified at the inequality of the status quo and focus our attention in that direction, unless the receiving—governments, households and individuals—share the same priority, at the same time then our efforts become largely misdirected. It is not uncommon therefore, in households with limited income or resources, to sacrifice sufficient food for other priorities of schooling, rents, land acquisition, security of tenure, etc. In this way assuming sufficient resources to acquire food, supplemental to Sen's and others' entitlement theories is the notion that food just might not be the top priority. However, far from justifying any contraction of the food security concept, this has implications for policy and intervention and serves to promote or support at least, the wider societal aspects of the concept; that is security of tenure, education, employment and health services among others. Moreover, understanding that food security is but one element in a full complement of livelihood securities is indicative of the complexities of isolating the food security concept vis-à-vis individual choices and national policy motivation for that matter. For this reason, coping or adaptive strategies need to be considered not so much as a catchall for 'getting by' but rather to be isolated and studied so that livelihoods policies or social safety nets are in place to offset the need to make such choices. Further, in this sense, coping strategies are a bit of an oxymoron as by 'coping', people are left with unsustainable choices that effectively limit future coping strategies, and that these in themselves, ultimately become failure strategies.

TABLE 19.4

Worst Famines of the Twentieth Century

Date	Location	No. of Deaths	Additional Information	Cause
1913–14	W. Africa (Sahel)	125,000	Altered cropping patterns, changing fiscal policy of colonial powers and seasonal hunger was exacerbated by drought.	Drought and poor governance
1914–18	Armenia, Belgium Mount-Lebanon, Germany	Millions	Famine and associated death from persecution, tuberculosis, rickets, influenza, dysentery, scurvy, keratomalacia and hunger oedema during World War I. Reports too of cannibalism and exhumed bodies used for food.	Famine excerbated by WW
1917–19	Persia	Quarter of the population	Famine. As much as 1/4 of the population living in the north of Iran died in the famine exacerbated by the Russian revolution.	Famine and Conflict
1921–22	Soviet Union	5–9 million	Massive crop failures were greatly magnified with failures of Soviet central planning and Stalin's decision to withhold food.	Drought
1927	N.W. China	3–6 million	The famine was initially triggered by drought then exacerbated by local warlords and excessive taxation.	Drought
1928–29	China (Hunan)	2 million	Not much has been recorded about this famine.	Drought and conflict
1932–34	Soviet Union (Ukraine)	7–8 million	Stalin's embarked on his massive collectivisation programme in which the government seized grain for export earnings to buy industrial equipment. In response, revolts by peasants led to all collective land, agricultural produce and implements being declared state property.	Government Policy
1936–38	Sichuan, China,	5 million	Not much has been recorded about this famine.	Famine
1940–44	Warsaw, Leningrad, Greece, Netherlands	70,000	Famine in Warsaw Ghetto. Leningrad blockade by Nazi's and the Finnish. Famine worsened by freezing temperatures, dysentery, bronchopneumonia, tuberculosis and typhus.	Famine, d seas and conflict
1943	China (Henan)	5 million	Combination of Japanese invasion and by Chinese government grain seizures to feed its troops and finance the war.	Conflict and Policy
1942–44	India (Bengal)	2.1–3 million	Combined crop failure and the exportation of foods by India's British administration to Allied soldiers. End of rice imports from Burma after Japanese invasion and a lack of food price control.	Conflict and policy
1943	Rwanda-Burundi	300,000	Famine in Rwanda causing migrations to the Congo, Burundi in 1943–44.	Conflict and drought
1946–47	Soviet Union (Ukraine and Belorussia)	2 million	Drought and government policy—the reinforcement of agricultural collectivisation policies during WW2.	Drought and government policy
1957–58	Ethiopia Tigray	100,000–397,000	Famine in Tigray, Ethiopia.	Drought and Locusts

				Political action
1958–62	China	30–33 million	Mao Zedong's (Mao Tse-Tung) 'Great Leap Forward' plan to modernise agriculture and increase grain production. Collectivisation and the order to give the state a large percentage of their crops allowed China to double its grain exports and cut imports of food.	
1967–70	Nigeria (Biafra)	1 million	Biafran famine caused by Nigerian blockade.	Conflict
1969–1974	West Africa (Sahel)	101,000	Sahel drought.	Drought
1972–74	Wollo, north-eastern Ethiopia	40,000–80,000	Recurrent crop failures and continuous food shortage.	Crop failure
1974	Bangladesh	1.5 million	Famine in Bangladesh.	Flood and market failure
1979	Cambodia	1.5–2 million	Resulting from a decade of conflict—first the civil war from 1970 to 1975 then the Khmer Rouge, which ended with the Vietnamese invasion that ended Khmer Rouge rule in 1979.	Conflict
1980	Uganda (Karamoja)	30,000	Drought and a breakdown of civil order caused famine in Karamoja.	Conflict and Drought
1982–85	Mozambique	100,000	This famine was triggered by drought but exacerbated by the conflict between the Frelimo government and the south African backed Renamo.	Conflict and Drought
1983–85	Ethiopia	590,000,000–1 million	Caused in part by the 1982–1983 El Nino Southern Oscillation.	Conflict and Drought
1983–85	Sudan (Darfur, Kordofan) +Sudan South	250,000	Drought and economic crisis combined with government denials of the crisis exacerbated by conflict.	Drought, policy and conflict
1991–93	Somalia	300,000–500,000	Rival warlords battled for control of the southern coast and hinterland. This brought war and devastation to the grain-producing region between the rivers, spreading famine throughout southern Somalia.	Conflict and Drought
1995–99	North Korea	2.8–3.5 million	Between 2.8 million and 3.5 million people died because of a combination of flooding and government policy.	Flood and government policy
1998	Sudan (Bahr El Ghazal)	70,000	Drought and war.	Conflict and drought

Sources: Fisher, H. H., *The Famine in Soviet Russia 1919–1923 the Operations of the American Relief Administration*, Books For Libraries Press, Freeport, New York, 1927; Chamberlin, W. H., *Foreign Affairs*, 12(3), 495–507, 1934; Gantt, W. H., *British Medical Journal* 2(3939), 19–22, 1936; Bennett, M. K., *The Journal of Political Economy*, 57(3), 185–198, 1949; Scrimshaw, N. S., *Annual Review of Nutrition* 7(7), 1–22, 1987; Waal, A. D., *Evil Days: Thirty Years of War and Famine in Ethiopia*, Human Rights Watch, New York, 1991; Kalayjian, A. S., Shahinian, S. P., Gergerian, E. L. and Saraydarian, L., *Journal of Traumatic Stress* 9(1), 1996; Watson, F., *Field Exchange* 8, 20–23, 1999; Devereux, S., *Famine in the Twentieth Century*, Institute for Development Studies, Brighton, 2000; Ellman, M., *Cambridge Journal of Economics* 24, 603–630, 2000; Dyson, T. and Ó Gráda, C. O., *Famine Demography: Perspectives from the Past and Present*, Oxford University Press, Oxford, 2002; Hionidou, V., *Population Studies*, 56(1), 65–80, 2002; Patenaude, B. M., *The Big Show in Bololand: The American Relief Expedition to Soviet Russia in the Famine of 1921*, Stanford University Press, Stanford, CA, 2002; Peterson, M. D., *Starving Armenians: America and the Armenian Genocide, 1915–1930 and After*, University of Virginia Press, Charlottesville, London, 2004; Leo Lucassen, D. F. and Oltmer, J., *Paths of Integration: Migrants in Western Europe (1880–2004)*, Amsterdam University Press, Amsterdam, 2006; Oltmer, J., *Nationalities Papers 1465–3923*, 34(4), 429–446, 2006; WIT, World Ecology Report, World Information Transfer, New York, 2008; Ó Gráda, C., *Famine: A Short History*, University Presses of California, Columbia and Princeton, Princeton, 2009.

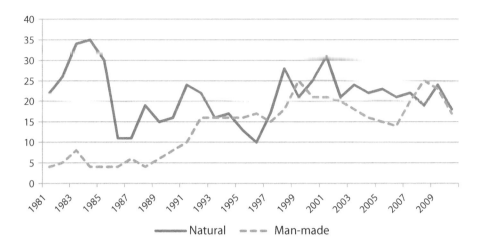

FIGURE 19.9 Countries requiring external food assistance. (Compiled from various FAO Crop Prospects and Food Commodity Reviews.)

19.6.5 Emergency Aid

In total, various estimates of the number of deaths by war and oppression from about 1900 to the present day have been put at anywhere between 165 and 262 million with many favouring the 190–200 million range. This figure however incorporates all deaths from war, and this includes crossover of some of the 70 million people that died as a result of famines during the same period (White 2010; Rummel 2011). Also, in annual terms, hundreds of thousands continue to perish in conflicts while anywhere between 35 and 43 million refugees and IDPs are forcibly displaced. Furthermore, in any given year, as many as a further 50 million people are estimated to be displaced through natural disasters such as earthquakes, floods, tsunamis and landslides (WHO 2010; OCHA 2011).

As a result, it can be seen that countries with emergencies and crises that required humanitarian assistance and food aid, apart from a few wild spikes, have been averaging between 15 and 25 over the last 30 years or so (Figure 19.9). How these are dealt with in connection with food security are looked at in detail in Section 22.2.4; however, for now we are concerned with the frequency and effects these emergencies have on the population and preparedness of individuals and states.

What is immediately obvious with this picture is the slow but steady increase in man-made food emergencies compared to a more static, albeit erratic, number of natural emergencies that require external assistance. With this in mind, it can be suggested that even though extreme natural disasters, irrespective of assistance required, are on the increase (Figure 19.5), the fact that external food assistance remains fairly static is suggestive that in countries where natural disasters occur without the complication of conflict they are perhaps in a better position to cope. This is indicative of good stable political environments, established emergency procedures, easy access for the humanitarian community and a more palpable community comradery. This in turn, collectively better offsets the food insecure position. On the other hand, increased emergency assistance due to human-made disasters through armed conflict or complex emergencies* are perhaps more immediately vulnerable. This is because such conflict often occurs within a devastating social breakdown in all areas from weakened governmental infrastructure, tied up resources, restricted freedom of mobility and a difficult humanitarian environment, among other things. Such a brutally unforgiving landscape places enormous pressures on those in such situations. This is not

---

* Complex emergencies were initially characterised in the late 1980s onwards by armed conflict and political instability; nowadays however complex emergencies refer more so to humanitarian crises where disasters may result from multi-causal events or which results in considerable or total breakdown of legitimate authority. At other times complex emergencies are considered those requiring international action that goes beyond the mandate or capabilities of any single agency. This also includes war or conflict and associated difficult political and security environments.

helped either when year on year the humanitarian sector struggle to fund existing projects let alone finding the much needed capital for other slow burning, non-emergency situations where limited resources often see only the worst affected being aided.

19.7 Changing Diets

Long-term trends in dietary changes at the global level has seen every continent on Earth change their pattern of consumption. By way of illustration, Figure 19.10 (a and b), taken from the FAOSTAT database in 2011, shows that all except Africa are consuming markedly less cereals over the 40-year period till 2007. Asia, in particular, has seen a dramatic drop in cereal and pulse consumption, which has been largely offset by a huge increase in diets containing meat, fruit and vegetables. In fact, this trend, albeit not to the same extent and with the exception of Africa, is largely replicated throughout the world. Africans, on the other hand, continue to eat a staple diet of starchy roots and cereals and although there has been a recent increase in the amount of vegetable oil consumption, most of the other food groups are fairly reflective of a similar diet of 40 years ago.

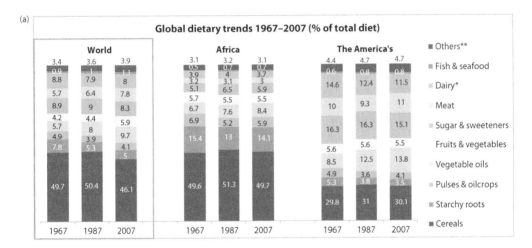

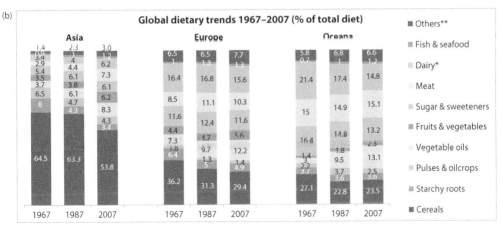

FIGURE 19.10 (a, b) Global dietary trends by region, 1967–2007. (Compiled from *FAOSTAT Food and Agriculture Statistics*, Food and Agriculture Organization, http://faostat.fao.org/, 2011.) *Notes*: Others** includes all other categories not mentioned, i.e. offal, sugar crops, tree nuts, spices, stimulants and alcoholic beverages. Dairy* has been loosely grouped with animal fats, milk, butter and eggs.

Overall though, the trend is clear and shows a 7.2% decrease in global cereal consumption; a drop in starchy roots from 7.8% to just 5% (36% drop) and increases in meat (37%); vegetable oils (70%); and fruits and vegetables (40%). Fuelling this dietary transition are many factors. In the first instance, the developed world has witnessed a doubling of incomes while the average per capita income in the developing countries grew nearly threefold. Paralleling this was the drop in real terms of food commodity prices, which although have moved up in recent times, are still below 1970 levels (Section 22.5.6) Combine this with, the advent of globalisation and improved processing and storage technologies as well as better global transportation systems and the movement and trading of food commodities is now better facilitated around the world. Additionally, these changing diets are also stimulated by a certain cultural globalisation which threatens to homogenise world diets through the encroachment of multinationals and promotion by the media. All the while urbanisation adds another degree of separation vis-à-vis the food on the table and traditional local production. Further, elaborating on urbanisation aspects of food culture it becomes that much easier for some to make the transition from low to higher incomes, which in turn further stimulates changing dietary demands. Moreover, in urban and peri-urban societies, there has almost been a break from the realities of seasonal produce with many consumers demanding choice irrespective of the time of year (Foresight Project 2011).

In combination then, growing incomes and increased availability of marketable goods is competing with traditional diets. This relationship between higher incomes and changing diets has been eloquently established by Engel's law, which postulates the elasticity of demand compared to incomes; that is, the more we earn the more we spend in real terms on food (albeit less overall percentage) (Eifert et al. 2002; SOFI 2008). In fact, the correlation between dietary consumption and incomes is set to further fuel changes in future food diversity. This is because with approximately 70% of the global population living in developing countries, and with a rate of population growth that is double that of the developed countries, much of this change will be seen to take place in these regions.

In a separate but related note, the worldwide increase in calorie consumption in general is a worrying trend. Excess food consumption of high-fat, high-sugar commodities as well as a propensity for processed foods is becoming commonplace in urban environs. And that's not all; this urbanisation is further accompanied by changes in physical activity patterns too. It is also a trend that is not confined to the developed world either, and as the developing countries benefit from economic growth so greater urbanisation and changing lifestyles will see more and more people exercising less and eating more of the wrong foods. An interesting aside here sees a prevalence for obesity and overweight people in developed countries largely among the poorer income groups; this is in contrast to the trend in developing countries that see more overweight problems among the higher income groups (USDA/ERS 2008b; Nellemann 2009). As a result, we are now faced with the unenviable situation that sees both extreme malnutrition and overnutrition existing side by side in many developing countries. Consequently, nutritional concerns take on a duality of concerns that needs to be balanced with the delicate policies targeting specific sectors of the society (WHO 2010).

19.7.1 Up the Trophics: From Wheat to Meat

There are believed to be approximately 10,000–50,000 edible plant species in the world, yet less than 200 make up the total variety of food we eat. Concentration and mechanisation has seen even this paltry number being reduced to just 15 on which we rely on for 90% of our food. Further, over 65% of this is provided by just three such species: rice, wheat and corn. In the animal world, just a couple of dozen livestock species are currently used for food, with practically 99.9% of all animal protein coming from just 9 animals: buffalo, cattle, sheep, pigs, chickens, ducks, geese and turkeys (FAO 1995a; Joseph 2009).

With this in mind, one particular question which continues to surface concerns the use of meat over grains and other less resource intensive foods. Health issues aside, few disagree with the potential logistical problems of the dietary trend that sees increasing numbers eating more meat and dairy products. This is in part because rearing livestock is a very grain-intensive process and this has implications not only for the obvious food security aspect but also other factors, not least of which concerns the sustainability and management of resources. The main issue arises in the resource-hungry inputs of meat. Globally we consume annually about 210 million cattle, 418 million sheep or goats, 1.1 billion pigs and 55 billion chickens.* That works out at about 10, 2, 15 and 13 kg of beef, lamb, pork and poultry, respectively, or about 40 kg of meat in total for every person for the whole year. To provide for all this, 33% of cropland and one-third of all grain grown is currently dedicated to the production of these animals. The grain-drain becomes clearer when we take into consideration the cost of inputs on a per-animal basis. As a simple guide, it takes approximately 3–4 kg of grain, 2.5 kg of roughage and 13,000–16,000 L of virtual water[†] to produce just 1 kg of meat (on-the-bone) (Nellemann 2009[‡]; WFN 2011). The worry here is that not only is this seen as an inefficient use of inputs that could be put to better use, the expected increase in demand for meat would also place upward pressures on cereals and grains. More than this though, it would result in corresponding increases in water, crop and land resources, which we frankly might find difficult to maintain. Moreover, at the industrial level of meat production, the practice is well understood to be both energy inefficient and environmentally damaging, although for the sake of a balanced view. Industrial chicken rearing is somewhat less intensive but still continues to use more energy inputs than cereals. In light of these findings, there are growing campaigns today aimed at reducing meat consumption in the industrialised world whilst attempting to slow or restrain the changing dietary patterns in developing countries (Brown 2005; Dyer 2006). However, faced with such ideas it is also recognised that the choice we make regarding the food we eat, where choice exists, is thought to be a

> ... complex mix of traditions, religion, culture, availability and not the least, financial constraints. (Nellemann 2009)

It is an often quoted, but no less truthful, analogy that suggests the less meat we eat the more food (cereals) becomes available for human consumption. The question arises then, with sufficient food being grown to meet global requirements even after the livestock has been fed, would this not then just become a moot point? It has to be said the answer to this would be a resounding yes. However, this does not take into account future demands and, to cease discussion there, would be to miss other wider implications of the simple notion of eating lower on the food chain. In understanding something of the debate in such ideas a quick foray into the concept of the food chain is given. The trophics describes the ecological level of an organism's position in the food chain, the pecking order if you like and another way of looking at issue of sustainable resources relates to the energy and food equivalents as we go up the food chain. Nearly all life on Earth is driven by solar radiation. Plants, algae and photosynthetic bacteria photosynthesise solar energy, which in turn, are consumed by bacteria, insects and animals. As animals eat plants and other animals eat yet other animals, the transfer of energy up the trophic levels diminish. It must be noted that this is not a reflection on the assimilation of eaten foods;[§] instead the energy-diminishing concept refers to the 10% rule of ecological efficiency. This is a rule of thumb that loosely suggests that only about 10% of the original energy consumed at one trophic level, after consumption, metabolic processes and energy expenditure in general, is converted into stored biomass which is then available to the next trophic level (Russell et al. 2007; Kling 2010). By way of example, only 10% of the food energy ingested by the chicken is stored as biomass. In turn only 10% of the chicken, that is eaten by the fox is converted (after energy expenditure, etc.) to stored fat and muscle. This idea continues up the food chain

---

* Calculation based on FAO food balance sheets 2007 and industry standard carcass weights. Bovine meat 63.2 million tonnes (mt) with a slaughtered carcass weight of approximately 300 kg = 210.6 million equivalent cattle; Mutton and goat meat 12.6 mt with a slaughtered carcass weight of approximately 30 kg, yielding 418.6 million equivalent sheep; Pigmeat, 99 mt with a slaughtered carcass weight of approximately 90 kg, yielding 1.1 billion pigs; and poultry meat at 83 mt with a slaughtered carcass weight of approximately 1.5 kg, yielding 55.4 billion chicken equivalent.

† Virtual water is amount of water calculated to produce a good or a service throughout the course of its growth or production.

‡ Calculations vary depending on who one reads and what are considered average carcass weights. Some, like the Water Footprint Network calculate kg beef weights off-the-bone which can drastically affect such calculations.

§ The assimilation of eaten foods is the amount of food digested and biologically used by another organism. For organisms eating plants about 15%–50% of this is made available whereas between 60% of this is 90% is assimilated if the food is from animal material. From this 98% and 80% is used for metabolism (in vertebrates and invertebrates respectively) and only 2% and 20% for growth and reproduction.

TABLE 19.5

Primary Commodity Equivalent of Annual Human Food Requirements

	Share	kcal/ day	gram/ day	kg/ year	Food Equivalent	Total Wheat Equivalent	Virtual Water
Carbohydrate	60%	1440	350	127.75	176 kg wheat	176 (total incl. wheat fat & protein) kg	
Fat	25%	600	66	24.06	17.02 kg (wheat fat) 5.4 kg (beef fat) 1.64 kg rapeseed oil	3.16 kg (rapeseed[a])	
Protein	15%	360	87	31.76	65.51 kg beef (lean beef, e.g. sirloin) 17.95 (wheat protein)	262.04 kg wheat	
Total equals wheat equivalent to feed one person for a year						441.2 kg	595,620 L or 596 m³

[a] In the absence of any conversion rates of rapeseed to wheat a direct 1:1 ratio was used, also in this calculation rapeseed was used as a balancing item. kcal conversion rates were based on Atwater's general conversion factors. Wheat was calculated using 72.5% carbohydrate, 10.2% protein, 9.7% fat ratio's; and beef calculated using 21.08% protein, 8.24% fat. Grain equivalent beef was converted at the ratio of 4 kg/1 kg beef. Virtual water of wheat equals 1350 L per kg.

and as a result the more links in the chain the less efficient the end user's are seen vis-a-vis the original, first trophic-level-inputs.

Building on this, there have been many anecdotes as well as scientific attempts regarding how much it takes to feed humans in terms of trophic energy (that which is accumulated up the chain). One such regularly quoted observation appeared in Miller's book entitled *Energetics, Kinetics and Life* in 1971 when he observed that it took

> Three hundred trout are needed to support one man for a year. The trout, in turn, must consume 90,000 frogs, that must consume 27 million grasshoppers that live off of 1000 tons of grass. (Miller 1971, p. 233)

Other examples are numerous and often anecdotal; in this book I have made my own back of the envelope calculation to give an example of the primary production wheat equivalent of a recommended average dietary requirement of 2400 kcal for one person for 1 year. Bearing in mind the average make-up of a recommended diet is in the region of 45%–75% carbohydrate, 15%–35% fat, and difference by protein, I have chosen 60% carbohydrate, 25% fat and the remainder 15% in protein for the calculations. From here, Table 19.5 highlights the various aspects of the calculation and arrives at approximately 440 kg, or nearly half tonne of wheat, which in turn requires 596 m³ of virtual water just to feed one person for one year. This estimate is not based on any nutritionally balanced diet and is used solely for illustrative purposes; however, it is interesting to note that this estimate falls somewhat short of the virtual water footprints calculations of both the vegetarian diet at 800 m³/year and the 1500 m³/year meat-based diet (next chapter). This is because my calculations are solely the virtual footprints of the products themselves and do not factor in any storage, process or movement of goods.

It is also worth taking into consideration the energy values along the production line, not just in the harvesting and production costs but also in the primary phase. Although this is beyond the remit of this book, essentially a more thorough calculation would take into account the conversion efficiency of photosynthesis or the free energy stored as carbohydrate (chemical energy) in the wheat. This would also have implications for wheat straw or the biomass that is often rejected. On a harvest index of between 0.5 and 0.6 (grain weight to biomass yield), this would mean that for each kilogram of wheat grain used approximately, the same amount of wheat straw is discarded. Of course, this might be used in other applications from fertiliser, silage and potential second-generation biofuel uses but collectively this all relates to the energy efficiency of the total biomass material that supports all living things on the planet in both food and energy terms.

In short, the changing diet movement indicates we could do more, support more people and live healthier lives by thinking more efficiently and more sustainably. Even the World Health Organization recommends eating lower on the food chain, although in fairness this is less to do with energy efficiency or the ecological issue but rather the health aspects of eating less animal fats. (WHO 2010).

One last note regarding the difficulty in changing dietary practices begins with a little known paper in 1968, which looked at malnutrition and national development. In his paper, Alan Berg enshrines the cultural diet in a romantic notion of immovable and transcendental importance:

> ... food habits also have deep psychological roots and are associated with love, affection, warmth, self-image and social prestige. As a result, there is perhaps no aspect of personal life less flexible than one's eating pattern. (Berg 1968)

Yet, on balance, while this plausible and even intuitive view invokes a sense of truth in the reader, trends in the interim 40 years have shown just how fickle the notion of a traditional or cultural dietary identity has become.

19.8 Discussion

The social capital or human development concept has gained much currency of late, indeed human rights, social justice and ultimately the right to food have all been enshrined in some form or another in legally binding multilateral instruments. Yet too often these seemingly important social advances are met with backpedalling and the much softer language of 'que sera' politics. Reconciling this dualism in food security is going to require strong collective social and political will in order that progress can maintain a solid forward looking path.

One challenge of the ever desirable social balancing act of food and population pressures sees many resources being stretched to their limits. This was also a very real fear for many in the past too and historically, recurrent population pressures have tested many of the greatest thinkers over the centuries. Out of this murmurings, the ideas of a sustainable limit quickly grew becoming a highly contentious perennial debate often open to opinion and subjective interpretation (Ehrich and Ehrlich 1990; Johnson 2000). However, while Malthus maintained that limits on populations correlated to the food supply, a fundamental flaw in his proposition underestimated humanity's ability to raise agricultural production to compensate (Kracht and Schulz 1999). Still, after much discourse on population and sustainability over the last three centuries there is growing universality of opinion that continuing on the present course without consideration of the future is not an option (GEO 2007; PAP 2008). This view was confirmed in 1992 in a succinct declaration which was reaffirmed at the Rio Earth summit by the Union of Concerned Scientists when they declared

> Human beings and the natural world are on a collision course ... The Earth is finite. Its ability to absorb wastes and destructive effluent is finite. Its ability to provide food and energy is finite. Its ability to provide for growing numbers of people is finite. And we are fast approaching many of the Earth's limits. (UCS 1992)

This view also reflects a noteworthy and current growing fear that the population is, or is soon to be living way beyond its current means. This suggests too, that in doing so humankind is possibly inflicting irreversible damage on the environment (GEO 2007; PAP 2008). This raises many concerns; although many cannot agree on just what these limitations turn out to might be both in terms of numbers or limiting factors. Whatever they turn out to be, without doubt there is a finite resource base and the extent to which man's innovation can manage these resources most efficiently and beneficially might ultimately determine this ceiling. In general though, in the absence of any concrete theories, it can be said that through a multi-pronged approach populations are best sustained with a combination of population control, or at least an awareness of our expansion rates, an understanding of our limitations, and managing such resources so as not to threaten the continuity of the species or the productivity of the environment.

These are most definitely challenging times and one that is without doubt going to occupy policy makers, philosophers and reductionists for years to come. One of the benefits too of this long overdue research will ultimately furnish a better understanding of the mainly closed loop system of the natural and biological spheres of Earth.

Living within these boundaries, one consideration that is very topical at the moment is the marginalisation of the many millions of smallholder farmers, pastoralists and fisher folk globally. Many such people work and reside on environmentally challenged areas often without adequate support services. Difficulty too is found with access to markets because of poor infrastructure in transport, storage or representation and while some effort is concentrated in this direction, with community-based rural development policies; more work is needed as such policies are often small scale and relatively ineffectual (Windfuhr and Jonsén 2005). Within this development picture too, other concerns regarding gender parity are also brought up. Women in their important roles as workers, caregivers and providers still suffer in those cultures in which women still lack equal access to food, health care, support, education and social standing. This is a difficult issue, for while numerous multilateral instruments have been advanced addressing these issues over the years, ingrained cultural traditions are hard to influence and change. This becomes a slow process that requires patience and relentless championing. Reflecting their unique roles, progress needs to be made too in bringing more women into key decision-making processes, not just in food security issues but at all levels of society (IFAD 2009).

The lack of progress in poverty reduction is also called into question as it is still understood to be one of the most fundamental underlying causes of food insecurity. Lower incomes mean a greater proportion of that money is spent on food, which in turn makes putting food on the table more susceptible to commodity price fluctuations and economic downturns. What is more, people living in poverty are more vulnerable to natural and man-made disasters, making it very difficult for this group to recover. Many, for instance, because of their financial status, might have been forced to cultivate or settle on cheaper or unwanted land prone to flooding, storms and landslides. Once in this situation, common, usually manageable events such as heavy rains can ruin a season's crops or force the family to become reliant on the state. Another worrying face of poverty describes the link between poverty, undernutrition and the full mental development of children in which undernourishment ensures they grow unable to fulfil their full potential in life. This often translates into more menial, lesser paid jobs, culminating in insecurity that continues to perpetuate a cycle of hunger and poverty from one generation to the next (WFP 2011).

The sad truth is that issues of poverty have been on the social agenda for centuries. The almost euphonious re-realisation in the 1980s that food insecurity was exacerbated by poverty is a somewhat poor excuse for the lack of progress seen in our collective efforts. Indeed we understood poverty to be at the heart of the food issue long before Malthus, yet Malthus, despite his most often quoted food population conundrum, was acutely aware of the plight of the poor vis-à-vis food security. Even back then, just as it has been in the interim period, it seems the rights of the poor were specifically and conveniently disposed of. A similar story is reflected in the right to food. Students of history are well aware of the moral and ethical imperative of human rights and the right to food. However, if it has taken us nearly three centuries to translate the idea's of Locke, Sherlock and Burke among others into voluntary codes of international conduct, then one has to wonder whether this is any real progress at all. While we sit and pat ourselves on the back for our many accomplishments, the reality is that the so-called right to food is having something of an identity crisis, with many believing the new voluntary codes are a whitewash, others continue to hail it as a real step forward. In this regard, it might be difficult to understand the full implications of the voluntary code, as on face value; it seems quite at odds with the original, legally binding Article 2 of the Covenant on Economic, Social and Cultural Rights, which obliged states to ensure legislative measures for the progressive realisation of the rights contained in the Covenant.

Further considerations of food security are brought to the attention of the reader in regards to emergencies and disasters. All nations are exposed to natural or human-made shocks at some point or another. Yet just how well a country copes with such adversity seems to be largely a measure of its economic and social development (USAID 2007). However, having acknowledged that, it is also to be noted that this can work in reverse too. Frequent and often protracted natural and human-made disasters in developing countries, continues to pose a key threat to the economic and social development progress and long-term food security of these countries. As a greater occurrence of both is anticipated, disaster preparedness, response and mitigation need to be addressed. Key here too are the monitoring of appropriate variables and competent early warning systems that facilitates proper analysis and rapid response. Implicitly bound up in this consideration too are important international networks of responders who can be called upon in times of crisis (FAO 2003a; Hartmannshenn 2004; Harvey 2010). On a similar note with regard to famines, over the years, they have variously been viewed as the whims of gods, the punishment of kings, acts of nature and more recently of political, economic and social consequences (Mehta 1929; Hutchinson 1998; Ó Gráda 2009). In reality though it is this last notion that reflects the changing trend and one that sees modern human-made famines rather than 'naturally' caused starvation taking over and becoming the predominant cause of hunger-related deaths (Ó Gráda 2009). Further, more recent realisations that famines are complex phenomena challenges the prevailing understanding of famine and hunger-related deaths. As it turns out, death by starvation is a relatively low occurrence; yet more frequent are the many associated opportunistic diseases or deficiencies such as diarrhoea, pneumonia and the like.

This brings us to the foods we eat. Food choice, where is exist, is succinctly summed up in the words of Nellemann, as

> ... a complex mix of traditions, religion, culture, availability and not the least, financial constraints. (Nellemann 2009)

Yet this traditional mix is changing too. Traditional cultural dietary diversity is losing out to a converging homogenous global diet. A diet that took hundreds of years to evolve in the developed world with all its associated benefits and problems is changing the face of developing nations diets in a matter of decades. Apart from the health connotations discussed in the next chapter, the effect of these changes places greater stress on an increasingly burdened supply chain. As more and more livestock is being consumed so more of the annual grain stock is being diverted to feed them. More animals too take up more land and increase dangerous greenhouse gases (GHGs) in the environment. However, changing dietary habits is going to be a difficult process to challenge as more and more of us enthusiastically sign up to the global homogenous dietary palate.

One of the many ways we can help to address the issues of increased production is by becoming less wasteful. On this score taking into account the sheer volume of waste at every stage of the food chain becomes cringeful, considering much of this is down to human inefficiency and worse, human profligacy: serious questions that need to be addressed in regards to the percentage of food lost pre- and post-harvest. In particular reducing post-harvest losses in processing, transport, distribution, retail as well as decreasing wasteful habits from consumers (particularly in richer countries) has the potential to substantially increase the availability of food whilst encouraging the more efficient use of resources. This ultimately has the potential to increase existing and future availability even before we consider expanding the lands or increasing the yields of our crops

References

AEZ (2000). Inventory of Land Resources. FAO and the International Institute for Applied Systems Analysis (IIASA). http://www.fao.org/ag/AGL/agll/gaez/index.htm (accessed 2 June 2009).

Albala, C., K. Yr and R. Uauy Dagach (2002). Nutritional Programmes for Enhanced Equity in Health in Chile. *Food, Nutrition and Agriculture* 30: 40–52.

Arcand, J. L. (2001). Undernourishment and Economic Growth: The Efficiency Cost of Hunger. European Media Seminar on Global Food Security, Royal Swedish Academy of Agriculture and Forestry. Stockholm: Food and Agriculture Organization.

Bennett, M. K. (1949). Food and Agriculture in the Soviet Union, 1917–48. *The Journal of Political Economy* 57(3): 185–198.

Berg, A. D. (1968). Malnutrition and National Development. *The Journal of Tropical Pediatrics* 14(supp3): 116–123.

Blössner, M. and M. d. Onis (2005). Malnutrition: Quantifying the Health Impact at National and Local Levels. *Environmental Burden of Disease Series, No. 12*. Geneva, Switzerland: World Health Organization.

Brown, L. (2005). *Outgrowing the Earth: The Food Security Challenge in an Age of Falling Water Tables and Rising Temperatures*. New York: W.W. Norton and Co.

Burke, L. (1790). *Reflections on the Revolution in France, and on the Proceedings in Certain Societies in London Relative to That Event. In a Letter Intended to Have Been Sent to a Gentleman in Paris*. Pall Mall, London: Printed for J Dodsley.

Cao, S., L. Chen and Z. Liu (2007) Disharmony between Society and Environmental Carrying Capacity: A Historical Review, with an Emphasis on China. Ambio. *A Journal of the Human Environment* 36(5): 409–415.

Cattell, C. C. (1874). *Co-Operative Production*. Birmingham: G. H. Reddalls.

CFS (2007). Assessment of the World Food Security Situation. Committee On World Food Security: Thirty-third Session. Rome: Food and Agriculture Organization.

Chamberlin, W. H. (1934). The Ordeal of the Russian Peasantry. *Foreign Affairs* 12(3): 495–507.

Clay, E. (2002). Food Security: Concepts and Measurement: Paper for FAO Expert Consultation on Trade and Food Security: Conceptualising the Linkages. Rome: Food and Agriculture Organization.

Cohen, B. (2006). Urbanization in Developing Countries: Current Trends, Future Projections, and Key Challenges for Sustainability. *Technology in Society* 28(1–2): 63–80.

Cohen, J. (1995). *How Many People Can the Earth Support?* New York: W.W. Norton.

Constantino, R. and L. Constantino (1988). *Distorted Priorities, the Politics of Food. Foundation for Nationalist Studies*. Quezon City: Philippines.

DeRose, L., E. Messer and S. Millman (1998). Case Study: The Importance of Non-Market Entitlements. In Derose, L., Messer, E., and Millman, S.In *Who's Hungry? And How Do We Know? Food Shortage, Poverty, and Deprivation*. Tokyo, New York, Paris: United Nations University Press.

Devereux, S. (2000). Famine in the Twentieth Century. Brighton: Institute for Development Studies.

Dollahite, J., C. Olson and M. Scott-Pierce (2003). The Impact of Nutrition Education on Food Insecurity among Low-Income Participants in Efnep. *Family and Consumer Sciences Research Journal* 32: 127–139.

Dyer, G. (2006) How Long Can the World Feed Itself? *Energy Bulletin*. http://www.energybulletin.net/node/21736.

Dyson, T. and C. Ó Gráda, Eds. (2002). *Famine Demography: Perspectives from the Past and Present* Oxford: Oxford University Press.

Ehrich, P. and A. Ehrlich (1990). Overpopulation. In *The Population Explosion*. P. Ehrich and A. Ehrlich, eds. New York: Simon & Schuster.

Eifert, B., C. Galvez, N. Kabir, A. Kaza, J. Moore and C. Pham (2002). The World Grain Economy to 2050: A Dynamic General Equilibrium, Two Sector Approach to Long-Term World-Level Macroeconomic Forecasting. *University Avenue Undergraduate Journal of Economics (Online)* www.econ.ilstu.edu/uauje/PDF's/CarrolRound/eifertpost.pdf (accessed 13 February 2011).

Ellman, M. (2000). The 1947 Soviet Famine and the Entitlement Approach to Famines. *Cambridge Journal of Economics* 24: 603–630.

ER (1996) A Conversation with Joel Cohen. *Environmental Review Newsletter* 3, (1): http://www.environmentalreview.org/archives/vol03/cohen.html.

Ericksen, P. J. (2008). Conceptualizing Food Systems for Global Environmental Change Research *Global Environmental Change* 18(1): 234–245.

ESA (2007). World Population Prospects: The 2006 Revision. New York: Population Division, UN Department of Economic and Social Affairs.

FAO (1995a). Dimensions of Need: An Atlas of Food and Agriculture. Rome: Food and Agriculture Organization.

FAO (1995b). Report of the First External Programme and Management Review of the International Irrigation Management Institute (IIMI). Rome: Food and Agriculture Organization.

FAO (2002) World Food Summit: Five Years Later Reaffirms Pledge to Reduce Hunger. *Summit News*: http://www.fao.org/worldfoodsummit/english/newsroom/news/8580-en.html.

FAO (2003a). Commodity Market Review 2003–2004. Rome: Food and Agriculture Organization.

FAO (2003b). Trade Reforms and Food Security: Conceptualizing the Linkages. Rome: Food and Agriculture of the United Nations.

FAO (2006). Policy Brief: Food Security. Rome: Food and Agriculture Organization 4.

FAO (2011). Website of the Food and Agriculture Organization of the United Nations. FAO. http://www.fao.org/hunger/faqs-on-hunger/en/ (accessed 15 February 2011).

Ferguson, A. R. B. (2008) Food and Population in 2050. *Optimum Population Trust* 8, (1): http://www.opti-mumpopulation.org/opt.journal.html.

Fisher, H. H. (1927). *The Famine in Soviet Russia 1919–1923 the Operations of the American Relief Administration* Freeport, New York: Books For Libraries Press.

Food Summit (1996). Rome Declaration on World Food Security World Food Summit. Rome: Food and Agriculture Organization.

Foresight Project (2011) Foresight: The Future of Food and Farming: Final Project Report. London: The Government Office for Science 202pp.

Freedom from Hunger (2010). Website of the Freedom from Hunger. Freedom From Hunger. http://www.freedomfromhunger.org/ (accessed 15 June 2010).

Gantt, W. H. (1936). A Medical Review of Soviet Russia: Results of the First Five Year Plan. *British Medical Journal* 2(3939): 19–22.

GEO (2007). Global Environment Outlook 4: Environment for Development. Malta: United Nations Environment Programme.

Gerster-Bentaya, M. and N. Maunder (2008a). *Food Security Concepts and Frameworks: Concepts Related to Food Security*: EC-FAO.

Gerster-Bentaya, M. and N. Maunder (2008b). *Food Security Concepts and Frameworks: What Is Food Security?* EC-FAO.

Gilland, B. (2006). Population, Nutrition and Agriculture. *Population & Environment* 28(1): 1–16.

Gottlieb, M. (1945). The Theory of Optimum Population for a Closed Economy. *The Journal of Political Economy* 53(4): 289–316.

Grebmer, K. v., B. Nestorova, A. Quisumbing, R. Fertziger, H. Fritschel, R. Pandya-Lorch and Y. Yohannes (2009). 2009 Global Hunger Index the Challenge of Hunger: Focus on Financial Crisis and Gender Inequality. *The Global Hunger Index*. Washington, DC: International Food Policy Research Institute.

Halhed, N. B. (1776). *A Code of Gentoo Laws, or, Ordinations of the Pundits : From a Persian Translation, Made from the Original, Written in the Shanscrit Language*. London: [sn].

Hall, D. O. (1998). Food Security: What Have Sciences to Offer? Paris, France: The International Council for Science

Hartmannshenn, T. (2004). Food Security: Guidelines for the Promotion and Execution of Food Security Projects by German Agro Action. Bonn: German Agro Action.

Harvey, D. (2010). An Abbreviated UK/EU Agricultural Policy History. Newcastle University. http://www.staff.ncl.ac.uk/david.harvey/AEF372/History.html (accessed 23 November 2010).

Hasan, M. A. (2001). Role of Human Capital in Economic Development: Some Myths and Realities. Bangkok, Thailand: Development Research and Policy Analysis Division, ESCAP.

Hausmann, R., L. D. Tyson and S. Zahidi (2010). The Global Gender Gap Report. Geneva, Switzerland: World Economic Forum, 334pp.

Hionidou, V. (2002). Why Do People Die in Famines? Evidence from Three Island Populations. *Population Studies* 56(1): 65–80.

Hurlbut, E. P. (1847). *Essays on Human Rights and Their Political Guaranties*. Edinburgh: Maclachlan, Stewart and Co.

Hutchinson, C. F. (1998). Social Science and Remote Sensing in Famine Early Warning. In *People and Pixels: Linking Remote Sensing and Social Science*. D. Livereman, E. F. Moran, R. R. Rindfuss and P. C. Stern, eds. Washington, DC: National Academy Press.

Hyder, A. A., S. Maman, J. E. Nyoni, S. A. Khasiani, N. Teoh, Z. Premji and S. Salim (2005). The Pervasive Triad of Food Security, Gender Inequity and Women's Health: Exploratory Research from Sub-Saharan Africa. *African Health Science* 5(4): 328–334.

IDMC (2011). Global IDP Estimates (1990–2009). The Internal Displacement Monitoring Centre (IDMC). http://www.internal-displacement.org/8025708F004CE90B/(httpPages)/10C43F54DA2C34A7C12573 A1004EF9FF?OpenDocument (accessed 12 March 2011).

IFAD (2009). Food Security: A Conceptual Framework. International Fund for Agricultural Development. http://www.ifad.org/hfs/thematic/rural/rural_2.htm (accessed 15 August 2009).

ILO (2009). Global Employment Trends for Women. Geneva, Switzerland: International Labor Organization, 78pp.

INSCR (2011). *The Integrated Network for Societal Conflict Research (INSCR)*. Center for Systemic Peace.

Ishay, M. R. (2008). *The History of Human Rights: From Ancient Times to the Globalization Era*. Berkeley and Los Angeles: University of California Press.

Johns, C. H. W. (1904). *Library of Ancient Inscriptions: Babylonian and Assyrian Laws, Contracts and Letters*. New York: Charles Scribner's Sons.

Johnson, D. G. (1997). Agriculture and the Wealth of Nations. *American Economic Review* 87(2): 1–12.

Johnson, D. G. (2000). Population, Food, and Knowledge. *The American Economic Review* 90(1): 1–14.

Joseph, B. (2009). *Environmental Studies*. New Delhi: Tata McGraw Hill.

Kalayjian, A. S., S. P. Shahinian, E. L. Gergerian and L. Saraydarian (1996). Coping with Ottoman Turkish Genocide: An Exploration of the Experience of Armenian Survivors. *Journal of Traumatic Stress* 9(1): 87–97.

King, M. and C. Elliott (1996). Education and Debate: Averting a World Food Shortage: Tighten Your Belts for Cairo Ii. *British Medical Journal* 313: 995–997.

Kling, G. (2010). *The Flow of Energy: Primary Production to Higher Trophic Levels*. Ann Arbor, MI: University of Michigan.

Kracht, U. and M. Schulz (1999). *Food Security and Nutrition: The Global Challenge*. New York: St. Martins Press.

Leo Lucassen, D. F., J. Oltmer (2006). *Paths of Integration: Migrants in Western Europe (1880–2004)*. Amsterdam: Amsterdam University Press.

Locke, J. (1714). *The Works of John Locke Esq: In Three Volumes*. Paternoster, London: Printed for John Churchill at the Black Swan.

Louria, D. B. (2000). Emerging and Re-Emerging Infections: The Societal Determinants. *Futures* 32(6): 581–594.

MacAuslan, I. (2009). Hunger, Discourse and the Policy Process: How Do Conceptualizations of the Problem of 'Hunger' Affect Its Measurement and Solution? *European Journal of Development Research* 21: 397–418.

Mehta, V. N. (1929). Famines and Standards of Living. *The Annals of the American Academy of Political and Social Science* 145: 82–89.

Mettrick, E. F. (1929). Population, Poverty, and Ethical Competence. *International Journal of Ethics* 39(4): 445–455.

Miller, G. T. (1971). *Energetics, Kinetics and Life: An Ecological Approach*. Belmont, CA: Wadsworth Publishing Company.

Mukudi, E. (2003). Education and Nutrition Linkages in Africa: Evidence from National Level Analysis. *International Journal of Educational Development* 23: 245–256.

Murai, S. (1994). Global Environment and Population Carrying Capacity. In *Population, Land Management, and Environmental Change*, J. I. Uitto and A. Ono, eds. Tokyo: UN University.

Nellemann, C. (2009). The Environmental Food Crisis: The Environment's Role in Averting Future Food Crises. Geneva, Switzerland: United Nations Environment Programme (UNEP) GRID-Arendal.

O'Niel, O. (2002). Second Reith Lecture. London: BBC Radio 4.

Ó Gráda, C. (2009). *Famine: A Short History*. Princeton: University Presses of California, Columbia and Princeton.

OCHA (2011). Website of the UN Office for the Coordination of Humanitarian Affairs. United Nations. http://ochaonline.un.org/AboutOCHA/ (accessed 23 July 2011).

ODI (1997). Global Hunger and Food Security after the World Food Summit. London: Overseas Development Institute.

OFDA/CRED (2010). *The Centre for Research on the Epidemiology of Disasters*. Brussels, Belgium: Universite catholique de Louvain.

Oltmer, J. (2006). The Unspoilt Nature of German Ethnicity: Immigration and Integration of 'Ethnic Germans' in the German Empire and the Weimar Republic. *Nationalities Papers 1465–3923*. 34(4): 429–446.

OPT (2009) Earth Heading for 5 Billion Overpopulation? *Optimum Population Trust*. http://www.optimum-population.org/releases/opt.release16Mar09.htm.

Orr, J. B. (1939). Diet and Nutrition: Nutrition Problems, Dietary Requirements for Health. *The Canadian Medical Association Journal* 41(1): 78–80.

Orr, J. B. (1940). Diet and Nutrition: Nutrition Problems, Dietary Requirements for Health. *The British Medical Journal* 1(4146): 1027–1029.

PAP (2008) Population and Human Development—the Key Connections. *People and the Planet Online*: http://www.peopleandplanet.net/doc.php?id=199andsection=2

Patenaude, B. M. (2002). *The Big Show in Bololand: The American Relief Expedition to Soviet Russia in the Famine of 1921*. Stanford, CA: Stanford University Press.

Peterson, M. D. (2004). *Starving Armenians: America and the Armenian Genocide, 1915–1930 and After*. Charlottesville, London: University of Virginia Press.

Poleman, T. T. (1972). Food, Population, and Employment: Ceylon's Crisis in Global Perspective. *Staff Paper*. New York:Department of Agricultural Economics 31, Cornell University.

Prentice, A. M. (2001). Fires of Life: The Struggles of an Ancient Metabolism in a Modern World. *Nutrition Bulletin* 26(1): 13–27.

Ravallion, M., S. Chen and P. Sangraula (2007). New Evidence on the Urbanization of Global Poverty Washington, DC: World Bank.

Rosegrant, M. W. and S. A. Cline (2003). Global Food Security: Challenges and Policies. *Science* 302(5652): 1917–1919.

Ross, E. B. (2003). Malthusianism, Capitalist Agriculture, and the Fate of Peasants in the Making of the Modern World Food System. *Review of Radical Political Economics* 35: 437–461.

Roughgarden, J. (1979). *Theory of Population Genetics and Evolutionary Ecology: An Introduction*. New York: Macmillan.

Rummel, R. J. (2011). Freedom, Democracy, Peace; Power, Democide, and War. http://www.hawaii.edu/powerkills/20TH.HTM (accessed 12 March 2011).

Russell, P. J., S. L. Wolfe, P. E. Hertz and C. Starr (2007). *Biology: The Dynamic Science*. Belmont, CA: Thompson Brooks/Cole.

Ruxin, J. N. (1996). *Hunger, Science, and Politics: FAO, WHO, and UNICEF Nutrition Policies, 1945–1978, Chapter II the Backdrop of Un Nutrition Agencies, by Joshua Nalibow Ruxin*. London: University College London.

SAC (1967). The World Food Problem: A Report of the President's Science Advisory Committee, Volume LII Report of the Panel on the World Food Supply. Washington, DC: The President's Science Advisory Committee (PSAC), The White House.

Save the Children (2010). Website of Save the Children Alliance. http://www.savethechildren.net/alliance/index.html (accessed 15 June 2011).

Scrimshaw, N. S. (1987). The Phenomenon of Famine. *Annual Review of Nutrition* 7(7): 1–22.

SD (2001). Gender and Nutrition: Gender and Development Fact Sheets. Rome: FAO/Sustainable Development Department.

Sherlock, T. (1718). *A Vindication of the Corporation and Test Acts: In Answer to the Bishop of Bangor's Reasons for the Repeal of Them. To Which Is Added: A Second Part, Concerning the Religion of Oaths*. London: Printed for J Pemberton at the Buck and Sun.

SOFI (2005). State of Food Insecurity. Rome: Food and Agricultural Organization.

SOFI (2008). The State of Food Insecurity in the World 2008. Rome: Food and Agriculture Organization 59.

SOFI (2009). The State of Food Insecurity in the World 2009. Rome: Food and Agriculture Organization 59.

Sulistyowati, C. (2002) Can Gmo Ensure Food Security Amid the Population Growth. Institute of Development Studies and Technological Assistance, *BIC News*: http://www.bic searca.org/news/2002/oct/indo/28.html

UCS (1992). 1992 World Scientists' Warning to Humanity. http://www.ucsusa.org/about/1992-world-scientists.html (accessed 21 July 2009).

UIS (2010). Adult and Youth Literacy: Global Trends in Gender Parity. *UIS Fact Sheet No. 3*. Geneva, Switzerland: UNESCO Institute for Statistics (UIS).

UN (1948). *Universal Declaration of Human Rights*: United Nations. A/RES/217 A (III).

UN (2002/3). Fighting Poverty through Better Health and Education. *Bulletin of the Eradication of Poverty* 2002/3 Edition: 14.

UN (2004). Voluntary Guidelines to Support the Progressive Realization of the Right to Adequate Food in the Context of National Food Security: Adopted by the 127th Session of the FAO Council November 2004. Rome: Food and Agriculture Organization.

UN (2011). Website of the United Nations. United Nations. http://www.un.org/ (accessed 23 March 2011).

UN ISDR (2006). Website of the UN International Strategy for Disaster Reduction. United Nations. International Strategy for Disaster Reduction. http://www.unisdr.org/ (accessed 13 February 2011).

UNEP/GRID Arendal (2005). Maps and Graphics Library. UNEP/GRID-Arendal. http://maps.grida.no/go/graphic/trends-in-natural-disasters (accessed 21 February 2011).

UNFPA (2007). State of the World's Population: Unleashing the Potential of Urban Growth. U.N. Population Fund.

UNHCR (2009). Global Trends: Refugees, Asylum-Seekers, Returnees, Internally Displaced and Stateless Persons. Geneva, Switzerland: Office of the United Nations High Commissioner for Refugees.

UNICEF (2009). The Convention on the Rights of the Child. United Nations International Children's Emergency Fund (UNICEF). http://www.unicef.org/rightsite/sowc/ (accessed 30 January 2011).

USAID (2007). Food for Peace. Fy 2008 P.L. 480 Title Ii Program Policies and Proposal Guidelines Washington, DC: United States Agency for International Development.

USDA/ERS (2008a). Food and Consumer Price Index (CPI) and Expenditures. *Food Security Assessment.* Washington, DC: USDA/Economic Research Service.

USDA/ERS (2008b). Gfa-19: Food Security Assessments Reports. *Agriculture and Trade Reports.* USDA Economic Research Service.

Van Den Bergh, J. C. J. M. and P. Rietveld (2004). Reconsidering the Limits to World Population: Meta-Analysis and Meta-Prediction. *Bioscience* 54(3): 195–204.

von Braun, J., Ed. (1995a). *Employment for Poverty Reduction and Food Security.* Washington DC: International Food Policy Research Institute.

von Braun, J. (1995b). A New World Food Situation: New Driving Forces and Required Actions. International Food Policy Research Institute. Washington, DC: The International Food Policy Research Institute.

Vos, F., J. Rodriguez, R. Below and D. Guha-Sapir (2009). Annual Disaster Statistical Review 2009: The Numbers and Trends. Brussels, Belgium: CRED: Centre for Research on the Epidemiology of Disasters 46.

Waal, A. D. (1991). *Evil Days: Thirty Years of War and Famine in Ethiopia.* New York: Human Rights Watch.

Walker, P. (2008). The Origins, Development and Future of the International Humanitarian System: Containment, Compassion and Crusades. ISA's 49th Annual Convention, Bridging Multiple Divides, San Francisco, CA: International Studies Association (ISA).

Watkins, S. C. and J. Menken (1985). Famines in Historical Perspective. *Population and Development Review* 11(4): 647–675.

Watson, F. (1999). One Hundred Years of Famine —A Pause for Reflection. *Field Exchange* 8: 20–23.

WFN (2011). Website of the Water Footprint Network http://www.waterfootprint.org/?page=files/productgallery&product=beef (accessed 13 July 2011).

WFP (2011). Website of the World Food Program. World Food Programme. http://www.wfp.org/ (accessed 25 February 2011).

White, M. (2010). Historical Atlas of the Twentieth Century. Matthew White. http://users.erols.com/mwhite28/warstat8.htm (accessed 21 March 2011).

WHO (2010). Website of the World Health Organisation. http://www.who.int/ (accessed 2 April 2010).

Windfuhr, M. and J. Jonsén (2005). Food Sovereignty: Towards Democracy in Localized Food Systems. Warwickshire: ITDG/FIAN 70.

WIT (2008). World Ecology Report. New York: World Information Transfer.

World Bank (2009). Understanding Poverty. World Bank. http://web.worldbank.org/WBSITE/EXTERNAL/TOPICS/EXTPOVERTY/0,,contentMDK:20153855~menuPK:373757~pagePK:148956~piPK:216618~theSitePK:336992,00.html (accessed 5 June 2009).

Zhang, W. (2008). A Forecast Analysis on World Population and Urbanization Process. *Environment Development and Sustainability* 10(6): 717–730.

20

Health and Nutrition

20.1 Nutritional Knowledge and Policy: Through Golden Ages

While the relationship between diet and health had long been noted in history, no definitive contemporary rationale could account for such linkage, and by the close of the nineteenth century, the aetiology of some common diseases was still thought to result from bacteria, mould and toxins exacerbated by poor hygiene and sanitation. The idea that some of these common diseases such as pellagra, beriberi and infantile scurvy might just be caused by nutritional deficits was beyond the grasp of most contemporary physicians and scientists. Still there were moments of epiphany as one example by Wilbur Olin Atwater illustrated when he had intuitively noted in 1889 that health and diet were directly related. At water prophetically suggested that, the intake of food and expenditure of energy should always be in equilibrium if obesity or poor health were to be avoided (Atwater 1889). Although this had important implications for the later progression of food security, such thoughts were largely the preserve of a few elite scientists before their time. However, until such knowledge became widespread, ongoing efforts to reduce nutritional disease and prolong the shelf life of food was increasingly being consigned to things like sterilisation, milling and polishing of rice. Unfortunately, as we know today this can cause degradation of vital nutrients and as a result nutritional diseases continued, largely unabated. At about this time there were perhaps two branches of conventional wisdom both of which sought recognition for reduced mortality rates; nutrition and health objectives. Although, as research later confirmed, the relationship between the two was too close to call to allow one camp victory over the other.

Nevertheless, propelled by the pioneering research of Casimir Funk and others, the discovery of the role of vitamins and minerals in maintaining good health was realised. With this new focus on nutrients and eating the right foods as well as supplements and the fortification of some foods, incidents of goitre (iodine), rickets (vitamin D), beriberi (thiamin), pellagra (niacin) and dental caries (fluoride) were all reduced. Consequentially by the mid-1920s, a paradigm shift in the understanding of nutrition was well underway (Funk 1922; Bendich 1992). From this point onwards, nutritional research took two separate but related routes. The first sought to build on the previous work of researchers and attempted to more fully understand the energy components of food, while the second branch occupied itself with finding an optimal constituency and balance of the nutritional trinity of carbohydrate, protein and fat (McMillan 1946).

At the international level also, nutrition made its first appearance on the worldwide stage first, to a lesser extent through the Office International d'Hygiène Publique (OIHP), but then more prominently with the creation of the League of Nations Health Organization (HOLN) in 1920. This was an exciting time for nutrition, with much research and understanding adding to a growing body of knowledge in the basic physiological requirements for health as well as an improved understanding of the body's requirements for vitamins, proteins and amino acids. It is also worthy of note that already by the late 1920s and early 1930s there was a growing disequilibrium research orientation with many concentrating on protein at the expense of more holistic nutritional views. This was in large part fuelled by the prevalence of kwashiorkor at the time. However, this emerging obsession with protein was a situation that was to prevail for decades. This fascination was perhaps first addressed officially in the Health Organization's Quarterly Bulletin of 1936 in which Ruxin (1996) quotes Professor Terroine as denouncing such lopsided research and advocated that in nearly all cases:

> . . . if you take care of the calories, the protein will take care of itself. (Ruxin 1996)

Out of this period though perhaps, one of the most striking of the League's innovations was the articulation of some of the fundamental causes of poor nutrition: poverty and social inequity. Although this had been noted in many previous reports, the real coup of the League was in its attempt to unite agricultural and economic policies in an attempt to address these underlying conditions (Gillespie 2003). At about this time with the effective politicisation of issues of health and hunger, it seemed that nutrition was now becoming central to policy and development paradigms of many national governments. Child feeding programmes rolled out over the industrialised world while in the United Kingdom, Europe and the United States such policies also became entrenched in legislation (Gunderson 2003). From here, large-scale international cooperation on nutritional health issues seemed to be to moving ahead for the first time. As nutritional science encompassed more and more disciplines such as pharmacology, biology, biochemistry, physiology, medicine and organic chemistry, it began to develop its own autonomous scientific identity. In this new field of nutrition, public nutritionists sprang up extolling the virtue and benefits of adequate and proper nutrition to the health and vitality of a nation's population. Nutrition and nutritionists, it now seemed, had now finally taken their rightful place and occupied a prominent and influential role in both public and official capacities.

As quickly as the nutrition movement was taking hold, the advent of World War II slowed progression somewhat. Although, having said that, what it did achieve during this time was to highlight the importance of the international dimension of nutrition and food security. Large disparities were noted among indigenous population's food and nutritional habits with concomitant widespread hunger and disease. Nevertheless it was not till after the war with the creation of the United Nations and its family of agricultural, economic and health bodies, did global nutrition really take shape. At this time both the HOLN and the OIHP were transferred to the UN and restructured as the World Health Organization (WHO). Simultaneously the International Labour Organization (ILO) as well as the Permanent Court of International Justice, renamed the International Court of Justice, became UN institutions. For the meantime though the previous nutritional work of the League of Nations Health Organization during the 1930s heavily influenced the work of the main Allied reconstruction agencies. This in turn ultimately helped shape the policies of future UN agencies involved in health projects. Interestingly too though, the poverty and economic aspect of the nutrition debate, brought up in numerous previous reports of the League and others at the time, seemed to move into the shadows. As a result, collective nutritional thinking at this time was translated into a development paradigm of expanded agricultural production, one which lacked structural underlying measures to tackle poverty and social inequity. However, all was not lost as one of the staunchest advocates of improved nutritional standards and food supplies within the UN family was John Boyd Orr, a nutritionist and first Director-General of the FAO. His enthusiasm and dogged determination is legendary in the field and it is through his dedication that much forward movement was made in the ideology of food security.

Previous to the war, nutritional issues in developing countries, despite some notable researchers' efforts had taken a backseat to what seemed like more domestic or industrialised nation's concerns. The close of the war changed this perspective and many now focused on the whole, global picture of food deprivation and undernutrition. Children too began to occupy more prominence in targeted nutritional policies. Widespread Applied Nutrition Projects (ANP) at the time too, met with some measure of initial success however, these were later panned by many as ineffectual and wasteful. By the 1960s, work was also finally being undertaken that would scientifically link what many had suspected for a long time: that of the relationship between nutrition and health. About this time too there was a break in the established dominance of scientists in influencing policy. Before the Bellagio conference, nutritionists, economists and other researchers would determine the problem, and after further research, the extent of the problem was identified and plans would be drawn up and executed. However, despondent over misleading or ambiguous statistics, the emphasis on quantitative goals was rationalised in favour of more qualitative social targets (Ruxin 1996). These and other new ideological developments at the time meant the closer collaboration of a multitude of disciplines from social theorists, nutritionists, food economists and policy makers, to name but a few. It was a collaboration that also meant that many institutions came to overlap in their mandates and a new period of political infighting erupted. Duplication of effort, inefficiency of leadership and misaligned objectives ensued and subsequently hindered real progress on the international governance front. All the while the nutritional status of the majority of the developing world witnessed painfully slow progress in undernutrition programmes, although some small improvement was seen in supplementation and fortification projects. Institutional reform was in the offing, the need for which was further reiterated at the 1974 World Food Conference. The conference took place amid a jubilant sense of enthusiasm. At in the conference, we saw the first real coordinated global initiative to end hunger within a decade. Unfortunately it was not to be; in fact, this was a time when the already cavernous gap between developed and developing countries became more pronounced.

Collectively the 1980s and 1990s became a time of intellectual reflection. By now the link between health and nutrition had long been established and it was clear that while great advances in nutritional science had precipitated an upward trend in health and vitality, it was just as clear such advancement was predominantly benefitting the industrialised world. Nutritional pioneers looked further afield. With sufficient food being produced, the issue of poverty and access once again came to the fore. Poverty, which had been highlighted as one of the fundamental obstructions to good nutrition back in the 1920s, was back on the agenda a full 60 years later. This is not to say the issue had been forgotten; it is just that it did not feature as prominently in the mainstream international development paradigms of the day. What was needed was an expansion of the policy of health and nutrition to denote security of access and utilisation. This involved a rethink that broadened the goal of nutritional attainment through reducing levels of poverty, through education, economic development and empowerment.

Partly as a result of changing characteristics of consumption, patterns of disease were shifting from nutrient deficiency to increasing rates of cardiovascular disease, diabetes and cancer. This shift effectively placed nutritional diseases, at both ends of the spectrum of malnutrition and at the centre of much health and nutritional policy.

20.2 Nutritional Science

The modern era of the science of nutrition was initiated by Lavoisier and pioneered by the likes of Frankland, Smith, Playfair, Voit, Rubner, Atwater, Chittenden and Lusk among others and by the close of the nineteenth century the three major energy-yielding food components—carbohydrate, protein and fat—had all been identified. Also by this time, the general gram caloric conversion rates calculated by Atwater, often rounded to carbohydrate 4 kcal, protein 4 kcal and fat at 9 kcal, were in place and are still being used today. Building on Lavoisier's calorimeter studies on animals, Atwater built the first human equivalent in 1892, which was later modified by Francis Benedict. The Atwater-Rosa-Benedict calorimeter was used extensively to calculate the energy expenditure of healthy individuals and later these experiments paved the way for the creation of the Harris-Benedict formula; which allowed for the prediction of basal metabolic rates. Collectively the previous breakthroughs in turn provided the underpinning knowledge that allowed dietary standards to finally be calculated with accuracy.

From here, the stage was set, and the impetus for the first real universal nutritional standards evolved out of the Boer War and the World War I. It was postulated that by pooling resources, the benefits of common standards would be beneficial in future nutritional planning. And so it was that the first real attempt at a semi-international dietary standard came from the Inter-allied Scientific Food Commission of 1917/18; which at the time adopted the energy requirements proposed by Lusk. Incidentally the primary purpose of this exercise was to determine the food requirements to be shipped from North America to the Allied forces in Europe. Out of this though came the principles of planning food allowances for basic needs and welfare intervention. After this in 1933, Hazel Stiebling, working for the USDA, proposed the first dietary standards that took into account vitamin and mineral requirements. Also following hot on these heels in 1934 and although not necessarily aimed so much at an international audience, the British Medical Association (BMA) and the Ministry of Health (MoH), after much wrangling, provided a joint memorandum of recommended calorie requirements. In the interim period, the League of Nations in 1932, alarmed by the general living conditions throughout the world, had already commissioned a report by Doctors Burnet and Aykroyd to answer essential questions of the nutritional requirements of humans. Their report was discussed at length, and the League's Technical Committee's eventual report, the Problem of Nutrition, was published in 1937 detailing, in part two, the physiological bases of nutrition. In the United States, similar initiatives were underway and another body, the Food and

Nutrition Board of the National Research Council, first devised their dietary recommendations in 1941 and published them in 1943.

These early pioneering recommendations were of the time, yet other issues were becoming clear. In the post-war period the expansion of nutritional scientific knowledge allowed for more investigation into individuals needs. The problem was that up till then (1945) individual requirements were based on population averages and the difficulty arose because such recommendations were diagnostic in nature and not intended to be prescriptive. For example, the rationing during the war provided (energy) entitlements at the individual level, that was based on average or mean requirement estimates of individuals within that group or class. The problem was clear; bearing in mind the variability of individuals how could a single estimate of needs be applicable to all individuals within the group? It could not, and confounding any solution was the further realisation that science and nutritional thinking were not yet advanced enough to deal with such issues. Somehow inherent variability of requirements had to be factored in and the need for a statistical element to describe this potential variation was needed. It was a dilemma that was touched upon by Pett, Morrell and Hanley in 1945 and then by Pett again in 1955, although the nutritional community had to wait several years before any answers were forthcoming. In fact, it took the 1971 seminal paper by Lorstad to finally address these concerns of inter-individual variation. In his paper he proposed a statistical solution to the problem that related group means to individuals on a basis of probability. From this point onwards, recommendations routinely utilised a statistical algorithm to calculate the probability of the proportions of individuals within a particular group meeting their daily requirements and by 1985 the expert meeting of the WHO/FAO had largely accepted this approach. At the same meeting, the expert group also finally rejected the use of a reference man or woman in order to allow more flexibility in specific groups' recommendations. Another innovation of the 1985 report was the improved use of the factorial approach in the calculation of energy requirements for adults, which used multiples of BMR to represent total energy expenditure. So it was when the UN FAO/WHO expert group met again in 2001 all requirements such as those of adults, children and infant could now be calculated from expenditure rather than intake. Although there has been further research regarding the use of BMR and the inter-variability of individuals, the latest data is still relevant today. A summary of the committee's average recommendations for each age grouping is outlined in Table 20.1.

20.3 Nutrition, Mortality and Disease

During the past decade and a half, the role of diet in preventing or at least controlling morbidity and mortality among both communicable (CD) and non-communicable diseases (NCD) has been strengthened through the research of numerous studies. People's diets, to a large extent, define their health, growth and development potentials and the responsibility of nutrition from the foetus to the grave is a shared one. As a result, for nourishment from breastfeeding to old age and palliative care, the individual relies on himself or herself, the family and the state. Food security in this sense comes with responsibility from all parties.

With regard to the poor or the overnourished, rapid changes in diets and lifestyles that have accompanied economic growth, globalisation, industrialisation and urbanisation has accelerated over recent years. Traditional cultural dietary diversity seems to be losing out to a converging homogenous global diet. What is more, as standard of living have improved, so changes that took hundreds of years to evolve in the developed world, along with all its co-morbidities, is being transposed onto the developing regions in the space of mere decades. This in turn is being reflected in a rapid and significant decline of the general nutritional and health status of these populations (Figure 20.1). This is happening in no small measure because inappropriate dietary patterns as well as reduced physical activity are increasing the incidents of diet-related chronic diseases including obesity, diabetes, cardiovascular diseases, hypertension and stroke.

This increasing per capita wealth has varied and somewhat mixed implications for both the national economy and food security. On the one hand, growing incomes ensure more options and better access to foods; yet on the other hand, the extra costs associated with overnutrition of, workdays lost to ill

TABLE 20.1

Energy Requirements of Successive Bodies

Organisation	Weight[a] (kg)	Inter-allied Scientific Food Commission	USDA Hazel Stiebling	BMA/ MoH	LoN	FAO/WHO	FAO/WHO	FAO/WHO	FAO/WHO[b]	FAO/ WHO/ UNU[c]
Meeting date		1918	1933	1934	1935	1949	1956	1971	1981	2001
Report published in					1936	1950	1957	1973	1985	2004
Age (years)						(kcal/day)	(kcal/day)	(kcal/day)		
Children										
<1	7.3		1200			803	803	820	820	620
1–3	13.4		1200	1150	1010	1200	1300	1360	1250	1050
4–6	20.2	1650	1500	1550	1560	1600	1700	1830	1700	1413
7–9	23.1	2100	2100	1850	2060	2000	2100	2190	2100	1738
Male adolescents										
10–12	35.9	2500	2500	2550	2560	2500	2500	2600	2200	2350
13–15	51.3		2500	2900	2850	3200	3100	2900	2500	2975
16–19	62.9		3–4000	3200	3000	3800	3600	3070	2850	3363
Female adolescents										
10–12	33			2900	2460	2500	2500	2350	1950	2138
13–15	49.9		2500	2900	2700	2600	2600	2490	2125	2438
16–19	54.4				3000	2400	2400	2310	2150	2500
Adult men	65	3000	3000		3000	3200	3200	3000	3325	2550–3600[d]
Adult women	55	2400	2500		3000	2300	2300	2100	2625	2000–3050[e]

Source: Starling, E.H., *The Feeding of Nations*, Longmans, Green & Co., London, 1919; Stiebling, H.K., *Food Budgets for Nutrition and Production Programs*, USDA, Washington DC, 1933; League of Nations, *The Problem of Nutrition: Volume I Interim Report of the Mixed Committee on the Problem of Nutrition*, League of Nations, Geneva, Switzerland 1936a; League of Nations, *The Problem of Nutrition: Volume II Report on the Physiological Bases of Nutrition*, Technical Commission of the Health Committee of the League of Nations, Geneva, Switzerland 1936b; FAO/WHO, *FAO/WHO Ad Hoc Committee of Experts on Energy and Protein: Requirements and Recommended Intakes*, Food and Agriculture Organization, Rome, 1973; Périssé, J., Joint *FAO/WHO/UNU Expert Consultation on Energy and Protein Requirements: Past Work and Future Prospects at the International Level*, Food and Agriculture Organization, Rome, 1981.

Notes: Data are based on average kcals note too the original might not fit neatly in the age ranges specified here as such these data are approximations based on original reports.

[a] From the Report of the 1971 Committee.

[b] Data for children is averaged across both male and female recommendations; adult data is averaged across rural and urban recommendations.

[c] Data for children is averaged across both male and female recommendations.

[d] Based on moderate work of 1.75 × BMR plus weight ranges for adults 50–90 kg.

[e] Based on moderate work of 1.75 × BMR plus ranges for adults of between 45 and 85 kg.

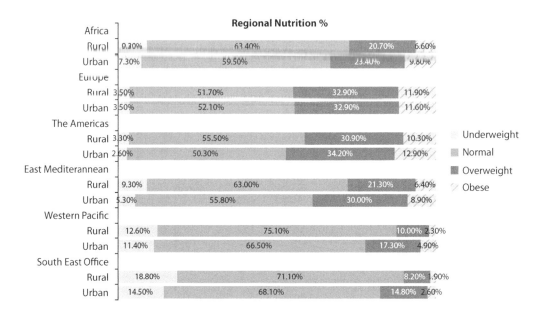

FIGURE 20.1   Prevalence of underweight and overweight 2002/3. (Re-created from Moore et al., *Journal of Obesity* 2010. With permission from Hindawi Publishers.)

health, revenue loss and early death have repercussions beyond the individual to their families and the ongoing security of food. With a large percentage of deaths already attributable to chronic diseases in developing countries, the difficulties arise not only with the expected increasing trend but also as a consequence of a cultural and political environment that singularly aims to create 'more': more wealth, more food, more convenience. This is no less true of the other side of the equation either; that of the undernourished. It has already been established that a perpetuated vicious cycle of malnourishment, poverty and disease needs to be broken if food security and better health are to prevail. In this sense, the morbidity and mortality of undernourishment is closely associated with opportunistic diseases like malaria, measles and gastro-intestinal infections. By linking the two, the burden of chronic and opportunistic disease-associated malnutrition is ultimately borne by the state as well as the individual.

Combined, the rapidly increasing burden of nutritional disease and disorders now become a key aspect of a country's public health profile. This has enormous physical and economic costs both at the country as well as the individual levels. The physical loss or debilitation of its workforce represents an opportunity lost to a nation's economic growth potential; either through sickness or mortality figures (Figure 20.2 and Table 20.2), or through economic loss which is perhaps that much more tangible (next sections). Focusing on the mortality trend for the moment, it can be seen in the bar charts that while figures for 2030 reflect a steep decline in communicable diseases from 30.6% to 13.8%, there is a considerable resultant gain in projected deaths due to non-communicable diseases and injuries from 69.4% to 86.1%. In some instances, particularly in cardiovascular deaths, there is considerable projected increase. In short, the picture is very telling. On the one hand, with so much focus on the nutritional and disease burdens of malnutrition as well as wider societal goals of social welfare, the picture for these ailments is looking optimistic. On the other hand, dietary diversity and growing per capita wealth alongside urbanisation look to change the face of primary health care in the foreseeable future. Out of the frying pan into the fire perhaps? In short, from whichever way the nutrition continuum is viewed, it is set to remain the primary concern for the decades to come.

While we have discussed many aspects of the burden of disease in previous sections, not too much attention has been paid to HIV and AIDS with regard to food security. The next section seeks to restore that balance.

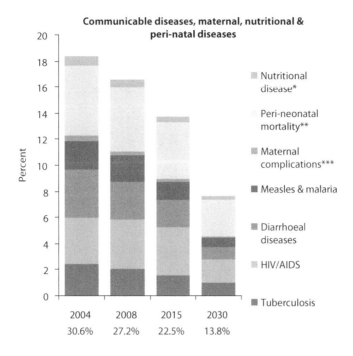

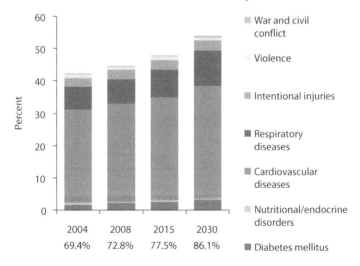

FIGURE 20.2 Mortality figures projected to 2030. (From the Global Burden of Disease: 2004 update, World Health Organization, Geneva, Switzerland 2008. With permission.) Note this graph does not include all such diseases in the represented categories. More detailed data can be found in Appendix B. *Notes:* *Nutritional Disease includes: Protein-energy malnutrition, Iodine deficiency, Vitamin A deficiency, Iron-deficiency anaemia. **Peri-neonatal Mortality includes: Prematurity and low birth weight, Birth asphyxia and birth trauma, Neonatal infections and other conditions. ***Maternal Complications includes: Maternal haemorrhage, Maternal sepsis, Obstructed labour.

TABLE 20.2

Mortality Figures Projected to 2030

Summary: Deaths (000s) by Cause, in WHO Regions, Estimates for 2004	2004	2004	2008	2008	2015	2015	2030	2030
	(000s)	% total	(000s)	% total	(000s)	% total	(000s)	% total
Population (000s)	6,436,826		6,700,628		7,186,888		8,110,599	
Deaths Unpop Division 98 Rev	51,949		51,949		51,949		51,949	
Total deaths	58,772	100	58,766	100	60,856	100	67,790	100
I. Communicable diseases, maternal and perinatal conditions and nutritional deficiencies	17,971	30.6	15,980	27.2	13,705	22.5	9370	13.8
Infectious and parasitic diseases	9519	16.2	8427	14.3	7044	11.6	4216	6.2
Respiratory infections	4259	7.2	3816	6.5	3427	5.6	2871	4.2
Maternal conditions	527	0.9	424	0.7	316	0.5	180	0.3
Perinatal conditions	3180	5.4	2913	5	2606	4.3	1898	2.8
Nutritional deficiencies	487	0.8	401	0.7	313	0.5	205	0.3
II. Non-communicable conditions	35,017	59.6	37,124	63.2	41,193	67.7	51,619	76.1
Malignant neoplasms	7424	12.6	8097	13.8	9259	15.2	11,928	17.6
Other neoplasms	163	0.3	178	0.3	201	0.3	253	0.4
Diabetes mellitus	1141	1.9	1294	2.2	1656	2.7	2229	3.3
Nutritional/endocrine disorders	303	0.5	310	0.5	331	0.5	395	0.6
Neuropsychiatric disorders	1263	2.1	1320	2.2	1429	2.3	1757	2.6
Sense organ disorders	4	0	5	0	5	0	6	0
Cardiovascular diseases	17,073	29	17,890	30.4	19,388	31.9	23,578	34.8
Respiratory diseases	4036	6.9	4426	7.5	5220	8.6	7373	10.9
Digestive diseases	2045	3.5	2010	3.4	2015	3.3	2146	3.2
Diseases of the genitourinary system	928	1.6	980	1.7	1089	1.8	1376	2
Skin diseases	68	0.1	71	0.1	80	0.1	103	0.2
Musculoskeletal diseases	127	0.2	131	0.2	142	0.2	175	0.3
Congenital abnormalities	440	0.7	408	0.7	373	0.6	294	0.4
Oral diseases	3	0	3	0	4	0	5	0
III. Injuries	5784	9.8	5663	9.6	5957	9.8	6801	10
Unintentional injuries	3906	6.6	3977	6.8	4181	6.9	4786	7.1
Intentional injuries	1642	2.8	1685	2.9	1777	2.9	2015	3

Source: World Health Organization, *WHO's Global Burden of Disease: 2004 Update*, World Health Organization, Geneva, Switzerland, 2008.

20.3.1 HIV/AIDS

At the outset, it is worth noting that HIV positive or AIDS status does not automatically signify food insecurity. The growth of HIV/AIDS is thought to have peaked in 1999 and while death rates are now showing a 19% reduction among people living with HIV between 2004 and 2009. However, in total since the emergence of the epidemic, over 60 million people have been infected with HIV while nearly 30 million have died. Currently, of the estimated 33.3 million people living with HIV, 15 million of these live in low- and middle-income countries while globally, only approximately 5 million people were receiving HIV treatment in 2009. In addition, just a little over half of all sufferers are found to be women and girls. There is promising progress though as shown in 33 countries, 22 of which are in sub-Saharan Africa. In these countries the incidence of HIV has fallen by more than 25% between 2001 and 2009. However, this is tempered by the fact that figures in seven countries, five in Eastern Europe and Central Asia have shown a collective increase of more than 25% over the same period. Collectively, though the figures are optimistic and it can be seen that annual newly infected numbers dropped in 2009 standing at approximately 2.6 million compared to 3.1 million in 1999. Death rates too were also slowing from an estimated 2.1 million in 2004 to about 1.8 million in 2009 (UNAIDS 2010).

The way in which the disease is viewed to is changing too. In Africa, the initial early stages of the evolution, HIV/AIDS was primarily thought of as a public health crisis whereas now it is largely accepted as having had a disproportionate effect on the deterioration of,

> … human, financial, social, political and cultural resources at the household, community, regional and national levels throughout much of Africa. (USAID/AED/WFP 2007)

As a result, growing emphasis is given over to multi-objective food insecurity responses incorporating both short- and long-term objectives.

Nonetheless, for all sufferers the nutrition and HIV/AIDS link is a complex one. As with diseases in general, adequate nutrition is vital to maintaining a healthy, responsive immune system, which in turn provides strength in managing opportunistic infections. With HIV or AIDS sufferers, this health-nutrition barrier takes on extra importance as it helps slow the progression of the disease while giving optimal quality of life. Inherent in the disease profile is significant weight loss, this has been positively associated with an increased risk of opportunistic infection, complication and early mortality, especially in the absence of anti-retroviral therapy (ART). Further, reduced food security affects the capacity of HIV sufferers to work, reducing their productivity and as a result are often unable to adequately provide income and resources for their household. There is also the question of care and households tend to spend more (up to 50% of poorer household's annual incomes) on tending to those infected.

Another aspect of the problem is the ongoing stigma and discrimination of HIV sufferers. As social networks within communities are fragmented, the affected find it more difficult to offer help to others or receive it themselves. Children suffering HIV find it especially hard with the early loss of an adult and subsequent lost indigenous knowledge transfer among generations. Second, other emotional and psychological traumas are faced as they are ostracised or forced to quit school or perhaps are disinherited by their parents. There are differences too in the urban/rural dynamic of HIV/AIDS. While initially it was seen as a predominantly urban problem, the resultant shift in current thinking suggests that it has a disproportionate effect on rural life. This is because in poor HIV-affected urban households, householders tend to have less personally destructive options of access to food through such things as informal lending, migrant work or perhaps prostitution. In rural settings however, households often try to meet expenses by selling their assets, or sending needed family members to urban areas in search of work or worse still they might just come to rely on child labour. All of these measures have the same effect of reducing long-term security of food, livelihood and opportunities.

While HIV populations continue to cite food as one of their greatest and most significant needs, it is shameful that despite high-level acknowledgement of the food security aspect of the problem, it has hitherto not been given adequate attention. With so many different variables at play regarding how HIV affects individuals and households in different communities, there is no 'one-size-fits-all' solution vis-à-vis targeting or programmes. What is agreed however is the need for more funding and expertise as well as culturally specific situation analyses (CFS 2001; USAID 2007; USAID/AED/WFP 2007; FANTA2 2008; Greenway 2008; Munzara 2010; WHO 2010e).

20.3.2 The Cost of Malnutrition

> What would be the cost/benefit ratio of a $10 million investment in food enrichment, for example, as compared to other forms of development expenditure, e.g. spending the $10 million for fertiliser or dams or roads or schools? What, in-fact, is the relationship of malnutrition to development—or, more specifically, what are the economics of malnutrition? (Berg 1968)

These were the questions brought up in 1968 by Alan Berg and they are still as relevant today as they were then. The costs of hunger and malnutrition in society manifests itself in many distinct ways. The immediately obvious one that comes to mind are the direct costs of dealing with its consequences. These might include the medical treatment costs of problem pregnancies, nutritional diseases and associated co-morbidities or early mortality. Then there are the days of lost labour, the decreased productivity or the shortfall of the national workforce. In turn, lower productivity means less national income, and reduced per capita spending power, poverty and static or diminishing human development. In fact, the cost implications become clearer as the list gets bigger. The obvious answer to all this, is investment but by how much and by whom?

Economically, the benefits of nutritional investment are often unseen and unappreciated yet malnutrition in its many forms have serious economic consequences for individuals, households and societies in general. This is perhaps among the strongest and most effective argument for investing in food security and national nutrition programmes. Consider for the moment that many currently malnourished children for instance will not survive through to retirement; in fact, many will die before they have been able to earn their first wage. Moreover, of those that do survive, not all will be employed.

As to the question who should be investing, there are cases to be made for both the state and the individual. Individuals who are at most risk of malnutrition are also far too often the poorest in society; on this basis, investing their own monies is a low take-up option. Alternatively, individuals have a moral obligation to themselves and their family's well-being in which case the investment in education is primary. As for the state, while it can be suggested that the immediate benefits of any intervention go to the individual and not the state, the country ultimately benefits

> … from a healthier more robust economy that generates government revenues through taxation and provides employment opportunities for other individuals. (Ross 2003)

Bearing all this in mind, difficulties arise as to how to calculate the cost of malnutrition against these factors. A good place to start is looking at the opportunity lost to individuals and nations of malnutrition over a lifetime. In this endeavour starting, as many studies do, with a low birthweight child (LBW) the FAO's Academy for Educational Development (AED) has developed a statistical model to asses such factors. In their work the model looks at the economic costs based on risk analysis of nutritional status, mortality ratios, learning performance and mental development as well as lost future income potential discounted to the present day. On top of this, their calculations take into account productivity ratios and nutritional disease prevalence. In short, the FAO's calculations of 2006 based on this model are illustrated in Figure 20.3. In total, it is estimated that to remove one LBW infant out of malnutrition reaps almost $1000 in benefits from reduced neonatal care to improved productivity. Meanwhile, a separate World Bank report estimated the losses in the developing world of productivity alone due to malnutrition over the lifetime of individual could equate to as much as 10% of lifetime earnings or, considering total accumulated losses of all malnourished individuals, it could end up equating to 2%–3% of GDP (World Bank 2008). In perspective, there are approximately 20 million LBW infants born every year and in yet another study, a total conservative estimate of losses for the developing world's existing and newborns is given at approximately 5%–10% of GDP or in the region of $500 billion to $1 trillion (Mason et al. 2001). A similar report however by the World Business Council for Sustainable Development in 2006 calculated the cost of malnutrition over the next ten

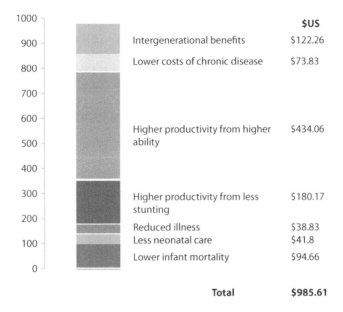

	$US
Intergenerational benefits	$122.26
Lower costs of chronic disease	$73.83
Higher productivity from higher ability	$434.06
Higher productivity from less stunting	$180.17
Reduced illness	$38.83
Less neonatal care	$41.8
Lower infant mortality	$94.66
Total	**$985.61**

FIGURE 20.3 Economic benefit of shifting one infant out of LBW status. (Based on data in the State of Food Insecurity in the World 2004, Rome, Food and Agriculture Organization, 2004.)

years to the global economy of $186 billion while in the United Kingdom, a separate study concluded the cost of malnutrition to the United Kingdom alone amounted to a whopping £13 billion every year (WBCSD 2006; Elia and Russell 2008).

With such wildly differing quotes, it is difficult to grasp any sort of consensus beyond the shared notion that the costs are vast. The economics of malnutrition regarding the cost of opportunity lost is tremendous, especially when factoring this against two simple facts, that to treat a single case of severe acute malnutrition (SAM) costs around $200 and to address the causes and prevent it, a mere $40–80 (Reinhardt 2004). Whichever figures one chooses to latch onto, the case for investment in preventative programmes have enormous cost–benefit ratios.

20.4 Care and Feeding Practices

20.4.1 Breastfeeding

Birthweight has come to be a telling indicator of a mother's existing health and nutritional status. Moreover, it also gives a good indication of the newborn's chances when it comes to survival and long-term health and psychosocial development potentials. Optimal infant feeding is best started with breast-feeding within one hour of birth. This provides both essential nutrients and protective antibodies against disease. This is also best continued exclusively for 6 months, after which slow weaning over the next two years with a mix of nutritionally adequate soft and semi-solid foods, whilst continuing to breastfeed is seen by the professionals as, nutritionally speaking, the best option for the newborn child. Also there is growing evidence that this early breastfeeding reduces a mother's risk of post-partum haemorrhage; a leading cause of maternal mortality. Unfortunately however, poor nutritional education, poverty or necessity all too often see commercial breast milk substitutes, weaning foods or needless supplementation continuing to be introduced too early, to the detriment of the health and nutritional status of the child. In fact, the optimal feeding regimen mentioned above is so important that it has been suggested that such practices alone have the potential to prevent 1.4 million annual under-5s deaths in the developing world (ChildInfo 2009).

20.4.2 Safe Water and Sanitation

The importance of water, sanitation and hygiene in the promotion of good health and food security cannot be underestimated. Ensuring people have access to safe drinking water as well as adequate sanitation improves the well-being and overall quality of life. Moreover, good clean water helps to reduce the prevalence of vector-borne diseases, is necessary for the preparation of nutritious food, supports the digestion, the transportation, use and ultimately adsorption of food. Moreover, it also aids in the elimination of wastes and toxins from the body; it is fundamental for the hygiene of the individual and not least is a valuable social resource in general recreational use.

20.4.2.1 Water

Water requirements for human consumption are based on the need to avert dehydration. While there is no universally agreed index of hydration, some estimates have been put forward. The US National Institutes of Health for instance in 2002, defined mild dehydration as a loss of body weight equal to 3%–5%, moderate dehydration as 6%–10% loss in body weight and severe dehydration resulting in a 9%–15% loss of body weight. On the other hand, Howard and Bartram (2003), defined dehydration in much lower terms with mild forms being equivalent to 1%–2% loss of body weight and severe dehydration existing beyond 2% losses.

In mild forms of dehydration, simple fluid replacement involving salt solutions usually suffice to address the problem, but in severe cases the process can take up to 24 hours involving rehydration strategies of water intake as well as osmolar intake through food and the like (Howard and Bartram 2003). As a guide, the World Health Organization suggests an intake of about 2 litres per 60 kg adult per day (excluding food water absorption). The WHO-UNEP-ILO International Programme on Chemical Safety, on the other hand, refers to more specific needs based on reference body weights of 70 kg for adult males and 58 kg for adult females with a combined average of 64 kg. These can be seen in Table 20.3.

Water for human use in terms of human development, is thought of as belonging to one of three access levels: improved water sources, consisting of piped water to a yard, dwelling or plot; other improved drinking water sources such as boreholes, wells, springs, rainwater, taps or standpipes used by the public; and lastly unimproved drinking water sources such as unprotected wells, springs, bottled water or from surface water sources including rivers, dams, lakes, ponds, streams, canals and even irrigation channels (UNICEF/WHO 2008). Water needs too, involve more than just consumption and the World Health Organization suggests that people with access to less than 5 litres of freshwater per person per day (pp/pd) cannot be assured of sufficient personal water requirements. This is considered no access as per WHO standards. The next level, basic access of 20 litres pp/pd, ensures that daily water requirements are met for personal consumption, hand washing and perhaps basic food hygiene. Beyond this though, things like household laundry and bathing are still unattainable. The next level of 50 litres pp/pd is considered intermediate and guarantees consumption and most other needs are taken care of; yet there is little room for manoeuvre. Ultimately the optimal level of 100 litres pp/pd gives sufficient assurance that all personal water needs are met and it is a level of service recommended by the WHO that denotes the least level of health concern (Howard and Bartram 2003). In the absence of optimal availability, reasonable

TABLE 20.3

Daily Water Intake Requirements

	Volumes (litres/day)		
	Average Conditions	**Manual Labour in High Temperatures**	**Total Needs in Pregnancy/Lactation**
Female adults	2.2	4.5	4.8 (pregnancy) 5.5 (lactation)
Male adults	2.9	4.5	
Children	1.0	4.5	

access then is considered to be at least 20 litres pp/pd from an improved water source no more than 1 kilometre from the user's dwelling (GEP 2011).

Today, 87 % or 5.7 billion people around the world have access to drinking water from improved sources: 54% from piped connections and 33% from other improved sources leaving a further 884 million without access to any form of improved water sources at all. Nevertheless, disturbing though these figures maybe, this does in fact represent improvements on figures over the last 15 years or so as Figure 20.4 attests. As an aside, when water is not piped into or near the home, the brunt of responsibility for fetching the household requirement falls on the women's shoulders who are considered twice as likely to land this job.

20.4.2.2 Sanitation

Sanitation can be described using the joint UNICEF/WHO analogy of a ladder in which the four rungs consist of the differing levels of achieved sanitation. The first rung called open defecation still sees people defecate in fields, forests, bushes, lakes and rivers and other open spaces as well as the disposal of human faeces along with solid waste matter. The next rung, unimproved sanitation amenities, might include pit latrines or other similar facilities that do not facilitate the hygienic separation of excreta from human contact. The third rung deals with the sharing of sanitation facilities often consisting of shared household or public toilets where there is an increase in this separation. Finally, improved sanitation facilities ensure the proper hygienic separation of human excreta and include flushing or poured flush toilets leading to a piped sewer or septic tank system. Another improved facility might also be considered to include well constructed and ventilated pit latrines or even a well maintained composting toilet (UNICEF/WHO 2008). In 2006 there were still 2.5 billion people around the world without improved sanitation. Most of these could be found in Asia and sub-Saharan Africa which combined, represented a total of 1.8 billion or 70% of those affected. Importantly too there is a large rural urban divide that saw 7 out of 10 people without improved sanitation in 2006 originating from rural habitats. This is despite good gains over the last 15 or so years (Figure 20.4).

Both safe water and sanitation, combined, still contributes to enormous mortality figures and suffering annually. For instance, every year alone 50% of malnutrition cases are associated with severe forms of diarrhoea alone. Then there are the other water and hygiene-led diseases and infections such as malaria, lymphatic filariasis, intestinal nematode infections, schistosomiasis and trachoma (Prüss-Üstün et al. 2008). In fact, ingested pathogens from unsafe drinking water, contaminated food or from poor hygiene practices resulting in diarrhoea and to a lesser extent cholera, typhoid and dysentery are collectively responsible for approximately 1.5 million deaths a year. Indeed, a World Health report in 2008 stated that as much as 10% of the annual disease burden could in fact be prevented simply by improving the water

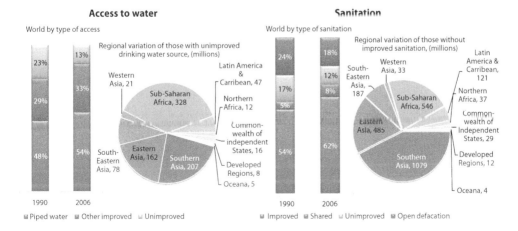

FIGURE 20.4 Water and sanitation by type and region. (Derived from data in the UNICEF/WHO report Progress on Drinking Water and Sanitation: Special Focus on Sanitation, UNICEF/WHO, New York/Geneva, 2008.)

situation. Combined then, taking care of water and sanitation could not only relieve the annual disease burden by 10% but could in reality prevent 6.3% of all deaths annually (ibid).

20.5 Discussion

The various manifestations of nutritional deficiencies had been well known and well documented for centuries; however, improvement had to wait for science to catch up before any real progress could be made. Now that the scientific knowledge is in place, it seems we once again have to wait for political and social will to catch up. When it comes to undernutrition in general, poor access to things like health services is still seen as contributory in maintaining the status quo and as with many development issues, this relates to continuing underinvestment of domestic health care services particularly in developing countries. Of note too is the fact that diet and health have now been inextricably linked, which now brings overnutrition problems into the food equation. The issues related to being overweight or obesity in developing countries is a relatively new phenomenon, already though, these are set to eclipse food insecurities within a generation. As incidents of food related non-communicable diseases like heart disease, strokes, etc. are set to rise, the already over-stressed health sector is in danger of having to dilute its nutritional spending to accommodate. As has been shown too, the cost of doing nothing however is large and is set to increase. Also from the health perspective, the idea of tackling undernutrition in isolation without regard to the problems of overweight and obesity would be short-sighted and counterproductive. Yet the dilemma of this all inclusive approach itself would raise important questions such as at what point would food security objectives cease and food health begin? This issue aside though, this places clear urgency on changing current and expected dietary and/or physical activity trends. Educational efforts that aim to inform and challenge the dietary trends would clearly be of measurable benefit. There are options open to us although some solutions would not be politically attractive. One solution would be to follow the tobacco industry's example which would see the sharing of health responsibility between the food processing companies, governments and the individual. I acknowledge there are massive hurdles with this approach and one that evokes the chicken and egg response with companies claiming they are only providing what the consumers want. In truth however, this is not the case with many if not most for-profit organisations seeking to maximise revenue by finding financially cheaper and technically easier ways of producing mass market food irrespective of health considerations or implications. This was a similar defence of the tobacco and alcohol industries; yet the cost to the health and economic situation of countries led to radical changes in the way these industries were allowed to conduct themselves. Of course, as individuals, we must take no small measure of responsibility ourselves, but it will be interesting to see, as the economic costs of health care spiral, just how policy will react.

In the area of micronutrients, as health and nutrition programmes are rolled out so malnutrition and nutritional disorders are being addressed. This is reflected in the successes of several stand-alone micronutrient intervention policies targeted at micronutrient deficiencies (or hidden hunger) such as iodine, vitamin A and iron deficiencies. While however such measures have reaped great returns, this almost exclusive focus on the single micronutrient intervention policy tends to divert attention away from the fundamental holistic dietary approach, which has to be the ultimate goal. While in response, proponents might offer that such measures were only ever meant to be a short-term objective until such times as the underlying fundamentals could be addressed, the more cynical among us could argue that these programmes, because of their cost to success ratios, are politically more rewarding. That said, projections favour a decreasing prevalence of many such deficiencies over the coming years. Yet in spite of all this, unless we intend to reposition the holistic goal of improving dietary quality and diversity to the periphery of the development process we must reconsider the intervention approaches in favour of replacement strategies. One way of achieving this would be to tie the policies of short-term goals to replacement programmes aimed at the longer-term objectives. However, another pessimistic caveat here suggests that by continuing to rely on the interventionist stance, we continue to depend on predominantly Western nutrition manufacturing interests.

In other areas, things like breastfeeding practices will benefit from continued education of the advantages of mother's milk over commercial substitutes. This goes hand in hand with best practice when it comes to weaning and other child caring considerations. With HIV too, new technologies are improving longevity and quality of life; however, increased community acceptance and social integration will further improve the lives of sufferers and their families. More understanding and acceptance improves employment and by extension food security prospective. Lastly the advances made in accessible safe water needs to be maintained and perhaps speeded up so that the whole package of social improvements can best be maximised.

References

Atwater, W. O. (1889). Public Health Papers and Reports. 27th Annual Meeting of the American Public Health Association. Minneapolis, MN: The Berlin Printing Company.

Bendich, A. (1992). Symposium: History of Nutritional Immunology: Vitamins and Immunity. *The Journal of Nutrition* 3: 601–604.

Berg, A. D. (1968). Malnutrition and National Development. *The Journal of Tropical Pediatrics* 14(supp3): 116–123.

CFS (2001). Assessment of the World Food Security Situation. FAO Committee on Food Security, Twenty-seventh Session. Rome: Food and Agriculture Organization.

ChildInfo (2009). A World Fit for Children. *Child Nutrition*. United Nations Children's Fund (UNICEF). http://www.childinfo.org/idd.html (accessed 12 February 2011).

Elia, M. and C. A. Russell (2008). Combating Malnutrition: Recommendations for Action. Redditch, UK: British Association for Parenteral and Enteral Nutrition (BAPEN) 7.

FANTA2 (2008). Nutrition, Food Security and HIV: A Compendium of Promising Practices. Washington, DC: Regional Centre for Quality of Health Care (RCQHC) and the Food and Nutrition Technical Assistance (FANTA) Project, managed by the Academy for Educational Development (AED).

FAO/WHO (1973). FAO/WHO Ad Hoc Committee of Experts on Energy and Protein: Requirements and Recommended Intakes. Rome: Food and Agriculture Organization.

Funk, C. (1922). *The Vitamines*. Baltimore, MD: Williams & Wilkins Company.

GEP (2011). Human Conditions. The Global Education Project. http://www.theglobaleducationproject.org/earth/human-conditions.php (accessed 11 January 2011).

Gillespie, J. A. (2003). International Organizations and the Problem of Child Health, 1945–1960. *Dynamis* 23: 115–142.

Greenway, K. (2008). Inter-Agency Task Team on Children and HIV and AIDS: Working Group on Food Security and Nutrition. Paris: IATT Food Security and Nutrition Working Group.

Gunderson, G. W. (2003). *The National School Lunch Program: Background and Development*. New York: Nova Science Publishers.

Howard, G. and J. Bartram (2003). Domestic Water Quantity, Service Level and Health. WHO/SDE/WSH 39

League of Nations (1936a). The Problem of Nutrition: Volume I Interim Report of the Mixed Committee on the Problem of Nutrition. Geneva, Switzerland: League of Nations.

League of Nations (1936b). The Problem of Nutrition: Volume II Report on the Physiological Bases of Nutrition. Geneva, Switzerland: Technical Commission of the Health Committee of the League of Nations.

Mason, J. B., J. Hunt, D. Parker and U. Jonsson (2001). Nutrition and Development Series Improving Child Nutrition in Asia. *Food and Nutrition Bulletin* 22(supplement 3).

McMillan, R. B. (1946). "World Food Survey"—a Report from F.A.O. *Review of Marketing and Agricultural Economics* 14(10): 372–378.

Moore, S., J. N. Hall, S. Harper and J. W. Lynch (2010). Global and National Socioeconomic Disparities in Obesity, Overweight, and Underweight Status. *Journal of Obesity (Online)* http://www.hindawi.com/journals/jobes/2010/514674/abs/ (accessed 21 February 2011).

Munzara, M. A. (2010). Land Grabbing and Nutrition: Challenges for Global Governance. *Right to Food and Nutrition Watch*. Germany: Brot für die Welt (Bread for the world) 90.

Périssé, J. (1981). Joint FAO/WHO/UNU Expert Consultation on Energy and Protein Requirements: Past Work and Future Prospects at the International Level. Rome: Food and Agriculture Organization.

Prüss-Üstün, A., R. Bos, F. Gore and J. Bartram (2008). Safer Water, Better Health: Costs, Benefits and Sustainability of Interventions to Protect and Promote Health. Geneva, Switzerland: World Health Organization 60.

Reinhardt, E. (2004). Vitamin and Mineral Deficiency, a Global Progress Report. *UN Chronicle*, September–November.

Ross, E. B. (2003). Malthusianism, Capitalist Agriculture, and the Fate of Peasants in the Making of the Modern World Food System. *Review of Radical Political Economics 2003* 35: 437–461.

Ruxin, J. N. (1996). *Hunger, Science, and Politics: Fao, Who, and Unicef Nutrition Policies, 1945–1978, Chapter II the Backdrop of Un Nutrition Agencies, by Joshua Nalibow Ruxin*. London: University College London.

Starling, E. H. (1919). *The Feeding of Nations*. London: Longmans, Green & Co.

Stiebling, H. K. (1933). *Food Budgets for Nutrition and Production Programs*. Washington, DC: USDA.

UNAIDS (2010). UNAIDS Report on the Global Aids Epidemic 2010. Geneva, Switzerland: UN Inter-agency Project on AIDS.

UNICEF/WHO (2008). Progress on Drinking Water and Sanitation: Special Focus on Sanitation. New York/Geneva, Switzerland: UNICEF/WHO 58.

USAID (2007). USAID P.L. 480 Title II Food Aid Programs and the President's Emergency Plan for Aids Relief: HIV and Food Security Conceptual Framework. Washington, DC: USAID Bureau for Democracy, Conflict & Humanitarian Assistance, Office of Food for Peace and the US President's Emergency Plan for AIDS Relief.

USAID/AED/WFP (2007). Food Assistance Programming in the Context of HIV. *Food and Nutrition Technical Assistance (FANTA) Project*. Washington, DC: United States Agency for International Development (USAID); Academy for Educational Development (AED); United Nations World Food Programme (WFP).

WBCSD (2006). DSM: Fighting Hidden Hunger. Geneva, Switzerland: The World Business Council for Sustainable Development (WBCSD).

WHO (2010). World Health Statistics. Geneva, Switzerland: World Health Organization 149.

World Bank (2008). World Development Report 2008 Agriculture for Development. Washington, DC: World Bank.

21

Environment and Natural Resources

Today, more than any other time over the last 50 years or so, environmentalism and a growing awareness of the finite nature of Earth's 'apparent abundances' is becoming more and more inextricably linked with the issues of food security. This is not surprising really considering that the natural resource base provides everything that we humans eat, wear, use or live in. Moreover, through the ages our reliance on the environment around us has increased exponentially and nowadays, as population growth places greater stresses on these resources, the sustainability of this relationship comes into question. As a consequence, the last few decades has seen growth in not just environmental movements, charities and organisations but a fundamental shift in the way many humans view themselves and their relationship with Earth. This is particularly manifested in the way many have chosen to no longer view themselves solely as consumers or end users of the Earth's products but rather as custodians of an equal and mutually beneficial symbiotic relationship. Thus, it this relationship and in particular the way in which it is viewed has huge implications in every aspect of our lives and not least in almost every facet of what we grow and eat. In short, from the way we cultivate our fields to the inputs we use and the emissions we generate; from the food processing plants through to the storage, the consumption patterns and the disposal of all our waste, environmentalism and sustainability is helping to redefine the food security concept (Steck 2008).

Perhaps the best way to examine the relationship between food security and the specific challenges facing us is to explore such issues through the multitude of biomes, or rather the various spheres (lithos, atom, bio, hydro) as well as other aspects such as energy and environmental accounting.

21.1 Biomes and Classifications

The total surface area of Earth is just over 510 million square kilometres, of which 70.8% (361 million square kilometres) is water and the remaining 29.2% (149 million square kilometres) is land (including inland water bodies) (NASA 2011). Biomes and classifications aim to categorise this simple division into more meaningful data. At its most basic level, the world can be divided into sub-divisions. These can be based on the regional or global sharing of common properties such as biotic community or climate, or perhaps using flora and to some extent fauna. These regions such as grassland, desert and marine are known as biomes, yet the exact number varies according to the extent and type of classification used. Used as a general classification, these generally range in number from anywhere between 4 to more than a dozen, although just how many specific instances of each there are is also up for debate with some suggesting there may be as many as 200 plus (Encyclopedia 2002; Woodward 2003; NASA 2011). As mentioned, biomes have been identified using vegetation type differences as well as regional climatic variations. At its simplest, five terrestrial biomes are often given, comprising forest, desert, grasslands, shrubland (chaparral) and tundra with the optional inclusion of wetlands; although in practice many other examples are also available (Table 21.1). More often these days though there is a propensity to categorise such biomes on the basis of land cover classification systems where land in this sense is used as a catchall term to encompass the physical Earth including water bodies. The classification of land cover and its usage is important in understanding the extent of the resources available to humanity and once again as with much else, information tends to be non-consensual and proprietary. The FAO for instance make use of 16 such international classification systems. On this note, the FAO's recent land cover classification system is currently in its second version, yet this tends to focus more on the vegetal index for obvious reasons (Gregorio and Jansen 2005; FAO 2009b). The most common source of land cover data these days is created through remote sensing. This includes data gathered by aerial photography and satellites. Of the satellite data, there has been several major sources over the years, which have included:

- 1992: IGBP-DISCover (AVHRR USGS)
- 1992: University of Maryland (UMD) (AVHRR)
- 2000: MODIS 1-km LC MODIS (Boston University)
- 2000: GLC2000 SPOT/VEGETATION JRC
- 2003: GLCNMO MODIS ISCGM
- 2005: GLOBCOVER ENVISAT/MERIS ESA

Each one of these however operate their own distinct categorisations and among the confusing array of classifications and interpretations none of the leading systems have been singularly adopted as a global standard (McGinley 2008; Tateishi 2010). Striking a balance between the many systems, Table 21.1 combines the major elements of many of these and gives an idea of the author's own basic hierarchical list of biomes.

TABLE 21.1

Biomes of the World

Biomes			
Terrestrial	Forest	Coniferous forests	Taiga/Boreal forest
			Temperature: –40°C–20°C, average summer temperature about 10°C; coniferous forest regions have cold, long, snowy winters, and warm, humid summers; well-defined seasons, at least four to six frost-free months. *Precipitation* (mm p/y): 300–900. Biology: Coniferous-evergreen trees (trees that produce cones and needles). Location: Canada, Europe, Asia and the United States.
		Temperate deciduous forests	*Temperature*: –30°C–30°C, yearly average is 10°C, hot summers, cold winters, most notable because they go through four seasons. *Precipitation* (mm p/y): 750–1500. *Biology*: Broadleaf (oaks, maples, beeches), shrubs, perennial herbs and mosses. *Location*: Eastern United States, Canada, Europe, China and Japan.
		Rainforests	*Temperature*: 20°C–25°C warm and frost-free. *Precipitation* (mm p/y): 2,000–10,000. *Biology*: Vines, palm trees, orchids, ferns. There are two types of rainforests: tropical and temperate. Tropical rainforests are found closer to the equator and temperate rainforests are found farther north near coastal areas. *Location*: Between the Tropic of Cancer and the Tropic of Capricorn.
	Desert	Desert/arid land	Temperature: Average of 38°C (day), average of –3.9°C (night). *Precipitation* (mm p/y): 250. *Biology*: Cacti, small bushes, short grasses. Perennials survive for several years by becoming dormant and flourishing when water is available. Location: Between 15° and 35° latitude (North & South of equator); examples are Mojave, Sonoran, Chihuahua, and Great Basin; Sahara; Negev and Gobi.
	Grasslands	Grasslands	*Grasslands*: Prairies in America & Steppes in Eurasia. *Temperature*: –20°C–30°C. *Precipitation* (mm p/y): 500–900. *Biology*: Grasses (prairie clover, salvia, oats, wheat, barley, coneflowers). *Location*: The prairies of the Great Plains of North America, the pampas of South America, the veldt of South Africa, the steppes of Central Eurasia, and surrounding the deserts in Australia. Every continent except the Antarctica.
		Tropical-Savanna	*Temperature*: Variable 20°C+. *Precipitation* (mm p/y): 760–1020. *Location*: Primarily Africa, Savanna grasslands a product of the tropical wet-dry climate found between tropical rainforests and desert. *Location*: Central Africa (Kenya), America and the North and East of South America (Brazil).

TABLE 21.1 (Continued)

Biomes of the World

Biomes			
Terrestrial *(Cont.)*	Shrubland/ Chaparral		*Temperature*: Hot and dry in the summer, cool and moist in the winter. *Precipitation* (mm p/y): 200–1000. *Biology*: Aromatic herbs (sage, rosemary, thyme, oregano), shrubs, acacia, chamise, grasses. *Location*: West coastal regions between 30° and 40° North and South latitude.
	Tundra	Arctic and Alpine	*Temperature*: –40°C–18°C. *Precipitation* (mm p/y): 150–250. *Biology*: Almost no trees due to short growing season and permafrost; lichens, mosses, grasses, sedges and shrubs dominate. *Location*: Arctic, Northern Hemisphere encircling North Pole, extending to south of the coniferous forests of the Taiga, also south of the ice caps of the Arctic. Alpine—extending across North America, Europe, and Siberia (high mountain tops).
Freshwater	Ponds Lakes Rivers Streams		*Temperature*: Varied. *Precipitation*: Varied. *Biology*: Characterised by rich biodiversity; ponds, lakes and rivers are fed through the rainwater system that often starts life in the oceans. A rich and varied ecosystem of both flaura and fauna. *Location*: Global.
	Wetlands	Marshes Bogs Fens Swamps (including mangrove swamps)	*Temperature*: Varied. *Precipitation*: Varied. *Biology*: wetlands are saturated with water at or near the surface. Marshes dominated by soft-stemmed vegetation, swamps mostly woody plants, bogs are freshwater wetlands often formed in old glacial lakes, consist of spongy peat deposits, evergreen trees and shrubs and a floor covered by thick sphagnum moss. Fens are freshwater peat-forming wetlands covered by grasses, sedges, reeds and wildflowers. Mangrove swamps are coastal tropical and subtropical wetlands regions with salt-tolerant trees, shrubs and others able to stand brackish to saline tidal waters. *Location*: Global.
Marine	Oceans, Continental shelves, Estuaries, Coral reefs		*Temperature*: Varied. *Precipitation*: Varied. *Biology*: Covering about three-quarters of the Earth's surface marine algae supply much of the world's oxygen supply while acting as a sink for atmospheric carbon dioxide. Photosynthesis and phyto- and zoo-planktons support a multitude of plant and marine animal growth. Evaporation of the seawater provides rainwater for the land. *Location*: Global.
Anthropogenic	Urban and built-up areas		*Temperature*: Varied. *Precipitation*: Varied. *Biology*: This type of biome creates its own significant ecological patterns caused by sustained direct human interaction with ecosystems. It supports all manner of plant and animal life. Land is primarily covered by buildings and other human-made structures as well as recreational open areas. *Location*: Global.

Source: UoM, *Global Change: Human Impact—the Changing Biomes*. University of Michigan, 2007; McGinley, M., *Biomes*, Environmental Information Coalition, National Council for Science and the Environment, Washington, DC, 2008; IGBP, The International Geosphere-Biosphere Programme. Royal Swedish Academy of Sciences, 2010; NASA, Earth Observatory. NASA, 2011.

Bearing all these in mind, there still exists some confusion though over the use of the terms land use and 'land cover' (Gregorio and Jansen 2000). There are, however, some fundamental differences which are worth pointing out; land-cover for instance is generally concerned with the bio-physical cover of the Earth's surface whereas land-use on the other hand is more concerned with the functional dimension of lands used in different activities and economic use (ibid). So why is land cover usage and classification important? Apart from the obvious need to classify biomes, the simple answer is that with the unprecedented magnitude and pace with which humanity is altering its environment. There is a need to quantify usage and change over time for all manner of reasons ranging from identification and conservation through to soil fertility. More too, this data gives us baseline measurements for the calculation, monitoring and measurement of ecosystems, water and energy cycles as well as crop production information (Lambin et al. 2001; DeFries et al. 2002). Also this knowledge is helping with understanding climate change, as most climate models now use digital land cover data in land surface parameterisations (LSPs), which aim to assist in calculations of albedo, surface roughness, evapotranspiration and respiration among others (LCLUC 2009). Finally, from a food security standpoint, accurately assessing land

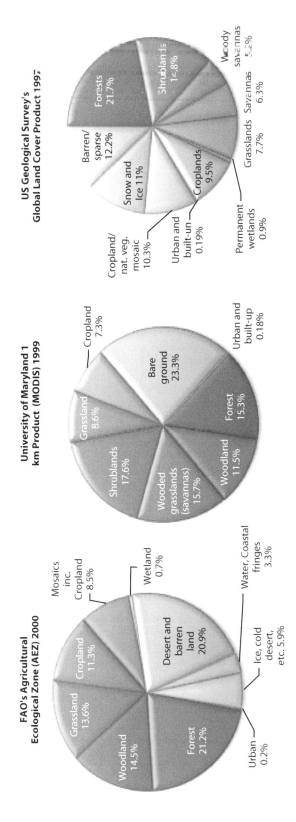

FIGURE 21.1 Comparison of global biome land cover using various classifications. (Compiled from data available in the FAO and the IIASA Inventory of land resources, 2010, http://www.fao.org/ag/AGL/agll/gaez/index.htm).

cover and change over time helps scientists determine future potential availability of food, fuel, timber as well as shelter and other resources.

As alluded to earlier, there are major differences in classifications too, and the lack of consensus is not confined to forest cover alone either (a hotly debated topic in its own right). Figure 21.1 demonstrates the complexity of land cover semantics where differing standards lead to varying results.

Thus, it can be seen that land cover and usage differs according to whichever dataset one chooses and having a brief overview of such systems helps contextualise some of the food security issues discussed in the following sections. Caution then is needed to ensure that data sources and classification systems are identified by whichever boundaries or classifications on which the information is based. It is also prudent to ensure that like for like data is used in any comparative studies.

The following sections look at various aspects of these resources and views any impacts or considerations it might have on, inter alia food availability, sustainability or degradation.

21.2 Lithosphere: Land Resources

Of the 149 million square kilometres of land globally, there are about 130 million square kilometres or 13 billion hectares if we exclude inland water bodies (rivers, lakes, etc.). As has also been documented in this book, fears about land availability as a constraint in feeding an ever-increasing population has a long history. This debate however is still not settled. The difficulty relates to the finite land resource, which with the exception of forest cover, has not changed drastically in over 45 years. For example, in Figure 21.2, it can be seen that in nearly half a decade the overall arable land used has only increased marginally from 1.28 to 1.38 billion hectares (9.8%–10.6%). Therefore, bearing in mind the more than doubling of the population during this period, it is not surprising that arable land per capita has dropped drastically from 0.48% to 0.2% hectares per person.

With such land restrictions, these figures serve to highlight the remarkable achievements that have been made in feeding more than twice the 1962 population with a mere 8% increase in land under cultivation. This is clear testament to the growing importance of technological advances within the agrarian sector. The argument then becomes two-fold: where will future production increases come from—productivity, land expansion or both? Furthermore, what is the potential to increase our land capacity and how can we be expected to feed a further 2.5 billion people by 2050 under such constraints? The

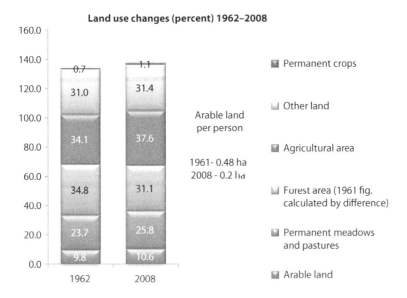

FIGURE 21.2 Land use changes 1962–2008. (Compiled from *FAOSTAT Food and Agriculture Statistics*, Food and Agriculture Organisation, 2011, http://faostat.fao.org/).

former has been addressed to some extent in other sections, whereas the latter, land expansion, is worth exploring a little more.

So what is the future potential of land? All else being equal, further expansion of agricultural land* will be based on multiple factors including soil constraints, erosion hazards, salinity, hydrology and suitable available land for expansion, not to mention the social acceptance of convertible lands, in particular current forested lands for example. A study by the FAO in 2000 calculated that in total between 3 and 3.3 billion hectares (approximately 25% of total land area), could in fact be made available to agriculture under the right conditions (AEZ 2000). Table 21.2 gives an indication of where this extra rain-fed cultivatable land might come from. It is also worth noting too that while these figures offer room for optimism, the expansion is likely to be uneven with some regions having nearly reached, or having already exhausted, their cultivatable potential (ibid). Such inequitable distribution of potential is likely to further aggravate regional food insecurity.

With this in mind, based on the range of possibilities and from a static existing arable land base of 1.3 billion hectares to an extra 1.7 billion hectares (based on the above figures minus the author's generous allowance for lands for pasture), this would equate to a per capita arable land projection of between 0.2 and 0.45, thus reversing some of the losses over previous decades. Over-optimistic? Perhaps. In contrast, another view suggests much of this so-called potential agricultural land is likely is to be geographically inaccessible, marginal and in direct competition with other social and economic demands (Muir 2010).

Despite the optimism, it is figures like these that are giving rise to many concerns that the Earth's natural resources are being stretched beyond their sustainable limits or carrying capacity (GEO 2007a; PAP 2008). The United Nations Environmental Program's (UNEP) fourth Global Environmental Outlook report has compared current population consumption levels with the planet's available resources and concluded that the human population is already living way beyond its sustainable means and thus possibly inflicting irreversible damage on the environment. The problem was not population per se, suggested Achim Steiner of the UN Environment Programme, but rather the per-capita impact on the planet (GEO 2007a). This has led some to call for more drastic population control measures such as forced family planning. On this, Steiner refutes such draconian measures, yet he does favour a more realistic, proactive and ethical approach in addressing the problem of finite resources (GEO 2007a). Having said that, this particular report, like so many others before it also offered its conclusions as providing:

... a final wake-up call for the international community. (GEO 2007b)

TABLE 21.2

Potential Agricultural Land

Region	Total Land	Total Potential Land for Cultivation	Underforest	Housing, Infrastructure	Not suitable
North America	2138	384	135	9	1637
South and Central America	2049	858	346	16	1048
Europe and Russia	2259	511	97	21	1645
Africa	2990	939	132	26	1909
Asia	3113	516	47	83	2407
Oceania	850	116	17	1	694
Developing	8171	2313	527	124	5383
Developed	5228	1012	247	33	3956
World Total	13,400	3325	774	156	9338

Source: FAO (2000) *Global Agro-Ecological Zones: Inventory of Land Resources*, 2000, http://www.iiasa.ac.at/Research/ LUC/GAEZ/index.htm.

---

* Agricultural land refers to land under temporary and permanent crops, meadows, fallow and land under kitchen or market gardens (World Bank 2007).

Indeed, and not to sound pessimistic but the work of the first food and agriculture Conference in 1943 had already recognised that in many cases the most advantageous foods required to improve people's general nutritional status were often those same foods that were produced in an environmentally sympathetic manner (UN 1943). In other words, the sentiments that there is no separation between food production and sustainability is not exactly new.

21.2.1 Land Cover and Land Use Changes

The unprecedented change in human land usage has implications too beyond that of security of food alone. The pace and reach of human-induced change in the biophysical make-up of Earth is not only affecting food production it also has consequences for the climate as well as the Earth's biota and biodiversity. In return, these changes further impact on the Earth's ability to continue to support human endeavours (Lambin et al. 2001). This symbiotic relationship is at the heart of a fundamental dynamic equilibrium that has to our shame been ignored or exploited for far too long. In a very real sense, then the need to enumerate the changes in land use and land cover are pivotal in understanding the health and productivity of the Earth's natural resource base. Such changes, however, as shown above are difficult to determine, especially considering that many people are singing from different hymn sheets when it comes to the standardisation of classification systems. That said, while such changes in land cover have been difficult to discern with any degree of accuracy there is widespread agreement that the change has been considerable (Oldeman 1994; Lambin et al. 2001; Kniivila 2004; Foley et al. 2005). The consequences of this change have been considerable too. One estimate from Ramankutty and Fole in 1999 suggests that as much as 6 million km^2 of forests and woodlands as well as 4.7 million km^2 of grasslands/savannas/steppes, for example, have been lost to global cropland expansion since 1850. Moreover of these figures about 1.5 and 0.6 million km^2 respectively has since been abandoned (Ramankutty and Foley 1999). Land use changes have also been accused of leading to soil and water pollution (next section) and subsequent fertility/productivity issues as well as being major release mechanisms of CO_2 (Section 21.4) (IPCC 2001). On top of this, changes in use of established land areas have diverse impacts on ecosystems, possibly affecting both its geological and biotic dynamics. Once again, considering the symbiotic relationship mentioned earlier, changes derived from inconsideration and ignorance are clearly unsustainable both in terms of equilibrium and food security.

One such strain comes from the relentless sprawl of urbanicity. Population increases have already seen huge urban centres tirelessly springing up over the recent past. This has culminated in the present day that sees over 50% of the population now living in such habitats. Moreover, as present trends continue, competition with agriculturally productive lands will only become more burdensome (World Bank 2009; WSF 2009). In China alone, as an example, they have been losing arable land at the rate of one million acres per year while in the United States, over 25 million acres have been conceded to urban sprawl since 1967 (equivalent to approximately 0.6 million hectares annually) (Muir 2009c). Such expansion of land for both agriculture and urban development comes at the expense of existing forest cover, grasslands and other land cover types. While this is something worth considering when factoring any future population-food dynamic other land cover and land use, changes are equally as consequential. Perhaps the most prevalent of these changes is the degradation of the productive lands themselves.

21.2.2 Pollution and Degradation

There are two types of degradation: natural and human-made. It is often overlooked for instance that soil erosion is a natural process of the Earth's geological cycle. The loss of topsoil through wind and water in fact rates above those of anthropogenic origins (Alexandratos 1995; DeRose et al. 1998b; Eswaran et al. 2006; Freedom 21 2008; Muir 2009a; World Bank 2009). In this respect, soil erosion, desertification and deforestation are all natural processes that continually reshape the geological and chemical composition of much of the Earth's surface. However, for millennia, when natural resources were abundant relative to population size, humanity has managed to live in harmony within this context. Yet, as this changing dynamic unfolded, highlighted over the last half century so, the need to understand and address these issues assumed greater importance. Anthropogenic degradation and erosion have been blamed on many practices. Specifically within the industrialisation of agriculture, this might be in the form of a general lack of understanding vis-à-vis erosion, excessive or ill-managed agricultural inputs, poor land management practices, overgrazing, improper irrigation leading to salinisation, waterlogging, or simply a laissez-faire attitude to the environment among others. In the developing countries these are all prevalent. However, there is also another dimension: one of need. When you are hungry and you are working the land that does not belong to you for instance, it is difficult to incentivise the farmer in conservation practices beyond his immediate needs. In this respect, people tend to optimise working practices for the benefit of the short term. Ironically, those with abundant plentiful and secure access to land might also be discouraged from conservation practices for when the land no longer yields there are options to simply abandon it in favour of pastures new (Morvaridi 1995; DeRose et al. 1998b).

Land degradation comes in many forms including soil erosion, overgrazing, waterlogging, excessive chemical inputs, leaching, acid rain and salinisation, among others. At the very least, such degradation can lead to significant lowering of biotic diversity, altered ecosystems and reduced agricultural productivity. At its worst however, it can also lead to desertification and subsequent barren lands unable to provide continuing sustenance for its inhabitants. While there is a distinct lack of up to date land degradation data, it is likely that as much as 14% of all land (19 of the Earth's 134 million square km) has been affected by human practices, with about 5.5 million of those existing in Asia and the Pacific alone (Oldeman et al. 1991; GEO 2000). A good proportion of this too is agricultural land (Oldeman 1994; Kniivila 2004; ECOSOC 2008; Muir 2009d; UOR 2009; World Bank 2009). As Figure 21.3 highlights, the principal regional sources of this degradation are seen in both Africa and Asia, with deforestation, poor agricultural practices and overgrazing. Once again, such data has its opponents with opposing views who charge that while there is indeed widespread evidence of soil erosion, the impact of this on agricultural production has not been well established, a point to bear in mind (Freedom 21 2008).

Looking more specifically at the type of soil degradation as opposed to the source, there are four types of human-induced activities: water, wind, chemical and physical. Each of these too are recognised to occur at four degrees of severity: light degree of soil degradation resulting in reduced productivity; a moderate degree of soil degradation showing greatly reduced productivity; strongly degraded soils which are difficult to reclaim and considered virtually lost without extensive reparation; and extremely degraded soils which are considered completely irreclaimable and beyond restoration. It can be seen from Figure 21.4 that water, totalling 56% (1094 million hectares), is the biggest culprit of human-induced degradation. This occurs through agricultural practices such as poor irrigation and salination. Next, at 38% (548 million hectares) is wind erosion from deforestation or the uprooting

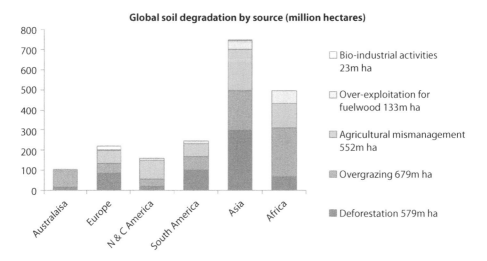

FIGURE 21.3 Causes of human-induced total overall land degradation. (Adapted from Oldeman, L.R., Hakkeling, R.T.A. and Sombroek W.G. (1991) *World Map of the Status of Human-Induced Soil Degradation.* Global Assessment of Soil Degradation (GLASOD.)

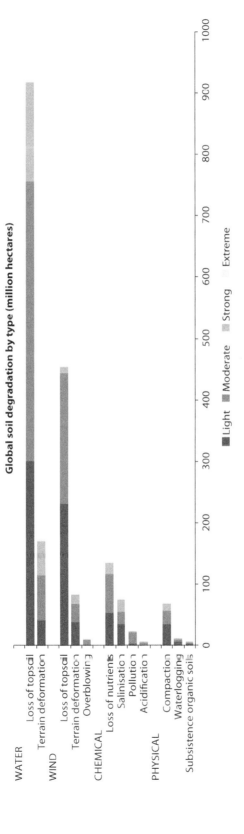

FIGURE 21.4 Total overall human-induced global soil degradation by type. (Adapted from Oldeman, L.R., Hakkeling, R.T.A. and Sombroek W.G. (1991) *World Map of the Status of Human-Induced Soil Degradation*. Global Assessment of Soil Degradation (GLASOD).)

of protective species of plants. Following this at 12% (239 million hectares) is chemical erosion and finally at 4% (83 million hectares) is the physical abuse of the soil with compacting and waterlogging, among others. (1991)

21.2.2.1 Soil Erosion

Such degradation has a huge impact on productivity and if unchecked, this could drastically affect future food security. In Africa, for instance, land is one of the most critical resources that the rural poor can access and if the soil is not adequately maintained, such losses could amount to as much as 50% of productivity over the next few decades (GEO 2000). Top soil is the first to be eroded and available evidence indicates that while it is unclear just how much is lost through wind, rain and irrigation runoff, it is likely to only be a small fraction of the total estimated soil moved (Trimble 1999). This is important as topsoil contains most of the nutrients and organic matter available to plants and organisms; therefore any excessive erosion above the rate of soil formation degrades the soil's structure and water-holding capacity resulting in less available moisture and nutrients (Muir 2009a). Much of this eroded soil ends up in rivers and streams where it settles as sediment, which in turn might negatively influence local ecosystems and further degrade water quality.

Since about 1940/1950s however, there has been increasing awareness of these problems as understanding of both the soil typology, its needs and constraints have collectively aided in a fuller more rounded awareness of such issues (Salter 1950). Nowadays, through increasing use of conservation tillage, the use of cover crops that bind the soil or act as windbreaks as well as contour ploughing that follows the lay of the land among other simple and practical measures are helping to add stability in soil fertility, livelihoods and food security.

21.2.2.2 Overgrazing

Historically, many pastures or rangelands used for livestock production have been dominated by perennial grasses with their extensive root systems and year-round ground cover. These plants are a useful soil stabilisation resource, yet continuous or rotational grazing, heavy grazing from large herds of livestock or from longer grazing periods reduce plant leaf areas and ultimately growth. Over time this leads to plants becoming weakened with reduced root length and a decreased ability to protect the soil from wind and rain erosion as well as risking shorter more vulnerable life spans of the plants themselves. The weakened sod then allows for the invasion of short-lived perennial grasses, herbs and other non-native weeds making it, in the severest cases, difficult for the native perennial grasses to re-establish themselves. Furthermore, some of these replacement weeds might also be harmful, poisonous or indigestible to the grazing herd (Muir 2009b).

21.2.2.3 Irrigation

Significant improvements in irrigated agricultural croplands have increased productivity ultimately allowing more food to be grown on the same amount of land (SOFI 2008). Irrigated lands too have also helped boost productivity whilst reducing volatility of crop yields. In fact, so effective is this technique that over the last century irrigated agriculture has expanded nearly fivefold to reach the present 16%–17% of all global cropland (FAO 2003b; Muir 2010). However, it is not all good news. Already irrigation is the largest user of water worldwide, and many such systems can be very inefficient with water diverted for irrigation seeping out of unlined canals or perhaps delivered to plants on a schedule that does not match a plants' needs (Muir 2010). Furthermore, over-exploitation of this increasingly scarce resource is rife with problems of eutrophication,[*] chemical runoff as well as the aforementioned unsustainable practices. Altogether the result can often exhaust or contaminate existing aquifers and other water sources (SOFI

---

[*] Eutrophication is a process that occurs as lakes, groundwater, streams and rivers age whereby the growth of algal blooms is a natural part of the ecosystem cycle (Eifert et al. 2002). Excess nitrogen and phosphate runoff from fertiliser overuse or misuse percolate into these bodies and accelerate the growth of these algae. This accelerated or 'cultural' eutrophication has a devastating effect as oxygen is depleted from the waters in turn creating large areas of 'dead' zones (Muir 2010). It is estimated runoff from fertiliser or animal waste account for 50%–70% of all nutrient runoff reaching these waters (Muir 2010).

2008). Indeed as it stood in 2008, 25% of the world's irrigated lands continuously drew more water than was actually renewed. This undoubtedly will be further complicated by the continuing effects of climate change and the resultant increase in irrigation needs as well as the rising demand for irrigated crops. As a result, irrigation costs are rising and as this and natural scarcity begins to have its effect it is the agricultural sector that stands to lose out (Rosegrant and Cline 2003).

21.2.2.4 Salinisation and Waterlogging

While freshwater might be considered salt free by many, this in fact is not the case. All freshwater dissolves salts, albeit in low concentrations, as it passes through the land. In the dry months or in naturally arid and semi-arid areas, these salts can become concentrated, which can then seep up back through the soil and evaporate, leaving salt deposits behind. While this build up can often be flushed away through plentiful natural rainwater in the wetter months, it can remain a problem for the poorly drained drier regions. The process of salinisation refers to this build up of salts in soil and in worst cases can reach concentrations of between 3000 and 6000 ppm. When this happens, most cultivated plants either struggle to take up water from the soil or in extreme cases find the water directly toxic (Muir 2009e).

Approximately 20% of the world's irrigated acreage is estimated to be affected by salinisation, reducing productivity and in some cases leading to the abandonment of lands. In fact so bad is the problem that it threatens to adversely affect productivity and affect food security in developing countries (Postel 1999; Khan and Hanjra 2009). However, while the cure for salinisation is to flush the soil with lots of fresh water, consideration needs to be given to nearby rivers and groundwater where the flushed water percolates (Muir 2007, 2009e).

Another associated problem of excessive irrigation is waterlogging. As with salinisation, this occurs primarily on poorly drained soils, affecting about 10% of all global irrigated land. Waterlogging occurs as excess water, either through irrigation or poor drainage effectively raises the water table in turn affecting soil moisture content. Unfortunately, this is often undetectable until it is too late. When this happens, the soil structure is damaged and the soils natural air pockets tend to fill with water, which can lead to the suffocation of plant roots (Sundquist and Broecker 1985).

21.2.2.5 Fertiliser and Pesticide Degradation

With fertiliser use and sustainability, there are two issues: one relates to the energy aspect of its manufacture, which has already been covered, while the other concerns itself with usage and pollution discussed here. The three major nutrients—nitrogen, phosphorus and potassium—are all required for good plant growth and of these nutrients, it is quite often nitrogen becomes the limiting factor. As has been discussed, plants cannot utilise atmospheric nitrogen directly. Instead it is provided in one of two ways: natural and inorganic. Natural processes involve the decaying of organic matter by microbial or bacterial action, which releases biologically available forms of nitrogen such as ammonium into the soil. The other way this can be achieved is by adding inorganic nitrogen fertiliser directly into the soil. Fertilisers then are used to supplement a soil's natural nutrient supply and to compensate for plant use and from nutrient leaching. In assuring greater food security, modern agriculture relies to a much greater extent on the use of these improved fertilisers (GEO 2007b).

Pesticides on the other hand is a general term incorporating fungicides and insecticides. These are needed to tackle the huge problem that sees insects, mammals, birds, bacteria, fungi, viruses and weeds consuming or destroying in the region of between 30% and 40% of the world's crop production every year. Originally, many first-generation pesticides were found to be highly toxic. However, while these were replaced by second-generation synthetic organic compounds such as DDT, ultimately these too were discovered to be just as toxic. In fact, DDT was one of the original persistent organic pollutants (POPs) which were eventually banned in most Western countries (Muir 2010).

With such large quantities being used on a global scale, there are big concerns regarding the use of fertilisers and pesticides in agriculture. While few doubt their usefulness, there are issues with the overuse or improper use of such chemicals, which can result in loss of soil fertility and pollution of surrounding soils and water tables and of course contamination of the food supply. Excessive or improper application reduces the soil's ability to retain valuable moisture and as water passes through the land, it carries with it these chemicals into nearby streams and rivers which itself is also absorbed in the water table. This has numerous consequences from localised poisoning of biota to the eutrophication of water bodies (next sections). The dilemma for modern agricultural practices is that it can create a cycle of dependency that in order to continually achieve greater yields more and more energy inputs from chemical fertilisers and pesticides are needed. This has serious environmental consequences for local ecosystems and as forecasts of increased fertiliser use is made so the sustainability of future productivity gains need to be protected through careful and practical considerations of these issues (Anderson and Cook 1999; Eifert et al. 2002).

21.2.3 Deforestation

Typical perceptions of deforestation is often associated with the consequences of the 'push' of population growth and poverty and while there is some element of truth to this, Lambin et al. suggest this is too much of a generalisation. In their well-researched study of tropical deforestation, Lambin et al. argue that as with much of land use or cover change, causation is all too often based on oversimplification attributed solely to poverty or overpopulation. In reality however, according to Lambin, there are wider more intricate forces at play and as the study suggests, variables such as changing economic opportunities linked to social, political and infrastructural changes are underpinned by migration, settlement, plantation and extractive industries. These in turn seem to follow deeper governmental desires to secure territorial claims and facilitate market opportunities or to promote interests of specific groups (Lambin et al. 2001).

What is more there is also a tangible element of weak governance where certain nation states depend heavily on natural resource income from exported timber. In turn, weak regulation, corruption and lack of enforcement sees large forest areas continually being logged. Of course, this is not the only element. There are the conversion of forest areas for the cultivation of crops, settlements and other large forest development projects (Lambin et al. 2001). There is however, some movement in the right direction with what appears to be a slowing of deforested areas. In a 10-year rolling average the FAO calculated that the annual rate of deforestation between 1990 and 2000 equated to about 832,000 square km. Compare this to the recent 10-year period 2000–2010, of about 521,000 square km, and this shows a considerable reduction (SOWF 2011). At work in this reduction is a collective growing awareness of the value of global forests in many aspects of society. Furthermore, the introduction of replanting programmes, the development of sustainable forest industries as well as more harmonious agricultural practices are all helping to encourage a global shift in the way this natural resource is being viewed.

21.3 Hydrosphere: Global Water Resource

21.3.1 Water Resources and Usage

Clean freshwater is crucial to all life; yet it represents only 2.5% of all readily available water on Earth (Figure 21.5). On a global geographic scale, water resources are small and randomly distributed relying heavily on precipitation patterns, climatic variability and whether ground, surface or desalinated water is used. Water underpins improved living standards, agricultural development and urbanisation and as a result is increasingly seen as a potential source of conflict (Rijsberman 2010; Tindall and Campbell 2010). Moreover, this finite resource is gaining more and more attention as a possible limiting factor in the growth, development and sustainability of populations. Indeed in the bleak words of the triennial publication the World Water Development Report they argue that,

> Water is linked to the crises of climate change, energy and food supplies and prices, and troubled financial markets. Unless their links with water are addressed and water crises around the world are resolved, these other crises may intensify and local water crises may worsen, converging into a global water crisis and leading to political insecurity and conflict at various levels. (WWDR-2 2006)

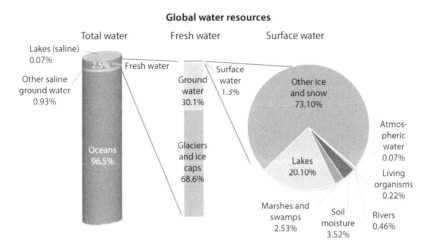

Global water resources

FIGURE 21.5 Global water division. (Compiled from Tindall and Campbell's *Water Security—National and Global Issues: US Geological Survey Fact Sheet.* US Geological Survey 6. 2010.)

On top of existing basic sanitation and water supply needs, currently the most important drivers of increasing freshwater withdrawals are population growth and rapid economic development. This development is creating pressure through population growth and increasing standards of living, which are then reflected in increased agricultural consumption. This has important connotations for food security, especially as water stresses begin to manifest. Moreover, as climate change exacerbates the changing dynamics of the hydrological cycle, freshwater distribution is also changing (WWDR-2 2006).

Per capita freshwater availability is determined by this hydrological cycle which provides the precipitation and the meltwater that collectively replenishes surface and groundwater resources. Total global annual freshwater resources in the form of precipitation over land are calculated to be about 110,000 km^3 per year. This comprises 70,000 km^3/year of green water (water in the soil) and 40,000 km^3/year of blue water (water in rivers and groundwater) (WWDR-2 2006; Lopez-Gunn and Llamas 2008). While some of this water goes towards replenishing the water stocks, some is withdrawn for human and animal use and some is effectively 'lost' through evaporation from the soil, evapotranspiration from vegetal matter, fatally polluted or simply when it joins the sea or another saltwater body (WWDR-2 2006). Within this cycle then there are thought to be about between 43,000 and 55,000 km^3 of global annual renewable freshwater depending on source (Rivera 2008). Of this, total global withdrawals are estimated to be about 3800 km^3, i.e. about 600 m^3 (571) per person per year or just under 9% of renewable freshwater. This figure of course is a global average, and the range of regional withdrawals is staggering, ranging from a mere 20 m^3 in Uganda to more than 5000 m^3 in Turkmenistan.

While agriculture consumes about 70% of all annual freshwater withdrawals, the remaining 30% is used for domestic needs—about 20%, 10% for industry and 1%–2% for the energy industry. Of the 70% water withdrawal for agricultural use, 29% goes to producing meat, 17% for meat products and 23% is used for the irrigation of mainly cereal crops. This irrigated land, while utilising 23% of the water, actually only nourishes about 5% or so of global agricultural lands (17/18% of croplands) while the largest majority of the rest (80%) is predominantly rainfed (FAO 2003b; Rockström et al. 2007; Lopez-Gunn and Llamas 2008; Muir 2010). Put in perspective, various estimates of total water requirements to produce the food needed to feed the current population every year ranges between 5200 and 6000 km^3. Thus, the shortfall in the 70% (2,663 km^3) of annual agricultural freshwater withdrawals is made up mainly from precipitation (Lopez-Gunn and Llamas 2008).

The problem is the increasing scarcity of available supplies. This belies the obvious face value statistics, which suggest that only 9% of the renewable 43,000–55,000 km^3 of water is being used. The difficulty is that water as in all natural resources is not distributed evenly with many regions now considered to be water-stressed. Perhaps 50 years ago, growth in water needs was met by things like increasing the number and sizes of dams that stored and channelled surface water where it was needed. By the late 1960s however, much of the world and Asia in particular began turning to underground water sources (38%) and the diversion of rivers (62%) to satisfy increasing irrigation needs. Today the trouble is many-fold for while there are indeed problems of efficiency, misuse and pollution are all taking its toll the over-whelming pressure comes from increased agricultural demand from growing populations and increased demands from better standards of living (Molden et al. 2007; WWDR 3 2009). This extra demand sees millions of irrigation wells around the world tapping into aquifers and the like operating nearly at, or beyond their renewable sustainable yield (Brown 2005).

With regard to aquifers in particular, these represent a major source of groundwater and as such it is worth knowing a little more. There are two types (replenishable and non-replenishable) and by far the largest source of aquifers, providing 75% of all this type of water, is the non-replenishable aquifers, also known as fossil aquifers. These are remnants of ancient wetter climes such as melt-ice from the Pleistocene ice ages and not being replenishable when they are depleted, they are gone (Brown 2005). Replenishable aquifers face their own set of challenges too in that water is often being withdrawn faster than is being recharged (Muir 2010). This is also becoming true of many river basins and lakes. So large in fact is the problem that the Aral Sea in Central Asia, once the world's fourth largest freshwater lake has to date lost 75% of its volume, owing largely to the diversion of water for irrigating cotton and rice fields (Muir 2010). Other examples of the global reach of this problem are highlighted in Table 21.3. As a

TABLE 21.3

Reduced Water Tables around the World

Global Water Deficits
United States: The Ogallala Aquifer in central United States runs from northwest Texas to South Dakota and supports about one-fifth of the irrigated farmland in the United States. Falling water levels and increased pumping costs has caused places like the Texas panhandle to revert back to dryland farming.
China: The world's largest grain producer is being forced to drill up to a 1000 m in some areas into the region's non-replenishable aquifers. Surveys have shown that aquifers in the Hebei Province in the heart of the North China are falling at an alarming rate with the World Bank estimating an annual deficit of as much as 37 billion tons.
India: Water tables are dropping by more than 1 m a year in Punjab (India's breadbasket) and several other states like Rajasthan, Gujarat, Haryana, Andhra Pradesh and Tamil Nadu. As a result, practices like double cropping wheat and rice may be restricted and may possibly force switches from water-hungry crops like rice to crops like millet or sorghum.
Pakistan: Shares the fertile Punjab plain and the fall in water tables in the region are comparable to those in India. Observations around Islamabad and Rawalpindi show falls of between 1 and 2 m a year from 1982 to 2000.
Iran: In the north-eastern agriculturally rich Chenaran Plain, the water table has been shown to be falling by 2.8 m a year.
Saudi Arabia: A 1984 Saudi national survey suggested that deep-well irrigated agriculture could last for another decade or so before the wells largely disappear. At this point, they will be forced to return to the shallow aquifers that are replenished by the kingdom's natural rainfall.
Yemen: The country's largest water table is falling by roughly 2 m a year while the Sana'a basin in western Yemen is falling by 6 m per year. The World Bank observes that groundwater is being mined to such an extent that parts of the rural economy could well be under threat within just a few years.
Israel: Is depleting the coastal and mountain aquifers that it shares with the Palestinians. Conflicts between Israelis and Palestinians over the water allocation are not uncommon. Depletion has reached such an extent that Israel recently discontinued the irrigation of wheat.
Mexico: In Mexico City's agricultural state of Guanajuato, the water table is falling by 2 m a year while at the national level more than half of all the water extracted from aquifers is not sustainable.
Egypt: Shortages of water in Egypt, which is entirely dependent on the Nile River, are well known.

Sources: Shah, T., Molden, D., Sakthivadivel, R. and Seckler, D. The Global Groundwater Situation: Overview of Opportunities and Challenges, International Water Management Institute, Colombo, 2000; Moench, M. Overcoming Water Scarcity and Quality Constraints: Groundwater—Potential and Constraints. *IFPRI 2020 Focus Brief no 9.* Washington DC: International Food Policy Research Institute, 2001; Brown, L., Water Deficits Growing in Many Countries: Water Shortages May Cause Food Shortages. *Great Lakes,* 2002; Brown, L., *Plan B: Rescuing a Planet under Stress and a Civilization in Trouble,* W.W. Norton and Company, New York, 2003; INTERFAIS, *Food Aid Flows:* World Food Program, 2003; Muir, P., *Bi301: Human Impacts on Ecosystems,* Oregon State University, 2010.

result of this inefficient and over use, scores of countries are running up regional water deficits to satisfy their huge demand.

As countries like Central Asia, the Middle East, North Africa, India, Pakistan and the United States continue to run up water deficits, farming practices are beginning to change on a global scale (Brown 2002, 2005). Indian farmers, for instance, are finding it difficult to expand their grain harvest while shortages in the Texas panhandle, Oklahoma and Kansas in the United States are forcing farmers to return to lower-yield dryland farming (Brown 2005). It can be seen then that huge unsustainable global freshwater withdrawals are placing a hefty burden on Earth's resources. In total, 80% of China's, 60% of India's and approximately 20% of the United States' grain harvest is cultivated on irrigated land. This translates to a total global grain harvest on irrigated lands of about 40%, in this context a water shortage threatens to become a food shortage. In fact, it has been suggested that since it takes approximately 1000 tons of water to produce one ton of grain, the importation of grain might be the most efficient means of offsetting a region's water shortage. As a result, future international grain trade could by proxy, become an international water market (Wade et al. 2002; Brown 2003, 2005). Lastly, to echo the words of Ban Ki Moon, Secretary-General of the United Nations, at the 2007 Asia-Pacific Water Summit,

> The consequences for humanity are grave. Water scarcity threatens economic and social gains and is a potent fuel for wars and conflict. (Lewis 2007)

Indeed areas of potential tension are already surfacing as China plans to use dams along the Mekong diverting water away from Myanmar, Thailand, Laos, Cambodia and Vietnam while separately India plans to divert water from the Ganges to its water-poor south, which will affect water received in Bangladesh. In the Middle East, Iraq is feeling the water stress caused by the damming of the Euphrates by Syria and Turkey (GFS 2011).

21.3.2 Sustainable Use

There are solutions to the problems of water scarcity, many of which are related to inefficiencies in irrigation, wastage and pollution. Shortages might be addressed in several ways. Irrigation efficiency, for instance, can be improved by transporting water in lined, covered irrigation canals or by using dripper systems or simpler still by matching crops to the ecological nature of the land more closely, thus requiring less irrigation. Other measures might see farmers raising animals on relative drylands and feeding them on a diet of grass and roughage instead of crops, which could help to considerably reduce the water burden. Desalination is also considered an effective and contributory solution to the problem of water shortages and presently about 1% of drinking water is produced by this method globally. At the moment desalinated drinking water produced in Ashkelon, Israel, costs around 52.7 cents per cubic meter and provides 13% of the country's domestic consumer demand, equivalent to about 6% of the countries needs. In Singapore, desalination plants currently satisfy about 10% of their water demand. Thus, with suitable advances and reduced economic costs of desalination, seawater could very well become one of the main sources of freshwater in the decades ahead (EJP 2005; EC 2006).

In this way, the lack of freshwater rather than land may end up being the principal constraint in efforts to secure food security and expand world food output. What this means in practical terms is that as competition for water among countries becomes increasingly intense so there is likely to be a shift in the way we view water. Water usage needs to become more efficient while new ways of tapping into the renewable hydrological cycle might also be found; however, perhaps more fundamentally water costs will need to go up, reflecting not only the increasing costs of developing new water sources but also reflecting the new inherent philosophical and economic value of this resource (Rosegrant and Cline 2003; Brown and Lall 2006; Khan and Hanjra 2009). This is similarly echoed by Lester Brown when he suggests that if we were to eliminate the artificially low prices of water by scrapping subsidies and raising water prices, this would result in water extraction to a point that ultimately becomes sustainable (Muir 2010).

TABLE 21.4

Water Footprints for Various Items and Areas

Item	Quantity	Virtual Water Usage (litres)
Bottle of beer	250 mL	75
Cotton T shirt	500 g	4100
A4 sheet of paper	80 g/m^2	10
Beef hamburger	150 g	2400
Pair of leather shoes	Each	8000
Rice	1 kg	3000
Wheat	1 kg	1350
Maize	1 kg	900
Meat—beef	1 kg	15,000–16,000
Meat—chicken	1 kg	6000
Country	**Total Water Footprint (WF) 10^9 m^3/yr**	**Per Capita WF m^3/cap/yr**
United States	696.01	2483
Greece	25.21	2389
Malaysia	53.89	2344
Afghanistan	17.29	660
Botswana	1.03	623
Yemen	10.70	619
Global Average	7452	1243

Sources: Hoekstra, A. and Chapagain, A., Water Footprints of Nations: Value of Water Research Report Series No.16, 2004; Water Footprint, Water Footprints of Nations: Value of Water Research Report Series No. 16, University of Twente, 2004; Hoekstra, A. and Chapagain, A., *Globalization of Water: Sharing the Planet's Freshwater Resources*, Blackwell Publishing, Oxford, 2008.

21.3.3 Water Footprint

A useful aid in quantifying water use is called the water footprint concept. It was first introduced in 2002 by Hoekstra and Hung as a consumption-based indicator of water usage. The central tenet is twofold: the first calculates virtual water use or the virtual water footprint of an item or object. This is the total amount of water used by plants or animals in order to grow and provide us with food while the same logic is also applied in the manufacturing and service sectors where water equivalents are calculated in the production or delivery processes. An example might be the water used to produce a gallon of beer or perhaps the bottle in which it is held. The second aspect of the water footprint is its ability to calculate the sum of all water used domestically as well as all imported water (virtual water) required to meet the needs of the goods and services for a specific area (Hoekstra and Chapagain 2004). This becomes the water footprint of an area and its introduction was a revolutionary new way in which environmental accounting empowered people to take stock of their actions and resources. In this way, water footprints can be calculated for items, individuals and areas as shown in Table 21.4 (Lopez-Gunn and Llamas 2008.

With this kind of data, it is now possible to calculate water footprints with a good degree of accuracy; for example an average vegetarian diet might represent about 800 m^3/year while a diet of red meat can reach the equivalent of about 1500 m^3/year (Lopez-Gunn and Llamas 2008). Also very telling in these calculations is the annual water footprint metric of 1243 m^3 per capita; compare this to the annual water withdrawals of approximately 600 m^3 and it becomes immediately clear that we are consuming more than twice as much water as we are withdrawing annually and while in water-rich areas this may not be a problem, in water-stressed areas this is often at the expense of diminishing resources.

21.4 Atmosphere

With regard to the atmosphere and food security, the main issues concern changing climatic conditions. The basic premises on which the phenomenon exists and whether it is largely due to natural climatic

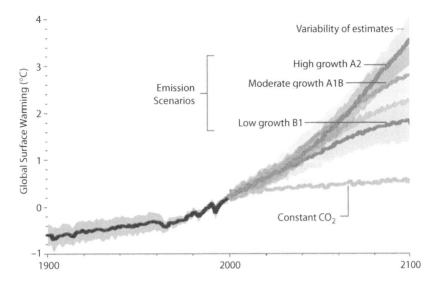

FIGURE 21.6   Projected global climate change. (Courtesy of Climate Change 2007: The Physical Science Basis. Working Group I Contribution to the Fourth Assessment Report of the Intergovernmental Panel on Climate Change, Figure 10.4. Cambridge University Press.)

variability or anthropogenic causes is debated in detail in Appendix A. Suffice to say current consensus leaves little doubt as to the veracity of the underlying assumptions and conclusions, which experts at the Intergovernmental Panel on Climate Change suggest could see global average estimated temperature rises of between 1.1°C and 6.4°C by 2100 (Figure 21.6) (IPCC 2007; Lynn et al. 2007; Roach 2007; Zhang et al. 2008; EIA 2011). Much concern these days focuses on the potential consequences of this warming. Such changes have possible grave implications for food security and in general the situation s forecast to create new food insecurities in the coming decades (von Braun 1995b; Rosegrant and Cline 2003).

Water vapour, not withstanding the top three contributing gases, to global warming are carbon dioxide, methane and nitrous oxide. Most greenhouse gases have both natural (bio- and geochemical processes) and anthropogenic (human) sources. Of the anthropogenic sources, Figure 21.7 highlights the different sectors responsible for the varying amounts. It can be seen from this that agriculture plays a considerable role in the accumulative total; this ultimately has consequences for the future sustainability of the sector especially as future projections see these figures rising (EIA 2011).

Of these figures, it is the agricultural livestock that accounts for a considerable proportion (9%) of the agricultural sector's overall greenhouse gas emissions. Between the various species, mainly ruminants and to a minor extent monogastrics, these represent a full 37% (86 million tonnes) of all human-induced methane, mostly in the form of enteric fermentation (gastric gases) and some 65% of all human-related nitrous oxide and 64% of ammonia gases (FAO 2000, 2006; Foresight Project 2011; UN 2011).

21.4.1 Consequences

So what are the consequences of these projected rises? Temperature and rainfall are critical determinants of cropping seasons worldwide, and any extremes in heat or floods, early frosts or late thaws serve to test the limits of many crop varieties. In some extreme cases, changes could even transform entire regional cropping patterns (DeRose et al. 1998b). Presently, summer drying and associated risks of drought and floods are now the two single most common causes of food production shortages in the world and any increase therefore in global warming would further negatively affect agricultural and rangeland productivity (Wolfson and Schneider 2002; WFP 2011). The bad news is that this is likely to happen; based on a range of models it is likely that tropical storms will become more frequent and more intense and following observed trends, move poleward with resultant changes in wind, precipitation and temperature patterns.

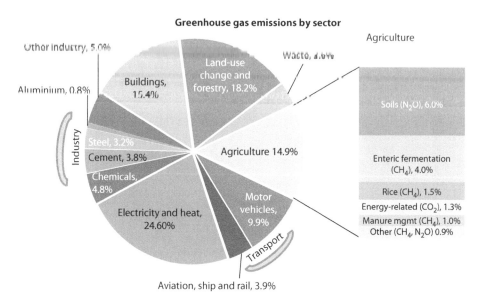

Greenhouse gas emissions by sector

Uther Industry, 5.0%

Aluminium, 0.8%

Buildings, 15.4%

Land-use change and forestry, 18.2%

Waste, 3.6%

Agriculture

Soils (N_2O), 6.0%

Industry

Steel, 3.2%

Cement, 3.8%

Chemicals, 4.8%

Agriculture 14.9%

Enteric fermentation (CH_4), 4.0%

Rice (CH_4), 1.5%

Energy-related (CO_2), 1.3%

Manure mgmt (CH_4), 1.0%

Other (CH_4, N_2O) 0.9%

Electricity and heat, 24.60%

Motor vehicles, 9.9%

Transport

Aviation, ship and rail, 3.9%

FIGURE 21.7 GHG emissions by sector. (Compiled from data in Baumert, Herzog et al., *Navigating the Numbers Greenhouse Gas Data and International Climate Policy*, World Resources Institute, Washington, DC, 2005.)

Playing to the worst case scenario offered in some publications, such geographically displaced heavy precipitation events are likely to have devastating consequences. This would be nowhere more true than in the low-income countries where extant poverty could further aggravate limited adaptive capabilities (von Braun 1995b; Easterling et al. 2007). Furthermore, the recent global mean sea level rises of 1.7 mm/year are likely to increase, mainly through thermal expansion of the seas but also from more glacial and polar ice melt. This has implications for global oceanic circulation as well as the snowmelt-dominated watersheds. Coastal erosion and flooding would also increase with the resultant loss of wetlands, mangroves, reefs and other important coastal ecosystems. The increased risk of storm surges would also almost certainly threaten coastal settlements, which could see tens of millions of people displaced. Moreover, rising sea levels too would threaten to seep into groundwater, affecting irrigated lands in many of the densely populated low-lying countries of the world. It has been suggested that a rise as small as 1°C above the historical average could result in the displacement of India's monsoon rains, for which it relies on for 70% of its water supplies (Muir 2010). This same 1% increase could also see grain yields in certain area's decline by as much as a 10% (Brown 2005; Dyer 2006).

Other extremes too involving shifting La Niña and El Niño weather events, heat waves, droughts and decreases in precipitation in many subtropical land regions may very well exacerbate desertification, forest fires and land degradation. All this would place an enormous strain on wildlife, plants, ecosystems and natural resources in general. Significant extinctions of plant and animal species too would be expected as they find themselves unable to migrate or keep up with a fast changing climate. Alpine glaciers and large permafrost tundras would come under strain too and could ultimately disappear forever.

Although some crops would benefit from modest warming, warmer climates could also accelerate the invasion of alien weeds in some areas. In sub-Saharan Africa in particular there is also concern that warmer climes might actually shorten the growing seasons of staple crops, further reducing food availability in certain regions while in South and East Asia, retreating Himalayan glaciers threaten to adversely affect the water cycle (FAO 2003a; Rosegrant and Cline 2003; Gregory et al. 2005; Ingram et al. 2008).

Finally, the world would be a different place to visit too with possible changes in traditional summer tourist and winter ski destinations. Vector-borne infectious diseases could witness alterations in their geographical distribution increasing previously unexposed populations to diseases such as malaria, dengue fever and cholera. All in all, the worst case scenario is one that grips many with fear and this does not even begin to consider the economic consequences of more frequent extreme weather events, leave alone their effect on the food security of nations.

All very scary however, alternative views balance this argument by accusing the IPCC, Brown and others of scaremongering and creating something out of science fiction (Verweij et al. 2006; Kininmonth 2008; Booker 2009). Indeed many debate whether we are in fact underestimating agriculture's capacity to adapt to environmental stimuli and in this point it has been illustrated that the Vikings for instance, once farmed the lush meadows of the now frozen Greenland. Other practices of deliberate translocation of crops and rapid substitution are often cited too, along with the ability to adapt to resource substitutions like dryland farming instead of irrigation (Easterling et al. 2007).

In weighing up the evidence, the potentially deleterious effects of climate change on food security is proving difficult to anticipate. Compounding this is the fact that global average temperature rises characterised as a single number, while useful can be misleading as in reality global warming will vary substantially from one region to another. Climate change will ultimately be both good and bad for food production depending on region and farming practices; however, of more concern perhaps will be the rate of change. The goal here would be to establish new agricultural zones faster than others are lost (GFS 2011). Thus in the end only preparation and adaptability are going to give us the tools to mitigate the worst effects of future climate changes on our daily lives but also on food security in particular (Wolfson and Schneider 2002; Rosegrant and Cline 2003; IPCC 2007; World Bank 2009; NOAA 2011).

21.5 Biosphere

21.5.1 Biodiversity

The main preoccupation concerning the biosphere and food security revolves around issues of biodiversity. Biodiversity belongs to us all; it is a global, regional and individual asset. Indeed, it is biodiversity that supplies the plant, animal and microbial genetic resources we need for food and agricultural production. It powers the many ecosystem processes such as nourishing and fertilising the soil and regulating pests and disease. It also helps control erosion, provides energy for our modern world and pollinates many of our crops. It is this relationship that in the words of Chappel and LaValle:

> The two problems of biodiversity loss and food insecurity are global in scope and cannot be viewed independently: in a world with limited resources the methods used to address one necessarily involve choices affecting the other. (Chappell and LaValle 2011)

Unfortunately though in our ignorance and the collective choices we have made over the centuries, we have contributed to a huge and consequential loss of biodiversity. Estimates of this impact currently range from several hundred times the natural rate of loss to between 1000 and 10,000 times. One of the underlying drivers of such biological genocide is the veracious capacity of us humans in commandeering large proportions of Earth's natural resources for our exclusive benefit. Currently, we have seen that about 34% of Earth's available land is being used for agriculture, with a result that a good proportion of this is to some extent degraded; moreover, we are currently running up freshwater deficits from nonrenewable aquifers and polluting others, further aggravating the situation. On top of this, net deforestation takes place to the tune of about 5 million hectares a year to accommodate things like agriculture while marine life has been severely overfished to the point that it threatens future capacity. Furthermore, as communities relinquish traditional lifestyles and agricultural practices in favour of intensification, we are proactively and forcefully attempting to manage immediate cropping ecosystems with huge inputs of fertilisers and pesticides. What is more, of the many thousands of species available to humans for food we are concentrating most of our eggs in one basket of less than two dozen species at the expense of wider dietary variety.

Wild estimates aside, there is some consensus as to the numbers of species of living organisms on our planet. These total approximately 1.75 million known species with perhaps as much as 14 million as yet unidentified (Appendix G). Indeed, currently over

> ... 80% of all endangered birds and mammals are threatened by unsustainable land use and agricultural expansion. (Nellemann 2009)

The present day loss rates mentioned above would take tens of millions of years to fully recover, yet we sometimes fail to realise that as a species our own future surely depends on our ability to conserve this rich biological diversity. We serve them as they serve us. Fundamental to this symbiotic relationship once again is the recurring notion of the carrying capacity of Earth and a sustainable limit to population numbers.

So far reaching is this loss that a global Convention on Biodiversity (CBD) was adopted at the UN Conference on Environment and Development in Rio de Janeiro in 2002. This effectively provided a road map that required all nations to subject development programmes to a biodiversity impact analysis in order to ensure that 'progress' was not linked to further biotic loss. Interestingly too, with regards to food security, one viewpoint from Chappel and LaValle convincingly argue that while agriculture is an important driver of biotic losses, a solution sees alternative rather than conventional means of agriculture as an under tapped resource that could simultaneously provide sufficient calories to achieve global food security whilst offsetting the worst effects of this loss of biodiversity (Chappell and LaValle 2011).

21.6 Energy

Energy, its use and availability, its cost, the possible environmental aspect as well as a number of other closely related issues ensure that it remains of major consideration in terms of food security. After all, energy is utilised at every level of food production, processing, transport, storage, retail and eventual cooking. Ensuring a good, stable, cost effective, environmentally acceptable source of this energy is a challenge beyond the scope of this book. However, an introduction to some of the issues is outlined here. Currently 86% of the world's energy needs are provided for by fossil fuels; yet the question is frequently raised regarding the continuity of these existing sources with regard to availability and pollution (WEO 2008, 2010). The debate is fairly divisive among both camps. With regards to availability, one camp suggest that fears of peak oil* production have come and gone with such frequency that oil deposits still confound sceptics; in this argument, newer more economically viable deposits are continually being found to offset such fears. The other side of the debate focuses on the positive aspects of shortages as giving impetus and momentum in the quest for alternative energy supplies such as renewables. The former argument tends to miss the longer term issues of sustainability and ignores the issues of pollution altogether; yet the latter argument is often predisposed to ideological persuasions that are sometimes emotively charged rather than reality driven. In time peak oil will undoubtedly come to a pass (if it has not already) and more emphasis will indeed be placed on these renewables. Until then however, fossil fuels are intrinsically woven into the economic fabric of the modern world that even the very real threat of global warming has had, or is projected to have little effect on its daily usage (see Figure 21.8).

One of the difficulties with reducing fossil fuel dependency is the sheer flexibility that it affords us. Take Figure 21.9 for instance and it can be seen that between both oil and gas the sheer number of uses these products can deliver becomes apparent. Moreover, sectors using oil and gas as feedstocks for such products are already heavily invested in terms of both technological and infrastructural. In this way, weaning industry off these feedstocks and into alternative sources is going to be a long and slow transition. Of course, this is an oversimplification and indeed there are many alternatives and substitutes

---

* Peak oil is the concept that attempts to look at the big picture of oil production. Peak oil is said to have been reached when we cannot sustain existing production of easy accessible oil and supplies begin to decline due to more difficult access, more complicated extraction processes. It is a contested maximum with some suggesting we have already peaked while others insist we still have many years yet to go.

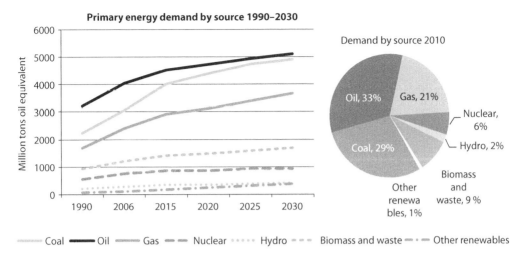

FIGURE 21.8 Global energy demand by type 1990–2030. (Compiled from WEO (2008). *World Energy Outlook 2008.* The International Energy Agency (IEA), Paris, 578, 2010.)

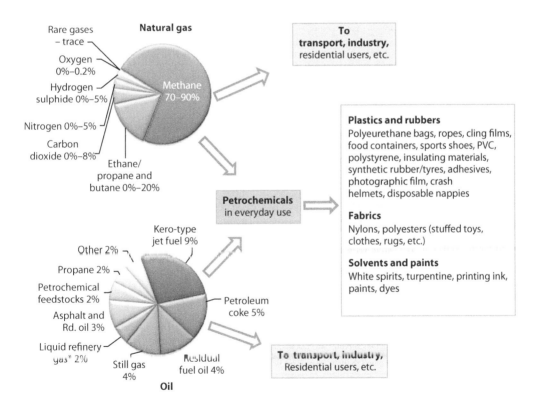

FIGURE 21.9 Oil and gas typical components and uses. (*Source:* Created from multiple datasets (APPE 2011; EIA 2011; Naturalgas 2011).) *Notes:* *Distillate fuel oil includes heating oil and diesel fuel. Liquid refinery gases include ethane/ethylene, propylene, butane/butylene, and isobutane/isobutylene.

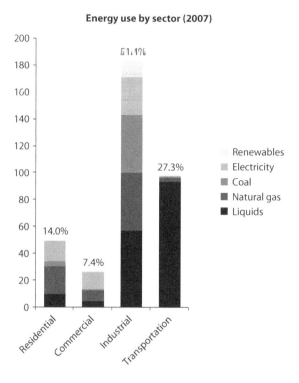

Energy use by sector (2007)

FIGURE 21.10   Percent of global energy consumption by sector. (Compiled from WEO (2008). *World Energy Outlook 2008*, The International Energy Agency (IEA) 578, Paris, 2010.)

available for many if not all of these processes; yet as we have already mentioned, the ties to these resources have far-reaching roots. For further discussion on energy (Appendix J) looks at the different sources of energy and their relative availability, strengths and limitations.

On a per sector basis, it can be seen that by far the biggest consumer of oil in 2007 was the transport sector utilising about 57% of all global production of liquid fuel. Industry on the other hand used the bulk of the coal produced in the same year as well as a good proportion of the natural gas and electricity supplies. Considering the varying energy values of oil and coal, etc. and calculating all this on British Thermal Unit (BTU) equivalents (Figure 21.10), it can be shown too that industry, consisting of manu-facturing, agriculture, mining, and construction and agriculture by itself consumes just over half the total global energy budget at 51.4%. Of this figure, energy used directly in agricultural activities has been variously estimated at between 3% and 4%; however, once we factor in indirect usage this figure can rise to as much as 26% (Ho 2008).

21.6.1 Energy in Agriculture

Although not immediately obvious at first the extent to which energy costs permeate the whole of the agricultural process is extensive, especially once we factor in this indirect usage. As an example, the production of ammonia for fertiliser consumed about 5% of global natural gas in the 1990s representing a little over 1.5% of total world energy production at the time. In total such fertilisers are generally the biggest energy costs within direct agricultural use as is shown in Table 21.5 (Sundquist and Broecker 1985; McLaughlin et al. 2000; Muir 2008).

Furthermore, factoring in things like food processing, packaging, storage and distribution the extent to which these figures are multiplied becomes a little clearer (Ho 2008). Affordable food production by industrial agriculture then relies on constant access to inexpensive energy. Energy-hungry livestock, feedlots, industrial fuel and 'food mile' costs together with inefficient irrigation systems suggest this

TABLE 21.5

Direct Agricultural Energy Consumption Broken Down
by Agri-Processes

Energy Consumption
31% for the manufacture of inorganic fertilisers
19% for the operation of field machinery
16% for transportation
13% for irrigation
8% for raising livestock (not incl. livestock feed)
5% for crop drying
5% for pesticide production
8% miscellaneous

Source: McLaughlin, Hiba et al., *Canadian Agricultural
Engineering*, 42(1), 2000.

level of energy intensiveness has implications beyond the idea of food security and agricultural sustainability, instead it transcends into issues of climate change and energy conservation as well (Eifert et al. 2002; HSUS 2008). With regard to energy creation, renewables like solar, wind, water and biofuels have had a long history. Indeed, as existing reliance on fossil based technologies becomes ideologically or economically unacceptable, we humans are a creative breed and we may even revert once again to simpler earlier ideological times of the 1930s as this example from Brandt highlights:

> It is within man's capacity to make adjustments and to invent new solutions. For a more adequate production of foodstuffs it is possible to intensify agriculture and horticulture; for power and fuel it is possible to shift from 'deposit resources' to 'flow resources' and harness the latent energies of lakes, rivers and tidal sites. (Brandt 1939)

However, even despite the financial and geopolitical risks inherent in fossil based energy, it is at the moment here to stay and as such intrinsically affects agricultural costs. One worrying concern both from a cost and environmental concern involves today's ever-increasing industrialisation of agriculture (CFS 2007). As more and more inputs are used in the farming methods, the health of the soil becomes strongly dependant on continuing inputs, without which the soil quickly degrades. This creates a cycle of dependency that in order to continually achieve greater yields characterised by the green revolution, more and more energy inputs from fossil fuels becomes the norm (Eifert et al. 2002). This is especially disconcerting when we consider that crop production uses approximately between 2% and 5% of the commercial energy in nearly all countries irrespective of their individual level of development (Vlek et al. 2004; Modi et al. 2007).

There are also other areas of energy creation that need consideration too.

21.6.1.1 Biofuels

One particular concern that has been publicly raised over recent years regards the relationship of biofuels vis-à-vis fossil based energy sources, climate change, food security and competition. While bioenergy production accounts for about 10% of global energy supply biofuels by themselves account for about 0.2%; this equates to about 1.5% of all road transport fuels. Moreover, in the last few years, biofuel cultivation occupied 2% of global cropland and utilised between 5% and 7% of all global coarse grains as well as 9% of all vegetable oil produced (CFS 2008; FAO 2009a). While nearly every country in some measure uses biofuels, it is Brazil that currently leads the world in this field. In 2005, all gasoline sold in Brazil contained at least 26% ethanol (Morgan 2005) and as of January 2008 all forecourt diesel in Brazil was mandated to contain at least 2% biodiesel (Biopact 2008). The United States follows with biofuel satisfying about 3% of the national supply while the European Union and others are ratcheting up production in the face of increasing oil prices (BBC 2007; Fritsch and Gallimore 2007; Brown 2008; EERE 2008; EIA 2011).

More is discussed about the creation of biofuels and the various considerations in Appendix J.6; however, one of the main questions being asked is that if biofuels are so good for both the economy and the environment, why are they not more widely available? The answer is twofold: first, given the costs of raw materials and energy use, coupled with forecourt duties (in many countries) actually producing biofuels costs nearly twice as much as those from fossil fuel (DEFRA 2003, Muir 2007). Second, there is much contention over the emission benefits of current biofuel products. On face value, biofuels are a great solution to a worsening problem of greenhouse gas emissions and high oil prices with many people, governments and NGOs espousing their benefits (Roubanis 2004; Stauffer 2007; National Geographic 2008; FAO 2009a). However, there are those who contend that biofuels in practice offer only small or negligible benefits in cost reduction on a life cycle basis. As highlighted in the appendices, first generation biofuel production is at present fairly inefficient and costs more in some areas to produce than conventional fuels. Also in question is the basic premise that biofuels are far superior in terms of CO_2 emissions with some suggesting such environmental gains are, by the numbers negligible. Indeed Searchinger (2008) concludes that CO_2 reduction benefits of first generation biofuels have not only been shown to be negligible but in some instances negative (Dyer 2006; Brown 2008; Smith and Edwards 2008). Also, as an aside, it has been suggested that because much of the energy used in biofuel production comes from fossil fuels, biofuels do not actually replace, in energy terms, as much oil as they use. Moreover some people fundamentally object to the diversion of food and land resources and see it as morally at odds especially in view of the globally malnourished. Others however, particularly concerning next generation technologies expected to be commercially available by 2020, see the combustion, pyrolysis, hydrolysis and gasification of residual crop biomass as becoming more and more clean and efficient and consequentially its potential as a valuable and sustainable source of carbon-neutral or even carbon-reducing energy is a given (CFS 2007; Mateos and González 2007).

In sum, then as more and more countries incentivise the expansion of the bioenergy sector, the effect on food security will depend closely on the types of technologies and feedstock used. How this compares to traditional fossil based energy sources depends to a large extent on the biofuel policy environment as well as petroleum prices (CFS 2007). This interplay was noted in 2008 where it has been posited that as much as one-third to perhaps three-quarters of food price rises seen in recent years have been attributable to biofuels (Chakrabortty 2008; Lovgren 2010). Interestingly too, as an aside, in the food fuel conflict, one quoted statistic is worth recounting here and that is the grain required to fill the tank of a 4 × 4 vehicle with approximately 100 L of ethanol could provide enough maize (240 kg) to feed one person for a year (Dyer 2006). Although this does not quite fit with the author's calculations detailed earlier, it is nevertheless a telling metric.

21.7 Environmentalism and Sustainability

21.7.1 Environmentalism

Environmentalism and sustainability have become buzzwords over recent decades and together they are collectively fostering a new self-awareness regarding humanity's impact on its surroundings. Technically, the two things are separate; yet they share considerable overlap that both environmentalism and sustainability are often considered together. Environmentalism tends to focus on the immediate physical surroundings with the well-being of the various elements that make up the numerous properly functioning ecosystems. It is concerned with the land, the air, the water and all its biota in respect of the Earth's overall health vis-à-vis pollution, degradation, optimisation and other social benefits for all. In short, the environmentalist movement involves taking a balanced holistic and philosophical approach to its continuing maintenance and conservation. Sustainability on the other hand looks at the environment from a natural resource usage perspective. In this sense, sustainability aims to ensure that resources are used in a sympathetic way that neither over stresses their future potential for provision nor degrades them to the point of collapse. In this respect, sustainability seeks to maintain the continuity of resources for both the present and future populations.

In response to increasing population figures, the second half of the twentieth century saw the doubling of global food production. However, this initial advancement was achieved without a great deal of consideration to the environment or to sustainability; indeed we have already discussed the various impacts of agriculture and other industries on the environment with the overreliance on fossil fuels, the overuse of chemical inputs as well as or from improper water management among many other issues. Consequentially all this left a significant environmental footprint on the ecosystem (Khan and Hanjra 2009).

Of course, such issues were tentatively picked up early on in the twentieth century as we have discussed with Rudolph Steiner's organic bio-dynamics in the 1920s as well as Rachel Carson's attack on questionable agricultural practices of the 1960s. However, it was perhaps not until the 1970s and the advent of the United Nations 1972 UNEP Conference in Stockholm that the issues were readily received by a more receptive audience. By the 1980s the idea that humanity's ever-expanding food needs had negative consequences for the environment were now firmly established and the introduction of newer more sustainable agricultural practices as well as conservation programmes began to be accepted in the mainstream. Out of this the idea that, as a species, we needed to modify our attitude towards the environment was an easy sell. In this way, it can be said that both environmentalism and sustainability can be thought of as an ideological movement firmly grounded in the practicalities of common sense. At this point in time too food security, through agricultural practices was already becoming increasingly linked to wider environmental considerations, climate change and also sustainability (Maxwell and Frankenberger 1992).

As a consequence of such progress, environmentalism and sustainability now enjoy widespread consensus as a priority in policy decisions which aim to improve standards of living, food security and numerous other goals (Nathoo and Shoveller 2003; Ericksen 2008). However, this balance between progress and sustainability becomes a juggling act with many disagreeing over the extent of governance and solutions required. While the ideas of the movement are generally and widely accepted, poor practices and financial considerations still prevail. In light of this, recent years have witnessed a backlash against the high energy, high-intensity agricultural methods that collectively characterise much modern conventional agricultural systems. This in itself is leading to ideas of a modern renaissance in regards to a broad category of alternative agricultural systems. These range from traditional or indigenous style farming to organic and include any number of options in between such as low-input, agro ecological, integrated pest management (IPM), no-till/conservation agriculture and the resurgence of bio-dynamism among others (Chappell and LaValle 2011). These holistic systems are espousing the new 'evergreen' revolution, which aims to be more sustainable and longer lasting than the previous green revolution (GHI 2010). Fundamental to this new ideology is the recognition of the interrelatedness of agriculture and the environment and within this new paradigm practitioners look to reduce costs, protect the environment as well as enhance natural processes through recycling and other complementary practices. Moreover while some see the many millions of smallholder farms as central in this quest this does not preclude the many modern industrial farmers, who with more consideration of environmental consequences would also contribute to a more lasting solution in the quest for global food security (IAASTD 2009; Chappell and LaValle 2011).

It is within these broad considerations that consumers operate with duality of purpose with many adopting a low or zero mile carbon footprint lifestyle while many more choose to feast on a diet rich with non-seasonal produce irrespective of geographical origin. Others still regard country of origin or a particular brand name as a marquee of quality and so link their purchasing habits to these perceived traits. This is the polychotomous decision tree we all encounter and is one that is going to play an increasingly important role. This will be particularly important as growing wealthier consumers' demands for out of season produce, convenience and quality tend to have consequences beyond the till and which effectively helps shape the landscape of food production on a global scale. What is needed in this respect is to build on existing educational efforts for more widespread awareness of the impacts of our purchasing decisions.

One method of looking at the Earth's limitations with regard to population and its natural resources is the concept of environmental or ecological accounting. We have already discussed the use of water footprints and virtual water, yet there are other methods too which include more footprinting methodologies of carbon and ecological varieties, as well as National Environmental Accounts or satellite accounts to the main National Accounts.

21.7.2 Environmental/Ecological Accounting

21.7.2.1 Ecological Footprint

The present global population of 6.7 billion is growing by about 80 million people per year and is set to increase to 9.2 billion by 2050 (ESA 2007; GEO 2007a). The demographic split however will see less developed regions absorbing more of this growth (from 5.4 billion in 2007 to an estimated 7.9 billion in 2050) compared with more developed regions, which are expected to remain static at around 1.2 billion (ESA 2007). The rural-urban shift is on the increase too, for while close to 50% (3.3 billion) of people currently (2008) live in urban areas, this is set to grow to approximately 61% within the next three decades (FAO 1995; von Braun 1995a; Cohen 2006; UNFPA 2007). As has been discussed, the Earth's carrying capacity is difficult to estimate with any degree of scientific consensus. Instead, Cornell University offers an alternative and more promising approach that calculates the Earth's ecologically productive areas holistically and combines them to arrive at an ecological footprint. One overriding advantage of this system is that it recognises multiple constraints on the Earth's productive ability. Put another way, the ecological footprint can be seen as an analysis of human demand on the Earth's natural resources via an ecological accounting system that determines the footprint against the bio-capacity (or biologically productive area) of the world (WWF 2006).

In methodological terms, an ecological footprint can be determined for each person or nation. By aggregating the total biologically productive land and water areas used in various ecosystems needed by a nation to generate all the resources it requires and to absorb all the waste products it generates. These measures can then be converted to an equivalent land area and can be compared or measured against all the ecologically productive land and water on the planet. Such measures are then suitable for comparison on a per-caput, national or global basis. Using this method, it was calculated that globally in 2003, approximately 11.2 billion hectares of biologically productive area was available (Kitzes et al. 2007). This represented nearly 1.8 hectares for every person on Earth (Table 21.6); in the same year however, humanity it transpires asked too much of the Earth requiring products and services equivalent to 14.1 billion hectares or 2.26 hectares per person. Others have gone further and have suggested that, for many years now, populations have been sustained at the cost of the of the depletion of the Earth's natural capital resources (Myers 1984; Jones and Wigley 1989; Postel 1989; Ehrlich and Ehrlich 1990; Schneider 1990; Jacobs 1991; Kitzes et al. 2007).

The concept of ecological footprinting however has been challenged on its use of crude estimates and the idea that the model generally seems not to factor in multiple uses of the same land. Despite these criticisms, environmental footprinting remains a popular metric and one that is continually being improved upon.

An alternative type of environmental accounting runs parallel to national economic accounting systems (NA) and is referred to as satellite accounts (SEEA 2003). These are increasingly appended to many countries' NAs and seek to detail metric, and where possible, monetary aspects of the environmental

TABLE 21.6

Earth's Total Biological Productive Capacity

Earth's Total Biocapacity	
11.2 billion global hectares of biologically productive area comprising:	2.4 billion hectares of water (ocean shelves and inland water)
	8.8 billion hectares of land comprising:
	1.5 billion hectares of cropland
	3.4 billion hectares of grazing land
	3.7 billion hectares of forest land
	0.2 billion hectares of built-up land

The amount of biocapacity available per person globally is calculated by dividing the 11.2 billion global hectares of biologically productive area by the Earth's population (6.3 billion in 2003). This ratio gives the average amount of biocapacity available on the planet per person:

= 1.8 global hectares per person

Source: Kitzes, Peller et al., *Science for Environment and Sustainable Society,* 4(1), 1–9, 2007.

system. These accounts still sometimes use some of the footprinting methodologies as well as proprietary methods of their own; however, combined the accounts seek

> to measure the contribution of the environment to the economy and the impact of the economy on the environment (UN/DESA 2011)

Whichever accounting system is employed, the real achievement here is that the environment is being monitored and accounted for. This can only be a good thing and one that helps determine humanity's effects and perhaps limits our abuse of the natural habitat.

21.8 Discussion

It should not come as a surprise that Earth around us is changing. Global warming, loss of biodiversity, land degradation and pollution are all symptoms of 'progress'; yet this progress is clearly unsustainable and threatens to swipe the rug from under our feet. In terms of energy use, we have raped and pillaged the Earth's stock of carbon based fossil fuels; fuels which took many millennia to form are being squandered in a few short centuries. Another interesting metric, this time from Zhang suggests that as much as two-thirds of the entire natural resource base on the planet has already been exhausted through over-exploitation. Of this figure, it seems 46% has been attributed to population growth and over half, 54% from better living standards (Zhang 2008). As a result, one of the greatest challenges facing humanity in the twenty-first century is how to feed and maintain a projected population of 9.2 billion by 2050 in a continuing and sustainable fashion (IAASTD 2010). Equally when it comes to the environment, the natural resource base and food security the issues are widespread and tap into many aspects of the agricultural, societal, economic and political debates. Many of these issues are also often highly interdependent and should where possible be addressed together.

The agricultural industrialisation, which has taken place in the past among open optimism of the benefits of all, initially gave little nod to the consequences of environmentalism and sustainability. Moreover, the resultant drain on fresh water resources; the reduction in biodiversity through simplification of production; and the acres of erosion and pollution of the water supply have borne witness, according to many, of the negative impact of the global open market (Windfuhr and Jonsén 2005). In the meantime, the pressure to produce more for less is continuing the trend of agricultural concentration in the hands of the big commercial interests, from large-scale landowners to multinational processors and retailers. In fact, all along the food chain, Windfuhr and Jonsén contend it is not the smallholders but rather these big interests that are the worst culprits in the woeful destruction of the environment, although conversely, debate in this research has already shown that in fact it is a charge that applies equally to those, big and small.

In terms of the social and economic costs to the environment, the continuation of business as usual does not bode well, and indeed is now widely understood to be unsustainable. In fact, as the literature attests, few people doubt there needs to be an alternative more sustainable way of providing food security that does not negatively impact on the future productivity of the natural resource base. In this way, a symbiotic relationship is encouraged whereby protection of the various ecosystems will ultimately positively contribute to future security. As it stands, if little changes one UNEP report *The Environmental Food Crisis* of 2009 highlighted the extent of the problem when it suggested that a shortfall of as much as 25% of food production could result by 2050 as a result of the combined consequences of climate change, land degradation and cropland losses, water scarcity as well as pest infestations (Nellemann 2009). This poses dilemmas of hitherto unimaginable proportions and if we are not careful, compensating for such losses could see agricultural practices becoming more intensified or more land coming under the plough (Ericksen 2008). Worse, as a result of these increased pressures, future food prices might well be raised by as much as 50%, increasing the vulnerability of the poorest people. Of course, the challenge here are multiplied when we consider the much talked of production increases needed to feed future populations. In this respect, all areas of production from soil management, pollution and degradation to storage and transport with freight miles, losses and emissions to consumerism with choice, profligacy and dietary habits needs to be challenged. Only by the continual challenging of current practices can sustainability issues be readily addressed.

In balancing economic growth and ecological security, we must encourage food security at the lowest possible environmental impact. In terms of land use and land change, there is a trade-off with constantly changing land needs for development, agriculture and leisure. This means that making such decisions needs to be based on the whole environment, that is to say every country must recognise their role in the feedback system of the whole planet. Any decisions to alter land usage must surely be reciprocated with concomitant decisions that replenish say the albedo or vegetation index This is grossly oversimplified but for illustration purposes if each country were to take responsibility for its part in the overall equilibrium of the planets radiative, greenhouse gas and atmospheric balance for instance, then the worst effects of climate change might be mitigated. The overriding difficulty in such attempts though is one of the unequal distribution of the natural resource base vis-à-vis land resources and needs. One answer would possibly see a system similar to that of the carbon trading market whereby states alter land usage, deforest and afforest in a sort of international trading system.

In addressing pollution and degradation, education and legislation are already in place in many industrial countries; this needs to be rolled out throughout the world. International standards must be set and the testing and enforcement of these standards must be maintained and expanded.

When it comes to agricultural methane it can be seen that globally, we consume just over 200 million cattle, a small proportion (less than 20%) of total actual numbers annually; however, considering the numbers of cattle are set to increase to perhaps 2.6 billion by 2050 this has consequences for future greenhouse gas emissions (Foresight Project 2011). Indeed as has already been observed the livestock sector is currently responsible for the biggest source of anthropometric methane emission in the world. Changing diets place upward pressure on this situation, so in this situation such things as methane sinks woodland soils or changing diets take on greater importance. Another aspect of emissions concerns the overreliance of fossil-based fuels in agriculture. While there is much talk of renewables and energy efficiency, the truth is fossil-based technologies are still cheaper than many of the leading alternatives. Sadly I am all too aware that only when oil becomes either toxic in the political sense or reserves start to run dangerously low that real alternatives will only then start to have a real impact. The question will be then can we afford, environmentally, to wait the 40, 50 or 60 years it might take for this to happen.

In terms of water management, we have already talked about the under-valued commodity and the possible solutions of bringing desalination within the zone of economic viability. On top of this though, the recent increase in land grabbing and import preferences over domestic production based on water stresses is turning the food chain into a proxy international water market. This might be seen in two lights. First, this signs up to the principles of comparative advantage economics and on paper is not necessarily seen as a bad thing. By importing water through food, this clearly is a more sustainable option for those water-stressed regions and just as importantly it also starts to address the economic undervaluing of water. On the other hand, placing both food and water at the mercy of international trading systems might be viewed as fraught with uncertainty and volatility. The market is already seen by many as an imperfect mechanism and outsourcing both food and water needs might just prove overreliance too far for the financially challenged, water-stressed regions of the world.

Environmentalism then, has been the spur for the modern notion of sustainability and one promising outcome of this new-found paradigm of accountability is the occurrence of environmental or ecological accounting. A good example of this approach is the footprinting network championed by the likes of Hoekstra and others. Other advances also include the satellite records to the National Accounts, detailing economical and virtual costs to the environment. Methodology this area though is relatively young and is being constantly refined and as this develops so more acceptable and widespread accounting techniques promises to revolutionise natural resource monitoring and accountability. This is no less true of the need to account for the loss of biodiversity. Monocultures would benefit from intercropping, integrated pest management systems and a other practices. Such measures too address multiple objectives that not only protect beneficial biological diversity but impacts on sustainability issues too. Marginal lands and agroforestry that are sympathetically cultivated also offer great opportunities to promote this new ethos. By seeking to live in harmony with diversity rather than attempting to managing it, the ultimate health of the planet and its ability to provide reduced vulnerability and more sustainable food source benefits both the environment and its populations.

On this last point in fact, solutions to a sustainable environmental policy vis-à-vis food security and all else, gives rise to challenges in every field of discipline, from biotechnologies to management of the food chain. In achieving the much noted growth in the food supply, the frontiers of all these disciplines are going to have to be pushed to come up with multiple solutions involving multiple stakeholders that needs to see everyone singing off the same hymn sheet.

References

AEZ (2000). Inventory of Land Resources. FAO and the International Institute for Applied Systems Analysis (IIASA). http://www.fao.org/ag/AGL/agll/gaez/index.htm (accessed 2 June 2009).

Alexandratos, N., Ed. (1995). *World Agriculture: Towards 2010.* Chichester: FAO & John Wiley & Sons.

Anderson, M. D. and J. T. Cook (1999). Community Food Security: Practice in Need of Theory? *Agriculture and Human Values* 16: 141–150.

APPE (2011). Association of Petrochemical Producers in Europe. The European Chemical Industry Council. http://www.petrochemistry.net (accessed 13 February 2011).

BBC (2007) Quick Guide: Biofuels. What Are Biofuels? *BBC News* 24 January. http://news.bbc.co.uk/1/hi/sci/tech/6294133.stm.

Biopact (2008) Brazil's Biodiesel Mandate Comes into Effect—Data Show Success of 'Social Fuel Program'. *Biopact* January 1. http://biopact.com/2008/01/brazils-biodiesel-mandate-comes-into.html; http://news.mongabay.com/bioenergy/2008/01/brazils-biodiesel-mandate-comes-into.html.

Booker, C. (2009). *Climate Change: This Is the Worst Scientific Scandal of Our Generation.* London: Telegraph Media Group Limited. 28 November 2009.

Brandt, K. (1939). Foodstuffs and Raw Materials.In *War in Our Time.* H. Speier and A. Kähler, eds. New York: W.W. Norton & Co.

Brown, C. and U. Lall (2006). Water and Economic Development: The Role of Variability and a Framework for Resilience. *Natural Resources Forum* 30(4): 306–317.

Brown, L. (2002) Water Deficits Growing in Many Countries: Water Shortages May Cause Food Shortages. *Great Lakes*: http://www.greatlakesdirectory.org/zarticles/080902_water_shortages.htm.

Brown, L. (2003). *Plan B: Rescuing a Planet under Stress and a Civilization in Trouble.* New York: W. W. Norton and Company.

Brown, L. (2005). *Outgrowing the Earth: The Food Security Challenge in an Age of Falling Water Tables and Rising Temperatures.* New York: W.W. Norton and Co.

Brown, L. (2008). Why Ethanol Production Will Drive World Food Prices Even Higher in 2008. Earth Policy Institute. http://www.Earth-policy.org/Updates/2008/Update69.htm (accessed 7 June 2009).

CFS (2007). Assessment of the World Food Security Situation. Committee On World Food Security. Thirty third Session. Rome: Food and Agriculture Organisation.

CFS (2008). Agenda Item II: Assessment of the World Food Security and Nutrition Situation. Committee On World Food Security: Thirty-fourth Session. Rome: Food and Agriculture Organisation.

Chakrabortty, A. (2008) Secret Report: Biofuel Caused Food Crisis—Internal World Bank Study Delivers Blow to Plant Energy Drive. *Guardian online* 3 July: http://www.guardian.co.uk/books/2001/mar/31/society.politics (accessed 15 March 2011).

Chappell, M. J. and L. A. LaVollo (2011). Food Security and Biodiversity. Can We Have Both? An Agroecological Analysis. *Agriculture and Human Values* 28: 3–26.

Cohen, B. (2006). Urbanization in Developing Countries: Current Trends, Future Projections, and Key Challenges for Sustainability. *Technology in Society* 28: 63–80.

DEFRA (2003). *Renewable Biofuels for Transport: The Facts on Biodiesel and Bioethanol.* UK: DEFRA.

DeFries, R., L. Bounoua and G. Collatz (2002). Human Modification of the Landscape and Surface Climate in the Next Fifty Years. *Global Change Biology* 8: 438–458.

DeRose, L., E. Messer and S. Millman (1998a). Food Deprivation. In *Who's Hungry? And How Do We Know? Food Shortage, Poverty, and Deprivation*. L. DeRose, E. Messer and S. Millman, eds. Tokyo, New York, Paris: United Nations University Press.

DeRose, L., E. Messer and S. Millman (1998b). The Relationship between Drought and Famine.In *Who's Hungry? And How Do We Know? Food Shortage, Poverty, and Deprivation*. L. DeRose, E. Messer and S. Millman, eds. Tokyo, New York, Paris: United Nations University Press.

Dyer, G. (2006). How Long Can the World Feed Itself? *Energy Bulletin*. http://www.energybulletin.net/node/21736 (accessed 17 June 2010).

Easterling, W., P. Aggarwal, P. Batima, K. Brander, L. Erda, S. Howden, A. Kirilenko, J. Morton, J. Soussana, J. Schmidhuber and T. F. In (2007). Food, Fibre and Forest Products.In *Climate Change 2007: Impacts, Adaptation and Vulnerability. Contribution of Working Group Ii to the Fourth Assessment Report of the Intergovernmental Panel on Climatechange*. M. Parry, O. Canziani, J. Palutikof, v. d. L. P. and C. Hanson, eds. Cambridge, UK: Cambridge University Press.

EC (2006). Environmental Technologies Action Plan Water Desalination Market Acceleration. Brussels: European Commission.

ECOSOC (2008). Land and Vulnerable People in a World of Change. New York: UN Economic and Social Council.

EERE (2008). Abc's of Biofuels. US Department of Energy: Office of Energy Efficiency and Renewable Energy (EERE). http://www1.eere.energy.gov/biomass/abcs_biofuels.html (accessed 12 February 2011).

Ehrlich, P. and A. Ehrlich (1990). *The Population Explosion*. New York: Simon & Schuster.

EIA (2011). Energy Information Administration. *Energy Information Administration Brochures*. US Department of Energy. http://www.eia.gov/ (accessed 14 February 2011).

Eifert, B., C. Galvez, N. Kabir, A. Kaza, J. Moore and C. Pham (2002). The World Grain Economy to 2050: A Dynamic General Equilibrium, Two Sector Approach to Long-Term World-Level Macroeconomic Forecasting. *University Avenue Undergraduate Journal of Economics (Online)*.

EJP (2005). *French-Run Water Plant Launched in Israel*. Brussels: European Jewish Press.

Encyclopedia (2002). Science of Everyday Things. *Encyclopedia.com*. http://www.encyclopedia.com (accessed 13 July 2011).

Ericksen, P. J. (2008). Conceptualizing Food Systems for Global Environmental Change Research *Global Environmental Change* 18(1): 234–245.

ESA (2007). World Population Prospects: The 2006 Revision. New York: UN Department of Economic and Social Affairs: Population Division.

Eswaran, H., P. Reich and F. Beinroth (2006). Land Degradation: An Assessment of the Human Impact on Global Land Resources. At the 18th World Congress on Soil Science.

FAO (1995). Report of the First External Programme and Management Review of the International Irrigation Management Institute (IIMI). Rome: Food and Agriculture Organisation.

FAO (2000). Global Impact Domain: 'Methane Emissions'. Rome: Food and Agriculture Organization.

FAO (2003a). Trade Reforms and Food Security: Conceptualizing the Linkages. Rome: Food and Agriculture of the United Nations.

FAO (2003b). World Agriculture: Towards 2015/2030. An Fao Perspective. Rome: Food and Agriculture Organization of the United Nations/Earthscan.

FAO (2006). Livestock's Long Shadow: Environmental Issues and Options Rome: Food and Agriculture Organization of the United Nations 416.

FAO (2009a). Climate Change Terminology. http://www.fao.org/climatechange/49365/en/ (accessed 2 June 2009).

FAO (2009b). Reference/international land use classification systems. http://www.fao.org/ag/agl/agll/landuse/clsys/index2.html (accessed 2 June 2011).

Foley, J., R. DeFries, G. Asner, C. Barford, G. Bonan, S. Carpenter, S. Chapin, M. Coe, G. Daily, H. Gibbs, J. Helkowski, T. Holloway, E. Howard, C. Kucharik, C. Monfreda, J. Patz, I. Prentice, N. Ramankutty and S. P (2005). Global Consequences of Land Use. *Science* 309: 570–574.

Foresight Project (2011). Foresight: The Future of Food and Farming: Final Project Report. London: The Government Office for Science 202.

Freedom 21 (2008). Meeting Essential Human Needs. 9th Annual National Conference, Dallas, Freedom 21.

Fritsch, A. J. and P. Gallimore (2007) Healing Appalachia: Sustainable Living through Appropriate Technology. *Earth Healing*. http://www.Earthhealing.info/biofuels.pdf; http://www.earthhealing.info/dailyfeb06.htm.

GEO (2000). Global Environment Outlook. Malta: United Nations Environment Programme.

GEO (2007a). Global Environment Outlook 4: Environment for Development. Malta: United Nations Environment Programme.

GEO (2007b) Planet's Tougher Problems Persist, Un Report Warns. *Global Environment Outlook 4*: http://www.unep.org/geo/geo4/media/media_briefs/Media_Briefs_GEO-4%20Global.pdf.

GFS (2011). Global Issues. Global Food Security Resource Centre. http://www.foodsecurity.ac.uk/issue/global.html (accessed 12 March 2011).

GHI (2010). Gap Report: The Global Harvest Initiative: Measuring Global Agricultural Productivity Washington, DC: The Global Harvest Initiative.

Gregorio, A. D. and L. J. M. Jansen (2000). Land Cover Classification System (Lccs): Classification Concepts and User Manual. Rome: FAO: Land and Water Development Division.

Gregory, P. J., J. S. I. Ingram and M. Brklacich (2005). Climate Change and Food Security. *Philosophical Transactions of the Royal Society B—Biological Sciences* 360(1463): 2139–2148.

Ho, M.-W. (2008). Organic Agriculture and Localized Food & Energy Systems for Mitigating Climate Change. *ISIS Reports*. London: The Institute of Science in Society.

Hoekstra, A. and A. Chapagain (2004). Water Footprints of Nations: Value of Water Research Report Series No.16.

Hoekstra, A. Y. and A. K. Chapagain (2008). *Globalization of Water: Sharing the Planet's Freshwater Resources*. Oxford: Blackwell Publishing.

HSUS (2008). The Impact of Animal Agriculture on Global Warming and Climate Change. Washington, DC: The Humane Society of the United States.

IAASTD (2009). IAASTD Global Report: Agriculture at a Crossroads. *IAASTD Synthesis Report*. Johannesburg, South Africa: International Assessment of Agricultural Knowledge, Science and Technology for Development (IAASTD) 6.

IAASTD (2010). *Agriculture and Development: A Summary of the International Assessment on Agricultural Science and Technology for Development*. Johannesburg, South Africa: GreenFacts.

IGBP (2010). The International Geosphere-Biosphere Programme. Royal Swedish Academy of Sciences. http://www.igbp.net/page.php?pid=369 (accessed 13 February 2011).

Ingram, J. S. I., P. J. Gregory and A. M. Izac (2008). The Role of Agronomic Research in Climate Change and Food Security Policy. *Agriculture Ecosystems & Environment* 126(1–2): 4–12.

INTERFAIS (2003). *Food Aid Flows*: World Food Program.

IPCC (2001). *Climate Change: The Scientific Basis*. Cambridge: Cambridge University Press.

IPCC (2007). Climate Change 2007: Working Group I: The Physical Science Basis, What Factors Determine Earth's Climate? Geneva, Switzerland: Intergovernmental Panel on Climate Change.

Jacobs, L. (1991). *Waste of the West: Public Lands Ranching*. Tuscon, AZ: Jacobs.

Jones, R. and T. Wigley (1989). *Ozone Depletion: Health and Environmental Consequences*. New York: John Wiley & Sons.

Khan, S. and M. A. Hanjra (2009). Footprints of Water and Energy Inputs in Food Production - Global Perspectives. *Food Policy* 34(2): 130 140.

Kininmonth, W. (2008). Illusions of Climate Science. *Quadrant Online* LII(10). http://www.quadrant.org.au/magazine/issue/2008/10/illusions-of-climate-science (accessed 21 April 2011).

Kitzes, J. A., A. Peller, S. Goldfinger and M. Wackernagel (2007). Current Methods for Calculating National Ecological Footprint Accounts. *Science for Environment and Sustainable Society* 4(1): 1–9.

Kniivila, M. (2004). Land Degradation and Land Use/Cover Data Sources. New York: UN Department of Economic and Social Affairs: Statistics Division 36.

Lambin, E. F., B. L. Turnerb, H. J. Geista, S. B. Agbolac, A. Angelsend, J. W. Brucee, O. T. Coomesf, R. Dirzog, G. Fischerh, C. Folkei, P. S. Georgej, K. Homewoodk, J. Imbernonl, R. Leemansm, X. Lin, E. F. Morano, M. Mortimorep, P. S. Ramakrishnanq, J. F. Richardsr, H. Skåness, W. Steffent, G. D. Stoneu, U. Svedinv, T. A. Veldkampw, C. Vogelx and J. Xuy (2001). The Causes of Land-Use and Land-Cover Change: Moving Beyond the Myths. *Global Environmental Change* 11(4): 261–269.

LCLUC (2009). *Land Earth System Data Records (ESDR)*: NASA Land-Cover and Land-Use Change Program.

Lewis, L. (2007) Water Shortages Are Likely to Be Trigger for Wars, Says Un Chief Ban Ki Moon. http://www.timesonline.co.uk/tol/news/world/asia/article2994650.ece.

Lopez-Gunn, E. and M. R. Llamas (2008). Re-Thinking Water Scarcity: Can Science and Technology Solve the Global Water Crisis? *Natural Resources Forum* 32(228–238): 12.

Lovgren, S. (2010). *Hardy Plant May Ease Biofuels' Burden on Food Costs*. National Geographic Society. October 28.

Lynn, B. H., R. Healy and L. M. Druyan (2007). An Analysis of the Potential for Extreme Temperature Change Based on Observations and Model Simulations. *Journal of Climate* 20: 1539–1554.

Mateos, E. and J. M. González (2007). *Biomass: Potential Source of Useful Energy*. Sevilla.

Maxwell, S. and T. Frankenberger (1992). Household Food Security: Concepts, Indicators and Measurements: A Technical Review. New York, Rome: UNICEF and IFAD.

McGinley, M. (2008). *Biomes*. Washington, DC: Environmental Information Coalition, National Council for Science and the Environment.

McLaughlin, A., G. J. W. Hiba and D. J. King (2000). Comparison of Energy Inputs for Inorganic Fertilizer and Manure Based Corn Production. *Canadian Agricultural Engineering* 42(1): 2.1–2.14.

Modi, V., S. McDade, D. Lallement and J. Saghir (2007). Energy Services for the Millennium Development Goals. Energy Sector Management Assistance Programme. New York: World Bank.

Moench, M. (2001). Overcoming Water Scarcity and Quality Constraints: Groundwater—Potential and Constraints. *IFPRI 2020 Focus Brief no 9*. Washington DC: International Food Policy Research Institute.

Molden, D., T. Y. Oweis, P. Steduto, J. W. Kijne, M. A. Hanjra, P. S. Bindraban, B. A. M. Bouman, S. Cook, O. Erenstein, H. Farahani, A. Hachum, J. Hoogeveen, H. Mahoo, V. Nangia, D. Peden, A. Sikka, P. Silva, H. Turral, A. Upadhyaya and S. Zwart (2007). Pathways for Increasing Agricultural Water Productivity. In *Comprehensive Assessment of Water Management in Agriculture, Water for Food, Water for Life: A Comprehensive Assessment of Water Management in Agriculture*. D. Molden, ed. London/Colombo: Earthscan/International Water Management Institute.

Morgan, D. (2005). *Brazil's Biofuel Strategy Pays Off as Gas Prices Soar. Washington Post Saturday*. Washington, DC: The Washington Post Company. June 18.

Morvaridi, B. (1995). Contract Farming and Environmental Risk: The Case of Cyprus. *Journal of Peasant Studies* 23(1): 30–45.

Muir, P. (2007) Study Finds Net Energy of Biofuels Comes at a High Cost. http://extension.oregonstate.edu/news/story.php?S_No=436&storyType=news.

Muir, P. (2008) Fossil Fuels and Agriculture. http://oregonstate.edu/~muirp/fossfuel.htm.

Muir, P. (2009a) Erosion from Inappropriate Agricultural Practices on Crop Lands. http://oregonstate.edu/%7Emuirp/erosion.htm

Muir, P. (2009b) General Concepts Related to Overgrazing: How Does Overgrazing Affect Rangelands and Lead to Degradation? : http://oregonstate.edu/%7Emuirp/genconce.htm

Muir, P. (2009c) Land. http://oregonstate.edu/%7Emuirp/landlim.htm

Muir, P. (2009d). Resource Limitations. w. http://oregonstate.edu/~muirp/reslimit.htm (accessed 5 July 2009).

Muir, P. (2009e) Salinization. Oregon State University. http://oregonstate.edu/%7Emuirp/saliniz.htm

Muir, P. (2010). *Bi301: Human Impacts on Ecosystems*. Oregon State University.

Myers, N. (1984). *Gaia: An Atlas of Planet Management*. New York: Anchor Press.

NASA (2011). Earth Observatory. NASA. http://Earthobservatory.nasa.gov/Laboratory/Biome/ (accessed 20 February 2011).

Nathoo, T. and J. Shoveller (2003). Do Healthy Food Baskets Assess Food Security? *Chronic Diseases in Canada* 24(2–3): 65–69.

National Geographic (2008). *Biofuels: The Original Car Fuels*. National Geographic Society.

Naturalgas (2011). Overview of Natural Gas. Natural Gas Supply Association. http://www.naturalgas.org/overview/background.asp (accessed 13 February 2011).

Nellemann, C. (2009). The Environmental Food Crisis: The Environment's Role in Averting Future Food Crises. Geneva, Switzerland: United Nations Environment Programme (UNEP) GRID-Arendal.

NOAA (2011). *National Climatic Data Center*. Boulder: National Oceanic and Atmospheric Administration. 2011.

Oldeman, L. (1994). *The Global Extent of Soil Degradation. In Greenland and Szabolcs*. Wallingford, UK: CAB International.

Oldeman, L. R., R. T. A. Hakkeling and W. G. Sombroek (1991). World Map of the Status of Human-Induced Soil Degradation. Global Assessment of Soil Degradation (GLASOD).

PAP (2008) Population and Human Development—the Key Connections. *People and the Planet Online*: http://www.peopleandplanet.net/doc.php?id=199andsection=2

Postel, S. (1989). Halting Land Degradation. In *State of the World 1989*. L. Brown, A. Durning, C. Flavin, L. Heise, J. Jacobson, S. Postel, M. Renner, C. Shea and L. Starke. New York: Norton.

Postel, S. L. (1999). *Pillar of Sand: Can the Irrigation Miracle Last?* New York: W.W. Norton.

Ramankutty, N. and J. A. Foley (1999). Estimating Historical Changes in Global Land Cover: Croplands from 1700 to 1992. *Global Biogeochemical Cycles* 13: 997–1027.

Rijsberman, F. (2010) Energy & Climate: Water and Food Security. *America.Gov* March: http://www.america.gov/st/cncrgy-cnglish/2010/March/20100310143830fsyelkaew0.4131739.html.

Rivera, A. (2008). International Year of Planet Earth 3. Groundwater Sustainable Development in Canada—Emerging Issues. *Geoscience Canada* 35(2): 73–87.

Roach, J. (2007) Global Warming 'Very Likely' Caused by Humans, World Climate Experts Say. *National Geographic News* February 2, 2007: http://news.nationalgeographic.com/news/2007/02/070202-global-warming.html.

Rockström, J., M. Lannerstad and M. Falkenmark (2007). Assessing the Water Challenge of a New Green Revolution in Developing Countries. *Proceedings of the National Academy of Sciences of the United States of America (PNAS)* 104 (15): 6253–6260.

Rosegrant, M. W. and S. A. Cline (2003). Global Food Security: Challenges and Policies. *Science* 302(5652): 1917–1919.

Roubanis, N. (2004). Biofuels and Their Growing Importance. Energy Statistics, Working Group Meeting: Special Issue Paper 6. Paris: OECD.

Salter, R. M. (1950). Technical Progress in Agriculture. *Journal of Farm Economics* 32(3): 478–485.

Schneider, S. (1990). *Global Warming*. New York: Random House.

Searchinger, T., R. Heimlich, R. Houghton, F. Dong, A. Elobeid, J. Fabiosa, S. Tokgoz, D. Hayes and T. Yu' (2008). Use of US Croplands for Biofuels Increases Greenhouse Gases through Emissions from Land-Use Change. *Science* 319(5867): 1238–1240.

SEEA (2003). The Handbook of National Accounting: Integrated Environmental and Economic Accounting (Seea). New York: United Nations, Eurostat, IMF, OECD, World Bank 598.

Shah, T., D. Molden, R. Sakthivadivel and D. Seckler (2000). The Global Groundwater Situation: Overview of Opportunities and Challenges. Colombo: International Water Management Institute.

Smith, K. and R. Edwards (2008) The Year of Global Food Crisis. *Sunday Herald Online*: http://www.sunday-herald.com/news/heraldnews/display.var.2104849.0.2008_the_year_of_global_food_crisis.php.

SOFI (2008). The State of Food Insecurity in the World 2008. Rome: Food and Agriculture Organisation 59.

SOWF (2011). State of the World's Forests. Rome: Food and Agriculture Organization of the United Nations 179pp.

Stauffer, N. (2007). *Ethanol Analysis Confirms Benefits of Biofuels*: Massachusetts Institute of Technology. January 8.

Steck, T. L. (2008). Human Population Explosion. http://www.eoEarth.org/article/Human_population_explosion (accessed 25 March 2011).

Sundquist, E. and W. Broecker (1985). *The Carbon Cycle and Atmospheric CO_2: Natural Variations Archean to Present*. Washington, DC: American Geophysical Union.

Tateishi, R. (2010). Global land cover mapping and geospatial data sharing system. Japan, Chiba University http://www.csrsr.ncu.edu.tw/08CSRWeb/ChinVer/C7Info/2010RSDMM/PPTpdf/12.pdf (accessed 2 June 2011).

Tindall, J. A. and A. A. Campbell (2010). Water Security—National and Global Issues: US Geological Survey Fact Sheet. US: US Geological Survey 6.

Trimble, S. (1999). Decreased Rates of Alluvial Sediment Storage in the Coon Creek Basin, Wisconsin, *Science* 285(5431): 1244–1246.

UN (1943). United Nations Conference on Food and Agriculture, May 18–June 3, 1943 : Final Act and Section Reports / Department of State, United States of America. Washington, DC: US G.P.O. United Nations Conference on Food and Agriculture. Washington, DC: United Nations Government Printing Office.

UN (2011). Website of the United Nations. United Nations. http://www.un.org/ (accessed 23 March 2011).

UN/DESA (2011). Integrated Environmental and Economic Accounting 2003 (Seea 2003). UN Department of Economic and Social Affairs (DESA) (accessed 12 March 2011).

UNFPA (2007). State of the World's Population: Unleashing the Potential of Urban Growth. U.N. Population Fund.

UoM (2007). *Global Change: Human Impact—the Changing Biomes*. University of Michigan.

UOR (2009). University of Reading: Agriculture, Policy and Development, History of Agriculture. University of Reading. http://www.ecifm.rdg.ac.uk/history.htm (accessed 4 June 2008).

Verweij, M., M. Douglas, R. Ellis, C. Engel, F. Hendriks, S. Lohmann, S. Ney, S. Rayner and M. Thompson (2006). Clumsy Solutions for a Complex World: The Case of Climate Change. *Public Administration* 84(4): 817–843.

Vlek, P. L. G., G. Rodríguez-Kuhl and R. Sommer (2004). Energy Use and Co2 Production in Tropical Agriculture and Means and Strategies for Reduction or Mitigation. *Environment, Development and Sustainability* 6(1–2): 213–233.

von Braun, J., Ed. (1995a). *Employment for Poverty Reduction and Food Security.* Washington DC: International Food Policy Research Institute.

von Braun, J. (1995b). A New World Food Situation: New Driving Forces and Required Actions. International Food Policy Research Institute. Washington, D.C: The International Food Policy Research Institute.

Wade, J., A. Branson and G. Qing (2002). Grain and Feed Annual Report: China Beijing: US Department of Agriculture.

Water Footprint (2004). Water Footprints of Nations: Value of Water Research Report Series No.16. University of Twente. http://www.waterfootprint.org/?page=files/home (accessed 13 February 2011).

WEO (2008). World Energy Outlook 2008. Paris: The International Energy Agency (IEA) 578.

WEO (2010). World Energy Outlook 2010. Paris: The International Energy Agency (IEA).

WFP (2011). Website of the World Food Program. World Food Programme. http://www.wfp.org/ (accessed 25 February 2011).

Windfuhr, M. and J. Jonsén (2005). Food Sovereignty: Towards Democracy in Localized Food Systems. Warwickshire: ITDG/FIAN 70.

Wolfson, R. and S. Schneider (2002). Understanding Climate Science.In *Climate Change Policy: A Survey.* S. Schneider, A. Rosencranz and J. Niles, eds. Washington DC: Island Press.

Woodward, S. (2003). *Biomes of Earth, Terrestrial, Aquatic, and Human-Dominated.* Westport, CN: Greenwood Press.

World Bank (2007). World Development Report 2007: Development and the Next Generation—Agricultural Inputs. Washington, DC: World Bank.

World Bank (2009). Land Resources Management. Washington, DC: World Bank. http://web.worldbank.org/wbsite/external/topics/extard/0,,contentmdk:20452620~menuPK:1308455~pagePK:148956~piPK:216618~theSitePK:336682,00.html (accessed 21 April 2009).

WSF (2009). Fresh Water: More Precious Than Oil. http://www.whole-systems.org/water.html (accessed 2 April 2009).

WWDR-2 (2006). Water: A Shared Responsibility. *World Water Development Report (WWDR).* Geneva, Switzerland: UNESCO 601.

WWDR-3 (2009). Water in a Changing World *World Water Development Report (WWDR).* Geneva, Switzerland: UNESCO.

WWF (2006). Living Planet Report 2006. Switzerland: World Wide Fund for Nature.

Zhang, W. (2008). A Forecast Analysis on World Population and Urbanization Process. *Environment Development and Sustainability* 10(6): 717–730.

Zhang, X., F. Zwiers and T. Peterson (2008). The Adaptation Imperative: Is Climate Science Ready? Geneva, Switzerland: World Meteorological Organisation.

22

Governance, Politics and Economics

22.1 Governance

22.1.1 Trends in Global Governance

The notion of governance in food security matters is a complex one and one that has tendrils going back centuries. Governance can be thought of as responsibility, guidance and oversight. In this sense, modern notions of governance have evolved from an eclectic mix of social accountability, political ideology, economic development paradigms as well as the many more subtle moral and ethical influences. This was highlighted in Part Two with a look at the many multifarious trains of thought and their indelible criss-crossing paths that culminated in many international agreements and instruments of peace and social rights. In Part Three, the notion of multilateral governance properly evolved and with the advent of World War One, food was seen as a right in many quarters. This was further entrenched with the advent of the League of Nations where the idea of food security took on a global remit. However, the League ultimately dwindled in influence and was eventually replaced after World War Two with the creation of the United Nations. This period saw three UN agencies the FAO, WHO and UNICEF given the mandates to tackle the growing problem of hunger and malnutrition. Initially though it was a difficult time politically, in so much as sovereign nations seemed uncomfortable with the notion of an international organisation beyond the control of national governments. Yet despite such circumspection, prudence reigned and the UN eventually received wide acceptance (Williams 2005). The FAO itself was born out the conference at Hot Springs in 1943 and on creation, it held its first session in Quebec City, which finally cemented the organisation as a specialised United Nations agency headed by Sir John Boyd Orr in 1945 (FAO 1946; Phillips 1981; FAO 2010). Since then the FAO has had more to do with the influence of the food security debate than any other agency or government (Shaw 2007).

Thus from the social and political detritus of two World Wars, the problem of hunger and malnutrition was firmly and finally fully politicised. Moreover, coupled with contemporary nutritional research and operating alongside an ideological righteousness the League and the UN set the precedent for the first international initiatives to address nutritional issues in all countries, both developed and developing. As the decades passed, so knowledge and research became invaluable tools in understanding hunger's many causes and potential solutions, and by the 1960s, the extent of the problem was becoming clearer. And it was shocking news. The international community responded to a resounding battlecry with new initiatives like the UN's Freedom from Hunger Campaign, the World Food programme, the International Grains Council's Food Aid Convention, as well as the US Food for Peace programme, all created in response. These too were complimented by new agencies and bodies such as the European Economic Community (later the EU), the UN's Research Institute for Social Development (UNRISD), the UK-based Institute of Development Studies (IDS) and the Overseas Development Institute (ODI) as well as numerous other institutions, charities and civil organisations.

Initially, co-operation among these groups was largely kept in-house; however, as needs arose cross-institutional collaborative efforts began to draw these bodies together. By the 1980s and 1990s, the UN had fully and firmly embraced a co-operative model of governance devolving much responsibility and policy direction to its many semi-autonomous specialised agencies as well as to the numerous governmental and non-governmental bodies it had finally opened its arms to. Collectively food security issues at this time had been institutionally defined and had grown to encompass wide and on face

value, divergent fields of interests including poverty, livelihoods, nutrition, health, politics and economics. As a result, other bodies such as the UN's Committee on Food Security (CFS), the Office of the
United Nations Disaster Relief Coordinator (UNDRO), the UN Environment Programme (UNEP), the
Washington Research Group, the Consultative Group on International Agricultural Research (CGIAR),
the World Resources Institute (WRI) as well as the International Food Policy Research Institute (IFPRI)
had collectively joined the foray and taken up the food security mantle (see Appendix K: for a map of
current food security stakeholders).

While these numerous seemingly unconnected bodies sought to analyse, research and ultimately
respond to these issues Pandora's box was finally fully opened revealing food security in all its complicated interconnectedness. This picture revealed the need for a more complete and overarching body of
macro-governance. However, for all its faults, there was only one real contender for this role: the UN.
Responding to these challenges the UN in consultation with multiple stakeholders implemented the
Millennium Development Goals (MDG) in 2000. Resulting from these events and the rationalisation of
mandates, objectives together with more than a modicum of common sense, the food aid organisation
of the UN, the World Food Programme has grown to become the world's biggest multinational food aid
organisation handling approximately 56% of all global food aid deliveries as well as 78% of emergency
food aid deliveries in 2006 alone (CCP 2005; Mousseau 2005; Rucker 2007; WFP 2008d).

Nonetheless, as is not uncommon in institutional behemoths like the UN, with political infighting, territorial disputes and duplication of efforts, further rationalisation was inevitable and ongoing attempts to
resolve these issues continue to provide a changing landscape of policy and direction. This has led many
to suggest that present global governance on food and nutritional security is fragmented and not providing clear guidance on hunger and malnutritional issues (Marzeda-Mlynarska and Curie-Sklodowska
2009; RTFN 2009). In support of such charges for instance, Schutter maintains that bodies like the IMF,
the World Bank and the World Trade Organisation often give conflicting policy guidance in respect
to food security issues. Further, it is suggested too that with so many stakeholders, it becomes practically impossible to develop common coherent policy objectives and as a consequence of unimproved
coordinated guidance, the underlying structural causes of food security are unlikely to be adequately
addressed. Even the FAO in this respect agrees with the notion of a lack of unified global governance
when, in a report to the then upcoming World Food Summit of 2009, it said that

> We recognize that there is a lack of coherence and efficiency in the current governance of world
> food security. The system is poorly organised and each institution operates to a large extent
> separately despite important progress in coordination. Responding to the global food insecurity
> crisis in an effective and sustainable way requires not only strong leadership and relevant poli-
> cies, strategies and programmes, but also coordinated implementation and monitoring capaci-
> ties. (FAO 2009c)

While this is a poor indictment of the current state of food security governance, it is a tacit and welcome
admission of the difficulties facing the various stakeholders. Indeed such difficulties were already recognised back in 1995 in the Commission on Global Governance's 1995 report 'Our Global Neighbourhood'.
Established to analyse global changes and to offer ways in which the international community could
better cooperate on global issues, the commission called for a new global ethic that would encourage
cooperation and collaboration as common goals. While the report called for a strengthened UN, based
on multilateral and multilevel hierarchical global governance, many people felt the language was perhaps
too authoritarian and drew images of a hegemonic world government at the expense of individual national
sovereignty (CGG 1995). Although good governance in today's understanding becomes a divisive topic;
only by such open and forthright self-awareness as shown in the above two examples will progress be
made with inroads into a phenomenon that has blighted humanity's moral landscape for far too long.

22.1.2 Food Security, Freedom and Democracy

An important but not so widely discussed area of food insecurity prevention centres around the type
of governance within a country. Many people and institutions recognise the impetus of political will

in tackling the issues of hunger and indeed there has been much theoretical posturing, yet for all this there is very little research that directly examines these principles. One study however by Stephen Scanlan in 2004 looked at these specific issues. In his study he aimed to quantitatively assess whether the political structure of a state or country affects the free passage of information and by extension the likely preparedness of its citizenry in addressing issues of hunger and malnutrition. His findings, using regression analysis, strongly supported the idea that democracy and the public's access to information were closely tied to the reduction of child hunger in less industrialised societies (Scanlan 2004). Scanlan posits several fundamental dynamics in this relationship, of which the most important seems to be the accountability of the democratic process. These sentiments echo others' thoughts including those of D'Souza and Birchall who also strongly believe the mechanisms and trappings of a properly functioning democratic country are such that the free press, the internet and other mediums of information exchange, facilitate openness and the collective empowerment of its people (D'Souza 1994; Birchall 2002). Indeed parallels can be drawn and in a wry comparison, Birchall offers too that just as turkeys do not vote for Christmas so oppressive regimes and dictatorial governance are not predisposed to abandoning the reigns of power, and indeed in these politically hostile environments, the future is at best risky or uncertain. Alternatively democratic or near democratic governed countries on the other hand tend to allow individuals the freedom to operate and to question the actions of government. In this sense, as democracy contributes to economic growth, stability and social development needs, so it can be seen that human development and democratic principles then go hand in hand. The encouraging news on this front is that the trend towards democracy is a positive one and as Figure 22.1 highlights the trajectory in global governance over the last 40 years or so shows a strong and continual move away from autocracy and toward democracy.

Once again, comparative data do not always agree nor is it always that straightforward. With democracies, for instance, a seemingly straightforward concept is interpreted differently by the various institutions. Freedom House, for instance, considers democracies based on the electoral principle of democratic society whereas the Economists Intelligence Unit uses a suite of indicators to determine true democracies from others. As a result, Figure 22.1 compares only loosely to Freedom House's 116 official electoral democracies in 2010, while the Economists Intelligence Unit, also in 2010, found around 79 full and flawed democracies in the world (representing 49.5% of the total population with a further 33 hybrid regimes representing another 14% of the population) (EIU 2010; Freedom House 2011). In sum, the differences reside in the official electoral tag compared to actual democratic governance indicators used by the EIU.

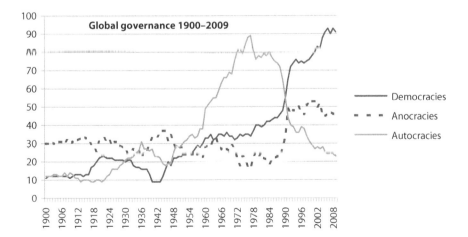

FIGURE 22.1 Trends in global governance 1900–2009. (Recreated from the Integrated Network for Societal Conflict Research (INSCR), Center for Systemic Peace, 2011, http://www.systemicpeace.org/inscr/inscr.htm. With permission.)*Note*: Anocracies are states with incoherent or inconsistent patterns of authority: partly liberal, partly authoritarian.

Governance in food security issues then can be thought of as more than just taking responsibility; it is a process of navigation through: competing political and social ideologies, conflicting government and institutional policies as well as commanding appropriate technologies and the concomitant dissemination of information (INSCR 2011; Mukherjee 2011). As the globe trends toward even greater democracy, it will be interesting to see how this translates into greater world food security. Another fascinating development currently unfolding in early 2011 are the protests and calls for democratic reform currently sweeping the Middle East and how these events eventually translate into improved financial and social developments will be one future historians can speculate on.

22.1.3 Globalisation: Hegemony, Multilateral Cooperation or People Power

Although perhaps taking a liberty with the above heading, this section briefly looks at the emergence and trappings of globalisation as a new orthodoxy against a background of changing political and economic dominance. Wild swings in political and economic dominance over the last couple of centuries, as defined by multilateral engagement and economic power, has been within the confines of and dominated by Western industrialised countries. In fact, of the 7000 or so multilateral treaties signed in the last 400 years, the majority were signed in the period between 1850 and 1960. According to Denemark and colleagues, this frenzy of activity relates to the political hegemony of prevailing dominant states vying for international supremacy (Denemark et al. 2007).

It is largely an historic given, that after the fall of Napoleon by about 1815, Great Britain emerged as the dominant global power up until, that is about, 1885 whereupon the rising powers of France, Austria, the United States and Germany threatened its lead role (Denemark et al. 2007). This was then followed by the meteoric rise of the United States after World War Two into the front running position; which it has largely dominated from an economic perspective ever since. Politically however, regional alliances like the European Union have given the United States a run for its money; as is China more recently. Without getting drawn too far into political ideology, the point worth making here is that global power fluctuates and whether political or economic, hegemony influences policy beyond its own borders. Dominant global power can be thought of as being achieved through either fear, as in Cold War USSR, or economic, as in the United States and China or through ideology as in the United Nations. This is a loose analogy and the irony of implicating the UN as a global dominant power is not lost on the author. However, liberties aside, a global, benevolent dominant power facilitates food security through multi-lateral agreement and by paving the way through beneficial legislation and terms of trade that promotes rather than hinders progress. Multilateralism in this way is an appropriate and cooperative means of addressing common goals, to manage problems of coordination and to resolve conflicts of interest. Well, that is the theory perhaps.

Meanwhile, as all this tooing and froing was taking place, another dominant global power was snaking its way to its present position of some merit: globalisation. Early globalisation it seems, while the progeny of emerging dominant states as instruments of economic policy, quickly outgrew its parents and took on a life of its own. As a result, the middle to late nineteenth and early twentieth centuries witnessed the connection of countries as never before. New railroads, telegraph, mass circulated newspapers and new trading routes allowed people access to goods and services from around the world at unprecedented levels; globalisation was coming of age. Its march has been relentless and by the late 1980s, globalisation and the advent of a global free market had emerged as the new orthodoxy (Vaidya 2006). Globalisation has brought much politics and social agenda into the fold; however, food is a relative newcomer. Only by the 1990s, and after much wrangling by the UN and others concerning agriculture and negotiations at the General Agreements on Trade and Tariffs (GATT) talks, was food finally and fully admitted into the global free-trading arena. In theory, this effectively allowed countries to buy the food they needed at the best price on the international market unimpeded by cost barriers. This economic and cultural liberalisation according to Kennedy et al. (2004) is presently spurred on by urbanisation and is effectively transforming the whole food chain from production, processing, retailing and marketing around the world (Kennedy et al. 2004). Direct foreign investment by large multinational food and supermarket chains is resulting in cheaper food, greater availability and more diversity. These practices, however, caution Kennedy et al. (2004), can bring fundamental changes to the production, procurement and distribution systems, often at the expense of smaller local agents and traditional food markets. On top of this it has been suggested that globalisation is bringing about a gradual shift towards a more universal food culture (Kennedy et al. 2004; Dimitri et al. 2005; Scheuerman 2008; Walker 2008). So, as societies become increasingly politically and culturally integrated through a global network of communication, technology and information; and as the reduction in barriers see the subsequent increased flow of commodities, services, labour and capital, so food security has taken on a global dimension. As with individual, household and regional food security issues, the global trade in food and agriculture brings with it its own challenges. As many countries aim to feed from the same trough, the many global inequalities in terms of trade benefit some more than others. Moreover, warn the FAO and others, this global integration comes at a price as potentially increased vulnerability of global agrarian policy convergence is reflected in production and price volatilities (McMichael 1994; FAO 2003e).

There are other economic and social benefits of globalisation too that are related to poverty and food security. International trade can raise domestic incomes while increased standards of livestock quality control might also translate into raising standards of animal welfare, which in turn has the potential to reduce disease and increase food safety. Globalisation to boot also has the inherent ability to reduce food prices to consumers, warn the FAO and others, although this does not always happen in reality. The downsides too are also worth noting; the same strict quality controls for instance that raise standards might also act as barriers. Moreover, international trade tends to favour big business and centralised procurement systems acting as exclusionary hurdles to smallholders or small producers (FAO 2005).

22.1.4 Free Trade versus Protectionism

Closely associated with globalisation is the notion of free trade. Free trade however is often a misnomer or more judgmentally an oxymoron, as there are frequently additional barriers to the free movement of goods and services across international or trading block borders. Moreover, as a consequence of generations of such practices, presently some degree of national protectionism is seen as the norm throughout the world. In this world view, the fundamental ideologies of free trade and comparative advantage, two economic models often seen as complimentary however, are then viewed with mixed blessings. Originally pioneered by the likes of economists such as Adam Smith, Robert Torrens and David Ricardo, free trade is predicated upon the free movement of goods and services without outside influence from governments or other bodies. Free trade too pre-supposes certain comparative advantages whereby given different skill sets and natural resources, countries become more efficient in some industries than others. Given these conditions, the idea of free trade under these circumstances ensures that both sides mutually benefit from trade. In this freely functioning system, the supply and demand for food is said to be in equilibrium and subsequently reflects the true needs of populations. Moreover, there is considerable evidence too that globally integrated trade tends to grow both economically and through better standards of living, more so than those of more inward looking stances (IMF 2010). Global competition, as previously mentioned, also ensures lower prices and just as importantly the rational use of resources makes such trade more environmentally friendly and sustainable.

The downsides of free trade, however, are in its application. Free trade, as mentioned earlier, does not really exist and has not existed since its inception. This was recognised early on and as a result the UN initiated the GATT in 1947. This international treaty's purpose was to substantially reduce tariffs and other trade barriers as well as to eliminate trade preferences in a global, mutually reciprocal and advantageous way. Over the years, the GATT held many meetings or rounds of talks, and was eventually taken over by the World Trade Organisation (WTO). Its many rounds consisted of the Annecy Round (1949); Torquay Round (1951); Geneva Round (1955–1956); Dillon Round (1960–1962); Kennedy Round (1964–1967); Tokyo Round (1973–1979); Uruguay Round (1986–1994); and currently the Doha Round (2001, ongoing), which between them have tackled literally thousands of obstacles in the quest for true free trade. In respect of food incidentally, it was only by the time of the Uruguay Round that agriculture was finally admitted into the talks. While these talks have indeed reduced many such barriers, there still exist numerous preferential trading agreements, import tariffs, quotas and subsidies. This is no less true of agriculture. Instruments like the EU's common agriculture policy (CAP) or the US Farm Bill, for instance, continue to artificially distort the internal agricultural sector, providing generous subsidies and excess surpluses that the ideology of free trade would have precluded. Further, when trading externally, the EU's subsidy programme creates unfair advantages, which make it difficult for developing countries and others to trade on a level playing field. This has resulted in huge export growth for EU agricultural products over the past decades at the same time as developing countries witnessed considerable decreases. Others too like the United States also practice widespread protectionist policies, which proffers a competitive edge to their domestic farmers, yet all the while developing countries are encouraged to liberalise their markets in favour of free trade. This was perhaps one of the major failings of the Uruguay Round of negotiations. Before Uruguay Round, it had been hoped that the success of the EU CAP model in raising living standards and production within its borders could be mimicked by other less developed countries; however, the 1995 GATT round incorporated agriculture into the liberalisation process for the first time, disallowing this sort of protectionism. At the same time, the EU refused to unravel its own trade distorting policies (Warnock 1997; SED 2004; Barnes 2006; Harvey 2010; Murphy 2010). This is not to suggest that the developing countries are totally innocent themselves, as numerous countries, particularly in Africa and the Middle East for instance, also follow protectionist policies that marginalise their global participation in free trade (IMF 2010).

It is this dichotomy of 'do as I say and not as I do' politics that many see as one of the overriding barriers to real progress for developing countries in developing healthy trade balances. Indeed unless the status quo is challenged, many are likely to become sceptical of the ultimate benefits of so-called liberalisation. Hypocrisy and protectionism, suggest critics of such lopsided measures, says it favours rich countries as well as large corporations and producers while continuing to subjugate impoverished countries. Moreover, free trade powered by privatisation and fuelled by profit is charged with reinforcing short-termism and promoting a laissez-faire attitude towards environmentalism. Alarmingly this runs contrary to the good stewardship needed for the well-being of natural resources and sustainable production methods (Harvey 2010).

On top of these general free trade issues, the FAO also suggests that unequal trade practices between the North (developed countries) and the South (developing) countries is further exacerbated by major structural and government policy obstacles including inadequate transport and communication facilities, poor investment opportunities and marketing strategies (FAO 2004). It has also been said that considerable gains in South-South trade stands to gain a great deal from reducing barriers among themselves. Further suggestions indicate that another way for poorer countries to increase their bargaining power at the international level is to ally together and develop closer regional free trade systems similar to the EU example. Indeed many developing regions' existing regional trade agreements are seen as a good starting point for promoting trade among themselves.

Placed in perspective, globally, the total amount of food that crosses international borders is relatively small at just 9.6% of total merchandise traded in 2010; moreover, this traded food also represents a much smaller proportion of the total global food produced (WTO 2010). Even this small amount of trade however, is crucial in terms of global food security; especially when it comes to the low-income food deficit countries of the world (Murphy 2010). As domestic markets of less developed countries are potentially distorted from cheap food imports, so export markets for these same countries encounter considerable barriers to entry. Once again, for the sake of perspective, David Harvey quotes FAO as suggesting the total subsidies paid to Organisation for Economic Cooperation and Development (OECD) countries in agricultural protectionism in 2007 was in the region of $365 billion per annum while and IMF report estimated increased world trade from all traded merchandise globally, were all barriers removed, to be in the region of $250 billion to $680 billion per year (Harvey 2010; IMF 2010).

Thus, as a result of all this, it can be argued that it is perhaps not so much nation states' global political and economic hegemony that is the driving force behind societal change but rather the collective social and ideological values of billions of global consumers. In this picture of global dominant power then, globalisation seems to be an economic and political force all of its own and firmly in the hands of the people. The globalised community for better or worse is here to stay and it is difficult to see any attempt at rolling back of this increasing trend as anything other than a futile gesture. While some might see this Frankenstinian model taking on a will of its own to the detriment of global food security, others see it less of a monster and more of a benevolent force that can only be for the better.

22.1.5 Policies and Political Will

Already, extensively throughout this book, politics appears never to be far in the background. Indeed, it has been shown that poverty and hunger owe much of their existence and persistence to political causes from the erection of trade barriers and programmes affecting agriculture resources, war, conflict and social displacement through to the lack of capital investment and political instability. On the back of this, it has been suggested too that in order to improve food security, an enabling political environment is a must. This in turn is predicated on a conducive social and economic environment free of conflict, the promotion of equality as well as gender parity, among many others. Against this backdrop, politics has been implicated more than the whether in the way a government employs choices to address or ignore the causes of hunger (von Braun 1995; ActionAid 2003; CFS 2008b). Political will alone however, is perhaps insufficient and what is further needed is the expedient conversion of this will into actionable policy, incorporating both frontline programmes as well as macro-economic and structural adjustment programmes (SOFI 2008; FAO 2009a). Moreover this needs to be sustained in the face of setback after setback so as not to be worn down by failure, as noted by V. N. Mehta. Commenting on the will of the farmers in response to repeated food crisis in India in the early part of the twentieth century, Meta insightly writes that

> The impotence of man and the omnipotence of the forces of nature could not have been more unerringly demonstrated. This terrible race experience has worked darkly into the warp of the cultivator's mind and the resultant 'will to action' has ever after acquired a fatal limp. His life has become a gamble in rain and he is a gambler not in the exuberance of plenitude but the despair of destitution. (Mehta 1929)

Indeed, some criticise existing political motivations suggesting that the current systems favour economic growth over equitable social and economic development (Schuftan 2010). Others also feel that if governments are to take Roosevelt's freedom from want enshrined in Article 25 of the 1948 Universal Declaration of Human Rights seriously, then they have an obligation to ensure equity or reasonably fair access of peoples to essential resources such as food, health care, housing and education among others (Roosevelt 1941; UN 1948; Latham 1997).

There are many ways in which policy decisions affect food security, more than can be addressed in this book. I concentrate on three: social safety nets, land rights and policy reform.

22.1.6 Safety Nets and Food Reserves

The ideas of food safety nets are as old as civilisation itself. In the words of Sophia Murphy, lands cannot be moved, harvests are unpredictable and consumption is neither elastic nor optional (Murphy 2009). As a result, since ancient times, people have sporadically stockpiled grain as a reserve for leaner times, although, having said that within the past few decades this practice has been drastically reduced. Murphy suggests four reasons for this decline. Firstly, the lack of efficiency in managing reserves is given as outweighing potential benefits. Secondly, a political and economic shift in the 1980s, which mirrored Milton Friedman's economic paradigm of less is more. A view which also suggests that market forces are inherently better at providing for our needs rather than government intervention. Thirdly, confusion over policy objectives in respect of such grains and whether they might be used to stabilise prices, aid in the development of markets, help to compensate shortfalls in foreign exchange reserves or of course to stave off hunger. Lastly, Murphy suggests that less optimal policies were favoured in place of the more difficult economical polices associated with reserves that are social, political and geographical in nature (Murphy 2009). Instead, in the place of many stockpiles around the world, a more pragmatic market-led system aims to take its place, which, as has been shown during times of peace, works relatively well as any emergencies are quickly offset to a noticeable degree by international aid. Or so we thought, Murphy highlights the inadequacies of the recent financial crisis in relying on the market as a major policy tool in the fight against hunger. The crisis of 2008 indeed prompted a revival of the idea of food reserves at the G-8 meeting in 2009; yet on closer examination such reserves would be difficult to administer in areas already least able to support their populations. In conclusion, Murphy suggests that grain reserves, particularly in vulnerable areas, are best complemented with other policies of support and provision.

Other safety nets aiming to enhance access to food are said to be more effective if they are ultimately embedded in wider, more general social safety net programmes (EC/FAO 2008). Such programmes would be of benefit to the widest possible vulnerable groups such as those unable to work for one reason or another or perhaps those affected by recession or natural disasters. These nets might include targeted direct feeding programmes such as soup kitchens, free school meals, food fortification, provisioning for expectant and nursing mothers as well as the under-5s. Alternatively, food-for-work programmes might also help support local community projects such as irrigation, roads or buildings such as schools and health centres. Yet another form might include income transfer programmes that may be in the form of cash or in kind payments such as food stamps, subsidised rations and other targeted measures. Lastly safety nets might also include measures like agricultural input subsidies or crop insurance. All in all, and as is evidenced, there are many initiatives responsible governance can take to ensure adequate food provisions to the most vulnerable of its citizenry (SOFI 2008).

22.1.7 Land Rights

Throughout the developing world, land is one of the most critical resources that the rural poor can access. The UN generic land policy model is predicated on five equitable constructs of land management: land distribution, land utilisation, land tenure security, land administration and land adjudication. While other numerous factors of food security are acknowledged, unequal or insecure access to land and land tenure have the most profound effect on determining a person's or a household's access to food security (ECA/ SDD 2004).

Land tenure or the system of rights, both social and institutional, governs access to land and other resources. Such rights might exist as a result of law, statutory or customary, through marriage, inheritance or through power; they might also exist under freehold use, leasehold or by mutual agreement with no contractual basis. The difficulty in many countries is that these rights are not always backed up in law nor are they always static. Land under customary tenure is effectively ownership through habitual use; however, such land is rarely recognised in law and is not protected nor is it tradable collateral, denying its users access to credit and other pecuniary investments. Rights to land too can change relatively quickly, affecting the security of those working in it. This also affects perceived future benefits and affects decisions of investment, whether for instance such investment might result in benefits for existing land users or not. Moreover, communal or common land, not protected by institutional or community governance, might well suffer degradation as users, lacking ownership, often lack the motivation to invest in its future fertility. This last charge, however, has been levelled for years although there is some question as to its continuing relevance as many including Maxwell and Wiebe oppose this belief that community land is automatically doomed to degradation (Maxwell and Wiebe 1999).

Unfortunately too, in many post-colonial areas, there still exists residual policies of pseudo-feudal systems, which see vast tracts of land owned by a privileged few and the rent or leasing of which has led to long-term tension and animosity. Added to this are failed structural adjustment policies and a laissez-faire attitude to change in the face of enormous political pressures of land titling, particularly in Africa, which have collectively compounded to leave a modern legacy of land vulnerability in many developing countries. Things are beginning to change, however, with the slow reform of land titling that sees more and more private ownership. One of the major driving forces behind such privatisation and promotion of individual land rights and a keen area of debate is the commercialisation of agriculture. On the one hand, commercialisation, especially for export purposes, sees the reduction of subsistence food at the household level and a concomitant increase in cash income. This, however, has been said to increase market vulnerability and potential food insecurity. On the other hand, it is argued that such integration into the exchange or globalised economy is a pre-requisite for future sustained growth and development. These two opposing arguments have fuelled much literature and has been tackled to some degree in the preceding sections.

The salient point of this brief look at land tenure highlights the fact that there is a clear and direct linkage between land policies and food security issues. With tenure insecurity stemming from people having too few rights, lack of duration of these rights, a lack of support in exercising such rights or perhaps prohibitive high costs of enforcement, land reform and land registration then, is an important policy goal in attaining security of land and food (ECA/SDD 2004; Roth and Myers 2010).

22.1.8 Reform: Traffic Lights or Roundabouts?

It can be seen from the above that achieving food security must see policy and investment reforms on multiple fronts. In the medium to long term, pressures on the existing agricultural industry will come from increased competition in agricultural markets as a result of various WTO reforms. As the CAP, the Farm Bill and other advantages are eroded as surely these must eventually come, so a new dynamic in global free trade will ensue. Agriculture must also respond to the changing demand and supply dynamics as new technologies, sympathetic environmental policies and more powerful consumers exercise increasing control and choice over practices, quality and traceability (UOR 2009). Policy reform must also consider investment in human resources, agricultural research, rural infrastructure as well as water and natural resources. Further, development policies must also increase agricultural production and international trade in efforts to boost incomes and reduce poverty (Rosegrant and Cline 2003; UOR 2009). One major difficulty in all this is that in many countries the policy environment is protective and inward-looking. In light of this, the burning question then arises, as individual nations do we, as a collective, legislate all the way in the form of traffic lights dictating access or denial via policy and rights, or do we provide enabling environments in the form of roundabouts where structures and recommendations are in place that enable the national and individual self determination of food security. Although this is perhaps a simplistic summary, the implications of policies of governance, in the multilateral sense versus self-determination, is one that continues to bubble beneath the surface.

22.2 International Humanitarian Aid

Despite the growth of humanitarianism in the late nineteenth century or so, aid agencies on the other hand are of fairly recent origin (Paras 2007; Riddell 2007; Walker 2008). The formal institutionalisation of international aid effort began after World War Two with the institutions like the International Labour Organisation (ILO) and the UN's Relief and Rehabilitation Administration's (UNRRA) efforts to raise funds to improve the standard of living in poorer countries. This was immediately followed with the unprecedented Marshal plan (Riddell 2007). Although surpluses had in fact been used in 1896 and during World War One, it was not until the rolling out of the Marshall Plan that surpluses as aid in bolstering recovery and development was being fully embraced (Hjertholm and White 2000; Walker 2008).

With progress in the fledgling green revolution, government policy and programmes during the early 1950s agricultural surpluses once again returned. Previous ideas that such surpluses could be used to alleviate hunger abroad seemed to offer a single solution to the dual problems of hunger and domestic interests. And as history shows us, few countries embraced this notion more enthusiastically than the United States and as the Marshall Plan; which was so lucrative for US domestic farmers, came to an end, so US farmers lobbied for the continuation of food aid resulting in the 1954 Agricultural Trade Development and Assistance Act PL480 (Salter 1950; Hjertholm and White 2000). In the meantime, as a result of concerns that surpluses and aid could act to distort local markets, the first discussions on food aid were held at the FAO Conference in November 1953. This meeting sought to address difficulties in absorbing the commodity surpluses accumulating in North America and disposing of them in a way that was not trade distorting. Reviewing the situation, the conference concluded that possible international repercussions of commercial export markets of donor countries could indeed threaten the orderly development of international trade (CCP 2005). Out of this, by 1954, the FAO had formulated its principles of surplus disposal (Walker 2008). Despite FAO's warnings, however, the expansion of the US surplus policy and the power of the US agricultural industry of the 1950s and 1960s, the world increasingly saw domestic agricultural interests being put above foreign policy interests (Sullivan 1970). This resulted in foreign disposal programmes becoming less related to international aid and cooperation and more to do with the needs of domestic American agriculture in the creation of export markets and the dumping of surpluses. Initially, the aid agencies blissfully turned a blind eye as the expansion of such surpluses enabled voluntary agencies to greatly expand their humanitarian activities and reach ever more needy people (Sullivan 1970). However, this was to have unforeseen implications for the independence of these relief agencies and in particular the nature of their relationship with the government. Other issues too arose with regard to the church–state dilemma inherent in American domestic politics and the fear of possible repercussions with agencies' ability to remain neutral and focused. This in turn led to paralysis in the 1960s of the voluntary agencies' coordinating agency, the American Council (Sullivan 1970).

At about this time, relief operations had come to an end in Europe and new frontiers were being opened up in the less developed countries of Asia, Latin America and Africa. These new areas of opera-tion absorbed more surpluses than previous programmes. One reason for this was that the new experi-mental feeding or developmental programmes such as the school lunch programmes and food-for-work projects began to be linked with less short-termism and more closely associated with longer-term aid development programmes (Sullivan 1970). However, it was not until 1967 with the establishment of the Food Aid Convention that the first truly multilateral instrument with the express purpose of dealing with food aid as a phenomenon in its own right was established (Barrett and Maxwell 2005).

Today, there is a profusion of organisations and agencies involved with food security, so much so that it is sometimes not clear who is responsible for what and accountable to whom.

22.2.1 The Architecture of the International Humanitarian Food Aid System

Structurally, there are a number of stakeholders and international instruments that make up the humani-tarian sector. In a little over 100 years too, a remarkable accompanying body of international humanitar-ian treaty law has also been established. This collective inertia has culminated in a broad field containing a multitude of stakeholders from single issue non-governmental organisations (NGO) to complex insti-tutions responsible for all manner of rescue and relief efforts. Despite such proliferation, it can be said however, that the frontline of international humanitarian food aid is populated by only a few select bilat-eral and multilateral organisations. These include the OCHA, WFP, FAO, FAC, IFAD, UNDP, WHO, UNICEF, USAID and the WTO. On top of this, a strong band of international NGOs, trusts and foun-dations also operate in partnership with these institutions or if need be can also act unilaterally. For a comprehensive diagram of the many stakeholders, see Appendix K. All these organisations have differ-ent ideological mandates that provide the impetus for the provision of aid (Global Policy Forum 2005; Wahlberg 2008). In fact, such is the importance of these NGOs that the WFP co-opts more than 2800 NGOs in its activities across the globe (WFP 2009b).

22.2.2 The Rise of the NGO

On this point, it is interesting to note the rise in numbers and political leverage of NGOs (also known as civil society organisations (CSO)) and private voluntary organisations (PVOs) and their memberships in recent decades (Bloem 2001; Hall-Jones 2006). Numbers of international NGOs, for instance, currently operating in more than one country nowadays currently range anywhere from between 30,000 and 37,000 (Bloem 2001; GPF 2002; YIO 2002; Hall-Jones 2006). In recognising the importance of such groups the United Nations first established a channel for civil society participation at the Freedom from Hunger Campaign (FFHC) established by the 1959 FAO Conference (CFS 2008a). However, many consider the 1992 Rio Earth Summit as the defining watershed of the NGO power base, a time when enough public pressure was applied to persuade governments to adopt greenhouse gas emission agreements (Bloem 2001; Hall-Jones 2006). The ability to wield such direct political leverage and their capacity to confront government policy directly and overtly has given NGOs and their ilk overriding advantages over other politically sensitive institutions like the World Bank, IMF and UN agencies (Hall-Jones 2006).

In terms of size, influence and financial strength, some of the biggest NGOs are found in the area of humanitarian work (Table 22.1). In fact Karajkov (2007) suggests as much as 70% of available relief funding goes direct to the eight biggest NGOs: World Vision, Save the Children, CARE, Catholic Relief Services, Oxfam, Doctors Without Borders, International Rescue Committee and Mercy Corps (ibid).

TABLE 22.1

International NGOs and Private Foundations/Organisations Ranked by Asset Value

International NGOs	
Action Against Hunger (AAH)	Crescent Societies (IFRC)
Action Aid	International Rescue Committee (IRC)
American Jewish World Service (AJWS)	Islamic Relief
Cafod, the Catholic Agency for Overseas	Lutheran World Federation
CARE	Mercy Corps
Caritas Internationalis	Oxfam International
Catholic Relief Services	Refugees International
Christian Aid	Relief International
Doctors Without Borders	Salvation Army, USA
Feed The Children	Save the Children
Food For The Hungry International (FHI)	The Salvation Army
International Committee of the Red Cross (ICRC)	US Committee for Refugees (USCR)
International Federation of Red Cross & Red	World Jewish Aid
	World Vision International

Organisation/Foundation	**Assets**
Bill and Melinda Gates Foundation	$29 billion (2005)
Wellcome Trust (UK)	$24.5 billion (2002)
Ford Foundation	$11.6 billion (2005)
Fondazione Cassa di Risparmio delle Province Lombarde (Savings Bank Foundation of Lombardy)	$7.8 billion
YMCA	$5.3 billion (2005)
The American Red Cross	$3.8 billion (2005)
Goodwill Industries International	$3 billion (2005)

Sources: Karajkov, R., *The Power of N.G.O.'s: They're Big, but How Big? World Press*, 2007; ESA, European Space Agency: Land Cover Implementation Team Project Office, 2008.

It can also be seen from Table 22.1 that some private organisations financially dwarf some of biggest NGO agencies; however, in this, a cautionary note by Karajkov (2007) offers that it is not always about financial power. As financially miniscule in comparison, human rights and advocacy groups like Human Rights Watch, Amnesty International and Freedom House can, when needed, pack a weighty political punch.

22.2.3 Emergency Response

The ultimate role of responsible governance then is to prevent these food disasters or emergencies; however, some emergencies do happen. What is needed in these cases are measures that expediently deal with such emergencies as they arise. Food emergencies can be complex phenomena, which require equally complex solutions. In response, the global humanitarian community has developed an extensive experiential base for responding quickly and effectively (FAO 2007; Wahlberg 2008). In this sense, food aid, although not a sustainable solution to food insecurity, has nonetheless a vital humanitarian role to play. Food assistance, as food aid is more and more often referred to these days, is not solely about emergencies either. Aid is also an important non-emergency preventative measure. For whichever scenario it is used, food aid can include direct food-based provisions such as general rations, supplementary feeding, school feeding and food-for-work projects; food subsidies such as cash transfers and vouchers; or agricultural support. Strictly speaking food subsidies are not considered true food aid assistance although on occasion it can form part of the package. Other concerns include cash transfers and whether indeed these should be included as food aid. Despite these many considerations, food aid or assistance has been credited with saving millions of lives and improving the lives of many more (CFS 2007a; Harvey et al. 2010). In 2009 humanitarian assistance, including food aid, was estimated to be around $15.1 billion, of which $11 billion came from government donations (Figure 22.2) (representing a drop of 11% from the previous year). Collectively too, this aid formed only a small part of the overall $130 billion Overseas Development Assistance (ODA) given out in 2009 (Kellett et al. 2010).

In 2009 global food aid saw the delivery of 5.7 million tonnes of food reaching over 100 million people, which cost in the region of $4.4 billion (Figure 22.3) (OCHA/FTS 2006; WFP 2008a). Up to 60% of this was used in humanitarian emergency situations; yet despite such huge numbers however, this was still somewhat inadequate and fell short of actual needs. The IMF illustrated the full scale of the problem when they calculated in 2003 that food aid provided only about 7 kg of every metric ton of food aid needed; and things have not changed greatly since then (ActionAid 2003; Barrett and Maxwell 2005). As can be seen from Figure 22.3, food aid formed the largest part of the overall humanitarian assistance

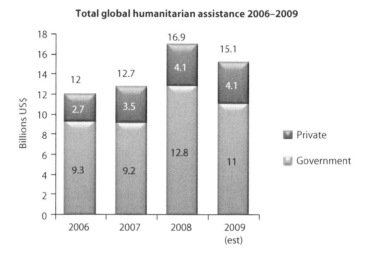

FIGURE 22.2 Total humanitarian assistance 2006–2009. (Adapted from Kellett, J., Malerba, D., Brereton, G., Walmsley, L., Sweeney, H., Keylock, J., Smith, K., Poole, L., Stoianova, V., and Sparks, D. *Global Humanitarian Assistance Report 2010*, Global Humanitarian Assistance (GHA), Shepton Mallet, UK, 2010.) *Note*: Does not include Overseas Development Assistance.

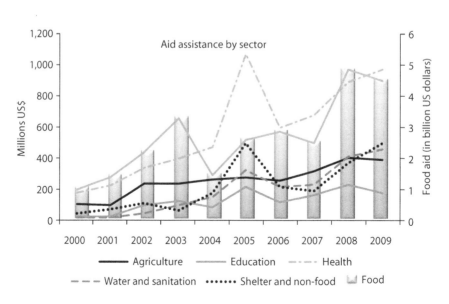

FIGURE 22.3 Humanitarian assistance by sector. (Courtesy of OCHA/FTS (2006). Financial Tracking Service (FTS) of OCHA, 2006.)

package worldwide while health care and education are clearly important contributors together with water and sanitation considerations.

Largely driven by donors and international institutions, aid agencies govern the allocation, utilisation and reporting of aid. Yet, according to many, with so many organisations operating within competing ideologies and under poor coordinated leadership, the sector has been and continues to be in disarray (ODI 1997, 1999; Walker 2008; Karunakara 2010). The problem is seen to be that institutions remain largely unaccountable with existing food aid-related codes of conduct, all being voluntary and applying solely to agencies that distribute food aid rather than the donors, governments and operational NGO agencies themselves (EuronAid 1995; Project 2004; Wahlberg 2008). Having said that, Maxwell and others (Darcy and Hofmann 2003; Maxwell 2006; Wahlberg 2008) inform us that governments in all countries are facing increased public pressure to be more open, accountable and transparent. In response to these concerns too, there are proposals underway to replace the current Food Aid Convention in an effort to prevent the manipulation of aid recipients and to ensure the operational independence of humanitarian agencies against geopolitical or commercial considerations (ibid). Moreover, with such concerns of oversight, the debate was taken to the WTO and while technically it does not operate a mandate to oversee humanitarian operations, given the history of humanitarianism and food aid as a mechanism of surplus disposal it would seem, according to Maxwell at least, inevitable that the WTO be drawn in (ActionAid 2003; Clay 2006; Maxwell 2006). Separately in response to these charges, efforts in 2006 by the NGOs, the International Red Cross and the likes of the United Nations, sought to create a Global Humanitarian Platform (GHP) bringing together coordination of UN and non-UN humanitarian organisations on an equal footing.

Despite this the charge of a sector in disarray was once again brought to bear in 2010. The prime example of this supposed failing was given in the recent earthquake and cholera disasters in Haiti. In response to this emergency, it has been estimated that over 12,000 humanitarian agencies including NGOs, the UN and others were present with over 420 participating in the UN health cluster alone. Despite this, and with over 2500 deaths due to cholera in the face of this overwhelming humanitarian presence, the concern has been raised that by legitimising NGOs with responsibilities, the fear is some may not have the infrastructure, capacity or know-how to carry out the necessary work entrusted to them (Karunakara 2010).

22.2.3.1 Emergency Response Procedures

So what are these procedures, who organises them and how? When it comes to emergencies or disasters, on the ground, each of the stakeholders involved, either working independently or in conjunction with others, has its own emergency response procedures. This section briefly looks at the various responses from different sections of the United Nations system. In their system, the moment an emergency is detected, a chain of events aims to ensure rapid and effective deployment of firstline humanitarian responders. It is worth noting from the outset that under the charter of the UN, while the sovereignty of the state is of primary in any emergency, any breakdown or weakening of leadership may result in responsibility of the victims superseding those of the state (see Section 22.2.6) (UNDAC 2006).

With this in mind, emergencies are coordinated by the UN Office for the Coordination of Humanitarian Affairs' Disaster Assessment and Coordination (UNDAC) team. Initially, in each disaster-prone country, an existing UN Disaster Management Team (UN DMT) is convened and chaired by the UN's resident coordinator (RC)/humanitarian coordinator (HC) (quite often the UNDP representative). This team consists of country level representatives of

- Food and Agricultural Organisation (FAO)
- United Nations Development Programme (UNDP)
- United Nations Children's Fund (UNICEF)
- World Food Programme (WFP)
- World Health Organisation (WHO)
- United Nations High Commissioner for Refugees (UNHCR), when present in the country
- Any other agency representative that is deemed necessary for the specific emergency type

TABLE 22.2

UN Cluster System of Emergency Responsibility

Cluster	Lead Agency
Food	WFP
Refugees	UNHCR
Education	UNICEF
Agriculture	FAO
Nutrition	UNICEF
Water and sanitation	UNICEF
Health	WHO
Camp coordination and management	UNHCR—complex emergencies IOM—natural disasters
Emergency shelter	UNHCR—complex emergencies IFRC—natural disasters
Protection	UNHCR—complex emergencies. UNHCR/UNICEF/OHCHR—jointly determined under the leadership of the RC/HC in natural disasters
Logistics	WFP
Telecommunications	OCHA for emergency telecommunication (tc) UNICEF for common data services WFP for common security and tc services
Early recovery	UNDP

Source: United Nations Disaster Assessment and Coordination (UNDAC) Handbook, Office for the Coordination of Humanitarian Affairs, New York and Geneva, 2006.

Usually the RC remains as lead coordinator but if one agency ends up with the lion's share of the work, the Inter-Agency Standing Committee (IASC) may designate this agency as lead agency and its representative RC/HC. The RC also liaises with local government or local authority officials and any other voluntary or aid agencies in strategic coordination of efforts. After the Asian tsunami of 2005, a 'cluster' system was developed to improve aid coordination. These are areas of responsibility with clearly defined boundaries with each cluster being represented by a UN agency. The designated clusters are highlighted in Table 22.2 and although responsibility for food, refugees, education and agriculture are not strictly clusters, they are included for completeness.

With regards to outside agencies, there may be a large collection of NGOs in any emergency. These might already be organised under umbrella organisations such as the International Council of Voluntary Agencies (ICVA), InterAction, the Steering Committee for Humanitarian Response (SCHR) or similar organisations; however, the activities of such groups are also integrated within the overall OCHA's coordination effort.

Often in these situations too, food is rarely the only resource needed by those suffering; typically, affected persons also need access to complementary non-food resources like clean water, reasonable sanitation, shelter and health care. So in terms of food aid and associated assistance what happens and how does it work?

22.2.4 Humanitarian Aid in Action: Modern Food Aid

Food aid in particular is a large part of the humanitarian aid effort and getting food aid to recipients in the developing world can be a daunting and controversial task. Pulling off the complicated journey from fields to feeding centres calls into play a number of disparate players including international bodies, the agriculture industry, NGOs and advocacy groups (Rucker 2007). Each of these organisations has specific mandates and there is considerable overlap in terms of their functions and responsibilities. Moreover, while there exists little formal means of coordination among these various interests, a handful of major organisations provide guidance on the process (Rucker 2007). Figure 22.4 illustrates just how important

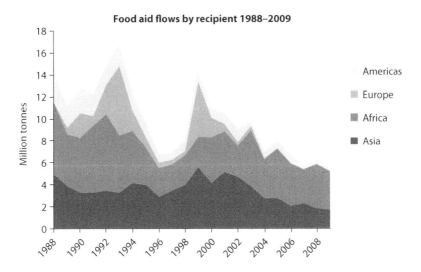

FIGURE 22.4 Food aid flows by recipient 1988–2009. (Courtesy of the World Food Programme, 2011.)

these flows have been and continue to be over the last two decades or so, especially in the African and Asian continents.

In 2009, eight countries requiring assistance received 55% of all food aid for that year. These included Ethiopia at 17%, the Sudan at 9%, Somalia and the Democratic People's Republic of Korea (DPRK) at 6% each, Kenya at 5% and Afghanistan, Pakistan and Zimbabwe all at 4% (WFP 2011).

So where does this aid come from and what form does it take?

22.2.4.1 Types and Sources of Food Aid

In the Food Aid Convention (FAC) in 1960 the world's wealthiest countries committed to providing about five million tonnes of food aid per year and as of 1999 donors were able to make equivalent contribution in cash instead of food. However, the food aid process is quite complex whereby some of this food is delivered to the needy; other times it is sold to fund development and food security projects and other times it is sold to pay NGO costs of transport, etc. At other times too, food is donated to be sold in the recipient countries as part of concessional domestic fiscal policy objectives such as balance of payments, etc.

Food aid then is provided in many different types and through many differing channels. Where physical food aid is delivered, it targets vulnerable groups and is provided in the form of

> *General rations*—providing a complete basket of commodities in quantities sufficient to meet requirements; *supplementary feeding*—targeting groups especially at risk of malnutrition, such as pregnant women and small children; *therapeutic feeding*—in feeding centres or clinics or *food-for-work programmes*—when wages are paid in food. (WFP 2009c)

These are provided in three categories either in the form of grants (gifts) or sold or loaned on concessional terms more favourable than world markets (Figure 22.5). The three categories include the following:

1. *Relief, or Emergency Food Aid*: Typically for emergency situations, in cases of war or natural disasters, where emergency food aid from government and NGO agencies is intended for direct, free distribution (ActionAid 2003; Barrett and Maxwell 2005; Mousseau 2005).
2. *Programme Food Aid*: This is a form of 'in-kind' aid whereby food is grown in the donor country for distribution or sale abroad. This is typically bilateral on a government-to-government basis. Some programme food aid is donated to recipients while the rest is sold on concessional

terms. Programme Food Aid, or in-kind food aid, makes up the majority of aid from the United States (Mousseau 2005) and at times too, this type of food aid is sold by the recipient central government who then uses the proceeds for budgetary and balance of payments relief in a practice called monetisation (discussed in the proceeding sections) (Barrett and Maxwell 2005).

3. *Project Food Aid*: This is food aid delivered as part of a specific project related to promoting agricultural, social or economic development as part of a programme of poverty reduction or disaster prevention.

Deliveries of food (usually free) can either be used directly or monetised whereupon the proceeds are used for project activities (Barrett and Maxwell 2005). Most food aid is channelled through multilateral agencies, mainly the WFP and NGOs and are categorised by the way the food aid is used, i.e. project, programme or emergency (Figure 22.5), or by the way it is sourced, i.e. direct transfers or triangular and direct purchases (Figure 22.6).

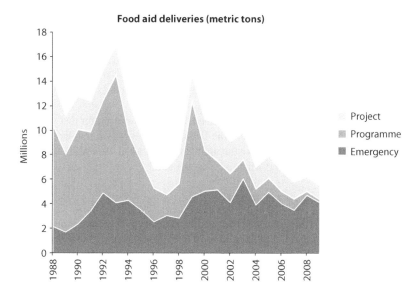

FIGURE 22.5   Food aid by type (1988–2009). (Courtesy of the World Food Programme, 2010.)

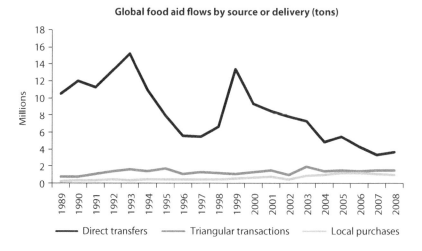

FIGURE 22.6   Aid flows by delivery source (1989–2008). (Courtesy of the World Food Programme, 2010.)

Direct transfers are food aid donations that originate in the donor country and are a form of 'tied aid' in the sense that they are limited to food sourced in the donor country. That is to say, rather than donating money, the governments or agencies purchase food in their own countries to be shipped out to where it is needed. Often in addition, directly transferred food aid is tied to additional conditions such as the use of donor-country contractors. For instance, the United States and Canada require that at least 70% and 90% respectively of the procurement, packaging and shipping be handled by firms of their own countries.

Triangular Transactions: Triangular purchases describe food aid purchased in a country outside of the donors for use as food aid in another. Triangular purchases are usually financed by cash contributions from donors.

Local Purchases refers to the procurement of food in the recipient country and is one of the most cost-effective ways to source food aid (WFP 2009c).

Although not strictly in the above category there is one other consideration worth mentioning here and that concerns monetisation.

Monetisation is the practice whereby donated food is sold in the recipient country's market to increase the local supply of food whilst at the same time providing funds to cover the costs of these deliveries by NGOs. It is also a practice used to generate cash for development projects such as road building, nutrition education and microcredit, among others. Originally monetisation derived from divergent political and organisational interests and nearly all of this practice today comes from the United States. It is also prudent to mention that not all food sold is considered monetisation. This exception usually relates to food deliveries used for programme food aid (often provided as balance of payments support), which is then sold but not considered the same as monetised food aid. Recent years has seen increasing criticism of monetised aid and is discussed in the following sections (Harrell 2007; Simmons 2009).

The extent of each of these types and sources of aid for 2009 are highlighted in Table 22.3. It can be seen that aid over the years is decreasing, while the WFP's role in delivering this aid is increasing to the point in 2009 when it was responsible for about two-thirds of all food aid or about 95% of all multilateral aid. It can also be seen that monetisation, or more precisely food sold, is also on the decrease while more and more food aid is being received through multilateral donations.

Interestingly 100% of the last two years' donations have been on a full grant basis with no concessional sales; before that 8% was concessional in 2007. Also of note is the increasing prevalence of multilateral aid up from 29% in 1999 to the 70% mentioned earlier. This highlights a more cooperative and centralised role of aid, at the centre of which is the World Food Programme That said, while the WFP remains the most significant emergency food aid organisation, some governments still provide large amounts of food directly to NGOs either for emergencies, development projects or also directly to governments themselves (bilateral). Note too that while local and triangular purchases of food have remained largely static, all of the reduction in overall food aid tonnage (12 million tonnes in 1990 to 2.8 million tonnes in 2009) has resulted from a significant decrease in direct transfers (food purchased by donors in their home countries). And lastly, of particular note is that total tonnage of food aid in 2009 of 5.7 million mt is the lowest since records began in 1961. This declining food aid trend has huge knock-on effects for the numbers of insecure people that can be targeted and is reflected in the types of aid given. As a result of declining donations, both programme and project aid has fallen as proportion of the aid needs so as to maintain delivery of aid in emergencies (WFP 2009a).

22.2.4.2 Daily Rations

As direct transfers decline and cash donations take prominence, sourcing food becomes more important. The WFP aims to purchase food aid as close to the source of the emergency as possible. This cuts down on transport costs and speeds up the delivery process while at the same time supports local economies (WFP 2011). Consideration is given to the rationing and nutritional content of this food and the WFP operates on a basic daily ration of 2100 kcal of energy. While diets are different all over the world, where possible cultural dietary needs are also considered. However, in emergency situations the main focus is to provide sufficient calories and other nutrients. By way of example, a generic but appropriate WFP food basket might contain a staple, pulses, vegetable oil (fortified with vitamin A and D), sugar and iodised salt. These might also be complemented with special blended foods like Corn Soya Blend that also help

TABLE 22.3

Supply and Distribution of Food Aid 2005/9

	2005		2006		2007		2008		2009	
	million tonnes	%	million tonnes	%	million tonnes	%	million tonnes	%	million tonnes	%
Global Food Aid	8.2		6.7		5.9		6.3		5.7	
By category										
Emergency	5.2	63%	4.2	63%	3.7	63%	4.8	76%	4.3	75%
Project	2.1	26%	1.6	24%	1.4	24%	1.2	19%	1.2	21%
Programme	0.9	11%	0.9	13%	0.9	15%	0.3	5%	0.2	4%
By food type				0%		0%		0%		0%
Cereals	7.1	87%	5.7	85%	5.1	86%	5.4	86%	4.9	86%
Non-cereals	1.1	13%	1	15%	0.8	14%	0.9	14%	0.8	14%
Global food aid deliveries										
WFP share of total	4.5	55%	3.8	57%	3.1	53%	4	63%	3.8	67%
Procurement in developing countries	2.4	29%	2.3	35%	2.3	39%	2.1	33%	1.8	32%
By mode										
Local purchase	1.2	15%	1.3	19%	1	17%	1.1	17%	0.9	16%
Triangular purchase	1.3	16%	1.3	19%	1.4	24%	1.5	24%	1.9	33%
Direct transfer	5.7	70%	4.2	63%	3.5	59%	3.7	59%	2.8	49%
By sale										
Sold	1.8	22%	1.6	24%	1.4	24%	0.5	8%	0.4	7%
Distributed	6.4	78%	5.1	76%	4.6	78%	5.7	90%	5.3	93%
By channel										
Multilateral	4.4	54%	3.8	57%	3.3	56%	4.2	67%	4	70%
Bilateral	1.8	22%	1.1	16%	1.3	22%	0.6	10%	0.4	6%
NGOs	2	24%	1.8	27%	1.4	24%	1.5	24%	1.4	24%

Source: WFP, Food Aid Flows, World Food Program (WFP), Rome, 2005, 2006, 2007, 2008a, 2009a.
Note: Missing data has been calculated from given values. Also values for different years are often
reestimated in updated publications. These values are all from original contemporary
publications.

to improve an individual's micronutrient uptake, as illustrated in Figure 22.7, providing a total energy count of 2100 kcal including 58 g of protein and 43 g of fat. There are also alternative rations for those without access to cooking facilities made up of pre-cooked ready to eat foods.

In other non-emergency situations, supplementary rations aimed at the more vulnerable groups might also be given out to address specific nutritional needs or as a preventative measure in children and pregnant women (WFP 2008b).

While considering the relatively small emergency ration weight of about 500 g per person per day, this might seem very manageable. Yet scale this up to say a relatively small emergency of 100,000 people who need to be fed over the next month, the logistics become somewhat daunting. In fact, over the month this would require daily deliveries of nearly two 26-ton trucks full of food rations, as well as 500 cubic metres or 500 tonnes of freshwater every day (based on 5 L per person). Then there are the issues of shelter, sanitation, water for washing and personal hygiene as well as health and disease risks and much more and all these challenges begin to stack up.

As the humanitarian sector matures the availability of fortified foods increase as does the types of projects and programming at agencies' disposal. One of the overriding requirements of this complicated mix

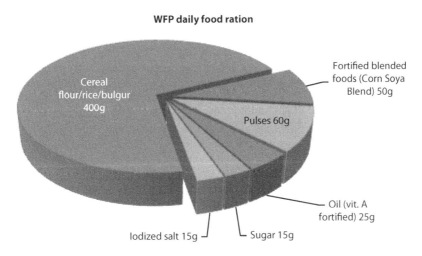

WFP daily food ration

Cereal flour/rice/bulgur 400g

Fortified blended foods (Corn Soya Blend) 50g

Pulses 60g

Oil (vit. A fortified) 25g

Iodized salt 15g

Sugar 15g

FIGURE 22.7 Daily WFP rations in food emergencies. (Courtesy of the World Food Programme, 2010.)

of choices is the need to choose appropriate commodities and programming approaches in combination with suitable non-food interventions such as water and sanitation in a holistic manner that is both efficient and culturally acceptable (WFP 2008b). For example, in considering whether to bring cash or food into non-emergency food programming areas, careful thought must be given to whether food for direct distribution might upset the local balance of supply and demand. Market demand for instance may suffer, which might disincentivise local producers; on the other hand cash transfers effectively increase personal incomes and of course more purchasing power. This could also have inflationary pressures and end up pushing up food prices countering any cash benefits (Bonnard 2008a). Such considerations plague all levels of food aid decision making with much criticism frequently aimed at many aspects of the sector. The following highlights many other those criticisms simultaneously illustrating the extent of the difficulties.

22.2.5 Problems of Aid

The many issues outlined here illustrate the breadth of consideration needed in considering appropriate food aid and best practice. Perhaps two of the most often quoted charges levelled at aid is that firstly it creates dependency and secondly it largely serves political self-interests. In the first charge, it has been suggested that aid is not sustainable in that it tends to undermine local agricultural production by lowering prices of the commodities supplied. This has been said to further discourage local production and possibly prolonging the dependence on food aid deliveries (Clay et al. 1998; ActionAid 2003; Cohen 2005). These are criticisms that have emerged perennially. In retaliation, the FAO's 2006 State of Food Agriculture (SOFA) report introduces empirical evidence that shows aid is unlikely to cause dependency simply because of its sporadic and unpredictable nature. Moreover, the levels of aid tend to be too small and insufficient to be relied upon (FAO 2006; CFS 2007a). With regards to the second charge of the politicisation of aid, there are perhaps two distinct elements: the domestic situation and favourable trading practices. This has been prevalent in the long-standing charge that the United States is guilty of using aid to promote domestic interests' terms of trade (Rowe 1949; Anon 1953; Murphy 1956; Kristjanson 1960; Sullivan 1970; Goldin and Mensbrugghe 1993; Reutlinger 1999; ODI 2000; ActionAid 2003; Subramanian 2003; Clapp 2005). The wide practice of direct transfers, albeit decreasing considerably over the years nevertheless, still accounts for the majority of aid given today. Of this, 82% was supplied by the United States. This practice of aid donation uses government money to buy commodities on their local markets often but not always surpluses. This food is then further 'tied' with other conditions, which see the United States conditionally offering this aid on the proviso that 70%–75% of packaging and transport costs are spent on US registered companies. The Canadians too are guilty of this and ensure that 90% of their direct transfer aid is tied this way. In consequence, it has also been said that as a direct result of these significant financial flows, three interest groups—the shipping companies, agribusiness

firms including farmers and processors and a small but powerful group of non-governmental organisations known to some analysts as the 'iron triangle'—cooperate through intense lobbying and pressure to
perpetuate the status quo (Mousseau 2005; Johnson 2007).

Other cash donations are rarely 'politically free' either; in this practice donors often impose conditions on how the money is to be spent or in which countries the money must be targeted. By contrast,
donors give only a small percentage of aid in unconditional 'multilateral' aid (Mousseau 2005; WFP
2008c, 2011). Further, according to Barrett and Heisey (2002) this suggests that bilateral food aid tends
to mainly be supply-driven and motivated by domestic interests whereas multilateral trends to food aid
is more recipient-oriented. The second form of politicisation is the practice of favourable donations that
go, rather than where it is perhaps most needed, but more so to ingratiate or curry favour in the political
sense (Subramanian 2003). Others critics too are of the view that political antagonism often sees aid
used as political leverage (DeRose et al. 1998; ActionAid 2003; Poole-Kavana 2006). However, perhaps
the most heinous charge levelled at the aid sector is the continual political hegemony exercised by the
United States. In this charge, Mousseau (2005) and Wahlberg (2008) criticise several key players such
as the WFP and some of the major US-based NGOs for being under the influence of, and following the
priorities of, US foreign policy with regards to areas of aid intervention. This is a strong charge, and one
that is difficult to prove one way or the other, although in fairness though, there has been considerable
change in such practices over the last few years. Organisations like CARE International, for instance, are
now refusing to accept direct transfers of food aid from the US government, as are others as it has been
largely discredited as inefficient and time consuming with some transfers taking weeks if not months to
arrive at their destinations. By contrast given the right tools the WFP can deliver food aid in as little as
48 hours (ActionAid 2003; Barrett and Maxwell 2005, 2006).

The next main concern receiving a lot of attention recently is the practice of monetisation. This is
mainly practiced by large NGOs to raise money for shipping and handling of food aid as well as generating funds for development projects by the sale of donated food (Murphy and McAfee 2005). In 2006,
NGOs monetised 68% of all their project food aid (Wahlberg 2008). However, it has been found that
monetisation severely distorts local markets and production. In fact so deep was the concern that it was
fully expected that the WTO Doha round of trade negotiations would prohibit this practice (ActionAid
2003; FAO 2006; Harrell 2007; Simmons 2009). This never happened and some fully expect the practice
to continue at the very least and perhaps increase over the next few years.

Article 11 of the International Covenant on Economic, Social and Cultural Rights states that the right
to food implies food 'free from adverse substances, and acceptable within a given culture.' (UNHCR
1999; Wahlberg 2008). In this respect, food aid has been charged at times with changing consumption
patterns in recipient countries away from locally produced crops in favour of those supported by aid
imports (Goldin and Mensbrugghe 1993; Subramanian 2003; SOFA 2006; Wahlberg 2008). Moreover,
while the WFP accepts donations of GM food, they are required to respect recipient countries' right to
reject GM food aid. However, that said it has been argued that there have been cases where the WFP has
given countries in crisis no choice but to accept GM crops (Wahlberg 2008).

Another concern of aid is that national interests and media attention, rather than need, sometimes
determine how governments and private donors prioritise crises. High media profile crisis like the Asian
tsunami in December 2004 or the Haiti crisis of 2010 received plenty of funds while many hunger emergencies went unreported. In the worst cases, repeated failures or repeated appeals risks donor fatigue
(Wahlberg 2008; WFP 2008c). Without these valuable donations as well as lack of full funding governments, the WFP is forced to reduce international food aid to crisis management rather than through
preventive projects (Weingärtner 2004). This calls into question the various costs of the different types of
aid with Barrett and Maxwell (2005) arguing in 2005 that for every $1.00 of food procured in the United
States, it actually cost $2.13 while the European programme food aid was said to cost $1.33 for every $1
of food provided. Similar calculations by the OECD in 2005 determined, too that in-kind food aid was at
least 30% more costly than food aid that was purchased directly in recipient countries.

Lastly it has been suggested that in its current form, the Food Aid Convention (FAC) has numerous
shortcomings described as outdated and ineffective (Mousseau 2005; Barrett and Maxwell 2006; Carter
and Barrett 2006). Criticisms are also levelled at that the FAC for being housed in the International
Grains Council, a commercial trade promotion body based in London. What is more are suggestions

that, like the CCSD and many others, the FAC lacks enforcement mechanisms and transparency in its functioning (Morrison and Sarris 2007). Godoy however further queries whether mismanagement of aid is being used at the expense of longer term objectives and wonders whether in the long run this might actually undermine local democracy and sovereignty (Godoy 2007). Couple these concerns with other outdated institutional practices, duplication and ineffectiveness and it should come as no surprise that repeated calls for the substantial overhaul of the aid system in favour of more transparent accountable and outward looking mechanisms is being made (Carter and Barrett 2006; Morrison and Sarris 2007).

This long list of problems might indeed leave some thinking well why bother? The truth is for all its faults the evidence is clear that some aid, albeit imperfect, is better than no aid. Moreover, such concerns are not new and as more pressure is brought to bear so the sector tends to re-invent itself on a regular basis. Concerning the many charges laid at the feet of the United States it is worth pointing out that for all the problems discussed the United States still remains the biggest single donor by a long shot regularly contributing more than half of all annual aid (Darcy and Hofmann 2003; Murphy and McAfee 2005).

Before we leave aid, there is one last aspect that requires some attention and this is the right to protect or the right of humanitarian intervention.

22.2.6 The Right to Protect

Earlier I mentioned that international law dictates a nation's sovereignty supersedes all else with the caveat that in situations of weakened or failing states, victims of humanitarian disasters take precedence. This is in reality a grey area and one that I will explore further here.

The foundations of current national sovereignty date back to the Treaties of Westphalia in 1648 that ended the 30 years of war in Europe. Out of this melee, Europe emerged from the Middle Ages into a world of sovereign states (SEP 2003). Importantly, sovereignty established itself as the substantive constitutional authority of the time and one in which no other state had the right to intervene in domestic matters. In other words, Westphalia established the authority of the head of state in the governance of its own matters. While these treaties were initially in response to the Holy Roman Empire's attempts to establish a Christian world empire in Europe, they are fundamental to the modern notion of sovereignty. Building on this the UN's Charter also embodied the notion that as long as it remains within international law, what happened within a state is the state's own affair.

That still remains the case today; well, largely, since about the 1990s there has been some glaring cases of state ignored, and sometimes state sanctioned human rights violations. As a result, a growing feeling within the international community began to emerge that the state was not always the benevolent or responsible entity that perhaps it should be. Consequentially, more and more people began to question whether the concept of non-interference should still prevail in the face of such violations. In this scenario, faced with a humanitarian emergency—whether through conflict, natural disaster or otherwise—and if the government or ruling parties—through impotence, ineptitude or worse complicity—fail to adequately protect is people, then the argument goes that surely the international community has a moral obligation to intervene. In this case, the authority of the international community to protect the citizenry should supersede those of sovereignty. Indeed, the so-called right of 'humanitarian intervention', or the Kouchnerian* notion was used to intervene on humanitarian grounds by the UN in Iraq in 1991 and Somalia in 1992, and then by France in Rwanda in 1994. The trouble was, this laid dangerous precedents for intervention without the UN Security Council's express permission. With little guidance on this matter and with shockwaves rippling through traditional humanitarian agencies, Kofi Annan, then Secretary General of the UN, posed this dilemma to the General Assembly in 1999 and again in 2000. Far from being opposed to the idea, Annan was seeking an internationally agreed and acceptable way in which such action could be taken. In response, Canada's then prime minister Jean Chrétien established the International Commission on Intervention and State Sovereignty (ICISS) to effectively promote international debate on the relationship between intervention and sovereignty. A year later, after extensive

---

* Kouchnerian notions refer to the term 'devoir d'ingérence' (duty of intervention), which was coined in the late 1980s by Professor Mario Bettati and French politician Bernard Kouchner. The idea is that another state has a moral right to intervene in another state's affairs in the name of humanitarianism.

research and consultation, the report, 'The Responsibility to Protect', rubber-stamped the notion shifting the emphasis of sovereignty from one of rights and control to one of responsibility. In this shift the state was seen to be responsible for the protection of its people rather than in control of it and if, for one reason or another, the state failed in its duty then the international community would be empowered to act. Following the report in 2001 the idea was unanimously adopted by world leaders and further enshrined in the outcome document of the 2005 World Summit High-level Plenary Meeting of the 60th Session of the General Assembly. This provided a framework in which the ideals, variously known as the right to humanitarian intervention, or the right to protect (R2P), finally crystallised the obligation to act (UN 2005). Moreover, it was also established by this outcome document that the legal framework was in fact already in place in the forms of well-established principles of international law (Rosenberg 1994).

While predominantly aimed at conflict and genocide, the same framework has been said to be applicable in times of natural disasters. Take, for instance, the poor government response by the Myanmar authorities in the face of Cyclone Nargis in 2008. In this case the government limited the amount of aid entering the country to a fraction of what was actually needed and indeed went as far as rejecting aid from certain countries. While the UN and others criticised Myanmar's response as inadequate and inhumane, some senior politicians discussed whether the situation justified invoking the 'responsibility to protect' (Ford 2009). As it was a middle ground was found however, just by provoking the discussion, many fundamental questions of independence were posed. The difficulty here is that up till now, the tacit notion that humanitarianism operated inside a political vacuum would no longer apply. While perhaps morally unimpeachable the consequences of this type of uninvited intervention are far reaching. Current humanitarian intervention that moves so freely in a political no-man's land, operating within the doctrine of 'silent neutrality', to borrow a phrase, runs the risk of partisan humanitarianism. How this pans out for the workers in the field and the policy makers at home is difficult to imagine. In the event of forced entry are aid workers to be considered combatants? Will they require training, perhaps protection? What of failed or limited successes with interventions such as seen in Haiti are we then, by the same rules to be held accountable? Many interesting issues are raised in such a scenario; however, since the ideas are in its infancy the repercussions for global food security are, as yet unknowable (Evans et al. 2001; Abbott 2005).

So, how are all these food security issues monitored considering, as we have in this book, that there is no single measure of food security? Good governance determines that a suite of indicators is the best approximation but how and what are these proxy measures?

22.3 Measuring, Classifying and Monitoring Food Security

Measuring and monitoring food security, as has been discussed throughout the book, is a complex process. Different types of food security information are needed by different stakeholders and inherent in monitoring and assessment of these situations are baselines from which deviation can be assessed. In all cases however, regardless of needs, access to reliable and timely data is cardinal (DeRose and Messer 1998; Hoddinott 1999). As discussed in Section 1.4, many indicators are monitored besides the core groupings and also, as previously discussed, the various stakeholders will have varying perspectives and may interpret and view food insecurity differently (Clay 2002; Maunder 2006a). A few examples of the different stakeholders are illustrated:

- *Policy-makers*: as well as their advisors and line ministries, such as agriculture, finance and environment are responsible for directing funds and/or services to the food insecure or vulnerable;
- *Government officials and elected representatives*: nationally or locally elected representatives;
- *Local government authorities*: local government staff might be responsible for the administration of food security and nutrition on a day-to-day basis;
- *Private sector and parastatal organisations*: private and parastatal (owned wholly or partly by the government) organisations providing services to the insecure;
- *The donor community*: unilateral, bilateral or multilateral agencies such as UN, international NGOs and civil societies involved in food security;

- *Researchers*: statisticians, academics and the like at universities, national statistics offices, research institutes and other stakeholders interested in food and nutrition issues;
- *Training institutions*: those involved in the management of food insecurity information systems, analysis and dissemination of collected data (ibid).

Each in turn will require indicators for monitoring and the analysis of different aspects of food security. These might include

- *Market transparency*: providing information on current market data concerning movement and prices of food commodities;
- *Early warning*: monitoring food security situations in the event of an impending crisis;
- *Emergency needs assessment*: short-term analysis for effective response;
- *Vulnerability analysis*: aiming to determine risk analysis and management of vulnerable groups and situations;
- *Household food security*: determining the micro level incidence of food security;
- *Policy and programme development*: analysis is conducted in respect of government or institutional policy and perhaps longer term development planning;
- *Advocacy*: analysis might be required to promote understanding of underlying causes and outcomes of food insecurity.

Under these circumstances, food security is a complex concept that no one measure can adequately describe. As such decision makers rely on a pool of appropriate and complimentary set of indicators and applied analytical tools at both the national and international levels (Hoddinott 1999; Devereux *et al.* 2004; Gerster-Bentaya and Maunder 2008b; Sibrián et al. 2008). However, data collection and monitoring is an expensive exercise (Hoddinott 1999; Maxwell et al. 2003). Perhaps the first to collect global data was the League of Nations; however, since then, the United Nations FAO has been routinely collating regular hunger statistics dating back to its inception (FAO 2011). Moreover, changing requirements have meant more volume and frequency of data needs, as in the example of the malnourishment metrics, or the MDG targets and others are being demanded. In response, there is concern that scarce national resources are being utilised without offering commensurate return from the information collected (FAO 2008b). This has led to calls for rationalisation of data needs in terms of efficient use of existing data as well as taking account of duplication (FAO 2002c).

So how are institutions measuring and monitoring food security? The international community relies on a suite of indicators and composite indices like that of the Global Hunger Index (GHI) (Appendix F). In this way,

> Developing indicators to measure the different facets of food security presents numerous conceptual and methodological challenges. (Nathoo and Shoveller 2003)

For a long time too, primacy was given to measuring the quantitative aspect of food insecurity; however, increasingly of late stakeholders have been seeking techniques for food insecurity and hunger that are simple to use and easy to analyse. Consequentially many have moved toward the use of qualitative food security measures to supplement this quantitative data.

22.3.1 Classification Systems

Classification systems of food security are not new; as long ago as the 1880s the Indian Famine Codes sought to categorise famines according to severity. Nowadays there are numerous ways in which food security situations are classified. Agencies and institutions such as Oxfam, WFP, FAO GIEWS, MSF, FEWS NET and others have over time developed a multitude of food security classification systems. There are important differences in each which depend to a large extent on the institutions involved. In spite of this proliferation, once again consensus is lacking and as yet there is no commonly used scale. This has implications for comparability of the different food crises among stakeholders.

The Integrated Food Security and Humanitarian Phase Classification Framework (IPC) is one such model. This classification, just like many others, is not an assessment methodology in itself; instead it integrates information from diverse sources to classify food security according to reference outcomes drawn from what they feel are accepted international standards (IPC 2007). First developed by the Food Security Analysis Unit (FSAU) of Somalia, it is now being developed and the aim is to roll it out as a WFP/FAO industry standard; to date, the IPC has been introduced in 14 countries (Gerster-Bentaya and Maunder 2008a; WFP/VAM 2010). The IPC scale starts from a generally food secure situation and extends to a humanitarian catastrophe; incremental in five phases: Phase 1, generally food secure; Phase 2, moderately/borderline food insecure; Phase 3, acute food and livelihood crisis; Phase 4, humanitarian emergency; Phase 5, famine/humanitarian catastrophe. Each phase relies on a collection of indicator cut-off points that mark the severity of the situation (IPC 2007).

Another such model is the USAID's FEWSnet, which aims to provide early warning and vulnerability information on emerging and evolving food security crises. In their system, crises are based on indicator cut-off thresholds and divided into: No Acute Food Insecurity; Moderately Food Insecure; Highly Food Insecure; Extremely Food Insecure; and Famine. While also a partner in the IPC system previously mentioned, the USAID still uses their own classification (FEWSNET 2010). Another example of a food security classification system is the US Food Security Core Module (FSCM). Over the last decade or so, the United States has developed new methods for monitoring food security at the national and sub-national levels. The Food Security Core Module (FSCM) is an 18-question module providing a means of measuring the prevalence and severity of food in security in the United States. While individually these questions do not provide great insight, collectively they significantly correlate with more traditional quantitative measures. The qualitative scale used in their assessment highlights four levels ranging from security or insecurity:

1. *Food secure*: little or no evidence of food insecurity;
2. *Food insecure without hunger*: food insecurity concern is evidenced by households' adjustments to food management;
3. *Food insecure with moderate hunger*: adults food intake is reduced and experience hunger;
4. *Food insecure with severe hunger*: adults and children experience hunger and show evidence of more severe hunger (FAO 2002c; Kennedy 2002).

This survey has been administrated every year since 1995 and is now being adopted by other countries. Other examples of disparate systems include Oxfam, Médecins Sans Frontières (MSF), the Famine Magnitude Scale of Howe and Devereux (2004) and the Food Insecurity Classification developed by Darcy and Hoffman (2003) (Darcy and Hofmann 2003; Howe 2004). The main point illustrated in the above highlights the various proprietary classification systems that as yet are non-convergent. There is growing consensus of the need for cross-comparability and the IPC in this area is one to watch.

22.3.2 Monitoring and Early Warning Systems

The line between classification of a situation and monitoring is not always that distinct. Some of the monitoring systems also classify food security situation and vice versa. Much attention today is focused on the monitoring and surveillance of food and nutrition and once again taps in to the indicator debate, of which ones and how many. It began after the 1970s famine crisis of Africa when much credence was given to shortfalls in supply as a result of production failure. This was further highlighted at the 1974 World Food Conference, which ultimately was instrumental in improving information systems, particularly in the area of vulnerability mapping and assessments. Notably during this time it was believed that as a result of reduced regional food supplies and access, individual or household (micro) insecurity manifested in increased prevalence of undernutrition; ergo supply deficits related directly to nutritional status (Maxwell and Frankenberger 1992). This model was influential in utilising macro food balance sheets and nutritional surveillance programmes as dominant indicators of the time (Maxwell and Frankenberger 1992). As understanding of the problem increased so too did the realisation of the need to develop forward-looking analyses aimed, not only at the currently food insecure, but at those who were likely to become food insecure in the future. As a result, while traditional early warning systems concentrated on monitoring agricultural production, the newer models expanded, taking in the wider societal influences of environment, water and natural resources (Scaramozzino 2006). This meant that early warning systems solely concentrating on potential food crisis were oftentimes no longer sufficient to address decision makers' needs. Today stakeholders seek cross-national, forward-looking, early-warning analysis to ascertain humanitarian crises and to support resource allocations and other planning needs (CFS 2000).

These changes also mean that many early warning systems of today, now also monitor a broader set of indicators including access, health and environmental considerations (Maunder 2006b). Change has also been witnessed with the advent of technological advanced data gathering techniques and many food security early warning systems now regularly involve the use of satellite based data. Data from NASA, the Moderate Resolution Imaging Spectroradiometer (MODIS); or Landsat satellites from the National Oceanic and Atmospheric (NOAA) or the joint European Space Agency (ESA) are all used to regularly monitor repetitive coverage of continental Earth surfaces in the visible, near-infrared, short-wave, and thermal infrared regions determining vegetative indices and crop coverage.

Several of the more important monitoring systems are briefly looked at here.

22.3.2.1 FAO's Global Information and Early Warning System (GIEWS)

One such early warning system is the GIEWS. Located in the Commodities and Trade Division of FAO, the Global Information and Early Warning System (GIEWS) began life in 1975. Over time, it has become one of the leading sources of information regarding food production and security for all countries around the world regardless of FAO membership. It works closely with the WFP and other UN bodies as well as many external humanitarian agencies and the resultant information consists of regular situation reports, which are disseminated through its many publications. The scheme also maintains an integrated information system known as the 'GIEWS Workstation', this is a software platform similar to that of the KIDS system (see below) that aims to bring together much of the information in a GIS interface for ease of visualisation (FAO 2002b). The GIEWS publications include Crop Prospects and Food Situation, Food Outlook, CFSAM Reports and Alerts, the Sahel Report, Food Crops and Shortages as well as the Africa Report.

22.3.2.2 Food Insecurity and Vulnerability Information and Mapping Systems (FIVIMS)

As a result of the need to formulate and implement suitable policies and programmes to realise the objectives of the World Food Summit in 1996, the FIVIMS initiative was launched. The Food Insecurity and Vulnerability Information and Mapping System (FIVIMS) promotes the development of a framework specific to individual country needs. It is an inter-agency initiative including over 25 members comprising multilateral, bilateral and NGOs. In principle, the FIVIMS system is loosely defined to include any information system or network that monitors the food security situation of people. It recognises the uniqueness of each country's situation and as such recommends a selection of indicators based on needs. These indicators in turn are based on the notion of a common global standard aimed at comparability and are taken from a pool of 15 domains. Outcomes of this analysis are tables, maps and charts detailing information about the food security situation of homogenous vulnerable groups as well as that of the national and subnational populations. As with the GIEWS this can also be aided by the use of the key information database system (KIDS) and allows for a certain standardisation of information, which contributes enormously to the resources of information needs (Devereux et al. 2004; CFS 2007b; FIVIMS 2008).

22.3.2.3 Food Insecurity Information and Early Warning Systems (FSIEWS)

In 2000 the FAO produced a handbook for defining and setting up a food insecurity information and early warning system (FSIEWS). Its aim was to take the GIEWS approach to food security and translate this into a national or regional tool in the fight against hunger. As a framework it is adapted to the strengths of each country relying heavily on existing monitoring systems and its particular special strength is its application of the four areas of food security (availability, stability, access and utilisation). FSIEWS aims to give reliability of analysis whilst at the same time give sufficient time for decision-makers to act.

22.3.2.4 The Key Indicator Data System (KIDS)

Expediting the dissemination of much of the UN information is the KIDS framework. It has been developed by the World Agriculture Information Centre (WAICENT) of the FAO for the purpose of mapping and disseminating food security information relevant to FIVIMS. Essentially the Key Indicator Data System (KIDS) is a software framework that allows for the visualisation and dissemination of thematic information. This is achieved through a combination of quantitative data being overlaid on geographic or geospatial information systems (GIS).

22.3.2.5 The World Food Programme (WFP) Vulnerability
Analysis and Mapping (VAM)

The World Food Programme (WFP) has a wide remit and one of its main functions is food security analysis. As information requirements differ depending on circumstances so the WFP conducts a wide range of different analysis commonly referred to as VAM (Vulnerability Analysis and Mapping). This service was created in 1994 to analyse and map food insecurity by producing reports that describe the vulnerable of exposed population groups (Maunder 2006b). Several tools at its disposal include the following:

- *Comprehensive Food Security and Vulnerability Analysis (CFSVA)*: referred to as the precrisis baseline study, it provides a picture of the food security situation during normal times (non-crisis years);
- *Crop and Food Security Assessment Mission (CFSAM)*: conducted jointly with FAO are the agricultural production related emergency reports;
- *Emergency Food Security Assessment (EFSA)*: a report is undertaken following a disaster and covers areas affected to determine the impact on households and their livelihoods and to provide response recommendations;
- *Food Security Monitoring System (FSMS)*: is an ongoing activity to monitor changes in food security conditions. It provides advanced notice of a an impending crisis triggering an EFSA;
- *Joint Assessment Mission (JAM)*: conducts investigations in collaboration with the UNHCR to further understand the needs of the vulnerable;
- *Market analysis*: is undertaken to provide information on the functioning of markets and availability of food;
- *Economic Shock and Hunger Index (ESHI)*: developed as a result of the global financial and economic crisis to better understand the interactions between economics and food insecurity (WFP/VAM 2010).

Before intervening in a country, it becomes important to first analyse the food security situation. VAM's strong emphasis on geospatial mapping of data ensured it became a significant factor in providing food insecurity information. More importantly, these reports established a baseline vulnerability for numerous countries, which quickly caught on. As well as these traditional services, VAM collect information on emergency needs assessments and early warning (Mock et al. 2006; WFP/VAM 2010).

Other early warning systems include USAID's FEWSnet. The joint EC/JRC MARS project and others which are highlighted in Table 22.4.

Many of these monitoring systems were traditionally organised around agricultural production or crop simulation models although more often of late complex modelling with multiple information is overlaid using GIS software that better enables cause and effect to be observed. Generally speaking, food security monitoring systems engage in four main areas: agricultural production monitoring (APM), the market information system (MIS), the monitoring of vulnerable groups (MVG) and the food and nutritional surveillance systems (also called food and nutrition monitoring) (FNSS) (FAO 2000a).

As well as monitoring of the food security situation, important data for monitoring health and nutrition comes from several major sources including nutrition and consumption surveys, anthropometry, agricultural surveys, environmental assessments, household budget (income and expenditure) surveys, poverty

TABLE 22.4

Early Warning Systems

System Acronym	Parent or Joint Body	Full System Name	Description	Year
RCMRD	UNECA UN Commission Africa	Regional Centre for Mapping of Resources for Development	Natural resources assessment and management in Africa	1975
FAO/GIEWS	UN	Global Information and Early Warning System	Continuously reviewing the world food supply/demand situation	1975
FEWS NET	USAID	The Famine Early Warning Systems Network (FEWS NET)	An international, regional and national collaboration providing early warning and vulnerability information on emerging food security issues	1985
MARS project	EC/JRC	Monitoring Agriculture through Remote Sensing (MARS) Techniques	Comprising four dimensions: GeoCAP (ex-MARS PAC), AGRI4CAST (ex-MARS STAT), FOODSEC and CID. Offers a variety of information about seasons of the important global agricultural areas in the world	1988
REWS	Southern African Development Community (SADC)	The Regional Early Warning System	Analyses and monitors food crop production prospects, food supplies and requirements	1990s
WFP/VAM	WFP	The WFP Vulnerability Analysis and Mapping VAM	Provides assessments to aid in understanding the nature of food insecurity as well as monitoring emerging food security problems	1994
ISDR	UN	The International Strategy for Disaster Reduction (ISDR)	Comprises numerous organisations which work together and share information to reduce disaster risk.	2000
GMFS	The Global Monitoring for Food Security	European Space Agency (ESA) funded project	Part of the Global Monitoring for Environment (GMES) GMFS provides crop monitoring services	2003
HEWSweb	IASC (Inter-Agency Standing Committee)	The IASC Humanitarian Early Warning Service	Primary mechanism for inter-agency coordination of humanitarian assistance involving key UN and non-UN humanitarian partners concentrates on early warning of impending humanitarian disasters	2004

Sources: FAO, GIEWS. The Global Information and Early Warning System on Food and Agriculture, Food and Agriculture Organisation, Rome, 24pp, 2002; UN ISDR, Website of the Un International Strategy for Disaster Reduction. United Nations. International Strategy for Disaster Reduction, 2006; GMES, Global Monitoring of Environment and Security. European Space Agency ESA/EU, 2007; USAID, What Is the Famine Early Warning Systems Network (Fews Net)? USAID, 2010; HEWSweb, Humanitarian Early Warning System. World Food Programme, 2011; MARS, The Monitoring Agricultural Resources (Mars) Unit Mission, EC Joint Research Commission (JRC), 2011; WFP, Website of the World Food Program, World Food Programme, 2011.

mapping, vulnerability assessments, food balance sheets, market analysis and demographic monitoring, among others. These are examined in detail in Appendix F. Overridingly, one of the difficulties with all this data collection and analysis is that, not only are both data collection and analysis tasks carried out by different bodies, but oftentimes the analysis stage is divided among diverse and disparate departments of the same institution or government (Maunder 2006b). With such organisational structures and different information needs and end products, integrated analysis can be hard to achieve and at its worst this can result in contradictory or inconsistent messages for decision makers (Maunder 2006b). Moreover such institutional division can and does result in fragmented understanding of the fundamentals of food security (Maunder 2006b). While progress is being made then, it is recognised that any multi-dimensional analysis is difficult to achieve in departments or ministries that are organised separately around agriculture, health, and so on for instance (CFS 2000; Devereux et al. 2004).

22.4 Development Paradigms

Development paradigms or development economics as they are interchangeably referred to have no universally accepted principles or set of guidelines. However, even before a cursory examination of the issues at hand, we must first understand what development is and how it is viewed. Development is a complex term acting sometimes as a focal point of growth in different sectors such as economic, agricultural and educational. Growth however is also problematic in that it to is also viewed differently; sometimes it may refer to economic goals and sometimes it might mean social well-being. Of agreement though is the general idea that historically, development ideals were ultimately bound up in the demise of feudalism and the evolution of capitalism. Development can be thought of as three interlinked but separate stages. Firstly, survival; this aspect of development ensures the essentials of food, shelter, good health and protection against threats are provided. The second stage, raising these basic standards of living, is concerned with creating more jobs, better education and increasing personal income. The last stage is occupied with increasing personal and social choices or freedom of choice. In a nutshell, this skeleton view condenses the current perceived wisdom of development and its attainment, naturally suggests a continuum ranging from underdevelopment to development.

Underdevelopment then can be used to describe the poor, illiterate and ill-educated people of a nation who are essentially under-resourced and possess limited choices. Economies in this situation often see people living subsistently, with basic essentials of food and shelter taking up most of the financial and physical resources of the individual. In this situation underdeveloped regions are characterised by over-population, low productivity, lack of technology and training as well as general inequality. Thus the classic view of achieving development entails increasing GDP; structural transformation of the economy, including a decline in the reliance on agriculture and the further boosting of socio-economic diversity and differentiation; increasing trade and geo-political relationships as well as engendering wider social sophistication encompassing integration, equality and choice. Of course, this is all predicated on stable government; good macroeconomic management resulting in a healthy trade and fiscal policy; personal security; increased education and literacy; good basic health and adequate infrastructure, among other things. Furthermore, while there is no standard inside-track to good development guidelines, development theory too is at odds. The debate takes several approaches but two common themes regarding underdevelopment commonly recur: inherent and dependence. Inherent failures suggest that underdevelopment is a consequence of poor economic efficiency. This suggests that resource allocation and market failure are caused by, or at least exacerbated by, government or policy failures, corruption and tyranny. The other protagonist of underdevelopment is the notion of dependence. This view offers that the inequality seen between the developed and underdeveloped nations was the result of external exploitation or colonisation by wealthy foreign nations. This would have been further compounded by inappropriate and inequitable distribution of wealth through an imbalance of the terms of trade between the rich and the poor.

22.4.1 The Beginnings of Modern Development Theory

In a paper written by Contreras exploring the widening economic gaps of richer and poorer nations he explains that by the 1950s the world could be divided into two groups: the rich and the poor countries (Contreras 1999). On the one side was the collection of Western European countries, Canada and the United States and on the other, Latin America, Asia and Africa. However, long before this gap became fully pronounced, the widening imbalance was the focus of debate during the 1940s and many people began looking for reasons for the increasing divide and ways of closing this gap (Contreras 1999). Development economics became the prevailing model and it received a lot of attention around this time. It was posited that as with other economic theories, development economics too was concerned with the sustainable and efficient allocation of resources. More than this though as an idea, development economics was seen as a notion that not only promoted economic growth but social growth too through the enactment of wide-ranging initiatives such as health and education to workplace policies in poorer countries, etc. (Mikesell 1968).

As a result a prevailing development theory emerged whose ideology was firmly embedded in mercantile ideology and nationalist economics; the national control of the economy, capital and labour. Also within this paradigm, Mikesell, outlining many of the prevailing economic theories of development economics by the late 1960s, suggested that many preconditions were needed for the steady, equitable economic growth. These included, but were not limited to social and labour mobility, education and incentive; economic freedom; political stability as well as well developed capital markets. He also saw aid as a necessary means of allowing developing countries to assume a critical mass in the effort to achieve self-sustained economic development. Furthermore, because of the inherent nature of underdeveloped economies he suggested that most of this capital would need to come from external sources (Mikesell 1968). Indeed this is exactly what happened in the aftermath of World War Two. Modern development economics grew up amid a growing sense of humanitarianism. With the perceived problems of the reconstruction of Eastern Europe after World War Two, authors such as Rosenstein-Rodian, Kurt Mandelbaum, Ragnar Nurkse and Sir Hans Wolfgang Singer all suggested solutions of aid and structural transformation for economic and social development (Pleskovič and Stern 2001; Meier and Stiglitz 2002; Riddell 2007). Although first aimed at a war-torn Europe, the practice was subsequently extended to Asia, Africa and Latin America as the next beneficiaries of this expanding international aid effort (Pleskovič and Stern 2001).

From these beginnings, much modern aid development theory and ideology emerged. Several notable development models have been cited by Contreras as being instrumental in shaping modern thought although not all were seen as beneficial; these are briefly outlined.

Structural-change theory emerged in Latin America in the 1940s. Structuralists urged economic development through an expanded domestic industrial sector. It also focused on changing the economic structures of developing countries, emphasising increased diversification into an industrialised and service economy.

Linear stages of growth model was first formulated in the 1950s by W. W. Rostow. The idea focuses on the accelerated accumulation of capital through investment to promote economic growth measured by a rising per capita income.

International dependence theory, this theory is more of an observation than an actionable development theory. It aimed to characterise a model of dependency that was already underway and was an important yet controversial contribution to development economic ideology in the 1970s. This theory postulates a political and economical dependence of weaker countries on their more powerful counterparts; the developed and developing countries. Such a relationship too is predicated on maintaining the status quo with dominant or developed partners continually siphoning off surplus value from developing countries.

Neoclassical theory, first gaining prominence in the West during the 1980s, it completely dismisses previous neo-Marxist (international dependence) theory as flawed and unrealistic. Instead the idea promotes a less interventionist approach to development. Ultimately it was this notion, the idea of a free markets uninhibited by excessive government regulation that ultimately became the prevailing paradigm for achieving rapid and successful development from the 1980s onwards (Contreras 1999).

22.4.2 Agriculture as Development

More specific to the food security debate has been the unfolding of the agricultural development paradigm. Following post-war structural rebuilding of physical infrastructure and industry, and in keeping with the emerging development paradigm, in terms of the agricultural sector the idea was to increase productivity allowing for the security of food and concomitant growth to ripple through the economy (Goodrich 1947). This, it was thought, would be achieved through the innovation of technology, financial and physical inputs, appropriate and adequate institutional frameworks, suitable infrastructure, a good information network and the right incentives. The idea of enlarging a country's international trade as a development tool, at this point had not yet emerged; in fact at the time this notion was seen as exploitative. Instead, protectionism and quotas were seen as legitimate policy goals with imports supplying the shortfall (Goodrich 1947; Warnock 1997). This view, according to McCalla (2007), enforced inward-looking policy objectives and a paradigm of self-sufficiency towards food security (Warnock 1997). However, change was on the horizon and the then recently formed Bretton Woods institutions created the international architecture, which helped open up international markets for goods, services and capital. More than this though these institutions also aimed to create a sense of stabilisation through outward looking programmes of advice and monetary assistance. This helped modernise Western agricultural development agenda's moving them towards a global perspective and the resultant idea of international agricultural trade as a development tool (Belshaw 1947). In the West, this meant radical changes in the way farming was viewed. Self-sufficiency was being replaced with cheap imports where comparative advantage became the underlying agricultural philosophy. Agricultural development in developing countries too was encouraged to follow the Western model and to collectively unite technological advances and development theory of the time into the now familiar industrialisation of agriculture model. Through this growth in agricultural industrialisation and the subsequent occupational redistribution of farm workers, it was hoped developing countries would not only help themselves in food security, but also allow them to grow economically and close some of the wealth gaps throughout the world (Belshaw 1947). This model became very successful in the West; however it had mixed successes in the developing regions with many failing to fully adopt the ideals.

Nonetheless, development ideology is not static and by the late 1980s the idea of a global free market had firmly taken hold. Countries, building on the model of comparative advantage, were encouraged to export whatever they could produce most efficiently in turn earning the foreign exchange to buy the food they needed. Importantly in the evolution of development, the 1970s and 1980s saw the focus of growth shifting from the sole aim of agricultural expansionism, with its implicit goals of social advancement, to one of explicitly reducing poverty and food insecurity. This would be achieved through broad-based strategies of land reform, investment in human capital (education, health and nutrition), improving gender equality as well as rural non-farm economic development. The 1990s also introduced a new focus of development adding environmental sustainability into the mix. This effectively gave priority to resource poor areas and invited a much broader approach in the ideology of development economics. Sustainable land and environmental considerations too involved the recognition of the notion of secure individual and community property rights vis-à-vis land and natural resources. It also included an array of educational initiatives and the monitoring of pollution and degradation. In short, this new approach to economic and social growth involved the full realisation of the wider environmental impacts and effects of food security on developmental issues (Harvey 2010).

22.4.3 Questioning Agricultural Free Trade

The current popular market-oriented agricultural model advocates the consolidation of smaller farms to form larger industrial units able to benefit from cost savings as well as effectively being able compete in the export market. Common in most of the recent paradigms and contrary to much of the post-war government-managed market economy, was the notion that global trade would naturally grease the wheels of development. Scathingly however, critics have argued that this belief in the private sector, has been wholly and inappropriately misplaced. It has been argued that by relying on the private sector to replace those important functions that had previously been provided for by governments would in fact result in more insecurity rather than less (Murphy 2010). This view gained further acceptance with a World Development Report in 2007 that essentially acknowledged this very shortcoming. Following on the heels of this growing realisation was the 2007–08 food price crisis, which saw widespread civil unrest and in some cases death. This firmed up the realisation in policy maker's minds of the basic need to manage the volatility inherent in global commodity markets and to further understand how to protect people from escalating food costs. This marked a shift away from the laissez-faire free market mechanism of food security and a return to the more cautious approach involving government managed intervention and social safety nets (Murphy 2010). Out of such feelings yet another paradigm emerged to challenge prevailing trends in development economics and food security; that of food sovereignty. This approach marks a departure from the hitherto large-scale industrial agriculture and liberalised trade-based development or food security model in favour of more localised, small-scale sustainable solutions (Windfuhr and Jonsén 2005).

22.4.3.1 Food Sovereignty

Food sovereignty is not the same as self-sufficiency, instead food sovereignty is an alternative view of food security that bucks the prevailing globalisation of the food chain. Introduced in 1993 in Mons

(Belgium) by a group of farmers' representatives called La Via Campesina, the idea of food sovereignty was finally launched at the World Food Summit in 1996. The movement brings together millions of small and medium-size farmers, landless people, peasants, indigenous people and others from around the world with the goal of actively engaging in the decision-making processes that affects all their lives. La Via Campesina defends small-scale sustainable agriculture by promoting food sovereignty. This alternative view of food security is opposed to large, impersonal, profit-hungry transnational companies and corporate-driven agriculture. Further, the idea favours less multilateral intervention and interference in favour of national self-determined food security policies. At the core of this belief is the notion, which has already been touched upon earlier, in that the millions of smallholder farmers are able to efficiently and sustainably provide adequate food security for the world. Almost every aspect of agriculture nowadays is influenced, governed or legislated at the international level. In this respect, one of the main concerns of the food sovereignty movement is this onward march of globalisation, which sees more and more multilateral instruments decreasing national decision-making abilities in such areas as trade, agriculture and land policies. This is witnessed in the Uruguay Round of negotiations where liberalising agricultural policies in favour of freer more open trade international norms become prescriptive at the national level. Further compounding this situation is the many developing countries' external debt positions which sees them having to instigate further World Bank and the IMF structural adjustment programmes. While such programmes may on the one hand appear economically beneficial, such aid comes with much policy guidance and in many instances mandatory conditions in such areas as the opening of international markets for agricultural products, among others. Also, as hinted at earlier the strong negotiating position of large foreign multinationals wield significant power beyond their political weight, and it is this power that is brought to bear in influencing terms of trade and sometimes by default, domestic policy.

In favour of change, the food sovereignty framework involves the devolution of this centralised decision-making process away the global and back to the national. In keeping with their fundamental ideals too, food sovereignty prioritises local food production and consumption whilst at the same time ensuring that the balance of power resides in the many people who produce the food rather than in the hands of the corporate behemoths (Windfuhr and Jonsén 2005; Clements 2009; La Via Campesina 2011).

22.5 Economics

Economics is fundamental to food security. It dictates the price of food commodities, its elasticity and substitution as well as the flow of investment and development agendas. In order to understand how food security responds to economic variables, the following looks at some basic economic principles before looking at some of the issues in more detail.

Economics is way in which a society uses its resources to satisfy its wants and needs. Every society relies on some kind of economic system. The three basic economic systems can be defined as traditional, command and market economies, while a fourth is perhaps the mixed economy. A traditional economy tends to rely on the status quo; what has been will be, so to speak. As an economic system, it often relates to regions relying on subsistence farming or other areas outside the mainstream command or market economies. It is also largely an economy based on custom and tradition and has been used in the past to denote primitive economic systems. A command economy is where a central government takes control of the economic decisions. This is particularly common in communist social groups such as Korea and Cuba. Lastly, the market economic system is based on capitalist values and seeks to leave the decisions of supply and demand to the market, which by its nature encourages change and growth. It is seen as the most efficient type of economy and its most famous proponent, Adam Smith, saw it as a flexible way in which people could trade based on their comparative advantages. A mixed economy tends to be one in which both the command and market forces are at work. This socialist approach sees the government owning the basic productive resources (although even this is changing) while the rest of the businesses are privately owned. Large government involvement in the mixed model has more say in the economy than it does in capitalism but less so than in communism. This allows for everyone to be provided for, particularly those who are not able to earn sufficiently to share in society's benefits; in such circumstance this is achieved through such things as health care, welfare and other social safety nets. While there are few downfalls to socialism, it is usually not as productive or efficient as capitalism; as a result, another form of mixed economy sees a mixture of capitalism and socialism. This is perhaps considered the best as it allows certain freedom of market forces while ensuring government intervention takes care of the welfare side of society.

22.5.1 Macro- and Microeconomic Principles

Irrespective of the different economic styles, any economy can be sub-divided into two parts: macroeconomics and microeconomics. In macroeconomics, analysts are concerned with the overall growth of the economy, the business cycle, inflation, monetary policy and policy regulation, development economics, and the overall balance of payments among others. In microeconomics on the other hand analysts are concerned with the operations of the components of a national economy, such as markets, individuals or firms. There are several economic areas relevant to food security ranging from the market mechanism to development economics.

22.5.2 Macroeconomics

22.5.2.1 Balance of Payments

The balance of payments (BoP) is a record of all the monetary transactions between country's consumer's, its businesses as well as government activity and the rest of the world. It is the balance sheet of a country. These might include payments for a country's exports and imports (goods, services and financial capital), as well as financial transfers. The BoP is made up of three accounts: the current, capital and reserve accounts and by definition, between them they must balance, i.e. Balance on current account + Balance on financial account + Balance on official reserve account = 0. These accounts are recorded using the regular double entry book-keeping methodologies, that is, to say any transaction has both a positive and negative effect on the accounts, i.e. as exports, etc. create a financial outflow of goods so it also creates a financial inflow in the form of payment for such goods, etc.

22.5.2.2 Current Account

The current account tracks all monies going out of and into the country. It itself is made up of three accounts: the balance of trade, net investment income and other transfers (Figure 22.8). When a country talks of a trade deficit, this essentially means more goods and services are being imported than are being exported, with the resultant net outflow of financial monies. A trade surplus on the other hand equals more exports than imports. The same is true of net investment; this can be negative with net investment outflow, or positive with net investment inflow.

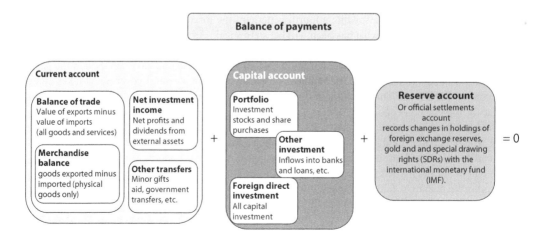

FIGURE 22.8 Balance of payments accounts.

22.5.2.3 Capital Account

The other half of the BoP account is the capital account. While the current account tracks changes in the income of a country the capital account can be thought of as tracking the assets of a country. It is made up of four items: foreign direct investment (FDI); portfolio investment; other investment; and the reserve account. The FDI refers to long-term capital investment such as the purchase of machinery, buildings and businesses. While this is recorded in the capital account, if the business for instance is not sold after the initial investment, any profits not re-invested are recorded in the current account and not the capital account. The portfolio investment refers to the purchase of shares and bonds; however as with FDI, any derived income is recorded in the current account. Next to these two accounts, other investment includes net capital flows into banks or that provided as loans. Large short-term financial flows between different nations is commonplace and aims to take advantage of fluctuations in interest rates or the exchange rates between currencies.

22.5.2.4 Reserve Account

The reserve account is also sometimes known as the official settlements account. It is the balancing mechanism of the BoP and is operated by a nation's central bank. With all else being equal including the terms of trade between two countries, any imports are financed by a corresponding drop in the country's reserves. Conversely any exports result in inflows of foreign exchange. By this principle, the BoP account should be balanced; however, there are always discrepancies. It is these discrepancies that set, or interfere with things like, inflation and exchanges rates, etc. This could be due to any number of reasons. One of these might be the result of an imbalance in the terms of trade; if for instance one grain exporting country has a bumper harvest while others are slack, this may result in market disequilibrium. While initially the net inflow of funds from favourable export prices is beneficial, if too much money flows into a country then foreign exchange, gold and other reserves quickly build up. This has the effect of increasing a country's assets and places pressure on the domestic currency. This currency appreciation is expressed through inflation and exchange rates and a strong local currency results. While this might be good for imports, this becomes overall unattractive with the terms of trade tipping out of balance and both the FDI and exports suffering as a result. The central bank, in this instance, is under pressure to intervene by stabilising or devaluing the currency; if it fails to do so the financial markets effectively adjust rates accordingly through foreign exchange rates instead.

22.5.3 Exchange Rates

Trade is the cornerstone of any economy. Early civilisation's traditional bartering for goods was of the time; however, as we progressed so this became inefficient and inflexible. Eventually coinage based on gold and silver and other perceived precious metals was used in trade exchanges; however, this had fundamental limitations in that such metals were scarce and any expanding economies could not always guarantee sufficient gold for expansion. One solution was the introduction of fiat money. Like gold, the value of the money was intrinsic, that is, built on the inherent belief in its worth. This allowed governments more flexibility and leverage to manage the money in circulation. Both gold coins and paper money became common place, although by the eighteenth century, paper money had become the dominant form of transactions, in Britain at least. This brought with it problems of its own and through leverage, too much credit was being introduced that had the undesirable effect of destabilising the economy. One way to instil the necessary controls on this money was to introduce gold convertibility.

Simultaneously international trade was booming and facilitating this was the global acceptance of gold's worth with most countries using physical gold or silver for the payment of such goods and debts. By 1821, England was the first country to adopt a gold standard and by 1871, a new international standard was introduced where much of the developed world pegged their currencies to gold. This underpinned the notion of the exchange rate system; the cost of exchanging one currency for another, this effectively allowed both country's currenies to be determined against their respective gold values and in turn against each other. This was known as the fixed exchange rate mechanism.

By the twentieth century, all the major industrial powers had converted to the new gold standard and by linking respective currencies to gold, nations could now use printed paper currency backed by gold in reserve for the payment of debts. This ensured a good deal of stability and heralded a new era of international trade among nations. However, by the outbreak of World War One, there was insufficient gold to back the amount of money needed to finance hostilities and many countries abandoned or suspended the gold standard. This brought about currency speculation as people no longer defined the values of their currencies in terms of gold. Instability resulted and it was not until the advent of the Bretton Woods Agreement that a version of the gold standard was brought back. This time it was based on countries pegging their domestic currency to that of the dollar, which in turn was pegged to gold. In this way, a new roundabout gold standard was introduced that allowed for renewed stability until inflationary pressures in 1971 saw the devaluation of the dollar and the eventual breakaway from the agreed convertibility standard. This allowed for a new floating rate exchange system where the private markets determine exchange rates through supply and demand. A floating exchange rate that is constantly changing is now in place and can be seen as a self-correcting mechanism that in the long term efficiently determines the worth of a currency. While today, the use of gold as a peg has been completely abandoned by all major currencies, in reality no currency is either fully fixed or fully floating. Some governments might choose to have a 'floating' peg in which the government reassesses its value periodically and adjusts it accordingly. There are inherent benefits and weaknesses in both fixed and floating exchange rate systems. The main reason to peg a currency is stability. However, this can sometimes lead to financial crises since a peg is difficult to maintain in the long run. Moreover, pegging is also seen nowadays in less sophisticated financial capital markets characterised by weak regulatory institutions. On the other hand, a floating currency is seen as the most efficient exchange rate mechanism these days and is associated with strong mature economies.

22.5.4 Debt and Structural Adjustment

The scale of structural debt in the developing world is enormous and has a significant bearing on these countries ability to achieve national food security. For instance, in 2008, the extent of the developing world's indebtedness was as follows:

- Total debts of the developing world in 2008: $3.425 trillion
- Total debt service of the developing world in 2008: $516 billion
- Total official development assistance in 2008: $128 billion

This means that for every $1 worth of assistance developing countries received in 2008, they paid $4 in debt repayments (Figure 22.9) (World Bank 2010a). So what is this debt and where has it come from? A country's government debt, also known as public or national debt, are the monies borrowed to finance

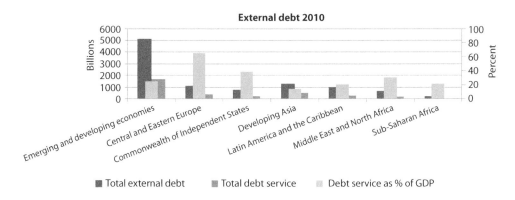

FIGURE 22.9 Regional external debt, 2010. (Compiled from World Bank and IMF data (IMF. *International Monetary Fund Website. International Monetary Fund*, 2010. Available at http://www.imf.org/external/; World Bank, *Worldbank: World Development Indices*. World Bank Group, 2010b.)

a nation's ongoing business. It can be divided into two groups: internal debt, which is owed to lenders within its own borders, and external debt, which is owed to foreign lenders, governments or multilateral institutions. Internal debts can be repaid by raising taxes and other domestic fiscal policies; however, external debts are paid for out of a country's international reserves, which can often place great demands on governments. External debt burdens are a common feature of modern economics. In 2009 the world's external debt reached $56.9 trillion against a world total GDP of $69.98 trillion; that's a total debt borrowing of just over 81% of GDP (CIA Factbook 2010).

The developing world debt or the so-called third world debt crisis arose out of a number of factors. After a period of decolonisation following World War Two, the World Bank and other multilateral creditors such as the IMF lent heavily to developing world governments for the purpose of expansion and development projects. This continued into the 1960s and 1970s; meanwhile, following the 1973 oil price shock was an international banking system that found itself flush with capital inflows from the OPEC countries. This new found liquidity was quickly funnelled through loans and other disbursements into developing countries and sparked global inflationary pressures. On top of this, odious debt was rising as the result of illegitimate or dictatorial governments, corruption and mismanagement, which in turn further impoverished nations, leading to additional borrowing needs. By the early 1980s, a sharp rise in world interest rates greatly increased the interest burden on debtor countries. This was coupled with a decline in export receipts of developing countries as commodity prices too began to fall. At about this time (1982), the Mexican government found itself in difficulty and unable to service its debts. At the risk of allowing Mexico to default and causing major havoc in the international financial markets, governments and multilateral institutions stepped in to re-finance the debt. However, the debt crisis continued with several other countries threatening to default on their loans too. This led to further widespread re-financing, with lenders proposing a variety of structural adjustment programmes (SAPs) to reorient and liberalise developing economies. Essentially, SAPs aimed to address domestic fiscal imbalances through restructuring of the national economic model. In the process one aim of the SAP's was to open up debtor country's trade to international markets by promoting more market-oriented policies. The rationale behind this move was that countries could generate export earnings that could be used to ensure repayment of the initial loans. Such programmes soon became a pre-requisite of lending and the conditions imposed had the effect of restricting expenditure on infrastructure, health and education and maintaining the prioritisation of debt repayments. There is much debate surrounding these issues with many critics arguing against the financial prudence of the lenders. Others still see the continuing debt problem as the ongoing and relentless raping of weaker countries' financial and natural resources. In recent years, there has been much talk of debt relief while several countries have also adopted aggressive domestic policies in the hope of retiring public external debt in favour of replacing it with domestically issued debt. How this pans out remains to be seen.

While no single, or set of indicators, give a thorough analysis of a country's ability to carry these debts, analysts use various debt ratios to assess the external situation of countries and to assess or analyse their creditworthiness. In some measures, a country's total indebtedness can be classified as the present value of debt to gross national income (GNI) while another measure might look at the external debt as a proportion of the GNI. However, perhaps the two most important measures include the debt service ratio and the total external debt ratio. The debt service is an important indicator of debt sustainability and is measured as the total debt service as a proportion of total exports of goods and services (plus workers' remittances). In short, it is the debt in relation to foreign exchange earnings. The total external debt ratio on the other hand highlights the external debt as a proportion of the international reserves held, which indicates a country's capacity to meet debt service payments from existing stocks (World Bank 2010a).

The massive external debt incurred by the governments of developing countries has had and continues to have enormous impact on policy and food security. This has left many in crippling poverty unable to afford the basic right of food security. It also creates economic fragility, for if a country cannot afford to keep up with its external debt repayments, it is forced to re-evaluate its domestic economy and in extreme circumstances perhaps devalue its currency. The need to keep the national accounts and the balance of payments in order is not in dispute either. However, by prioritising this debt as previous SAPs had done, it pushes the domestic economy, expansion and development agenda into positions of secondary importance. That said debts are an important international instrument and as discussed earlier servicing external debts is usually paid for out of foreign export earnings. Thus, any shortfall needs to be raised through the stimulation of export earnings and there are perhaps two important ways of doing this. The first involves diverting local economic industries to satisfy the needs of external markets, which often entails cash cropping and the like; and the second way is to address the terms of trade by reducing its currency value, thus making exports more attractive (Van Esterik 1984). It can be argued that by devaluing a countries exchange rate it brings with it mixed blessings. On one hand domestic export industries expand, earning valuable foreign currency while on the other, producers become vulnerable to world market prices. On the domestic front too, devaluing a currency can aggravate internal inflation, making any reliance on international markets for imports and primary inputs that much more fraught with vulnerability. The choices are difficult and far more complex than this simple summary reveals. Yet a country's external debt position is an important but under-appreciated factor in the overall food security arsenal; get this wrong and the domino effect might certainly cripple all efforts of progress.

With regards to SAPs and their use, it is worth noting that on balance and, following the train of thought of DeRose and colleagues, any SAP that aims to cut social welfare, currency devaluation and insist on greater cash cropping to service such debts, however well intentioned, is ultimately to blame for much increase in developing countries' poverty and food insecurity status. This was a point echoed by the FAO when it pointed out that practices like this forced on developing countries have subsequently and radically changed some country's economic trajectories with concomitant repercussions on food security (FAO 2000a). To be fair to a balanced view though, it has not all been bad; in some long-term structurally adjusted economies, food production and security have actually been seen to have increased (Meller 1992; Weeks 1995). Moreover, a review of developing countries' external debt in 2008 found that the total external debt of developing countries in 2006 was equally matched with the total international reserves held by these countries. Of course, this hides the vast regional disparities inherent in such aggregates but it shows at least that things are heading in the right direction (UNCTAD 2008).

22.5.5 Microeconomics

22.5.5.1 Markets

Whichever type of economy is in use nationally, there is a global market system that facilitates trade among buyers and sellers and it is a system that works fairly efficiently. Trade in this manner can be thought of as a circular flow of income between consumers and producers (Harvey 2010). Oversight of the global market in which free trade is favoured is achieved through the World Trade Organisation and just how efficient these markets are depends on one's viewpoint. Free market systems based on the capitalist model tend to be least regulated, relying instead on little oversight and efficiency of the supply and demand pulls and pushes. These markets are favoured by democratic and social democratic societies operating market or mixed economies while the more command or centrally planned economies tend to be more active in the markets. Markets are an integral part of everyone's lives. As a consumer, whether urban, rural, rich or poor, we all rely on market systems for essential goods and services. Markets too provide employment and a means of selling our goods; in this way many livelihoods are affected by markets and associated industries. Markets also contribute to several of the pillars of food security: availability, access and stability and in areas of deficit; where not enough food is produced, inhabitants rely on the markets to supply the shortfall. Conversely where there is a surplus, markets facilitate movement of stocks to areas of need. This commodity flow is regulated by price movements and generally speaking, supply is often seen to fall as the price falls and when the price rises, the demand is seen to fall.

How well the markets respond and facilitate the movement of commodities then is an important concept. This is known as market integration and it is the measure of how well the markets translate prices from one market to another. If a market is well integrated, price movements in one is quickly mirrored by price movements in another; this is usually determined through statistical analysis and correlation measure. The resultant level of market integration in turn heavily depends on government regulation, commodity types, location, market infrastructure and communication networks. Further affecting these flows is the civil stability of a country or region including corruption and banditry. Such analysis of the market system is also prudent as, excluding aid, market supplies tend only to flow towards effective demand (Bonnard 2008b). In its absence, households and individuals' food needs can only be met through their own production, social transfers or government and humanitarian intervention.

Markets rely on demand and supply. Demand is of two types: primary and derived. It is critical to agricultural markets and the relationship between primary and derived demand is an important concept. The primary demand for any commodity is the demand made by the final user or consumer whereas all other demand along the supply chain is considered derived. An example of derived demand is the demand for grain to feed livestock or for the biofuel sector. Another term used in connection with demand is effective demand; this makes the distinction between those who have the will and the means to buy a product and those who have the will but not the means to buy the product. Derived demand, as in the example above, is important in that in many instances, it can directly compete with primary demand and force an upward trend in prices.

22.5.5.2 Elasticity

Further affecting demand of food is the notion of elasticity. Elasticity of demand can be thought of as the strength of demand compared to price movements, or in other words, the responsiveness of demand versus price. If a commodity is elastic, a small increase in the price will cause a large decrease in demand; conversely if another product was inelastic a small increase in the price would result in either zero movement or only a small decrease in demand. Knowing the elasticity of commodities helps the food security analyst estimate any effective shifts in supply or demand may have on market prices (Bonnard 2008b). An important note here is too is that irrespective of elasticity, a scarcity of food commodities in general will result in price rises as the market responds to the fundamentals of supply and demand economics. There are a number of things that can affect the elasticity of demand of a commodity and understanding this can help analysts determine how households might respond to rises in basic food prices.

22.5.5.3 Terms of Trade

At the macro level, the terms of trade (ToT) is a measure of how much is exported per unit of import; that is, to say exports divided by imports or ToT = 100 × Average export price index/Average import price index. An improvement in a nation's terms of trade is experienced when export prices rise faster than import prices. This means that fewer exports have to be given over in exchange for a given volume of imports. Conversely, deterioration in a nation's terms of trade arises when import prices rise faster than export prices. A nation's terms of trade then, fluctuates in line with changes in export and import prices. In turn, export and import prices are not only the product of supply and demand but of exchange and inflation rates too. A big fall in the terms of trade indicates a real reduction in living standards with imports of goods and services becoming relatively more expensive.

On the micro level, the same applies, the terms of trade help to determine the purchasing power of individuals, households and businesses. In this case, it is not imports or exports but goods, services and finance. Consider the pastoralist who sells his livestock in the dry season to buy cereal. At this time of year, his stock is of lesser quality and he receives less income per animal. If this happens at the same time as cereal prices rise, his terms of trade is said to have shrunk.

22.5.5.4 Parity Pricing

Parity pricing is used to compare commodity and other prices across borders. It is used to determine or quantify the incentives to trade and to produce goods and services. It is a useful comparative metric in times when local producers are in competition with producers and suppliers from outside the country. It is built on the principles of purchasing power parity (PPP) (see Section 4.3.2.1) and in the market sense, there are two types of parity prices: import and export parity. Import parity price is the value of an imported commodity from another country including associated cost, insurance and freight charges;

it is effectively its CIF (Cost, Insurance and Freight), that is the landed cost of an import on the dock or other entry point in the receiving country. In this way if this imported good's price compared to the same product locally is cheaper then it becomes an attractive alternative to buying locally. This determines the comparative costs of each item and helps decision makers on where to source goods from. The export parity price conversely, is the value of a commodity sold in a foreign location but valued at local (location of origin) prices. Unlike imports, exports are valued as free on board (FOB) and helps determine whether a country's exports are competitive with the same commodity produced in other countries (Ward 1977).

22.5.5.5 Market Distortions

This is how markets operate in the very general sense yet many governments and institutions interpret market economics differently. While some sign up to the idea of a capitalist market economy, others feel private ownership is misguided and prefer market mechanisms along the lines of social democracies with some form of intervention. Mixing these different views in a single global market is difficult, which accounts for the currently stalled WTO-Doha round of free trade talks.

In the absence of true free trade, an example of how the markets can be distorted is that of the EU CAP. According to David Ricardo's comparative advantage theory, this would suggest that the because of the EU's relative high land and labour costs, they would not naturally have engaged in agriculture to the extent they have done so. In fact by introducing subsidies that have, in effect, artificially stimulated the markets, the EU has managed to create a largely self-sufficient agricultural sector. Moreover, while policies were ostensibly designed to increase production and guarantee farmers a reasonable living, this in fact was achieved at considerable cost. Further, this practice involved the introduction of external customs tariffs for non-EU members and routinely destroying or dumping of agricultural surpluses on the international market (Windfuhr and Jonsén 2005).

22.5.6 Investment

Economic investment in sectors like agriculture, education and health services is of vital significance in the fight against food insecurity. In fact, overall this general and systematic lack of investment in agriculture has been cited as one of the biggest obstacles to achieving lasting food security today (UN News Centre 2011). Yet this investment is still conspicuously lacking. It was once considered that poor farmers, bound by tradition and a backward mentality, were largely to blame for relatively lower productivity in the developing countries rather than a lack of investment per se (Johnson 1997). However, a study by Schultz back in the 1960s dispelled this myth and by recognising that such inefficiencies were due to technological constraints, lack of opportunity and economic incentives, it was soon realised that vital domestic investment was much needed (Johnson 1997). Indeed, today, as a result of this realisation and with the majority of the poor living in rural areas, it is finally recognised that a sustained reduction in hunger is only possible with this much needed investment in agricultural and rural development (CFS 2007a). Sadly however, in spite of what has been learned and despite World Bank estimates that rural economic growth reduces poverty by as much as four times faster than other sectors, the proportion of foreign aid allocated to agriculture in Africa actually fell from 17% in 1980 to 3% in 2005 (ARI 2009). As a means of escaping poverty and increasing food security then, it follows that investments in agriculture both from government and private streams are needed to facilitate modernisation (Rosegrant and Cline 2003; CFS 2008b; UN News Centre 2011).

An associated issue regarding investment in research and development, suggests Pardey and colleagues, is that in 1945 perhaps as much as 90% of global agricultural research was funded by the public sector. By the 1980s this division had changed with investment then representing 70% public funded and 30% private. By 2006 Pardey and colleagues further proposed that private sector investment in research and development had reached about 40% and in some instances as much as 70% in some European countries (Pardey et al. 2006). This is particularly prevalent in the fledgling biotechnologies where genetically modified crops see huge potential returns. Already privately owned patents in this area see GMO seeds being dominated by a handful of companies. What this means for the future privatisation of agricultural research in this sense will also be one to watch (Pray et al. 2007).

22.5.7 Commodity/Food Prices

The escalating prevalence of food price volatility and frequent food insecurity in recent years has been attributed, by some, to the consequences of a global industrial development model based on competitive export-orientated agriculture, self interest and speculation. In 2007/08, it has been estimated that price hikes alone pushed a further 200 million (or 133 or 115 million depending on who you read) people globally into food poverty (SOFI 2008; World Vision 2009; GFMG 2010). There appears to be much speculation and difference of opinion as to the root causes of current upward price pressures on food commodities. For some, it is the rapid expansion of the developing economies, especially China and India, coupled with a growing change in dietary habits while others cite increased general demand owing to growing wealth as well as combined high fuel prices (Shah 2007; Watts 2007; FAO 2008a; Smith and Edwards 2008). The Agricultural Outlook 2007–2016 report however suggests a combination of factors including the lowest grain stocks held for 25 years together with drought or scarcity of water in wheat-growing regions. In addition, the report also postulated that increased demand and use of food crops like cereals, sugar, oilseed and vegetable oils for biofuels too were playing a central role in food price spikes (OECD-FAO 2007; Shah 2007; FAO 2008a). Indeed, an interesting observation regarding this last point is the apparent strong correlation between selected agricultural foodstuffs and oil prices (Figure 22.10). It is not the first time this has been suggested either and if this convergence were to continue then the possibility of oil prices acting as a price-floor for agricultural commodities is certainly not inconceivable (von Braun 1995; Schmidhuber 2003; Brown 2008; SOFI 2008).

There are many widely accepted consequences of these erratic movements in price. First and foremost, it is generally agreed that it will be the poorer sections of the community who will suffer the most. Speculation aside, on a fundamental level the demand for food is price inelastic, that is, the quantity people demand varies little with changes in prices (Lee 1993). Despite this though, people still exercise food choices. Rising food prices in developed countries for instance, will force many to substitute expensive foods rather than reducing the quantity consumed; by contrast, in poorer low-income countries, people are more responsive to price, reducing the demand for meat, dairy and vegetable products and increasing staples like bread and cereals. As such, food's continuing price volatility threatens to reverse the gains made in the prevalence of undernourished people in the world in recent years. Furthermore rising food prices are also translating into social unrest. Pasta protests in Italy, tortilla demonstrations in Mexico, and maize protests in Kenya are just a few of the many demonstrations

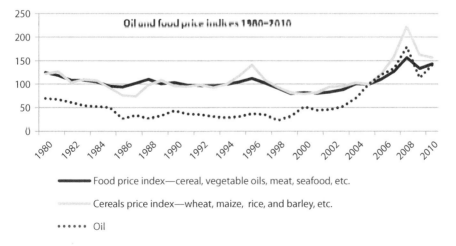

FIGURE 22.10 Selected agricultural foodstuffs and oil prices 1980–2010. (Compiled from data from the IMF *World Economic Outlook 2010*, International Monetary Fund, Washington, DC, 2010.) *Note*: Indices set at 2005=100.

around the world, which according to von Braun, are set to increase (Lee 1993; von Braun 1995; King and Elliott 1996; Regmi et al. 2001; Eifert et al. 2002; Watts 2007; BBC 2008a; Brown 2008; Delva 2000, Clapp 2009; FAO 2009b).

An interesting statistic that puts this dilemma in perspective was made by Regmi in 2001, who suggested that for every 1% increase in the price of food, food consumption expenditure in developing countries decreased by about 0.75% while the caloric intake was reduced by 0.5% (Regmi et al. 2001, Brown 2008). With this in mind, responding to growing concerns about the implications of the soaring prices of 2007/08, governments recently began by first: removing quotas and tariffs on imports; banning or introducing export duties; buying food at preferential prices; increasing grain stockpiling and introducing food subsidy interventions. However, this has largely been seen as a short-term fix, which was neither beneficial or sustainable in the longer term (Dupont and Thirlwell 2009). In a more holistic approach, the UN, in recognising the seriousness of the crisis, held a special session of the Economic and Social Council (ECOSOC) members to identify the causes of and measures it could undertake to bring prices down. Its response was a three-fold combination of short- and long-term plans. In the short term, this meant seeking full funding for life-saving projects and to use its huge logistics capacity to support distribution networks while in the longer term, to support governments engaging in agricultural development programmes (SOFI 2008).

Of course, high food commodity prices too are not all always necessarily bad in and of itself either. Higher prices are good for sellers or farmers who are net sellers of food. This also translates into increased agricultural labour as well as better wages. Furthermore, such increased revenue can foster or at least encourage increasing agricultural investment (SOFI 2008). In sum there are inherent winners and losers in food price increases. However, and in the event of continued increased price volatility, international efforts in this matter need to focus on those countries or segments of society we know are most adversely prone to such shocks.

22.5.7.1 Financial Speculation

A separate but related phenomenon, that of commodity speculation, has also been accused of instigating further food price rises. It has been argued that increasing food prices as a result of financial speculation is adding to the upward pressures of inflation in turn further affecting the stability of food security for the poor (SOFI 2008; Clapp 2009). By way of example, as the global food price bubble overheated in 2007/08, escalating tension and social unrest spread around the world. Consequently people began looking for answers (Shah 2007; BBC 2008b; Brown 2008; ECOSOC 2008; SOFI 2008). However, unlike the previous cereal price rises of the early 1970s, which was attributable to the OPEC oil embargo among other things, the 2007/08 commodities situation saw price rises everywhere and across the commodity board. As many drivers of change were identified so too did commodity trading come under the spotlight (Figure 22.11) (Watts 2007; Brown 2008).

According to Michael Masters' (Phillips 2008) testimony before a US senate committee, financial speculative commodity transactions by institutional investors were to blame for increasing commodity-price volatility and contributing to food and energy price inflation. His unequivocal position suggested that the situation led to increased panic, further hoarding and price hikes bigger than would otherwise have been the case. In this case, Masters clearly indicates a strong correlation between the rapid increase in commodity investment instruments over the preceding 4 years with an equivalent increase in commodity spot price over the same period. Indeed this is a trend borne out in terms of commodity's volume traded in the preceding 5 years alone, which saw combined global trading in futures and options more than double, while globally traded agricultural financial instruments in the first 9 months of 2007 alone rose by 30% over the previous year (SOFI 2008). Yet not everyone shared Masters' definitive view, balancing this opinion, Paul Krugman of the *New York Times* convincingly argued in his article entitled 'The Conscience of a Liberal' against the effects of commodity speculation on actual food prices (Krugman 2008). Krugman's fundamental premise was that futures were simply bets about commodity outcomes that had no bearing on actual spot prices. Yet this position oversimplifies the argument for speculation of this sort, as futures contracts are not simple bystanders in the speculative process. They are in fact legally binding agreements to buy or sell commodities in a designated future date at a price

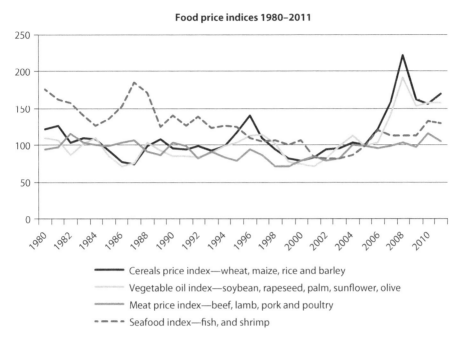

Food price indices 1980–2011

Cereals price index—wheat, maize, rice and barley

Vegetable oil index—soybean, rapeseed, palm, sunflower, olive

Meat price index—beef, lamb, pork and poultry

- - - - Seafood index—fish, and shrimp

FIGURE 22.11 Global commodity prices 1980–2011. (Compiled from data from the IMF *World Economic Outlook 2010*, International Monetary Fund, Washington, DC, 2010.)

agreed upon today. Do they affect real prices? Of course. Just read Kolb and Overdahl's book as well as numerous others' work on the subject of futures and it becomes enlightening just how real (spot) prices are affected and indeed just how easy it is to manipulate the market to do so (Kolb and Overdahl 2006 ; Torero and Braun 2009).

22.5.8 Corporate Control: Transnational Corporations

In the field of food security, a lot has been written about ever growing more powerful, market distorting corporations. Indeed the FAO notes that large transnational companies increasingly dominate world agricultural commodity markets, which leave them wielding direct and increasing power over what is produced and distributed (FAO 2004; Mousseau 2005) Just to picture the scale of the economic power these businesses wield, the top 500 worldwide businesses by market value in 2010 included 23 food producers or retailers, 20 biotechnology companies and 13 oil and gas giants (Figure 22.12). The top four food companies were Walmart (United States), Nestle (Switzerland), Unilever (Netherlands/UK) and Tesco (UK) with a combined market value of $533.2 billion and a turnover of $637.1 billion in 2010, of which Walmart alone held $405 billion. Interestingly while Walmart was only seventh in the league, it had by far the biggest turnover beating its next rival, the oil giant Exxon Mobil, by over $100 billion. So great is the food giants' economic power that in turnover alone, the big four—Walmart, Nestle, Unilever and Tesco—earned more in 2010 than the 2009 GDP earnings of 171, 137, 127 and 133 countries, respectively. In fact, the combined turnover of the 23 top food companies nearly equalled as much as the Russian Federation earned in 2009 (Financial Times 2010; World Bank 2010b).

Of course, the influence of multinational agri-food-retail businesses in the global food system does not differ significantly from other sectors; however, other sectors rarely share the same inelastic captive markets enjoyed by the agri-business sector. Everyone needs food, we need to grow it, move it, process it and retail it; and of concern are the fundamental relationships in this chain from farmer to consumer. This becomes especially important as the trend to a larger more integrated global food chain increasingly sees fewer larger multinational or transnational firms concentrating the lion's share of the market

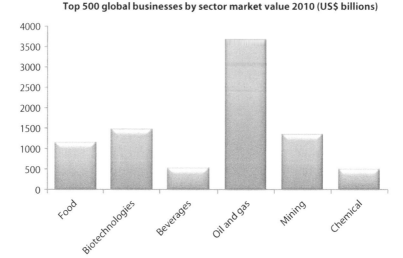

FIGURE 22.12 Top 500 businesses by market value 2010. (Courtesy of Financial Times' *World Top 500 Companies by Market Value*, http://media.ft.com/cms/66ce3362–68b9–11df-96f1–00144feab49a.pdf, 2010.)

into the hands of the few. In this debate there has been an increasingly vocal backlash against what is seen as the commodification of the food chain. Private corporations are concerned with profits and some argue there is an elementary dichotomy in the food-as-profit and food-as-a-right ideologies (ActionAid 2010; La Via Campesina 2011). While corporations were small to medium-sized, the argument goes that the many and varied stakeholders enabled a more equitable division of power between the producers, the processors, the retailers and the consumers and as a result the benefits were evenly shared. However, with a concentration of the agri-business sector both in traditional horizontal[*] integration and the more outwardly looking vertical[†] integration strategies the democratic free market is being replaced by a de facto oligopoly.[‡] From this point, it does not take an economic genius to know that a market concentrated with fewer companies sees their effective economic power base exponentially increase. This tends to be the argument of the anti-globalisation, anti-GMO and anti-capitalist movements; moreover, the juxtaposition of the two stances is increasingly being polarised. On the one hand, as the onslaught of relentless capitalist economies demand ever growing profits, the companies respond with increasing innovation, ingenuity and novelty. On the other side of the debate, some traditionalists dismiss GMOs out of hand in Frankenstinian horror while anti-capitalist movements descend on G-8 summits and food sovereignty proponents aim to roll back WTO negotiations in favour of a return-to-Earth approach to food production.

Another charge of this increasing corporate innovation is all too often profits are derived at the expense of the primary producers. One example of this is the advent of genetics and the growing proclivity of biologically based intellectual property rights (IPR) systems. Combined, genetic engineering and IPRs are promoting monopolistic privilege over material that many believe should be common property. Effectively, what this concentration and power endorses is the proprietary ownership of seed and livestock that effectively ensures farmers' continuing servitude. Furthermore IPRs become an obstacle of technological transfer too, a barrier of common social development if you like. This is especially so when

---

[*] Horizontal integration is a microeconomic principle that sees expansion of the firms at the same level in the value chain as in retailers buying up other retailers.

[†] Vertical integration is a microeconomic principle that sees expansion of the firms into other areas of the supply chain as in retailers buying up or merging with food processing industries or investing in primary producing activities.

[‡] Oligopoly is a market form in which a market or industry is dominated by a small number of sellers.

you consider that developing countries often lack the sophisticated patenting infrastructure enjoyed by more industrialised countries. Interestingly in this point Windfuhr and Jonsén observe that

> Whereas more than 90% of genetic resources for food and agriculture are from biotopes in the South, corporations in developed countries claim 98% of the patents on genes and living organisms. (Windfuhr and Jonsén 2005)

This is but one example and there are many more in the full length of the food chain that include both producer and retailer side concentrations. As more and more of our food is being innovated, transacted, processed and retailed by a smaller band of transnational or multinational companies many are questioning the sustainability of continuing trends. In this sense, one observer, Olivier De Schutter, special rapporteur on the right to food at the 17th Session of the UN Commission on Sustainable Development commented that

> Trade is mostly done not between States but between transnational corporations … The expansion of global supply chains only shall work in favor of human development if this does not pressure States to lower their social and environment standards in order to become 'competitive States', attractive to foreign investors and buyers. (CSD 2009)

On balance, De Schutter advocates multilateral-based incentives and regulation, encouraging the global agri-food chain to take responsibility in their role on sustainable food development and individual's right to food.

22.6 Discussion

In the twentieth century, the governance of peace and food security has been largely a corroboration of governments, international agencies and NGOs (FAO 2009c) and its importance cannot be underestimated

> Democratic peace, cosmopolitanism, and global governance are among the most powerful conceptual frameworks in contemporary world politics. (Aksu 2008, p. 368)

The world is constantly evolving, democracy has become the world's most common form of governance with 116 official electoral democracies (with caveats, see Section 22.1) (Freedom House 2011) while globalisation continues to integrate the world's markets for goods, services and finance. This research clearly highlights that democracy is a key ideal, although not necessarily a pre-requisite, for sustained development that aims to reduce poverty and increase food security in the poorly performing countries. Democracies facilitate change and with the right balance of institutional governance embracing notions of human rights, development philosophies, greater public resources, improved physical infrastructure, education, health and international cooperation, more inroads can be made into many of the issues surrounding food insecurity. While it is noted that democracy is not necessarily a requisite, good governance most definitely is. As funds and resources are misappropriated for the benefit of the ruling elite, their supporters and other cronies, external assistance and limited domestic resources are lost to the true grass-roots reformers on the ground. Some might argue that you cannot separate good governance from democracy, yet history shows us there have been numerous benevolent despots (if that's not an oxymoron) throughout the ages unless of course you sign up to Lord Acton's dictum:

> Power tends to corrupt, and absolute power corrupts absolutely. (Engel-Janosi 1941)

So, benevolent despots aside, democracies are the preferred political persuasion. Democracies too, it has been shown, empowers its citizenry through associations, movements and the media to monitor and comment openly on the conduct of governance and its public officials ensuring, in theory at least, full and open accountability. Yet for all this there is still a lack of coherent and efficient governance in current world food security efforts and one that is recognised at the highest levels (FAO 2009c). Lack of investment, a shortage of a code of conduct for foreign direct investment (FDI), difficult and unwieldy legislation, poor coordination at the international level coupled with inadequate or powerless bodies, underequipped and underfunded are just some examples of the essence of the problems that need addressing from the perspective of governance.

In this respect, leadership is needed and as the world's only superpower and largest donators of humanitarian aid, the United States is uniquely positioned to set the tone of and provide examples of best practices in international governance and multilateral cooperation. Yet in this guidance the United States is sadly lacking. US hegemony is being exercised in many areas yet the majority of this is motivated by self-interest. This has led some to ask why, as leading proponents of human rights, has the United States not yet ratified the UN Convention on the Rights of the Child nearly ten years after it came into force, or the Convention to Eliminate All Forms of Discrimination Against Women (CEDAW), or for that matter, why would it not ratify the Kyoto Protocol? The reasons are especially nonsensical as the United States actively participated in guidance on many of these matters. In short many reasons have been given for these shortcomings from partisan politics to conflicts of domestic interests; however the UN, in their humble opinion, feels the biggest single reason still remains the simple lack of political will (UN News Centre 2010). It is easy to be philosophical about such events, especially if one were to read Ayn Rand who opposed altruism (but not kindness) on the grounds that people should not sacrifice themselves for the greater good. Indeed in her view,

> Capitalism was destroyed by the morality of altruism. Capitalism is based on individual rights—not on the sacrifice of the individual to the 'public good' of the collective. Capitalism and altruism are incompatible. It's one or the other. It's too late for compromises, for platitudes, and for aspirin tablets. There is no way to save capitalism—or freedom, or civilization, or America—except by intellectual surgery, that is: by destroying the source of the destruction, by rejecting the morality of altruism. (Rand 1967)

For Rand, capitalism and altruism were philosophical opposites that could not

> ... co-exist in the same man or in the same society. (Rand 1967)

These are fair points and a train of thought that the author finds hard to fault. Moreover, from the governmental perspective, it is a difficult line to enforce or for that matter to take the moral highground as the United States like any government has an obligation first to its own citizens. In truth we cannot, nor should we enforce altruism at gunpoint, after all most of us are not politicians. Yet this leaves a vacuum in international guidance and one that the United States resists all efforts to fill. The harsh reality is that if the United States cannot, or is not, prepared to support the much needed change beyond lip service, then their global hegemony needs to be tempered with a healthy measure of scepticism and common sense and the reigns of influence needs to be passed to those that can and will. Such needed governance was exemplified in the 2008 food price crisis that placed huge strain on international commodity markets. Indeed, for better or worse, good governance was also witnessed in the recent 2010/11 financial crisis which saw governments rally global support in pursuit of a single common goal-stability.

Governance aside for the moment, as the food price drama continues to play out in the international markets, there is an immediate recognised need to restore faith in the capitalist ideology. For some though, this is a moot argument as the failings of a system that left hundreds of millions of people unable to afford the food they needed is indicative of the fundamental flaw in free trade economics itself. Even Jacques Diouf, Director-General of the FAO, recognises the problems of doing nothing to effect change in the system. In his view, the principles of comparative advantage hold and that the world food markets need to become more reliable if it is to provide

> ... a credible alternative to the politically more secure, but economically more costly system of national food self sufficiency. (Christiaensen 2009)

In the field of policy, policy choices can often have indeterminate effects on other policies while others might be used to serve multiple objectives. Getting the right balance is difficult and can sometimes lead to thorny trade-offs. Raising interest rates for instance to combat inflationary pressures such as higher food prices tends to reduce investment and places upwards pressure on exchange rates. This in turn affects exports, growth and investment. Shielding the vulnerable too, from food shocks through export restrictions, subsidies and safety nets can turn out to be expensive and reduces exports earnings. The point being that there is no one policy that promotes food security, multiple policies aiming to shore up security need to be considered in tandem and across multiple government ministries or institutional divisions. This is no less true of interventionist policies, particularly in food commodities. There is a long history of rival opinion as to the validity of intervention in food prices. On the one hand, some people feel that food, as a commercially traded commodity, it needs to reflect fair market price determined by free trade. Others however, because of the vulnerability of stocks and the volatility of prices, feel that basic foodstuffs should be placed 'outside the area of private or commercial gain' so as to prevent unjust profiteering and more equitable access by the population (Clynes 1920, p. 147).

In considering development strategies to lift developing countries out of poverty this is seen as an area beset with complex considerations and one that has repeatedly defied international efforts of improvement over the decades. While the Western model of economic growth was predicated on among other things, technological innovation, investment in human capital, knowledge acquisition, the expansion of the population as well as political, and some say Western, predatory exploitation, the success of these events have proven difficult to elucidate let alone replicate in developing nations. As a result, much posturing over the successes of this Western model leave only marginal agreement over what needs to be done in the poorer regions. Many too also suggest that it is just not possible to transplant the Western model of economic and social development across to the developing world as this was achieved in a different time faced by many different challenges. Fundamentally, in light of these contradictions, many disagree over the very nature of development strategies themselves. While some argue for the status quo, that is development should continue to be built on economic principles, others feel a more socialist model would best serve developing countries' overall development agendas. It is this latter view that has been repeatedly interpreted in UNDP's annual Human Development Index (HDI) where some nations are regularly seen to achieve greater social development with less economic growth. Indeed, this counter-intuitive outcome is one that strongly challenges traditional assumptions that economic development is necessary for social development to follow. This then begs the question: progress in what? Economic growth or social growth, or both? If however, as is the present tendency the aim is to combine both, then the challenge involves the best way of integrating these objectives into one overarching development model.

In short, development paradigms are the models or ideas that underpin progress. In this way, an overall development model is often made up of many independent models all pointing in the same direction. The many models can include inter alia economic, social, environmental as well as agricultural. The difficulty is that such models are rarely truly aligned and in some instances we may even see competing or incompatible ideologies. Considering agriculture development paradigms, it can be seen then that over time agriculture paradigms have shifted, sometimes quite radically. At times, the agricultural sector has at times been treated as little more than a global allotment in which food is drawn out at will with little or no regard to its developmental potential in the wider economy. At other times, it has been suggested that by investing in agriculture, rural education and industry more labour can become mobile and readily transposed from the agricultural sector to the so-called 'more effective' or prosperous industrial sectors. Later, further shifts recognised the fact that poverty was predominantly a rural problem and one that could be effectively addressed through social and economic investment. And finally, it seems after a long and circular route, agricultural development paradigms have brought us to the current realisation that indeed the agricultural sector can have a valuable role to play in the overall strategy of both the economic and social development of a country or region. In this scenario, McCalla (2007) suggests that the agricultural sector

> ... becomes much more complex, more interdependent with the rest of the economy and charged with meeting multiple goals. [Where] Increasing food production is no longer a goal in itself. (McCalla 2007, p. 16)

This goes to the heart of the problem as many developing countries were encouraged to follow the comparative advantage model and base their food security not on self sufficiency but on the global free trade model. While in principle this might not seem an unreasonable idea, in practice, it has not worked out that well for developing countries. Initially because of cheap labour and available land, the comparative advantage of many developing countries lay in their ability to produce vast quantities of food cheaply. As a result, they were encouraged to concentrate on cash crops and plantations for export while importing the food they needed for domestic consumption. This was also reinforced through the structural adjustment policies of the IMF. In time however, the West's agricultural innovation and technology were translated into greater yields and lower agricultural costs. Combined with the protectionism prevalent in many Western agricultural policies it meant that food from the West was now cheaper than many foods coming out of the developing countries. This effectively depressed local production in the poorer countries whose natural comparative advantage had been undermined by artificial trade-distorting policies of the industrialised world. Not only are the developing countries now struggling to compete on the global market but also the continuing imbalance in the terms of trade ensure the continuing cheap supply of food runs in one direction. What is left is a dualistic or dichotomous approach to developing countries. On the one hand, there is much effort by the Western (or north) industrialised countries to keep the status quo and on the other, the same industrialised countries along with multilateral organisations seek to offer aid and assistance in changing their economic fortunes for the better. Of note too in this debate is that agriculture by its nature will only ever give limited returns. Just looking at the West's model of economic growth and their comparative economic sectors, it becomes obvious that as a proportion of GDP, industry and the service sectors create huge added value in the economy, yet primary agricultural production has limited value added. Value-added agriculture comes with industrial food processing and to a much lesser extent some differentiation of hard to find or luxury commodities. Diversifying out of agriculture into the luxury or value added sectors is one approach to increasing economic growth. Yet there are two immediate issues with this approach. Firstly, as Western food processing technologies increases up the chain, the barriers to international trade in such areas are increased, further discouraging investment in the developing countries. Secondly, if we are suggesting a model based on the Western example of out-of-agriculture and into industry, what of the hundreds of millions of smallholder farmers in developing countries? Of particular note here too is the introduction of the WTO Agreement on Agriculture (AoA). In 1995, agriculture became the target of trade barrier reductions and developing countries were bound to the opening up many of their agricultural markets. However, there still exists a fundamental imbalance in the level of liberalisation for different groups of countries with many developing countries' smallholder farmers still facing challenges competing with subsidised exports from industrialised countries. There are bigger problems however; for while export subsidies have been targeted by the WTO, there is a gray area that still allows vast domestic subsidies in agriculture such as the CAP. With poor countries unable to afford to subsidise their farmers and with the binding AoA agreements developing countries once again face an unequal playing field.

When considering economic growth, further disagreements as to which model might best serve policy objectives are common and indeed necessary in a democratic development model. Some suggest capitalism and free trade is the only model worth developing while others, as mentioned earlier, feel the whole capitalist system itself is at fault. Some others still suggest that control of economic growth is a myth and that development is simply a case of collative adaptation. However, talking of capitalism, the much vaunted notion of free trade is a bit of a misnomer and one that does not actually exist. Even despite the best efforts of the GATT negotiations and the WTO, this has generally been the case for many years. As has been noted, the competing notions of economic development paradigms from different camps might go a long way to explain this melee. This however does not detract from those hailing free trade and the market economy as the panacea for the equitable distribution of social and economic benefits. Sadly though in reality instead, governments and private firms talk freely of open market mechanisms whilst at the same time seeking to protect themselves from market volatility via protectionism. In this way, any government, transnational corporation or other entity that colludes to distort free and fair market pricing through off-market agreements, contracts or plain protectionist policies are in direct contravention of the principles of comparative advantage and fair trade: those same policies incidentally for which most signed up for. As a result, perhaps the idea of global free trade is like democracy; it is fine in theory but in practice a more social democratic approach along the lines outlined by Einstein in 1949 is perhaps warranted (Einstein 1949). Otherwise if left to its own devices, it seems free trade, like democracy, favours the stronger party. As a result, capital is further concentrated in fewer hands and as history shows us, capital becomes economic, becomes political power. This perpetuates a cycle familiar to us all where power begets power and harking back to Einstein's social ideals, in such situations, individuals become

> … more conscious than ever of his dependence upon society. But he does not experience this dependence as a positive asset, as an organic tie, as a protective force, but rather as a threat to his natural rights, or even to his economic existence. (Einstein 1949)

Alas, this new social democratic approach to agriculture and free trade, rather than promoting a new social order, borrows from socialist democracy's ideological base and directs goals towards a more considered social-ethic. In this way collective barriers of trade need to be forcibly reduced and mechanisms put in place that allow countries to determine their own food security priorities should they wish. After all, and tapping into certain ideals promoted by the food sovereignty movement, a country's food priorities should not be subject to the whims of others. The difficulty of these two seemingly mutually exclusive points of view (the status quo versus Einstein's socialist democracy) is not lost on the author and I can offer no answer in its solution, not at least in this limited space.

Modern economic and development paradigms are not the only ideologies under fire. The benevolence of food aid is also continually being questioned and the argument is perhaps more complicated than it at first may seem. Long-held concerns of dependency have been answered to some extent by the FAO, yet charges of disruption to recipient countries markets persist (FAO 2006). Whether food aid indeed disincentivises recipients, encourages governments to postpone or divert needed reforms or worse, relinquish social responsibility of the food security problem is constantly being raised. Underpinning a lot of these issues is the fundamental use of aid; the question is whether food aid should be used only as an emergency measure as suggested by the EU or whether it should be used in the development projects as promoted by the United States and the WFP. In some instances humanitarian aid and development projects clearly overlap, yet where there are purely development objectives the case is not so clear cut. Other more serious issues are raised regarding the 'tying' of aid as well as direct transfers. Tying remains a controversial practice and imposes significant restrictions on donor organisations in its use. Simultaneously, the United States is still the biggest donor of direct transfers of aid; although thankfully it is on the wane (90% in 1980s to 50% presently). However direct aid still represents a significant dollar investment and the largest source of aid within the overall humanitarian package. The continual conditions attached to aid by the United States sees large domestic procurement requirements as well as further legislation that sees between 50% and 75% of commodities being processed and packed in the United States before shipment. On top of this a further 75% and 50% of the aid is required to be transported by US-registered vessels by USAID and the USDA respectively (FAO 2006; Harvey et al. 2010).

It's not just the United States. In fact the whole idea of aid as a political tool or an extension of domestic foreign policy is one that receives a lot of attention and is completely at odds with the whole humanitarian ethos. Addressing such practices is one thing however, what is more difficult perhaps is the difficulty in addressing a structural ideology that still places humanitarianism below that of national self-interests. A recent example from the 2002 US Agency for International Development report entitled 'Foreign Aid in the National Interest' rather tellingly signs up to the notion that what is good for the United States is good for the World. To be fair though despite this less than altruistic opening gambit the document is a rather enlightening, partly self-critical, appraisal of the many challenges ahead and well worth a read (USAID 2002).

So despite the many less than optimal practices and the much open criticism of US policies in this field, it is perhaps worth remembering that in spite of these limitations, US food aid still remains important source of much needed assistance.

An interesting and pertinent complaint from the Committee on World Food Security (CFS) suggests that when food security crises are treated as purely transitory phenomena as they are now, it promotes tunnel vision and effectively inhibits much needed broader development analysis and integrated emergency and non-emergency strategies (CFS 2007a). This view also taps into the wider societal malaise and another important component of humanitarian assistance: donor fatigue. With increasing incidents of emergencies and disasters and an average 30% annual deficit in requested funding year on year, it seems the public wanes and wanes in its support. Yet this is nothing now as it seems a half a century ago we were plagued by the same nonchalance and it makes one wonder whether Norman Cousins's (editor-in-chief of the *Saturday Review*) insightful understanding of half a century ago could ever be reversed:

> A nation conditioned by affluence might possibly be suffering from compassion fatigue, or from conscience sickness, the peril of narrowing our field of vision to leave out the unpleasant view of life disfigured by hunger. (Cousins 1961)

Of note too is the fact that the reduction in direct transfers mentioned earlier has not resulted in equivalent increases in cash contributions for local or triangular purchases. While overall food assistance (spending and direct transfers) has increased, the direct food donations or food aid flows have decreased. This could be seen as a good trend with less use of inefficient direct transfers; however; it might also mask other important changes that are difficult to examine as the data does not lend itself to detailed analysis of this sort (Harvey et al. 2010). In any event, the difficulty in such circumstances is that any shortfalls in crisis situations has to be found somewhere and these are often found by reallocating resources from valuable development programmes (FAO 2002d; Kellett et al. 2010).

Moving on from aid and considering corporate concentration both up and down the food chain, it is clear that such businesses effectively offer the consumer choice and convenience. The burning question here is at what price? In principle for every consumer there is a producer on the other side of the equation, who, all things being equal, benefits from mutual exchange. However, this equation is being distorted by longer food chains and corporate middle-men. Indeed customers want more choice; they tell the producers that they need better quality, uniformity, no blemishes on their food, but above all they want cheaper goods. And we do. So with an increasing captive audience, we empower these corporations to act on our behalf who in turn are said to artificially distort the equilibrium for greater profit. As has been seen, many of the biggest companies in the world turnover many more times the GDP of numerous states or countries and of course these are the benefits of the market system. This gives them enormous economic, and more importantly, political power in influencing financial and policy decisions, not only in their own countries but in others too. As the likes of Walmart turns over just under 25% of the UK's annual $2 trillion GDP, just what influence is brought to bear in negotiations with suppliers and how well equipped is host government politically, legally and morally to balance the inflow of foreign direct investment with free and fair trade? Another example using the British government again sees them being held to ransom over increased taxation of the banking sector, with many banks currently threatening to flee the capital in favour of more economically friendly waters. As multinationals acting as merchants of the capitalist system increasingly challenge the sovereign state head on, are we not in danger of reaching a point where sovereign entities become somewhat subservient to both multinational interests and the capitalist economy? Do not misunderstand me, I am not anti-capitalist, but more of a moderator or devil's advocate, if you like. So I ask again: At what price? Of course, we want cheap food as we want cheap housing, water and cars; in fact, we want cheaper everything, it is the doctrine within which the modern world flourishes. But do we want cheap food if it means our fellow man goes hungry? Or if the environment suffers? Or at the expense of our future legacy? I suspect not, but once again guilt and altruism at gunpoint is not the answer. What we need is a radical re-evaluation of our current lifestyle choices. After all capitalism and the pursuit of money over everything else is but 500 years old, yet, by the same token I for one am not naive enough to think that we can, or should, turn the clocks back. Progress I feel, has to be seen in the rebalancing of our collective moral compass. Perhaps one answer would be social groups to equal the giants of industry that in turn advocate solid moral guidance predicated on socially just policies, good governance and a properly working ethical compass. Indeed, we are talking about the ever growing NGO base, growing in numbers, financial, moral as well as political might. These bodies of social conscience do what they can and should be applauded; yet all too often we hear similar stories of poor governance, mismanagement, tunnel vision and on occasion self-interests placed over the public good. Where to go from here?

Who knows? But it does leave one questioning whether there is fundamental conflict of interest in collective national policies that attempt to steer global development paradigms through multilateral agreements that ultimately return to negatively affect domestic interests. Once again this harks back to the global versus local debate and whether or not national self interests trump globalisation. We cannot have it both ways, in too many instances nations outwardly and evangelically preach of the benefits and desirability of a global village: benevolent, interconnected and equitable. Yet inwardly, and unashamedly non-discreetly, preach nationalism in terms of buying local, protecting jobs and generally preserving domestic integrity in all things capital. Is this hypocrisy, misguided or simply wrong?

References

Abbott, C. (2005). Rights and Responsibilities: Resolving the Dilemma of Humanitarian Intervention. Oxford Oxford Research Group 17pp.

ActionAid (2003). Food Aid: An Actionaid Briefing Paper. ActionAid

ActionAid (2010). Act!Onaid. Actionaid. http://www.actionaid.org.uk/100041/our_history.html (accessed 23 January 2011).

Aksu, E. (2008). Perpetual Peace a Project by Europeans for Europeans? *Peace & Change: A Journal of Peace Research* 33(3): 368–387.

Anon (1953). Food Surpluses a Weapon for the Cold War. *Journal of Agricultural and Food Chemistry* 1(10): 655.

ARI (2009). Green Revolution. *Briefing Note 0902*. London: Africa Research Institute.

Barnes, D. G. (2006). *A History of English Corn Laws: From 1660–1846*. London: Routledge.

Barrett, C. and K. Heisey (2002). How Effectively Does Multilateral Food Aid Respond to Fluctuating Needs? *Food Policy*(5–6): 477–491.

Barrett, C. and D. Maxwell (2005). *Food Aid after Fifty Years: Recasting Its Role*. London: Routledge.

Barrett, C. and D. Maxwell (2006). Towards a Global Food Aid Compact. *Food Policy* 31(2): 105–118.

BBC (2008a) Clashes at Nairobi Food Protest. *BBC Online*. http://news.bbc.co.uk/2/hi/africa/7429303.stm.

BBC (2008b) The Cost of Food: Facts and Figures. *BBC Online*. http://news.bbc.co.uk/1/hi/in_depth/7284196.stm.

Belshaw, H. (1947). The Food and Agriculture Organization of the United Nations. *International Organization* 1(2): 291–306.

Birchall, J. (2002). Evolutionary Economics. John Birchall. http://www.themeister.co.uk/economics/evolutionary_economics.htm (accessed 15 January 2010).

Bloem, R. (2001) The Role of Ngos in the Age of a Democratic Civil Society. http://www.globalpolicy.org/component/content/article/177/31613.html.

Bonnard, P. (2008a). *Markets Assessment and Analysis: Assessing Markets*. EC-FAO.

Bonnard, P. (2008b). *Markets Assessment and Analysis: Markets and Food Security*. EC-FAO.

Brown, L. (2008). Why Ethanol Production Will Drive World Food Prices Even Higher in 2008. Earth Policy Institute. http://www.Earth-policy.org/Updates/2008/Update69.htm (accessed 7 June 2009).

Carter, M. and C. Barrett (2006). The Economics of Poverty Traps and Persistent Poverty: An Asset-Based Approach. *Journal of Development Studies* 42(1): 178–199.

CCP (2005). A Historical Background on Food Aid and Key Milestones. Committee on Commodity Problems: Sixty-fifth Session. Rome: Food and Agriculte Organisation.

CFS (2000). Assesment of the World Food Security Situation: Suggested Core Indicators for Monitoring Food Security Status. Committee On World Food Security: Twenty-sixth Session. Rome: Committee On World Food Security.

CFS (2007a). Assessment of the World Food Security Situation. Committee On World Food Security: Thirty-third Session. Rome: Food and Agriculture Organisation.

CFS (2007b). Report on the Development of Fivims. Committee On World Food Security: Thirty-third Session. Rome: Food and Agriculture Organisation.

CFS (2008a). Agenda Item V: Participation of Civil Society/Non-Governmental Organizations (Csos/Ngos) Committee On World Food Security: Thirty-fourth Session. Rome: Food and Agriculture Organisation.

CFS (2008b). Follow-up to the World Food Summit: Report on Progress in the Implementation of the Plan of Action. Committee On World Food Security: Thirty-fourth Session. Rome: Food and Agriculture Organisation.

CGG (1995). *Our Global Neighborhood the Report of the Commission on Global Governance*. Oxford: Oxford University Press.

Christiaensen, L. (2009) 2009 World Summit on Food Security. *Our World* (30 January 2011): http://ourworld unu.edu/en/2009 world-summit-on-food-security/

CIA Factbook (2010). The CIA World Factbook. https://www.cia.gov/library/publications/the-world-factbook/geos/xx.html (accessed 12 October 2010).

Clapp, J. (2005). The Political Economy of Food Aid in an Era of Agricultural Biotechnology. *Global Governance*, (Oct–Dec 2005), 11: 467–485.

Clapp, J. (2009). The Global Food Crisis and International Agricultural Policy: Which Way Forward? *Global Governance* 15(2): 299–312.

Clay, E. (2002). Food Security: Concepts and Measurement: Paper for Fao Expert Consultation on Trade and Food Security: Conceptualising the Linkages. Rome: Food and Agriculture Organisation.

Clay, E. (2006). The Post Hong Kong Challenge: Building on Developing Country Proposals for a Future Food Aid Regime. London: Overseas Development Institute.

Clay, E., N. Pillai and C. Benson (1998). *The Future of Food Aid: A Policy Review*. London: Overseas Development Institute.

Clements, R. E. (2009). Scoping Study into the Impacts of Bioenergy Developmenton Food Security. Nairobi, Kenya: Policy Innovation Systems for Clean Energy Security (PISCES).

Clynes, J. R. (1920). Food Control in War and Peace. *The Economic Journal* 30(118): 147–155.

Cohen, D. (2005). Achieving Food Security in Vulnerable Populations. *British Medical Journal* 331: 775–777.

Contreras, R. (1999). Competing Theories of Economic Development. *Transnational Law and Contemporary Problems* 9: 93–108.

Cousins, N. (1961). *Confrontation*. New York: McCall's Publishing Company.

CSD (2009). Contribution of Mr. Olivier De Schutter Special Rapporteur on the Right to Food; 17th Session of the UN Commission on Sustainable Development. New York: UN Commission on Sustainable Development.

D'Souza, F. (1994). Democracy as a Cure for Famine. *Journal of Peace Research* 31: 369–373.

Darcy, J. and C. Hofmann (2003). According to Need? Needs Assessment and Decision Making in the Humanitarian Sector. London: Overseas Development Institute.

Delva, J. G. (2008) Uneasy Calm in Haiti after Food Price Protests. Reuters. http://uk.reuters.com/article/latestCrisis/idUKN0932761420080410.

Denemark, R., M. Hoffman, L. Twist and H. Yonten (2007). Dominance and Diplomacy: Trends in the Utilization of Multilateral Agreements by Global Powers. ISA 48th Annual Convention. Chicago: International Studies Association.

DeRose, L. and E. Messer (1998). Food Shortage. In *Who's Hungry? And How Do We Know? Food Shortage, Poverty, and Deprivation*. Tokyo, New York, Paris: United Nations University Press.

DeRose, L., E. Messer and S. Millman (1998). Food Deprivation.In *Who's Hungry? And How Do We Know? Food Shortage, Poverty, and Deprivation*. L. DeRose, E. Messer and S. Millman, eds. Tokyo, New York, Paris: United Nations University Press.

Devereux, S., B. Baulch, K. Hussein, J. Shoham, H. Sida and D. Wilcock (2004). Improving the Analysis of Food Insecurity: Food Insecurity Measurement, Livelihoods Approaches and Policy: Applications in FIVIMS. Rome: FAO-Food Insecurity and Vulnerability Information and Mapping Systems (FIVIMS) 52.

Dimitri, C., A. Effland and N. Conklin (2005). *Economic Information Bulletin: Number 3: The 20th Centurytransformation of US Agriculture and Farm Policy*. USDA.

Dupont, A. and M. Thirlwell (2009). A New Era of Food Insecurity? *Survival* 51(3): 71–98.

EC/FAO (2008). An Introduction to the Basic Concepts of Food Security. Rome: EC-FAO Food Security Programme 3.

ECA/SDD (2004). Land Tenure Systems and Their Impacts on Food Security and Sustainable Development in Africa. Addis Ababa, Ethiopia: UN Economic Commission for Africa.

ECOSOC (2008) Global Food Crisis. http://www.un.org/ecosoc/GlobalFoodCrisis/index.shtml

Eifert, B., C. Galvez, N. Kabir, A. Kaza, J. Moore and C. Pham (2002). The World Grain Economy to 2050: A Dynamic General Equilibrium, Two Sector Approach to Long-Term World-Level Macroeconomic Forecasting. *University Avenue Undergraduate Journal of Economics (Online)*.

Einstein, A. (1949). Why Socialism? *Monthly Review* 1(May).

EIU (2010). Democracy Index 2010: Democracy in Retreat. London: Economist Intelligence Unit 46pp.

Engel-Janosi, F. (1941). Reflections of Lord Action on Historical Principles. *The Catholic Historical Review* 27(2): 166–185.

ESA (2008). European Space Agency: Land Cover Implementation Team Project Office. http://www.gofc-gold. uni-jena.de/sites/background.php (accessed 27 June 2009).

EuronAid (1995). Ngo Code of Conduct on Food Aid and Food Security. Brussels: EuronAid.

Evans, G., M. Sahnoun, G. Côté-Harper, L. Hamilton, M. Ignatieff, V. Lukin, K. Naumann, C. Ramaphosa, F. Ramos, C. Sommaruga, E. Stein and R. Thakur (2001). The Responsibility to Protect: Report of the International Commission on Intervention and State Sovereignty. Canada: ICISS.

FAO (1946). Report of the First Session of the Conference Held at the City of Quebec, Canada, 16 October to 1 November, 1945. Washington, DC: FAO 89.

FAO (2000a). *Handbook for Defining and Setting up a Food Security Information and Early Warning System (Fsiews)*. Rome: Food and Agriculture Organisation.

FAO (2002b). GIEWS: The Global Information and Early Warning System on Food and Agriculture. Rome: Food and Agriculture Organisation 24pp.

FAO (2002c). Measurement and Assessment of Food Deprivation and Undernutrition: International Scientific Symposium. International Scientific Symposium: FIVIMS—An Inter-Agency Initiative to Promote Information and Mapping Systems on Food Insecurity and Vulnerability. Rome: Food and Agriculture Organisation.

FAO (2002d) World Food Summit: Five Years Later Reaffirms Pledge to Reduce Hunger. *Summit News*. http:// www.fao.org/worldfoodsummit/english/newsroom/news/8580-en.html.

FAO (2003e). Trade Reforms and Food Security: Conceptualizing the Linkages. Rome: Food and Agriculture.

FAO (2004). The State of Agricultural Commodity Markets 2004. Rome: Food and Agriculture Organisation.

FAO (2005). The Globalizing Livestock Sector: Impact of Changing Markets. *Committee on Agriculture Report*. Rome: Food and Agriculture Organisation.

FAO (2006). The State of Food and Agriculture 2006: Food Aid for Food Security? Rome: U.N. Food and Agriculture Organisation, 183pp.

FAO (2007). FAO's Role & Effectiveness in Emergencies. Rome: Food and Agriculture Organisation 208 pp.

FAO (2008a). Crop Prospects and Food Situation No. 1 2008: Global Cereal Supply and Demand Brief. Rome: Food and Agriculture Organisation.

FAO (2008b). FAO Methodology for the Measurement and Assessment of Food Deprivation: Updating the Minimum Dietary Energy Requirements. Rome: FAO Statistics Division.

FAO (2009a). Address by Jacques Diouf Director-General of the Food and Agriculture Organization of the United Nations (Fao). Food Security for All. Madrid: Food and Agriculture Organization.

FAO (2009b). Food Outlook. Rome: Food and Agricultural Organisation.

FAO (2009c). Secretariat Contribution to Defining the Objectives and Possible Decisions of the World Summit on Food Security on 16, 17 and 18 November 2009. Rome: Food and Agriculture Organisation 10.

FAO (2010). The Food and Agriculture Organization of the United Nations. UN Food and Agriculture Organisation. http://www.fao.org/about/en/ (accessed 26 January 2010).

FAO (2011). Website of the Food and Agriculture Organisation of the United Nations. FAO. http://www.fao. org/hunger/faqs-on-hunger/en/ (accessed 15 February 2011).

FEWSNET (2010). Special Brief: Revising the Fews Net Food Insecurity Severity Scale. Washington: US Agency for International Development (USAID) 7.

Financial Times (2010). World Top 500 Companies by Market Value. *Financial Times*. Financial Times. http:// media.ft.com/cms/66ce3362-68b9-11df-96f1-00144feab49a.pdf (accessed 21 March 2011).

FIVIMS (2008). The FIVIMS Initiative: Food Insecurity and Vulnerability Information and Mapping Systems. http://www.fivims.org/index.php?option=com_content&task=blogcategory&id=20&Itemid=37 (accessed 6 June 2009).

Ford, S. (2009) Is the Failure to Respond Appropriately to a Natural Disaster a Crime against Humanity? The Responsibility to Protect and Individual Criminal Responsibility in the Aftermath of Cyclone Nargis. *Social Science Research Network (SSRN)*.

Freedom House (2011). Website of Freedom House. Freedom House, Inc. http://www.freedomhouse.org/template.cfm?page=1 (accessed 21 January 2011).

Gerster-Bentaya, M. and N. Maunder (2008a). *Food Security Concepts and Frameworks: Concepts Related to Food Security*. EC-FAO.

Gerster-Bentaya, M. and N. Maunder (2008b). *Food Security Concepts and Frameworks: Food Security Analysis*. EC-FAO.

GFMG (2010). The 2007/08 Agricultural Price Spikes: Causes and Policy Implications. London: DEFRA/ The Global Food Markets Group.

Global Policy Forum (2005). How Does Food Aid Work? . Global Policy Forum. http://www.globalpolicy.org/socecon/hunger/general/2005/0916fact.htm (accessed 18 June 2008).

GMES (2007). Global Monitoring of Environment and Security. European Space Agency ESA/EU. http://www.gmfs.info/ (accessed 12 March 2011).

Godoy (2007) G8-Africa: Farm Subsidies a Taboo Subject? *IPS*: http://www.ipsnews.net/news.asp?idnews=37966.

Goldin, I. and D. v. d. Mensbrugghe (1993). Trade Liberalization: Whats at Stake? *Policy Brief No. 5*. Paris: OECD Development Centre 32pp.

Goodrich, L. M. (1947). From League of Nations to United Nations. *International Organization* 1(1): 3–21.

GPF (2002). Global Policy Forum: Growth of International NGOs between 1990 and 2000. http://www.global-policy.org/ngos/tables/growth2000.htm (accessed 14 June 2008).

Hall-Jones, P. (2006). *The Rise and Rise of NGOs*: Global Policy Forum.

Harrell, E. (2007) The Power of One: How to End the Global Food Shortage. *Time Magazine Online*. Wednesday, August 15: http://www.time.com/time/nation/article/0,8599,1653360,00.html.

Harvey, D. (2010). An Abbreviated UK/EU Agricultural Policy History. Newcastle University. http://www.staff.ncl.ac.uk/david.harvey/AEF372/History.html (accessed 23 November 2010).

Harvey, P., K. Proudlock, E. Clay, B. Riley and S. Jaspars (2010). Food Aid and Food Assistance in Emergency and Transitional Contexts: A Review of Current Thinking. London: Overseas Development Institute 102.

HEWSweb (2011). Humanitarian Early Warning System. World Food Programme. http://www.hewsweb.org/home_page/default.asp (accessed 13 February 2011).

Hjertholm, P. and H. White (2000). Survey of Foreign Aid: History, Trends and Allocation. Copenhagen: Institute of Economics University of Copenhagen.

Hoddinott, J. (1999). Choosing Outcome Indicators of Household Food Security. Washington, DC: International Food Policy Research Institute 29.

Howe, J. (2004). *The End of Fossil Energy and a Plan for Sustainability*. Fryeburg, US: McIntire Publishing.

IMF (2010). International Monetary Fund Website. International Monetary Fund. http://www.imf.org/external/ (accessed 4 January 2010).

INSCR (2011). *The Integrated Network for Societal Conflict Research (INSCR)*. Center for Systemic Peace.

IPC (2007). The Integrated Food Security Phase Classification: Technical Manual Version 1.1. Rome: FAO, IPC Global Partners.

Johnson, D. G. (1997). Agriculture and the Wealth Of. Nations. *American Economic Review* 87(2): 1–12.

Johnson, S. (2007) Op-Ed: Let Them Eat Pie. *The Stanford Daily Online* Monday, 29 October 29: http://www.stanforddaily.com/2007/10/29/op-ed-let-them-eat-pie/.

Karajkov, R. (2007) The Power of N.G.O.'S: They're Big, but How Big? *World Press*. http://www.worldpress.org/Americas/2864.cfm

Karunakara, U. (2010). *Haiti: Where Aid Failed*. London: Guardian News and Media Limited. 28 December 2010.

Kellett, J., D. Malerba, G. Brereton, L. Walmsley, H. Sweeney, J. Keylock, K. Smith, L. Poole, V. Stoianova and D. Sparks (2010). Global Humanitarian Assistance Report 2010. Shepton Mallet, UK: Global Humanitarian Assistance (GHA).

Kennedy, E. (2002). Measurement and Assessment of Food Deprivation and Undernutrition: Keynote Paper: Qualitative Measures of Food Insecurity and Hunger. International Scientific Symposium. Rome: FAO.

Kennedy, G., G. Nantel and P. Shetty (2004). Globalization of Food Systems in Developing Countries: Impact on Food Security and Nutrition: A Synthesis of Country Case Studies. Rome: FAO.

King, M. and C. Elliott (1996). Education and Debate: Averting a World Food Shortage: Tighten Your Belts for Cairo Ii. *British Medical Journal* 313: 995–997.

Kolb, R. W. and J. A. Overdahl (2006). *Understanding Futures Markets*. UK: Blackwell Publishing.

Kristjanson, R. L. (1960). Discussion: Impact of Surplus Disposal on Foreign Competitors and the International Perspective on Surplus Disposal. *Journal of Farm Economics* 42(5 Proceedings of the Annual Meeting of the American Farm Economic Association): 1081–1083.

Krugman, P. (2008) The Conciense of a Liberal: Speculative Nonsense, Once Again. *The New York Times Online*: http://krugman.blogs.nytimes.com/2008/06/23/speculative-nonsense-once-again/ (accessed 2 March 2011).

La Via Campesina (2011). La Via Campesina: The International Peasant's Voice. *News & Views*. http://www.viacampesina.org/en/index.php?option=com_content&view=frontpage&Itemid=1 (accessed 21 February 2011).

Latham, M. C. (1997). Human Nutrition in the Developing World. *Food and Nutrition Series No. 29*. Rome: Food and Agriculture Organisation.

Lee, J. (1993). Observations on Agricultural Policy, Policy Reform and Public Policy Education. 43rd Annual Conference of the National Public Policy Education Committee, Florida.

MARS (2011). The Monitoring Agricultural Resources (Mars) Unit Mission. EC Joint Research Commission (JRC). http://mars.jrc.ec.europa.eu/mars/ (accessed 1 January 2011).

Marzeda-Mlynarska, K. and M. Curie-Sklodowska (2009). Multilevel Governance of Food Security: Theoretical Model or New International Practice. 4th Annual Conference of the GARNET network. Rome: GARNET Network of Excellence.

Maunder, N. (2006a). *Food Security Information Systems and Networks: Lesson 1. Food Security Information Systems*. Rome: EC FAO.

Maunder, N. (2006b). *Food Security Information Systems and Networks: Lesson 2. The Institutional Context*. Rome: EC FAO.

Maxwell, D. (2006). Global Trends in Food Aid. Khartoum Food Aid Forum. Khartoum: The World Food Programme.

Maxwell, D., B. Watkins, R. Wheeler and G. Collins (2003). The Coping Strategies Index: A Tool for Rapidly Measuring Food Security and the Impact of Food Aid Programmes in Emergencies. Rome: Food and Agriculture Organisation.

Maxwell, D. and K. Wiebe (1999). Land Tenure and Food Security: Exploring Dynamic Linkages. *Development and Change* 30(4): 825–849.

Maxwell, S. and T. Frankenberger (1992). Household Food Security: Concepts, Indicators and Measurements: A Technical Review. New York, Rome: UNICEF and IFAD.

McMichael, P. (1994). *The Global Restructuring of Agro-Food Systems*. Ithaca, NY: Cornell University Press.

Mehta, V. N. (1929). Famines and Standards of Living. *The Annals of the American Academy of Political and Social Science* 145: 82–89.

Meier, G. M. and J. E. Stiglitz (2002). *Frontiers of Development Economics: The Future in Perspective*. New York: Oxford University Press.

Meller, P. (1992). Adjustment and Equity in Chile. Development Centre Studies. Adjustment and Equity in Developing Countries. Paris: Organisation for Economic Development.

Mikesell, R. F. (1968). *The Economics of Foreign Aid*. Chicago: Aldine Publishing Company.

Mock, N., N. Morrow, S. Aguiari, X. Chen, S. Chotard, Y. Lin, A. Papendieck and D. Rose (2006). Comprehensive Food Security and Vulnerability Analysis (CfSVA): An External Review of WFP Guidance and Practice. Rome: Development Information Services International.

Morrison, J. and A. Sarris (2007). Wto Rules for Agriculture Compatible with Development. Rome: Food and Agriculture Organization

Mousseau, F. (2005). *Food Aid or Food Sovereignty? Ending World Hunger in Our Time*. Oakland, CA: Oakland Institute.

Mukherjee, S. (2011). Blame the Rulers Not the Rain: Democracy and Food Security in Zambia and Zimbabwe. South Africa: South African Institute of International Affairs (SAIIA).

Murphy, S. (2009). Strategic Grain Reserves in an Era of Volatility. Minneapolis, Minnesota: Institute for Agriculture and Trade Policy.

Murphy, S. (2010). Trade and Food Reserves: What Role Does the Wto Play? Minneapolis, Minnesota: Institute for Agriculture and Trade Policy.

Murphy, S. and K. McAfee (2005). US Food Aid—Time to Get It Right. Minneapolis: Institute for Agriculture and Trade Policy.

Murphy, W. (1956). Editorial: Food Surpluses and Energy Shortages. *Journal of Agricultural and Food Chemistry* 4(2).

Nathoo, T. and J. Shoveller (2003). Do Healthy Food Baskets Assess Food Security? *Chronic Diseases in Canada* 24(2–3): 65–69.

OCHA/FTS (2006). Financial Tracking Service of Ocha. UN OCHA Financial Tracking Service (FTS).

ODI (1997). Global Hunger and Food Security after the World Food Summit. London: Overseas Development Institute.

ODI (1999). What Can We Do with a Rights-Based Approach to Development? London: Overseas Development Institute.

ODI (2000). Reforming Food Aid: Time to Grasp the Nettle. *Briefing Paper*. London: Overseas Development Institute.

OECD-FAO (2007). Agricultural Outlook 2007–2016. Paris/Rome: OECD-FAO.

Paras, A. (2007). 'Once Upon a Time . . . : Huguenots, Humanitarianism and International Society'. ISA's 48th Annual Convention. Chicago: International Studies Association.

Pardey, P. G., N. Beitema, S. Dehmer and S. Wood (2006). Agricultural Research: A Growing Global Divide. Washington, DC: IFPRI.

Phillips, L. (2008). Senators Hear Testimony on Legislation to Reduce Food and Energy Prices by Curbing Excessive Speculation in the Commodity Markets. Washington, DC: Senate Committee on Homeland Security and Governmental Affairs.

Phillips, R. W. (1981). FAO: Its Origins, Formation and Evolution 1945–1981. Rome: Food and Agriculture Organisation.

Pleskovič, B. and N. Stern (2001). Annual World Bank Conference on Development Economics. Annual World Conference on Development Economics, Washington DC: World Bank.

Poole-Kavana, H. (2006) 12 Myths About Hunger. http://www.foodfirst.org/en/12myths.

Pray, C. E., K. O. Fuglie and D. K. N. Johnson (2007). Private Agricultural Research.In *Handbook of Agricultural Economics: Vol. 3—Agricultural Development: Farmers, Farm Production and Farm Markets*. R. E. Evenson, P. Pingali and T. P. Schultz. Amsterdam: Elsevier.

Rand, A. (1967). *Capitalism: The Unknown Ideal*. New York: New American Library.

Regmi, A., M. Deepak, J. Seale and J. Bernstein (2001). Cross-Country Analysis of Food Consumption Patterns.In *Changing Structure of Global Food Consumption and Trade*. A. Regmi. Washington, DC: United States Department ofAgriculture Economic Research Service.

Reutlinger, S. (1999). Viewpoint: From 'Food Aid' to 'Aid for Food': Into the 21st Century. *Food Policy* 24(1): 7–15.

Riddell, R. C. (2007). *Does Foreign Aid Really Work?* Oxford: Oxford University Press.

Roosevelt, F. D. (1941). *Xxii Annual Message to the Congress, 1941*: United States Government Printing Office.

Rosegrant, M. W. and S. A. Cline (2003). Global Food Security: Challenges and Policies. *Science* 302(5652): 1917–1919.

Rosenberg, I. H. (1994). Nutrient Requirements for Optimal Health: What Does That Mean? *Journal of Nutrition* 124(9 Suppl): 1777S–1779S.

Roth, M. and G. Myers (2010). Issue Brief: Land Tenure, Property Rights, and Food Security: Emerging Implications for USG Policies and Programming. Washington DC: United States Agency for International Development.

Rowe, H. B. (1949). Issues in American Foreign Food Policy. *Journal of Farm Economics* 31(1 (Part 2: Feb)): 281–290.

RTFN (2009). Who Controls the Governance of the World Food System? *Right to Food and Nutrition Watch*. Germany: Brot für die Welt (Bread for the world).

Rucker, A. (2007) Key Players in Food Aid. *eJournal USA*: http://www.america.gov/st/health-english/2008/Jun e/20080616003149xjyrrep0.4639551.html.

Salter, R. M. (1950). Technical Progress in Agriculture. *Journal of Farm Economics* 32(3): 478–485.

Scanlan, S. J. (2004). '*Feast or Famine? Food Security, Democracy, and Information Technology in Less Industrialized Countries*' Paper presented at the annual meeting of the American Sociological Association, Hilton San Francisco & Renaissance Parc 55 Hotel, San Francisco, CA, August 14, 2004.

Scaramozzino, P. (2006). Measuring Vulnerability to Food Insecurity. Rome: FAO Agricultural and Development Economics Division (ESA) 26.

Scheuerman, W. (2008). *Globalization*. Stanford, CA: Stanford University.

Schmidhuber, J. (2003). The Outlook for Long-Term Changes in Food Consumption Patterns: Concerns and Policy Options. FAO Scientific Workshop on Globalization of the Food System. Rome: FAO.

Schuftan, C. (2010). Land Grabbing and Nutrition: Challenges for Global Governance. *Right to Food and Nutrition Watch*. Germany: Brot für die Welt (Bread for the world) 90.

SED (2004). European Trade Barriers and Developing Countries. Netherlands Ministry of Foreign Affairs Sustainable Economic Department.

SEP (2003). *Sovereignty*. Stanford: Stanford University.

Shah, A. (2007) Food Aid. http://www.globalissues.org/food/aid/#Whatisfoodaid.

Shaw, J. (2007). *World Food Security: A History since 1945*. Basingstoke: Palgrave Macmillan.

Sibrián, R., S. Ramasawmy and J. Mernies (2008). Measuring Hunger at Subnational Levels from Household Surveys Using the FAO Approach. Food and Agriculture Organisation: Statistics Division.

Simmons, E. (2009). Monetization of Food Aid: Reconsidering US Policy and Practice. Washington DC: Partnership to Cut Hunger in Africa.

Smith, K. and R. Edwards (2008) The Year of Global Food Crisis. *Sunday Herald Online*: http://www.sunday-herald.com/news/heraldnews/display.var.2104849.0.2008_the_year_of_global_food_crisis.php.

SOFA (2006). The State of Food and Agriculture 2006: Food Aid for Food Security? Rome: Food and Agriculture Organisation.

SOFI (2008). The State of Food Insecurity in the World 2008. Rome: Food and Agriculture Organisation 59.

Sphere Project. (2004). *Humanitarian Charter and Minimum Standards in Disaster Response*. Oxford: Oxfam Publishing.

Subramanian, S. (2003). An Analysis of the Relationship between the Agricultural Policies of the European Union and the United States, and Their Implications on the World Food Economy. *Journals of the Stanford Course: Engr 297b* (Controlling World Trade): 32.

Sullivan, R. R. (1970). Politics of Altruism—Introduction to Food-for-Peace Partnership between United-States-Government and Voluntary Relief Agencies. *Western Political Quarterly* 23(4): 762–768.

Torero, M. and J. v. Braun (2009). Alternative Mechanisms to Reduce Food Price Volatility and Price Spikes. CGIAR/IFPRI.

UN (1948). *Universal Declaration of Human Rights*: United Nations. A/RES/217 A (III).

UN (2005). *Draft Resolution Referred to the High-Level Plenary Meeting of the General Assembly by the General Assembly at Its Fifty-Ninth Session: 2005 World Summit Outcome*: UN. A/60/L.1

UN ISDR (2006). Website of the Un International Strategy for Disaster Reduction. United Nations. International Strategy for Disaster Reduction. http://www.unisdr.org/ (accessed 13 February 2011).

UN News Centre (2010). *Somalia and US Should Ratify UN Child Rights Treaty—Official*: UN News Service.

UN News Centre (2011). *UN Food Experts Call for Increased Agricultural Investment to Offset Soaring Prices*: UN News Service.

UNCTAD (2008). Review Session on Chapter V of the Monterrey Consensus: External Debt. Geneva: United Nations Conference on Trade and Development (UNCTAD).

UNDAC (2006). United Nations Disaster Assessment and Coordination (UNDAC) Handbook. New York and Geneva: Office for the Coordination of Humanitarian Affairs (OCHA).

UNHCR (1999). Substantive Issues Arising in the Implementation of the International Covenant on Economic, Social and Cultural Rights: General Comment 12—the Right to Adequate Food. *CESCR: Committee on Economic, Social and Cultural Rights - Twentieth Session*. Geneva: UNHCR.

UOR (2009). University of Reading: Agriculture, Policy and Development, History of Agriculture. University of Reading. http://www.ecifm.rdg.ac.uk/history.htm (accessed 4 June 2008).

USAID (2002). Foreign Aid in the National Interest Promoting Freedom, Security, and Opportunity. Washington, D.C.: US Agency for International Development 169pp.

USAID (2010). What Is the Famine Early Warning Systems Network (Fews Net)? USAID. http://www.fews.net/ml/en/info/Pages/default.aspx?l=en (accessed 30 June 2010).

Vaidya, A. K., Ed. (2006). *Globalization: Encyclopedia of Trade, Labor, and Politics*. Santa Barbara: ABC-CLIO.

Van Esterik, P. (1984). *Intra-Family Food Distribution: Its Relevance for Maternal and Child Nutrition*. Ithaca, NY: Cornell University Press.

von Braun, J. (1995). A New World Food Situation: New Driving Forces and Required Actions. International Food Policy Research Institute. Washington, DC: The International Food Policy. Research Institute.

Wahlberg, K. (2008) Food Aid for the Hungry? . http://www.globalpolicy.org/socecon/hunger/relief/2008/01wahlberg.htm.

Wallun, R (2009) The Origins Development and Future of the International Humanitarian System: Containment, Compassion and Crusades. ISA's 49th Annual Convention, Bridging Multiple Divides. San Fransisco, CA: International Studies Association (ISA).

Ward, W. A. (1977). Calculating Import and Export Parity Prices: Training Material of the Economic Development Institute, CN-3. Washington D.C: World Bank.

Warnock, K. (1997). Third World: Feast or Famine? Food Security in the New Millennium (Commentary). *Journal of Race and Class* 38(3): 63–72.

Watts, J. (2007) Riots and Hunger Feared as Demand for Grain Sends Food Costs Soaring. *Guardian Online*: http://www.guardian.co.uk/world/2007/dec/04/china.business

Weeks, J., Ed. (1995). *Structural Adjustment and the Agricultural Sector in Latin America and the Caribbean.* New York: St. Martin's Press.

Weingärtner, L. (2004). Food and Nutrition Security, Assessment Instruments and Intervention Strategies: Background Paper No. I. *Food Security Information for Decision Making.* Rome: EC-FAO Food Security Information for Action Programme.

WFP (2005). Food Aid Flows. Rome: World Food Program (WFP).

WFP (2006). Food Aid Flows. Rome: World Food Program (WFP).

WFP (2007). Food Aid Flows. Rome: World Food Program (WFP).

WFP (2008a). Food Aid Flows. Rome: World Food Program (WFP).

WFP (2008b). *Ten Minutes to Learn About ... Improving the Nutritional Quality of WFP's Food Basket—an Overview of Nutrition Issues, Commodity Options and Programming Choices.* World Food Programme. 2008 (S).

WFP (2008c). WFP's Operational Requirements, Shortfalls and Priorities for 2008. Rome: World Food Programme.

WFP (2008d). The World Food Programme: The First 45 Years. World Food Programme. World Food Programme. http://www.wfp.org/aboutwfp/history/index.asp?section=1&sub_section=2 (accessed 25 April 2010).

WFP (2009a). Food Aid Flows. Rome: World Food Program (WFP).

WFP (2009b). World Food Program: Non-Governmental Organisations. World Food Programme. http://www.wfp.org/aboutwfp/partners/ngo.asp?section=1andsub_section=4 (accessed 11 April 2008).

WFP (2011). Website of the World Food Program. World Food Programme. http://www.wfp.org/ (accessed 25 February 2011).

WFP/VAM (2010). *VAM Understanding Vulnerability Food Security Analysis.* Rome: World Food Program.

Williams, A. J. (2005). 'Reconstruction' before the Marshall Plan. *Review of International Studies* 31(3): 541–558.

Windfuhr, M. and J. Jonsén (2005). Food Sovereignty: Towards Democracy in Localized Food Systems. Warwickshire: ITDG/FIAN 70.

World Bank (2010a). Website of the Worldbank. World Bank. http://www.worldbank.org/ (accessed 13 July 2010).

World Bank (2010b). *Worldbank: World Development Indices.* World Bank Group.

World Vision (2009). Website of World Vision. World Vision. http://www.worldvision.org.uk/ (accessed 23 May 2010).

WTO (2010). International Trade Statistics 2010. Geneva: World Trade Organisation.

YIO (2002). *Yearbook of International Organizations: Guide to Global and Civil Society Networks/2001/2002.* Brussels: Union of International Associations.

Part V

The Final Analysis—Food Security

He who has never hoped can never despair.

George Bernard Shaw
Caesar, Act IV

Parts I–IV provided us with what might have seemed at times, an overwhelming amount of information. While Part IV, to some extent, aimed to summarise the main elements worthy of consideration within the food security debate, this sections looks to bring all of this information and place it in context. We begin by finally and definitively dispelling some of the myths of the origins of food security that, without doubt in my opinion, have contributed greatly to the loss of valuable past learned lessons. Following this we continue by amalgamating all of the current ideas of food security before attempting to firm up as much of the causality of the phenomenon as is possible. At this juncture there is also some small attempt to redefine the food security definition to encapsulate all that has been learned, while the final pages reflect on some of these findings and challenges with a view to stirring up healthy and positive discourse on the topic.

23

Origins: Aetiology and Etymology— Dispelling the Myths

For myself, one of the most striking observations I regularly encounter in the field of food security knowledge is the sheer short-sightedness of those claiming to further the cause. In particular, I refer to the origins and evolvement of the concept in both general and specific terms. The myth that the modern notion of food security is a relatively new phenomenon is one that continues to be perpetuated at the highest levels. In fact, it is staggering just how widespread this misconception is. This has worrying implications, not least of which is the notion that past lessons have not been learned or debated, let alone heeded. Examples are abundant and while many believe the first official reference to world food security was in Resolution XVII of the UN's 'International Undertaking on World Food Security' at the World Food Conference in 1974, others firmly believe the actual concept itself also originated from this time (UN 1975; Maxwell and Frankenberger 1992; Maxwell and Smith 1992; Padilla 1997; FAO 2003a; Nehme 2004; FAO 2006; World Bank 2008; Forge 2009; IFAD 2009b; Lutz et al. 2010; SDC/ GPFS 2010; Tefera 2010; Yu et al. 2010; OCHA 2011). This list of references is nowhere near exhaustive; instead it represents just a cross-section of numerous people and multilateral bodies of global reach who still believe to this day in the relative infancy of the concept. I would indeed go as far as to say the idea that food security emerged fully formed as a concept in the mid-1970s is frankly laughable were it not for the pervasiveness of its many believers. Instead, research for this book has revealed a different more illuminating heritage. So where did this modern notion of food security, stripped of all its accumulated baggage and in its narrowest definition, originate? Did the idea emerge fully formed? Was it an accumulative process or has as it evolved out of something else? If not, what of its direct antecedents? In answer, there are two ways of looking at this: aetiology, the origin of the idea; and etymology, the origin of the term. Both are looked at here.

23.1 Etymology of Food Security

The etymology of the 'food security' catchall can be found in literature well before the 1970s. Indeed, the term itself has been used in all manner of ways representing such notions the literal physical food security as well as in its modern connotation. On page 87 of the well-known *Farm Journal* of 1964 for instance, a small ad proclaims to offer food safety in the guise of food security when it proudly proclaims

> BUY FOOD SECURITY!...BUY A FRIGIDAIRE FREEZER!...Food security—that's what you get—with every Frigidaire freezer. (Frigidaire 1964, p. 87)

Granted, it is not quite what Maxwell and Frankenberger had in mind when they were examining the multitude of definitions of food security in 1992, but it is illustrative of the point. On a more serious note however, it has also been used in a more literal translation to mean security of food supplies. When writing about his wartime experiences working for the Senate Subcommittee on War Mobilisation, Bela Gold bemoaned the lack of logistical oversight in the mobilisation of food when he tells of

> ... our deplorable ignorance about where our lines of food security were relatively most vulnerable. (Gold 1949, p. 370)

However, in its modern connotation, the food security terminology can be traced back further. Early examples for instance exist in the publication *Farm Journal* in 1940 in which talking of America's food preparedness compared to other nations it points out

> America's food security rests solidly on two assets which exist nowhere else in such numbers—progressive farmers and modern farm machinery. (*Farm Journal* 1940, p. 5)

Even when considering alcohol use in primitive societies, Donald Horton positively related the use of this libation as reducing anxiety in things like food insecurity. Horton suggested that cases of hunger anxiety partly reflected the means of access and observed

> ...the greater the danger of food shortage the more difficult the conditions of life generally. Under such conditions anxiety should be high. Where food-getting techniques are less primitive, the greater is the chance that conditions of food security may prevail. Under these conditions anxiety may be relatively low. (Horton 1943, p. 263)

In a similar vein, Selling and Ferraro in 1945, studying diet and nutrition from the psychological perspective, noted that fear and insecurity about food also acted on their behaviour noting that

> ...youngsters, who had not known food security in Germany or later in Austria or Italy, gorged themselves as long as they were unsure of the reception which they could expect in their foster homes... (Selling and Ferraro 1945, p. 31)

These last two examples offer great insight into how food security acted on the individual mindset in times of perceived shortages. There was also a macro level to these early ideas of food security and linking the concept to a country's overall food planning objective. Economist and notable author Dr. Baljit Singh, in his book *Population and Food Planning in India* (1947) wrote that

> ...if [the] extent of irrigation is an index of food security, the people of India are no more protected today than they were at the beginning of the century... (Singh 1947, p. 53)

Once again as in the previous three examples, the terminology in the above reference is clearly used in the familiar modern context. And it does not end there; also in 1947 the newly merged *United Nations World** magazine, talking of wartime food shortages and hunger prophetically opined

> ...it is difficult for well fed Americans to realise how desperately most of the people of the world need to be sure that they will never again starve. Peace will not be secure until it is based on food security. (Walsh 1947, p. 24)

On top of these and just as fundamental to the modern notion is the book by Percy Chew in 1948 entitled *Plowshares into Swords*, in which several references are made regarding certain country's precarious hold on food security in the face of growing industrialisation (Chew 1948, p. 217).

In fact, recent history is dotted with numerous instances of a concept cited as only arriving on the scene in the 1970s and I could go on, but I feel the point here has been largely made. However, before I do leave this idea, there is one particular incidence of the term's usage that is worth examining and it is one of the most important works uncovered in this research. Above and beyond those already cited, there is perhaps one man who stands head and shoulders above the rest. Even as far back as 1886, it would seem the modern notion of food security was achieving critical mass. His name was John Towne Danson and in a remarkable book, the *Wealth of Households*, Danson (1886), in my view, understood more of the integrated issues concerning the security of food than any others at the time and perhaps some since. In

---

* The United Nations World was formed in February 1947 out of a merger between Asia and the Americas and Inter-American (PSBI 2010).

his writings, we get more than a glimpse, more than a mere inkling of the nature and scale of the issues at hand as he uncannily elucidates many of the underlying assumptions of the current concept. Moreover, albeit intellectually and not empirically, he carefully ties in more than a few of the threads of food insecurity to causes of poverty, health, trade and the political climate. For this reason, it is worth teasing out some of his ideas in detail.

Born in Liverpool, John Towne Danson was very industrious; he worked as a journalist for the *Daily News* and the *Globe*; he became private secretary to (later *Sir*) Benjamin Hawes and was also a Barrister and even an insurance underwriter. Danson had many interests too he was an enthusiastic student of philosophy, economics, politics, archaeology and science as well as nurturing a lifelong affection for statistics. Lastly, J.T. Danson was an author and ardent chronicler (NML 1760–1976; RSS 1898; Danson 1906). These life experiences are evidently abundant throughout his book, and it is in these clearly articulated insights that he brings to bear an uncanny skill for elucidating much overview in the subjects of economics and social interconnectedness, and more importantly—food security. The core of the book itself is largely concerned with economic trade and growth; however, Danson ties all this in with an expansive preliminary text underscoring the needs for, inter alia, wealth creation and cooperation. This is complemented in the last part of the book with the political and philosophical justification that reinforces his underlying assertions as well as the advocation of empowerment and self-determination over dependency (Danson 1886). In the book, Danson sees society as a mutual enterprise, one where every member is of value and in turn to be valued as an equal. This in his view, Impels Us to Society (Danson 1886, p. 5), and in this endeavour, he acknowledges the primacy of children, especially with regard to health and nurturing. From his words, it would seem that Danson viewed society and its responsibilities to each other as nothing short of an intrinsic human right. Indeed, it is from this position that Danson offers us the first ideas of food security. In recognising the extent of society's reliance on agriculture and the shared responsibility of storing harvest as security against starvation, he suggested that

> Nature pays us our wages only once a year. This is an inexorable law. And he who does not make provision for complying with it must either do without the harvest which distinguishes the civilized from the savage man, or depend upon his more provident man to aid him. (Danson 1886, p. 7)

and further pondering on the equality of man he offers that

> They share the food-security of a well ordered community; and they share many other of its advantages [too]. (Danson 1886, p. 7)

This view, partially egalitarian but far from naïve, serves to highlight in his opinion, a logical natural and progressive society and one that is ordered and evolutionary. Yet in these ideals he was also all too aware of the realities:

> To say that men are born free and equal is to contradict all we know of them. In fact, they are born more unequal than any animal we know; and the only freedom they have yet had, or apparently can ever have, is such as they may fight for, out of society, or agree for, within. (Danson 1886, p. 6)

In fact, it is this individuals' societal struggle marred by inequality, both social and economic, that propels Danson's central tenet:

> Had all men been equal…had they all exercised the rights, incident to a state of society…[they] would have been equally entitled to the crop. (Danson 1886, p. 7)

Acutely aware of this disequilibrium, Danson goes on to offer that the disaffected, in order to raise themselves out of the cycle of poverty and hunger would do well to become capitalist. By extension

Danson suggests that, the poor, by employing his wage-earning potential to the maximum, by investing in himself, he is investing for his future and for his brethren and country alike. Furthermore, in reference to the crippling nature of poverty, Danson argues that

> The most urgent of all the duties of life must surely be that, the due performance of which alone enables us to do the rest; and before we can do, or even attempt to do, any other, we must provide for our lowest wants. The labour that provides food, clothing, and lodging must precede all other labour … (Danson 1886, p. 9)

Only then he contends can the evils of poverty be tempered (Danson 1886). In directly relating poverty, health and mutual cooperation, Danson promotes the mutually reinforcing and symbiotic notion that if we improve the health and wealth of the individual, we improve the health and wealth of the nation, or in Danson's words,

> All gain and all lose by the gain and loss of each. (Danson 1886, p. 2)

While others have indeed held discourse on many of themes addressed by Danson, none have given as much impetus to the inclusivity and specificity of food security as bundled within a wider package of societal needs; nor, more importantly in the context of this book had anyone else by the end of the nineteenth century properly articulated the now familiar modern concept of food security in this fashion. Indeed, in the food security debate, it is this legacy that empowers Danson as, in my research, one of the more progressive of the nineteenth century thinkers and one who is fully deserving of recognition as giving proper and solid foundation to the hitherto sometimes circumspect notions of human rights, poverty alleviation, societal cooperation, humanitarianism and ultimately food security.

Having noted all that however, and in the final analysis of the etymology of the 'food security' catchall, I would be surprised to find Danson as having been the first to elucidate the term in its entirety in this way, for no other reason than the ideas seem fairly well advanced and almost fully formed by today's standards. This leads me to believe he was either even more exceptional than I have previously mentioned or more likely he is, like many of us, building on ideas that came before him. Irrespective of whomsoever turns out to be the originator of the phrase, it is clear that it was certainly in full usage, in its modern connotative form nearly a full century before it is often credited.

23.2 Aetiology and Evolution of Food Security

Having identified the etymology of the term, the challenge in comprehending the origins of food security itself is an altogether more difficult one. This is made more challenging by the fact that the concept has not followed a direct linear path from the past to the present. More often, it has been a road beset with bumps, distractions and u-turns peppered with ignorance and controversy. The journey too has been further aggravated by frequent ideological and philosophical disagreements, which ultimately pulled the fledgling threads of food security in all directions. Despite such difficulties, it can be said that food security is an amalgamous concept embodying a number of elements from a diverse multitude of disciplines. If I were asked to highlight the major developmental periods in food security, I would loosely lean toward the following as illustrative purposes only as the reality is much more complex:

- 1850–1914: Social and scientific revolutions provided the foundation upon which modern notions of human rights, the right to food and ultimately food security would emerge.
- 1914–1945: Inter-war years where scientific, ideological and political principles of social welfare, poverty and nutrition, etc. began to converge.
- 1945–1974: Post-war reconstruction and development characterised by institutional and humanitarian evolution and supply side issues predominant of which was agricultural expansionism.

- 1974 to present: Global awakening where from about the 1970s food security issues created a groundswell of public and institutional momentum focusing on the individual with supply side issues of access, poverty, water and non-food inputs as well as embracing increasingly wider societal concerns of environment and sustainability.

These ideas are developed further in the following sections.

23.2.1 1850–1914

From the start, early socio-cultural dynamics ensured that pioneering civilisations were beset with frequent challenges of population and food supplies. Specialisation, trade and division of responsibility for a greater part held these two forces in relative equilibrium but the often severe threat of food shortages and famine were never far from the surface. Many like Malthus questioned the ongoing sustainability of the food and population issue, which sparked much polarised debate over the centuries; however, holding these worst prophecies at bay was a growing integration of the global society. Events such as the Muslim agricultural revolution as well as the Columbian exchange and ultimately the British agricultural evolution of around 1650 helped foster trade and knowledge exchange. These in turn introduced new agricultural techniques and products that further increased productivity that helped allay fears of food shortages.

   Separately in areas of health, primitive but strongly held cultural and spiritual beliefs of vitalism and spontaneous generation hindered scientific progress for centuries. This led the majority of people to treat health issues in isolation of food based on an ancient humoral system of the four humors. Consequently, early epiphanous insights by the likes of Hippocrates, Aristotle and others, that food was integral to overall health and vitality were not properly acted upon for centuries. Indeed, it took the work of Redi, Spallanzini and others to finally mark the beginning of the end of vitalism and allow a new scientific awareness of chemical and biological processes to grow in its place. Finally by the time Darwin had postulated his theory of evolution to the masses, scientific advancement was such that people were now beginning to look very seriously at the life maintaining forces of food and nutrition. Questions such as what did it take to maintain life, what kinds of foods and in what quantities emerged as an infant nutritional discipline was formed. From here people like Lavoisier, Thompson, Smith, Voit, Rubner, Atwater and many many more provided scientific knowledge that was to underpin future policy and understanding. Converging with nutrition too was the notion of foods' properties of health and well-being although it would be a while yet before the two were scientifically linked.

   All the while, in a separate but loosely related thread, political, economic and humanitarian forces were on a collision course that was to shape future ideological debates for generations. On the one hand, political and economic frontiers, via increased trade, colonialism and aggressive expansionist policies were embedding themselves in Western growth models. This effectively led to greater wealth and increased agricultural resources for the host nations; progress incidentally, which was facilitated by ever-increasing industrialisation. On the other hand, wanton destruction and the toll of human suffering brought on by ongoing conflict, particularly in Europe, over ideological differences as well as territorial disputes, witnessed perhaps the first true act of important multilateralism for the common good: the Westphalian Treaties of 1648. Yet although this effectively established the sovereign state system we recognise today it did not put an end to interstate conflict. So, as war continued to claim its toll an increasing local and global social divide gave much needed impetus to fledgling humanitarian interventionist ideals. By now the likes of Durant and the Red Cross (which eventually led to the Geneva Convention), sought oversight of common decency and fair play at the global level while grass roots social interventionism saw school feeding, poverty alleviation and education programmes grow at the more local levels.

   Ultimately this series of mini scientific, socio-cultural, political, industrial, chemical, biological, economic and agricultural revolutions ultimately ushered in a new philosophical revolution of its own: the Age of Enlightenment. This was an all-pervasive awakening that heralded a new collective awareness in all things scientific, ideological and political. Open fora were held on such diverse and collectively important societal issues like hunger, carrying capacity, political ideology and humanitarianism.

Meetings, discussions, debates and ideas were abundant and they were held frequently in all areas of concern and enjoyed the patronage of all levels of society.

This then provided the foundation upon which the genesis of food security would emerge. Up till now, many millions in each generation had led lives blighted by hunger and malnutrition and it was against this backdrop that the late nineteenth century and the turn of the twentieth century was perhaps the point at which many of the modern notions of food security began to converge. Collectively then, the rise of this new nutritional knowledge and the lifting of the value of life above the sordid was seen by one commentator, as one of the greatest events in human history (McCollum 1957). However, food security is more than just nutrition and it was in these early decades that several events conspired to elevate the role of food in society to the level of political consideration. In Britain, the advent of the Boer War and the poor general nutritional state of the soldiers was an embarrassment to the government. Simultaneously, concern too was being expressed at the increasing prevalence of poverty and disease. By this time tentative rumblings of food as a right were being heard louder and more frequently. However, it eventually took the growing industrialised democratic societies of the West, with all the trappings of free speech and the free flow of information to highlight these many scandalous situations and encourage the subsequent politicisation of hunger. By now, Smith and others had already calculated individuals' needs in terms of energy intake while others had identified the main food components of good dietary habits. Still, even in spite of these advances, before any great social undertaking could be advanced, the advent of the World War I concentrated minds on more immediate matters.

23.2.2 1914–1945

As the war progressed, however, it became increasingly obvious to those in charge that the logistics of war also involved food considerations; and not just of the warring soldiers either, but of the domestic populations too. Government intervention in agriculture was ratcheted up and new inter-allied War Commissions aimed to satisfy food needs based on a more scientific footing were set up. Fear of a repetition of the World War I and its associated problems were deeply felt by many at the time (Tanner 2004) and, after the end of hostilities, the birth of the League of Nations gave further credence to multilateralist goals of commonality. Agriculturally and politically, this was also a time of collective social stocktaking. As a result, several things happened over the subsequent years that impacted strongly on the fledgling food security concept. Rationing had raised the spectre of famine and agricultural insufficiencies while more research into health and nutrition allowed for the development of new recommended dietary allowances. Moreover, social science and political interests were also introspectively examined and which ultimately led to greater profile of food security issues in general. Importantly too at this time, the many challenges of food and malnutrition were seen, not just as a scientific quandary, but just as much a problem of fundamental social values as doctors and health officials began to speak out at what they saw as a prevailing injustice within the social environment (de Onis 2003; Pemberton 2003). It seems that health and nutrition was finally, in Britain and America at least, was further entrenching itself in the political agenda.

In terms of advancing the food security concept, one overriding lesson learned from the Great War was the justifiable government market and societal intervention policies for the greater good. Also from Britain's example, it became clear to governments that in cases of emergencies or extreme food shortages, a voluntary code of conduct was of limited use. This advanced the important precedence of market intervention beyond that of simple production, supply or protectionist policies to overarching goals of provision and equitable access for all peoples (Blainey 2000, p. 88). Up till about the early 1930s, then the issues of food security were still largely considered in local or national terms. However, this was about to change. At about this time the Health Organization of the League of Nations was publishing much research that re-oriented this narrow focus toward an increasingly global perspective. Although action generally continued to be coordinated on the local or national levels several reports, namely, the Burnet and Aykroyd and the Mixed Committee's Physiological Basis of Nutrition at last began to open the world's eyes to the extent of the food insecurity problem (Burnet and Aykroyd 1935; League of Nations 1937). Once again however, before the information was properly digested and could be acted upon, World War II broke out. Once more the war fully tested nations' abilities to feed themselves as the beginning of hostilities again saw price increases on both sides of the Atlantic exacerbated by uncoordinated government food purchases on a commercial competitive basis; expanding food buying by businesses in anticipation of government purchases, as well as other speculative activity (Reid and Hatton 1941). Once again too, as war progressed, so the Americans in particular responded to national calls for increased agricultural production and once in overdrive, overproduction proved difficult to control, especially as new technologies greatly increased productivity (Hardin 1943; Barton and Cooper 1948; Salter 1950).

23.2.3 1945–1974

The end of the war was a pivotal time for food security and several things happened which ultimately was to shape the debate for decades. Without doubt though, the most important event at this time was the replacement of the League of Nations with the creation of the United Nations as a world multilateral organisation. After the devastation of the World War II, the new body, building on its forerunner's humanitarian roots and its multilateral ethos' sought perpetual peace, social equality and many other social freedoms in the 1945 Charter of the United Nations. In terms of food security however, no other single instrument outlined the importance of an individual's right to food, to this point and perhaps since, was the Universal Declaration of Human Rights' (1948) freedom from want embodied in Article 25. Other social goals in both this declaration and the UN's Charter echoed the spirited four freedoms speech by Roosevelt in 1941. These freedoms of worship, of speech and expression, from want and from fear all collectively sought to finally put an end to the misery and suffering of untold millions in the hope of a more equitably shared investment in human social capital (UN 1945, 1948).

The second major event was the emergence of a new 'development economics' discipline. In order for the new ideals of the UN to take shape, a new way of looking at economic principles that specifically targeted these social and economic objectives in developing countries was needed. Up till now, domestic paradigms were national in scope and largely driven by economic principles whereby social development was seen as a consequence of this economic growth. Unfortunately, growth in these countries was often erratic, slow and the trickle-down effect of economic wealth into social development seemed to be lacking. As a result, the new branch of development economics emerged and began to question the efficiency of neoclassical economic principles in addressing social and economic inequities in developing countries. Questioning the role of liberalised markets and advocating the state take more responsibility in the development process, economists like Rodan, Mandelbaum, Nurkse and Singer set out new principles for economic progress. These principles applied as much to the industrialised nations as they did to the developing regions as the new development paradigms was meant to be seen as an inclusive global model of mutual benefits.

Another important consideration for food security at this time was the recurring issue of food surpluses. Spurred by innovation, technological advances and government intervention policies of parity and protectionism, surpluses was fast becoming a perennial embarrassment for the West (McMillan 1946). Something more had to be done, this took the form of aid. The notion of surpluses as aid had already been established by the United States in 1896; however, the most famous example of this practice was the Marshal Plan, which when combined with other needed resources, seemed to work very well (Shaw 2007). By now many nations saw opportunities in combining domestic interests with these international objectives and the usually separate but interconnected economic, social and political paradigms converged on a multithreaded global development paradigm. As a result, food security through surpluses was now seen as a legitimate foreign policy goal and one the fulfilled dual objectives with little downside. The symbiotic helping-you–helping-me relationship that developed around this time was also one that prevailed for decades, and indeed still continues in no small measure today. Besides these measures though and as a result of the aforementioned goals and development paradigms, agriculture was seen as pivotal in achieving such aims. As a result, the prevailing agricultural paradigm of the day advocated increased domestic production and self-sufficiency and by extension a concomitant increase in global aggregate food production (FAO 2006; IFAD 2009b). This, at a time of recurring surpluses! Consequentially, agricultural production increased across the world; however, in the Western world at least, this was coupled with inadequate growth in purchasing power and with the international markets attempting to supply cheaper produce, these markets were effectively closed off, leaving the foreign agricultural industry to languish in restrictive international trade agreements, unemployment and depression (McMillan 1946). What started out as an ideological development strategy, it seemed, was fundamentally unworkable if national self interests were to be protected.

Despite how this played out (discussed in previous sections) another piece of the jigsaw that helped cement the ideology of food security and development paradigms at the time was the accumulation of nutritional knowledge. Abutting these developments was the seemingly concrete and united international political will, at about this time, that enabled grand gestures and targets to be elucidated for the eventual elimination of hunger and malnutrition. Thus, armed with this new invigorating enthusiasm the question in many minds was not if or how we could achieve such laudable aims, but by when. While progress was slow, the concept did begin to take on a life of its own and during the 1950s, food security began to include associated notions beyond that of supply alone. Of note was the recognition of the disproportionate vulnerability of certain groups of people, in particular pregnant women and children under 5. In recognising this, and while increased production remained the overall agricultural goal, agencies of the UN such as UNICEF, the FAO and the WHO began to administer applied nutrition projects (ANP) focusing on medical, educational and economic factors as well as other facets of a more holistic food security model. Unfortunately, while these measures are seen today as vital in the overall goal of food security, in the 1950s, these programmes represented only a small portion of development budgets and were largely discredited at the time as ineffectual and of little use. Continued surpluses too began to be scrutinised in their role of development with some questioning their ultimate benefit (Goodrich 1947). This led to oversight being created in the form of the UN's Consultative Sub-Committee on Surplus Disposal (CSSD). In the meantime, the International Freedom from Hunger Campaign launched by the FAO in 1960 was an attempt by the UN to enable countries to move away from reliance on surpluses and to foster more self sufficiency (Argeñal 2007). Also at about this time (1963), the World Food Program (WFP) began operations in support of food security while developing countries eagerly embraced the notion of self-sufficiency and development in general. This was also the case with many of the new found independent countries, which incidentally were also largely agricultural economies.

Collectively during this period many began to voice concerns about international price stability in the commodity markets as well as the perceived unfair agricultural policies of industrialised countries. In response, the UN Conference on Trade and Development (UNCTAD, 1964) sought to find ways of promoting the smooth and equitable integration of developing countries into the world economy. As an aid in this endeavour, large debt was created in many developing countries attempting to mimic the success of the industrialised countries (Munasinghe 1998; Slusser 2006). Growth, development and food security, it seemed, was within their grasp. Unfortunately, a series of setbacks including the poor translation of the basic principles of development, corruption and general mismanagement meant that further borrowing was sought. This extra borrowing came with conditions: structural adjustment policies (SAP). However, despite their best intentions, these policies effectively hindered rather than helped with overall food security. Indeed, many such policies included conditions requiring fundamental changes to recipient country's domestic fiscal and economic policies while development paradigms were imposed encouraging the liberalisation of domestic markets, opening them up to international trade. Effectively, it seemed, dual competing development objectives were now being introduced. On the one hand, agricultural expansion and self-sufficiency were being encouraged in the spirit of earlier paradigms and on the other, self-sufficiency was redefined to suggest the provision of food through economic means rather than domestic production. Considering the rural poor and the agricultural sector at the time made up the majority of the hungry and malnourished throughout the world an unhealthy and unsympathetic division of investment followed. As SAPs diverted development funds away from domestic improvement projects to restructure export-oriented trade, vitally needed investment in local infrastructure and agriculture became scarce.

Simultaneously protectionist policies of the industrialised countries ensured imperfect terms of trade that resulted in outward net capital flows for many developing countries. It is interesting to note too that economic growth at this time was continuing unabated in most of the industrialised countries. What was needed was some real inroads into the legacy of years of protectionist policies that would see a true free global market economy. While the GATT negotiations began in 1947 with this express purpose, agriculture was vehemently protected and still excluded from all talks to this point. This protectionism was noted by many as a real barrier to equality in international trade and one that worked almost exclusively in the favour of industrialised countries. In some regards too, many developing countries did not help themselves at this time as some remained inward looking and operated import subsidies of their own. Some nations, on the other hand, particularly in East Asia, abandoned their long-held inward-looking import-substituted industrialisation model and concentrated on rapidly growing their expanding export markets. At the same time, international institutions, private foundations and NGOs began investing in rural development projects including education, which was once again now seen as crucial to development. On the whole though while this period undoubtedly saw the progression of the food security concept, it still seemed like richer countries were becoming richer while poorer countries were becoming poorer.

23.2.4 1974 to Present

Not surprisingly, development paradigms, which were at best competing and at worst diametrically opposed, ensured that global food security could not find its feet. In the face of this apparent dilemma, optimism was waning and was being quickly replaced by pessimism. The advent of the 1974 World Food Conference (WFC) aimed to re-invigorate this flailing momentum. While the genesis of food security is firmly grounded in the late nineteenth and early twentieth centuries and while the immediate post-war era was pivotal in providing the means of achieving these objectives, so the 1974 WFC was a seminal moment in food security's history. Not least for the reason that it gelled the disparate strands of the notion into a single formal concept. As a result of this first World Food Conference, while food production remained the overriding goal other macro-level, instruments were introduced for consideration including measures concerned with providing buffer mechanisms against production shortfalls or balance-of-payments support for countries facing temporary food shortages. Other important insights elucidated at this meeting were not new and echoed some of the much earlier League and early United Nations realisations, that the causes of food insecurity were not so much failures of food production but more importantly structural problems relating to poverty and the uneven distribution of wealth. The fact that it took another 40 years for this realisation to sink in is perhaps the single greatest tragedy of food security efforts to date. However, also emerging out of this meeting were new multilateral instruments and organisations like the World Food Council, which sought to provide coordination for the growing number of food security stakeholders, as well as the International Fund for Agricultural Development (IFAD) as a financial institution to fund development projects in developing countries. Largely as a result of the conference, the international goal of food security generated enormous worldwide attention. This attention translated into a wider appreciation of the global situation and culminated in more concentration of effort and a broader understanding of the issues of poverty, access and investment, among other things. Of note too, at this juncture, is the fact that this meeting not only elucidated the macro-level variables believed to be at play but it also helped to refocus the individual as the ultimate beneficiary of global or national food security measures. As a result, predominantly following the conferences many initiatives, many can be forgiven for believing that food security might indeed have emerged out of this time as it was an era of frenetic that food security activity (FAO 2003b; Ericksen 2008; IFAD 2009b, 2009a).

Importantly too for food security, the convergence of years of experiential institutional learning came together at the World Food Conference and marked the first official multilateral definition of the concept. The first attempt was very firmly aimed at the production value of the concept and defined food security in terms of a continuous food supply that provided sufficient and stable priced foods, food security in these terms were described as

> Availability at all times of adequate world food supplies of basic foodstuffs to sustain a steady expansion of food consumption and to offset fluctuations in production and prices. (UN 1975)

In 1976 food security received a further boost when the International Bill of Human rights consisting of: The Universal Declaration of Human Rights; the International Covenant on Economic, Social and

Cultural Rights; and the International Covenant on Civil and Political Rights, finally became legally binding. Importantly while other instruments also enshrined the right to adequate food, the bill via the Covenant on Economic, Social and Cultural Rights (Article 11.1) dealt more exclusively and comprehensively than any other international undertaking to the present day. The article specifically recognises

> ...the right of everyone to an adequate standard of living for himself and his family, including adequate food, clothing and housing, and to the continuous improvement of living conditions. (UN 1966)

Thus in this article, the human right to adequate food that we all now enjoy was crucially established in law. It also applied to everyone even, despite the use of the non-pc term—'himself', those female headed households. Just as importantly the Covenant also bound states to

> ...take appropriate steps to ensure the realization of this right.... (UN 1966)

However, before this was fully accepted by all member states further clarification was sought at the later 1996 World Food Summit. In essence though, instruments were now in place that facilitated and indeed compelled states to act. Yet, despite progress this was not the end of the journey and food security concepts were to undergo several more changes before its final modern incarnation.

By the end of the decade though, the 'FAO Plan of Action for World Food Security' adopted in 1979 was still firmly based in the macro perspective of grain reserves, import and export quotas, food aid, agricultural techniques as well as increased productivity initiatives. However, Amartya Sen's work was about to challenge this. During the 1980s, issues of poverty and access were receiving increasing attention and finally in 1981 Amartya Sen published what many believed was the groundbreaking notion of 'food entitlement'. This idea took the notion of poverty and inadequate access to food a stage further by outlining a suite of entitlements over which an individual had command and any of which could be traded or used to acquire food. Such entitlements included monetary means such as wages, savings or access to credit; land on which to grow food; community or families that could offer support in times of need; as well as other social safety nets. This new way of looking at hunger highlighted some important considerations which were implicit in much literature but which were hitherto rarely elucidated to such degree. Perhaps the main issue in Sen's writings was that adequate food at the aggregate level did not necessarily ensure adequate access at the household or individual level. It was a simple and a somewhat obvious deduction but it turned out to be one that required a stubborn shift in contemporary thinking (Maxwell and Frankenberger 1992; Anderson and Cook 1999; IFAD 2009b). Shifting emphasis like this reoriented the focus of measuring food security not just of the supply side variables but towards those of demand and issues of poverty. Included in this widening analysis were growing financial consideration of different income groups including metrics like the $1-a-day, the Gini coefficient for income distribution and the like. Further, it now became necessary to consider non-financial allocation of resources like access to land, property rights, secure tenure as well as understanding the local community social safety net systems. A plethora of research followed using both quantitative and qualitative studies, which attempted to highlight the complex interplay of an ever growing band of variables.

Meanwhile, blighted by decreases in living standards due to among others including SAPs and the debt crises, many of the poorest countries resorted to removing the few social safety nets afforded to their poorest citizens (Argeñal 2007). It took the resurgent interest culminating from the 1984–1985 famines in Africa to once again rally international momentum and focus minds (Maxwell and Frankenberger 1992). People began to apply Sen's theory of entitlements to famines, further reinforcing the poverty aspect of food insecurity. This was reflected in the FAO's food security redefinition to include the important notion of access whereby all people had 'both physical and economic access to the basic food that they need' (FAO 1983). In 1986 the World Bank too added to the concept by suggesting that food security was also temporal and could result from either transitory or chronic conditions. Transitory or temporary conditions tended to refer to the onset of food insecurity through sudden or slow onset shocks such as drought, illness, conflict or other temporary situations. These were often associated with acute symptoms of malnutrition where a sudden withdrawal of food resulted in rapid loss of weight and other associated problems. Chronic conditions on the other hand was seen as resulting from a slow burning, underlying condition of food shortages, which, while not immediately catastrophic, entailed a long and slow decline in nutritional status. This might be associated with protracted conflicts, prolonged natural disasters or perhaps underlying structural problems among others (World Bank 1986).

At this time, just as food security proponents were becoming more optimistic in their analysis, so ecologists and demographers were not. Population growth once again emerged as a potential threat to ongoing attempts of ensuring a sufficient food supply. On top of this there was also a small but growing band of environmentalists who were becoming more vocal in their opposition to the unsustainability of existing growth models. Many were criticising the relentless march of capitalism as a catalyst in increasing the severity of environmental problems. In response, the 1987 report of the World Commission on Environment and Development *Our Common Future*, also known as the Brundtland report (after Norwegian prime minister and chairwoman of the commission Gro Harlem Brundtland), offered an alternative term, 'sustainable development', to embody a more longer term view of relationship between provision and the Earth's natural resources (WCED 1987). Meanwhile, advances in understanding meant that the current notion of food security was soon considered incomplete and subsequently the International Conference on Nutrition, held at FAO in 1992, broadened the definition to incorporate a component of food utilisation. This in turn was understood to be governed by such criteria as hygiene, sanitation, clean water and any physiological condition affecting food absorption (Ericksen 2008). As a result, the FAO now developed and promoted a broad framework of food security identifying the now familiar four key dimensions of availability, access, utilisation and stability. Moreover, by the late 1980s and into the 1990s, and building on Sen's entitlement theory, a new food security paradigm sought to view the subject from an individual livelihoods perspective. This ultimately cemented a people-centric view of the concept and while national and global objectives were still in play, by now the individual had largely achieved primacy (Chambers and Conway 1991; Maxwell 1996; FAO 2003b; Ericksen 2008). The benefits of this new approach were self evident, in analysing food security, the livelihood system seemed to lend itself to better interpretation of 'coping strategies' and nutritional analysis. Insights from the livelihoods approach further challenged prevailing theories on food security and ultimately led to the recognition that households actually had multiple objectives beyond solely achieving and maintaining security of food. In this realisation, it was understood that people may in fact go hungry in order to preserve other household assets. Beyond this, it was also hoped that the livelihoods approach would also generate more sensitive and appropriate policy responses tailored to local situations. In fact so intuitive was this new perspective that nowadays, its applicability is widely recognised and is increasingly being applied in emergency contexts including in important areas of vulnerability, risk coping and risk management (Devereux et al. 2004; FAO 2006).

During this period, two more definitions, one new and one redefined, came to greatly influence contemporary analysis. The first was the USAID's definition:

> Food security. Access by all people at all times to enough food for an active, healthy life. Food security includes at a minimum: (1) the ready availability of nutritionally adequate and safe foods, and (2) an assured ability to acquire acceptable foods in socially acceptable ways (e.g., without resorting to emergency food supplies, scavenging, stealing, or other coping strategies). (Andersen 1990; USDA 2009)

The other was a reworking by the United Nations, which by 1996 was refined to address concerns about food composition and nutritional balance. Further, food preferences were also now considered in respect of a social and culturally specific context ensuring increased complexity in translation (FAO 2003b). In short, food security definitions had lost their simplicity and came to represent an ideal rather than a fundamental need:

> Food security, at the individual, household, national, regional and global levels [is achieved] when all people, at all times, have physical and economic access to sufficient, safe and nutritious food to meet their dietary needs and food preferences for an active and healthy life. (FAO 1996, 2003a)

Also at the 1996 World Food Summit, members whilst reaffirming the 1966/76 right to adequate food sought clarification or more precise definition in its scope. Consequently in May 1999 the UN's Committee on Economic, Social and Cultural Rights (CESCR) released the landmark report or 'General Comment 12'. This elucidated the rights in principle but then further action was required to produce guidelines on the best way of achieving these. In the follow-up 1996 World Summit meeting held in 2002, members finally proposed an International Alliance Against Hunger to fulfil the objectives of the 1996 summit. Moreover, an Intergovernmental Working Group was to report to the Committee on World Food Security within 2 years with a voluntary set of guidelines of achieving the right to adequate food (FAO 2004). However, as these new guidelines were rolled out, some saw the word 'voluntary' as watering down of the legal binding right and indeed this was mirrored in the US stance, which, while agreeing to the proposed rights in principle, did not see these rights as actually inferring any legal obligation, either internationally or domestically (UNHCR 1999; Claude and Weston 2006). As an aside and to dispel a common misconception in the application of the 2004 Voluntary Guidelines, it is interpreted that the right to food is not to be misinterpreted as the right to be fed. Some assume that the right to food is tantamount to governments handing out free food whereas in reality the right to food infers the obligation of governments to provide a conducive environment that allows individuals to either produce food or to buy it for themselves. In essence, the right to food is the right to feed oneself in dignity (UNHCR 1999).

Just after the turn of the millennium, FAO's State of Food Insecurity 2001 report further refined the UN's definition of 1996 to add a social dimension suggesting that food security exists when people have

…physical, social and economic access to sufficient, safe and nutritious food…. (SOFI 2001)

23.3 Food Security Today

When considering current notions of food security against the vast body of information covered in this book, it helps to break down the concept into its constituent components of access, availability, utilisation and stability. On top of this, though it is also helpful to consider the policy, social and political as well as the physical environments.

In terms of access to food, there are several considerations: economic, physical and social, although the overriding concern is economic; that of poverty. Poverty is perhaps the single biggest obstacle to achieving food security and much past and current policy aims to reduce its prevalence in multiple ways including job creation schemes, improved equity distribution and to encourage savings and the building up of capital assets among others. However, before any long-lasting measures can be assured at the individual level first and foremost the country itself must be in comparative economic health. This means dealing with macro-economic issues of multilateral trade agreements, industry composition, healthy balance of payments accounts, attracting foreign direct investment and importantly an appropriate economic and social development model. Simultaneously, micro-economic and social considerations build on these macro issues to facilitate the movement and investment of capital with things like improving the market infrastructure to ensuring easy and reliable access, strong employment law (protecting both workers and employers), regional incentives for firms to invest in poorer areas and the diversification of industry to protect against vulnerability. Progress can at times be self perpetuating and successes in these areas can also bring with it increased investment, growth and employment opportunities. Another important consideration of access is physical access to land; yet this only delivers profitable food security if there is strong land legislation that ensures worker and owner rights enforceable by legislation. That means land reform in reducing traditional land claims through custom and marriage and replacing or enhancing these with recognised instruments of law. This allows people to consider the future gains of investment in the lands over which they have legitimate control. Other aspects of access include social and community support, which entails family networks, friends and community-based social institutions that can step in to help in the event of individual or family crises or in times of vulnerability. Access to food then requires investment in these areas especially for the more vulnerable community members.

When considering availability in the food security concept, we are traditionally talking of availability or production of food. This means different things to different people whether they are looking to be self-sufficient in food or be supplied through the global market; on the whole however, availability is considered the aggregate supply of food. Currently, sufficient food is being produced although many governments have their eye on future population growth and the concomitant demand that is going to require a considerable increase in productivity. In keeping with social objectives too, such increases need to be met sustainably and for many this means the substantial rethinking of existing practices. In this, policy makers are being pulled in several directions; on the one hand the trend of intensification is attractive for its efficiency and economies of scale, this allows for greater returns on investment and also allows it to compete on the world stage providing valuable export earnings. On the other hand, such intensification is attracting a popular backlash from environmentalists, small farmers and the general public who fear for the social fabric of society and the environment. In response to these concerns, proponents of change offer an alternative development ideology that suggests small-scale farming, permaculture and a more harmonious relationship with the land that benefits millions of rural workers. Both camps are attractive in their own ways, and it becomes a difficult and politically charged policy choice. On top of these decisions, considerations of the environment and sustainable availability are also widely recognised as an issue of priority and as such, modern notions of food security need to take into account many aspects of the physical land. Indeed it is now widely acknowledged that if left unaddressed poor farming practices, whether through intensification or small-scale farmers, can degrade vast tracts of land, destroy biodiversity and drain an area of valuable freshwater and other natural resources. Such practices also threaten to erode existing gains by reducing yields and placing future food security in jeopardy. In juggling the issues of availability and sustainability, policy makers must also consider technologies such as genetically modified organisms (GMOs), agricultural energy costs, foreign land grabbing investments, climate change effects and more. Availability then is a complex interwoven set of considerations that see many trade-offs and compromises between many competing factions, policies and objectives.

The next aspect of the concept deals with stability. In this sense, stability refers to the environment surrounding those of the other pillars of access, availability and utilisation. Continued food security rests on any threats to these instabilities being addressed quickly and decisively. Two particular dimensions of note are those of political and economic stability. Weak government reduces confidence in investors who might defer capital inflows in favour of more stable environments while poor economic governance or unstable international markets can be reflected in poor terms of trade and currency pressures. Furthermore, mismanagement and corruption see domestic investment suffer and good fiscal and social policy beset with contradiction, poor execution or complete absence. Social unrest and conflict too tends to divert much needed resources away from domestic development. All the while, the solidity of social support or livelihood protection programmes for vulnerable groups depends on the continued benevolence of stable regimes. In this regard, free and open political models such as democracy or near democracies provide stable long term benefits irrespective of who is in power.

Biological utilisation, as implied in the modern notion of food security, connotes safe and nutritious food. In achieving this, governments aiming to implement food security enact and enforce legislation in food processing, storage and transport systems. Of interest here is also the education and dissemination of standards, recommendations and best practices not only for industry but also at the individual level. So ensuring individuals are equipped with this knowledge entails addressing things like illiteracy, access to learning and appropriately skilled personnel. In food security terms, utilisation also infers good health, this requires a whole network of support services from hospitals to clinics as well as a good complement of health professionals such as doctors, nurses and nutritionists. Such services too need to be accessible and affordable to all. Lastly, incumbent on good biological utilisation of food is water and sanitation requirements. Without water for drinking, washing, cooking and cleaning, all the food security in the world means nothing. Relatedly, poor sanitation sees people taking on infection and disease that become counterproductive to food security goals.

When considering the social and political as well as the physical environments, I mentioned earlier the need for an appropriate economic and social development model. Getting the right balance between objectives that are not always aligned is perhaps one of the most difficult challenges facing policy makers. Policies aimed at reducing and maintaining a low-wage, low-resource cost base and attracting foreign direct investment for instance might see an initial growth in industrial jobs but in the long term may also see the breakdown of local traditional industries unable to compete. Other policies too that secure much needed income from abroad might also see governments selling off vast tracts of valuable agricultural land well below market value and perhaps (as yet unknown) also at the expense of local livelihoods and potential food security. This is not to suggest these issues cannot be balanced but simply to point out that such considerations are there. Moreover, there are other considerations in choosing the right development model for a country. Modern development paradigms are not so clear-cut and nor are they all predicated on economic growth or neo-liberalism. This is highlighted year after year in the UNDP's Human Development Index (HDI) whereby some countries achieve great social advancement with small amounts of economic growth. In this argument, one does not now necessarily follow from the other and indeed the two objectives can now be separated.

All taken in consideration then, this is the modern notion of the food security concept and it can be seen that it is one beset with fundamental challenges of breadth and of scope, of competing ideologies, of disagreement and a constantly evolving remit. Furthermore, such notions can be culturally specific, making food security a difficult political, social and economic animal to tame. Complex and unwieldy or in the words of the FAO-politically acceptable and morally unimpeachable?

References

Andersen, S. A. (1990). Core Indicators of Nutritional State for Difficult to Sample Populations. *The Journal of Nutrition* 120 (Supplement 11): 1559–1600.

Anderson, M. D. and J. T. Cook (1999). Community Food Security: Practice in Need of Theory? *Agriculture and Human Values* 16: 141–150.

Argeñal (2007). History of Food Security in International Development: Background Paper Developed for Christian Children's Fund Honduras. Honduras: Christian Children's Fund.

Barton, G. T. and M. R. Cooper (1948). Relation of Agricultural Production to Inputs. *The Review of Economics and Statistics* 30(2): 117–126.

Blainey, G. (2000). *A Short History of the World*. Ringwood, Victoria: Viking.

Burnet, E. and W. R. Aykroyd (1935). Nutrition and Public Health. *Quarterly Bulletin of the Health Organization* IV(2): 152.

Chambers, R. and G. R. Conway (1991). Sustainable Rural Livelihoods: Practical Concepts for the 21st Century. *Discussion Paper 296*. Brighton: Institute of Development Studies.

Chew, A. P. (1948). *Plowshares into Swords: Agriculture in the World War Age*. New York: Harper.

Claude, R. P. and B. H. Weston (2006). *Human Rights in the World Community: Issues and Action*. Pennsylvania, PA: University of Pennsylvania Press.

Danson, J. T. (1886). *The Wealth of Households*. Oxford: Clarendon Press.

Danson, J. T. (1906). *Economic and Statistical Studies, 1840–90*. London: T. Fisher Unwin.

de Onis, M. (2003). Commentary: Socioeconomic Inequalities and Child Growth. *International Journal of Epidemiology* 32: 503–505.

Devereux, S., B. Baulch, K. Hussein, J. Shoham, H. Sida and D. Wilcock (2004). Improving the Analysis of Food Insecurity: Food Insecurity Measurement, Livelihoods Approaches and Policy: Applications in FIVIMS. Rome: FAO-Food Insecurity and Vulnerability Information and Mapping Systems (FIVIMS) 52.

Ericksen, P. J. (2008). Conceptualizing Food Systems for Global Environmental Change Research. *Global Environmental Change* 18(1): 234–245.

FAO (1983). World Food Security: A Reappraisal of the Concepts and Approaches. *Director General's Report*. Rome: FAO.

FAO (1996). Rome Declaration on World Food Security and World Food Summit Plan of Action. Rome: Food and Agriculture Organization.

FAO (2003a). Trade Reforms and Food Security: Conceptualising the Linkages: Food Security: Concepts and Measurement. Rome: Food and Agricultural Organization.

FAO (2003b). Trade Reforms and Food Security: Conceptualizing the Linkages. Rome: Food and Agriculture of the United Nations.

FAO (2004). Voluntary Guidelines to Support the Progressive Realization of the Right to Adequate Food in the Context of National Food Security. Rome: Food and Agriculture Organisation.

FAO (2006). Policy Brief: Food Security. Rome: Food and Agriculture Organization 4.

Farm Journal (1940). Poultry. *Farm Journal and Farmers Wife* 88, p. 5.

Forge, F. (2009). Food Security in the Americas: Situation and Policy Options. *Background Paper prepared for the Inter-Parliamentary Forum of the Americas*. Ottawa: Parliament of Canada.

Frigidaire (1964). *Food Security*. Philadelphia: Farm Policy.

Gold, B. (1949). *Wartime Economic Planning in Agriculture: A Study in the Allocation of Resources*. New York: Columbia University Press.

Goodrich, L. M. (1947). From League of Nations to United Nations. *International Organization* 1(1): 3–21.

Hardin, C. M. (1943). The Food Production Programs of the United States Department of Agriculture. *The Annals of the American Academy of Political and Social Science* 225: 191–200.

Horton, D. (1943). The Functions of Alcohol in Primitive Societies: A Cross Cultural Study. *Quarterly Journal of Studies on Alcohol* 4(2): 199–320.

IFAD (2009a). About the International Fund for Agricultural Development The International Fund for Agricultural Development http://www.ifad.org/governance/index.htm (accessed 24 February 2010).

IFAD (2009b). Food Security: A Conceptual Framework. International Fund for Agricultural Development. http://www.ifad.org/hfs/thematic/rural/rural_2.htm (accessed 15 August 2009).

League of Nations (1936a). The Problem of Nutrition: Volume I. Interim Report of the Mixed Committee on the Problem of Nutrition. Geneva, Switzerland: League of Nations.

League of Nations (1936b). The Problem of Nutrition: Volume II. Report on the Physiological Bases of Nutrition. Geneva, Switzerland: Technical Commission of the Health Committee of the League of Nations.

League of Nations (1937). Nutrition: Final Report of the Mixed Commission of the League of Nations on the Relation of Nutrition to Health, Agriculture and Economic Policy. Geneva, Switzerland: League of Nations.

Lutz, A. E., M. E. Swisher and M. A. Brennan (2010). Defining Community Food Security. Tallahassee, FL: University of Florida, The Institute of Food and Agricultural Sciences (IFAS).

Maxwell, S. (1996). Food Security: A Post-Modern Perspective. *Food Policy* 21(2): 155–170.

Maxwell, S. and T. Frankenberger (1992). Household Food Security: Concepts, Indicators and Measurements: A Technical Review. New York and Rome: UNICEF and IFAD.

Maxwell, S. and M. Smith (1992). Household Food Security: A Conceptual Review. In *Household Food Security: Concepts, Indicators, Measurements: A Technical Review*. S. Maxwell and T. R. Frankenberger, eds. New York/Rome: UNICEF/IFAD.

McCollum, E. V. (1957). *A History of Nutrition*. Boston, MA: Houghton Mifflin Company.

McMillan, R. B. (1946). 'World Food Survey'—A Report from F.A.O. *Review of Marketing and Agricultural Economics* 14(10): 372–378.

Munasinghe, M. (1998). Special Topic I: Structural Adjustment Policies and the Environment: Introduction. *Environment and Development Economics* 4: 9–18.

Nehme, N. (2004). Proceedings No. 11 of the Agricultural Policy Forum on Food Security and Development. Damascus, Syria: Ministry of Agriculture and Agrarian Reform: National Agricultural Policy Center.

NML (1760–1976). *The Danson Family Archive*. Liverpool: National Museums Liverpool, Maritime Archives and Library.

OCHA (2011). Website of the Un Office for the Coordination of Humanitarian Affairs. United Nations. http://ochaonline.un.org/AboutOCHA/ (accessed 23 July 2011).

Padilla, M. (1997). Food Security in African Cities—The Role of Food Supply and Distribution Systems. Rome: Food and Agriculture Organization.

Pemberton, J. (2003). 'Malnutrition in England' University College Hospital Magazine 1934 Some Reflections in 2003 on the 1930s. *International Journal of Epidemiology* 32: 496–498.

PSBI (2010). Archive: Guide to the Papers of Pearl S. Buck and Richard J. Walsh. Pearl S. Buck International (PSBI). http://www.psbi.org/site/PageServer (accessed 23 November 2010).

Reid, R. and E. H. Hatton (1941). Price Control and National Defense. *Northwestern University Law Review* 36(255): 283–284.

RSS (1898). John Towne Danson. *Journal of the Royal Statistical Society* 61(2): 372–374.

Salter, R. M. (1950). Technical Progress in Agriculture. *Journal of Farm Economics* 32(3): 478–485.

SDC/GPFS (2010). Sdc Global Programme Food Security (Gpfs) Strategic Framework 2010–2015. Berne: Swiss Agency for Development and Cooperation: Federal Department of Foreign Affairs FDFA.

Selling, L. S. and M. A. S. Ferraro (1945). *The Psychology of Diet and Nutrition*. New York: W.W. Norton & Company.

Shaw, J. (2007). *World Food Security: A History since 1945*. Basingstoke. Palgrave Macmillan.

Singh, B. (1947). *Population and Food Planning in India*. Bombay, India: Hind Kitabs.

Slusser, S. (2006). *The World Bank, Structural Adjustment Programs and Developing Countries: A Review Using Resource Dependency Theory*. Montreal: Paper presented at the annual meeting of the American Sociological Association, Montreal Convention Center, Montreal, Quebec, Canada, August 10, 2006.

SOFI (2001). The State of Food Insecurity in the World 2001. Rome: Food and Agriculture Organization

Tanner, J. (2004). *Incorporated Knowledge and the Making of the Consumer: Nutritional Science and Food Habits in the USA, Germany and Switzerland (1930s to 1950s)*. Bielefeld, Germany: ZIF (Zentrum für Interdisziplinäre Forschung/Centre for Interdisciplinary Research).

Tefera, M. M. (2010). Food Security Attainment Role of Urban Agriculture: A Case Study from Adama Town, Central Ethiopia. *Journal of Sustainable Development in Africa* 12(3): 223–249.

UN (1943). United Nations Conference on Food and Agriculture, May 18–June 3, 1943 : Final Act and Section Reports / Department of State, United States of America. Washington, DC: US Government Printing Office. United Nations Conference on Food and Agriculture.

UN (1945). *Charter of the United Nations, 24 October 1945, 1 Unts XVI*. United Nations.

UN (1948). *Universal Declaration of Human Rights*. United Nations. A/RES/217 A (III).

UN (1966). *International Covenant on Economic, Social and Cultural Rights*. United Nations: A/RES/2200 A (XXI).

UN (1975). Report of the World Food Conference 1974. New York: United Nations.

UNHCR (1999). The Right to Adequate Food Factsheet No.34. Office of the Commissioner of Human Rights. Geneva, Switzerland: UNHCR.

USDA (2009). Food Security in the United States: Measuring Household Food Security. United States Department of Agriculture. http://www.ers.usda.gov/Briefing/FoodSecurity/measurement.htm (accessed 15 October 2009).

Walsh, R. J. (1947). *World Food Shortage*. New York: U.N. World, Inc. Vol 1 No 1.

WCED (1987). Report of the World Commission on Environment and Development: Our Common Future. New York: United Nations.

World Bank (1986). Poverty and Hunger: Issues and Options for Food Security in Developing Countries. Washington, DC: World Bank Group.

World Bank (2008). World Development Report 2008 Agriculture for Development. Washington, DC: World Bank.

Yu, B., L. You and S. Fan (2010). Toward a Typology of Food Security in Developing Countries. *IFPRI Discussion Paper*. Washington, DC: IFPRI.

24

Causality and Future Research

As has been shown, there are an enormous number of variables at play that influence to one degree or another the food security of individuals, regions, nations and the world. Also, as has been stated in the introduction, the goal of this book was not necessarily to provide solutions to the many issues but highlight as much of the debate as was possible. This would then provide a platform from which further research and analysis can be brought to bear on future overarching strategies. Indeed, the many possible integrated solutions could ultimately become the subject of another book; however, for now with so much information coming together in one place, it would be remiss of me not to at least remark on some of the glaring findings that this research has uncovered. With this in mind, the following section highlights certainly not all but many of the issues and potential considerations that may eventually contribute towards possible solutions as raised throughout this book.

When considering causes of food security and, as this book attests, there are numerous, context-specific variables that may act independently or in conjunction with others. To what extent each variable influences overall food insecurity is a cause of much discussion; however, there does exist a general consensus of the main protagonists involved, and these overridingly include, persistent poverty, lack of social and economic development as well as weak governance. Of course, this is the simple version and the solutions are that much more complex. While the following highlights some of the findings that have emerged from this research, bear in mind too this list is not exhaustive nor is it an attempt to analyse potential solutions or the benefits or pitfalls of each. Instead, these are observations that on closer analysis might just provide the seedlings of real effectual solutions. Bear in mind too that the following only loosely aggregates the issues into sections, and no attempt is made to separate out the individual and the state; the micro and the macro; or lastly the short-, medium- or long-term consideration.

24.1 Agriculture, Forestry and Fisheries

Considerations of agriculture that impact on food security issues might include the physical and institutional structure of the agricultural sector as well as prevailing agricultural development paradigms. This might concern whether or not collectively they promote growth through targeted investment, condu cive legislation, adequate institutional infrastructure as well as good access to markets. On the production side too, food security is dependent on numerous issues: technology transfer and appropriate farm practices; the availability and equitable access to resources, both natural and social; the cost of inputs such as energy, labour and fertilisers, etc.; pest control measures; yield and productivity ratios; the soil typology, its quality and fertility; governance and sustainability; land tenure issues; land management programmes; agroforestry as well as issues of sustainability, etc. With all this in mind, the potential for advancement in any number of these issues, with regards to food security concerns, is huge. A few potential areas of consideration might include the following:

- Agricultural development paradigms: We must endeavour to change the way agriculture is viewed in its role as a driver of social and economic development through a more holistic agricultural overview. A new vision perhaps that encourages greater understanding of agriculture's importance in both the overall make-up of a country's economic potential as well as its role in the social fabric of the community. In this way, any push for increased productivity in the sector must bear in mind the consequences of altering the physical and social make-up of the existing industry. After all what's development, if not for the people, by the people?

- Temper the growth of large-scale farms to reverse the declining employment trend and slow the migration to urban cities. Help to increase rural development projects that encourage the young into agriculture.
- Improve agriculture in the developing world through technological transfer and investment in agricultural infrastructure, roads, markets, water harvesting devices, institutions and credit, among others.
- Increase smallholder access to seeds, fertilisers, animal feed, technical assistance and other inputs.
- Introduce hitherto under-used crop varieties back into mainstream diets.
- Actively allow and encourage alternative farming techniques such as organic and permaculture increasing urban and peri-urban agriculture as well as small-scale animal husbandry.
- Allow those so minded the space to develop food sovereignty or at least incorporate its principles into present practices.

24.1.1 Land

- Improve agricultural practices with more consideration of alternative methods to monoculture. Where monoculture persists, more consideration of inputs, soil health, pollution and wastage is needed. Also encourage conservational practices with appropriate tilling, land cover and integrated pest management, etc.
- Better land management involving afforestation, care of marginal lands.
- Land ownership and tenancy rights that encourage stability and protection through legislation, also increase the rights of access to land for landless people, pastoralists and smallholder farmers.
- Ensure that land grabbing is not at the expense of local food security and that transactions are transparent and fair.
- Perhaps an international trading system of land use/change, trading on such things as albedo, radiative forcings, etc. similar to that of the carbon trading market.

24.1.2 Water

- Better overall water management taking into account needs, availability and usage as well as safety, and pollution, etc.
- Desalination of sea and brackish water is widely practiced and rapidly growing as a viable source of new fresh water in the world; this needs to be further encouraged particularly in the water stressed regions.
- The monitoring of plant needs perhaps introducing more practices like drip irrigation, while soil water reserve status avoids water loss by drainage and waterlogging.
- Improving soil structure and increasing the organic matter in the soil helps with fertility and water retention, maximises soil infiltration and minimises efflux of surface waters.
- Reduce and prevent soil salinisation.
- Crop according to water availability: avoid crops with high water needs in low-availability regions.
- Restore and maintain wetlands such as marshes, bogs and mangroves, etc. for biodiversity, water retention and cropping.
- Introduce water harvesting by digging catch pits, crescent bunds, etc. and recycle whenever possible.
- Invest in new and existing aquaculture and artisanal fisheries as well as processing techniques and greater emphasis on common sense that does not see over-quota fish being dumped already dead back into the sea.

24.2 Science and Technology

With regard to science, technology and food security issues there is much to be considered such as whether the green revolution continues to provide increased yields or whether new technologies are within farmers' budgets, are of concern. Further, the cost of ongoing biotechnologies and whether these too are accessible or not or whether subsequent license conditions are reasonable. Other issues concerning genetic engineering depend to a large extent on the public's acceptance and the impact of the ongoing trend that sees the privatisation of agriculture research compared to public ownership. Further consideration might include the following:

- Green or evergreen revolution technologies need to be widely disseminated with more careful consideration of environmental issues together with chemical and fossil inputs are needed.
- Increased research and development spending, especially public-funded research can offset some of the concerns over the increasing private sector patented biologically based intellectual property rights (IPR) systems.
- The use of novel research like biomimetics can help emulate natural processes that aid in more sustainable and healthier human technologies and designs that illuminate new solutions to current problems.
- Numerous public and private concerns over GMO health and safety issues need to be tackled if there is to be a meaningful adoption of these technologies beyond the marginal. Of benefit too would be the promotion of market-assisted selection (MAS) technologies as supplementary to transgenic GMOs.
- Further consideration can be given too to underutilised existing advances in single cell proteins or perhaps newer technologies such as in vitro meat as viable sources of alternative proteins.
- Consider alternative livestock feed (currently primarily wheat, barley, maize and soya), research new feeds that do not compete with food crops and look into alternative under utilised feed sources such as algae, seaweed, SCP and second-generation biofuel technologies that effectively release biomass glucose.
- Second-generation biofuel technologies can be used to provide alternatives to using food crops.

24.3 Socio-Cultural

Food security and natural disasters whether floods, droughts, hurricanes and earthquakes depend to a large degree on the preparedness and emergency procedures of both the nation, the individual and the effectiveness of humanitarian responses. Population growth too, is an important food security dynamic that determines the available per capita resources while sufficient employment opportunities, social safety nets and prevalence of poverty influence overall access to food. Poor or inadequate infrastructure, inequitable access or ownership of land as well as meager connection to markets further impede a fluid food security dynamic. Moreover, increasing urbanisation removes an important and direct means of access to home-grown produce becoming is reliant instead, on wage and market efficiency. In turn social safety nets take on greater importance as does the realisation of the right to food; which sadly remains largely insufficient. Important gender inequalities too often see individual low levels of education and poor or inequitable intra-household distribution of food impact on women's security of food. On top of this, unequal wage-earning potential sees female-headed households continually placed at a disadvantage. Further, women's roles in the provision for children and their effective roles as household carers would benefit from nutritional and disease education. In other issues such as war or conflict for instance, the challenge to provide food security is made more difficult as disruption along the food supply chain is increased. Refugees and displaced people also lose their livelihoods while lands are abandoned or destroyed. Further consideration in the area of socio-cultural issues might include the following.

24.3.1 Poverty

- With poverty found in three-quarters of the rural population and one-quarter in slums or shanty towns, political and social conscience need to be prodded and poked continually to maintain momentum in the fight against this widespread scourge.
- Create employment through investment, education and policy environments.
- Enhance income and other entitlement options to food.
- The promotion of social welfarism where the collective supports the individual either through state-sponsored programmes or insurance initiatives.
- Help to establish stable democracies or at least political stability where economic and social goals can take root and grow.
- Empower more entrepreneurship through humanitarian and governmental projects.

24.3.2 Population

- Credible research into optimum carrying capacity of Earth in terms of population and the potential limiting factors otherwise without knowing our limits much progress is just academic.
- Population growth can be slowed by investing heavily in female literacy and family planning services.
- Continue to disseminate information on per capita effects on Earth's resources of water, land, greenhouse gases, etc.

24.3.3 Education and Information Dissemination

- Invest in education and continue to roll out education initiatives in the developing world focusing both on school age and adult learning.
- Advertising the problems of food security and not just headline grabbing hunger statistics in order to raise the issue's profile.

24.3.4 Waste Management

- Looking at the management of waste at all levels of the food chain from pre- and post-harvest losses to human profligacy is likely to free up considerable resources.
- Improved processing and food storage techniques.
- The active recycling of food wastes.
- Reduce the fickle notion of straighter cucumbers, unblemished carrots and the wrong colour of red apples, etc.
- Reduce the artificial mechanisms that sees the industrialised nations' foods kept artificially low and to raise food costs to reflect their true value, thus providing incentive to be more mindful.

24.3.5 Human Development

- Right to food needs to be taken beyond a voluntary code and have its authority reinstated in the original spirit of the legally binding obligation of the Covenant on Economic, Social and Cultural Rights.
- Encourage, legislate and force a change in gender equality. This will help close the gender gap and free up valuable time, create opportunities and a level-playing field in food security matters and encourage more women to take up positions of decision-making influence.
- Tailored social safety nets like food-welfare programmes, direct food supplements, food stamps and food-for-work programmes prove to be reasonably effective; these need to be up-scaled with more investment and commitment.

- Improve disaster and emergency preparedness with more, better-trained personnel, improved humanitarian infrastructure, increased liaison with international organisations, better vulnerability analysis and pro-active strategies for dealing with refugees and the displaced.
- Regular updated database of key responders in emergencies whether local government, charities and organisations or international agencies.
- Improve peace, conflict resolution, strengthen international arbitration and enforcement.

24.3.6 Diet

- Encourage a more balanced diet that takes into account the health of the person as well as the health of the planet. Educate people on the benefits and pitfalls of different eating patterns that tackle both under- and over-nutritional issues.
- Influence change in dietary habits for the reasons mentioned above but also to cash in on the benefits of eating lower on the food chain as well as alleviating some pressures of demand whilst lowering the resource footprint and greenhouse gas emissions.

24.4 Health and Nutrition

Epidemics and pandemics such as cholera, influenza, typhus and HIV/AIDS as well as more common or localised incidents of diarrhea affect the uptake of water and nutrients, further increasing the likelihood of malnutrition in food-deficit areas. Also where there is poor access to health professionals or health clinics, these can have major impacts on otherwise easily treatable diseases. The cost of this care also impacts on the poor and competes with income needed for food. Immunisation and food fortification policies that tackle easily treatable deficiencies that if left untreated divert more national resources away from much needed social improvements. The provision of safe water and sanitation services is also a major factor in the fight against food in security. Further consideration might include the following.

24.4.1 Health

- Education of nutritional issues that help disseminate wider understanding of good and bad nutrition; this will also help reverse the trend of non-communicable diseases like heart disease, etc.
- Community support for HIV/AIDS sufferers that promote food security and inclusivity so as to encourage integration and better support mechanisms.
- Highlight the cost of malnutrition to spur the call to action. The transparent financial cost of malnutrition is huge yet not always clear; by improving the analysis of economic costs to nations will encourage these nations to act.
- Continue to promote good healthcare and breastfeeding practices.
- Increase access to affordable health service professionals.
- Increase access to safe water, improve piped water and reduce distances needed to travel. The same is required of sanitation, reduce occurrence of open pit wastes and increase infrastructure allowing for flushing or other hygienic practices.

24.4.2 Nutrition

- Micronutrient deficiency programmes should be examined and brought into a holistic programming strategy that sees the phasing out and replacement of supplements and fortification programmes with underlying structural changes. Research into how and in what sort of time frame these goals can be filled is a good step in this direction.
- Nutritional status and the re-introduction of dietary diversification through the production and consumption of micronutrient-rich foods.

24.5 Environment and Natural Resources

An important consideration of existing and future food security is closely connected to the natural resource base. This directly affects Earth's carrying capacity and the food security of nations. Land resources including land cover and land use changes as well as pollution affect the global environment. Moreover, ruminant livestock emissions are the largest source of methane from human-related agricultural activities. This further exacerbates changing weather patterns, increasing incidents of natural disasters, storms, flooding and droughts. Pollution and degradation of soils through waterlogging, salinisation, imbalanced nutrient withdrawals, loss of balanced biodiversity in pest control all collude to reduce the productive ability of these lands. Valuable water resources too are at risk; overuse or inefficient use of water leads to shortages while fertiliser runoff threatens localised rivers and streams with ecological dead zones. Vast energy reliance in agriculture makes it costly and further taps into the climate change debate. There is much overlap here with other sections; however, other considerations include the following:

- Improve sustainable agricultural practices, conservation tillage, more natural integrated pest control systems, manageable and sympathetic chemical inputs.
- Support local farmers and consumers simultaneously cutting down on 'food miles' reducing fossil fuel use and CO_2 emissions.
- Better control and accountability for pollution of lands, water and the atmosphere.
- Encourage footprinting to account for our effect on the lands and its environs and improve methodologies in its calculation.
- Increase and protect existing biodiversity with the re-introduction of crops and careful consideration of other flora and fauna.

24.5.1 Energy

- Energy—Promote the use of second-generation biofuel technologies, as well as introduce more renewables into the energy mix including wind, water, solar, etc., to reduce reliance on fossil fuels.
- De-carbonise the energy sector—Either tax it to the extent that renewables become economically more viable or wait until peak oil starts to push up prices. Of course, the second option is more costly in terms of the environment, while the first is more costly to political and business interests.

24.6 Governance

Ultimately, individual, national and multilateral governance has to be exercised if food security is not going to suffer. This is particularly dependent upon strong political will at both the national and international levels. National policies that promote agricultural development or the sustainable management of resources through adequate institutional infrastructure, education and enforcement serve to promote efforts to achieve security of food. On the other hand the lack of institutional support and the weak protection of agricultural workers in such things as land rights issues as well as the failure to adequately regulate the food chain has immediate and strongly felt consequences. Furthermore, international market trading is hampered by true free trade while ongoing protectionist policies serve to further imbalance a much sought level playing field. Poor governance also affects the effective provisioning of food and social safety net measures, while underinvestment in agriculture, health and development reduces long-term food security status. National debt and service payments reduce the available financial resources for domestic improvements while volatile food prices affect many people's ability to secure adequate food at reasonable prices. Lastly, the concentration of market-distorting power in the hands of fewer and fewer players tips the balance away from the majority into the hands of the few. Further consideration might involve the following:

- Translate more of the investment into economic growth through efficiency and decreased corruption.
- Need to facilitate social and political will, enhance it and direct it. Poor rudderless navigation is no excuse. If political hegemony of well placed nations is not forthcoming governance must be ceded to those that can and will.
- Overt national interests competing against multilaterally agreed international instruments cannot be tolerated. This hypocrisy needs to be addressed with extreme prejudice.
- Promote democracy as this promotes a conducive social infrastructure that further facilitates food security.
- Poor governance needs addressing, much duplication despite recent efforts continue to plague food security efforts. This can be aided by reducing complexity and bureaucracy in the lumbering food security sector.
- More inter-agency collaboration whilst ensuring inter- or cross-agency mandates are in alignment and are not competing.
- Voluntary codes need to be replaced with obligation and legislation.
- Policies affecting food security need to be coordinated across government ministries of agriculture, health, education and industry.
- Rationalise and clarify stakeholder frameworks by elucidating the global framework and highlighting areas of expertise in both UN, government and NGO spheres. Only then can the global coverage of the concept be measured, the skill base assessed and gaps filled. Further by inviting more strategic alliances and collaborative programmes among stakeholders, costs can be rationalised, duplication of effort can be minimised and food security benefits maximised.
- More responsive institutional policy—policy can sometimes take what seems to be an age from think tank to board rooms to frontline action. Reducing the think tank to frontline lead time and allowing re-orientation and re-direction mid-policy or mid-stream produces more reactive and proactive measures and cuts time and cost overruns on unproductive policies.
- Targets must be made and kept, while importantly, targets must be targets; backpedalling or the repositioning of goal posts mid-target to suit political needs must be avoided. Also set more targets, only by setting ourselves tangible goals rather than vague promises can we ever hope to address these issues. In the process we make ourselves accountable, which is severely lacking at present.
- Institutional weakness, if these bodies are not up to the job, strengthen or get rid of them.
- Look at corruption, not just of recipient governments but rules concerning deferred gifts and similar practices. Ensure openness and transparency so as not to invite abuse.

24.6.1 International Humanitarianism

- Enhance food provisioning to the most vulnerable.
- Continue the removal of conditions and ties inherent in multilateral aid, aim for all donations to be in cash.
- Food aid is not the most effective way of providing long-term support. Food assistance's role in the wider humanitarian assistance development strategy needs to be better coordinated so that all emergency, short- and long-term objectives are joined up.
- Improved co-ordination among stakeholders and more transparency in donor contributions.

24.6.2 Monitoring

- Improve the monitoring of the many aspects of food security variables to form a comprehensive picture and facilitate early warning including developing risk analysis management.
- Improve indicator methodologies. Standardise international measures and reduce the time lag for important statistics.
- Improve the proxy method of determining food security. Look more into forward-looking metrics or perhaps utilising more qualitative data such as community surveys, etc.

24.6.3 Development

- Ensure that rural, agricultural, economic, employment and social development policies are not fundamentally at odds with each other.
- Ensure that worker's, women's, land and other rights are enshrined in legislation to create stability and social equity.
- If this remains the preferred method of doing business, free trade needs to be fair and just by ensuring global fair trade based on equality through a mutual reduction in trade distorting barriers, equitable terms of trade, free, open and fair market pricing.

24.6.4 Economics

- Competent monetary and fiscal policies that create stable, investment friendly economies and healthy balance of payments.
- Push for further debt relief for developing countries and provide budgetary or balance of payments support.
- Simplify procedures of existing international financial mechanisms that aim to support not penalise users.
- Reducing barriers to free trade might bring with it restructuring of industries. If these artificial comparative advantages are withdrawn too quickly this may well prove difficult for developing countries unable to adapt quickly or efficiently to such change.
- Diversify the domestic economic base to increase stability and reduce vulnerability; this requires huge investment in both industry and education.
- Revitalise rural financial systems such as microcredit and other income-creating measures supported by low-interest loans for the poor.
- Improve rural access to credit to target poorer farmers unable to secure credit on normal commercial terms.
- If comparative advantage (CA) is continually distorted on the global level, consider a mixture of CA and self-sufficiency/reliance.
- Increase regulation regarding corporate control, redefine and update definitions of monopolies and oligopolies, and re-evaluate measures to curb excessive concentrations.
- More research is needed in the effects of speculation in agricultural derivates markets such as futures, etc.
- Consider excluding basic food commodities from the market mechanism. Furthermore, price regulation and government subsidies can be crucial safety nets that aim to combat price volatility. Perhaps re-introduce international commodity agreements (ICAs) or other transparent global institutional arrangements.
- Increase the holding of national, regional, and international grain reserves.

- Increase market efficiency through improved information and reporting techniques along the lines of the International Energy Agency that aims to appease the market in times of shortages providing up to date reliable information.
- Encourage the diversification of household's livelihood strategies.

24.6.5 Research

- More research into coping strategies to understand better how people prioritise their goals and objectives in relation to food security.
- Avoid paralysis by analysis—Increase institutional learning by reducing the roundabout way in which lessons are learned, past learning is ignored and mistakes are repeated. Allow more roundtable discussion and open fora where experts and non-experts alike can frequently and openly debate the multitude of issues and potential solutions; past present and proposed.

In summary the above quick, back of the envelope solutions, provide much to contemplate. Perhaps there is nothing new here; proponents of change may have elucidated many of these issues already; yet this exercise helps to bring together all of the issues under one roof so to speak. Clearly, some of these observations represent ideals, and some are simply just suggestions that have not been debated or tested empirically and will face much resistance. The idea of scrapping all tied and conditional aid for instance will be laughed out of the debate; encouraging people to change their dietary habits in the face of increased wealth and more choice will be challenging at best, while getting stakeholders to collaborate without letting institutional ego and petty power struggles getting in the way may prove to be insurmountable. Further, I confess some remarks might seem glib, and this was intentional as sometimes the most obvious solutions are perhaps the best. Suggestions too, of a trading market for land, or further taxation on oil are not an attempt at egalitarian utopia. It is just understood that only by aiming for the best approximation of these ideals can we step out of the current paralysing dystopia where real progress might surely be made.

Perhaps if we all adopt a 1% rule to effect change we may start to make a difference. Perhaps by donating 1% of our incomes to charity; by feeding cattle 1% less grain; use 1% more renewables; eat 1% less meat; exercise 1% more; increase NGO membership by 1%; cede 1% national sovereignty to the UN; etc. ad infinitum, then who knows, real progress might be made in more areas than just food security. Indeed, in the absence of a major sociological revolution perhaps millions or billions of these small measures may accumulate to tip the balance in favour of a better more equitable and sustainable future.

25

Redefining Food Security

Despite a title that suggests otherwise, at the outset of this project, my intention was not to attempt to redefine food security; the reference in the title was initially a philosophical allusion to the rediscovery of the concept through rigorous analysis of its underlying principles. However, in light of everything that has been learnt in this research, I found myself challenging existing conceptual definitions. In and of itself, the notion of food security seems straightforward, and this has been the trap that many have fallen into. This was even noted in an influential 1967 US Presidential report, which suggested that

> Despite its true complexity, the problem, at first glance, seems deceptively straightforward and is, therefore, unusually susceptible to oversimplification. (SAC 1967)

It should come as no surprise then that the modern concept of food security for all has since become infinitely more complex with numerous multifaceted dimensions that invites definition, interpretation and division from all quarters. In its simplest form, food security is described by many in the 1983 FAO definition as ensuring that all people at all times have both physical and economic access to the basic food they need. This often quoted, carefully scripted rubric however belies the vast and often contentious nature of food security itself. Since these early days, the concept has evolved and so have the interpretations. In fact, definitions since then have spiraled so much so in fact that by the turn of the twenty-first century there were over 200 definitions in circulation. Many too, it has been said, were often moulded to suit institutions, organisations and academics' own purposes. This continual redefinition of food security over the past four decades can also be seen as highlighting the considerable evolution of official thinking and one that has led to the present propensity for definitions to take a reductionist view that concentrates on the component parts of the whole. Indeed, so subjective is the concept that large bodies of literature attempting to interpret that which is implicit and explicit continues to surface regularly, although as IFAD commented in 2009 that

> While the focus on the disaggregated has now become common, the various definitions of food security still differ. (IFAD 2009)

In examining if and how to improve on current definitions, I have chosen two of the most often quoted of the current choices available, these belong to the UN and the USDA.

- UN: Food security is a situation that exists when all people, at all times, have physical, social and economic access to sufficient, safe and nutritious food that meets their dietary needs and food preferences for an active and healthy life. (FAO 1996; SOFI 2001)
- USAID: Access by all people at all times to enough food for an active, healthy life. Food security includes at a minimum: (1) the ready availability of nutritionally adequate and safe foods, and (2) an assured ability to acquire acceptable foods in socially acceptable ways (e.g., without resorting to emergency food supplies, scavenging, stealing, or other coping strategies). (Andersen 1990; USDA 2009)

In order to make any comparisons, we must consider what the modern notion of food security entails. To help in this endeavour the diagram in Figure 25.1 highlights the cycle of malnutrition from birth to adulthood and to the next generation. To be food-secure, policies aim to intervene at any and every stage of this cycle putting an end to the often self-perpetuating concept.

Using the lifecycle model, the notion of food security (Section 23.2) and the model of food security (Figure 25.2) we are able to tease out the main ideas of the modern concept. I have listed the four pillars as a place to start although I have added a fifth pillar simply to pop in the wider societal objectives considered as part of the overall food security goal. Loosely the main aspects of these pillars include the following:

- *Access*: Poverty; economic development; physical—land access/rights; social—community institutions, family networks.
- *Availability*: Production; farming practices; infrastructure; sustainability; natural resource base; sufficient/adequate food.
- *Stability*: Sustainable economic, political, social and physical environment; conflict and social unrest, vulnerability and risk management.
- *Utilisation*: Health and health services; food safety; nutritional food; clean water; good sanitation; education.
- *Environment*: Governance; cultural considerations; development ideology; physical environment considerations; sustainable policy.

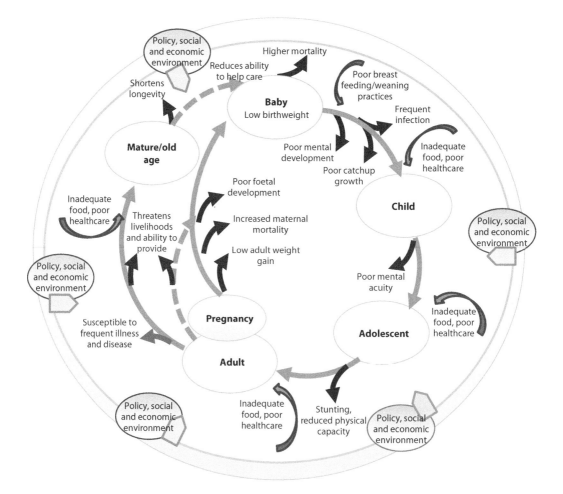

FIGURE 25.1   Malnutrition lifecycle analysis. (Based on the ideas of Nina Seres in the ACC/SCN-appointed Commission on the Nutrition Challenges of the 21st Century, 2000.)

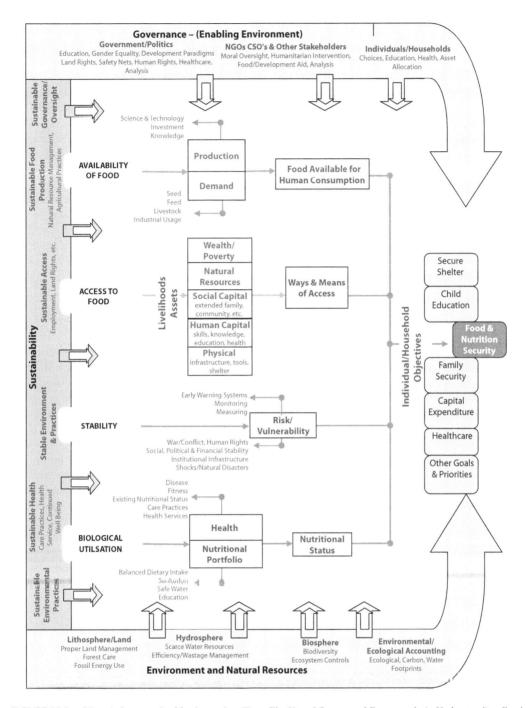

FIGURE 25.2 Gibson's framework of food security. (From *The Use of Conceptual Frameworks in Understanding Food Security*, Manchester Metropolitan University, UK, 2011.)

Looking at the ideas that propel the keywords and comparing them to the definition we can then make an assessment as to the continuing validity of the current meanings. In terms of both the UN and the USAID definitions, there is not really that much that can be said is wrong. They are both in agreement on the idea that food security is required by all people at all times. They both too agree on the pillars of access and availability although they are couched in different terms. Other similarities are reflected in the notions of adequate, safe and nutritious foods as well as the goal of active and healthy lives. In fact, it is quite difficult to separate the two meanings, although the USAID definition does, however, add the idea of sourcing such foods in socially acceptable ways, which is debatable whether it is implicit in the FAO's rendition or not. Of note however, are two key ideas that appear to be missing from both definitions, these are: stability and sustainability. However, while neither mention stability by word, it is quite clear that 'at all times' in both definitions is an implied reference to the risk management, or stability aspect of the concept. Sustainability, on the other hand, is a little more difficult to determine. Sustainability of access is perhaps implied by the same term 'at all times', yet it is lacking in any respect in connection with the environment. This is not bad considering these definitions are at least 10 years old and while I have read some who would consider changing the concept to food and nutrition security I believe the definition adequately reflects such a notion if not in title in principle and content. However, having said that, I believe food security by itself intuitively conjures up food related issues of supply, demand and access, whereas nutrition security places the emphasis on the individual thus linking the whole chain of food security from food acquisition to an individual's overall nutritional well-being. In light of this and recognising that the essence of a good definition needs to be short and to the point, I would recommend only two small changes to reflect these considerations as follows:

> Food and nutrition security ensures that all people, at all times, have physical, social and economic access to nutritionally adequate, safe and sustainably sourced food that meets their dietary needs and food preferences for an active and healthy life.

Whether this catches on as a new definition is not the goal of this book and neither the point; the exercise is simply used to illustrate a continually evolving concept that is slightly different today than it was 10 years ago and is likely to be different again in the next decade or so as new pressures and new objectives come to light. It is worth mentioning too that as with much else the devil is in the detail. Breaking the definition down, as we have already probably shown, the individual constructs of, say, sustainability, safe, nutritious, dietary needs, food preferences and the like are widely interpreted differently. It is these constant challenges that are set to push us to our limits in finding common ground from which we can act.

Lastly, an interesting observation that comes out of this exercise is the question of just how far can we push the food security issue to the limit. In its present form for instance, food security represents nothing short of what could be described as a social charter along the lines of the UN's Declaration of Human Rights or the EU's, the South Asian Association for Regional Co-operative Development's (SAAC) or Canadian's Social Charters. One could argue too that such an expanded definition threatens to overwhelm policy makers and add further confusion where already murky boundaries are becoming further blurred. In fact, glibness aside, perhaps the answer does indeed call for nothing short of a social revolution.

References

Andersen, S. A. (1990). Core Indicators of Nutritional State for Difficult to Sample Populations. *The Journal of Nutrition* 120 (Suppl 11): 1559–1600.

FAO (1996). Rome Declaration on World Food Security and World Food Summit Plan of Action. Rome: Food and Agriculture Organization.

IFAD (2009). Food Security: A Conceptual Framework. International Fund for Agricultural Development. http://www.ifad.org/hfs/thematic/rural/rural_2.htm (accessed 15 August 2009).

SAC (1967). The World Food Problem: A Report of the President's Science Advisory Committee, Volume LII Report of the Panel on the World Food Supply. Washington, DC: The President's Science Advisory Committee (PSAC), The White House.

SOFI (2001). The State of Food Insecurity in the World 2001. Rome: Food and Agriculture Organization.

USDA (2009). Food Security in the United States: Measuring Household Food Security. United States Department of Agriculture. http://www.ers.usda.gov/Briefing/FoodSecurity/measurement.htm (accessed 15 October 2009).

26

Food for Thought: Discussion and Considerations

26.1 Where Did It Come From?

With regard to the notion of food security, despite what has been suggested in other literature, the modern concept did not begin in the 1970s. Instead, it borrowed, regurgitated and built on numerous age-old ideological and philosophical foundations. As surely as modern civilisation bears the idiosyncrasies of a melting pot of cultural history, so too does the notion of feeding the masses draw on millennia of acquired lessons learnt, of altered philosophies and changing paradigms. The quest to address hunger and malnutrition provides a long and colourful history, dating back to the earliest civilisations. It has also been shown that while the concept reached its zenith in the twentieth century, the idea of food as an inalienable right; one of the central tenets of the concept, also had its genesis in the centuries leading up to it. Migrating workforces and the buildup of urban civilisations, increased poverty and suffering coupled with the almost continuous misery of conflict, frequent famines and social injustice shaped much reflective discourse. New political and economic theories emerged while intellectual and philosophical thought fuelled by eminent thinkers collectively galvanised a growing social conscience during an 'age of enlightenment'. This new collective social awakening was further fuelled by a new style of contemporary journalism where stories of hunger and starvation elicited a humanitarian outpouring among its readership. From here, increased philanthropy from individuals and humanitarian bodies solidified calls for perpetual peace and social justice while a growing international conscience led to the first real attempts at global governance. This convergence of political and social ideology became the driving force behind efforts to provide sufficient, safe and secure food for all. Thus, it can be said that the twentieth century stood apart as an era marked by extreme hardship, out of which considerable collective social growth effectively followed to ensure a future of social and political interdependence, promoting, among other things, social equality, accountability and fair trade.

26.2 So What of Food Security's Principles?

A quick glance behind the simple unassuming facade of the food security concept reveals something much more involved, infinitely more complex and perhaps somewhat overwhelming. It is this apparent paradox and duality that has invited much conjecture and subjectivity on the matter, often with as much over-simplification as unnecessary complication. At its core, food security is about enabling people to feed themselves; however, as has been seen throughout this book, such simplistic ideology misses the crux of the issue by a wide margin contributing to continued poor progress. Food security is about: poverty; good direction and governance; stoking and shaping political philosophy; understanding nutrition; agricultural science; environmental and natural resource considerations; economics; inclusionism—ensuring developed and developing countries are on ideological, visionary and moral parity; crisis management; aid; globalisation; science and technology; information dissemination; sustainability, personal and household choice; shelter; water and sanitation; rural/urban issues; women; vulnerable groups; issues of vulnerability and stability; monitoring; emergency preparedness; land rights; conflict management; internal and external trade; investment issues of agriculture, education and health services; and ultimately human rights. And the list goes on. However, as the concept was working its way towards full maturity by the 1990s, this evolution came with a certain amount of confusion where clarity should have been. By the 1990s, food security had firmly taken hold in the eyes of policy makers, nutritionists and a growing public, yet even if the terminology was still alien to most, they understood the burden of hunger and malnutrition. In the world of policy making, the new catchall was constantly being examined and defined in order that appropriate policy could be enacted. The difficulty by this time was that the concept had long shed its singular dimension of a lack of food to embrace more complex, collective notions. Consequently, encasing all of these considerations into the simple catchall of food security tended to cloud rather than clarify the situation. Not surprisingly, out of all this, the modern concept has taken on a duality of identity and purpose. On the one hand progress aims to satisfy the expectations of food security while on the other, the realisation that food insecurity itself, is as much a product of structural, social and political constructs dictates that there are push and pull dimensions to the modern concept that require wider societal considerations.

26.3 What Are Some of the Challenges and Difficulties?

In the absence of a one size-fits-all solution, answers require more than a scant understanding or singular perspectives. The main challenge here is that while such disciplines may have come along in leaps and bounds in past decades, a truly inter disciplinary mastery of the subject is sorely lacking by all but the very few. Compounding this dilemma is the frequent notion that a good body of literature attesting to new ideas on the subject hides the real fact that many such ideas have indeed been previously posited and explored in history. Not helping things in this aspect is the remarkably short-sighted propensity to concentrate on food security as a product of the 1970s. As a result, it is not surprising that so many lessons of the past have been ignored and are now being relearned, at great cost. Moreover, this paradox sits uncomfortably with the notion that as a result, poor conceptualisations, a lack of adequate definitions and often oversimplified constructs makes finding solutions fraught with ill-considered ideas and much subjectivity. This lack of a fundamental understanding of the multi-dimensionality and complex interplay of the many variables at play is reflected in the division and fragmentation of institutional efforts and painfully inadequate progress.

While poverty remains one of the single biggest obstacles in achieving lasting food security, so another major obstacle, social and political will, is found dilly-dallying in vague commitments and non-accountability: Millennium Development Goals (MDGs) notwithstanding. This is perhaps exacerbated, and I hide behind the veil of conjecture here, by economic and social development paradigms that appear at best to be at odds with each other and at worst diametrically opposed. This I believe comes about as a result of evolutionary rather than planned development strategies; and one that I believe is also borne out by the research. In this way, just as science and technology builds and adapts on previous successes and failures, so political ideology too is fluid and dynamic and benefits from an understanding of preceding concepts. In the field of social and development economics, for instance, it is clear that much groundwork emerged from centuries of conflict and rebellion. Successive theories led to the emergence of capitalism and a growing disparity in the twentieth century between the rich and poor nations. In turn, social development developed as an add-on to this basic model and one that was made to fit. It is not surprising then, that when such economic and social woes met with growing global unease, altruism and humanitarianism strengthened its resolve in an effort to redress the balance. Unfortunately, today, continuing from past practices, international multilateral will is increasingly being bullied, usurped and cajoled, with certain institutions and stakeholders at certain times becoming political agents where, instead of their long-held nod at neutrality, often end up bowing to sovereign state politics. An uncomfortable reality perhaps but one that needs addressing.

With regard to specifics, modern development goals demand long-term sustainable growth in every sphere, political, economic, social and cultural. Economic goals are predicated on an open economy that is market driven with underlying capitalist principles of privatisation. The role of government in this model of growth is simply to set and enforce appropriate rules ensuring a level, fair and open playing field whilst providing support in the form of necessary public goods. This economic model for a long time also doubled as the prevailing social development model too where one was thought to necessarily follow the other. However, while capitalism has served its masters well, the race for economic returns has created an unhealthy environment of competition. This has allowed the winners to adequately secure their everyday needs and wants with relative ease, while too many of those unfortunate losers of this economically led model of social advancement are left lacking even the basic dignity of sufficient food.

In this picture, economic growth is a primal force equated with bigger, better and stronger, yet it is a monster that needs to be tamed. Economically led social development is seen by many as a fundamentally sound concept, and why not? The trickling down through the economy of social advancement has been well documented in the West, but what happens if we take this model to its logical conclusion? Stronger more powerful corporate entities that question rather than serve the state; who knows? Certainly not I. Yet if this becomes the reality, the norm if you like, in whose hands do we place the social equity of the population, who then looks out for you and I? Such a model seems to be fundamentally at odds with the more equitable and ideologically pure social democracy or dare I say Marxist communism. Although, having said that, capitalism serves a purpose, a necessary purpose too, so long that is, as long as capitalism is subservient to a greater social objective. The moment capitalism becomes self-serving, the beast needs to be tempered by tighter legislative reigns so that the hitherto concentration of state managed, citizen-owned capital, does not become corporate capital.

Current models of agricultural development too have also been challenged as unsustainable, leading many to consider alternative models of growth. Agricultural development paradigms continue to evolve and have practically come full circle in the last 75 years or so. Initial agricultural expansion efforts of self-sufficiency were to a large extent replaced over the interim period by a push towards export-led growth favouring large farms reaping benefits from economies of scale. As understanding of the issues of food security developed so the current agricultural paradigm sees the distinction between food crops and market or export crops being blurred. As a result, there is a growing split in perceived future agricultural development needs. On the one side, there are those that feel that mimicking the Western model of concentration and intensification in developing worlds, relying on global free trade and comparative advantage is the best way forward. On the other hand, food sovereignty proponents argue that this is damaging to both the rural economy, to the social fabric and to the environment. Followers of this latter ideal suggest that by investing in and empowering existing small-scale farmers, not only can sufficient global food needs be met today and in the future but it can be done in a way that is sustainable, does not threaten the livelihoods of farmers and is sympathetic to the environment. This they argue too, also falls within the capitalist ideology; as the now widely held belief is that currently, the largest private sector economic activity comes from the millions of small-scale farmers themselves. Moreover, by incentivising farmers through investment in agriculture and technology, it has been suggested that this could provide the millions of small farmers the means to produce food and marketable surpluses that can be traded to further improve incomes and future investment.

Social development is also high on many governments and multilateral agencies agendas these days. The MDGs sought to finally raise the level of humanity out of the shame of deprivation and provide equality for all on many fronts. Couple this with the rights bestowed on individual's by the UN Charter from the freedom of speech, the right of political affiliation, as well as the freedom's of want and the right to food. Combined these lofty goals represent a high point of moral achievement for the global collective, albeit one that has not been easy. Beset by problem after problem, the challenge today is to translate these goals into tangible realities. However, on this front, there appears to be a disturbing pattern among frequently offered goals, targets and objectives; too often what starts amid a fanfare of publicity and enthusiasm follows a path of re-orientation, back-peddling and re-calibrated targets. Much lip-service is paid to many social goals, not least of which is global food security. Yet, relentless press releases, reports, meetings and conferences that continue to do little more than agree general acceptance of underlying ideals, that look for confirmation of commitment, repeat affirmations, re-commit, re-confirm and re-affirm targets and goals, in fact do little to actually put food in the mouths of the hungry. Sadly too, all these posturing and windless promises fail to fill the sails of action and instead serve only to increase the pile of ever-growing broken pledges. We know what needs to be done and frankly we ought to be ashamed. Bloated bureaucracy and political lily-livering is not the only blame on offer either. We, as individuals, seem to be suffering a long-term bout of compassion fatigue further compounded by practicing arm's length humanitarianism. In this endeavour, we might nod knowingly at the inequities on show yet, on the flip side, conveniently fail to embrace the realities for fear of the guilt-by-comparison that would send our moral compasses in a spin. In this respect, Ayn Rand was wholly correct when she suggested that we should not sacrifice ourselves for the greater good yet, on the flip side, what's a little sacrifice, if not a sacrifice.

Another rarely talked of challenge in current day understanding is to do with information overload. Any student or official for that matter of food security will be beset by frustration after frustration as literature can be constantly contradictive, plagued by inaccuracies, data lag, subjective interpretation and inconsistent terminology. This can often leave students feeling overwhelmed and confused and left with the constant nagging doubt that they are not in possession of the latest current thinking or relative data. In short, with so much information needed to comprehend and tackle the issues of food security by a number of disparate stakeholders, whether single or multiple issue institutions, governments and multilateral organisations, the collection, analysis and dissemination of relevant information is in disarray. This reflects no fault of any stakeholder in particular, more a general observation. While the UN, USAID, EU and numerous others all provide in-depth analysis on many issues, much valuable information is simply lost in the sheer unnavigable and proprietary compartmentalisation of a complicated and unclearly bounded subject. Moreover, reading two similarly 'objectively' focused reports from world class institutions might yield two very different conclusions, each with numerous supporting references of their particular view. In other words, it seems that whatever one's particular opinion, academic references can be sourced to not only support a particular viewpoint but positively reject credible alternatives. While I accept this is the scientific way and indeed a necessary academic development process, I would surely recommend more regular food security symposia along the lines of the climate change IPCC which seeks to search out scientific fact from conjecture. In this way, regular periodic analysis might trend towards tentative agreements on the many issues involved. The problem then is many-fold: the sheer volume of data, its organisation, the lack of consensus from an overarching perspective and the dissemination. What is required is the rationalisation of all this disparate information on current analysis, consensus, disagreement, stakeholders, programmes and ideologies into a single overarching one-stop shop—a clearing house, if you like, that serves to enlighten people, not confuse them. In this respect it is clear that some have tried, particularly the UN, although unfortunately not to any great success.

27

Closing Remarks

The primary aim of writing this book was to synthesise the enormous and disparate material on the subject of food security. This was embarked upon, not so much in attempt to provide solutions, but rather to present a solid foundation on which a full overarching understanding of the modern notion can be reasonably achieved. In this undertaking, the book has flirted with a multitude of diverse disciplines. These have included science, economics, politics and philosophy among others and, in this endeavour, it has been part theoretical, part conceptual, but all the time thoroughly pragmatic. It is worth noting too that in touching upon so many fields, I have sometimes only been able to offer small glimpses into multiple complex areas that surround an inherently difficult subject to navigate. However, despite these limitations, it is hoped that the underpinning knowledge provided in this book can act as a springboard, from which more specialised esoteric knowledge can be sought, to aid in further navigation of the various threads.

History provides us with a rich seam of knowledge in this area and one that does not require huge expense or vast expertise to mine. As a consequence, the historical aspect of the book was an important one and, apart from attempting to shed light on the origins and evolution of the subject, its ultimate aim was in offering a fresh perspective. One consideration borne in mind throughout the historical mining however, was the fact that it is perhaps all too easy to look retrospectively at the development of food security, from our present position offering a modern interpretation of events as they unfolded. I have attempted to avoid this by presenting where possible, contemporary motivational analyses that better underlines the cultural, political and social influences of the day. Ultimately by bridging the divide separating past lessons from today's food security merry-go-round, it becomes clear that history has already posed many of the questions facing modern day theoreticians. More than this though, and somewhat surprisingly, history has also posited many of the solutions that seemed to have later popped up as 'epiphanous' realisations, or 'new' knowledge, by individuals and institutions alike. In the end, it is envisaged that by reconnecting the past with the present, it is hoped valuable insight might provide for future original and forward-thinking analysis, based on progressive forward-looking objectives. In this sense, certain parallels can be drawn with past philosophical re-alignments: I refer of course to the 'Age of Enlightenment'. History too, it seems, often dwells on a problem for, what at times, must seem like an age. Yet it is encouraging to learn however, that if the problem sufficiently primes the social and political conscience, then a growing critical mass of opinion, can often become distilled into a targeted social objective. This is also no less true of today; momentum such as this, is crucial if the tendrils of change are to percolate into actionable policies that provide needed inertia in the continuation of social and ideological progression.

For my part, my role in this project was to play devil's advocate by highlighting some of the obvious contradictions, asking some difficult questions and challenging the status quo. In this, I mostly adopted an objective stance. However, it is not always possible to be truly objective in the face of certain glaring obstacles to food security like poverty, politics and ideology; especially when some of the solutions are self-evident. So where has it all gone wrong? Well, after what was a magnificent and serendipitous confluence of converging ideologies, science and social awakening over the last two centuries; much it would seem. Today's world is still a place of distortion and inequity. We continue to witness uneven global development, the unsustainable use of natural resources, increasingly polarised capital flows, worsening climate change and a growing disillusionment of humanities ability to deal with such problems. Furthermore, centuries-old traditions and patterns of life are being rapidly replaced by revolutionary globalisation, spearheaded by new political realities. All this too, continues alongside stubborn poverty and persistently high numbers of malnourished, only to be met with apathy, fatalism and indifference from the general population at large as well as the wider political establishment. In this environment, the modern notion of food security has lost some of its direction. Ultimately, poor governance, political will and hypocrisy are writ large in the many failings of food security. As a result, what started out as Western predatory development policies based on capitalism and liberalisation, has pervaded the global cultural psyche replacing traditional values with consumerism, materialism and the law of the jungle. Yet on the flip side of this argument; it is a fact that hard-fought capitalist tendencies of privatisation, naturally trends towards larger profits and bigger corporations. On this basis, we cannot, in all good conscience, actively encourage the free market system, then sit back and complain when it is working too well. We are masters of our own destinies and rebalancing our need for economic growth might better be served by a tempering of humanitarian social values.

Pre-Second World War, we might have been forgiven for not knowing better, as many countries were still firmly focused on the national perspective. Today however, with a global outlook, the advent of the UN and a growing awareness of the problems of the 'haves' and 'have-nots,' continued excuses do not really stand up under scrutiny. We have the know-how, the means and a multilateral will of sorts, that we need to stop tackling food security on a crisis footing but instead live up to our many promises and put an end to it once and for all. It is all very well too, dealing in what some might see as the abstracts of ideology and philosophy, but what about the reality on the ground? In answer, the intellectual separation and specialisation of the food security subject by economists, social historians, demographers, political scientists and the like, whilst having undoubtedly furthered the concept; have in fact done little for the overarching comprehensiveness of the notion. Instead, a surfeit of reductionist theoretical frameworks has left the concept in disarray, rudderless and without stewardship.

That said, a pre-requisite in any solution then is coming to grips with this underlying abstract, so to speak, in the hope of forging better more focused and inclusive ideological guiding principles. This is not to dwell on the ideological either; there are many numerous realities that if enacted would see real and rapid results. In short shrift, it has been said before, but the world really does have to rethink its relationship with food: the way we produce it and the way we consume it. In this regard, striking the right balance with a food security paradigm that falls somewhere between an oversimplified notion that is little more than a simple proxy for chronic poverty, or an overly inclusive one that requires nothing less than a sociological revolution, is difficult given the complexity of the notion and the aspirations of the international community. Moreover, if the world is to avoid a dystopian future based on the continued segregation of the 'haves' and 'have-nots' of food security, the continued subjugation of the poorer developing nations, in an increasingly wealthy global economy remains one of the major ethical, economic, and political challenges of our time. More than this though, in spite of everything that has happened over the last century or so, all the promises that have been and continue to be made; allowing hunger and malnutrition to persist is simply unaffordable in both social, moral and economic costs. In this endeavour, there is no substitute for benevolent political will and, although the once thought notion that food security goals could be achieved in isolation are long gone, embracing new ideas, such as women or smallholders and the like, might well turn out to be the key to food security for the future. Indeed, on this point of progress and not to sound trite, or draw unnecessary comparisons, but one obvious parallel is worth highlighting. This was the swift, decisive and expedient action with which the worldwide political glitterati responded to loud calls of global unity in the wake of the financial market crisis of recent years. All this too, in the name of collective social common good. If a fraction of this cooperation and governance was marshaled in the name of food security, then we might perhaps be able to place a feather in the cap of our achievements and set ourselves, just that little farther apart, from our poor past performances. Unfortunately, in this regard though, a planet so rich in possibilities, is apparently incapable of guaranteeing its inhabitants the basic minimum rights of food and good health. In this, double standards seem to be the unwritten guiding principle. It seems too, that all too often the rule of law is usurped, the moral conscience hijacked and the collective will sapped by sticky political fingers intent on self-interest. Indeed, while food as a right has long been the lofty goal of international agencies and stakeholders alike, pendulous political and philosophical ideologies are surely paralysing the industry. As a result, the sector is like a lumbering behemoth bloated by its own goals and objectives, rudderless by weak institutional direction and lack of political will that leaves a disenfranchised public wondering when we might just come to a consensus, roll our shirt sleeves up and get on with the job at hand.

Appendices

A map of the world that does not include Utopia is not worth even glancing at, for it leaves out the one country at which Humanity is always landing.

Oscar Wilde
The Soul of Man Under Socialism

Appendix A: Country Classifications

As noted throughout this book, a large number of official UN and other agency reports produce information using geographical, political or other country classifications as well as sub-regional aggregates. However, practices vary among agencies and literature and are oftentimes used interchangeably. Several standards have been used throughput this book and as far as was possible all references have conformed to one of more of these standards. These have included the UN Standard Country or Area Codes for Statistical Use (Series M, No.49/Rev 4), generally referred to as 'M49': the Millennium Development Goals (MDG), World Bank, Organization for Economic Co-operation and Development (OECD); and the World Health Organization classifications as can be seen in excerpt of Figure A.1. The full version of the table can be found on CRC's website at: http://www.crcpress.com/product/isbn/9781439839508

Country groupings header (two super-groups): **Geographical Based Groupings** and **Economic and Development Based Groupings**

ISO	UN M49 Country and Regions	MDG Regional Classification	WB Regions	WHO Regions	UN sub-Saharan Africa	MDG sub-Saharan Africa	FAO sub-Saharan Africa	FAO Near East and North Africa	UN Middle East (ME) and North Africa NA (ESCWA)	Oil Producing Exporting Countries (OPEC)	Organisation for Economic Cooperation and Development (OECD)	Commonwealth of Countries	European Union (EU)	Commonwealth of Independent States (CIS)	The League of Arab States (LAS)	MDG Developing a	MDG Developed a	WB Developing Economies c	WB high income Developed Economies c	Human Development Report (HDR) 2009	International Monetary Fund (IMF)	MDG Least Developed Countries (LDC)	MDG Landlocked developing countries (LLDC)	MDG Small island developing States (SIDS)	Commonwealth of Independent	MDG Transition countries of Southern Europe	UN Transition countries of South Eastern Europe b	FAO Industrialised Countries
	Africa (Afr)																											
	Eastern Africa (EA)																											
BDI	Burundi	SSA	SSA	AFRO	x	x	x									x		Low			Emerging	x	x					
COM	Comoros	SSA	SSA	AFRO	x	x	x								x	x		Low			Emerging	x		x				
DJI	Djibouti	SSA	MENA	EMRO	x	x									x	x		Low-Middle			Emerging	x						
ERI	Eritrea	SSA	SSA	AFRO	x	x	x									x		Low			Emerging	x						
ETH	Ethiopia	SSA	SSA	AFRO	x	x	x									x		Low			Emerging	x	x					
KEN	Kenya	SSA	SSA	AFRO	x	x	x					x				x		Low			Emerging							
MDG	Madagascar	SSA	SSA	AFRO	x	x	x									x		Low			Emerging	x						
MWI	Malawi	SSA	SSA	AFRO	x	x	x					x				x		Low			Emerging	x	x					
MUS	Mauritius	SSA	SSA	AFRO	x	x	x					x				x		Upper-Middle			Emerging			x				
MYT	Mayotte	SSA	SSA		x	x										x		Upper-Middle										
MOZ	Mozambique	SSA	SSA	AFRO	x	x	x					x				x		Low			Emerging	x						
REU	Réunion	SSA			x	x										x												
RWA	Rwanda	SSA	SSA	AFRO	x	x	x					x				x		Low			Emerging	x	x					
SYC	Seychelles	SSA	SSA	AFRO	x	x	x					x				x		Upper-Middle			Emerging			x				
SOM	Somalia	SSA	SSA	EMRO	x	x	x								x	x		Low				x						
UGA	Uganda	SSA	SSA	AFRO	x	x	x					x				x		Low			Emerging	x	x					
TZA	Tanzania, United Republic of	SSA	SSA	AFRO	x	x	x					x				x		Low			Emerging	x						
ZMB	Zambia	SSA	SSA	AFRO	x	x	x					x				x		Low			Emerging	x	x					
ZWE	Zimbabwe	SSA	SSA	AFRO	x	x	x									x		Low			Emerging		x					
	Middle Africa (MA)																											
AGO	Angola	SSA	SSA	AFRO	x	x	x			x						x		Low-Middle			Emerging	x						
CMR	Cameroon	SSA	SSA	AFRO	x	x	x					x				x		Low-Middle			Emerging							
CAF	Central African Republic	SSA	SSA	AFRO	x	x	x									x		Low			Emerging	x	x					
TCD	Chad	SSA	SSA	AFRO	x	x	x									x		Low			Emerging	x	x					
COG	Congo	SSA	SSA	AFRO	x	x	x									x		Low-Middle			Emerging							
COD	Democratic Republic of the Congo	SSA	SSA	AFRO	x	x										x		Low			Emerging	x						
GNQ	Equatorial Guinea	SSA	SSA	AFRO	x	x										x				High	Emerging	x						
GAB	Gabon	SSA	SSA	AFRO	x	x	x									x		Upper-Middle			Emerging							
STP	Sao Tome and Principe	SSA	SSA	AFRO	x	x	x									x		Low-Middle			Emerging	x		x				
	Northern Africa (NA)																											
DZA	Algeria	NA	MENA	AFRO				x	NA	x					x	x		Upper-Middle			Emerging							
EGY	Egypt	NA	MENA	EMRO				x	NA						x	x		Low-Middle			Emerging							
LBY	Libyan Arab Jamahiriya	NA	MENA	EMRO				x	NA	x					x	x		Upper-Middle			Emerging							
MAR	Morocco	NA	MENA	EMRO				x	NA						x	x		Low-Middle			Emerging							
SDN	Sudan	SSA	SSA	EMRO	x	x	x		NA						x	x		Low-Middle			Emerging	x						
TUN	Tunisia	NA	MENA	EMRO				x	NA						x	x		Low-Middle			Emerging							
ESH	Western Sahara	NA					x										x												
	Southern Africa (SA)																											
BWA	Botswana	SSA	SSA	AFRO	x	x	x					x				x		Upper-Middle			Emerging		x					
LSO	Lesotho	SSA	SSA	AFRO	x	x	x					x				x		Low-Middle			Emerging	x	x					
NAM	Namibia	SSA	SSA	AFRO	x	x						x				x		Upper-Middle			Emerging							
ZAF	South Africa	SSA	SSA	AFRO	x	x						x				x		Upper-Middle			Emerging							x
SWZ	Swaziland	SSA	SSA	AFRO	x	x	x					x				x		Low-Middle			Emerging		x					
	Western Africa (WA)																											
BEN	Benin	SSA	SSA	AFRO	x	x	x									x		Low			Emerging	x						
BFA	Burkina Faso	SSA	SSA	AFRO	x	x	x									x		Low			Emerging	x	x					
CPV	Cape Verde	SSA	SSA	AFRO	x	x	x									x		Low-Middle			Emerging	x		x				
CIV	Cote d'Ivoire	SSA	SSA	AFRO	x	x	x									x		Low-Middle			Emerging							
GMB	Gambia	SSA	SSA	AFRO	x	x	x					x				x		Low			Emerging	x						

FIGURE A.1 Country groupings: Geological, economic and developmental.

References

Arab League (2010). Website of the League of Arab States. Website of the League of Arab States. http://www.al-bab.com/arab/docs/league.htm (accessed 15 May 2010).

CIS (2010). Website of the Commonwealth of Independent States (Cis) Commonwealth of Independent States (CIS) http://www.cisstat.com/eng/ (accessed 10 May 2011).

CoN (2010). Website of the the the Commonwealth of Nations. The Commonwealth of Nations. http://www.commonwealth-of-nations.org/Member-Countries-Brief-Intro,38,30,1 (accessed 14 August 2010).

Cyberschoolbus (2010). UN Global Teaching and Learning Project. UN Global Teaching and Learning Project. http://cyberschoolbus.un.org (accessed 12 January 2011).

Europa (2010). Website of the European Union. *The EU at a Glance*. Europa. http://europa.eu/index_en.htm (accessed 21 June 2010).

FAO (2010). *Food Security Statistics: Country Groupings*. Rome: Food and Agriculture Organization.

HDR (2009). Human Development Report 2009 Overcoming Barriers: Human Mobility and Development. New York: United Nations Development Programme (UNDP).

IMF (2010). International Montetary Fund Website. International Monetary Fund. http://www.imf.org/external/ (accessed 4 January 2010).

MDG (2010). Millennium Development Goals Indicators. United Nations Development Group. http://unstats.un.org/unsd/mdg/ (accessed 12 July 2011).

OECD (2010). About the Organisation for Economic Development. Organisation For Economic Development. http://www.oecd.org/home/0,3305,en_2649_201185_1_1_1_1_1,00.html (accessed 30 March 2010).

OPEC (2010). The Organization of the Petroleum Exporting Countries. The Organization of the Petroleum Exporting Countries (OPEC) http://www.opec.org/opec_web/ (accessed 15 May 2010).

UN (2010). *United Nations Statistical Database*, United Nations.

UNPAN (ND). List of Country Groupings and Sub-Groupings for the Analytical Studies of the United Nations World Economic Survey and Other UN Reports. New York: United Nations Public Administration Network (UNPAN).

WHO (2010). Website of the World Health Organization. http://www.who.int/ (accessed 2 April 2010).

World Bank (2010). Website of the Worldbank. http://www.worldbank.org/ (accessed 13 July 2010).

Appendix B: Mortality Categories

Table B.1 contains the Full World Health Organization's breakdown of the Global Burden of Disease as of 2008 (2004 update).

TABLE B.1

WHO Mortality Category and Figures

Cause	World	
Population (000)	6,436,826	
	(000)	**% total**
TOTAL Deaths	58,772	100.0
I. Communicable diseases, maternal and perinatal conditions and nutritional deficiencies	*17,971*	*30.6*
Infectious and parasitic diseases	**9519**	**16.2**
Tuberculosis	1464	2.5
STDs excluding HIV	128	0.2
Syphilis	99	0.2
Chlamydia	9	0.0
Gonorrhoea	1	0.0
HIV/AIDS	2040	3.5
Diarrhoeal diseases	2163	3.7
Childhood diseases	847	1.4
Pertussis	254	0.4
Poliomyelitis (c)	1	0.0
Diphtheria	5	0.0
Measles	424	0.7
Tetanus	163	0.3
Meningitis	340	0.6
Hepatitis B (d)	105	0.2
Hepatitis C (d)	54	0.1
Malaria	889	1.5
Tropical diseases	152	0.3
Trypanosomiasis	52	0.1
Chagas disease	11	0.0
Schistosomiasis	41	0.1
Leishmaniasis	47	0.1
Lymphatic filariasis	0	0.0
Onchocerciasis	0	0.0
Leprosy	5	0.0
Dengue	18	0.0
Japanese encephalitis	11	0.0
Trachoma	0	0.0
Intestinal nematode infections	6	0.0

continued

TABLE B.1 (Continued)

WHO Mortality Category and Figures

Cause	World	
Ascariasis	2	0.0
Trichuriasis	2	0.0
Hookworm disease	0	0.0
Respiratory infections	**4259**	**7.2**
Lower respiratory infections	4177	7.1
Upper respiratory infections	77	0.1
Otitis media	5	0.0
Maternal conditions	**527**	**0.9**
Maternal haemorrhage	140	0.2
Maternal sepsis	62	0.1
Hypertensive disorders	62	0.1
Obstructed labour	34	0.1
Abortion	68	0.1
Perinatal conditions (e)	**3180**	**5.4**
Prematurity and low birth weight	1179	2.0
Birth asphyxia and birth trauma	857	1.5
Neonatal infections and other conditions (f)	1144	1.9
Nutritional deficiencies	**487**	**0.8**
Protein-energy malnutrition	251	0.4
Iodine deficiency	5	0.0
Vitamin A deficiency	17	0.0
Iron-deficiency anaemia	153	0.3
II. Non-communicable conditions	*35,017*	*59.6*
Malignant neoplasms	**7424**	**12.6**
Mouth and oropharynx cancers	335	0.6
Oesophagus cancer	508	0.9
Stomach cancer	803	1.4
Colon/rectum cancer	639	1.1
Liver cancer	610	1.0
Pancreas cancer	265	0.5
Trachea/bronchus/lung cancers	1323	2.3
Melanoma and other skin cancers	68	0.1
Breast cancer	519	0.9
Cervix uteri cancer	268	0.5
Corpus uteri cancer	55	0.1
Ovary cancer	144	0.2
Prostate cancer	308	0.5
Bladder cancer	187	0.3
Lymphomas, multiple myeloma	332	0.6
Leukaemia	277	0.5
Other neoplasms	**163**	**0.3**
Diabetes mellitus	**1141**	**1.9**
Nutritional/endocrine disorders	**303**	**0.5**
Neuropsychiatric disorders	**1263**	**2.1**

TABLE B.1 (Continued)

WHO Mortality Category and Figures

Cause	World	
Unipolar depressive disorders	15	0.0
Bipolar affective disorder	1	0.0
Schizophrenia	30	0.1
Epilepsy	142	0.2
Alcohol use disorders	88	0.1
Alzheimer and other dementias	492	0.8
Parkinson disease	110	0.2
Multiple sclerosis	17	0.0
Drug use disorders	91	0.2
Post-traumatic stress disorder	0	0.0
Obsessive-compulsive disorder	0	0.0
Panic disorder	0	0.0
Insomnia (primary)	0	0.0
Migraine	0	0.0
Sense organ disorders	**4**	**0.0**
Glaucoma	0	0.0
Cataracts	0	0.0
Refractive errors	0	0.0
Hearing loss, adult onset	0	0.0
Macular degeneration and other (g)	4	0.0
Cardiovascular diseases	**17,073**	**29.0**
Rheumatic heart disease	298	0.5
Hypertensive heart disease	987	1.7
Ischaemic heart disease	7198	12.2
Cerebrovascular disease	5712	9.7
Inflammatory heart disease (h)	440	0.7
Respiratory diseases	**4036**	**6.9**
Chronic obstructive pulmonary disease	3025	5.1
Asthma	287	0.5
Digestive diseases	**2047**	**3.5**
Peptic ulcer disease	270	0.5
Cirrhosis of the liver	772	1.3
Appendicitis	22	0.0
Diseases of the genitourinary system	**928**	**1.6**
Nephritis/nephrosis	739	1.3
Benign prostatic hypertrophy	39	0.1
Skin diseases	**68**	**0.1**
Musculoskeletal diseases	**127**	**0.2**
Rheumatoid arthritis	26	0.0
Osteoarthritis	7	0.0
Congenital abnormalities	**440**	**0.7**
Oral diseases	**3**	**0.0**

continued

TABLE B.1 (Continued)

WHO Mortality Category and Figures

Cause	World	
Dental caries	0	0.0
Periodontal disease	0	0.0
Edentulism	0	0.0
III. Injuries	*5784*	*9.8*
Unintentional injuries	3906	6.6
Road traffic accidents	1275	2.2
Poisoning	346	0.6
Falls	424	0.7
Fires	310	0.5
Drowning	388	0.7
Other unintentional injuries	1163	2.0
Intentional injuries	**1642**	**2.8**
Self-inflicted	844	1.4
Violence	600	1.0
War and civil conflict	184	0.3

Source: Reprinted from the WHO Global Burden of Disease: 2004 Update. World Health Organization, Geneva, 2008.

Appendix C: The Gibson Framework of Food Security

The Gibson model of food security was proposed as part of my PhD in answer to challenges of disseminating the food security concept. Unlike many other conceptual frameworks, the three models (Figure C.1a–c) are progressive and as a consequence do not suffer from oversimplification, assume linkages or presume esoteric fore-knowledge of the concept. They start with the simple (Figure C.1a) and increase in complexity with each model (Figure C.1b and c).

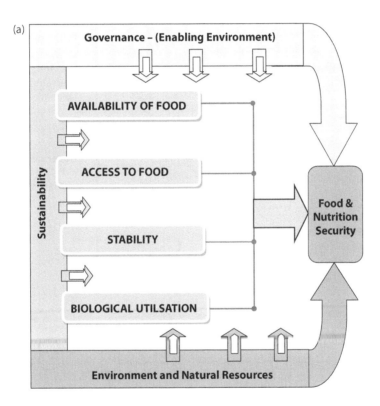

FIGURE C.1a Three-part conceptual framework of food security. (From Gibson, M., *The Use of Conceptual Frameworks in Understanding Food Security*, Manchester Metropolitan University, Manchester, pp. 285, 2011.) Part a—The first model is not too dissimilar from the early FAO four pillars framework. However unlike the FAO, this model's four pillars are bounded by three key aspects; the environment, sustainability and governance. These all explicitly collude to provide a conducive enabling environment of food security.

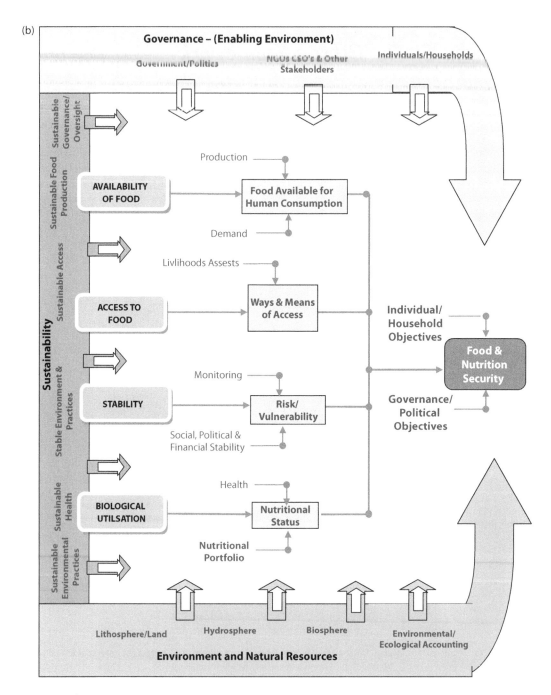

FIGURE C.1b The second model expands on the first by introducing some of the underlying variables of the enabling environment. Furthermore, the four pillars themselves are also expanded to accommodate more detailed analysis. Note too that the end outcome of food and nutrition security is influenced at both the individual micro and the national macro level. This is one of the key findings of this research; that is food security is the responsibility of both the individual and the collective.

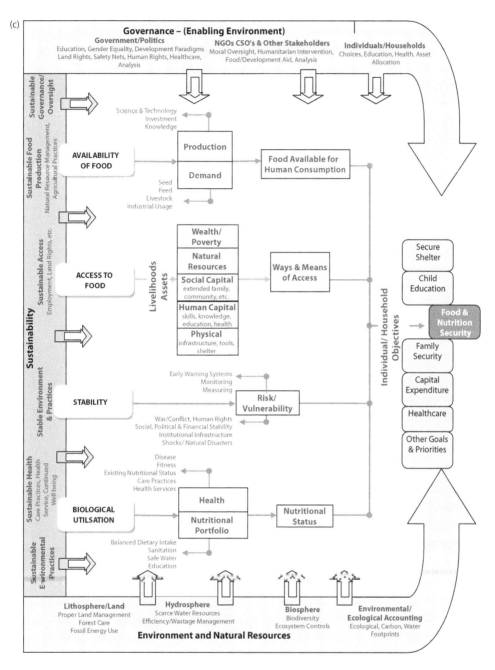

FIGURE C.1c In the full and final model the many variables are given a level of detail that is designed to engender a deeper and more holistic sense of understanding of the concept. Of note are two aspects in particular; that of the introduction of the livelihoods assets construct; and the idea that food security is not the only household or individual objective. Previous to this model both these items tended to be viewed in isolation of the bigger picture. By contextualising them in this way it is hoped a more rounded food security concept emerges.

Reference

Gibson, M. (2011). *The Use of Conceptual Frameworks in Understanding Food Security*. Manchester: Manchester Metropolitan University.

Appendix D: Metabolism

All animal cells require energy and building blocks. We obtain all that we need from the food we eat which contains all six of the major nutrient classes: carbohydrates, proteins, fats, vitamins, minerals and water. Between them, the nutrients not only provide the raw materials needed for building cells but they are also converted into the energy required for the body's processes through the chemical reactions called metabolism, sometimes referred to as cellular respiration. To understand this process, we need to take a quick dip into molecular biology and organic chemistry.

Simplifying the process, metabolism is concerned with the breaking down of complex molecules, a process known as catabolism. In this state the broken-down molecules are either used for energy or for building cells and other components, a process known as anabolism. These are all transported via the blood stream, which is produced in the bone marrow and is made up of two components: cells and plasma. The cellular portion is made up of three parts: red blood cells (RBCs) (also called enthrocytes), which transport oxygen to the cells and take away CO_2; white blood cells (WBCs), which help to fight infection; and platelets, which the body uses for clotting. The plasma is the liquid portion of the blood, and apart from being a vehicle to carry the cells, the plasma also transports dissolved electrolytes, nutrients, vitamins, hormones, clotting factors and antibody proteins such as albumin and immunoglobulins that helps to fight infection.

Unlike plants which convert sunlight into energy via photosynthesis (autotrophs), animal or human metabolism relies on organic molecules or compounds as their starting blocks (heterotrophs). Such metabolic reactions then are the chemical reactions in living organisms that are responsible for converting the chemical compounds (food) into usable energy or components. Primarily foods are broken down into available food molecules (nutrients) in the digestive tract. First, salivary glands secrete amylase that helps digest starch, then in the stomach, hydrochloric acid and pepsins help break down the structure of proteins into peptides. This concoction of food, saliva and gastric secretions collectively known as chyme then moves to the small intestine. From here, the pancreas produces proteolytic lipases and a host of other digestive enzymes and together with the liver's bile salts the chyme can be further broken down for easier ingestion. The resultant glucose, simple carbohydrates and other nutrients are then transported across the intestinal wall and absorbed into the blood. From here, the liver parenchymal cells and other tissues help to finally complete the initial catabolic process of breaking down carbohydrates into simple sugars, fats into fatty acids and proteins into amino acids. These are then made available to the cells (King 2010). The different molecules are further changed or broken down into numerous other chemicals in the cell depending on the body's needs, referred to as their metabolic pathway. Of the numerous pathways, there are a handful of important metabolic paths in human nutrition; these are shown in Table D.1.

D.1 Carbohydrate Metabolism

Of particular interest is carbohydrate metabolism, which uses several of the pathways described in Table D.1. Carbohydrate metabolism is important for two reasons: first, glucose ($C_6H_{12}O_6$), a monosaccharide or simple sugar (broken down from carbohydrate), occurs widely in most plant and animal tissues and as mentioned earlier is the major source of energy in the body; and second because carbohydrates are the main source of food for many of the world's poor often providing as much as 80% of their total diet (WFP 2000).

The digestive system treats all carbohydrates in essentially the same way. It attempts to break them down into their single sugar molecules and where possible it converts most digestible carbohydrates into glucose (known as blood sugar); the cell's universal energy source (HSPH 2010). In the carbohydrate pathway, the glucose is absorbed into the cell from the blood. From here it enters the cytosol, the aqueous liquid inside the cell where through the initial process of glycolysis (the oxidation of glucose and

TABLE D.1

Major Metabolic Pathways

Glycolic or glycolysis: glucose oxidation that produces ATP.
Krebs' or the citric acid cycle: acetyl CoA oxidation to obtain GTP and other metabolites.
Oxidative phosphorylation: disposal of the electrons released by glycolysis and citric acid cycle. Much of the energy released in this process can be stored as ATP.
Pentose phosphate pathway: synthesis of pentoses and release of the reducing power needed for anabolic reactions.
Urea cycle: disposal of NH_4^+ in less toxic forms.
Fatty acid b-oxidation: fatty acids breakdown into acetyl-CoA, to be used by the Krebs' cycle.
Gluconeogenesis: glucose synthesis from smaller precursors, to be used by the brain.

Source: Silva, P., A General Overview of the Major Metabolic Pathways. http://www2.ufp.pt/~pedros/bq/integration.htm, 2002.

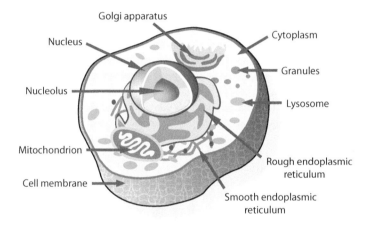

FIGURE D.1   Structures of the cell. (Courtesy of Patrick Newsham, © 2011.)

the first step in all cellular respiration) converts the glucose molecule to pyruvic acid (or pyruvate). Two options at this point depending on the cells needs and resources decides whether this pyruvic acid is then aerobically or anaerobically treated. Anaerobic glycolysis is vital for tissue or skeletal muscles that have high energy requirements or where there is insufficient oxygen supply in the cell to provide aerobic activity, or at times when there is an absence of oxidative enzymes. It is produced by lactic acid fermentation and is only effective during short periods of intense exercise or energy needs (Hunter 2008). In aerobic conversion however, pyruvic acid, after undergoing further metabolic processing, is transformed into acetyl-CoA before it enters the citric acid cycle (also known as the Kreb's cycle). Meanwhile oxygen from the lungs is diffused into the blood. This oxygenated blood is then fed to the cells, to the mitochondrion in particular (see Figure D.1). In the citric acid cycle, the oxygen in turn reacts with the acetyl-CoA, oxidises it and creates a small amount of energy, which is stored in nicotinamide adenine dinucleotide (NADH). Both hydrogen and waste carbon dioxide are also produced in this reaction (Hunter 2008). The carbon dioxide is transported back via the blood to the lungs where it is expired while the hydrogen continues on in other reactions to form adenosine triphosphate (ATP) molecules and water. The ATP molecule acts like a mini battery and subsequently allows one of two things to happen: molecular growth or energy release. If the cell requires it, the ATP molecule can recombine simple molecules into complex ones for growth and repair; this process is known as anabolism or biosynthesis (also called biogenesis) and this is how body tissue and cells grow. If on the other hand, energy is required for movement or the metabolic process itself, the ATP molecule is degraded back to adenosine diphosphate (ADP) and energy is released. When the organism is at rest and energy is not immediately required, the reverse reaction takes place, thus storing energy like a battery.

Any excess intake of carbohydrate (glucose) is stored as glycogen (a complex carbohydrate stored in the liver and skeletal muscle) or as fat in the form of triglycerides (in adipose tissue) providing a storage of energy for later use. If needed this glycogen then becomes is the first energy reserve to be tapped, then if further energy is required, the body's store of triglycerides (fat) is then raided (Albright and Stern 1998; Silva 2002; UoArizona 2004; Britannica 2009; King 2010; UoAkron 2010; UoBristol 2010).

D.2 Lipid (Fat) Metabolism

As with carbohydrates, the body needs lipids or fats for energy and growth. These are composed of fatty acids and glycerol and any intake in excess of needs as with carbohydrates is stored in the body as triglycerols (or triglycerides) for later use. In response to energy demands, ingested fatty acids or stored fat can be mobilised. Initially, triglycerols are catabolised in an enzymic process of lipolysis into their component parts, free fatty acid and glycerol. The free fatty acid and glycerol are treated separately. Free fatty acid, as with the process of carbohydrate metabolism, is oxidised in the mitochondrion organelles of liver cells in a process called β-oxidation. This leads to the formation of NADH and the energy battery, ATP. The glycerol on the other hand is converted to glucose in a process of gluconeogenesis and becomes available to the citric acid cycle and subsequently it too is metabolised into NADH and ATP (Albright and Stern 1998; Merck Manual 2005; King 2010).

D.3 Protein Metabolism

For use in the body, protein is catabolised into its amino acids. While amino acids usually act as substrates or inputs for the building of the body's own proteins, if needed, amino acids are converted to carbohydrates (glucose) by gluconeogenesis primarily in liver cells. These are then converted into pyruvate through glycolosis before entering the citric acid or Kreb's cycle for oxidation and ultimately the production of ATP. If intake is in excess of the body's requirements, proteins too are metabolised to glycogen or triglycerols (fat) and subsequently stored for later use. If however protein is converted for energy use, then less protein becomes available for growth, cell regeneration and other needs. In this regard, protein is especially important as far as children are concerned as they need extra protein for growth. Consequently, insufficient food intake or an imbalanced intake of the macro-nutrients means that important growth proteins are diverted for energy use (Latham 1997).

D.4 Energy Preferences

The body's first need is for energy. If needed, all macronutrient substances, protein, fat and carbohydrates with the exception of water can ultimately be catabolised into usable glucose, glycerol and free fatty acids (fats) for use as energy. However, this interchangeability of energy sources is not random; the body has preferences. Carbohydrate is the major source of energy for the body and with its relatively quick and easy conversion into glucose carbohydrates become the body's fuel of choice. In the absence of extra intake however, once the glucose runs low, the body first taps into its glycogen reserves. Once these are depleted or become dangerously low, the body then looks to the fatty deposits (the adipose cells) after which time, in the case of prolonged abstinence, the body then converts protein into its constituent amino acids and then into glucose for energy.

Adipose tissue is the body's storage system; it contains or can synthesise fat from either the dietary fat we eat or is converted from surplus carbohydrate or protein in our diets. The main reason the body converts all this extra carbohydrate, protein and fat into body fat is the relative energy profile of fat itself. Each type of molecule produces a specific amount of energy. The energy yield from one gram of fatty acid is approximately 9 kcal (37 kJ), compared to just 4 kcal/g (17 kJ/g) for carbohydrates and proteins. Thus because fats are such potent and efficient forms of energy, the body stores any excess carbohydrates, proteins and fats, as fats for use when needed (Wilson 2008).

References

Albright, A. L. and J. S. Stern (1998). Encyclopedia of Sports Medicine and Science. http://sportsci.org (accessed 12 April 2011).

Britannica (2009). Encyclopaedia Britannica Online. 2009.

HSPH (2010). The Nutrition Source. Harvard School of Public Health. http://www.hsph.harvard.edu/ (accessed 5 October 2010).

Hunter, A. D. (2008). *Chemistry 1506: Allied Health Chemistry 2 Section 12: Specific Catabolic Pathways, Molecular Destruction.* Department of Chemistry, Youngstown State University.

King, M. W. (2010). The Medical Biochemistry Page. IU Center for Regenerative Biology and Medicine. http://themedicalbiochemistrypage.org/home.html (accessed 6 November 2010).

Latham, M. C. (1997). Human Nutrition in the Developing World. *Food and Nutrition Series No. 29.* Rome: Food and Agriculture Organization.

Merck Manual (2005). *Inherited Disorders of Metabolism: Fatty Acid and Glycerol Metabolism Disorders.* Whitehouse Station, NJ: Merck Sharp & Dohme Corporation.

Silva, P. (2002). A General Overview of the Major Metabolic Pathways. University Fernando Pessoa. http://www2.ufp.pt/~pedros/bq/integration.htm (accessed 21 November 2010).

UoAkron (2010). *Department of Chemistry: Carbohydrate Metabolism.* Hardy Research Group, The University of Akron.

UoArizona (2004). *University of Arizona: The Biology Project.* The University of Arizona.

UoBristol (2010). *Acetyl Coenzyme A: The Molecule That Makes Fats, or Burns Them by Paul May.* University of Bristol.

WFP (2000). *Food and Nutrition Handbook.* Rome: World Food Programme (WFP).

Wilson, M. M. G. (2008). Disorders of Nutrition and Metabolism.In *The Merck Manual Online Medical Library.* R. S. Porter and J. L. Kaplan, eds. Whitehouse Station, NJ: Merck Sharp & Dohme Corporation.

Appendix E: Micronutrients

A closer look at the various vitamins and minerals will aid in the understanding of their different roles as well as giving us an appreciation of their relative importance.

E.1 Vitamins

E.1.1 Vitamin A

Vitamin A covers the group of mostly fat soluble vitamins including (1) retinoids, the most important and most useable of which retinal comes from animals and (2) pro-vitamin A (or vitamin A precursors), which can easily be converted into vitamin A. These are the carotenoids*; they are of plant origin and the most common is beta-carotene (Table E.1).

E.1.2 Vitamin B Complex

B-complex vitamins are grouped together and although chemically different they are loosely similar in many respects. Their physiological functions, predominantly coenzymic, involve working as catalysts alongside enzymes in metabolism where they are also useful cofactors in the body's chemical conversions.

TABLE E.1

Vitamin A: Function, Characteristics and Sources

Vitamin	Chemical Names	Function	Sources
Vitamin A	Retinol: various other retinoids (retinal, retinoic acid, retinyl esters), and some carotenoids (alpha-, beta- and gamma-carotenes plus beta-cryptoxanthin).	One of the most versatile vitamins. Vitamin A is essential for the visual cycle—the neural transmission of light to vision. It plays an important role in the body's immune system as well as in the reproductive system, particularly to the developing foetus. Vitamin A is also of importance to the vitality and integrity of healthy epithelial cells (skin, membrane and soft tissue) and lastly to the general growth and development of the body.	Vitamin A comes from both animal and vegetable sources. In animals, retinol is available in meat, eggs, milk, cheese, cream, liver, kidney, cod and halibut fish oil. In vegetables, vitamin A is available through carotenoids. These are pigments found in fruits and vegetables. The most commonly consumed form of pro-vitamin A is beta-carotene. This is both the yellow orangey pigment found in orange, cantaloupe, pink grapefruit, apricots, carrots, pumpkin, sweet potatoes, winter squashes as well as the dark green pigment found in broccoli, spinach and most dark green leafy vegetables. As a rule, the brighter the pigment the more vitamin A. These sources of beta-carotene include one extra advantage in that they are free of fat and cholesterol.

Sources: WHO/FAO, *Vitamin and Mineral Requirements in Human Nutrition (Draft Version)*, WHO/FAO, 2004; Britannica, Encyclopædia Britannica Online, 2009; FSA, Healthy Diet. UK Food Standards Agency. http://www.eatwell.gov.uk/healthydiet/, 2010; MedlinePlus, Medline/PubMed Database, US National Library of Medicine/National Institute of Health, 2010.

---

* Carotenoids are organic pigments used by plants, algae, and photosynthetic bacteria in the absorption of light energy in the photosynthetic process and to protect against photo damage. Carotenoids can be generally classified into carotenes (alpha-carotene, beta-carotene, and lycopene) and xanthophylls (beta-cryptoxanthin, lutein, and zeaxanthin). Four carotenoids—beta-carotene, alpha-carotene, gamma-carotene, and beta-cryptoxanthin—are particularly useful in the visual cycle of humans with beta-carotene being the most common. They are also act as antioxidants (LPI 2010) in this page as usual.

TABLE E.2

Vitamin B: Function, Characteristics and Sources

Vitamin	Chemical Names	Functions	Sources
Vitamin B$_1$	Thiamin	Thiamine, a coenzyme, aids in the body cells metabolisation of carbohydrates into energy. It is also essential for the proper functioning of the heart, muscle, and nervous system, and lastly it helps with the production of hydrochloric acid for use in the digestive process.	Thiamine is commonly found in all the main food groups. Sources include lean meat (especially pork), milk, eggs, cheese, fish, wholegrain breads, dried beans, peas, and soybeans and lastly, fruits and vegetables. While generally do not contain very high quantities of thiamine, consumed in large amounts they soon become an important source.
Vitamin B$_2$	Riboflavin	Riboflavin works alongside other B vitamins as a coenzyme in the metabolisation of carbohydrates. It also helps keep the nervous system as well as skin, eyes and the mucous membrane healthy. Lastly it also helps in the production of steroids and red blood cells.	Riboflavin is found in small amounts in many foods such as lean meats, eggs, legumes, nuts, green leafy vegetables, dairy products and milk.
Vitamin B$_3$	Niacin (nicotinic acid and nicotinamide)	As with many of the B complex biochemical interrelationships, Niacin (another coenzyme) aids in the process of energy conversion. It is also important for the well-being of the digestive system, skin and nervous system.	Niacin is found in dairy products, lean meats—beef, pork and chicken, fish, nuts and legumes. Niacin is also present in cereals—wheat and maize flour.
Vitamin B$_5$	Pantothenic acid	Pantothenic acid helps break down proteins, carbohydrates and particularly fatty acids in fats.	Pantothenic acid is found in very releasable forms in nearly all lean meats, kidneys, eggs, milk and milk products, fish, legumes, yeast, in cabbage family vegetables, white and sweet potatoes as well as whole-grain cereals.
Vitamin B$_6$	Pyridoxine, Pyridoxamine and Pyridoxal	Vitamin B$_6$ has a coenzyme function in and metabolism protein amino acids. It also helps the immune system produce antibodies and maintain normal nerve function as well as form red blood cells.	A good source of vitamin B$_6$ can be found in pork, chicken, turkey, cod, soya beans, nuts, legumes, eggs, milk potatoes, whole grains such as oatmeal, wheatgerm and rice.
Vitamin B$_7$	Biotin	Biotin (another coenzyme) is needed to aid the metabolisation of fatty acids and carbohydrates and proteins in the production of energy.	Biotin is found in many foods. Good sources include meats, kidney, liver eggs, salmon, sardines and some fruit and vegetables. It can also be found in brewer's and nutritional yeasts.
Vitamin B$_9$	Folic acid (folate)	Folic acid is the human-made form of folate (B$_9$'s natural form). Folic acid works with B$_{12}$ to form healthy red blood cells. It also acts to reduce neural tube defects such as spina bifida in unborn babies. It also acts with B$_{12}$ and vitamin C in the metabolisation and creation of protein molecules. Lastly folic acid helps tissue growth and cell activity.	Folate is found in small amounts in many foods. Sources include liver, poultry, pork, shellfish, chickpeas, beans, legumes, citrus fruits, wheat bran and other whole grains, dark green leafy vegetables as well as asparagus and peas.
Vitamin B$_{12}$	Cobalamin (cyanocobalamin, hydroxycobalamin, methylcobalamin)	Vitamin B$_{12}$ can be stored in the liver for years. It is important for metabolism, and to process folic acid. It also aids in the formation of red blood cells and keeps the central nervous system in good health.	Vitamin B$_{12}$ is found in some algae (seaweed), meat, poultry, salmon, cod, shellfish, milk, cheese, eggs and yeast extract.

Sources: WHO/FAO, *Vitamin and Mineral Requirements in Human Nutrition (Draft Version).* WHO/FAO, 2004; FSA, Healthy Diet. UK Food Standards Agency. http://www.eatwell.gov.uk/healthydiet/, 2010; MedlinePlus, Medline/PubMed Database, US National Library of medicine/ National Institute of Health, 2010.

TABLE E.3

Choline: Function, Characteristics and Sources

Name	Function	Sources
Choline	It has many functions: aiding in the synthesis of cell structure; helping to form the precursors of cell signalling; nerve impulse transmission as well as aiding the brain to produce neurotransmitters. Choline, a fatty acid also stops fat accumulating in the liver and also helps to remove toxins and wastes and slows the build-up of cholesterol.	Most choline in foods is in the form of phosphatidylcholine. Phosphatidylcholine in turn is a major constituent of lecithin; therefore foods high in lecithin are good sources of choline, these include: meat, liver, milk, eggs yolks, peanuts, soybeans, oatmeal, cabbage and cauliflower.

Sources: WHO/FAO, *Vitamin and Mineral Requirements in Human Nutrition (Draft Version).* WHO/FAO, 2004; FSA, Healthy Diet, UK Food Standards Agency. http://www.eatwell.gov.uk/healthydiet/2010; LPI, Micronutrient Information Center, Linus Pauling Institute, Oregon State University, 2010; MedlinePlus, Medline/PubMed Database, US National Library of medicine/National Institute of Health, 2010.

In this respect the B vitamins frequently overlap. They are also water soluble and often coexist in the same foods. Choline, though strictly not a vitamin, is often grouped alongside the B-complex vitamins and will be treated as such in this book (LPI 2010) (Table E.2).

Choline is not strictly a vitamin, but an essential nutrient. Although the body can synthesise choline in small amounts for proper maintenance of health, it must be consumed in the diet to maintain adequate levels (Table E.3).

E.1.3 Vitamin C

Vitamin C is a water-soluble vitamin that is needed for normal healthy growth and development. It is also easily destroyed. While some mammals and birds are able to synthesise vitamin C, humans are unable to do so, and as a result it has to be regularly supplied in the diet. It is also an electron donor and an important protective antioxidant[*] (Table E.4).

TABLE E.4

Vitamin C: Function, Characteristics and Sources

Vitamin	Chemical	Functions	Sources
Vitamin C	Ascorbic acid, ascorbate	Vitamin C is required for the growth and repair of tissues in all areas of the body. It is also essential in the formation of collagen (a protein in skin, scar tissue, tendons, ligaments and blood vessels). Vitamin C is used too in the healing of wounds (tissue repair) and the maintenance of bones, teeth and cartilage.	Sources of vitamin C include all fruits and vegetables contain some vitamin C. Some of the richer available are citrus fruits, kiwi, cantaloupe, strawberries, papaya, mango, watermelon, raspberries, blueberries, cranberries, pineapples and brussels sprouts, cauliflower, cabbage, winter squash, red and green peppers, tomatoes, broccoli, leafy greens and also sweet and white potatoes.

Sources: WHO/FAO, *Vitamin and Mineral Requirements in Human Nutrition (Draft Version)*, WHO/FAO, 2004; FSA, Healthy Diet. UK Food Standards Agency. http://www.eatwell.gov.uk/healthydiet/, 2010; MedlinePlus, Medline/PubMed Database, US National Library of medicine/ National Institute of Health, 2010.

[*] Antioxidants: Vitamin C is one of many that include vitamin E and beta-carotene. Antioxidants are nutrients that essentially reduce or block some of the damage caused by free radicals. Free radicals (missing one electron) are natural by-products of metabolisation, immune system (cells sometimes create them to neutralise threats like viruses and bacteria) and from environmental factors such as pollution, radiation, cigarettes and herbicides. Free radicals by their nature try to stabilise themselves by stealing an electron from other molecules (oxidisation), while the first radical is stabilised; this in turn creates another unstable free radical. This chain reaction changes chemical structures and can ultimately hinder or destroy cells. This process naturally contributes to the aging process but it can also promote the development of health conditions such as cancer, heart disease and several inflammatory conditions like arthritis. Antioxidants help reduce this damage.

TABLE E.5

Vitamin D: Function, Characteristics and Sources

Vitamin	Chemical Names	Functions	Sources
Vitamin D	Ergocalciferol, cholecalciferol	Vitamin D is needed to regulate and maintain normal balanced levels of calcium and phosphate in the blood. These in turn are needed for mineralisation of bone, nerve conduction, muscle contraction, skin and generally good cellular functioning of the body. On top of this, vitamin D modulates cell cycle proteins, which control cell proliferation and increase cell differentiation in certain specialised cells of the body.	Vitamin D is found in a small number of foods. Good sources include liver, oily fish, oysters, eggs, cheese, butter and cream.

Sources: WHO/FAO, *Vitamin and Mineral Requirements in Human Nutrition (Draft Version)*, WHO/FAO, 2004; FSA, Healthy Diet. UK Food Standards Agency. http://www.eatwell.gov.uk/healthydiet/, 2010; MedlinePlus, Medline/ PubMed Database, US National Library of medicine/ National Institute of Health, 2010.

E.1.4 Vitamin D

Vitamin D is important in many respects, and most of us obtain most of our vitamin D requirements directly from sunlight on the skin. It is fat-soluble and forms under the skin in reaction to sunlight, in particular in higher concentrations of ultraviolet light; for this reason the summer sun is the best source of vitamin D. However, there are good sources of dietary vitamin D too (Table E.5).

E.1.5 Vitamin E

Vitamin E is the major fat-soluble antioxidant. It encompasses a family of eight naturally occurring similar forms of vitamin E that are synthesised from homogentisic acid by plants (Table E.6).

E.1.6 Vitamin K

Vitamin K refers to a group of fat-soluble proteins synthesised by plants and the bacteria that line the gastrointestinal tract. Vitamin K is often referred to as the blood-clotting vitamin (Table E.7).

TABLE E.6

Vitamin E: Function, Characteristics and Sources

Vitamin	Chemical Names	Functions	Sources
Vitamin E	Family of eight, including four tocopherols (alpha, beta, gamma, delta) and four tocotrienols (alpha, beta, gamma, delta)	Vitamin E is important in the formation of red blood cells. It also helps the body to use vitamin K. However its primary role is that of an inhibitor of tissue cells oxidative processes. Just as importantly vitamin E is an antioxidant protecting the body from both endogenous (origin within the body) and exogenous (origin outside) free radicals.[a]	Vitamin E is found in animal fats and plants. Good sources include: vegetable oils, nuts and nut oils, margarine, seeds, egg yolk, chickpeas, soya beans, avocados, carrots, parsnips, red peppers, olives, tomatoes, sweet corn, sweet potatoes, watercress, spinach and other green leafy vegetables, parmesan, cheddar, wheat germ and oatmeal.

Sources: WHO/FAO, *Vitamin and Mineral Requirements in Human Nutrition (Draft Version)*, WHO/FAO, 2004; FSA, Healthy Diet. UK Food Standards Agency. http://www.eatwell.gov.uk/healthydiet/, 2010; MedlinePlus, Medline/ PubMed Database, US National Library of medicine/ National Institute of Health, 2010.

[a] Free radicals. See the footnote on Antioxidants.

TABLE E.7

Vitamin K: Function, Characteristics and Sources

Vitamin	Chemical Names	Functions	Sources
Vitamin K	K_1—phylloquinones (plants); K_2—menaquinones (bacteria); 14 in total designated MK_n, i.e. MK_1–MK_{14}; several synthetic vitamins K_3 (menadione), K_4 (menadiol diacetate), and K_5 (2-methyl- 4-amino-1-naphthol)	As mentioned, vitamin E has a haemostatic function. That is, it is essential for the synthesis of certain proteins not only necessary for the clotting of blood (procoagulants) but also for the prevention of clotting (anticoagulant). This coagulation cascade relies on the vitamin K-dependant proteins: factors II (prothrombin), VII, IX, and X and to some extent Z; this anticoagulant process is then regulated also by vitamin K-dependant proteins C and S, particularly in the female menstrual cycle. On balance. the vitamin K group is known as the clotting vitamin.	K_1: Phylloquinone forms the main dietary intake of vitamin K and is abundant in green vegetables, peas, lettuce, asparagus, brussel sprouts, broccoli, spinach, kale, some oils, sunflower, olive, soybean, avocado and kiwi. K_2: menaquinones—dairy products are particularly rich in vitamin K_2 species. Camembert, for example, contains about 40 ng/g of vitamin K_1 but about 600 ng/g of K_2. Vitamin K in general can be found in green leafy vegetables, vegetable oils and cereals while small amounts can be found in meat and dairy products.

Sources: Fain, J.N., *Mol. Pharmacol.*, 7, 465–479, 1971; WHO/FAO, *Vitamin and Mineral Requirements in Human Nutrition (Draft Version)*, WHO/FAO, 2004; Britannica, Encyclopaedia Britannica Online, 2009; FSA, Healthy Diet, UK Food Standards Agency. http://www.eatwell.gov.uk/healthydiet/, 2010; LipidBank, *Lipidbank Database*, University of Tokyo, 2010; LPI, Micronutrient Information Center, Linus Pauling Institute, Oregon State University, 2010; MedlinePlus, Medline/PubMed Database, US National Library of medicine/ National Institute of Health, 2010.

E.2 Minerals

Both minerals and trace elements are examined in Table E.8.

TABLE E.8

Minerals: Their Functions and Sources

Mineral	Function	Sources
Potassium	As an electrolyte and the principal positively charged ion, potassium helps maintain the body's fluid balance. Together with sodium and chloride (both negative ions), the three act as a sort of pump to regulate the cell's membrane electrical 'potential' used for nerve impulse functions. It also acts to regulate the heartbeat, aid in protein synthesis, and the metabolisation process as well as synthesising glycogen from glucose.	Many good sources of potassium include all meats, fish including salmon, cod, flounder and sardines; broccoli, peas, lima beans, tomatoes, potatoes, sweet potatoes squash, citrus fruits, cantaloupe, bananas, kiwi, prunes and dried apricots; milk yogurt and nuts.
Chloride	Chloride is also an electrolyte. In itself, chloride is ionized chlorine, that is, it is chlorine elements that have in this case picked up an electron to form an anion (negatively charged ion). In this form, chloride can join positively charged sodium ions (cations) to form sodium chloride (table salt) or with hydrogen cations to make hydrochloric acid or stomach acid. As an electrolyte alongside sodium, potassium, and calcium, chloride is important in helping to maintain the equilibrium of fluid both inside and outside the cell. Controlled by the kidneys it also acts to keep the blood pH levels in balance.	Most of the chloride in the body comes from table salt or sea salt as sodium chloride. It can also found in many vegetables including: seaweed, tomatoes, lettuce, celery, olives as well as cereals such as rye.
Sodium	Sodium, a soft metal, is an essential element for all animal and some plant species and as such plays a crucial role in a several life-sustaining processes. Working like potassium but with a negative charge, sodium ions (anions) help create the potential difference in cell membranes that allow for electrical impulses to generate muscle and nerve signals. Along with chloride, sodium is an electrolyte that makes up the principal ions in extracellular liquids that include interstitial that and blood plasma.	As with chloride, the most common form of sodium is table salt (sodium chloride), natural foods containing sodium are: seafood, beef, poultry; many vegetables, including celery, beets, carrots, artichokes, kelp and other sea vegetables.
Calcium	Calcium is the most abundant mineral in the body, and a major structural component required for bone (about 99% of intake), teeth and also a small essential amount found in the blood and soft tissue. Formation and absorption is aided by vitamin D. Teeth and other important functions require the remaining 1%, which is also needed for cell signalling and regulates muscle contraction (including the heart). Lastly calcium is a component in the good functioning of the nerve cells and promotes normal blood clotting.	Sources of calcium include dairy products, fish, prawns and sardines with bones, seaweed, olives, celery, carrots, sweet potatoes, cabbage all other greens, artichoke, many beans, spinach, broccoli watercress, parsley, dried figs, apricots, rhubarb blackcurrants, molasses, brewer's yeast, many nuts and some seeds, wheat germ, brown rice, barley and carob flour.
Phosphorous	Phosphorus, an essential mineral, is found in every cell; however, most (85%) is found as a structural component of bone and teeth. The majority of the phosphorus in the body is in the form as phosphate (PO_4), which aids in the metabolisation of carbohydrate and fats as well as the synthesis of protein. It is also instrumental in vitamin B complex functions as well. Phosphorous is active in assisting in nerve and muscle signals and contractions (including the regularity of the heartbeat), the functioning of the kidneys, and binding with haemoglobin to transport oxygen around the blood. Lastly it is also functional in the nucleic acids DNA and RNA.	Good sources are the protein food groups: red meat, fish, poultry, dairy products, bread, rice and oats. Fruits and vegetables, on the other hand, contain only small amounts of phosphorus. Although whole-grains contain ample phosphorous, it is in a form (phytin) not absorbed by the body.

TABLE E.8 (Continued)

Minerals: Their Functions and Sources

Mineral	Function	Sources
Magnesium	Magnesium is importantly involved in numerous essential metabolic reactions. Over 60% of the body's magnesium can be found in the skeleton, about 27% in muscle, the rest is in cells with about 1% existing outside of the cells. Apart from magnesium's structural role in bone, it aids in the production of metabolism of carbohydrates and fats in energy production, cell signalling and cell migration. It also ensures the proper working of the parathyroid glands (produces hormones for bone health).	The richest sources are green leafy vegetables (such as spinach) and nuts. Good sources also include bread, fish, meat and dairy foods.
Manganese	Manganese helps anabolise (create) some of the enzymes in the body. It also has important roles in catabolism, (the breaking down) of carbohydrates, amino acids and cholesterol for energy. In the cells, a number of manganese-activated enzymes act as antioxidants slowing down the oxidative process and the formation of free radicals, particularly in the mitochondrion, which consumes 90% of cell-available oxygen. Manganese is also indirectly involved in the wound healing process as well as promoting healthy cartilage and bone.	These include pineapple, nuts, green vegetables—peas, runner beans, spinach, brown rice, whole grains such as rye, spelt and oats as well as some breads. It is also found in tea.
Iron	There are two types of iron: heme, mainly of animal origin and easily absorbed, and nonheme, found in plants and less easily absorbed into the body; it has a number of important roles. Essential for metabolism, DNA synthesis, growth, healing, reproduction, the immune system as well as being a cofactor in many enzyme reactions making up the multitude of proteins in the body. Iron is particularly important in its role preventing anaemia assisted by Vitamin A. Iron makes the oxygen-carrying proteins haemoglobin (in blood) and myoglobin (in muscles), both transporters of essential oxygen around the body. Anaemia is the low level of red blood cells (or haemoglobin) needed in its oxygenation.	Good iron sources are liver, meat, beans, nuts, dried fruit, whole grains, soybean flour, whole grains, eggs, oysters, salmon, tuna and most dark green leafy vegetables, except spinach. Contrary to popular belief, spinach is not a good source of iron as it contains a substance (oxalic acid) that makes it difficult for the body to absorb spinach's iron.
Zinc	Zinc is present in all body fluids and tissues and is involved in numerous aspects of cellular metabolism. It acts as a stabiliser at the molecular structure of cellular components contributing to the general maintenance of cell integrity. As such zinc plays an important role in growth and development, the immune system response (wound healing), neurological function and reproduction.	Good food sources include meat, shellfish, milk and dairy foods, bread, and cereal products like wheatgerm.

Sources: WHO/FAO, *Vitamin and Mineral Requirements in Human Nutrition (Draft Version)*, WHO/FAO, 2004; FSA, Healthy Diet, UK Food Standards Agency, http://www.eatwell.gov.uk/healthydiet/ 2010; LPI, Micronutrient Information Center, Linus Pauling Institute, Oregon State University, 2010; MedlinePlus, Medline/PubMed Database, US National Library of Medicine/National Institute of Health, 2010; MIT, Optimizing Your Diet, Massachusetts Institute of Technology, 2010.

E.3 Trace Elements

As with micro and macro minerals, trace elements are found in varying classifications and can be just as important in the body's many processes (Latham 1997; MIT 2010). Following the convention of this book, these are considered those elements and chemical compounds needed by the body in amounts less than 1 mg (Table E.9).

TABLE E.9

Trace Elements: Functions and Sources

Mineral	Function	Sources
Copper	As with iron, copper aids in the formation of red blood and white blood cells. It also acts as a trigger in the release of iron to form oxygen-carrying haemoglobin. On top of this, copper has a function in energy production, neuromitter synthesis as well as central nervous system. It also has a role to play in antioxidation, keeping the blood vessels and bones healthy and lastly as a regulator of gene expression.	Good sources of copper: shellfish, particularly oysters, offal (kidneys, liver), beans, potatoes, dark leafy greens, dried fruits, cocoa, black pepper, yeast, nuts and whole grains.
Iodine	Most of the Earth's source of iodine is found in oceans, while what there is in soils tends to vary greatly by region. Dietary iodine is metabolised as iodide ion and boasts 100% bioavailability, that is to say, it is absorbed totally from food and water; dietary iodine is absorbed throughout the gastrointestinal tract; however, the only known physiological role for iodine in humans is in the synthesis of hormones by the thyroid gland.	Iodine source include seafood (particularly cod, sea bass, haddock and perch), kelp, dairy products and plants grown in iodine-rich soil; however fortified salt is the main food source of iodine.
Selenium	Selenium has a variety of functions. It helps make special antioxidants that helps to prevent cell damage. It is also important in the function of the immune system as well as thyroid hormone metabolism. Selenium has also been said to stimulate antibodies and help to boost fertility, especially among men.	A good source of selenium can be obtained from fish, shellfish, red meat, grains, eggs, chicken, liver, garlic, brewer's yeast, wheat germ, brazil nuts.
Chromium	Chromium functions in the metabolism of carbohydrates and fats. It aids in the maintenance of healthy insulin and blood sugar levels. As well as glucose metabolism, chromium also stimulates fatty acid and aids in cholesterol synthesis and levels, which are important for brain and other body processes.	Chromium is everywhere; in the air, water, soil as well as in plants and animals. For dietary intake however, good sources are meat, liver, eggs, oysters, green peppers, apples, bananas, spinach, romaine lettuce, onions, tomatoes, potatoes, whole grains, lentils and brewer's yeast.

Sources: WHO/FAO, *Vitamin and Mineral Requirements in Human Nutrition (Draft Version)*, WHO/FAO, 2004; FSA, Healthy Diet, UK Food Standards Agency, http://www.eatwell.gov.uk/healthydiet/ 2010; LPI, Micronutrient Information Center, Linus Pauling Institute, Oregon State University, 2010; MedlinePlus, Medline/PubMed Database, US National Library of Medicine/National Institute of Health, 2010; MIT, Optimizing Your Diet, Massachusetts Institute of Technology, 2010.

Numerous other elements, particularly trace elements, can be found in the human body besides those discussed above. These include cobalt, molybdenum, silicon, fluoride, vanadium, nickel, arsenic and tin. While these are nutritionally important, from the standpoint of food security and especially from the public health perspective and limitations of this book, there is little or no evidence that deficiencies in these are responsible for major health problems that might affect their food security (Latham 1997).

References

Britannica (2009). Encyclopaedia Britannica Online. 2009.

Fain, J. N. (1971). Effects of Menadione and Vitamin K5 on Glucose Metabolism, Respiration, Lipolysis, Cyclic 3',5'-Adenylic Acid Accumulation, and Adenyl Cyclase in White Fat Cells. *Molecular Pharmacology* 7(4): 465–479.

FSA (2010). Healthy Diet. UK Food Standards Agency. http://www.eatwell.gov.uk/healthydiet/ (accessed 15 August 2011).

Latham, M. C. (1997). Human Nutrition in the Developing World. *Food and Nutrition Series No. 29*. Rome: Food and Agriculture Organisation.

LipidBank (2010). *Lipidbank Database*. University of Tokyo.

LPI (2010). Micronutrient Information Center. Oregon State University: Linus Pauling Institute. http://lpi. oregonstate.edu/ (accessed 12 September 2010).

MedlinePlus (2010). *Medline/PubMed Database*. US National Library of medicine/ National Institute of Health.

MIT (2010). Optimizing Your Diet. Massachusetts Institute of Technology. http://web.mit.edu/athletics/sports-medicine/wcrminerals.html (accessed 12 January 2011).

WHO/FAO (2004). *Vitamin and Mineral Requirements in Human Nutrition (Draft Version)*. WHO/FAO.

Appendix F: Malnutrition—
Its Assessment and Measurement

Before we can properly classify the extent of any form of malnutrition, we must first be in a position to assess or measure it. This is not as easy as might be expected. Measuring nutritional status in individuals is a completely different process to measuring nutritional status of groups or populations. In the field of food security however, both are equally important as measures from the individual form the backbone of much extrapolation in regional and national assessments. In light of this, nutritional status can generally be assessed using one of several methods. For ease, these might be considered as direct or indirect. Three direct methods include clinical examination, biochemical or laboratory tests and anthropometry while two indirect methods include surveys and statistical analysis (Onis 2000). The following matrix summarises the many methods used in unison or complimentarity to help determine nutritional status (Table F.1).

As can be seen from the table, there are many methods in use. In reality, practitioners rarely rely on one technique and methods must be chosen for their appropriateness and applicability while multiple methods might also be used for triangulation and confirmation (Neithercut et al. 1987; Myatt et al. 2005). Also of note is the idea that there can be considerable overlap of methodologies, for instance, indirect analysis using statistics will usually be drawing on other methods of assessment such as aggregated laboratory results or accumulated direct anthropometric data. Having collated the data, there now needs to be some way of qualifying the extent of the problem. Back in the 1970s, Waterlow (1972) suggested complimentary systems of categorisation: one to quantifiably assess the situation and another more qualitative measure to distinguish between such measures of marasmus, kwashiorkor and its many intermediary forms (Waterlow 1972). Prophetically, this duality has come to pass although more likely by chance than by design. With this in mind, it can be said that such classifications can now be achieved both quantitatively and qualitatively. Moreover, they need not be mutually exclusive either; indeed some may be complimentary as in the use of clinical and sub-clinical classifications. Also some classifications will be more applicable to the individual than to the community as a whole while others might be better for use in the field as opposed to the laboratory. In this way, qualitative assessments can be made in respect of the type and severity of malnutrition such as kwashiorkor and marasmus using accepted clinical symptoms like visually determined stunting, moon face, oedema or pot-belly together with sub-clinical measures such as apathy and strength of appetite. Quantitative measures on the other hand might include anthropometrically determined stunting and wasting strengthened or complimented by such measures as biochemical markers like serum protein levels.

Independent of, and separate to specific biochemical or anthropometric classifications, there is one degree of classification that can be applied across the board and that is the clinical classification. At its simplest, this classification could best be described as one of degree.

Clinical malnutrition: It is a state characterised or determined by its apparent or overt signs and symptoms of nutrient deficiency. A person is considered clinically malnourished if the absence of nutrients causes a disease state or nutritional abnormality that can be reversed by the administration of appropriate essential dietary components. (WHO/FAO 2004).

Sub-clinical malnutrition: On the other hand is often a process that has started but has not yet manifested symptoms, that is, to say it might be present to a lesser degree but without overt or apparent signs (asymptomatic). Without apparent symptoms, it becomes more difficult to diagnose and oftentimes laboratory techniques such as biochemical markers and functional assays are used in support of its detection. In this way, malnutrition and many diseases, including diabetes, can be present sub-clinically before becoming clinical (WHO/FAO 2004).

TABLE F.1

Methods Used for Assessing Nutritional Status

Direct	
Clinical	Overall visual appearance, hair, mouth, eyes, nails, skin, thyroid gland and joints and bones.
Laboratory	Laboratory techniques might include: biochemical, haematological and microbiology tests. These can include: serum protein tests; urine dipstick and microscopy; tests for specific vitamin and mineral levels; tests for nutritional anaemias; detection of abnormal metabolites and stool samples for the presence of ova or intestinal parasites.
Anthropometric	Anthropometrics are an essential component of clinical examinations especially of infants, children and pregnant women. Measures are many and include: skin fold thickness measures, Circumferences measures and various Relational measures including: body mass index (BMI).
Indirect	
Statistical Analysis	Collating and assessing various national and regional statistics are an important approach in understanding nutritional issues—data from various sectors are used including those from health (mortality, morbidity), sociology (population density, rural urban shift), agricultural production, economics (income and expenditure), environmental and ecological (biodiversity, climate change, soil fertility, land availability).
Survey Methods	Surveys might include household food surveys, 24-hour dietary recall, food diaries, food frequency questionnaires and observational studies

Sources: Compiled from multiple sources: WHO/FAO 2004; Myatt et al., Technical Background Paper: *A Review of Methods to Detect Cases of Severely Malnourished Children in the Community for Their Admission into Community Based Therapeutic Care Programs*, WHO, Geneva. 2005; Elamin, A., *Protein Energy Malnutrition*, University of Pittsburgh, Pittsburgh, PA, 2008.

Bearing this in mind, specific assessment methods and classification standards are discussed in each category of direct and indirect methodologies shedding light on the different ways people and groups are nutritionally stratified.

F.1 Direct

Direct assessments often deal directly with the patient and can involve such investigations as body measures, oral histories and laboratory testing. These measures tend to be more objective dealing with a high degree specificity and can be compared with known standards; they are non-invasive (mostly) and are comparatively cheap to undertake. On the other side, problems of comparable reference standards (local versus international) can inject uncertainty and to some degree lead to arbitrary subjectivity. There are also questions of inter-observer measurement errors while lastly such direct assessment methods are fine at the individual level but are difficult to scale up to the national level for policy use or comparative analysis.

F.1.1 Clinical

Doctors can readily diagnose severe, long-term undernutrition and in some cases emerging nutritional deficits based solely on the person's appearance (Thomas 2007). Particular attention is paid to things like the hair, mouth, gums, nails, skin, eyes, tongue, muscles, bones and thyroid gland (Table F.2). General clinical examinations might be further complimented with oral enquiries looking at nutritional history such as diet, weight loss, the presence of other disorders and perhaps the ability to obtain and prepare food. Such complimentary questions might, in the absence of clear or obvious signs help confirm the diagnosis. Methods based on thorough physical examination and oral dietary evaluation have the specific advantage in that they are quick and easy to perform; however, by itself such diagnosis might easily miss early, or less severe cases of deficiency (Thomas 2007).

TABLE F.2

Signs Used in Clinical Examination

Clinical Examination
Overall visual appearance: Stunting, emaciation, weight loss.
Hair: Sparse, easily comes out, corkscrew or coiled hair.
Mouth: Glossitis, spongy and bleeding gums, cheilosis and angular cheilotitus, fissured tongue, leukoplakia.
Eyes: Exophthalmia, night blindness, blurring, conjunctival inflammation
Nails: Spooning and transverse lines.
Skin: Pale pallor, dermatitis, hypopigmentation, purpura, signs of haemorrhaging, follicular hyperkeratosis.
Thyroid gland: Goiter is a reliable sign of iodine deficiency.
Joints & bones: Helps to detect signs of rickets (vitamin D deficiency) and/or scurvy (vitamin C deficiency).
+ Oral history

F.1.2 Laboratory/Biochemical

Biochemical or laboratory assays measure specific aspects of a person's metabolism to help determine nutritional status. With regards to malnutrition, there are only a few useful laboratory or biochemical tests that yield sufficiently robust markers for analysis. Such tests might include biochemical, haematological and microbiology tests. However, used 'judiciously', analyses of blood, urine, adipose tissue, stools, nails and hair can aid in confirmation of nutritional status (Bowers 1999). The word judicious has been used but caution would have been just as adequate for there are many factors that can affect these tests. Non-nutritional factors such as drugs, time-frame (natural time-based variations in indicators) as well as inter-personal variability means that interpretation of these data from standard reference ranges becomes difficult. Even using different laboratory tests on the same biochemical markers has been shown to produce different results (Sullivana 2000). In this sense, no single measure ought to be used as a diagnostic, which in turn limits the usefulness of these tests as definitive markers for malnutrition.

Having said that, arguably the most important of these tests are those that measure the overall nutritional status in terms of calorie and protein. Serum albumin or albumin is a protein in the blood that is essential in transporting molecules and maintaining body fluid balance. Its concentration in blood plasma or serum* is determined by the balance among production (in the liver), distribution within extracellular fluid and degradation. As a result, blood serum albumin measures for some, are indicative of the body's overall protein status (Thomas 2007). As a test, albumin is not solely confined to the blood, it can also be detected in the urine using Urine Dipstick Chemical Analysis (Sullivana 2000; Merck Manual 2005). However while some clinicians still use this indicator, there is growing consensus in the face of overwhelming evidence as to albumin's ineffectiveness and indeed its appropriateness for use to identify malnutrition or even as an evaluation of any success of nutritional interventions. This is related to the fact that serum albumin levels are drastically altered by many processes, most notable of which is the relationship with the body's inflammatory response; that is, it is a negative acute-phase reactant in which serum albumin decreases as inflammation arises regardless of protein consumption (Shenkin 2006; Collins and Friedrich 2010). Another protein, transferrin, which primarily functions in transporting iron around the body as well as other proteins like thyroxine-binding prealbumin (transthyretin) and retinol-binding protein and lastly although not as common, insulin-like growth factor 1 (IGF-1) are also used as well. Other measures too, although not blood serum proteins, are sometimes used as indicators of malnutrition. These include: C-reactive protein (CRP), total lymphocyte count (TLC) and serum total cholesterol (Collins and Friedrich 2010). All these tests however, as with albumin, are sensitive but non-specific indicators of adequate nutrient intake. In this way, as cautioned above, these tests can show trends rather than be used as diagnostics per se and as such need to be interpreted with regard to other data and to clinical context (Salbe et al. 1995; Bowers 1999; Sullivana 2000; Swenne et al. 2007). Another important determinant of nutritional health are anaemias and when considering nutritional anaemias, blood screening still remains the best method to use. Tests might include red blood cell count,

---

* While blood plasma contains blood cells and clotting factors, blood serum or serum is the clear watery fluid left if both blood cells and clotting factors are removed from the plasma.

TABLE F.3

Laboratory and Biochemical Tests for Malnutrition

Macronutrients	
Protein	Albumin, transferrin, thyroxine binding prealbumin (transthyretin), retinol binding protein, insulin-like growth factor 1 (IGF-1), C-reactive protein (CRP).
Lipids	Serum cholesterol, triglcerides, lipoporoteins.
Others	Total lymphocyte count (TLC).
Vitamins	
A, Retinol	Serum retinol.
B, Calciferol	Serum 1.25 dihydroxycholecalcifer.
E, Tocopherols/Ttocotrienols	Serum trocopherol, enthrocyte, homolysis test.
K, Phylloquinones, etc.	Prothrombin time.
B_1, Thiamine	Enthrocyte transketolase activity.
B_2, Riboflavin	Enthrocyte glutathione reductase activity.
B_3, Niacin	Urinary N1-methylinicotinamide.
B_6, Pyridoxine	Serum pyridoxal phosphate, enthrocyte aminotransferase activity.
B_9, Folic Acid	RBC Folate.
B_{12}, Cobalamin	Serum B_{12} assay.
C, Ascorbic Acid	Serum ascorbic acid, leukocy ascorbate.
Minerals	
Iron	Serum ferritin, red blood cell count, haemoglobin.
Zinc	Serum zinc, RBC, WBC.
Magnesium	Serum magnesium, RBC magnesium.
Copper	Ceruloplasmin.
Calcium	Serum alkaline phosphate.

Sources: Compiled from multiple sources: Bowers, L.J., In Mootz R.D. and Vernon, H.T. eds., *Best Practices in Clinical Chiropractic.*, Aspen Publishers, Frederick, MD, 1999; Sullivana, D.H., *J. Gerontol A*, 56, M71–M74, 2000; Elamin, A., *Protein Energy Malnutrition*, University of Pittsburgh, Pittsburg, PA, 2008.

haemoglobin, mean corpuscular volume (MCV) and red cell distribution width (RDW). With regards to specific vitamin and mineral deficiencies, biochemical analyses are many and varied; some of the assessments or markers are listed in Table F.3.

On the whole, biochemical metrics used carefully are a useful and valuable aid in assessing malnutrition, particularly in the early subclinical stage and as indicators of marked changes in nutritional profile over time. In this sense, they are objective, accurate and reproducible. On the downside however, laboratory or biochemical analysis can be time-consuming, expensive, need trained personnel and cannot be reasonably applied on a large scale (Onis 2000; Udall Jr et al. 2002; Myatt et al. 2005; Shils et al. 2006).

F.1.3 Anthropometry

Another direct method: anthropometry uses, various body measurements either by themselves, as in skin fold thickness or as is common in relation to each other, like weight-for-height, etc., to determine normal healthy ranges. These can be performed on both adults and children and from these measurements assessments can be made as to an individual's nutritional status. Generally, in normal healthy adults, regardless of age, ethnicity or geographic location, measures such as the relationship between weight and height, for instance, is relatively constant. In normal healthy children too, growth profiles are similarly constant and progressive, making them calculable and therefore predictable (Shils et al. 2006; WHO 2006b; WHO/MGRSG 2006; Pérez-Escamillal and Segall-Corrêal 2008).

Anthropometry is very useful in the field of malnutrition. As well as various relational measures of weight and height, etc., other useful indicators might include skin fold thickness (triceps, biceps, below scapula, above iliac and upper thigh) or circumference (head, mid upper arm, waist, calf, and hip, etc.)

TABLE F.4

Anthropometric Assessment Methods

Skin fold thickness (triceps, biceps, below scapula, above iliac and upper thigh) or;

Circumferences (head, mid upper arm, waist, calf, and hip), and;

Various relational measures including: body mass index (BMI); waist and hip ratio (WHR); weight-for-age (W/A); height-for-age (H/A); weight-for-height (W/H); mid-upper-arm-circumference (MUAC); mid-upper-arm-circumference-for-age (MUAC/A); and mid-upper-arm-circumference-for-height (MUAC/H), etc.

(see Table F.4). In this way anthropometry has an important advantage over other nutritional indicators in that such body measurements are sensitive 'over the full spectrum' of severity of malnutrition and importantly, allows for the discrimination between short- and long-term forms of the phenomenon (Onis 2000; Myatt et al. 2005).

For these reasons, for many, the best technique for assessing undernourishment continues to be anthropometry and in continuing to enjoy universal acceptance, it remains one of the most important internationally comparable standards of nutritional assessment (Shils et al. 2006). Although popular, it is worth bearing in mind too that anthropometry does have its critics as discussed throughout the book.

In practice, undernourishment in adults is measured using a variation of the weight-for-height (W/H) metric: the body mass index (BMI).[*] This is because BMI figures in adults, as mentioned above, are comparably stable when considering age, sex, ethnicity and geographic location. These data in turn can then be independently assessed against standard international WHO BMI references (NHS/NOO 2009). On the other hand, when measuring undernutrition in children, BMI data varies too much between the sex and age of children therefore measures of weight-for-age (WFA), height-for-age (HFA) and weight-for-height (WFH) tend to be used. Individually, these three measures can separate (in general terms) the type of malnutrition suffered. For instance, a low height-for-age could suggest stunting, while low weight-for-height is indicative of wasting; and low weight-for-age is a combination of both linear growth and body proportions, which helps determine those underweight (Shetty 2002; NLIS 2010). For a quantitative assessment, these values can then compared to known growth curves or reference values to determine exact departures from the expected norm (WHO 1995; Onis 2000; Maynard et al. 2001; Udall Jr et al. 2002; Nandy et al. 2005; Checchi et al. 2007).

Despite such universal popularity of anthropometry, there are a few disadvantages. Firstly and most glaring perhaps is the specific inability to detect actual specific micronutrient deficiencies themselves such as zinc, thiamine and others, etc. Secondly, and just as importantly, while anthropometric data is extremely useful for establishing an overall nutritional profile, it is very inept at determining changes over the short term. This is related to the delay in malnutrition and the body's physiological responses. Another important shortfall of this method of data gathering is the inability to find causality, for while nutritional deprivation might easily be determined by such measures; its causes might indeed be wholly nutritionally based or might just as readily result from factors such as repeated infections or poor child care. Further, such anthropometric approaches are difficult, realistically, to scale up to the global level due to high cost of collecting representative data. Hence the argument is made for a need for indirect methodologies like those of the FAO's prevalence of undernourishment calculations. As in most other methods too, there is still arguably not one anthropometric measurement that is suitable in all situations (Gorstein et al. 1994; Shetty 2002; Myatt et al. 2005).

F.1.3.1 Anthropometric Classification

The general theory behind anthropometry above details the advantages and disadvantages of this method over others. The actual practice of anthropometry, particularly in the area of classification, is not surprisingly, more complicated than might at first appear. The following explores some of these main issues.

---

[*] BMI is predominantly used to determine weight ranges from under- to over-weight, other more accurate alternatives to this measure have included skinfold thickness measurements, underwater weighing, bioelectrical impedance, dual-energy x-ray absorptiometry (DXA) and isotope dilution. Unfortunately, such methods can be expensive, might not always be available or might require highly trained personnel in application (CDC 2010).

In adequately classifying malnutrition, three criteria are needed: that of appropriate reference values, an agreed or at least acceptable classification system and a method of quantifying one against the other, that is, to say a system of comparing observed data with reference values to give meaningful analysis.

F.1.3.1.1 Reference Values

When considering appropriate measurement types and reference values for children and adults with regard to anthropometrically assessed undernutrition, two different approaches are used. In adults (20 years and over) body Mass Index (BMI) is commonly used, while for children (0–5 and 5–19) growth standards are recommended[*] (CDC 2010; WHO 2010b). In the case of international standards, the World Health Organisation currently provides generally accepted measures of BMI for adults, the BMI for children 5–19 as well as standard growth reference charts for 0–5 year olds.

Body mass index (BMI or Quetelet Index): The relationships between height-and-weight and weight-and-health are widely acknowledged (Gilmore 1999). By measuring the former, it becomes possible to make judgments of the latter. To this end the BMI has become the most commonly used method of determining if an individual's weight falls within the accepted range attributed to good health. The BMI, or Quetelet index, was first introduced by Belgian astronomer and social statistician Adolphe Quetelet in his treatise on man's development in 1835 (Quetelet 1835). Although Quetelet first outlined the correlation between weight and height as an anthropometric indicator, it was not until 1972 with the publication of Ancel Key's 'Indices of relative weight and obesity' that the term 'body mass index' was coined (Keys et al. 1972).

The statistical measure of a person's weight divided by the square of their height given as kg/m^2 is a quick and easy proxy for underweight or overweight adults, although it does have certain limitations though (NHS/NOO 2009). Importantly, among these is that although weight and body fat are closely correlated, body fat is independent of height or stature in adults. Also BMI calculations do not discriminate between muscle and adipose and can lead to false-positives and false-negatives. For instance, as noted earlier, a person losing 10% of their body weight over a short period might indicate severe nutritional or health problems, but in using the BMI scale if a person 1.58 m tall and initially weighing 67 kg loses 10% of body weight, the equivalent BMI would drop from 27 to 24 kg/m^2, leaving them still within the healthy range. In this way, while BMI is best used to ascertain a general populations' body composition for reference and public health purposes, individuals should only use BMI as a comparative to that cohort. Although in reality its use is commonplace, and its misuse, as an absolute measure of an individual's health, is still, nevertheless, misleading. Lastly supporting evidence is not always clear as to BMI's appropriateness for use as a predictor of undernutrition (Cook et al. 2005). Nonetheless, despite these misgivings and Ancel's cautionary note that BMI is best used for referencing populations rather than individuals; as well as many offered alternatives, the measure remains a sufficiently acceptable, sensitive and convenient measure for underweight and overweight that is still widely used today (Micozzi and Albanes 1987; Gallagher et al. 1996; Bagust and Walley 2000; Merck Manual 2005; Stevens et al. 2008; CDC 2010).

BMI values for adults are age and gender independent and while further consultation with regard to ethnic populations is underway. The latest (2006) WHO body mass index tables are considered suitable for all ethnicities regardless of relative stature. The WHO 2006 BMI values then can be universally applied; this makes them an ideal and simple method when compared with the cut-off values in Table F.5 to assess nutritional status in adults across the world (WHO 2004; Cook et al. 2005; WHO 2010c).

For children however, BMI is a different story, for while it is calculated the same way as adults, it is not presented in the same way as a single matrix as in Table F.5. Instead reference data for child BMI are presented in formats based on projected growth curves. As mentioned earlier this is essentially because of the constant changes during childhood growth. This means that the relationship between weight-for-height, and height-for-age and weight-for-age is constantly changing, which is also, unlike adults,

---

[*] While the WHO offer BMI for all ages 0–5, 5–19 and 20+, there exist no international standard references for adolescents; this is because data used to construct the tables were based on US data from the early 1970s and thus is not representative. Although the WHO is working on a new standard, for now only the 0–5 and 20+ reference data can be universally applied (WHO 2010b).

TABLE F.5

The International BMI Classification with Cut-Off Points

Classification	BMI (kg/m²)	
	Principal cut-off points	**Additional cut-off points**
Underweight	<18.50	<18.50
Severe thinness	<16.00	<16.00
Moderate thinness	16.00–16.99	16.00–16.99
Mild thinness	17.00–18.49	17.00–18.49
Normal range	18.50–24.99	18.50–22.99
		23.00–24.99
Overweight	≥25.00	≥25.00
Pre-obese	25.00–29.99	25.00–27.49
		27.50–29.99
Obese	≥30.00	≥30.00
Obese class I	30.00–34.99	30.00–32.49
		32.50–34.99
Obese class II	35.00–39.99	35.00–37.49
		37.50–39.99
Obese class III	≥40.00	≥40.00

Source: Reproduced from the *WHO Global Database on Body Mass Index (BMI),* World Health Organisation, 2010. http://apps.who.int/bmi/index.jsp?introPage=intro_1.html.

dependent on sex (Birch 2009). As such, specific BMI-for-age-and-sex charts are presented alongside many child growth charts (not to be confused with each other) and are useful in gleaning data about general overnutrition and obesity and to some degree underweight. Importantly however, as a measure of undernutrition child BMI-for-age-and-sex charts are wholly inappropriate. This is because of the way BMI is calculated and the nature of the different types of undernutrition. As has been shown, BMI indices in adults are largely independent of such variables as height, weight, sex and age and therefore BMI calculations that determine the relationship between height and weight alone is a simple and adequate metric. In children however, height, weight, sex and age are intimately related and responsive to specific situations and accordingly such growth in turn is highly dependent on nutritional intake, making them very sensitive measures of nutritional status. For this reason, while child BMI charts are adequate for general growth measures, they are very insensitive to detailed child nutritional status (Shetty 2002; Onis et al. 2004b; WHO 2006b; CDC 2010). By elucidating this interaction further, the difficulties becomes clearer. In children, a low weight-for-height (or length) determines whether there has been severe acute calorie deficiency resulting in excess thinness known as wasting (Waterlow 1976; MacAuslan 2009; World Vision 2009). Short height-for-age, on the other hand, tends to reflect the more long-term linear growth failure associated with chronic calorie deficiency resulting in restricted growth or stunting (Waterlow 1976; WFP 2008; MacAuslan 2009; World Vision 2009). On top of these measures an appropriate weight-for-age is representative of both linear growth and body proportion and can be used in the identification of generally underweight children but without being able to distinguish between stunting or wasting (MacAuslan 2009; World Vision 2009).

Child Growth Reference Charts. By comparison, therefore the BMI's singular calculation of weight and height, while largely remaining a satisfactory (despite previous caveats) measure for adults does not sufficiently distinguish between the components of child growth necessary for adequate analysis. Instead measures of anthropometric undernutrition in children are calculated from international growth reference charts against which appropriate cut-off values are applied (Beaton and Bengoa 1976; Shetty 2002).

In this endeavour, anthropometric parameters—height-for-age; weight-for-height; and weight-for-age—among other useful metrics such as head circumference, etc. are plotted over time, with each one separately showing growth trajectories patterns as well as acceptable ranges over a specified period. This allows comparisons to be drawn in every major anthropometric indices to determine first whether there is cause for concern (nutritionally speaking) but also to ensure growth over time is adequate and progressing nicely. There have been several important standards over the years. The first real guidelines came out of the Harvard standard (or Boston Standard) based on measurements of children in Boston between 1930 and 1956 (Gueri et al. 1980). The charts ran simultaneously with the UK Tanner growth curves until the WHO in 1966 endorsed a simplified version of the Harvard charts effectively establishing a world standard (WHO 2006a). This lasted until 1977 when the WHO adopted the National Center for Health Statistics' (NCHS) growth charts, which despite known limitations, became the standard of choice for the international community. For three decades, these NCHS/WHO international growth reference standards prevailed until more recently the WHO launched a new set of standards, the WHO Multicentre Growth Reference Study (MGRS). Importantly, this new study sets 'standards' of growth rather than 'reference' values so characteristic of previous standards (WHO 2008b). This, it is anticipated, will make for better comparisons of actual versus expected values and allow for more informed value judgments to be made. Lastly it has also been recognised by the WHO that with the new standards there will be increases in those classified as stunting for infants 0–6 months compared to the old standards with decreases thereafter. It is also understood that an increase in those infants showing signs of wasting up to about 70 cm will also be noticed; these are due to new methodological considerations as well as reflecting revised understanding. (WHO/MGRSG 2006). Endorsed by many (WHO 2010a), the initial response seems encouraging although some advocate caution as with all new systems, until familiarity and its outcomes (anomalous or otherwise[*]) can be properly assessed (WHO 2006a; Garza and Onis 2007; Fergusson 2009).

Having established internationally acceptable reference standards, it remains to decide by which anthropometric classification criterion these measures should be compared. While there have been many such classifications over the years, there are some noteworthy standards that benefit from further exploration, not least because several of them are still in use today but also as an example of the different approaches that people have taken in addressing the challenge.

F.1.3.1.2 Classification

Each classification tends to focus on one or more anthropometric measurements and each chooses its own cut-off points to determine different categories or levels of undernutrition. With regard to cut-off points, there are generally three ways of statistically comparing children's anthropometric data to reference populations: Z-scores (standard deviation scores), percentiles and percent-of-median. Gomez (following discussion) used the percent-of-median as his cut-off point mainly because of the generally non-normalised[†] growth curves that were used (Flegal 1999; Gibson 2005). However, since the late 1970s WHO and more recently USAID advocate the use of Z-scores (standard deviations), which whilst useful at the population level need to be used with caution on an individual level. That said, the use of Z-scores tends to be the system most widely used today as its application is considered the simplest and best way to describe and compare the data (Cogill 2003).

Z- Scores or standard deviation[‡] score (SD) are statistical measures that determine how close or far away a value is compared to the average values of a group of reference values. The biggest advantage is this method allows for the comparison of groups of Z-scores within and cross-country at the population level (WHO 1995).

Percentile: This is the position a reference data point falls within the ranking of 100 percentiles, i.e. if something falls within the 10th percentile, then that value is equal to or greater than 10% of that group.

---

[*] Differences between the new WHO and the old WHO/NCHS reference standards are substantial, particularly regarding infancy, a crossover period where both datasets are used is recommended until familiarity and trends can be better analysed (WHO 2010a).

[†] This is based on the concept of normal distribution where a graph of the data is distributed normally, that is, a simple bell curve (or Gaussian function) where most of the data falls near the middle (average). If data, on the other hand, does not fall in such a smooth pattern (it has lots of outliers) then this can make for difficult analysis or interpretation. By smoothing out this data or 'normalising' this data by using different formulae, the process of interpretation and viewing trends can become easier.

[‡] The standard deviation is a statistic that determines how closely all the various data points are clustered around the average. When the examples are tightly bunched together and it produces a distribution curve (a bell-shaped curve) that is steep, this means the standard deviation is small. When the examples are spread wider apart and the curve is relatively flat, this suggests a relatively large standard deviation.

TABLE F.6

Gomez Anthropometric Classification of Undernutrition

Degree	Percent of Reference Weight for Age
First degree (moderate[a])	76–90%
Second degree	61–75%
Third degree	60% and less

Source: Gomez et al., *J. Trop. Pediatr.*, 2, 77–83, 1956.

[a] Gomez: usage not the author's. Also, although contrary to Gomez's original incarnation first, second and third degree terminology has become synonymous with mild, moderate and severe (Gueri et al. 1980).

TABLE F.7

Wellcome Trust Malnutrition Classification

Weight for Age	With Oedema	Without Oedema
60–80%	Kwashiorkor	Undernutrition
< 60%	Marasmic-kwashiorkor	Marasmus

Source: Based on the Wellcome Group, *Lancet*, 296: 302–303.

This method is simple to use and is often used in clinical settings where ease of use is an advantage. A major drawback with this method though is that data cannot be compared outside its own group of values. In this way unlike Z-scores, percentiles cannot be compared across the board.

The percent-of-median is simply the actual reference data point compared to the mean of all the data points expressed as a percentage. For example if a person weighs 8 kg and the average for the group of the same height is 10.2 kg, then $8/10.2 \times 100 = 78.4\%$.

Lastly, despite this seemingly detailed and purposeful level of analysis, it has to be borne in mind that such cut-off points remain to a large degree, arbitrary (Onis 2000).

In the following section, four classifications: the Gomez, Wellcome, Waterlow, WHO and the CIAF are examined here because they illustrate the different approaches used and because to some extent they are all still in use today.

Gomez classification: - In response to the non-standardisation of clinical features of undernutrition, the Gomez classification was first used in 1946 by Gomez and colleagues in Mexico (Gomez et al. 1956). The classification denoted malnutrition as an arbitrary percentage of the theoretical average (percent of the mean) weight for the child's age as presented by then Harvard reference growth standards. These were then presented as first, second degree and third degrees* (see Table F.6) Gomez et al. 1956; Onis 2000). The classification was a landmark in its simplicity and despite many criticisms of the method (Gueri et al. 1980; Onis 2000; Maleta 2003), the main being that it does not distinguish weight loss or small stature (i.e. discriminate between different biological processes of chronic or acute malnutrition), the Gomez classification, albeit modified is still used in some countries today (Onis 2000; Cogill 2003; Maleta 2003; Khan et al. 2008; Sepahi et al. 2010).

Wellcome Trust Classification: As with Gomez's system, the trust's classification was based on the weight-for-age metric (Table F.7). Perhaps the two main disadvantages with this system however, are that firstly, marasmic graded children could easily include those who were either stunted (and of normal body proportions) or wasted; while the second drawback was in the sole use of the metric—weight-for-age

---

* Unfortunately, the Gomez weight-for-age classification is often misquoted or more often its percentile ranges are commonly modified without reference to such adjustments. Of particular note in this regard is Waterlow's malnutrition classification assessment for the WHO where he apparently misquotes Gomez' first degree as 75–90%, and his second degree as 60–75% in an otherwise excellent analysis of classification systems (Waterlow 1976). I say apparent as there appears to be no other literature that picks up on this fact. Even if this turns out to be the case such criticism is a minor and one might argue trivial in light of the small adjustments/errors involved. Yet something of such seemingly little import has consequences; when assessments by people in key positions contain such errors, these errors are passed on blindly to the point where even today, much academic literature believes Gomez' classification to be that (miss)quoted by Waterlow and others (Cogill 2003; Maleta 2003; Tetanye 2004; Chowdhury et al. 2008; Kulsum et al. 2009).

measures alone. Also the same limitations with regards to Gomez exist in that such measures take no account of stunting and hence of the history of malnutrition.

In spite of the shortfalls in those classifications, it would seem that the Wellcome Trust categorisation is still in use and enjoys just as much wide appeal as both Gomez' and Waterlow's (Oshikoya and Senbanjo 2009).

Waterlow's Classification. Another widely used and just as well-known classification system comes from Waterlow (Waterlow 1973, 1976; Waterlow et al. 1977; Briones et al. 1989; Avencena and Cleghorn 2001). In response to meta-analysis of existing classifications, Waterlow attempted to reconcile the many limitations with his own classification. In a series of publications, Waterlow elucidated and modified his ideas over the years. This resulted in a cross-tabulated matrix, a summary of which can be seen in Figure F.1. This was by far the greatest contribution, albeit somewhat arbitrarily, to the classification of malnutrition to that point (Waterlow 1973; Avencena and Cleghorn 2001; Edelstein and Sharlin 2009).

This new system importantly attempted to distinguish between stunting and wasting, which previously, were often 'lumped' together in classifications based solely on one anthropometric measure; usually weight-for-age. In Waterlow's matrix, two measures were used: weight-for-age and weight-for-height. This compensated for the lack of distinction between long-term stunting (chronic malnutrition) and short-term wasting (acute malnutrition) in previous indices. By effectively allowing the cross-comparison of height and weight-for-age with the degree of severity, Waterlow offered a useful matrix of outcomes that remains in use today. These are: (1) normal; (2) wasted but not stunted, (acute PEM); (3) wasted and stunted (acute and chronic PEM); and (4) stunted and not wasted (past PEM with present adequate nutrition a.k.a. 'nutritional dwarfs') (Waterlow 1976; Udall Jr et al. 2002).

WHO: By the late 1970s, the FAO/UNICEF/ WHO Expert Committee on Nutritional Surveillance had recommended the use of height-for-age and weight-for-height as preferable primary indicators of nutritional status in children (WHO 1976; Briones et al. 1989). At about the same time, it was also recommended that the use of Z-scores (Table F.8) be used as the international cut-off points when compared to the international reference standards. This system remains WHO's position today and has consequently been readily accepted by many users. The WHO also usually present regular figures on stunting, wasting and underweight. The only difference is that the reference tables have been updated from the previous WHO/NCHS charts to use the WHO Multicentre Growth Reference Study (MGRS). Although that said, a long gestation period is anticipated for these new tables to be wholly tested and accepted. As a result, the MDG goals and others that rely on measure of child undernutrition will for the immediate future, continue to use the WHO/NCHS tables (WHO 1995; Blössner and Onis 2005; WHO 2010a).

			Height-for-Age %	1972	>95	95–87.5	87.5–80	<80
				1976/7	>95	95–90	90–85	<85
Grade			Stunting		0	1	2	3
		Severity			Normal	Mild	Moderate	Severe
			Standard Deviation (approx.)			−1 More than −2.00	−2 (−2 to 2.99)	−3 (−3 to 3.99)
Wasting	Weight-for-Height (%)							
0	>90	Normal						Stunted
1	90–80	Mild	−1	More than −2.00				Stunted
2	80–70	Moderate	−2	(−2 to 2.49) (−2.5 to 2.99)				Stunted
3	<70	Severe	−3	−3.00 or less	Wasted	Wasted	Wasted	Stunted & Wasted

FIGURE F.1 Waterlow's cross-tabulated undernutrition matrix. (From Waterlow, J.C., *Br. Med. J.*, 3, 566–569, 1972; Waterlow, J.C., *Lancet*, 302, 87–89, 1976; Waterlow et al., *Bull. World Health Organizat.*, 55, 489–498, 1977.) *Note*: Percentages are based on the Harvard Reference Standards.

TABLE F.8

WHO's Z Score: Standard Deviation (SD) Cut-Off Points

	Overnutrition	Undernutrition		Prevalence in a population (%)			
		Moderate	Severe	Low	Medium	High	Very high
Weight-for-height (wasting)	>+2 SD	<–2 SD	<–3 SD	<10	10–19	20–29	30+
Height-for-age (stunting)		<–2 SD	<–3 SD	<5	5–9	10–14	15+
Weight-for-age (Underweight)		<-2 SD	<–3 SD	<10	20–29	30–39	40+

Source: Compiled from the WHO Global Database on Child Growth and Malnutrition, World Health Organisation. http://www.who.int/nutgrowthdb/about/en/, 2010.

By 1983 the World Health Organisation had endorsed Waterlow's 1976/77 undernutrition classification system contributing to its worldwide adoption (Waterlow 1976; Waterlow et al. 1977; WHO 1983; Briones et al. 1989; NLIS 2010).

Composite index of anthropometric failure (CIAF): More recently, in response to existing shortfalls of anthropometric classifications using weight, height and age, Svedberg proposed a new classification: the composite index of anthropometric failure (CIAF) in 2000 (Svedberg 2000). In particular, Svedberg was concerned over the use of the single weight-for-age metric often used by the UNICEF and WHO and in the headline media as a measure that encompasses all the stunted and wasted children. In fact, Svedberg stresses that far from being inclusive of stunting and wasting, figures of underweight are not the sum of the other two and in fact importantly missed cases of undernutrition. In response and building on the interconnectedness of these measures, he proposed to disaggregate the components of weight-for-age, height-for-age and weight-for-height into six sub-groups based on singular and multiple deviations from the norm (stunting, stunting plus underweight, stunting plus wasting and underweight, etc.). From here, Svedberg constructed an all-inclusive algebraic equation, which was then used to determine a comprehensive total of undernourished (Svedberg 2000; Bhattacharyya 2006).

F.1.3.1.3 Cut-Off Points

With regard to appropriate cut-off points, any classification would only work if there was universal (or at least wide) agreement of a single standard (Waterlow 1976). The challenging question is then at what points should these individuals be considered mildly, moderately or severely malnourished and at what gradients (percentages or Z-scores) are these to be determined? The problem is there is no standard with many such recommendations relying on arbitrary cut-off points reflecting convenience, standard deviations (Z-scores) and practical experience (Waterlow 1976; WHO 1995; Pérez-Escamillal and Segall-Corrêal 2008). The difficulty here is that despite these best-'guestimates', such reference cut-off points have no proven 'physiological validity as indicators of the severity of undernutrition' and as such, Gopalan suggested any judgment based solely on current anthropometric measurements would be misleading. Further, Gopalan wondered back in 1984 whether such a classification was even possible:

> … it is doubtful if a single convenient index which can take into account the complex and multiple facets of undernutrition can ever be developed. (Gopalan and Kamala 1984)

F.2 Indirect

Indirect measurements, it can be said, are largely survey and statistically based. Many survey types including the 24-hour dietary recall, household food surveys, food diary techniques as well as observational type surveys all combine to add valuable data to the collective pool. Other methods used to assess nutritional status include compiling statistical databases on a multitude criteria from food balance sheets to birth and death rates, fertility, sanitation and education among many others. All of these aid in the analysis, although not always directly, of nutritional standing. Perhaps the most well known of the indirect statistical analysis methods used are those of the FAO's prevalence of undernourishment figures.

In execution, much of today's statistical data, as well as that gathered from databases and the like, often rely to some degree on data gathered using a wide variety of surveys regarding nutrition and hunger. Many of these, particularly, the household food surveys also tend to be broad brush approach types, which aim to elicit much information that might include all manner of qualitative data or statistics of income, expenditure, household size and individual anthropometrics as well as numerous other data.

F.2.1 Survey's

Of the wide variety of data and surveys collected, here we are interested in those surveys that specifically try to determine nutritional intake. These dietary intake surveys can be very expensive, time-consuming and frustrating due to lack of accuracy, detail or simply incomplete (Bowers 1999; Shetty 2002; Donovan 2009).

F.2.1.1 Household Income and Expenditure or Budget Surveys (HBS) (HIES)

Conventional household budget (or income expenditure) surveys can provide a great deal of relevant data for food security analysts. This method is based on interviewing households regarding the amount of money spent on food and other necessities over different time reference periods of perhaps weeks or month preceding the survey. The amount of calories consumed are then estimated on an average per-member, per-day basis (Pérez-Escamillal and Segall-Corrêal 2008).

F.2.1.2 24-hour Dietary Recall

One of the most popular methods of obtaining data the 24-hour dietary recall survey method uses trained interviewers that ask probing questions to determine previous 24-hour intake. To reduce errors and aid, recall interviewers try to tease out information to determine among other factors such as, total food intake, portion size, atypical intakes (due to special occasions, etc.). All the time trying to break down barriers in the face of possible reluctance for full disclosure, perhaps because of potentially embarrassing dietary habits or excesses. This makes this type of survey difficult and unreliable (Bowers 1999).

F.2.1.3 Food Frequency Questionnaire

To determine food intake over a period of time, the food frequency questionnaire (FFQ) orders food individually or by grouped and aims to determine daily weekly and sometimes monthly intakes. It has been said that one of the general pitfalls of this type of survey is that longer term questionnaires tend to overestimate intake whereas shorter ones tend to underestimate intake.

F.2.1.4 Food Diaries

By keeping a record over 3-, 4- or 7-day periods (over 7 days decreases validity) the food diary can give an oversight of overall dietary habits. There are benefits and drawbacks to how this is carried out; by writing down food intake as and when it is eaten the diary holder might become self-conscious, in turn affecting actual intake. If on the other hand, if it is filled out later at the end of the day, the danger is that of memory recall and the possibility that intake in inaccurately detailed (Bowers 1999). There are certain advantages with this type of data collection however, the most obvious is accuracy, in that the food measured is actual intake and not just availability (errors of estimation and interpretation aside). It also addresses the quality issue of dietary intake, able to better determine the makeup and therefore macronutrient and micronutrient profiles. In these respects, surveys allow for the up scaling or application of this data to the national level (Pérez-Escamillal and Segall-Corrêal 2008).

One of the main disadvantages with the survey methods is directly related to respondents' memory recall and accuracy of dietary portion estimates. This can lead to substantial error reporting, especially if interviewers are insufficiently trained to tease out the relevant information. Also the 24-hour recall survey as well, to be useful would need to be repeated over different 24-hour periods to even out intra-day variabilities. Such errors if multiplied at the national level might markedly magnify distortions (Pérez-Escamillal and Segall-Corrêal 2008).

F.2.2 Statistical Assessment

Statistical assessments of malnutrition attempt to estimate the numbers of undernourished either at regional, national or global levels. This might be the relatively simple up-scaling and extrapolation of data recorded from representative surveys or anthropometrics, to indicate prevalence of undernutrition as in the WHO and UNICEF examples. There is also more involved and complicated models as used by the FAO and USAID; or even perhaps a combination of both as in the IFPRI hunger index. These are explored in detail in the Section F.3.

F.3 Calculating the Prevalence of Undernourishment

F.3.1 FAO Prevalence of Undernourishment

The FAO calculation of undernutrition is a popular indirect measure relied upon by the international community. After much revision and criticism, the methodology was updated in 2002 (Naiken 2002). while individual parameters of energy requirements, population and BMI tables, were revised in 2004 and 2006, respectively (FAO/WHO/UNU 2004; FAO 2008; SOFI 2008). Essentially, this metric aims to determine a cut-off point below which calculations can be made as to the number of people are deemed to be undernourished. However, achieving this goal requires complex calculations that take into account several variables. Three key elements are involved in the framework these include the total availability of food calculated as per caput average daily food intake or dietary energy supply (DES); the inequality of access/consumption of that food (the coefficient of variation); and the energy requirement cut-off point for the average person (Naiken 2002; SOFI 2004; FAO 2008).

Generally speaking, in these calculations the FAO ideally need access to representative household dietary information on a per-member basis. The first choice in this respect would be the specialised food consumption or dietary surveys, unfortunately and directly related to the precise nature of the survey types, this involves complicated procedures, which include weighing food items used in the preparation of each meal (FAO 2002b). This makes this style of data relevant but costly especially on a nationally representative scale. The second choice would be the Household Income/Expenditure (HIES) or the Household Budget Surveys (HBS). Unfortunately again, the HIES/HBS cannot be used as a basis for estimation of distribution at the country or national level because of the imprecise nature of the data, the variable timeframes of each country's statistics and the limited number of countries with sufficient data (FAO 2002b). However there are other ways in which suitable data can be gathered.

Dietary energy supply (DES) is also known by several other names including average dietary intake and confusingly[*] the average daily consumption. However, in light of the difficulties of the data collection considerations just described, there are two ways in which DES can be calculated at the sub-national or national levels. At the sub-national level, derivation of average intake can be garnered from the survey methods for use in regional or group DES calculations. Otherwise in the prevalence of undernourishment figures at the national levels, the FAO employs the FBS (FAO 2002). National DES is generally considered a good representation of food supply firstly because it is gathered regularly and secondly this data by its nature is assumed to be log-normal, that is, the data is distributed in a way that clusters around the

---

[*] Confusingly the FAO often refers to availability or supply as a proxy for consumption because of its assumed close correlation. This is particularly evident in the FAO's calculation of undernourishment figures when DES (availability) is calculated on the basis of being the consumption equivalent. (FAO 2002d).

average and strengthens its representability. The DES calculation involves the a national food supply, taking into account total production, trade (import and export), changes in stocks, wastage and types of usage such as seed, animal feed and human use. From here, total supplies can be converted into per-person calorie equivalents (Naiken 2002; FAO/WHO/UNU 2004; FAO 2008).

Inequality of access: Next attempts are made to quantify household differences in consumption and distribution; that is, their inequality of access to food. This is estimated using the 'coefficient of variation' (CV) calculation and the 'Gini coefficient'. The CV determines the distribution of income across populations (household survey data are usually sufficiently detailed for this measure) and is calculated using a function of the Gini coefficient. Secondly, there is a need to establish the effect of income on the amount of foods purchased. This is a well understood phenomenon that suggests a positive correlation between income and food purchases. This is described as the elasticity of the demand, or the product of the Engel curve (FAO 2002b, 2008; MDG 2010). An important note at this point is that so far in the calculations households are treated as if everyone were 'average' and no distinction is made of the different age-sex groups and thus requirements of individuals. Therefore to this point the FAO have calculated and variously distributed a 'per capita dietary energy consumption' for the figurative average person in that country (Naiken 2002). The next steps needed are to determine an appropriate requirement cut-off point and to make allowances for the different age-sex demographic (FAO 2002b; Naiken 2002; Svedberg 2002b; FAO/WHO/UNU 2004; FAO 2008; Wiesmann et al. 2009; MDG 2010).

Energy requirement cut-off point: While the discussion on energy requirements in Table 3.4 focused on 'average' daily energy intake (the desirable level of achievement for users and policy makers alike), to calculate those that fall into the category of undernourished or hungry there needs to be a further measure—the minimum energy requirement or cut-off point below which people are considered undernourished. To achieve this the FAO treats adults and children separately whereby reference body weights are calculated as the lowest acceptable weight for heights and secondly the relevant energy requirements are then ascribed.

Reference body weights are treated differently for under and over 10-year-olds. These are taken as the median (or middle range) as dictated in the new BMI reference tables (WHO, 2006). With adults and children over 10, it can be seen from the BMI tables that given a specific height, there exist a range of weights that fall within the healthy band. Unlike energy recommendations though the minimum acceptable calorie cut-off requirements are considered the lowest acceptable weight for height, as that of the fifth percentile of that particular category (Naiken 2002).

Minimum energy requirements for children and adolescents under 18 years old: Minimum energy requirements are then calculated on a weight basis by multiplying the above reference weight by energy recommendations per kilogram of body weight. For adults, the above reference weight is multiplied by the lowest acceptable physical activity level (PAL); for men, it is equivalent of 1.55 BMR and females 1.56 BMR. This then constitutes the minimum dietary energy requirement (MDER), the of minimum calorie cut-off point (WHO 1995; NHLBI 1998; FAO 2008b; WHO 2010c). As noted, dietary energy requirements differ by gender, age and different activity levels.

Accordingly, to aid in a more precise estimate of the undernourished, the challenge then is to determine a single MDER that represents the entire population demographic. This is achieved by calculating the weighted average of the MDERs of the different gender and age groups that make up the population. It should also be noted here that because of inter-year demographic variability this measure needs to be constantly revised (Naiken 2002; SOFI 2004; FAO 2008; SOFI 2008). This calculation then sets a cut-off point and a nutritional gap between what is available, general access and individual needs. This can be expressed as a headcount or as a percentage of the population (SOFI 2008; Wesenbeeck et al. 2009).

There are enormous advantages and disadvantages of the FAO methodology. First and foremost, perhaps is the fact that the FAO is the only the only organisation to collect and provide statistics on the prevalence of undernourishment on a worldwide basis. The FBS for instance cover nearly all countries providing a rich and easily accessible source of information. The DES, as calculated from the FBS as well as contributing to the undernourishment figures, also provides a short-term measure of aggregate food availability (FAO 2008; Pérez-Escamillal and Segall-Corrêal 2008). Yet despite this, many difficulties are encountered when the calculation is broken down into its constituent parts. Not least of which is the notion that the formula relies on assessing a good number of variables that by their nature are inherently difficult to measure (Svedberg 1999). Indeed, much criticism has been levelled over the years at the FAO's formula on many fronts. It has been questioned from all angles from the conceptual framework to its statistical methodology and indeed FAO's handling of such criticism. In its conceptualisation, some have charged that the framework is unnecessarily indirect and arguably very sensitive to assumptions (Haddad 2002; Klasen 2008; Pérez-Escamillal and Segall-Corrêal 2008). While others, as early as the late 1970s questioned the appropriateness of using food availability as the determinate of hunger, offering instead, as many have since, that measures poverty in fact might be a better metric (Reutlinger 1977). This methodology has also been accused too of being a blunt instrument, aggregating information in a way that is unable to pinpoint hunger 'hotspots' so to speak (Wesenbeeck et al. 2009).

When it comes to such criticisms, one of the more vocal and perhaps longest running critics, Svedberg has provided much commentary over the years on the many shortfalls of the FAO's figures (Svedberg 1987, 1999, 2000, 2002a, 2002b). Various charges have generally suggested that the FAO statistics point policy makers in the wrong direction with erroneous and inappropriate data, wholly unsuitable for measuring or monitoring the prevalence of undernourishment. The complaints are long and of a technical nature, too numerous and esoteric to add any more to this discussion, although important and relevant none-the-less. A common thread within Svedberg's criticisms regard inherent biases of internal parameters that deliver false-positives and false-negatives that place either upward or downward biases and which in turn reflect subsequent overestimation or underestimation of figures. Of particular concern is the distributional impact of poverty on calorie consumption (ibid). Although in the case of the latter charge, in his defense Loganaden Naiken, former chief statistician of FAO/ESSA Statistics Division, denies this, suggesting that Svedberg and others have in fact misinterpreted or perhaps misunderstood some of the correlations involved (Naiken 2002; FAO 2008). Relatedly, on the note of FAO's acceptability of criticism, Ian MacAuslan, social policy consultant of Oxford Policy Management, recently suggests that FAO representatives simply play lip service to the notion of criticism and alternative methodologies. Instead, he opines that FAO has attempted to discredit such measures and accuses them of plainly ignoring people's concerns (MacAuslan 2009). Whether one sees or agrees with this or not is perhaps a subjective call and one on which the author is divided; besides, referring to the protein-energy fiasco, it would not be the first time the FAO have buried their head in the sand.

The FAO is certainly not surprised by such criticisms either. After all they suggest that given the nature and attention of such headline statistics, it is almost to be assumed, par-for-the-course even (SOFI 2004). With an eye on the balanced view however, credit has to be given to FAO for their unwavering determination to improve their own methodology and to encourage alternative, complementary measures. One such important International Scientific Symposium was convened in 2002 to review current methods of measurement and analysis (FAO 2002b). Among the frank and open exchanges between champions of different methodologies, advantages and disadvantages, as well as relative strengths were examined and discussed. Noting its own limitations, the FAO too made particular reference to the need for more information regarding the situation at the sub national levels and suggested that household surveys were well placed to fill this gap. It was also agreed that importance should be given to those metrics that correctly identify trends over time; however perhaps one of the most encouraging outcomes was the recognition that each method contributed something of the whole (FAO 2002a). That is, by formally re-establishing the fact that there still does not exist, one 'individual measure [that] suffices to capture all aspects of food insecurity'. In this way, a complementarity of metrics to aid in the measurement and classification of malnutrition. Collectively for instance, the FAO's undernourishment figures would provide information on availability and access; anthropometric data offers useful metrics on utilisation; and stability of access can be provided through the use of qualitative data methodologies.

F.3.2 WHO Method

The WHO calculates the prevalence of undernutrition, stunting, wasting and underweight of under-5s for numerous (over 100) countries (MDG 2010). In a drive to standardise anthropometric data, the WHO initiated the Global Database on Child Growth and Malnutrition in 1986. This helped promote the analysis over time and space and importantly enabled cross-border comparability and dissemination of nutritional surveys worldwide (Onis and Blössner 2003). Initially a database is compiled using results from numerous WHO, UNICEF (UN Children's Fund) and other countrywide household, nutrition, health and demographic as well as cluster surveys, which are then appropriately validated and tested for quality control. Such data, using the relevant cut-off Z-scores is then compared against WHO's current reference population to denote –2 standard deviations (SD) as moderate, and –3 SD as severe undernutrition. This calculation is applied to metrics that determine stunting, wasting and underweight. The method of standardising the data in this way allows the accumulation of valuable data over the years and allows for trends to be discerned over time. In fact the current database holds data on more than 31 million under-5s data collected since 1965 (Onis et al. 2004a).

Despite the precise and consistent nature with which anthropometric data can be measured, the WHO's method of determining undernutrition suffers the same limitations as all anthropometric data methodologies described previously (Shetty 2002). On top of this, these measures in themselves tell little of the specific causes of such stunting or wasting, as many other factors such as infection, previous episodes of malnutrition, stress, and genetics might all play a part in the child's physical morphology. Further, when undernutrition is identified, there are limitations in what some of the measures can tell us as well:

> Its composite nature complicates its interpretation. For example, the indicator fails to distinguish between short children of adequate body weight and tall, thin children. (MDG 2003, 2010; Scaramozzino 2006)

Lastly, measures of child undernutrition are taken to be representative of the population as a whole, while there is much anecdotal evidence to support such empirical correlations, extrapolating such data to the adult population has its limits too (Shetty 2002).

F.3.3 US Method

Global figures of undernourishment from the United States are calculated in the US Department of Agriculture's Economic Research Service (ERS) annual food security assessment. These are based on early work of Reutlinger and Selowsky in 1978, which still forms the backbone of ERS's approach (Reutlinger and Selowsky 1978). However, while US agencies refer to these calculations as 'food security', they are in fact measures of absolute undernutrition and undernutrition by income groups. The first thing to note in the ERS's calculation is the use of minimum calorie requirements. Whereas the FAO calculates regional average minimum requirements of approximately 1800 kcal, the ERS takes FAO's developing countries' regional average requirements as between 2000 and 2100 kcal, often simplified as 2100 kal, and uses this figure to act as their average minimum requirement over all developing countries (USDA/ERS 2008a, 2008b; Meade 2010).

Using this then, two measures of food gaps (previously food needs) are calculated: firstly the national average nutrition gap. This is, the difference, at national levels, between a per capita daily requirement of 2100 kcal and overall supply or availability. Secondly, a distribution gap is calculated, whereby the shortfall of each income group in relation to the 2100 calorie standard is calculated and presented as equivalent tonnage grain shortfall.

Nutrition gap: In the first calculation, the national nutrition gap, the US and FAO methodologies are similar in their defined objectives, that is, calculating food availability, measuring inequality in distribution and the measured shortfall between production and supply. However, this is where the similarity ends. The ERS computes their figures using grain equivalent measures; to do this, they determine three groupings that collectively encompass 100% of supply: grains, root crops and 'other' calculate each and combine and present them in a single grain equivalent measure. The first part of the equation is total food availability, that is, domestic supply minus non-food use. Domestic supply in turn is seen as the sum of domestic production, commercial imports, changes in stock and any food aid received. These four parameters are based on many assumptions (see earlier critique) and complex functions. The first, domestic production, is seen as the product of the area planted and yield response. This is in turn is based on considerations of rural labour, fertiliser, technology, capital use and real domestic prices (itself a product of income, substitute products and exchange rates) and substitute commodity prices. Secondly commercial imports are seen as a function of real-world and substitute food prices, foreign exchange availability and import restriction policies. From here, allowances are made in respect of changes in stock and food aid and in respect of non-food use (less, seed, livestock feed, exports and other uses). Next, as in the FAO model, calculations are then made of the relationships between income and calorie consumption (Engels curve) and variations in income distribution. Estimations can then be made as to those falling below the threshold of 2100 kcal and presented in national shortfalls of tons of grain, equivalent to caloric value (Reutlinger and Selowsky 1978; USDA/ERS 1997; Senauer and Sur 2001; USDA/ERS 2008a).

Distribution gap estimates of food gaps alone fail to identify access by different income bands. In response, the ERS's projections of food availability or consumption by different income groups determine this distribution gap. This involves indirectly calculating consumption of five different income groups (quintiles) by assuming a declining relationship between consumption and reducing income quintiles. This calorie-income relationship is then used to calculate the amount of food needed to raise consumption of each income quintile to meet per capita nutritional requirements of 2100 calories.

When concerned with the income, ERS's distribution gap calculation is its greatest strength. By way of example, in very simplistic terms and assuming £2 daily was sufficient to satisfy daily nutritional requirements: Imagine then, five people, earning between them a total of £10 daily. In the first of the ERS's calculations, this would suggest that total income (£10) divided by total people (5) would equate to everyone (100%) being satisfactorily fed. Imagine next that income distribution was skewed with one person earning £5 and the other four only £1. Although it immediately becomes obvious that four people or 80% of the population would be hungry, ERS's first calculation (the nutrition gap) would still register a 100% sated population, whereas their second calculation (the distribution gap through the use of quintiles) though would determine that only 20% of the population was actually properly fed. With income being an important determination of food security, this is a valuable tool. Another strength of the US methodology is their recognition of associated shortfalls with their calculations. Although sadly, while this was recognised a long time ago, the methodology has not changed a great deal over the years. (USDA/ERS 1997, 2008b). On the downside, certain employed assumptions are made with regard to commercial imports being products of domestic and world commodity prices as well as foreign exchange (FE) availability. Other assumptions too suggest FE reserves are constant, i.e. FE availability equals FE use, while the idea that import commodity prices are exogenous or outside the influence of individual countries is also employed. On top of this, while much historical data on which these estimates are based are updated using relevant growth parameters with food aid, etc., assumptions are also made regarding each country receiving static levels of food aid throughout a 10-year period (ibid).

F.4 International Food Policy Research Institute's (IFPRI) Global Hunger Index (GHI)

Launched in 2006 IFPRI hopes to mobilise support and draw attention to the plight of the hungry with their new multidimensional hunger index. A composite index incorporating other institutional data, the GHI combines three key dimensions of undernutrition. These include the FAO's proportion of undernourishment, the WHO's prevalence of underweight and the UNICEF's under-5 mortality rates (Wiesmann 2006). It was felt that by combining the above indices a valuable tool could be employed as an international tool for advocacy and one that would aid in the determination and analysis of the hunger problem. In calculating the index, sees only those countries considered developing and in transition being included in the metric. Developed countries are treated as having largely taken care of the problem of hunger. Also GHI are not calculated for countries supplying more than 2900 kcal (average 1995–97) and countries with under-5 mortality rate below 1.5% (15 per 1000 live births). After these exclusions the GHI index then is simply calculated giving equal weight to each metric using the following equation:

$$GHI = \frac{PUN + CUW + CM}{3}$$

where GHI = Global Hunger Index; PUN = proportion of the population undernourished %; CUW = prevalence of underweight in children under five %; and CM = % of under-5 child mortality.

The overriding advantage of this method is its ability to integrate related multidimensional nutritional metrics into one user friendly tool. By combining these indices random errors in individual methodologies are reduced, further strengthening or triangulating extant data. Lastly, a single Index from 0 to 100 is simple to use tool that gives hierarchical ranking and an ability to spot trends over time. Limitations in this methodology however relate to the individual measures and have been adequately covered. By strengthening each institutions metrics the GHI would benefit.

References

Avencena, I. T. and G. Cleghorn (2001). The Nature and Extent of Malnutrition in Children. In *Nutrition in the Infant: Problems and Practical Procedures*. V. R. Preedy and R. Watson, eds. London: Medical Media.

Bagust, A. and T. Walley (2000). An Alternative to Body Mass Index for Standardizing Body Weight for Stature. *Quartely Journal of Medicine* 93: 589–596.

Beaton, G. H. and J. M. Bengoa (1976). Some Concepts and Practical Considerations in Planning and Evaluation. In *Nutrition in Preventative Medicine: The Major Deficiency Syndromes, Epidemiology and Approaches to Control*. G. H. Beaton and J. M. Bengoa, eds. Geneva: World Health Organisation.

Bhattacharyya, A. K. (2006). Composite Index of Anthropometric Failure (CIAF) Classification: Is It More Useful? *Bulletin of the World Health Organization* 84(4): 335.

Birch, L. (2009). *An Evaluation to Review the Effectiveness of an Established Residential Weight Management Intervention on Short Term Health Outcomes in Overweight and Obese Children and Adolescents.* Chester: University of Chester. Doctor of Philosphy thesis.

Blössner, M. and M. d. Onis (2005). Malnutrition: Quantifying the Health Impact at National and Local Levels. *Environmental Burden of Disease Series, No. 12.* Geneva: World Health Organization.

Bowers, L. J. (1999). Assessment of Nutritional Status. In *Best Practices in Clinical Chiropractic*. R. D. Mootz and H. T. Vernon, eds. Maryland: Aspen Publishers.

Briones, E., E. Perea, M. P. Ruiz, C. Torro and M. Gili (1989). The Andalusian Nutritional Survey: Comparison of the Nutritional Status of Andalusian Children Aged 6–60 Months with That of the Nchs/Cdc Reference Population. *Bulletin of te World Health Organisation* 67(4): 409–416.

CDC (2010). Website of the CDC. US Centres for Disease Control. http://www.cdc.gov/ (accessed 15 June 2011).

Checchi, F., M. Gayer, R. F. Grais and E. J. Mills (2007). Public Health in Crisis Affected Populations a Practical Guide for Decision-Makers. London: Humanitarian Practice Network (HPN) at ODI.

Chowdhury, M. S. I., N. Akhter, M. Haque, R. Aziz and N. Nahar (2008). Serum Total Protein and Albumin Levels in Different Grades of Protein Energy Malnutrition *Journal of Bangladesh Society of Physiologist* 3: 58–60.

Cogill, B. (2003). Anthropometric Indicators Measurement Guide. Washington, DC: USAID Food and Nutrition Technical Assistance Project.

Collins, N. and L. Friedrich (2010). Using Laboratory Data to Evaluate Nutritional Status. *Ostomy Wound Management* 56(3): 14–16.

Cook, Z., S. Kirk, S. Lawrenson and S. Sandford (2005). Bapen Symposium 3 on 'from Beginners to Zimmers': Use of Bmi in the Assessment of Undernutrition in Older Subjects: Reflecting on Practice. *Proceedings of the Nutrition Society* 64: 313–317.

Donovan, C. (2009). *Availability Assessment and Analysis: Assessing Availability.* EC-FAO.

Edelstein, S. and J. Sharlin (2009). *Life Cycle Nutrition: An Evidence-Based Approach.* Massachusetts: Jones and Bartlett Publishers.

Elamin, A. (2008). *Protein Energy Malnutrition.* Pittsburg: University of Pittsburgh.

FAO (2002a). Directory of Fao Statutory Bodies and Panels of Experts: Section One—Statutory Bodies. Rome: Food and Agriculture Organisation.

FAO (2002b). Measurement and Assessment of Food Deprivation and Undernutrition: International Scientific Symposium. International Scientific Symposium: FIVIMS—An Inter-Agency Initiative to Promote Information and Mapping Systems on Food Insecurity and Vulnerability. Rome: Food and Agriculture Organisation.

FAO (2008). FAO Methodology for the Measurement and Assessment of Food Deprivation: Updating the Minimum Dietary Energy Requirements. Rome: FAO Statistics Division.

FAO/WHO/UNU (2004). Human Energy Requirements Report of a Joint FAO/WHO/UNU Expert Consultation: Rome, 17–24 October 2001. Rome: UNU/WHO/FAO.

Fergusson, P. L. (2009). *Severe Acute Malnutrition and Hiv in Children in Malawi.* Chester: University of Chester. Doctor of Philosphy thesis.

Flegal, K. M. (1999). Curve Smoothing and Transformations in the Development of Growth Curves. *American Journal of Clinical Nutrition* 70(suppl): 163S–165S.

Gallagher, D., M. Visser, D. Sepúlveda, R. N. Pierson, T. Harris and S. B. Heymsfield (1996). How Useful Is Body Mass Index for Comparison of Body Fatness across Age, Sex, and Ethnic Groups? *American Journal of Epidemiology* 143(3): 228–239.

Garza, C. and M. d. Onis (2007). Symposium: A New 21st-Century International Growth Standard for Infants and Young Children. *The Journal of Nutrition* 137: 142–143.

Gibson, R. S. (2005). *Principles of Nutritional Assessment.* Oxford: Oxford University Press.

Gilmore, J. (1999). Body Mass Index and Health. Health Reports 11(1).

Gomez, F., R. R. Galvan, S. Frenk, J. C. Munoz, R. Chavez and J. Vazquez (1956). Mortality in Second and Third Degree Malnutrition. *The Journal of Tropical Pediatrics* 2: 77–83.

Gopalan, C. and S. J. R. Kamala (1984). Classifications of Undernutrition—Their Limitations and Fallacies. *Journal of Tropical Pediatrics* 30(1): 7–10.

Gorstein, J., K. Sullivan, R. Yip, M. D. Onis, F. Trowbridge, P. Fajans and G. Clugston (1994). Issues in the Assessment of Nutritional Status Using Anthropometry. *Bulletin of the World Health Organisation* 72(2): 273–283.

Gueri, M., J. M. Gurney and P. Jutsum (1980). The Gomez Classification. Time for a Change *Bulletin of the World Health Organisation* 58(5): 773–777.

Haddad, L. (2002). 2001 Assessment of the World Food Security Situation. *ODI Food Security Briefings.* London: Overseas Development Institute 4.

Jelliffe, D. B. (1966). The Assessment of the Nutritional Status of the Community (with Special Reference to Field Surveys in Developing Regions of the World). *WHO Monograph Series NO 53.* Geneva: WHO 271.

Keys, A., F. Fidanza, M. J. Karvonen, N. Kimura and H. L. Taylor (1972). Indices of Relative Weight and Obesity. *Journal of Chronic Diseases* 25(6–7): 329–343.

Khan, M. R., I. Mahmud, S. Samiha, N. Islam, M. Rahman, S. Sohelin, I. Jahan, T. Sarram, R. Zereen, K. M. M. Islam, D. Hossain and A. Rouf (2008). Knowledge, Attitude and Feeding Practices among the Mothers Having under-5 Children in a Rural Community of Bangladesh. *Ibrahim Medical College Journal* 2(1): 35–36.

Klasen, S. (2008). Poverty, Undernutrition and Child Mortality: Some Inter-Regional Puzzles and Their Implications for Research and Policy. *Journal of Economic Inequality* 6(1): 89–115.

Kulsum, A., J. L. A. and J. Prakash (2009). Dietary Adequacy of Indian Children Residing in an Urban Slum— Analysis of Proximal and Distal Determinants *Ecology of Food and Nutrition* 48(3): 161 177.

MacAuslan, I. (2009). Hunger, Discourse and the Policy Process: How Do Conceptualizations of the Problem of 'Hunger' Affect Its Measurement and Solution? *European Journal of Development Research* 21: 397–418.

Maleta, K. (2003). *Growth in Undernutrition in Rural Malawi.* Medisiinarinkatu: University of Tampere.

Maynard, L. M., W. Wisemandle, A. F. Roche, W. C. Chumlea, S. S. Guo and R. M. Siervogel (2001). Childhood Body Composition in Relation to Body Mass Index *Paediatrics* 107(2): 344–350.

MDG (2003). *Handbook: Indicators for Monitoring the Millennium Development Goals.* New York: United Nations Development Group.

MDG (2010). Millennium Development Goals Indicators. United Nations Development Group. http://unstats. un.org/unsd/mdg/ (accessed 12 March 2011).

Meade, B. (2010). *Calorie Requirements.* Washington, DC: ERS/USDA.

Merck Manual (2005). *Merck and the Merck Manuals.* New Jersey: Merck Sharp & Dohme Corporation.

Micozzi, M. S. and D. Albanes (1987). Three Limitations of the Body Mass Index. *American Jornal of Clinical Nutrition* 46: 376–377.

Myatt, M., T. Khara and S. Collins (2005). Technical Background Paper: A Review of Methods to Detect Cases of Severely Malnourished Children in the Community for Their Admission into Community Based Therapeutic Care Programs. Geneva: WHO.

Naiken, L. (2002). Measurement and Assessment of Food Deprivation and Undernutrition: FAO Methodology for Estimating the Prevalence of Undernourishment International Scientific Symposium. Rome: Food and Agriculture Organisation.

Nandy, S., M. Irving, D. Gordon, S. V. Subramanian and G. D. Smith (2005). Poverty, Child Undernutrition and Morbidity: New Evidence from India. *Bulletin of the World Health Organization* 83(3): 161–240.

Neithercut, W. D., A. D. Smith, J. McAllister and L. F. G. (1987). Nutritional Survey of Patients in a General Surgical Ward: Is There an Effective Predictor of Malnutrition? *Journal of Clinical Pathology* 40: 803–807.

NHLBI (1998). Clinical Guidelines on the Identification, Evaluation, and Treatment of Overweight and Obesity in Adults: The Evidence Report. US: US National Institutes of Health: National Heart, Lung and Blood Institute.

NHS/NOO (2009). Body Mass Index as a Measure of Obesity. London: National Health Service National Obesity Observatory.

NLIS (2010). *Nutrition Landscape Information System (NLIS)*. Geneva: World Health Organisation.

Onis, M. d. (2000). Measuring Nutritional Status in Relation to Mortality. *Bulletin of the World Health Organization* 78 (10): 1271–1274.

Onis, M. d. and M. Blössner (2003). The World Health Organization Global Database on Child Growth and Malnutrition: Methodology and Applications. *International Journal of Epidemiology* 32(5): 18–26.

Onis, M. d., M. Blossner, E. Borghi, E. A. Frongillo and R. Morris (2004a). Estimates of Global Prevalence of Childhood Underweight in 1990 and 2015. *Journal of the American Medical Association* 291(21): 2600–2606.

Onis, M. d., C. Garza, C. G. Victora, M. K. Bhan and K. R. Norum (2004b). The Who Multicentre Growth Reference Study (Mgrs): Rationale, Planning, and Implementation. *Food and Nutrition Bulletin* 25(supplement 1): S1–S89.

Oshikoya, K. A. and I. O. Senbanjo (2009). Pathophysiological Changes That Affect Drug Disposition in Protein-Energy Malnourished Children. *Nutrition & Metabolism* 6(50): 7.

Pérez-Escamillal, R. and A. M. Segall-Corrêal (2008). Food Insecurity Measurement and Indicators. *Revista de Nutrição* 21(Supplemental): 15s–26s.

Quetelet, A. (1835). *Sur L'homme Et Le Développement De Ses Facultés: Ou, Essai De Physique Sociale*. Paris: Bachelier, imprimeur-libraire.

Reutlinger, S. (1977). Malnutrition. A Poverty or a Food Problem? *World Development* 5(8): 715–724.

Reutlinger, S. and M. Selowsky (1978). *Malnutrition and Poverty; Magnitude and Policy Options*. Baltimore, MD: Johns Hopkins University Press.

Salbe, A. D., D. P. Kotler, A. R. Tierney, J. Wang, R. N. Pierson and R. G. Campbell (1995). Correlation between Serum Insulin-Like Growth Factor 1 (IGF1) Concentrations and Nutritional Status in Hiv-Infected Individuals. *Nutrition Research* 15(10): 1437–1443.

Scaramozzino, P. (2006). Measuring Vulnerability to Food Insecurity. Rome: FAO Agricultural and Development Economics Division (ESA) 26.

Senauer, B. and M. Sur (2001). Ending Global Hunger in the 21st Century: Projections of the Number of Food Insecure People. *Review of Agricultural Economics* 23(1): 68–81.

Sepahi, M. A., B. Baraty and F. K. Shooshtary (2010). HDR Syndrome (Hypoparathyroidism, Sensorineural Deafness and Renal Disease) Accompanied by Hirschsprung Disease. *Iranian Journal of Pediatrics* 20(1): 123–126.

Shenkin, A. (2006). Serum Prealbumin: Is It a Marker of Nutritional Status or of Risk of Malnutrition? *Journal of Clinical Chemistry* 52: 2177–2179.

Shetty, P. (2002). Measurement and Assessment of Food Deprivation and Undernutrition: Keynote Paper: Measures of Nutritional Status from Anthropometric Survey Data. International Scientific Symposium. Rome: Food and Agriculture Organisation.

Shils, M. E., M. Shike, C. Ross and R. J. Cousins (2006). *Modern Nutrition in Health and Disease*. Philadelphia, PN: Lippincott, Williams and Wilkins.

SOFI (2004). The State of Food Insecurity in the World 2004: Monitoring Progress Towards the World Food Summit and Millennium Development Goals. *SOFI*. Rome: Food and Agriculture Organisation.

SOFI (2008). The State of Food Insecurity in the World 2008. Rome: Food and Agriculture Organisation 59.

Stevens, J., J. E. McClain and K. P. Truesdale (2008). Selection of Measures in Epidemiologic Studies of the Consequences of Obesityselection of Obesity Measures. *International Journal of Obesity* 32: S60–S66.

Sullivana, D. H. (2000). Guest Editorial: What Do the Serum Proteins Tell Us About Our Elderly Patients? *The Journal of Gerontology: Series A* 56(2): M71–M74.

Svedberg, P. (1987). Undernutrition in Sub-Saharan Africa: A Critical Assessment of the Evidence. *Working Papers*. UN World Institute for Development and Economics Research (WIDER).

Svedberg, P. (1999). 841 Million Undernourished? *World Development* 27(12): 2081–2098.

Svedberg, P. (2000). *Poverty and Undernutrition: Theory, Measurement, and Policy*. Oxford: Oxford University Press.

Svedberg, P. (2002a). Measurement and Assessment of Food Deprivation and Undernutrition: Part III: Parallel Contributed Papers Sessions—Fallacies in—and Ways of Improving—the FAO Methodology for Estimating Prevalence of Undernutrition. International Scientific Symposium: FIVIMS—An Inter-Agency Initiative to Promote Information and Mapping Systems on Food Insecurity and Vulnerability. Rome: FAO.

Svedberg, P. (2002b). Undernutrition Overestimated. *Economic Development and Cultural Change* 51(1): 5–36.

Swenne, I., M. Stridsberg, B. Thurfjell and A. Rosling (2007). Insulin-Like Growth Factor-1 as an Indicator of Nutrition During Treatment of Adolescent Girls with Eating Disorders. *Acta Pædiatrica* 968: 1203–1208.

Tetanye, E. (2004). Use of Anthropometry for Evaluation of Nutritional Status in Children. Geneva: Geneva Foundation for Medical Education and Research.

Thomas, D. R. (2007). Disorders of Nutrition and Metabolism: Undernutrition. In *The Merck Manual Online Medical Library*. R. S. Porter and J. L. Kaplan, eds. New Jersey: Merck Sharp & Dohme Corporation.

Udall Jr, J. N., Z. A. Bhutta, A. Firmansyah, P. Goyens, M. J. Lentze and C. Lifschitz (2002). Malnutrition and Diarrhea: Working Group Report of the First World Congress of Pediatric Gastroenterology, Hepatology, and Nutrition. *Journal of Pediatric Gastroenterology and Nutrition* 35: S173–S179.

USDA/ERS (1997). Food Security Assessment Gfa-9. *Situation and Outlook Series*. Washington, DC: USDA Economic Research Service.

USDA/ERS (2008a). Food Security Assessment, 2008–09: Overview—Food-Security Impact of the Financial Downturn, 2008–18. *Food Security Assessment*. Washington, DC: USDA/Economic Research Service GFA-20.

USDA/ERS (2008b). Website of the Economic Research Service. *Food Security Assessment*. USDA/Economic Research Service. http://www.ers.usda.gov/briefing/GlobalFoodSecurity/questions.htm#securityassessed (accessed 20 April 2010).

Waterlow, J. C. (1972). Classification and Definition of Protein-Calorie Malnutrition. *British Medical Journal* 3: 566–569.

Waterlow, J. C. (1973). Note on the Assessment and Classification of Protein-Energy Malnutrition in Children. *The Lancet* 302(7820): 87–89.

Waterlow, J. C. (1976). Classification and Definition of Protein Energy Malnutrition.In *Nutrition in Preventative Medicine: The Major Deficiency Syndromes, Epidemiology, and Approaches to Control*. G. H. Beaton and J. M. Bengoa, eds. Geneva: World Health Organisation.

Waterlow, J. C., R. Buzina, W. Keller, J. M. Lane, M. Z. Nichaman and J. M. Tanner (1977). The Presentation and Use of Height and Weight Data for Comparing the Nutritional Status of Groups of Children under the Age of 10 Years. *Bulletin of the World Health Organization* 55(4): 489–498.

Wellcome Group (1970). Wellcome Trust Working Party: Classification of Infantile Malnutrition. *The Lancet* 296(7667): 302–303.

Wesenbeeck, C. F. v., M. A. Keyzer and M. Nubé (2009). Estimation of Undernutrition and Mean Calorie Intake in Africa: Methodology, Findings and Implications. *International Journal of Health Geographics* 8(37): 1–18.

WFP (2008). *Ten Minutes to Learn About … Improving the Nutritional Quality of WFP's Food Basket—an Overview of Nutrition Issues, Commodity Options and Programming Choices*: World Food Programme. 2008.

WHO (1976). Joint FAO/UNICEF/WHO Expert Committee on the Methodology of Nutritional Surveillance. *Technical Report Series no. 593*. Geneva: World Health Organisation 24.

WHO (1983). *Guidelines for Assessing the Nutritional Impact of Supplementary Feeding Programme Nonserial Publication*. Geneva: World Health Organization.

WHO (1995). Physical Status: The Use and Interpretation of Anthropometry: Report of the Expert Committee. Geneva: World Health Organisation.

WHO (2004). Appropriate Body-Mass Index for Asian Populations and Its Implications for Policy and Intervention Strategies. *The Lancet* 363(9403): 157–163.

WHO (2006). *Who Child Growth Standards: Length/Height-for-Age, Weight-for-Age, Weight-for-Length, Weight-for-Height and Body Mass Index-for Age: Methods and Development.* Geneva: World Health Organisation.

WHO (2008). Transitioning to the WHO Growth Standards: Implications for Emergency Nutrition Programmes IASC Nutrition Cluster Informal Consultation. Geneva: World Health Organisation 34.

WHO (2010a). *Global Database on Child Growth and Malnutrition.* Geneva, Switzerland: World Health Organisation.

WHO (2010b). Website of the World Health Organisation. World Health Organisation. http://www.who.int/ (accessed 2 April 2010).

WHO (2010c). *The Who Global Database on Body Mass Index (BMI).* Geneva, Switzerland: World Health Organisation. 2010.

WHO/FAO (2004). *Vitamin and Mineral Requirements in Human Nutrition (Draft Version)*: WHO/FAO.

WHO/MGRSG (2006). Who Multicentre Growth Reference Study Group: Child Growth Standards Based on Length/Height, Weight and Age. *Acta Pædiatrica*(Suppl 450): 76–85.

Wiesmann, D., L. Bassett, T. Benson and J. Hoddinott (2009). Validation of the World Food Programme's Food Consumption Score and Alternative Indicators of Household Food Security. *IFPRI Discussion Paper.* Washington, DC: The International Food Policy Research Institute (IFPRI).

World Vision (2009). Website of World Vision. World Vision. http://www.worldvision.org.uk/ (accessed 23 May 2010).

Appendix G: Biological Systematics

There is an amazing diversity of life or biota on the planet. While approximately 1.8 million species have actually been described, the potential for undiscovered species is huge, with various estimates putting this figure anywhere between 5 and 100 million, although the figure is more likely to be in the region of 10–15 million. Describing, analysing and cataloguing this information is a behemoth of a task and in keeping of the high spirited nature of scientific debate is no less troubled by controversy. Systematic biology (systematics) is a field of study which attempts to describe, preserve and provide scientific names and classifications as well as understand the evolutionary interrelationships of all living things. Biological systematics also aims to study and provide new theories or mechanisms of evolution. Taxonomy, on the other hand, is that part of systematics that deals with providing scientific names and classifications although for many, taxonomy continues to be the source of much confusion with many terms being used interchangeably and out of context. The following helps clarify.

G.1 Taxonomy

While originally taxonomy referred only to the identification and classification of organisms, it is now becoming common to use it in contexts beyond biology. In this way, taxonomy and typology are sometimes used interchangeably. In this book however, a clear separation is used whereby taxonomy is used in its biological sense and typology for all other categorisational needs. Taxonomy specifically deals with the grouping of organisms sharing similar characteristics whether physical, genetic or based on other criteria. Taxon or taxa (plural) are a grouping of similar organisms judged to belong to a single unit or group. This group is then given a name and a rank. Taxonomy in this sense then is both a naming convention as well as a means of categorisation. It is a convention that has had a long and distinguished history beginning with Aristotle who suggested that all life fell into two taxonomic categories: bloodless (invertebrates) and blooded (vertebrates). From these early beginnings, this process of classifying life developed over the centuries and ultimately led to the Carolus Linnaeus (Karl von Linné)[*] classification as set out in his book *Systema Naturæ* in 1735 (NCBI 2011). The rank-based method popularised by Linnaeus arranged the taxonomy of all organic cellular life in a hierarchical structure. Although Linnaeus's rankings were slightly different from the modern day rankings, he did set the standard by which most followed. The hierarchical taxa ranging from domain to species essentially dictates a family structure with the lowest (species) being subsumed by the group above it (genus), as in Table G.1. Also within these divisions, there might in addition be many subdivisions of the main taxonomic levels such as subphylum, superclass and infraclass. It is also worth noting too that the ranks and hierarchical structure are not static either; they change over time and among disciplines or viewpoints as well as through necessity as new knowledge dictates (following discussion). That said, Table G.1 highlights an acceptable starting point.

While the above taxonomy then is the general naming and categorising of organisms or the framework so to speak, populating this framework is filled with equally diverse difficult decisions and depends largely on which particular methodology or paradigm is followed. As a result, classification systems display diversity of opinion and clashing guidelines and subsequent developments within the various doctrines can be thought of as belonging to three camps: Linnaean, phenetic and phylogenetic (cladistics) (UCMP 1996).

---

[*] Karl von Linné (or similar) published over 180 books and as his writings were mostly in Latin, he is known by scientific world today by the Latinized name he chose for himself, Carolus Linnaeus.

TABLE G.1

Taxonomic Ranking of Life

Main Taxonomic Ranks				Number of Taxa per Rank[a]
Latin Terminology		**English Terminology**		
Regio		Domain		3
Regnum		Kingdom		3–7 (see discussion below)
Phylum[b]	Division[c]	Phylum (pl. Phyla)[b]	Division[c]	111
Classis		Class		268[d]
Ordo		Order		1068[e]
Familia		Family		
Genus		Genus (pl. Genera)		
Species		Species		1.7–100 million[f]

Sources: UCMP, *Journey into Phylogenetic Systematics*, University of California Museum of Paleontology, Berkeley, CA, 1996; Britannica, Encyclopædia Britannica Online, 2009; Catalogue of Life, Catalogue of Life: 2010 Annual Checklist. Species 2000 and the Integrated Taxonomic Information System (ITIS), 2011; WRI, Relative Number of Described Species in Major Taxa, WRI, 2011.

[a] Approx. numbers based on catalogue of Life Checklist.
[b] Phylum is the Taxon (pl. Taxa) used in zoology.
[c] Division is the Taxon (pl. Taxa) used in botany.
[d] Includes 32 phyla with class not assigned.
[e] Includes 87 class with order not assigned.
[f] 1.7 million discovered but estimates vary as to those yet undiscovered.

G.1.1 Linnaean Taxonomy

As well as determining the structure and the basic principles of scientific naming,[*] Linnaeus, although not actually responsible for its inception, lends his name to the early practice of categorising organisms into the above framework. Before the advent of modern, genetically based evolutionary concepts of classification, this grouping of biological organisms into different categories was based largely on their morphological characteristics and presumed natural relationships. The leading proponent of this, as mentioned, was Linnaeus and as a result this type of classification can be called Linnaean taxonomy.

Since Darwin however, biological classification has come to be understood in terms of evolutionary relationships across space and time. This shift in perspective saw the beginning of an alternative classification system to Linnaeus's that attempted to categorise organisms not based solely on morphological characteristics and similarities but through the perceived relationships of organisms from an evolutionary perspective. Combined with the growing microbiological knowledge of the time, the classification system was in for a mini revolution. For a long time, distinctions among the top level rankings was simply made between plant and animal kingdoms; however, it was soon realised that microscopic life did not fit particularly well in either category and around the middle of the nineteenth century debate about the various existing kingdoms ensued. Although not fully settled, general consensus did arise as to the general grouping of this infusoria or microscopic life:

> … Protozoa (primitive animals), Protophyta (primitive plants), Phytozoa (animal-like plants), and Bacteria (primarily regarded as plants). (Scamardella 1999, p. 207)

---

[*] Perhaps Linnaeus's greatest legacy besides this was his use of binomial nomenclature or binomen, a two-part rank-based name. This is the practice of using the last two sub-groupings (the genus and species) as the Latin (or Latinized) name, more popularly known as its scientific name. For example, the scientific name for the human race is derived from the genus type Homo and the species type sapiens, thus the term *Homo sapiens* is unique to humankind. Note too, the convention for its notation, the first letter of the genus is capitalised while the species is not, moreover, convention also dictates that the name is italicised.

TABLE G.2

Evolution of the Kingdom Classification of Life

Carolus Linnaeus —1735	Richard Owen— 1860	Ernst Haeckel —1866	Edouard Chatton —1937	Herbert Copeland —1938	Robert Whittaker —1969	Woese and Fox—1977	Woese, Kandler and Wheelis—1990
2 kingdoms	3 kingdoms	3 kingdoms	2 empires	4 kingdoms	5 kingdoms	6 kingdoms	3 domains
			Prokaryota	Monera	Monera	Eubacteria	Bacteria
						Archaebacteria	Archaea
	Protozoa	Protista			Protista	Protista	
				Protista	Fungi	Fungi	
			Eukaryota				Eukarya
Vegetabilia	Plantae	Plantae		Plantae	Plantae	Plantae	
Animalia	Animalia	Animalia		Animalia	Animalia	Animalia	

Sources: Linnaeus, C., *Species Plantarum: Exhibentes Plantas Rite Cognitas Ad Genera Relatas Cum Differentiis Specificis, Nominibus Trivialibus, Synonymis Selectis, Locis Natalibus Secundum Systema Sexuale Digestas*, Impensis Laurentii Salvii, 1753a; Linnaeus, C., *Systema Naturae Per Regna Tria Naturae, Secundum Classes, Ordines, Genera, Species, Cum Characteribus, Differentiis, Synonymis, Locis*, Joannis Wilhelmi, 1753b; Owen, R. (1860). *Palaeontology*. Edinburgh: Adam and Charles Black. Haeckel, E., *Generelle Morphologie Der Organismen*, Reimer, Berlin, 1866; Copeland, H., *Q. Rev. Biol.*, 13, 383–420, 1938; Imprimerie, C.E., *Titres Et Travaux Scientifiques (1906–1937)*, S. Sottano, Sète, Italy, 1938; Whittaker, R., *Science* 163, 150–160, 1969; Woese, C.R. and Fox, G.E., *Proc. Natl Acad. Sci.*, 74, 5088–5090, 1977; Woese et al., *Proc. Nal Acad. Sci.*, 87, 4576–4579, 1990.

Within this microcosm however, there was no clear understanding of their nature or their evolutionary relationship to each other or even the existing kingdoms. As a result they ended up variously being classified and reclassified numerous times over the years depending on different camps' views (ibid). In 1937 however a breakthrough came when Chatton proposed a division of life based on the cellular level. Although not complete, it marked the beginning of modern understanding which separated prokaryotes (bacteria), which lack a cell nucleus; and eukaryotes, which essentially contain a cell nucleus and other membrane-bound complex cell structures—organelles. However, this was not the end of the matter and during the 1970s, Woese, dissatisfied with this phenetic or morphological classification alone, further divided prokaryotes into two domains: bacteria and archaea. Further clarification in 1990 saw the emergence of a domain system above the kingdoms that placed more emphasis on the different evolutionary and biochemical relationships at the molecular level (genes) between the two prokaryotic and the single eucaryotic domains. This evolution of kingdoms is summarised in Table G.2, which, not so incidentally, coincided with the rise of the phylogenic (cladistics) paradigms of classification.

While a good place to start, the above discussion and table clearly show that not everyone was in agreement. Although the new three-domain system has generally been accepted as replacing the traditional two-empire system there are still many biologists who prefer the old 5 kingdom ranking of Whittaker's (WRI 2011).

G.1.2 Phenetic Taxonomy

Although strictly speaking, the Linnaean categorisation is phenetic (physical similarity); phenetic taxonomy or numeric taxonomy, as it is sometimes known, gets its name through the practice of using correlations and algorithms to determine relationships and linkage. Developed in the 1930s in response to charges of speculation in general taxonomy, Sokal and Sneath used a method based on mathematical algorithms to generate similarity coefficients in an attempt to strengthen relationships among organisms. The trouble with this approach was that it was still fundamentally based on phenotypic characteristics or similarities and while ancestrally, some organisms might appear to share certain similarities; genetically speaking they might not be related at all. This tendency of classifying unrelated organisms together was, the critics argue, the methods biggest failing. While the system is not in widespread use today, it did however offer one major contribution to modern taxonomy, the widespread use of computational-based modelling.

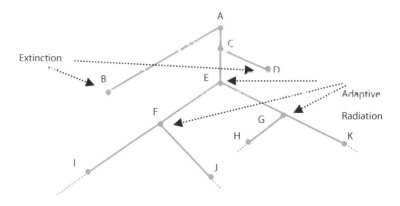

FIGURE G.1   Phylogeny: branching—adaptive radiation.

G.1.3 Phylogeny

Phylogeny on the other hand, is the branch of taxonomy that attempts to determine such groupings based on an organism's shared genetic (ancestral) lineage rather than physical similarities. It began when phylogeny or phylogenetic taxonomy (also known as cladistics) arrived around the time of the 1950s when modern genetic advances suggested that all organisms were genetically related. In this way, phylogenists tend not to use the traditional Linnaean hierarchical taxonomic ranking as the various levels are considered complex and indefinite. Instead phylogenists variously attempt to display relationships in an evolutionary tree. Based on the idea of common ancestry, or branching evolution (adaptive radiation). These are known as clades (also taxa) and is the origin of the term 'cladistics'.

There are three distinct ways of doing this:

1. *Monophyletic*: A grouping of any or and all descendents of a shared common ancestor. In the example in Figure G.1, all are descendents of letter A; however it also means that I & J are also monophyletic grouping of F. Other groups include G with H, K, or E with F,I,J,G,H,K, etc.
2. *Paraphyletic* descended from a common ancestor, that is, they are monophyletic groups. They are distinguished from monophyletic groups through the exclusion of the most recent common ancestor usually because of morphologic distinctiveness. In this way, in the above example, paraphyletic groups might be I & J, or H & K or C & D for instance.
3. *Polyphyletic* refers to the grouping of organisms based on common traits rather than direct ancestry (homoplasy). Consequently, such groups are often comprised of organisms that evolved separately, giving rise to groups with more than one common ancestor. As a branch of taxonomy, it is not considered in line with phylogentic principles and as such are constantly being rooted out in favour of monophyletic clades.

The process of phylogeny, after the many advances in genetic engineering, is becoming easier and the use of DNA/RNA mapping now allows for speedier and greater accuracy of the process.

G.1.3.1 Phylogenic Classification

With regard to the phylogenic classification, it can be seen that confusion once again prevails (Queiroz 2005). In reality, what is often heralded as a classification of taxonomy in strictness phylogeny is in fact a naming convention. The International Code of Phylogenetic Nomenclature (PhyloCode) is currently developing a formal set of rules governing the nomenclature of phylogeny. Importantly, unlike the rank-based codes of Linnaeus and others, the PhyloCode does not assign hierarchical groupings. Instead phylogenic classification uses cladograms (similar to Figure G.1 above), which elucidate ancestral relations between taxa. These diagrams often evolve into what is commonly known as the tree of life or family

TABLE G.3

Species Known to Currently Exist in the World

Category	Species	
	IUCN (2010)	GEO (2000)
Animals vertebrates	62,305	45,000
Animals invertebrates	1,305,250	1,250,000
Plants	321,212	250,000
Fungi & protists	51,563	70,000
Bacteria		4000
Protoctists (algae, protozoa, etc.)		80,000
Total	1,740,330	1,700,000

Sources: GEO, *Global Environment Outlook*, United Nations Environment Programme, Malta, 2000; IUCN, The IUCN List of Threatened Species, International Union for Conservation of Nature and Natural Resources (IUCN), Cambridge, 2010.

tree type structures and once several species are mapped can resemble a complex self-replicating fractal system. The clade system, rather than using a two-name identifier, which is then classified according to rank, employs several identifiers such as species, specimens or apomorphies to indicate actual organisms and ensure mutual exclusion. Thus this formulaic approach helps indicate lineage and in this way; a single defined taxon becomes this ancestor and all of its descendants. While this does not assign rank, it does allow for its optional use of a suitable rank-based ordering system (Queiroz 2005; PhyloCode 2011). Furthermore, Linnaean taxonomy does not preclude the monophyletic and paraphyletic groups of phylogeny as taxa either. Indeed since the early twentieth century, Linnaean taxonomists have widely endeavored to conform to this monophyletic convention, at least at the family- and lower-level taxa all the while other cladists advocate an all encompassing classification system of all organisms based on the three 'apparently' monophyletic kingdoms of Eukaryota (Eukarya), Eubacteria (Bacteria), and Archaebacteria (Archaea).

G.2 Biodiversity

So with all this in mind how are the current biotic forms reflected in the Earth's biodiversity? Table G.3 illustrates two perspectives averaging around 1.7–1.75 million known species with estimates of a further 12.5 million as yet un-described species.

References

Britannica (2009). Encyclopædia Britannica Online. 2009.

Catalogue of Life (2011). Catalogue of Life: 2010 Annual Checklist. Species 2000 & the Integrated Taxonomic Information System (ITIS). http://www.catalogueoflife.org/annual-checklist/2010/info/about (accessed 21 March 2011).

Copeland, H. (1938). The Kingdoms of Organisms. *Quarterly Review of Biology* 13: 383–420.

GEO (2000). Global Environment Outlook. Malta: United Nations Environment Programme.

Haeckel, E. (1866). *Generelle Morphologie Der Organismen*. Berlin: Reimer.

Imprimerie, C. E. (1938). *Titres Et Travaux Scientifiques (1906–1937)*. Sète, Italy: S. Sottano.

IUCN (2010). The IUCN List of Threatened Species. Cambridge, United Kingdom: International Union for Conservation of Nature and Natural Resources (IUCN).

Linnaeus, C. (1753a). *Species Plantarum: Exhibentes Plantas Rite Cognitas Ad Genera Relatas Cum Differentiis Specificis, Nominibus Trivialibus, Synonymis Selectis, Locis Natalibus Secundum Systema Sexuale Digestas*. Impensis Laurentii Salvii.

Linnaeus, C. (1753b). *Systema Naturae Per Regna Tria Naturae, Secundum Classes, Ordines, Genera, Species, Cum Characteribus, Differentiis, Synonymis, Locis*, Joannis Wilhelmi.

NCBI (2011). The Ncbi Taxonomy Homepage. The National Center for Biotechnology Information. http://www.ncbi.nlm.nih.gov/ (accessed 21 March 2011).

PhyloCode (2011). Phylocode. International Society for Phylogenetic Nomenclature. http://www.ohio.edu/phylocode/ (accessed 12 March 2011).

Queiroz, K. d. (2005). The Phylocode and the Distinction between Taxonomy and Nomenclature. *Systematic Biology* 55(1): 160–162.

Owen, R. (1860). *Palaeontology*. Edinburgh: Adam and Charles Black.

Scamardella, J. M. (1999). Not Plants or Animals: A Brief History of the Origin of Kingdoms Protozoa, Protista and Protoctista. *International Microbiology* 2: 207–216.

UCMP (1996). Journey into Phylogenetic Systematics. University of California Museum of Paleontology. http://www.ucmp.berkeley.edu/clad/clad4.html (accessed 21 March 2011).

Whittaker, R. (1969). New Concepts of Kingdoms of Organisms. *Science* 163: 150–160.

Woese, C. R. and G. E. Fox (1977). Phylogenetic Structure of the Prokaryotic Domain: The Primary Kingdoms. *Proceedings of the National Academy of Sciences (PNAS)* 74: 5088–5090.

Woese, C. R., O. Kandler and M. L. Wheelis (1990). Towards a Natural System of Organisms: Proposal for the Domains Archaea, Bacteria, and Eucarya. *Proceedings of the National Academy of Sciences (PNAS)* 87: 4576–4579.

WRI (2011). Relative Number of Described Species in Major Taxa. WRI. http://archive.wri.org/page.cfm?id=581&z (accessed 12 March 2011).

Appendix H: Millennium Development Goals

TABLE H.1

Official List of MDG Indicators

Millennium Development Goals (MDGs)	
Goals and Targets (from the Millennium Declaration)	**Indicators for Monitoring Progress**

Goal 1: Eradicate extreme poverty and hunger

Target 1.A: Halve, between 1990 and 2015, the proportion of people whose income is less than $1 a day.	Proportion of population below $1 (PPP) per day; poverty gap ratio; share of poorest quintile in national consumption.
Target 1.B: Achieve full and productive employment and decent work for all, including women and young people.	Growth rate of GDP per person employed; employment-to-population ratio; proportion of employed people living below $1 (PPP) per day; proportion of own-account and contributing family workers in total employment.
Target 1.C: Halve, between 1990 and 2015, the proportion of people who suffer from hunger.	Prevalence of underweight children under-five years of age; Proportion of population below minimum level of dietary energy consumption.

Goal 2: Achieve universal primary education

Target 2.A: Ensure that, by 2015, children everywhere, boys and girls alike, will be able to complete a full course of primary schooling.	Net enrolment ratio in primary education; proportion of pupils starting grade 1 who reach last grade of primary; literacy rate of 15- to 24-year-olds, women and men.

Goal 3: Promote gender equality and empower women

Target 3.A: Eliminate gender disparity in primary and secondary education, preferably by 2005, and in all levels of education no later than 2015.	Ratios of girls to boys in primary, secondary and tertiary education; share of women in wage employment in the non-agricultural sector; proportion of seats held by women in national parliament.

Goal 4: Reduce child mortality

Target 4.A: Reduce by two-thirds, between 1990 and 2015, the under-five mortality rate.	Under-five mortality rate; infant mortality rate; proportion of 1 year-old children immunised against measles.

Goal 5: Improve maternal health

Target 5.A: Reduce by three quarters, between 1990 and 2015, the maternal mortality ratio.	Maternal mortality ratio; proportion of births attended by skilled health personnel.
Target 5.B: Achieve, by 2015, universal access to reproductive health.	Contraceptive prevalence rate; adolescent birth rate; antenatal care coverage; unmet need for family planning.

Goal 6: Combat HIV/AIDS, malaria and other diseases

Target 6.A: Have halted by 2015 and begun to reverse the spread of HIV/AIDS.	HIV prevalence among population aged 15–24 years; condom use; proportion of population aged 15–24 years with comprehensive correct knowledge of HIV/AIDS; ratio of school attendance of orphans to school attendance of non-orphans aged 10–14 years.
Target 6.B: Achieve, by 2010, universal access to treatment for HIV/AIDS for all those who need it.	Proportion of population with advanced HIV infection with access to antiretroviral drugs.

continued

TABLE H.1 (Continued)

Official List of MDG Indicators

Millennium Development Goals (MDGs)	
Goals and Targets (from the Millennium Declaration)	**Indicators for Monitoring Progress**
Target 6.C: Have halted by 2015 and begun to reverse the incidence of malaria and other major diseases.	Incidence and death rates associated with malaria; proportion of children under-5 sleeping under insecticide-treated bednets; proportion of children under 5 with fever who are treated with appropriate anti-malarial drugs; incidence, prevalence and death rates associated with tuberculosis; proportion of tuberculosis cases detected and cured under directly observed treatment short course.

Goal 7: Ensure environmental sustainability

Target 7.A: Integrate the principles of sustainable development into country policies and programmes and reverse the loss of environmental resources. Target 7.B: Reduce biodiversity loss, achieving, by 2010, a significant reduction in the rate of loss.	Proportion of land area covered by forest; CO_2 emissions, total, per capita and per \$1 GDP (PPP); consumption of ozone-depleting substances; proportion of fish stocks within safe biological limits; proportion of total water resources used; proportion of terrestrial and marine areas protected; proportion of species threatened with extinction.
Target 7.C: Halve, by 2015, the proportion of people without sustainable access to safe drinking water and basic sanitation.	Proportion of population using an improved drinking water source; proportion of population using improved sanitation facility.
Target 7.D: By 2020, to have achieved an improvement in the lives of at least 100 million slum dwellers.	Proportion of urban population living in slums.

Goal 8: Develop a global partnership for development

Target 8.A: Develop further an open, rule-based, predictable, non-discriminatory trading and financial system. Includes a commitment to good governance, development and poverty reduction—both nationally and internationally. Target 8.B: Address the special needs of the least developed countries. Includes: tariff and quota free access for the least developed countries' exports; enhanced programme of debt relief for heavily indebted poor countries (HIPC) and cancellation of official bilateral debt; and more generous ODA for countries committed to poverty reduction. Target 8.C: Address the special needs of landlocked developing countries and small island developing states (through the Programme of Action for the Sustainable Development of Small Island Developing States and the outcome of the twenty-second special session of the General Assembly). Target 8.D: Deal comprehensively with the debt problems of developing countries through national and international measures in order to make debt sustainable in the long term.	Some of the indicators listed below are monitored separately for the least developed countries (LDCs), Africa, landlocked developing countries and small island developing states; Official development assistance (ODA); Net ODA, total and to the least developed countries, as percentage of OECD/DAC donors' gross national income; proportion of total bilateral, sector-allocatable ODA of OECD/DAC donors to basic social services (basic education, primary health care, nutrition, safe water and sanitation); proportion of bilateral official development assistance of OECD/DAC donors that is untied; ODA received in landlocked developing countries as a proportion of their gross national incomes; ODA received in small island developing states as a proportion of their gross national incomes; market access; proportion of total developed country imports (by value and excluding arms) from developing countries and least developed countries, admitted free of duty; average tariffs imposed by developed countries on agricultural products and textiles and clothing from developing countries; agricultural support estimate for OECD countries as a percentage of their gross domestic product; proportion of ODA provided to help build trade capacity; debt sustainability; total number of countries that have reached their HIPC decision points and number that have reached their HIPC completion points; cumulative); debt relief committed under HIPC and MDRI initiatives; debt service as a percentage of exports of goods and services.

TABLE H.1 (Continued)

Official List of MDG Indicators

Millennium Development Goals (MDGs)	
Goals and Targets (from the Millennium Declaration)	**Indicators for Monitoring Progress**
Target 8.E: In co-operation with pharmaceutical companies, provide access to affordable essential drugs in developing countries.	Proportion of population with access to affordable essential drugs on a sustainable basis.
Target 8.F: In co operation with the private sector, make available the benefits of new technologies, especially information and communications.	Telephone lines per 100 population; cellular subscribers per 100 population; internet users per 100 population.

Source: The Millennium Development Goals and targets from the Millennium Declaration, signed by 189 countries, including 147 heads of State and Government, in September 2000 UN (2000b) and from further agreement by member states at the 2005 World Summit (Resolution adopted by the General Assembly; A/RES/60/1).

Reference

UN (2008). Millennium Development Goals. http://www.un.org/millenniumgoals/ (accessed 15 July 2010).

Appendix I: Global Warming—The Basics

In order to fully appreciate the complex nature of global warming, it is useful to explore some of the fundamentals of the Earth's climatic processes. The following section takes a brief look at some of these basic ideas, which when combined helps to give an overview of the big picture in connection with food security.

I.1 Climate

To many, climate is often described in terms of the variability in the averages of temperature, precipitation and wind over a given period of time. This might be in timescales of millennia, decades or even just a few short months. The Earth's climate system is in constant fluctuation through changes in its own internal or external dynamics, called 'forcings'. Internal variabilities are interactions such as internal plate tectonics and oceanic circulations while external forcings need to be separately identified by their natural or anthropogenic (human induced) sources. Natural external forcings might include volcanic eruptions or perhaps solar variability while anthropogenic forcings might consist of changes in concentrations of greenhouse gases and land cover usage (IPCC 2007; NOAA 2011a, 2011b).

At the most basic level, the Earth's climate system is powered by solar radiation energy from the sun. It arrives in various wavelengths: longer wavelengths like microwave and radio waves and the more important high-energy short wavelengths like ultraviolet radiation. However because of the atmosphere, not all of this radiation gets through (Figure I.1). While some is absorbed by the Earth (about 30%) increasing the planet's temperature, some is reflected directly back into space; this reflection is called the albedo effect.

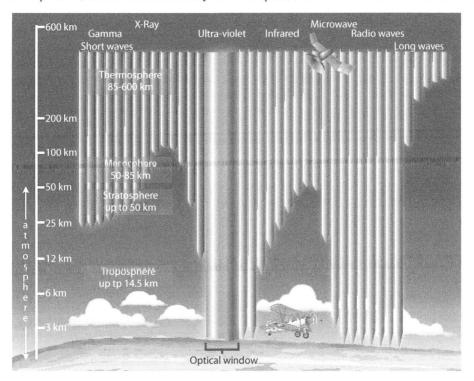

FIGURE I.1   Electromagnetic waves travelling through the atmosphere. (Based on an idea by Windows to the Universe, University Corporation for Atmospheric Research. Courtesy of Patrick Newsham, © 2011.)

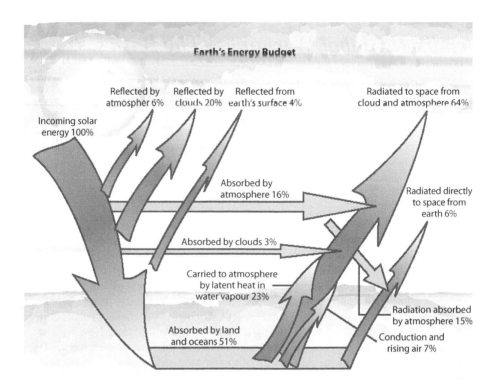

FIGURE I.2   Earth's radiative forcings. (Based on an idea by NASA Langley Research Center Atmospheric Science Data. Courtesy of Patrick Newsham, © 2011.)

Between incoming radiation, the albedo effect and the resultant thermal emissions via longwave radiation (infrared), an equilibrium or balance known as the radiation/energy budget or radiative forcings is created (NASA 2007). As part of the incoming radiation is absorbed by the Earth's surface (51%) the air is warmed and part of this radiative energy is then transferred back into the atmosphere via thermal currents, evapotranspiration and infrared radiation (Figure I.2).

In turn, some of this is then absorbed by clouds and greenhouse gases, which then re-emit part of this longwave radiation back to Earth and part out to space, hence the greenhouse effect (Kiehl and Trenberth 1997; IPCC 2007). This greenhouse effect is important for were Earth to rely solely on incoming radiation alone, based on our distance from the sun, then the average surface temperature would be $-18/19°C$; instead as a result of this interaction, the mean temperature is nearer to $+15°C$.

There are three ways in which this delicate balance can be upset. First, natural variations in the incoming solar radiation through aspects like Earth's changing orbit, or sunspots, etc. Second, by the changing the amount of solar radiation reflected (the albedo) through such events like changes in cloud cover, snow, ice and vegetative cover as well as atmospheric particles (volcanic eruptions, soot from burning coal etc). Third, by altering the longwave (heat) radiation emitted from the Earth through changes in for example, greenhouse gas concentrations (GCM 2005; IPCC 2007). Detecting and understanding such mechanisms has been the focus of a great body of research by scientists and policy makers.

Basics aside, getting to the heart of the debate of global climate change and the associated issues of contention needs further insight and clarification in two particular areas, namely, historical and current world temperature records as well as the nature and role of greenhouse gases themselves (GHG).

I.2 Global Warming: Is it Happening?

Accurate global direct temperature measurements do not exist before the mid-1800s; however, there are many climatologists assessing other variables that depend on temperature. For instance, just as marine

TABLE I.1

The Main Components of the Atmosphere

Gas	Pre-Industrial Values (1750[a])	2004/2005 Averages (In Parts per Million)	Mole Factor Percentage of Total Atmosphere (2004/2005)
Nitrogen N_2		780840	78.084
Oxygen O_2		209480	20.948
Argon		9340	0.934
Water H_2O[b]		30000	3
Carbon dioxide CO_2	280 ppm	385	0.0385
Methane CH_4	715 ppb	1.774	0.0001774
Nitrous oxide N_2O	270 ppb	0.319	0.0000319
Misc. gases[c]		0.03	0.000003

Sources: IPCC, Climate Change 2007: Working Group I: The Physical Science Basis, What Factors Determine Earth's Climate? Intergovernmental Panel on Climate Change, Geneva, 2007; CDIAC, Carbon Dioxide Information Analysis Center, 2011; NASA, Earth Observatory, NASA, 2011.

[a] Approximate pre-industrial date.

[b] Varies considerably with location, altitude and climate from 0% to 4%, with a global average of around 3%.

[c] Miscellaneous gases include CFCs, PFCs, HFCs and SF_6.

fossils are a reliable method of measuring temperatures, the oxygen isotopes in the Antarctic Vostoc ice cores are equally as important. Furthermore, important paleoclimate data can also be found in such varied sources as tree rings, corals reefs and lake sediments. In this way, scientists are able to provide an accurate and valuable picture of global temperatures, precipitation and volcanic activity enabling climatologists to reconstruct approximate near-global temperature records stretching back thousands of years (Petit et al. 2000; NASA 2005; IPCC 2007). With all this information, there is general agreement that the global average temperature has increased from about 0.7 to about 1.4 degrees Fahrenheit since the late 1800s. Further, there is widespread consensus that this warming has been unprecedented over the last 1000 years (Mastrandrea and Schneider 2005; EIA 2011; NOAA 2011b). There is less agreement however, over whether this falls outside of the natural range of variability or indeed whether a normal range in temperatures can actually be adequately quantifiable (Stouffer and Ya 1994; Ghil 2002; GCM 2005). On top of this, it is the difficulties of ascribing such changes to either natural or anthropogenic causes that strikes at the heart of the global warming debate. So where do these gases come from, how are they measured and what links temperature rises to activities of the greenhouse effect and in particular rises in CO_2?

The atmosphere comprises several gases (Table I.1); however, because of the density of air at different temperatures and altitudes it makes it awkward to measure gases by weight (mass fractions). Instead measurements use the WMO standard 'mole fraction' (or mixing ratio), which represents the relative number of molecules in a given air sample. This can be reliably measured in both the troposphere and stratosphere as chemicals (species) through the process of convection and turbulent mixing (eddy diffusion) are said to be 'well mixed'.

Greenhouse gases then are any gases that contribute to the absorption and reflection of the radiative forces of the Earth's energy budget. Many occur naturally in various degrees (Table I.2); these include water vapour (the most potent of the GHGs) followed by carbon dioxide, methane, nitrous oxide, ozone (not to be confused with the ozone layer) and a few trace gases. These gases as well as contributing to the Earth's thermal balance also protect life by absorbing UV radiation and some of the infrared radiation emitted by the Earth as mentioned earlier. Although they are produced through both biological and geochemical processes, they have remained in relative balance for millennia; that was up until about 200 years or so ago (Houghton et al. 2001; IPCC 2007; WMO 2009; CDIAC 2011).

TABLE I.2

Natural and Anthropogenic Sources of Greenhouse Gases

Gas	Natural Sources	Anthropogenic Sources
Carbon dioxide	Evaporation from the oceans, carbon dioxide released through natural decay; respiration of animals and plants	Fossil fuel combustion, forest clearing, cement production, deforestation, urban landscapes
Methane (CH_4)	Bacteria producing methane (swamp gas) in wetlands; natural gas; termites; oceans; vegetation	Landfills; production and distribution of natural gas; raising ruminant animals (cattle and sheep); fossil fuel combustion: rice paddies: biomass burning
Nitrous oxide (N_2O)	Tropical soils; evaporation from the oceans	Fertiliser use; fossil fuel and biomass burning; raising animals; nylon production, other industrial activities; manure
Ozone in the troposphere	Lightning	Carbon monoxide; hydrocarbons and nitrogen oxide chemically react to produce ozone; fossil fuels; ozone from motor vehicle emissions is a potent component of smog
Hydrofluorocarbons (HFCs)	No natural sources	Refrigeration gases, aluminium smelting, semiconductor manufacturing, etc
Perflourocarbons (PFCs)	No natural sources	Aluminium production, semiconductor industry, etc.
Sulfur hexafluoride (SF_6)	No natural sources	Electrical transmissions and distribution systems, circuit breakers, magnesium production, etc
Halocarbon[a]	Biomass burning, macroalgae, oceanic algae	Chlorofluorocarbons; refrigerants, propellants and solvents aerosols; foam manufacture; air conditioning

Sources: IPCC, Climate Change 2007: Working Group I: The Physical Science Basis, What Factors Determine Earth's Climate? Intergovernmental Panel on Climate Change, Geneva, 2007; Muir, P., *Bi301: Human Impacts on Ecosystems*, Oregon State University, 2010; ICBE, Calculating GHG Emissions. International Carbon Bank and Exchange, 2011; Weart, S., *The Discovery of Global Warming: Other Greenhouse Gases*, Harvard University Press, Cambridge, MA, 2011.

[a] A group of gases containing fluorine, chlorine and bromine.

Interest in the greenhouse effect is not new. As early as the mid-nineteenth century, scientists like John Tyndall realised that gases in the atmosphere, particularly carbon dioxide (CO_2) and water vapour, caused a 'greenhouse effect', affecting the planet's temperature. Building on this, Svante Arrhenius, at the turn of the century, suggested that emissions from industry could someday affect global warming. Others too like G.S. Callendar in the late 1930s championed the cause but were largely dismissed, and it was not until the 1960s with the rising measures of CO_2 by Keeling and others that researchers really began to take an interest. Subsequently, by discovering that CO_2 really did play a key role in the greenhouse effect, so carbon cycle studies proliferated. However, it took until the mid-1980s when a French-Soviet team looking at ice cores at Vostok Station in Antarctica found that CO_2 levels had risen and fallen in close tandem with temperatures that consensus really began to converge. Later, it was discovered too that these temperature changes in fact preceded CO_2 changes by several centuries prompting some to suggest that warming cycles can take thousands of years to be fully realised (Shackleton and Pisias 1985; Shackleton 2000).

Meanwhile, interest in chlorofluorocarbon (CFCs) effect on the stratospheric ozone layer, which blocks harmful ultraviolet rays from reaching the Earth's surface combined with other, weightier arguments, provoked a few scientists to a look at how the upper atmosphere was affected. Their concerns were discussed in Kyoto in 1973 and provided the impetus for Molina and Rowland to study the chemical emissions from human activities which indeed confirmed a threat to the ozone layer. This was followed by a scientist from the National Aeronautics and Space Administration (NASA), Veerabhadran Ramanathan, who discovered that a single molecule of CFC could be 10,000 times as potent as a molecule of CO_2

and that together with 30 other trace gases studied could collectively trigger as much global warming as CO_2 itself (Barrett and Maxwell 2005). This galvanised the scientific community and led to the ground-breaking1987 Montreal Protocol. In light of this, other gases including nitrous oxide (N_2O) and methane (CH_4), previously overlooked because of their minuscule quantities compared with CO_2, began to take on new momentum. Although jointly, these studies did not prove that by itself the greenhouse effect was responsible for the global warming. What it did do was firm up a growing consensus that the greenhouse effect was an important component in the system of climatic feedbacks that would now have to be taken seriously (Lorius et al. 1990; Pälike et al. 2006). This led to the 1992 United Nations Framework Convention on Climate Change (UNCCC) agreement in Rio de Janeiro and ultimately to the Kyoto Protocol.

I.3 Global Warming Potential

Because different gases have potentially disproportionate effects on warming through differing radiative or heat-trapping properties, a way had to be found to compare the potential harmful effects (including longevity in the atmosphere) on a singular comparative scale. This was achieved by converting all gases into CO_2 equivalents using conversion factors (Table I.3). However, one point worth mentioning is considering the importance of water vapour as the most important of GHGs; one glaring omission from this

TABLE I.3

Comparison of GWPs and Lifetimes Used in the SAR[a] and the TAR[a]

	Atmospheric Lifetime (Years)		GWP (100 year)	
Gas	SAR[a]	TAR[a]	SAR	TAR
Water vapour (H_2O)	NG	NG[b]	NG[b]	NG[b]
Carbon dioxide (CO_2)	50–200	5–200	1	1
Methane (CH_4)	123	8.4/12	21	23
Nitrous oxide (N_2O)	120	120/114	310	296
Hydrofluorocarbons				
HFC-23	264	260	11,700	12,000
HFC-125	32.6	29	2800	3400
HFC-134a	14.6	13.8	1300	1300
HFC-143a	48.3	52	3800	4300
HFC-152a	1.5	1.4	140	120
HFC-227ea	36.5	33.0	2900	3500
HFC-236fa	209	220	6300	9400
HFC-4310mee	17.1	15	1300	1500
Iodocarbons				
FIC-1311	<0.005	0.005	<1	1
Fully Fluorinated Species				
SF6	3200	3200	23,900	22,000
CF4	50,000	50,000	6500	5700
C2F6	10,000	10,000	9200	11,900
C4F10	2600	2600	7000	8600
C6F14	3200	3200	7400	9000

Source: Adapted from Climate Change 2001: The Scientific Basis. Contribution of Working Group I to the Third Assessment Report of the Intergovernmental Panel on Climate Change, Table 6.12. Cambridge University Press.

[a] SAR—Second Assessment Report, IPCC 1996; TAR—Third Assessment Report. IPCC 2001.

[b] NG = Not given.

table is its global warming potential (GWP); indeed nowhere in the IPCC's literature does it supply a figure for water vapour's GWP (Houghton et al. 2001; IPCC 2007; EPA 2011; ICBE 2011; Watkins 2011).

It is also worth mentioning that while the relative compilation of the two largest sources of greenhouse gases, water and CO_2 in the atmosphere (troposphere) is approximately 98.7%–1.2%, respectively, this makeup is reversed in the upper atmosphere (the stratosphere). In this rarefied atmosphere, the relative mix becomes about 80% carbon dioxide and about 20% water vapour; this also has implications for their respective roles in global warming.

I.4 Historic Greenhouse Records

Through the ice cores, it has been noted that for about a thousand years previous to the industrial revolution, defined by the IPCC as 1750, the amount of greenhouse gases present in the atmosphere had not changed appreciably (CDIAC 2011). On this basis, it was considered safe to assume that, in the absence of human intervention, this would have remained constant over the 150 years from 1850 to 2000. However, concentrations of various greenhouse gases have been shown to have increased. Moreover during the same period the rate of warming is increasing too, and according to Earth Policy indicators, the last two decades of the twentieth century were the hottest for nearly half a millennia (National Geographic 2007; EPI 2011). There is no doubt then that temperatures and CO_2 levels are rising but can the two be correlated? The IPCC and others think so: observations in a recent report suggest the increase in global average temperatures since the mid-twentieth century are very likely (90% probability) due to anthropogenic increase in greenhouse gas concentrations (Roach 2007; Zhang 2008; EIA 2011). Yet this issue, of whether the current warming is wholly natural or is increasing as a result of the cumulative effects of increased greenhouse gases exacerbated by humanity is the crux of many differing opinions (World Climate Report 2006; Brahic and Page 2007). On a brief note of causality, it is prudent to mention that while scientists are still unsure whether temperature increases cause CO_2 concentrations to increase or vice versa, at the very least their interaction is generally accepted. Lastly then, as the twenty-first century enters its second decade more corroborating research (ocean temperatures and fossil leaves) has left only the hardened critic doubting that CO_2 and other greenhouse gases are at least somewhat responsible for the present climate change (Wolfson and Schneider 2002; Brown 2005; Weart 2011).

I.5 Aerosols

While not a gas, aerosol particulates do have a warming impact worth mentioning. Essentially, they are small particles of varying size, concentration and chemical composition. As with greenhouse gases, these too originate from both natural and human activities. Fossil fuel and biomass burning increase aerosols in the atmosphere containing sulphur, soot (black carbon), nitrate, dust and organic compounds among others (IPCC 2007). These gases and aerosols create feedback mechanisms in the climate system called positive or negative feedback respectively that can either amplify or diminish greenhouse effects. These in turn also effect changes in climate forcings (ibid).

So, accepting that greenhouse gases are in part at least, responsible for global warming the international response was the Kyoto Protocol, which had its roots in the 1992 Earth Summit and the UN Framework Convention on Climate Change. However, it took a further 13 years before the treaty was entered into force in 2005. Section I.6 looks at the protocol in more detail.

I.6 The Kyoto Protocol

I.6.1 The Agreement

The Kyoto Protocol was adopted in Kyoto, Japan, in December 1997 and formally came into effect in 2005 to cover the emissions of six primary greenhouse gases: carbon dioxide, methane, nitrous oxide, hydrofluorocarbons, perflurocarbons and sulphur hexafluoride. The Kyoto Protocol involved a long drawn out process of inter-governmental negotiations over a 13-year period. Beginning in 1992, 154 countries signed the United Nations Framework Convention on Climate Change in Rio de Janeiro. The central element of the Convention was a commitment to stabilise greenhouse gas levels offsetting further potentially dangerous anthropogenic additions to climate change.

The 1992 Convention was only a general agreement in principle and the Convention did not provide any specifics or precise emission reduction targets. These issues were addressed at subsequent Conference of Parties (COP), most notably the Berlin Conference in 1995, which outlined specific targets, timeframes and policies in achieving the necessary reductions in greenhouse gases. A year later in COP2, Geneva, nations agreed the Ministerial Declaration that accepted the science of climate change as compelling, and thirdly, in 1997 at the Kyoto COP3, member countries signed the Protocol. Following the signing of the Protocol, participating nations held a further string of COP meetings to work out the Protocol's details in Buenos Aires (1998),The Hague (2000), Bonn (2001), Marrakech (2001), New Delhi (2002), Milan (2003), Buenos Aires (2004) and Montreal (2005).

I.6.2 How Does It Work?

Essentially governments are separated into two general categories: developed countries, referred to as Annex I countries and developing countries, referred to as Non-Annex I countries. Between January 2008 and 2012, Annex I countries have to reduce their greenhouse gas emissions by a collective average of 5% below their 1990 levels while Annex 2 countries such as China and India, are not obliged to limit their emissions at all. Under the Treaty, countries must meet their targets primarily through national measures. However, the Kyoto Protocol offers an additional means of meeting targets by way of three market-based mechanisms: Emissions trading (carbon trading), the clean development mechanism (CDM) and the joint implementation (JI). Effectively this allows countries to trade emissions and implement emission-reduction projects in developing countries to trade saleable units.

I.6.3 Pros and Cons

Since its signing in 1997, the Kyoto Protocol has been the centre of controversy for while some applauded its environmental benefits, others expressed concerns over its effectiveness and potential economic consequences and several key countries still, either refused to sign or ratify the agreement at all. In 1997 for instance while the Clinton Administration committed the United States to the Kyoto agreement, the election of George W. Bush witnessed a reversal of policy citing the highly critical nature of the Protocol's components, in particular the exemption granted to China. Later in 2002 the Australian government pulled out too expressing concern over the potential effectiveness of the protocol without the participation of the likes of the United States, China and India.

Proponents point out however that even without the United States and developing countries, the protocol will result in a slower rate of increase in greenhouse gas emissions offering that something is better than nothing. Further, it is often cited is as an important first step in a truly global emissions reduction regime of the future (Maich 2005; Mendelsohn 2005; UNFCCC 2011).

References

Barrett, C. and D. Maxwell (2005). *Food Aid after Fifty Years: Recasting Its Role*. London: Routledge.

Brahic, C. and M. L. Page (2007) Climate Myths: Ice Cores Show CO_2 Increases Lag Behind Temperature Rises, Disproving the Link to Global Warming. *New Scientist* (16 May 2007): http://environment.newscientist.com/channel/Earth/climate-change/dn11659.

Brown, L. (2005). *Outgrowing the Earth: The Food Security Challenge in an Age of Falling Water Tables and Rising Temperatures*. New York: W.W. Norton and Co.

CDIAC (2011). Carbon Dioxide Information Analysis Center. http://cdiac.esd.ornl.gov (accessed 20 January 2011).

EIA (2011). Energy Information Administration. *Energy Information Administration Brochures*. US Department of Energy. http://www.eia.gov/ (accessed 14 February 2011).

EPA (2011). Environment Protection Agency. US Environment Protection Agency. http://www.epa.gov (accessed 15 January 2011).

EPI (2011). Earth Policy Institute. Earth Policy Institute. http://www.earth policy.org/ (accessed 12 February 2011).

GCM (2005). *Natural Climatevariability*. Arlington, VA: George C. Marshall Institute

Ghil, M. (2002). Natural Climate Variability.In *Encyclopedia of Global Environmental Change*. M. MacCracken and J. Perry, eds. Chichester: John Wiley.

Houghton, J., Y. Ding, D. Griggs, M. Noguer, P. van der Linden, X. Dai, K. Maskell and C. Johnson (2001). *Ipcc: Climate Change 2001: The Scientific Basis*. Cambridge: Cambridge University Press.

ICBE (2011). Calculating Ghg Emissions. International Carbon Bank and Exchange. http://www.icbe.com/0.asp (accessed 23 February 2011).

IPCC (2007). Climate Change 2007: Working Group I: The Physical Science Basis, What Factors Determine Earth's Climate? Geneva, Switzerland: Intergovernmental Panel on Climate Change.

Kiehl, J. and K. Trenberth (1997). Earth's Annual Global Mean Energy Budget. *Bulletin of the American Meteorological Society* 78: 197–208.

Lorius, C., J. Jouzel, D. Raynaud, J. Hansen and H. Le Treut (1990). The Ice Core Record: Climate Sensitivity and Future Greenhouse Warming. *Nature* 347: 139–145.

Maich, S. (2005). *Kyoto Protocol's Shortcomings*. Canada: Historica Foundation

Mastrandrea, M. and S. Schneider (2005). *Global Warming: Nasa World Book Online*: NASA

Mendelsohn, R. (2005) An Economist's View of the Kyoto Climate Treaty. *National Public Radio*. http://www.npr.org/templates/story/story.php?storyId=4504298.

Muir, P. (2010). *Bi301: Human Impacts on Ecosystems*. Oregon State University.

NASA (2005). Paleoclimatology: The Ice Core Record.

NASA. http://Earthobservatory.nasa.gov/Study/Paleoclimatology_IceCores/ (accessed 20 August 2010).

NASA (2007). Earth's Radiation Budget Facts. NASA Atmospheric Science Data Center. http://eosweb.larc.nasa.gov/EDDOCS/radiation_facts.html (accessed 13 February 2011).

NASA (2011). Earth Observatory. NASA. http://Earthobservatory.nasa.gov/Laboratory/Biome/ (accessed 20 February 2011).

National Geographic (2007). *Global Warming Fast Facts: Global Warming, or Climate Change, Is a Subject That Shows No Sign of Cooling Down*. National Geographic Society.

NOAA (2011a). Climate Attribution, What Is Attribution? National Oceanic and Atmospheric Administration. http://www.cdc.noaa.gov/CSI/whatis/ (accessed 12 March 2011).

NOAA (2011b). *National Climatic Data Center*. Boulder: National Oceanic and Atmospheric Administration. 2011.

Pälike, H., R. Norris, J. Herrle, P. Wilson, H. Coxall, C. Lear, N. Shackleton, N. Tripati and B. Wade (2006). The Heartbeat of the Oligocene Climate System. *Science* 314: 1894–1898.

Petit, J., D. Raynaud, C. Lorius, J. Jouzel, G. Delaygue, N. Barkov and V. Kotlyakov (2000). *Trends: A Compendium of Data on Global Change: Historical Isotopic Temperature Record from the Vostok Ice Core*. U.S.A: Oak Ridge.

Roach, J. (2007) Global Warming 'Very Likely' Caused by Humans, World Climate Experts Say. *National Geographic News*, 2 February 2007. http://news.nationalgeographic.com/news/2007/02/070202-global-warming.html.

Shackleton, N. (2000). The 100,000-Year Ice-Age Cycle Identified and Found to Lag Temperature, Carbon Dioxide and Orbital Eccentricity. *Science* 289: 1897–1902.

Shackleton, N. and N. Pisias (1985). Atmospheric Carbon Dioxide, Orbital Forcing and Climate. In Sundquist, E. And Broecker, W/ (Eds.) the Carbon Cycle and Atmospheric CO_2: Natural Variations Archean to Present. *Geophysical Monograph 32*. Washington, DC: American Geophysical Union, pp. 303–317.

Stouffer, R. and M. Ya (1994). Model Assessment of the Role of Natural Variability in Recent Global Warming. *Nature* 367: 634–636.

UNFCCC (2011). The Kyoto Protocol. United Nations Framework Convention on Climate Change. http://unfccc.int/kyoto_protocol/items/2830.php (accessed 13 February 2011).

Watkins, T. (2011) Saturation, Non-Linearity and Overlap in the Radiative Efficiencies of Greenhouse Gases. *University Online Magazine*. http://www.sjsu.edu/faculty/watkins/radiativeff2.htm.

Weart, S. (2011). *The Discovery of Global Warming: Other Greenhouse Gases*. Cambridge, MA: Harvard University Press.

WMO (2009). The State of Greenhouse Gases in the Atmosphere Using Global Observations through 2008 *WMO-GAW Annual Greenhouse Gas Bulletins*. Geneva, Switzerland: World Meteorological Bulletin.

Wolfson, R. and S. Schneider (2002). Understanding Climate Science.In *Climate Change Policy: A Survey*. S. Schneider, A. Rosencranz and J. Niles. Washington, DC: Island Press.

World Climate Report (2006). *Co2 Emissions Link to Temperature Trends: A Quandary? World Climate Report*.

Zhang, W. (2008). A Forecast Analysis on World Population and Urbanization Process. *Environment Development and Sustainability* 10(6): 717–730.

Appendix J: Energy Sources

Considering a big part of the costs involved in agriculture, industry and transport are energy costs, cheap fuel is essential to the economic viability of food security. With this in mind it is worth understanding a little more of the whole energy market.

J.1 Fossil Fuels

When it comes to the global energy mix, the total world supply of energy primarily consists of 86% fossil fuels: that is commonly made up of oil (petroleum), coal and natural gas, and to a lesser extent, oil shale and peat. This is because in terms of current, economically viable energy sources, fossil fuel costs continue to remain lower than most other energy sources (WEO 2008; OPEC 2010; GEO 2011; GEP 2011; IEA 2011).

J.1.1 Coal

Coal is formed through coalification whereby fossilised plant debris forms in conditions deficient of enough oxygen to breakdown the organic matter. There are three main types of coal: anthracite, bituminous and lignite. Each contains varying amounts of carbon (the energy content) which in turn dictates its physical properties. Anthracite, the hardest of the three, contains the most carbon while bitumous and lignite decrease in that order (Energyquest 2011). Continuing existing usage and under present economic and technological conditions it has been estimated that with reserves of about 826 billion tonnes there is about 119 years supply of available coal (Figure J.1) (BP 2010; WCA 2011).

J.1.2 Oil and Natural Gas

Oil and natural gas, are often found together and are formed through a chemical process called catagenesis. They are formed from tiny plants (phytoplankton) and animals (zooplankton) collectively called plankton that settled to the bottom of ancient seas and rivers. Differences in temperature and depths usually determine the extent to which oil results over gas (Energyquest 2011; Naturalgas 2011). Out of all the alternative conventional and non-conventional energy sources, oil in particular has a significantly greater energy content than in any other currently available source. For this reason and considerations of relative costs, lifestyles of the developed and developing countries through industrial and economic growth have grown to become dependent on inexpensive oil and gas. Affordable industrial agriculture needs low-cost natural gas and oil for fertilisers, pesticides, machinery, processing and transportation. So how much oil is there and how long will it last? Goldman Sachs said in 1999 that 90% of global conventional oil had already been found and through production histories and reserve figures a picture emerged that suggested the peak of all liquid hydrocarbons would come around 2010. However, 2010 has come and gone and production has still not peaked. In response the International Energy Agency's World Energy Outlook 2010 now suggests that peak oil will not be reached before 2035 (WEO 2010). In fact, although peak oil might be a way off, with estimated oil reserves at 1337.2 billion barrels (Figure J.2) the projected longevity of this resource at present demand is still only 46 years at most (OPEC 2009; BP 2010; WCA 2011).

Natural gas, on the other hand, is one of the cleanest, safest and cheapest of all the energy sources. It is composed mainly of methane, but it can also include ethane, propane, butane and pentane. It can be measured volumetrically or as with all other fuels by using its energy equivalent. Natural gas reserve estimates (Figure J.2) from end 2009 are approximately 189,712 billion cubic feet, so continuing at current usage levels that gives us a supply of just about 63 years (OPEC 2009; WCA 2011).

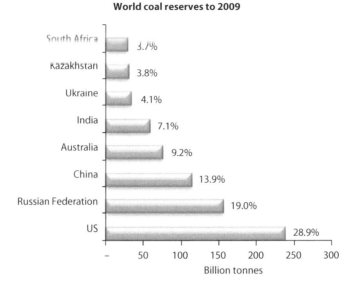

World coal reserves to 2009

FIGURE J.1   Proven coal reserves to 2009. (From the *BP Statistical Review of World Energy.* Statistical Review. UK: British Petroleum, 2010.)

J.1.3 Future Fossil Energy Supplies

While doomsayers forewarn of the collapse of world energy resources (Howe 2004; Lendman 2008), a conference report by Freedom 21 (2008) provides an alternative view suggesting that economically available supplies of oil and gas continue to confound the experts. In their view, 50 years ago many experts pessimistically estimated that only a 13-year supply of oil remained and that today's estimates of 40+ years, suggest Freedom 21, is equally misjudged. The report also goes on to suggest that as prices continue to increase and with more efficient technology, former uneconomically recoverable oil will become profitable to extract further increasing the world's supply by as much as 50%. The report also suggests that under these circumstances a 5000-year supply of shale oil becomes economically accessible.

J.2 Nuclear Energy

Natural mined uranium consists of different isotopes (atomic mass numbers), of which U-235 is only about 0.7%. Uranium-235 is the key ingredient in nuclear fusion, and the raw mined uranium needs to be enriched with these heavier (U-235) isotopes to a concentration of about 5% (NRC 2011). Once enriched, 1 kg of uranium-235 can theoretically produce as much electricity as 1500 tonnes of coal (Emsley 2003, p. 479). Although it has been said that, through excessive government regulations and political inexpediency, uranium as a feedstock for power generation tends to be expensive. At current rates of use, estimations of uranium deposits can supply nuclear reactors for the next 100 years or so (Freedom 21 2008).

J.3 Renewables

While renewable energy supplies are ethically desirable, it is observed that major technological breakthroughs and better policy environment considerations will need to be addressed if renewable energy is going to be anything more than a niche sector. At the moment with the exception of hydropower, all renewables—biomass, geothermal, wind and solar power—are expensive compared to fossil fuel alternatives (Freedom 21 2008).

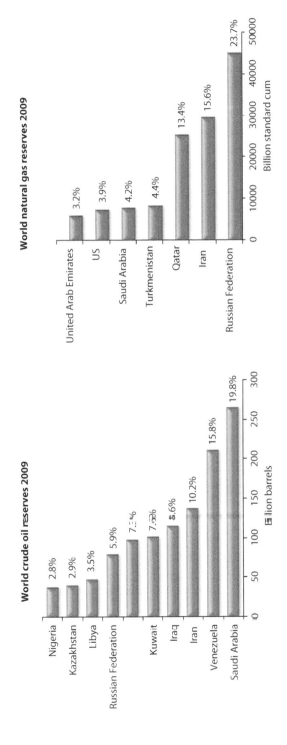

FIGURE J.2 Global crude oil and gas reserves end 2009. (From OPEC's Annual Statistical Bulletin. Vienna: Organization of the Petroleum Exporting Countries, 2009.)

J.4 Biofuels

Partly in answer to rising fuel prices and an over-reliance on imported oil together with Kyoto Protocol commitments to cut greenhouse gas-associated emissions, biofuels including, among others, ethanol and biodiesel have been thrown into the political limelight (DTE 2007; National Geographic 2008). Biofuels themselves are not new; indeed, they have been around since the start of the twentieth century. Pioneers like Henry Ford and Rudolph Diesel designed cars to use biofuels only to witness the discoveries of huge petroleum deposits, effectively putting biofuels on the back burner (Ostrolenk 1930). Although compared to petroleum, biofuel use in vehicles is still quite low, it does have the potential to displace a significant proportion of the world's transport fuels (Roubanis 2004). So what are biofuels?

J.4.1 What Are Biofuels?

For ease of analysis, bioenergy systems can be divided into three main categories: (1) traditional biomass (used in cooking and general heating), (2) modern biomass-based technologies (often used in electricity generation) and (3) liquid biofuels used predominantly in the transport industry (IBEP 2006; CFS 2007). The first category, traditional biomass, accounts for up to 90% of household energy consumption in developing countries. With four out of five people living in rural areas without access to electricity; fuelwood, charcoal and animal dung for cooking is the norm for perhaps 2.5 billion people around the globe (IBEP 2006; CFS 2007). Being one of the most common forms of generated energy, it also accounts for over half the wood harvested globally and represents approximately 15% of current world energy use (IBEP 2006; Fritsch and Gallimore 2007). Second, increased usage of wood and wood residues are being exploited in the generation of electricity alongside more traditional fossil fuels among the industrialised countries (CFS 2007). However, it is the third category that has direct implications for food security, that of liquid biofuels. Biofuels encompass a long list of fuels made from processed biomass (organic materials) which include: methane, ethanol, diesel or other liquid fuels made from plant and organic sources (IBEP 2006; US DOE 2008). Present first-generation technology uses food crops, which places it in direct competition with food and feedstock needs. Currently energy produced this way utilises about 1% of the world's arable land and supplies about 20 million tons of oil equivalent (MTOE) or about 1% of global annual road transport fuel demand. Further food security implications are invoked if, as some projections are realised, between four- and seven-fold increase in biofuel use in road transport is achieved (CFS 2007). Compared with projections from the EU that suggest that as much as 13% transport fuel could ultimately be displaced through existing cropland, the food fuel debate becomes more urgent.

Of the two main biofuel liquids bio-ethanol is by far the most abundant and accounts for more than 90% of the market (CFS 2007).

J.4.2 Bio-Ethanol

Ethanol (ethyl alcohol) is made through the fermentation of sugars and starch into alcohol from plants such as maize, sugar cane, sugar beet, sorghum, and wheat; the higher the natural content of sugar the better (CFS 2007). In a large, modern efficient plant, the yield from a bushel (25.5 kg) of maize is about 9.5 liters of ethanol. It is the most widely used biofuel today and can be used either as a stand-alone alternative fuel or as an additive to petrol/gasoline (Pimentel et al. 1998; US DOE 2008). It can be seen from Table J.1 that different crops produce different yields of ethanol.

Plants consist of cellulose (30%–70%), hemicellulose and lignin (which cannot be converted to sugar). Converting the cellulose sugars is quite a complex process, and the majority of plant cellulose material cannot be commercially viably converted at present. As such, conventional ethanol production currently utilises grain and sugar crops as the basic feedstock as these contain the most sugars. There is however substantial research underway in this area and if successful, a much larger potential feedstock base of cellulosic materials would become available (see Section J.6.4) (CFS 2007).

TABLE J.1

Ethanol Yields from Different Crop Varieties

Crop	Yield (%)
Sugar cane	80–85
Rapeseed	40–50
Sugar beet	40–45
Grains	20–30

Sources: von Braun, A New World Food Situation: New Driving Forces and Required Actions. International Food Policy Research Institute, The International Food Policy Research Institute, Washington, DC, 1995; DEFRA, *Renewable Biofuels for Transport: The Facts on Biodiesel and Bioethanol*, DEFRA, UK, 2003; IEA, The International Energy Agency, 2011.

J.4.3 Biodiesel

Biodiesel is made through a chemical process that reacts an oil or fat feedstock with methanol and potassium hydroxide in a process called esterification (CFS 2007). This can be vegetable oils (such as soy, sunflower, rapeseed and canola), or animal fat. By-products of this process including crushed 'bean-cake', an animal feed as well as glycerin used in cosmetics, medicines and foods, improve the economic benefits of producing biodiesel (US DOE 2008).

J.4.4 Second-Generation Biofuels

The limitations of first-generation biofuels with regard to technological efficiency has paved the way for research into new biofuel technologies of hydrolysis and gasification that extract more fermentable sugar. This effectively allows more of the woody ligno-cellulosic portions of biomass to be converted into ethanol, thus paving the way for more 'waste' biomass being used as opposed to food crops. Such second-generation technologies would also have the added advantage of effectively avoiding crop-usage conflicts (IEA 2011). Another associated advantage of future cellulose-to-ethanol fermentation systems is in the choice of fuels in its conversion. At present, the majority of grain-to-ethanol production processes in North America and Europe all use fossil fuel inputs whereas future cellulose processes will use cellulosic biomass including any unused parts of the plants being processed. However, while this technology currently exists, it is still expensive and somewhat inefficient and as a result is not expected to reach commercialisation until about 2020 (WEO 2008).

References

BP (2010). BP Statistical Review of World Energy. *Statistical Review*. UK: British Petroleum.

CFS (2007). Assessment of the World Food Security Situation. Committee On World Food Security: Thirty-third Session. Rome: Food and Agriculture Organization.

DEFRA (2003) *Renewable Biofuels for Transport: The Facts on Biodiesel and Bioethanol*. UK: DEFRA.

DTE (2007). Indonesia and Biofuel Fever. *Down to Earth* 74 (August).

Emsley, J. (2003). *Nature's Building Blocks: An a-Z Guide to the Elements*. Oxford: Oxford University Press.

Energyquest (2011). Fossil Fuels—Coal, Oil and Natural Gas. California Energy Commission. http://www.energyquest.ca.gov/story/chapter08.html (accessed 12 February 2011).

Freedom 21 (2008). Meeting Essential Human Needs. 9th Annual National Conference. Dallas, TX: Freedom 21.

Fritsch, A. J. and P. Gallimore (2007) Healing Appalachia: Sustainable Living through Appropriate Technology. *Earth Healing*. http://www.Earthhealing.info/biofuels.pdf http://www.earthhealing.info/dailyfeb06.htm.

GEO (2011). The Group on Earth Observations. The Group on Earth Observations. http://www.earthobservations.org/ (accessed 12 February 2011).

GEP (2011). Planet Earth: The Executive Summary. The Global Education Project. http://www.theglobaleducationproject.org (accessed 21 February 2011).

Howe, J. (2004). *The End of Fossil Energy and a Plan for Sustainability*. Fryeburg, US: McIntire Publishing.

IBEP (2006). Introducing the International Bioenergy Platform (IBEP). *International Bioenergy Platform*. Rome: Food and Agriculture Organization 18.

IEA (2011). The International Energy Agency. The International Energy Agency (IEA). http://www.iea.org/ (accessed 13 February 2011).

Lendman, S. (2008). Global Food Crisis: Hunger Plagues Haiti and the World. Montreal, Canada: Centre for Research on Globalization.

National Geographic Society (2008). *Biofuels. The Original Car Fuels*: Washington, DC: National Geographic Society.

Naturalgas (2011). Overview of Natural Gas. Natural Gas Supply Association. http://www.naturalgas.org/overview/background.asp (accessed 13 February 2011).

NRC (2011). Uranium Enrichment. The US Nuclear Regulatory Commission (NRC). http://www.nrc.gov/materials/fuel-cycle-fac/ur-enrichment.html (accessed 13 February 2011).

OPEC (2009). Annual Statistical Bulletin. Vienna: Organization of the Petroleum Exporting Countries

OPEC (2010). The Organization of the Petroleum Exporting Countries. The Organization of the Petroleum Exporting Countries (OPEC) http://www.opec.org/opec_web/ (accessed 15 May 2010).

Ostrolenk, B. (1930). The Surplus Farm Lands. *The Annals of the American Academy of Political and Social Science* 148: 207–211.

Pimentel, D., M. Pimentel and M. Karpenstein-Machan (1998). Energy Use in Agriculture: An Overview. *Agricultural Engineering International: CIGR Journal* 1.

Roubanis, N. (2004). Biofuels and Their Growing Importance. Energy Statistics, Working Group Meeting: Special Issue Paper 6. Paris: OECD.

US DOE (2008). Abc's of Biofuels. US Department of Energy. http://www1.eere.energy.gov/biomass/abcs_biofuels.html (accessed 5 July 2009).

von Braun, J. (1995). A New World Food Situation: New Driving Forces and Required Actions. International Food Policy Research Institute. Washington, DC: The International Food Policy Research Institute.

WCA (2011). Coal. World Coal Association. http://www.worldcoal.org/coal/ (accessed 13 February 2011).

WEO (2008). World Energy Outlook 2008. Paris: The International Energy Agency (IEA) 578.

WEO (2010). World Energy Outlook 2010. Paris: The International Energy Agency (IEA).

Appendix K: Stakeholders

Throughout the book I constantly refer to stakeholders of food security. While overall, stakeholders include all those with a stake in the concept, in the following map I provide an organisational chart of those institutions, charities and bodies that provide much needed governance, aid, finance and direction. In this sense the stakeholder map becomes a useful visual reference of the complex worldwide institutional framework highlighting the breadth and interrelatedness of those involved. While this is not exhaustive by any stretch of the imagination, it does provide a sold foundation from which to springboard. Figure K.1 too is only a partial representation of a larger map that can be found on CRC's website at: http://www.crcpress.com/product/isbn/9781439839508

While the above figure articulates and elucidates the current landscape of food security stakeholders it also highlights the sheer complexity of the phenomenon. This complexity however did not materialise overnight. By way of illustration, the Figure K.2 shows a timeline of the major organisations and events that unfolded over the last 100+ years that ultimately culminated in the present situation. Once again because of the size and complexity of this map, only a partial map is shown here while the full detailed version can also be found at the above CRC website.

Food Security

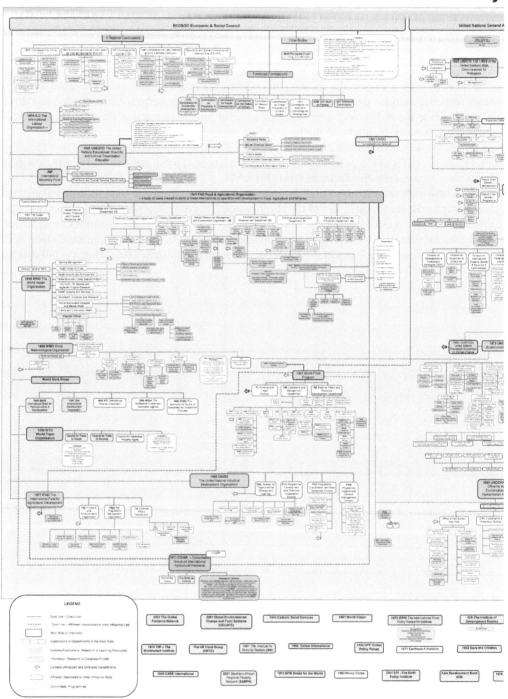

FIGURE K.1 Food security stakeholder map. (Reprinted with Courtesy of Mark Gibson ©.) The full size image available at CRC's website: http://www.crcpress.com/product/isbn/9781439839508

Stakeholders

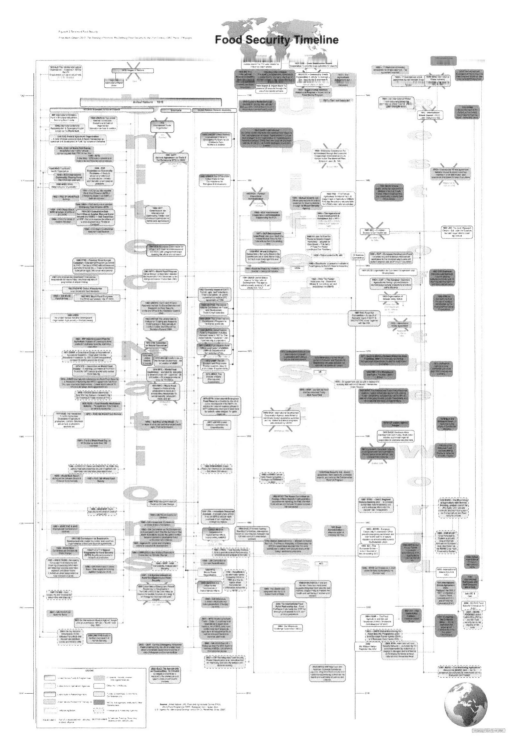

FIGURE K.2 Food security timeline map. (Reprinted with Courtesy of Mark Gibson ©.) The full size image available at CRC's website: http://www.crcpress.com/product/isbn/9781439839508

Figure K.2 Sources: ECOSOC Website 2009, WFP Organisational Chart 2008, ECA website 2009, ESCAP website 2009, ECE website 2009, ECLAC Organisational Chart 2009, ESCWA Organigram 2009, UNESCO Organigram 2009, ILO Organisational Chart 2009, IMF Organisational Chart 2007, FAO 2007 Organigram (director generals bulletin no. 2007/03), WHO Organisational Chart 2008, WMO Organisational Chart 2009, World Bank Group website 2009, WTO website 2009, UNIDO Organigram 2008, IFAD web 2009, UNICEF website 2009, UNDP Organisational Chart 2009, UNFPA Website 2009, UNCTAD Organigram 2009, UNEP Organigram 2009, UNOCHA Organigram 2009, OECD website 2009, DESA website 2009, USAID Organisational Chart 2008, UNHCR Organigram 2006, AU website 2009, SADC web site 2009, UNU website 2009, IGC Website 2009, ACF Action Against Hunger Website 2009, ActionAid Website 2009, Asia Development Bank ADB Website 2009, BFW Bread for the World Website 2009, CARE International Website 2009, Catholic Relief Services Website 2009, CGIAR Website 2009, Doctors Without Borders/Médecins Sans Frontières (MSF) Website 2009, Earthwatch Institute Website 2009, GPF Global Policy Forum Website 2009, IFPRI The International Food Policy Research Institute Website 2009, International Red Cross and Red Crescent Movement Website 2009, Mercy Corps Website 2009, ODI Overseas Development Institute Website 2009, Oxfam International Website 2009, Save the Children Website 2009, The Earth Policy Institute Website 2009, The Worldwatch Institute Website 2009, USAID Website 2009, USDA Website 2009, World Vision Website 2009, GECAFS website 2009, GRID-Arendal website 2009, WRI & Erthtrends website 2009, The Global Footprint Network website 2009, GreenFacts website 2009, IDS website 2009, CGFI website 2009, UKFG website 2009.

Appendix L: Conversion Rates

L.1 Energy

1 Btu [British thermal unit] = 1.055 kJ [kilojoules] = 252 cal [calories]

1 cal = 0.003967 Btu = 4.184 J

1 quad [quadrillion Btu] = 1015 Btu

1 Btu/lb = 0.556 kcal/kg

Energy Equivalents

	Ton Oil Equivalent (Toe)	Million British Thermal Units (Btu)	Million Kilocalorie (kcal)	Gigajoules	Gigawatt Hours (GW·h)
1 T Oil	1 Toe	39.7	10	41.9	11630
1 T Diesel	1.01 Toe	40.4	10.1	42.42	11746.3
1 m³ Diesel	0.98 Toe	39.2	9.8	41.16	11397.4
1 T Petrol	1.05 Toe	42	10.5	44.1	12211.5
1 m³ Petrol	0.86 Toe	34.4	8.6	36.12	10001.8
1 T Biodiesel	0.86 Toe	34.4	8.6	36.12	10001.8
1 M³ Biodiesel	0.78 Toe	31.2	7.8	32.76	9071.4
1 T Bioethanol	0.64 Toe	25.6	6.4	26.88	7443.2
1 m³ Bioethanol	0.51 Toe	20.4	5.1	21.42	5931.3

Sources: IEA (2011). The International Energy Agency. The International Energy Agency (IEA), http://www.iea.org/ (accessed 13 February 2011). WEO (2008). *World Energy Outlook 2008.* Paris: The International Energy Agency (IEA). WEO (2010). *World Energy Outlook 2010.* Paris: The International Energy Agency (IEA).

L.2 Volumetric

Volumetric Conversions

To	US Gallon (Gal)	UK Gallon (Gal)	Barrel (Bbl)	Cubic Foot (ft³)	Litre (L)	Cubic Metre (m³)
From: Multiply By:						
US Gallon (Gal)	1	0.8327	0.02381	0.1337	3.785	0.0038
UK Gallon (Gal)	1.201	1	0.02859	0.1605	4.546	0.0045
Barrel (Bbl)	42.0	34.97	1	5.615	159.0	0.159
Cubic Foot (ft³)	7.48	6.229	0.1781	1	28.3 0.	0.283
Litre (L)	0.2642	0.220	0.0063	0.0353	1	0.001
Cubic Metre (m³)	264.2	220.0	6.289	35.3147	1000.0	1

Sources: IEA (2011). The International Energy Agency. The International Energy Agency (IEA), http://www.iea.org/ (accessed 13 February 2011). WEO (2008). *World Energy Outlook 2008.* Paris: The International Energy Agency (IEA). WEO (2010). *World Energy Outlook 2010.* Paris: The International Energy Agency (IEA).

L.3 Bushels

To convert Bushels/Acre to kilograms/hectare, the conversion factor is 62.71 (assuming a bushel weighs 56 lb).

L.4 Oil

1 barrel oil = 42 US gallons/35 UK gallons = 159 litres.

Approx: 7–9 barrels = 1 ton of oil depending on the type of oil. Or: 294 American gallons = 1 ton of oil (256 US gallons per ton of heavy distillate to 333 US gallons/ton of gasoline, with crude oil at 272 US gallons/ton).

Appendix M: Glossary

It's easy enough to preach morality on
a full belly.

Erwin Sylvanus
Dr. Korczak and the Children

3 Cs of food security is the author's summary
of the existing state of understanding in the
field of food security: confusion, consternation and contradiction.

Abiota: the non-living component of an ecosystem.

Acid rain is rain or snow that has an acidity
content greater than a pH of 5.6. Acids form
when atmospheric gases, primarily carbon
dioxide, sulphur dioxide and nitrogen oxides,
combine with water in the atmosphere.

Acute hunger is the quick onset and severe
hunger that might manifest with sudden
food shocks as in natural disasters or harvest failure.

Aerosol is airborne particulate matter; these
may be of natural or anthropogenic and
influence the climate, the term has also
mistakenly come to be associated with the
propellant used in aerosol sprays. See **particulate matter**.

Afforestation is the establishment of a forest
through tree planting on land that has never
had or has lacked forest cover for a long
time.

Agenda 21: programme of action on sustainable development adopted at the un conference on environment and development in
1992.

Agribusiness are those businesses working in
agriculture.

Agricultural land refers to arable land, land
under permanent crops and land under permanent meadows or pasture.

Agricultural revolution: a name given to different periods of agricultural advancement
over the centuries.

Agricultural run-off: water that flows from
agricultural fields to streams, rivers and
estuaries usually containing pollutants.

Agricultural waste is waste produced as a
result of various agricultural operations,
including manure and other wastes from
farms, feedlots and slaughterhouses as well
as harvest waste, fertiliser run-off and pesticide pollutants in water, air or soils.

Agroecology: study of the relationships between
agricultural crops and the environment.

Agroecosystem: the system of plant, microbes,
and animals inhabiting farmed land, pastures or grasslands, which interact with each
other and their physical environment.

Agrology/agronomy: branch of agriculture
that deals with the origin, structure and classification of soils, especially in their relation
to crop production.

Aid and assistance refer to flows that qualify as
Official Development Assistance (ODA) aid
activities include projects and programmes,
cash transfers, deliveries of goods, training
courses, research projects, debt relief operations and contributions to non-governmental
organisations.

Air pollutants/pollution: substances in air
that may consist of solid particles, liquid
droplets or gases, or in concentrations that
can interfere with human health or produce
other harmful environmental effects.

Airborne particulates: suspended particulate
matter found in the atmosphere as solid particles or liquid droplets. Sources of airborne
particulates include dust, emissions from
industrial processes, the burning of wood
and coal.

Albedo: a measure of the degree to which a
surface or object reflects solar radiation.
Snow covered areas have a high albedo
(up to about 0.9 or 90%) due to their white
colour, while vegetation has a low albedo
(generally about 0.1 or 10%) due to the dark

colour and light absorbed for photosynthesis. Clouds have an intermediate albedo and are the most important contributor to the Earth's albedo. The Earth's aggregate albedo is approximately 0.3.

Algae: simple rootless plants that grow in sunlit waters. They can affect water quality adversely by lowering the dissolved oxygen in the water. They are food for fish and small aquatic animals. However, when algae exist in excess, they take away oxygen from the water, affecting the aquatic balance.

Algae blooms: excessive and rapid growth of algae and other aquatic plants. It takes place when there are too many nutrients in the water and is aggravated when accompanied by a rise in temperature, forming visible patches that may harm the health of the environment, plants, or animals.

Alpine: the biogeographic zone above the altitude of the timberline and characterized by the presence of herbaceous plants and low shrubby slow-growing woody plants.

Alternative energy: derived from non traditional sources for example compressed natural gas, solar, hydroelectric and wind.

Amazonia: the Amazon basin area of south America.

Amino acids—see **protein**

Anabolism—see **metabolism**

Anthropogenic: man-made; caused by human activity.

Anthropometric: measurements of the human body.

Antioxidant helps to fight free radicals. See **notes appendix A: macronutrients**.

Aquaculture: the farming of aquatic organisms including fish, mollusks, crustaceans and aquatic plants with some sort of intervention in the rearing process to enhance production, such as regular stocking, feeding and protection from predators.

Aquifer: an underground geological formation, or group of formations, containing water. These are valuable sources of groundwater for wells and springs; also described as artesian (confined) or water table (unconfined).

Arable land: can be cultivated to grow crops. Usually includes all land generally under rotation, temporary meadows for mowing or pasture, land under market and kitchen gardens and land temporarily fallow.

Ariboflavinosis: caused by deficiency of riboflavin marked by cheilosis.

Arid land: desert or semi-desert lands characterised by low annual rainfall of less than 250 mm or by evaporation exceeding precipitation and sparse vegetation. Agriculture is usually impractical without irrigation.

Aridity—see drylands.

Atmosphere: the air surrounding the Earth, described as a series of shells or layers of different characteristics. The atmosphere, composed mainly of nitrogen (78.1%), oxygen (20.9%) and argon (0.93%) with traces of carbon dioxide (0.035%), water vapour (highly variable but typically 1%), and other gases, acts as a buffer between Earth and the sun. The layers—troposphere, stratosphere, mesosphere, thermosphere, and the exosphere—vary around the globe and in response to seasonal changes. The troposphere extends to a height of 8–15 km, depending on latitude, contains the bulk of Earth's weather. The stratosphere stretches from here to a height of 50 km. The mesosphere reaches 80 km and the exosphere, beginning at 500–1,000 km above the Earth's surface, blends the atmosphere into space.

Atoll: a coral island consisting of a ring of coral surrounding a central lagoon.

Atomic weight: the average weight (or mass) of all the isotopes of an element.

Autoimmune disorders are conditions that sees the body's own immune system mistakenly attack and destroy its own healthy body tissue.

Balance of Payments (BOP): summarises the economic transactions of an economy.

Barley: more tolerant of poorer soils and lower temperatures than wheat, barley is used mainly for livestock feed, malt and in preparation of food.

Basal metabolic rate: the energy required to sustain basal metabolism.

Beans, dry: annual herb. Seeds are of different sizes and colour. Dry beans are cultivated all over the world and used primarily for human consumption.

Beet sugar, raw—see **sugar beets**.

Beriberi is a condition caused by deficiency of vitamin B_1.

Bioavailability: portion of food ingested that is absorbed and available to the body. See **notes** in this section.

Biocapacity refers to the capacity of a given biologically productive area.

Biocide: an agent that kills many organisms in the environment.

Biodegradable: capable of decomposing under natural conditions resulting from bacterial or other microbial action.

Biodiesel is produced from organically derived oils or fats combined with alcohol (ethanol or methanol) in the presence of a catalyst.

Biodiversity: the variety and variability of living organisms and the ecosystems in which they occur.

Bioenergy: biomass used in the production of energy.

Bioethanol: a biofuel produced by converting a plants sugar or starch content into ethanol.

Biofuels is a gaseous, liquid or solid fuel that contains energy derived from a biological source. Examples of biofuel include alcohol from fermented plants rich in sugar/starch; dry wood and other biomass products like peat; agricultural waste; straw, fish, seed and vegetable oils.

Biogas: derived from the anaerobic fermentation of biomass and solid wastes used to produce heat or power.

Biogenic: produced by natural processes often used in the context of emissions that are produced by plants and animals.

Biological amplification is the accumulation of elements and compounds in the tissues of living organisms, usually concentrated as food travels up the food chain.

Biological evolution: changes in the gene pool of a species over time.

Biological pesticides: composed of biological substances including bacteria, fungi, viruses and protozoa, as opposed to the chemical substances used in conventional pesticides.

Biological stressors: organisms in habitats in which they do not evolve naturally.

Biological waste: waste containing natural organic material including plants and animal remnants.

Biologically productive area: land and water that support significant photosynthetic activity and biomass accumulation.

Biomass fuel: liquid, solid, or gaseous fuel produced by conversion of biomass, commonly called biofuels.

Biomass: organic no fossil material of biological origin. For example, trees and plants are biomass. Often expressed as the total living weight.

Biome: a large ecosystem that has distinct climate, geology, and organisms. There are three types: marine, freshwater and terrestrial biomes. Terrestrial biomes are typically based on dominant vegetation structure. For example, desert, tundra, grassland, savanna, woodland, coniferous forest, temperate deciduous forest and tropical rain forests.

Biosphere: the part of the Earth system comprising all ecosystems of the Earth and the atmosphere capable of supporting living organisms.

Biota: all living plant and animal life of a region or area.

Biotechnology: any technological application that uses biological systems or living organisms for the purpose of developing products, improve animals or plants or improve biological processes.

Birth rate: the crude birth rate of an area is the number of births actually occurring in that area in a given time period, divided by the population of the area as estimated at the middle of the particular time period. Expressed as per 1000 of population.

Body mass index (BMI): the indicator of weight in relation to height. It is calculated as weight (kilograms) divided by height (metres), squared. The acceptable range for adults is 18.5 to 24.9, but varies for children with age.

Bog: a poorly drained wetland rich in accumulated plant material.

Bone meal: animal meal produced from bones.

Brackish waters: water bodies with salinity between seawater and freshwater.

Broadleaf: the leaves of trees associated with deciduous forests.

Bunker fuels—see **international marine bunkers**.

Cachexia: a fat and muscle wasting condition associated with cancer and HIV

Calorimetry: the process of measuring heat lost or gained—usually in a calorimeter.

Cane sugar, raw: manufactured from sugarcane.

Canopy: the layer formed naturally by the leaves and branches of trees and plants.

Capital expenditure: measures the value of purchases of fixed assets. Those assets that are used repeatedly in production processes for more than one financial period.

Carbohydrates are the main energy source of the body and globally provide two-thirds of total energy. Carbs consist of sugar, starch and fibre.

Carbon cycle: the process by which carbon, the chemical foundation of living organisms, circulates throughout the natural world. Usually thought of as a series of the four main reservoirs: the atmosphere, terrestrial biosphere; oceans; and sediments (includes fossil fuels). All carbon reservoirs and exchanges of carbon from reservoir to reservoir are by various chemical, physical, geological and biological processes.

Carbon dioxide: a colourless, odourless, non-poisonous gas, which occurs naturally in the air and also results from fossil fuel combustion. Plants absorb CO_2 during photosynthesis, and plants and animals produce it as an end product of respiration. It plays an important role in controlling the Earth's surface temperature.

Carbon dioxide equivalent: a measure used to compare the emissions from various greenhouse gases based upon their global warming potential.

Carbon market: a popular term for a trading system through which countries buy or sell units of greenhouse gas emissions to meet their national limits.

Carbon sequestration: the uptake and storage of atmospheric carbon or the process of removing carbon from the atmosphere and depositing it in a reservoir. Trees and plants, for example, absorb carbon dioxide and release the oxygen. Fossil fuels continue to store the carbon until it is finally burned. See **carbon sinks**.

Carbon sink: parts of the biosphere that absorbs more carbon dioxide than it releases.

Carotenoids: organic pigments that naturally occur in plants and some other photosynthetic organisms such as algae, some bacteria and some fungus.

Carrying capacity: the theoretical equilibrium of population size against the resources available in a particular area. In effect, it represents the total population that a particular area can support without reducing its resource base.

Carryover stocks: left in the bin at the end of the production year carried over into the following year.

Cartesianism: is the name given to the philosophical doctrine (or school) of René Descartes.

Cash crop: agricultural produce marketed for cash rather than for personal use.

Catabolism—see **metabolism**

Celiac Sprue: (celiac disease) gluten-induced enteropathy of the digestive tract.

Central tendency: the tendency of quantitative data to cluster around a particular value.

Cereals: annual plants, generally of the gramineous family, yielding grains or seeds used as is or processed for food, feed, seed and industrial purposes.

Cereals, NEC: an FAO classification, which includes cereal crops that are not identified separately because of their minor relevance at the international level.

Cheilosis: (angular cheilitis) the inflammation and cracking of the corners of the mouth.

Child/infant mortality: the infant mortality rate, expressed per 1000 live births, of a child dying before reaching the age of one.

Chlorofluorocarbons (CFC): any of the various compounds consisting of chlorine, hydrogen, fluorine and carbon. They have been widely used as refrigerants, as aerosol propellants, cleaning solvents and in the manufacture of plastic foam. Found to contribute to the depletion of the ozone layer. The Montreal Protocol worked to reduce and eliminate production of CFCs worldwide.

Cholera: an intestinal infection.

Cholera: caused by the bacteria often through contaminated drinking water, which causes extreme diarrhea and severe vomiting.

Chronic hunger involves long-term severe shortage of food, often associated with longer term disruptions of food supply.

Clean development mechanism (CDM): one of the three market-based mechanisms defined in the Kyoto Protocol where developed countries finance greenhouse gas emissions-saving projects in developing countries and receive credits towards meeting mandatory limits on their own emissions.

Clean technologies: industrial processes that reducing pollutants and environmental impacts of industrial production.

Climate change refers to any significant change in measures of climate (such as temperature, precipitation or wind) lasting for an extended period (decades or longer). Climate change may result from natural climate variability or from human activity that alters the composition of the global atmosphere.

Climate convention: the UN Framework Convention on Climate Change (UNFCCC) and the Kyoto Protocol. See United Nations Framework Convention on Climate Change (UNFCC)

Climate feedback: an interactive processes in the climate where the result of an initial process triggers changes in a second process that in turn influences the initial one. A positive feedback intensifies the original process, and a negative feedback reduces it.

Climate lag: the delay that occurs in climate change as a result of factors that changes very slowly. For instance, the effects of carbon dioxide into the atmosphere may not be known for some time because a large proportion is dissolved in the ocean and is only released to the atmosphere many years later.

Climate: the long-term average weather of a given area. Climatic elements include precipitation, temperature, humidity, sunshine, wind, fog, frost, hail storms and other measures of the weather. See also **weather**.

CO_2 equivalent—see **carbon dioxide equivalent**.

Coal: a black or brownish black solid, combustible substance formed by the partial decomposition of vegetable matter without access to air. Includes anthracite, bituminous coal, sub bituminous coal and lignite.

Coarse grains: coarse grains generally refer to cereal grains other than wheat and rice.

Coastal plains: the lowland area between the sea and the nearest hills representing a strip of seabed recently emerged in geologic time.

Codex Alimentarius Commission: an international body responsible for developing the standards, guidelines and recommendations that comprise the Codex Alimentarius.

Coefficient of variation: the standard deviation of a variable divided by the mean.

Coking coal: an industrial process that converts coal into coke, one of the basic materials used in the production of iron.

Composite indices of food security are the aggregated often formulaic model to classify the extent of food insecurity within a region or population

Concept: is a representational framework derived or inferred from an idea, a set of notions or even an abstract principle. Its use might be analogous to finding a new word, description or framework that encompasses such new or existing thoughts.

Concessional loans are extended on generous terms.

Conservation tillage: a term that covering a range of soil tillage practices that leave residue cover on the soil surface reducing the effects of soil erosion from wind and water.

Consumer Price Index (CPI): the consumer price index measures changes over time in the general level of prices of a fixed basket of goods and services that a reference population acquires, uses or pays for consumption.

Corn belt region: in the Midwestern United States: Illinois, Indiana, Iowa, Minnesota, Nebraska, and Ohio where maize is grown on a large scale.

Correlation coefficient is a measure of the degree to which two variables relate to each other.

Cottonseed: although rather low in oil content, the seed of the cotton plant is an important source of vegetable oil, rich in polyunsaturated fatty acid.

Cover crop: a temporary vegetative cover that is grown to provide protection for the soil.

Cretinism and Cretin's disease: a condition of severely stunted physical and mental growth due to untreated congenital deficiency of thyroid hormones (hypothyroidism) brought on by insufficient iodine can also lead to deaf-mutism and other physical disabilities.

Crop residue: remaining organic residue after the harvesting and processing of a crop.

Crop rotation: an agricultural practice of rotating two or more crops on the same land area from year to year to reduce nutrient depletion and reliance on pesticides.

Crude death rate: a vital statistic based on the number of deaths occurring in a population during a given period of time, per 1000 population.

Crude oil is a mineral oil consisting of a mixture of hydrocarbons of natural origin. Being yellow to black in colour, of variable density and viscosity, it remains liquid at atmospheric pressure.

Cultivar: a contraction of cultivated variety; a variety of plant maintained under cultivation.

Cultivated land: used for the growing of crops on a cyclical or permanent basis.

Cultivation: the practice of growing and plants outside of their wild habitat.

Current accounts: current accounts record a countries balance sheet of economic production.

Current value: the actual value of an aggregate: the quantities multiplied by the prices of the same period.

Death rate: the number of deaths in a year expressed as a percentage of the population per 1000 people.

Deciduous forest: areas dominated by trees that shed their leaves at the end of the growing season.

Deforestation: processes that result in the change of forested lands to non-forest uses.

Demography: the study of human populations; their size, composition, distribution, density and growth.

Dermatitis: inflammation and irritation of the skin.

Desalinisation: the process of removing salt and related minerals from water for human consumption.

Desertification: the progressive degradation of existing fertile land to form desert. This can occur due to overgrazing, deforestation, drought and the burning of extensive areas.

Detritus: accumulated debris, consisting of both inorganic material and organic remains of plants and animals, often heavily colonised by bacteria, usually an important source of nutrients in a food web.

Developed, developing countries: there is no internationally recognised convention for the designation of 'developed' and 'developing' countries or areas. However, a country's 'development' status is often associated with its economic and technological achievement; in terms of production, per capita income, consumption and level of industrialisation.

Development aid: economic and other support provided to developing countries to promote better living standards, infrastructure, agricultural and other aspects of an economy.

Dichlorodiphenyltrichloroethane (DDT): an organochloride used as an insecticide. It is highly toxic to biota, including humans and is a persistent biochemical, which accumulates in the food chain.

Dietary energy deficit: the difference between the average daily dietary energy intake of an undernourished population and its average minimum energy requirement.

Dietary energy requirement: the amount of dietary energy required by an individual to maintain body functions, health and normal activity.

Dietary energy supply: food available for human consumption calculated after the deduction of all non-food consumption (exports, animal feed, industrial use, seed and wastage).

Dioxins: chemicals used in the production of pesticides that have no industrial use but are hazardous to human health.

Drip irrigation: using pipes with small holes that release water near the roots of plants in an effort to reduce waste water and eliminate runoff.

Drylands: areas characterised by lack of water. Dryland subtypes include sub-humid, semi-arid, arid and hyperarid.

Dysentery: usually as a result of parasitic worms, which cause extreme bloody diarrhea.

Earth summit: the United Nations Conference on Environment and Development, which took place at Rio de Janeiro, Brazil, in June 1992.

Ecological Footprint (EF): introduced by William Rees in 1992; it is a calculation that estimates the productive land and water area required to supply an individual or groups resources as well as to absorb the wastes they produce.

Ecological zones: areas of land or water characterised by their species and communities. These major ecozones are further divided into many smaller ecoregions, which share similar natural communities, climate and other physical characteristics.

Ecology: the branch of science studying the interactions among living things and their environment.

Economies in transition (EITS): countries with national economies in the process of changing from a planned economic system to market-based economies.

Ecosphere: biosphere.

Ecosystem—plant, animal and microorganism communities and their non-living environment interacting as a functional unit within a defined physical location.

Elasticity: a measure of responsiveness of one variable to a change in another.

Electrolytes regulate body fluids to facilitate communication or electrical impulses between cells.

Emissions: the release of greenhouse gases into the atmosphere over a specified area and period of time.

Emissions trading: a market-based approach allowing countries to trade excess reductions to offset emissions at another source inside or outside the country.

Endemic: native and restricted to a specific geographic area, usually referring to plants or animals.

Energy balance: averaged over the globe and over long time periods, the energy budget of the climate system must be in balance. Because the Earth derives all its energy from the sun, globally, the amount of incoming solar radiation must on average be equal to the sum of the outgoing reflected solar and infrared radiation emitted by the climate system.

Enteropathy refers to any pathology (disease) of the intestine.

Entomology: study of insects.

Environmental Protection Agency: US agency that implements and enforces federal environmental laws both foreign and domestic.

Environmentally sound technologies (ESTS): technologies that protect the environment and that are less polluting.

Epidemiology: the study of the aspects, occurrence and distribution of a disease.

Estuaries: generally those broad portions of a river, stream, brook or torrent near its outlet influenced by the marine water body into which it flows.

Ethanol (C_2H_5OH): otherwise known as ethyl alcohol, alcohol or grain spirit. A clear, colourless, flammable oxygenated hydrocarbon, in transportation, it is used as a vehicle fuel by itself (E100), blended with petrol/gasoline (E85), or as a petrol/gasoline octane enhancer and oxygenate.

European commission: one of the principal organs of the European Union.

European Union (EU): an economic integration organisation.

Eurostat: statistical office of the European Communities.

Eutrophication: process by which bodies of water become enriched in nutrients through natural or man-made processes. Nutrients deplete the oxygen in the water by stimulating the growth of algae and other aquatic plant rather than animal life.

Evapotranspiration is combined loss of water to the atmosphere via the processes of evaporation and transpiration.

Ex officio: Latin phrase meaning by virtue of one's position or function.

Exosphere: the uppermost layer of the atmosphere where gases can, to any appreciable extent, escape into outer space.

Exponential growth: in which quantities such as population size increases by a constant percentage of the whole over a specific time period; when graphically plotted, this type of growth yields a curve shaped like the letter J.

Export subsidies: export subsidies consist of all subsidies on goods and services that become payable to resident producers when the goods leave the economic territory.

External forcing: influence on the Earth's radiative system by an external agent such as solar radiation or the impact of extraterrestrial bodies such as meteorites. See also **feedback mechanisms and climate feedback**.

Fallow land: arable land set aside for a period of time before it is cultivated again.

Fats are essential for the body; the different fats are decided upon their carbon or hydrogen arrangement. Depending on the configuration of these elements fats can take any number of forms including: saturated, unsaturated (mono- and polyunsaturated fats) and trans-fats.

Fatty acids—see **Fats**

Fauna: all animal life.

Feed grains: any of several grains most commonly used for livestock these include corn, grain sorghum, oats, rye and barley.

Feedback mechanisms: factors which increase or amplify (positive feedback) or decrease (negative feedback) the rate of a process. An example of positive climatic feedback is the ice-albedo feedback. See **climate feedback**.

Feedlot: an area containing a high density of reared animal livestock.

Feedstock: raw material directly used in the production of goods.

Fertility rate: the number of children that would be born per woman, assuming no female mortality at child-bearing ages and the age-specific fertility rates of a specified country and reference period.

Fibre: the other complex carbohydrate is generally indigestible in humans. Sometimes referred to as roughage or dietary fibre. It can be either soluble or insoluble and aids in the digestive process.

Flash flood: of short duration with a relatively high peak discharge.

Flood plain: a low area along a river considered to be at risk of flooding.

Flora: all plant life.

Fluorocarbons: carbon-fluorine compounds that often contain other elements such as hydrogen, chlorine, or bromine. Common fluorocarbons include chlorofluorocarbons (CFCs), hydro chlorofluorocarbons (HCFCs), hydro fluorocarbons (HFCs), and per fluorocarbons (PFCs).

Food aid represents a transfer of food commodities from donor to recipient countries sometimes free or on a concessional basis.

Food balance sheets: food balance sheets (FBS) are compiled every year by FAO, mainly with country-level data on the production and trade of food commodities.

Food insecurity: a situation where people lack secure access to sufficient safe and nutritious food for normal growth and development coupled with an active and healthy life.

Food insecurity typology: when referring to the underlying cause or structure of food security it can be distinguished between chronic (continuous and drawn out) temporal (temporary or sudden) and seasonal (cyclical). This helps determine the type of international policy response to best target the needy.

Food production index number: refers to the net food production after the deduction for feed and seed of a country's agricultural sector output relative to the base period 1999–2001.

Food security: food security exists when all people, at all times, have physical, social and economic access to sufficient, safe and nutritious food that meets their dietary needs and food preferences for an active and healthy life.

Forcing mechanism: a process that alters the energy balance of the climate system, that is to say, changes in the relative balance between incoming solar radiation and outgoing infrared radiation from the Earth. Such mechanisms include changes in solar irradiance, volcanic eruptions and enhancement of the natural greenhouse effect by the emission of greenhouse gases.

Fossil fuel: a general term for combustible geologic deposits of organic materials formed from decayed plants and animals that have been converted to crude oil, coal, natural gas, or heavy oils.

Four pillars of food security refers to the FAO's (and others) four dimensions of availability, access, utilisation and stability.

Free market: a market in which buyers and sellers are free to conduct business on whatever terms they choose without government interference and where scarcities are resolved through changes in relative prices.

Free radicals—see **note on antioxidants**—appendix A: macronutrients

G7: seven of the largest industrial countries comprising Canada, France, Germany, Italy, Japan the United Kingdom, and the United States.

G77: a large negotiating alliance of developing founded in 1967 under the auspices of the United Nations conference on trade and development.

G8: the group of eight (developed countries) comprising Canada, the US, France, Germany, Italy, Japan, Russian Federation, United Kingdom.

General Agreement on Tariffs And Trade (GATT): from 1947 until 1995, an international organisation with a mandate to reduce protection and promote free trade among nations. In 1995 Gatt was succeeded by the World Trade Organisation (WTO).

Genetic engineering: techniques for the alteration of the structure or the production of new genes by the substitution or addition of new genetic material.

Genetically Modified Organisms (GMO): a plant or animal micro-organism or virus that has been genetically engineered by means of gene or cell technologies.

Genomics/pharmacogenomics: the study of genes and their function.

Genotype: the genetic constitution of an organism, cell, individual or taxon, as distinct from its physical appearance (its phenotype)

Geo-engineering: efforts to stabilise the climate system by directly managing the energy balance of the Earth, in an effort to overcome the enhanced greenhouse effect.

Geophysical: relating to the study of the physical characteristics and properties of the solid Earth, its air and waters.

Geosphere: the physical elements of the Earth's surface crust, and interior.

Geothermal energy: available as heat emitted from within the earth's crust, usually in the form of hot water or steam. It is exploited for electricity or directly as heat for district heating, or in agriculture.

Giardiasis: an infection of the digestive system caused by parasites called giardia intestinalis.

Gini coefficient of variation: measures the extent to which the distribution of income among individuals or households within an economy deviates from a perfectly equal distribution. The Gini measures the area between the Lorenz curve and a hypothetical line of absolute equality, expressed as a percentage of the maximum area under the line. Thus a Gini index of 0 represents perfect equality, while an index of 1 implies perfect inequality.

Gini index: the Gini coefficient of variation times 100.

Global carbon budget: the balance of the exchange of carbon between the carbon reservoirs of the carbon cycle.

Global surface temperature: the global surface temperature is the area-weighted global average of (i) the sea surface temperature over the oceans (the sub-surface bulk temperature in the first few meters of the ocean) and (ii) the surface air temperature over land at 1.5 m above the ground.

Global warming: an increase in the near surface temperature of the Earth. Although warming has occurred naturally in the distant past, the term is often used today to refer to the warming predicted to occur as a result of anthropogenic influences.

Global warming potential (GWP): the index used to translate the level of emissions of various gases into a common measure in order to compare the relative radiative forcing of different gases. GWP is calculated as the ratio of the radiative forcing that would result from the emissions of one kilogram of a greenhouse gas to that from one kilogram of carbon dioxide over a period of time (usually 100 years).

Globalisation: a term generally used to describe an increasing internationalisation of markets for goods and services, the means of production, financial systems, competition, corporations, technology and industries, characterised by the increased mobility of capital, faster propagation of technological innovations and an increasing interdependency and uniformity of national markets.

Glycemic index (GI) is a way of grouping carbohydrates according to their absorptive capacity or how readily the body can metabolise and utilise different carbohydrates in blood sugar (glucose) levels compared to pure glucose (glucose=100).

Goiter: condition resulting in enlargement of the thyroid gland.

Grain crops: mostly grass crops, that are grown for their edible seeds, such as corn, wheat, rye, buckwheat, amaranth.

Great plains: a grassland area in the western United States, extending from north Dakota to south Texas and from the rocky mountains east to western Minnesota and Missouri.

Green revolution: the efforts in the 1960s to increase world food production by introducing high-yield plant varieties and new techniques of irrigation and use of pesticides.

Greenhouse effect: warming of the earth's atmosphere through build-up of heat in the troposphere, caused by heat flowing back toward space from the Earth's surface being absorbed by water vapour, carbon dioxide, ozone, and other gases in the atmosphere and then reradiated back toward the Earth's surface.

Greenhouse gas (GHG): greenhouse gases are those gaseous constituents of the atmosphere, both natural and anthropogenic that absorb infrared radiation in the atmosphere. Greenhouse gases include water vapour, carbon dioxide (CO_2), methane (CH_4), nitrous oxide (N_2O), halogenated fluorocarbons (HCFCs), ozone (O_3), perfluorinated carbons (PFCs), and hydrofluorocarbons (HFCs).

Gross Domestic Product (GDP): the value of all final goods and services produced in a country in one year.

Gross National Product (GNP): the value of all final goods and services produced in a country in one year.

Groundwater recharge: replenishment of groundwater supply by natural or artificial methods.

Groundwater: the supply of fresh water found beneath the Earth's surface, often in aquifers accessed through wells and springs.

Haematoma: small (<1 cm) pooling of blood outside the blood vessels.

Hard coal: a black, natural fossil organic sediment.

Heath land: uncultivated open land, covered with vegetation often consisting to a considerable degree of ligneous and semi-ligneous plants (fern, heather, furze, genista, etc.) As well as herbaceous cover.

Heavily Indebted Poor Countries (HIPCS): heavily indebted poor countries comprise a group of approximately 41 developing countries consisting of 32 countries with a 1993 GNP per capita of $695 or less and whose 1993 present value of debt to exports is higher than 200 percent or whose present value of debt to GNP is higher than 80% (world bank classification of severely indebted low-income countries). Also, the group includes nine countries that have received concessional rescheduling from Paris Club creditors (or are potentially eligible for such rescheduling).

Hemorrhagic disease: a disorder resulting in discharge of blood from the blood vessels; profuse bleeding from coagulation impairment.

Hepatic steatosis: is fatty liver, the collection of excessive triglycerides and other fats inside liver cells.

Hepatic: to do with the liver.

Herbaceous: green and leaf-like.

Herbicide: a chemical pesticide designed to control or destroy plants, weeds, or grasses.

Hidden or **silent** hunger was the vernacular name given to the less publicised hunger and malnutrition that did not fall into the previously known protein-energy malnutrition category; often associated with micronutrient deficiencies such as vitamin A, iodine, iron deficiency anaemia, etc.

High, middle, and low income countries: the World Bank's main criterion for classifying economies is the gross national income (GNI) per capita, which is similar to but not the same as the gross domestic product (GDP). Based on its GNI per capita, every economy is classified as low income, middle income (subdivided into lower middle and upper middle) or high income. In low income country's GNI per capita is $745 or less; in lower middle income countries, it is $746 to $2975; in upper middle income countries it is $2976–$9205; and in high income countries it is, $9206 or more.

Human Development Index (HDI): a composite of three indicators: longevity, as measured by life expectancy at birth; educational and combined primary, secondary and tertiary enrolment ratios; and the standard of living, as measured by real gross domestic product (GDP) per capita (in purchasing power parity).

Hunger: the physiological symptoms, hunger pains in the stomach. Of feeling hungry.

Hydrocarbons: substances containing only hydrogen and carbon. Fossil fuels are made up of hydrocarbons. Some hydrocarbon compounds are major air pollutants.

Hydrochlorofluorocarbons (HCFCs): compounds containing hydrogen, fluorine, chlorine and carbon atoms. Although ozone-depleting substances, they are less potent at destroying stratospheric ozone than chlorofluorocarbons (CFCs). They have been introduced as temporary replacements for CFCs and are also greenhouse gases.

Hydrogenation—see **trans-fats**

Hydrologic cycle—see **water cycle**

Hydropower: electrical energy produced by falling or flowing water.

Hydrosphere: the component of the climate system comprising liquid surface and subterranean water, such as oceans, seas, rivers, fresh water lakes, underground water.

Hyperthyroidism: is an overactive thyroid gland producing excess thyroid hormones.

Hypoglycaemia: (low blood glucose) is a condition in which the level of glucose in the blood drops below a certain point.

Hypotension: is abnormally low blood pressure.

Hypothermia: is the abnormally low body temperature.

Hypotonia is decreased muscle tone.

Import dependency ratio (IDR) is defined as imports $\times 100/$ (production imports − exports).

Import duties consist of customs duties, or other import charges, which are payable on goods of a particular type when they enter an economic territory.

Import quotas: are restrictions on the quantity or value of imports of specific commodities for some given time period.

Import subsidies consist of subsidies on goods and services that become payable to producers when the goods cross the frontier of an economic territory.

Import tariffs: taxes imposed on certain imported goods or services. May be levied as a percentage of the value of imports or as a fixed amount per unit and are often used to increase government revenue and protect domestic industries from foreign competition.

Industrial revolution: a period of rapid industrial growth which began in England during the second half of the eighteenth century and spread to Europe then the United States. In this book, the term pre-industrial refers to the period before 1750.

Infant mortality rate: the number of infant deaths (children between the ages of 0 and 12 months) per 1000 live births.

Infrared radiation: emitted by the Earth's surface, the atmosphere and clouds. Also known as long-wave radiation, or heat.

Insecticide: a pesticide compound specifically used to kill or prevent the growth of insects.

Intensive agriculture: refers to agricultural practices that produce high output per unit area, usually by intensive use of manure, agrochemicals, mechanisation and so on.

Intergovernmental Panel On Climate Change (IPCC): the IPCC was established jointly by the United Nations Environment Programme and the World Meteorological Organisation in 1988 with the purpose of assessing the scientific and technical literature related to all significant components of the issue of climate change.

International Energy Agency (IEA): Paris-based energy forum established in 1974. Linked with the OECD to enable member countries to share energy information and to coordinate energy policies.

International marine bunkers: a term referring to fuels consumed for international marine and air transport. Deliveries of oils to ships for consumption during international voyages (bunker oils) represent a special case of flows of oil from the country. These oils are used as fuel by the ship and are not part of the cargo. http://www.iea.org/textbase/nppdf/free/2004/statistics_manual.pdf.

International Monetary Fund (IMF): founded in 1944 together with the World Bank to promote international monetary cooperation and growth of trade by promoting exchange rate stability.

International poverty line: it is an income level established by the world bank and set at $1 a day per person in 1985 international purchasing power parity (PPP) prices to determine which people in the world are poor.

Intervention purchase: the act of purchasing a commodity once its market price drops below a set administered price (the intervention price) so as to raise or stabilise the market price to at least the level of the intervention price.

Intervention stocks: held by national intervention agencies as a result of intervention buying of commodities subject to market price support.

Irradiated food: subject to brief radioactivity, usually gamma rays, to kill insects, bacteria, and mould, and to permit storage without refrigeration.

Jet stream: rivers of high-speed air in the atmosphere formed along the boundaries of global air masses where there is a significant difference in atmospheric temperature. The jet streams may be several hundred miles across and 1–2 miles deep at an altitude of 8–12 miles. They generally move west to east, and are strongest in the winter with core wind speeds as high as 250 mph.

Joule: unit of measurement of energy consumption: 1 terajoule = 10^{12} joule = 2.78 × 10 5 kWh),1 terajoule = 23.88459 toe.

Kilocalorie (kcal): unit of measurement of dietary energy. In the international system of units, energy is usually measured in joules (J), but the customary usage of thermochemical energy units of kilocalories (kcal) is mostly used. 1 kcal = 4.184 kJ.

Kwashiorkor is the PEU associated (with caveats) with protein deficiency.

Kyoto mechanisms: three procedures established under the Kyoto Protocol to increase the flexibility of achieving greenhouse gas emissions cuts; they are the clean development mechanism, emissions trading and joint implementation.

Kyoto protocol: the Kyoto Protocol to the United Nations Framework Convention On Climate Change (UNFCCC) was adopted at the third session of the conference of the parties (CoP) to the UN Framework Convention on Climate Change (UNFCCC) in 1997 in Kyoto, Japan. It contains legally binding commitments for Annex 1 countries to reduce their anthropogenic emissions of greenhouse gases (CO_2, CH_4, N_2O, HFCs, PFCs and SF_6) by at least 5 % below 1990 levels in the commitment period 2008 to 2012.

Land cover corresponds to the bio-physical description of the Earth's surface. It enables various biophysical categories to be distinguished—these include: areas of vegetation (trees, bushes, fields, lawns); bare soil, hard surfaces (rocks, buildings) and wet areas and bodies of water (watercourses, wetlands).

Land degradation: is the reduction or loss of the biological or economic productivity of land resulting from natural processes, land use changes or other human activities as well as habitation patterns such as land contamination, soil erosion and the destruction of the vegetation.

Land tenure: is the right to the exclusive occupancy and use of a specified area of land.

Least developed countries—see **Developed, developing countries**.

Legumes: the fruit or seed of various bean or pea plants consisting of a pod that opens when ripe.

Lentils: erect annual bushy herb native to Asia widely cultivated for its edible flattened seeds.

Lignite: a brownish-black coal of low rank with high moisture content; it is used almost exclusively for electric power generation. Also referred to as brown coal.

Lipids: are commonly referred to as fats and oils but also actually includes waxes; sterols (including cholesterol); fat-soluble vitamins (A, D, E and K); and phospholipids among others.

Liquefied natural gas (LNG): natural gas converted to liquid form by cooling to a very low temperature.

Lupins: used primarily for feed, though in some parts of Africa and in Latin

America varieties are cultivated for human consumption.

Lymphatic system: the circulatory system of lymph.

Macronutrients are carbohydrates, lipids (fats and oils), protein and water, which provide the necessary daily intake of energy.

Maize: a grain with high germ content. Used largely for animal feed, commercial starch production and sweeteners. It also has wide industrial use in paper industries, adhesive, textiles, laundry, starch and ethanol.

Malnutrition: is a physiological condition caused by oversupply, deficiency and/or the imbalance in energy, protein and other nutrients including vitamins and minerals.

Manure: dung and urine of animals that can be used as a form of organic fertiliser.

Marasmic-kwashiorkor: is the condition of PEU that displays the combined symptoms of kwashiorkor and marasmus.

Marasmus: is the PEU associated with a general lack of energy nutrients.

Maxwell and Frankenberger's: key study in 1992 was a meta analysis of nearly 200 separate definitions of food security.

Meadow: a term usually used describing a field of permanent grass used for hay, but has also been applied to rich, waterside grazing areas that are not suitable for arable cultivation.

Megaloblastic (pernicious) anaemia: a condition where the body does not absorb enough vitamin B_{12} from the digestive tract characterised by very large red blood cells.

Metabolism: the body's system of breaking down (catabolism) or building up (anabolism) nutrients or chemical compounds for use in the various bodily processes.

Meteorology: study of the atmosphere and its phenomena.

Methane (CH_4): a non-poisonous, colourless and flammable gas created by anaerobic decomposition of organic compounds. Methane is a major component of natural gas used in the home.

Metric ton: common international measurement equal to 1000 kilograms, 2204.6 pounds or 1.1023 short tons.

Micronutrients: including vitamins, minerals and trace elements enable the body to produce enzymes, hormones and other substances essential for the metabolic processes of proper growth and development.

Millet: small-grained cereals that include a large number of different varieties. The plant can tolerate arid conditions and the grain is highly nutritious and stores well. Used principally for food and traditional beer brewing. Also used as a feed for birds.

Mineral: any naturally occurring inorganic substance found in the Earth's crust as a crystalline solid.

Mire, bog and fen habitats are saturated, with the water table at or above ground level for much of the year.

Monoculture: an agricultural practice in which single plant varieties are cultivated in an area.

Monounsaturated fats—see **fats**

Monsoon: the monsoon climate is characterised by long winter-spring dry seasons preceding the rains; a summer and early autumn rainy season, and a secondary warming immediately after the rainy season.

Montane: the biogeographic zone made up of relatively moist, cool upland slopes below the treeline characterised by the presence of large evergreen trees as a dominant life form.

Montreal protocol: a treaty signed in 1987 by 24 countries, which pledged to phase out use of all chlorofluorocarbons by 1999.

Morbidity refers to the poor health or diseased state of individuals or populations.

Morphology: a branch of biology that deals with the form and structure of animals and plants.

Naphtha is a distillate of oil and is a feedstock used in the petrochemical industry.

National accounts: a set of macroeconomic accounts providing a comprehensive accounting framework within which economic data can be compiled and presented for purposes of economic analysis, decision-taking and policy-making.

Natural gas: whether liquefied or gaseous includes both gas originating from fields producing only hydrocarbons in gaseous form, and gas produced in association with crude oil as well as methane recovered from coal mines.

Natural resources: are natural assets, the raw materials that occurr in nature that can be used for among other things, economic production or consumption.

Neologisms: are new or invented words.

Nitrate & nitrite: are naturally occurring ions that are part of the nitrogen cycle. Nitrate is used mainly in inorganic fertilisers, and sodium nitrite is used as a food preservative.

Nitrogen: an essential nutrient in the food supply of plants and the diets of animals.

Nitrogen fixation: conversion of atmospheric nitrogen gas into forms useful to plants and other organisms by lightning, bacteria, and blue-green algae.

Nitrogen oxide: (NO_2) is a product of fossil-fuel combustion; soil cultivation practices; and biomass burning. It is a major contributor to acid deposits and the formation of ground-level ozone in the troposphere.

No till: planting crops without prior seedbed preparation into areas with existing cover crop, sod or crop residues.

Non-governmental organisation (NGO): a non-profit group or association that aims to realise particular social objectives such as environmental protection or helping indigenous peoples, etc.

Non-point-source pollution: pollution from sources that cannot be defined. See also **point-source pollution**.

Nutrients: any substance that contributes to the growth and health of a living organism.

Nutrition: relates to the nutrients supplied by the food we eat to maintain a healthy functioning body. Inherent is the implication of quality fit for purpose. In this regard too, 'adequate' food does not necessarily equate with 'good' nutrition.

Nutritional status: is the indication of a person's nutritional health determined by the quantity and quality of foods consumed and by the ability of the body to use them.

Oases: fertile or green spots in a desert or wasteland made possible by the presence of water.

Oedema (edema) is the buildup of fluid in the body that results in swelling.

Official Development Assistance (ODA): also known as foreign aid. Consists of loans, grants, technical assistance and other forms of cooperation from developed to developing countries.

Ogallala aquifer: an underground water source that stretches from Texas to south Dakota that is used for irrigation in the great plains.

Oil—see **crude oil, petroleum, fossil fuel, hydrocarbons**.

Oilseeds/oil-bearing crops are annual and perennial plants whose seeds, fruits, nuts and kernels are used mainly for the extraction of oil for use in the culinary and industrial sectors.

Organic farming: a method of production, which puts a high emphasis on environmental protection and regard to animal welfare considerations.

Organic fertilisers are fertilisers derived from animal products and plant residues containing sufficient nitrogen.

Organisation for Economic Cooperation And Development (OECD): An organisation that coordinates policy among developed countries to maximize their countries' economic growth and help non-member countries develop more rapidly.

Osteoporosis: reduced bone mineral density can be brought on by vitamin D deficiency.

Overnutrition refers to regular over ingestion of nutrients.

Oxidize: to chemically transform a substance by combining it with oxygen.

Ozone: a colourless gas found in two layers of the atmosphere: the stratosphere and the troposphere. It is a form of oxygen found naturally in the stratosphere providing a protective layer against harmful ultraviolet radiation. Ground-level ozone (in the troposphere) is considered an air pollutant and major component of photochemical smog.

Ozone hole: a large area of stratospheric ozone depletion over the Antarctic continent that typically occurs annually between August and October. The phenomenon is the result of chemical mechanisms initiated by man-made chlorofluorocarbons.

Pacific rim: the 34 countries and 23 islands in and around the Pacific Ocean with an area of 70 million square miles and a population of 2.4 billion people.

Paresthesia: is the tingling of the hands and feet or 'burning feet' syndrome.

Particulate matter (PM): solid particles or liquid droplets suspended or carried in the air (for example, soot, dust, fumes and mist). See **aerosol**.

Pathogen: microorganism that can cause disease in other organisms.

Pathophysiology is the study of the changes of normal bodily functions either mechanical, physical or biochemical brought about by disease, infection or other abnormal conditions.

Peak oil is the point in time when the maximum rate of global petroleum extraction is reached, after which the rate of production enters terminal decline.

Peat bog: acidic peat environment typical of wetland and characteristically colonised by low vegetation including moss, rushes and heather.

Peat: combustible soft, porous and compressed, fossil sedimentary deposit of plant origin with high water content used for fuel.

Peat soil: predominantly organic soil derived from partially decomposed plant remains that accumulate under waterlogged conditions.

Pellagra is a condition caused by deficient vitamin B_3 niacin.

Per capita is defined as data for each person.

Percentiles: the set of values which divide the total frequency into 100 equal parts.

Perennial is a plant that becomes dormant after one growing season and sends up new shoots for the next.

Perfluorocarbons (PFCs): a group of human made chemicals introduced as alternatives, along with hydrofluorocarbons, to the ozone depleting substances.

Permafrost: a layer of soil or bedrock beneath the surface of the Earth in which the temperature has been below freezing continuously from a few years to several thousands.

Permanent crops: unlike temporary crops, permanent crops are sown or planted once then occupy the land for some years; examples include cocoa, coffee, rubber, fruit trees, nut trees and vines.

Persistent organic pollutants (PoP): organic pollutants that remain in the environment and do not break down.

Pesticides: refers to insecticides, fungicides, herbicides, disinfectants and any substance used to prevent, control or destroy pests.

Petechia: small purplish spots of blood rising to the skin.

Petrochemical feedstock are substances created or derived during the refinement of oil and are used principally for the manufacture of chemicals, synthetic rubber, a variety of plastics, paints, synthetic fibres, medicines, industrial chemicals, fertilisers and other products.

Petroleum & petroleum products: a generic term applied to oil and oil products such as crude oil, lease condensate, unfinished oils, natural gas and non-hydrocarbon compounds blended into finished petroleum products. See also **crude oil**.

PEU/PEM: protein-energy-undernutrition or malnutrition is the inadequate intake of energy giving nutrients such as carbohydrates, proteins and fats.

Phenetic system is a taxonomic grouping of organisms based on mutual similarity of phenotypic (physical and chemical) characteristics which may or may not correlate with evolutionary relationships.

Phenotype is any observable characteristic or trait of an organism: such as its morphology, development, biochemical or physiological properties.

Phosphorus is a mineral used in the manufacture of commercial phosphate fertilisers.

Photosynthesis: complex process that takes place in living green plant cells that use energy from the sun to combine carbon dioxide (CO_2) and water (H_2O) to produce oxygen (O_2) and simple nutrient molecules such as glucose ($C_6H_{12}O_6$).

Phytoplankton: the plant forms of plankton. Phytoplankton is the dominant plants in the sea and forms the basis of the entire marine food chain. See also **zooplankton**.

Point-source pollution: a single identifiable source that discharges pollutants into the environment. See also **non-point-source pollution.**

Polymerisation: the process where single monomers (single atom or molecules) are converted into 3D complex structures containing many monomers.

Polyneuropathy is a neurological disorder that occurs when peripheral nerves around the body malfunction or become impaired.

Polyunsaturated fat—see **fats**

Population density: number of persons in the total population for a given year per square kilometre of total surface area.

Post-industrialisation: the phase in a country's economic development following industrialisation characterised by the growing role of service sector in the national economy.

Poverty: a low standard of living in terms of other people (relative poverty) or in terms of basic needs (absolute poverty).

Prairies: characterised by fertile soil and covered with coarse grasses, a prairie is an extensive, predominantly treeless, grassland.

Price index: a price index reflects the average changes in the prices of a specified set of goods and services between two periods of time.

Primary crops: primary crops are those, which come directly from the land and without having undergone any real processing.

Primary goods: also called commodities; are goods that are sold just as they were found in nature. Examples include oil, coal, iron and agricultural products like wheat or cotton.

Proteins: are made up of amino acids, the building blocks of the body and are the most satiating macronutrient. There are many in existence but the body requires only 20, these are known as essential amino acids. Some of these (8–9) the body cannot synthesize and must be ingested.

Protocol: international legal instrument appended or closely related to another agreement.

Public Law 480: (a.k.a.) Food For Peace is properly known as the Agricultural Trade Development Assistance Act 1954. Consisting of four parts or titles; 1, 2, 3 and 5, titles 2, 3 and 5 are managed by USAID while title 1 is managed by the USDA. Collectively these are avenues by which funding and food aid can be used overseas.

Pulses are annual leguminous crops yielding grains or seeds. The term is limited to crops harvested for dry grains only and therefore exclude crops harvested 'green' for food such as green peas and beans.

Purchasing power parity (PPP) are the rates of currency conversions that equalise the purchasing power of different currencies by eliminating the differences in price levels between countries.

Quartile: values separated into four equal parts.

Quinoa: a minor cereal tolerant of high altitudes and cultivated primarily in Andean countries.

Quintiles: values separated into five equal parts.

Quorum: the minimum number of delegations that must be present for a meeting to start or decisions to be made.

Radiation: emission of rays from either natural or human-made origins, such as radio waves, the sun's rays, medical x-rays and the fall-out of nuclear wastes.

Radiative forcing: without any radiative forcing in the form of GHGS and albedo, solar radiation coming to the earth would continue to be approximately equal to the infrared radiation emitted from the Earth. Therefore radiative forcing can be described as the change in the balance between incoming and outgoing radiations.

Rain forest: evergreen woodland of the tropics characterised by a continuous leafy canopy and an average rainfall of about 100 inches per year.

Ranching: commercial raising of animals. (including fisheries), mainly for human consumption, under extensive production systems, within controlled boundaries and paddocks or in open space where they grow using natural food supplies.

Rangeland: including unimproved grasslands, shrublands, savannahs, and tundra, whose plants can provide food (i.e., forage) for grazing animals, etc.

Rapeseed oil: from the point of view of production, rapeseed oil ranks third most important among vegetable oils preceded only by palm and soybean oils.

Reforestation is the artificial or natural re-establishment of forest in an area that was previously forested.

Renewable energy: obtained from sources that are essentially inexhaustible. Renewable sources of energy include wood, waste, geo-thermal, wind, photovoltaic and solar thermal energy.

Rickets: lack of vitamin D can bring on a condition known as rickets in children and osteomalacia in adults, weak or softened bones, bow legged.

Rio conference/conventions: the United Nations Conference On Environment and Development (UNCED), held in Rio de Janeiro, Brazil, in 1992. Outcomes included: the UN Framework Convention On Climate Change (UNFCCC); the Convention On Biological Diversity (CBD); Agenda 21; the establishment of the Commission on Sustainable Development (CSD) and the Rio Declaration on Environment and Development. UNCED also led to the negotiation and adoption of the UN Convention to Combat Desertification.

Rural population: residual population after subtracting urban population from the total population.

Rye: a grain tolerant of poor soils, high latitudes and altitudes. Mainly used in making bread, whisky and beer.

Saccharides are any of a series of compounds of carbon, hydrogen, and oxygen. They are synonym for sugars. They can be simple monosaccharides, or polymerised disaccharides and polysaccharides.

Sahel: the arid region south of the Sahara desert in West Africa.

Salinisation: change in the salinity status of the soil.

Savannah: one of the Earth's biomes characterised by tropical or subtropical temperatures; similar to grasslands, but receiving more precipitation and containing more trees. An almost transitional biome between those dominated by forests and those dominated by grasses.

Scurvy: a conditioned brought on by lack of vitamin C resulting in general weakness, anaemia and/or gum disease (gingivitis).

Secondary PEU is malnutrition caused as a result of infection or other condition that interferes with the absorption of nutrients as opposed to a lack of nutrients themselves.

Sequestration: the process of increasing the carbon content of a carbon reservoir. These include direct removal of carbon dioxide from the atmosphere through land-use change, afforestation, reforestation and practices that enhance soil carbon in agriculture or through the separation of carbon dioxide from flue gases and from fossil fuels production et cetera. See also **sink**.

Shale oil: a fine-grained sedimentary rock containing kerogen, a solid, waxy mixture of hydrocarbon compounds that when heated is converted to vapour, which can then be condensed to form a slow flowing heavy oil called shale oil.

Short ton: common measurement for a ton in the United States. A short ton is equal to 2000 lbs or 0.907 metric tons.

Silage: a crop that has been preserved in a moist condition to encourage partial fermentation. The chief crops stored in this way are corn (the whole plant), sorghum, and various legumes and grasses. The main use of silage is in cattle feed.

Silt is fine particles that are finer than grains of sand and larger than clay particles that can be picked up by the air or by water and deposited as sediment.

Sink: any process which removes a greenhouse gas or aerosol from the atmosphere either by destroying them or by storing them in some other form. Forests and other vegetation are good examples of sinks as they remove carbon dioxide through photosynthesis.

Smog: a term used to describe a mixture of smoke and fog which occurs when high concentrations of moisture are combined with smoke in the presence of high temperatures or thermal inversions and the absence of wind.

Sodium-potassium pump (NA/K ATPase) is an electrogenic responsible for controlling the cell membranes electrical potential. It also acts as an intercellular nutrient transporter through the use of the sodium gradient (potential difference).

Soil conservation: protection of soil from erosion and other types of deterioration to maintain and promote soil fertility and productivity.

Solar radiation: energy from the sun. Also referred to as short-wave radiation, solar radiation includes ultraviolet, visible and infrared radiation.

Soot particles: formed during at the outer edge of flames of organic vapours consisting predominantly of carbon.

Sorghum is the common name applied to several annual grasses cultivated for grain.

Soybeans: the soybean is a legume whose seed is contained in a pod. Cultivated mainly in North and South America, soybean oil is the most important oil-crop for both human food (oil) and animal feed (seed cake).

Spatial data is any data with a direct or indirect reference to a specific location or geographical area.

Specialised agency: autonomous international organisation linked to the United Nations through special agreement.

Standard error: the positive square root of the variance of the sampling distribution of a statistic.

Starch: a digestible complex carbohydrate; a polysaccharide (complex sugar).

Stressors: physical, chemical, or biological entities that induce adverse effects on things like ecosystems or human health.

Stunting: low height for age reflecting sustained past undernutrition.

Sugar beet: sugar is the sweet product obtained from pressed sugary juice extracted from sugarcane and sugarbeets. Sugar content ranges from 13% to 18%. Beets are also cultivated as a fodder crop and grown as a vegetable called red or garden beets.

Sugar cane: plant whose: sugar content ranges between 10% and 15%.

Sugar crops: cultivated primarily for sugar and for the production of alcohol.

Sugar: obtained from the transformation of sugarcane and sugarbeets.

Summit: meeting at which the participants are high-level officials, such as heads of state or government.

Surface water: all water open to the atmosphere including rivers, lakes, reservoirs, streams, seas, estuaries and so on.

Sustainable development: that meets the needs of the present without compromising the ability of future generations to meet their own needs

Swamp: a type of wetland dominated by woody vegetation but without appreciable peat deposits. Swamps may be fresh or salt water and tidal or non-tidal.

Synthetic fertiliser: commercially prepared mixtures of plant nutrients such as nitrates, phosphates, and potassium applied to the soil to restore fertility and increase crop yields.

Synthetic natural gas (SNG): a manufactured product chemically similar to natural gas.

Taxonomy: naming and assignment of biological organisms to taxa.

Temperate climate: the climate of the middle latitudes; the climate between the extremes of tropical climate and polar climate.

Temporary crops are those which are both sown and harvested during the same agricultural year.

Third world: less developed countries that have low per capita incomes, often with large agricultural sectors, and a shortage of capital.

Timberline or treeline: the elevation in a mountainous region above which trees do not grow, also the northern or southern latitudes beyond which trees do not grow.

Time series: a set of observations of a quantitative characteristic taken at different points of time.

Topsoil: the top layer of the soil that contains large amounts of organic matter.

Trans fats (see also **fats**): trans fatty acids are partially hydrogenated fats that if incorporated into the cell can create denser membranes that interfere with the normal functioning of the cell.

Transgenic: organisms whose genetic makeup includes a gene or genes from another genus or species.

Transition economies/countries: countries moving from centrally planned to market-oriented economies.

Transnational corporation: a corporation that carries out business in several countries.

Transpiration: the process by which water is drawn through plants and returned to the air as water vapour. See also **evapotranspiration**.

Treaty: international agreement concluded between states in written form and governed by international law.

Triglycerides are the collective name for the lipids—fats and oils and make up the largest proportion of dietary fats consumed by humans.

Triticale: a minor cereal crossed between wheat and rye, combining the quality and yield of wheat with the hardiness of rye.

Trophic level: are the classification of natural communities or organisms according to their place in the food chain. Green plants (producers) can be roughly distinguished from herbivores (consumers) and carnivores (secondary consumers).

Tropical climate: a climate typical of equatorial and tropical regions that combines continual high temperatures with considerable precipitation.

Tropical Sprue: endemic to tropical regions is a condition that causes malabsorptive problems.

Tuberculosis (TB): is a potentially fatal disease affecting almost any part of the body but is often mainly an infection of the lungs.

Tundra: type of ecosystem dominated by lichens, mosses, grasses and dwarf woody plants. Found at high latitudes (arctic tundra) or high altitudes (alpine tundra).

Turbidity: hazy or cloudy condition of water due to the presence of suspended particles.

Ultraviolet radiation (UV): the energy range just beyond the violet end of the visible spectrum constituting the major energy source of Earth.

Undernourishment refers to the condition of people whose dietary energy consumption is continuously below a minimum dietary energy requirement. See also **malnourishment**.

Undernutrition refers to insufficient nutrient intake.

United Nations Framework Convention On Climate Change (UNFCC): the international treaty unveiled at the United Nations Conference on Environment and Development.

Urbanisation/urban sprawl is the expansion of an urban area to accommodate its growing population.

Vitamins are nutrients needed in the body for various metabolic or other co-factoring processes.

Wasting: low weight for height, generally the result of weight loss associated with a recent period of starvation or disease.

Water conservation: preservation, control and development of water resources, both surface and groundwater and prevention of pollution.

Water cycle: the cycle of water through its natural process of evaporation and precipitation from the sea, through the atmosphere, to the land and back to the sea.

Water is an important and often overlooked macronutrient needed in large quantities by humans.

Water vapour: the most abundant of the greenhouse gases; it is the water present in the atmosphere in gaseous form. Water vapour is an important part of the natural greenhouse effect.

Waterlogging: natural flooding and over-irrigation that brings water at underground levels to the surface affecting soil quality.

Wetlands: lands where water saturation determines the nature of soil development and the types of plant and animal communities living in the surrounding environment. Often wetlands occur in areas of low-lying land where the water table is at or near the surface most of the time. Wetlands include swamps, bogs, fens, marshes and estuaries.

Wheat: one of the most important of the cereals, wheat is used mainly for human food, particularly in the form of bread and pasta.

World Bank: an international lending institution that aims to reduce poverty by strengthening economies and promoting sustainable development.

World regions and countries are aggregated differently according to the institution and purpose.

World Trade Organisation (WTO): an international organisation established on 1 January 1995, to succeed the general agreement on tariffs and trade.

Xerophthalmia: a name given to a set of conditions affecting the eyes from night blindness to Bitot's spots, etc.

Zooplankton: microscopic animals that float freely in the open water.

Index

Note: n = Footnote.